MOLECULAR
NANOSTRUCTURES

Previous Proceedings in the Series of
International Kirchberg Winterschools

Year	Conference	Publisher	ISBN
2002	XVI	AIP Conf. Proceedings Vol. 633	0-7354-0088-1
2001	XV	AIP Conf. Proceedings Vol. 591	0-7354-0033-4
2000	XIV	AIP Conf. Proceedings Vol. 544	1-56396-973-4
1999	XIII	AIP Conf. Proceedings Vol. 486	1-56396-900-9
1998	XII	AIP Conf. Proceedings Vol. 442	1-56396-808-8
1997	XI	World Scientific Publishers	981-02-3261-6

Other Related Titles from AIP Conference Proceedings

678 Lectures on the Physics of Highly Correlated Electron Systems VII: Seventh Training Course in the Physics of Correlated Electron Systems and High-Tc Superconductors
Edited by A. Avella and F. Mancini, August 2003, 0-7354-0147-0

677 Fundamental Physics of Ferroelectrics 2003
Edited by P. K. Davies and D. J. Singh, August 2003, 0-7354-0146-2

671 Hydrogen in Materials and Vacuum Systems: First International Workshop on Hydrogen in Materials and Vacuum Systems
Edited by Ganapati Rao Myneni and Swapan Chattopadhyay, July 2003, 0-7354-0137-3

640 DNA-Based Molecular Construction: International Workshop on DNA-Based Molecular Construction
Edited by Wolfgang Fritzsche, November 2002, 0-7354-0095-4

590 Nanonetwork Materials: Fullerenes, Nanotubes, and Related Systems, ISNM 2001
Edited by Susumu Saito, Tsuneya Ando, Yoshihiro Iwasa, Koichi Kikuchi, Mototada Kobayashi, and Yahachi Saito, October 2001, 0-7354-0032-6

577 Density Functional Theory and Its Application to Materials
Edited by V. Van Doren, C. Van Alsenoy, and P. Geerlings, July 2001, 0-7354-0016-4

To learn more about these titles, or the AIP Conference Proceedings Series, please visit the webpage **http://proceedings.aip.org**

MOLECULAR NANOSTRUCTURES

XVII International Winterschool/Euroconference on Electronic Properties of Novel Materials

Kirchberg, Tirol, Austria   8-15 March 2003

EDITORS
Hans Kuzmany
Universität Wien, Austria

Jörg Fink
Institut für Festkörper- und Werkstoff-Forschung
Dresden, Germany

Michael Mehring
Universität Stuttgart, Germany

Siegmar Roth
Max-Planck-Institut für Festkörperforschung
Stuttgart, Germany

Melville, New York, 2003
AIP CONFERENCE PROCEEDINGS ■ VOLUME 685

L.C. Catalog Card No. 2003111782
ISBN 0-7354-0154-3
ISSN 0094-243X
Printed in the United States of America

CONTENTS

FULLERENES, ENDOHEDRALS, AND FULLERIDES

CARBON NANOSTRUCTURE SYNTHESIS AND PURIFICATION

PROPERTIES OF SINGLE-WALL CARBON NANOTUBES

CHARACTERIZATION OF CARBON NANOTUBES

FUNCTIONALIZATION OF CARBON NANOTUBES

NON-CARBONACEOUS NANOTUBES

THEORY OF NANOSTRUCTURES

APPLICATIONS

PREFACE

The present book contains the proceedings of the 17th International Winterschool on Electronic Properties of Novel Materials in Kirchberg, Tirol, Austria. The winterschool was held from 8th to 15th March 2003 in Hotel Sonnalp. The series of these schools started in 1985. Originally the school was held every second year and was devoted to conducting polymers. After the discovery of high temperature superconductors, the periodicity changed to an annual format and the topic alternated between conjugated polymers and superconductors. Since fullerenes are both conjugated compounds and in some cases superconductors, it was tempting to choose fullerenes as the topic of the Kirchberg schools. The evident extension of this topic is carbon nanotubes and so the title changed from fullerenes via Fullerene Derivatives and Fullerene Nanostructures to Molecular Nanostructures. This gradual change enables us to keep a fairly large interdisciplinary scientific community together and stimulate numerous international cooperations. A compilation of the previous Kirchberg Winterschool is presented in the table at the end of this preface.

The term "molecular nanostructures" implies the "bottom up" (synthetic) approach, as opposed to the "top down" (lithography and etching) techniques in nanostructure technology. As for the physics, we are in a field where solid state physics and molecular physics overlap. This is nicely illustrated with the example of carbon nanotubes. Perpendicular to their axis, nanotubes are molecular as their diameter is of the order of a few nanometers, and different diameters lead to different electronic structures, while along their axis they are extended solids.

Contributions to the 17th Winterschool focused on new nanostructured materials, with data presented on functionalized fullerenes and carbon nanotubes, non-carbon nanotubes such as BN and MoS_2 tubes, and new biological nanostructures. The direction of nanoelectronics research was explored in depth, and advancements in composite technology, and novel applications for nanotubes were discussed. Importantly, participants were updated on the theoretical and experimental determinations of structural and electronic properties, as well as on characterization methods for molecular nanostructures.

The meeting could not have taken place without the support of the Bundesministerium für Wissenschaft und Forschung in Wien and the Verein für Förderung der Winterschulen in Kirchberg, as well as from numerous industrial sponsors. Without their contribution, all the enthusiasm and dedication could be wasted and so we express our gratitude to the sponsors and supporters.

Finally, we are indebted, to the manager of the Hotel Sonnalp, Frau Edith Mayer, and to her staff for their continuous support and for their patience with the many special arrangements required during the meeting.
Special thanks to Viera Skakalova for her efforts in editing and compiling the Winterschool volume 2003.

H. Kuzmany, J. Fink, M. Mehring, S. Roth
Wien, Dresden, Stuttgart 2003

Table of Previous Kirchberg Winterschools

Year	Title	Published By
2002	Structural and Electronic Properties of Molecular Nanostructures	AIP Conference Proceedings 633 (2002)
2001	Electronic Properties of Molecular Nanostructures	AIP Conference Proceedings 591 (2001)
2000	Electronic Properties of Novel Materials – Molecular Nanostructures	AIP Conference Proceedings 544 (2000)
1999	Electronic Properties of Novel Materials – Science and Technology of Molecular Nanostructures	AIP Conference Proceedings 486 (1999)
1998	Electronic Properties of Novel Materials – Progress in Molecular Nanostructures	AIP Conference Proceedings 442 (1998)
1997	Molecular Nanostructures	World Scientific Publ. 1998
1996	Fullerenes and Fullerene Nanostructures	World Scientific Publ. 1996
1995	Physics and Chemistry of Fullerenes and Derivatives	World Scientific Publ. 1995
1994	Progress in Fullerene Research	World Scientific Publ. 1994
1993	Electronic Properties of Fullerenes	Springer Series in Solid State Sciences 117
1992	Electronic Properties of High-T_c Superconductors	Springer Series in Solid State Sciences 113
1991	Electronic Properties of Polymers – Orientation and Dimensionality of Conjugated Systems	Springer Series in Solid State Sciences 107
1990	Electronic Properties of High-T_c Superconductors and Related Compounds	Springer Series in Solid State Sciences 99
1989	Electronic Properties of Conjugated Polymers III – Basic Models and Applications	Springer Series in Solid State Sciences 91
1987	Electronic Properties of Conjugated Polymers	Springer Series in Solid State Sciences 76
1985	Electronic Properties of Polymers and Related Compounds	Springer Series in Solid State Sciences 63

ORGANIZER

Institut für Materialphysik
Universität Wien

PATRONAGE

ELISABETH GEHRER
Bundesministerin für Bildung, Wissenschaft und Kultur

Magnifizenz
Univ. Prof. Dr. GEORG WINCKLER
Rektor der Universität Wien

HERBERT NOICHEL
Bürgermeister von Kirchberg

SUPPORTERS

BUNDESMINISTERIUM FÜR BILDUNG, WISSENSCHAFT UND KULTUR

**VEREIN ZUR FÖRDERUNG DER INTERNATIONALEN WINTERSCHULEN
IN KIRCHBERG**

SPONSORS

BRUKER Analytische Meßtechnik GmbH, Wikingerstraße 13, D-7500 Karlsruhe 21, Germany
CREDITANSTALT BANKVEREIN, Nußdorferstraße 2, A-1090 Wien, Austria
ELECTROVAC GmbH, Aufeldgasse 37-39, A-3400 Klosterneuburg, Austria
JOBIN YVON GmbH, Neuhofstraße 9, D-64625 Bensheim
NANOCYL S.A., Rue de Seminaire 22, 5000 Namur, Belgium
OMICRON VAKUUMPHYSIK GmbH, Idsteinerstraße 78, D-65232, Taunusstein, Germany

The financial assistance from the sponsors and the supporters is gratefully acknowledged

FULLERENES, ENDOHEDRALS, AND FULLERIDES

Self-assembling of C_{60}-imidazole and C_{60}-pyridine adducts in the Langmuir and Langmuir-Blodgett films via complex formation with water-soluble zinc porphyrins

Renata Marczak,[a] Krzysztof Noworyta,[a] Wlodzimierz Kutner,[a]
Suresh Gadde,[b] and Francis D'Souza[b,*]

[a]Institute of Physical Chemistry, Kasprzaka 44/52, 01-224 Warsaw, Poland
[b]Department of Chemistry, Wichita State University, Wichita, Kansas 67260-0051, USA

Abstract. The C_{60}-pyridine, C_{60}py, and C_{60}-imidazole, C_{60}im, adducts were found to self-assemble in films floating onto aqueous solutions of zinc tetrakis (N-methylpyridinium)porphyrin cation, Zn(TMPyP), or zinc tetrakis (4-sulfonatophenyl)porphyrin anion, Zn(TPPS). This self assembling was due to axial ligation of the C_{60} adducts (acceptors) by Zn porphyrins (donors), which lead to the formation of relatively stable donor-acceptor dyads in the water-air interfaces. The films were compressed in a Langmuir trough and characterized by isotherms of surface pressure vs. area per molecule as well as by the Brewster angle microscopy imaging. All systems formed stable aggregated Langmuir films of the "expanded liquid" type. Extensive compression of the films resulted in two-dimensional phase transitions. The area per molecule at infinite dilution of the adducts in films increased in the order: water<0.1 mM Zn(TPPS)<0.1 mM Zn(TPMyP). Comparison of the determined and calculated values of area per molecule indicated that orientation of porphyrins in the complexes was parallel with respect to the interface plane.

The Langmuir films were transferred, by using the Langmuir-Blodgett technique, onto quartz slides. The UV-vis spectroscopic study of these films revealed that Zn porphyrins were transferred together with the C_{60} adducts and that the transfer efficiency increased in the order: C_{60}py-Zn(TPPS)<C_{60}py-Zn(TMPyP)<C_{60}im-Zn(TPPS)<C_{60}im-Zn(TMPyP), i.e., in accord with the increase of stability of the respective dyads in solutions.

INTRODUCTION

Covalent linking of metalloporphyrin donors and C_{60} acceptors leads to formation of photoactive donor-acceptor dyads capable of electron or energy transfer upon photoexcitation [1]. The structure and chemical properties of the covalent linker between the donor and acceptor moieties is known to influence the excited state properties of such dyads [2]. While a number of covalently linked porphyrin-fullerene dyads have been constructed [3], studies of non-covalently bound dyads remain limited in number. C_{60} bearing pyridine or imidazole substituents, coordinated in an axial position to metallated porphyrins, phthalocyanines, or porphycenes, form relatively stable complexes [4] and, therefore, have been successfully utilized for

CP685, *Molecular Nanostructures: XVII Int' l. Winterschool/Euroconference on Electronic Properties of Novel Materials,* edited by H. Kuzmany, J. Fink, M. Mehring, and S. Roth
© 2003 American Institute of Physics 0-7354-0154-3/03/$20.00

studying photoinduced electron transfer in self-assembled porphyrin-fullerene dyads in solution.

Here, preparation and properties of highly ordered molecular structures of the self-assembled dyads in the Langmuir and Langmuir-Blodgett (LB) films, shown by way of example for the $C_{60}py$-Zn(TMPyP) dyad in Scheme 1, are presented. Such films may be useful for probing vectorial photoinduced electron transfer and also for constructing light-harvesting molecular electronic devices.

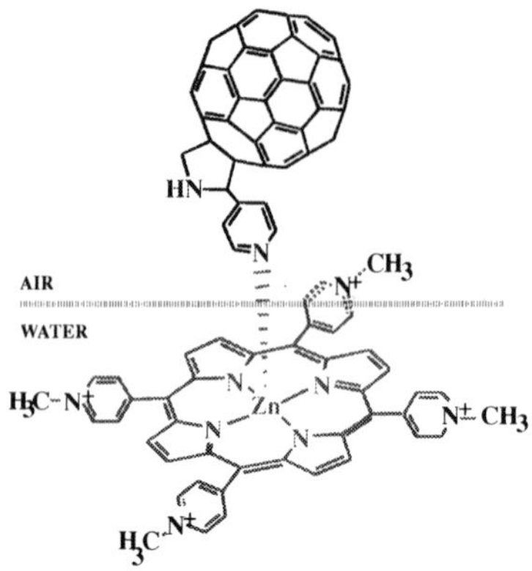

SCHEME 1. $C_{60}py$-Zn(TMPyP)

EXPERIMENTAL

Chemicals. The C_{60}-pyridine and C_{60}-imidazole adducts were prepared [4] according to a general protocol of the fulleropyrrolidine synthesis [5]. Zinc tetrakis (N-methylpyridinium)porphyrin chloride, Zn(TMPyP), and sodium salt of zinc tetrakis (4-sulfonatophenyl)porphyrin, Zn(TPPS), were from Mid Century Chemicals (Chicago, IL). Analytical grade chloroform from POCH (Gliwice, Poland) was used as received. Water used for preparation of subphase solutions of the Langmuir films was distilled and further purified (18.2 $M\Omega$ cm) with a Milli-Q filtering system of Millipore Corp. (Bedford MA, USA).

Instrumentation and procedures. A 601BAM trough of Nima Technology, Ltd. (Coventry, UK), equipped with a PS4 surface pressure sensor, was used for preparation of the Langmuir films. A 0.1 to 0.5 mL sample of 0.12 mM $C_{60}py$ or 0.11 mM $C_{60}im$ in chloroform was spread onto subphase of pure water, 0.1 mM Zn(TMPyP) or 0.1 mM Zn(TPPS). Then, chloroform was allowed to evaporate for 15 min and compression isotherms of surface pressure, π, vs. area per molecule, A, were recorded at 25 $cm^2 min^{-1}$, at 20 °C.

Area per molecule in the Langmuir films of the adducts and their Zn porphyrin complexes were calculated by using HyperChem (Release 5) of Hypercube, Inc., (Gainesville FL, USA), with PM3 parametrization.

RESULTS AND DISCUSSION

Relatively stable Langmuir films of the C_{60} adducts, spread onto surfaces of pure water as well as aqueous solutions of the Zn porphyrins, were obtained with no collapse points observed on the compression π-A isotherms (Fig. 1) in the attainable surface pressure range, $0 \leq \pi \leq 40$ mN m^{-1}, for all subphase solutions used as well as for both adducts spread. Most likely, the adduct molecules were pre-arranged in films by self-assembly due to axial ligation of the Zn porphyrins. The presence of inflection points in the isotherms suggests some re-orientation of molecules during compression. The area per molecule at zero surface pressure, A_1, determined from the isotherms increased with the decrease of the sample volume of the adduct spread indicating initial aggregation of the adduct molecules in films. Therefore, values of surface area per molecule at infinite adduct dilution, A_0, were determined for all systems by linear extrapolation of the A_1 values to the zero adduct amount in film [6]. The A_0 values were dependent on the composition of the subphase solution as relatively stable C_{60}py-Zn(TMPyP), C_{60}py-Zn(TPPS), C_{60}im-Zn(TMPyP), or C_{60}im-Zn(TPPS) complexes were formed in the (water solution)-air interface. Comparison of experimental areas per molecule at infinite dilution with those calculated indicates that adducts are vertically oriented in the films, i.e., their porphyrin planes are parallel to the interface, as shown for the C_{60}py films in Table 1. The determined values of compressibility, $\kappa = -(1/A_1)(dA/d\pi)_T$ (T is in Kelvins), indicate (Table 1) formation of the "expanded liquid" type films for all systems.

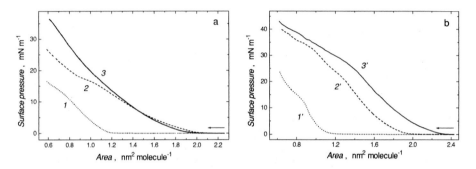

FIGURE 1. Langmuir trough compression isotherms for films formed from a sample of 0.3 ml of (a) 0.12 mM C_{60}py in CHCl$_3$ and (b) 0.11 mM C_{60}im in CHCl$_3$; water subphase - (*1* and *1'*), 0.1 mM Zn(TPPS) - (*2* and *2'*), and 0.1 mM Zn(TMPyP) - (*3* and *3'*).

TABLE 1. Compressibility, κ, and area per molecule, A_0, calculated and determined from the Langmuir π-A isotherms by extrapolation of area per molecule at zero surface pressure, A_1, to infinite adduct dilution, for films of C_{60}py floating on different subphase solutions.

Subphase composition	κ nm^2 molecule^{-1}	A_0 for C_{60}py-Zn(porphyrin) , nm^2			
		Calcd.[a] for orientation		Determined	
		Vertical	Horizontal	$A_{0,1}\pm$st.d.	$A_{0,2}\pm$st.d.
H$_2$O	0.026	0.90		1.1±0.6	1.60±0.02
0.1 mM Zn(TPPS)	0.025	2.23	3.87	2.3±0.1	2.4±0.1
0.1 mM Zn(TMPyP)	0.015	2.48	3.84	2.5±0.1	

[a]Semi-empirical calculations with PM3 parametrization for molecules in vacuum.

The BAM images of the C_{60} adduct films floating on the Zn porphyrin subphase solutions (Fig. 2) showed that the initially formed condensed phase domains approached each other during the compression and, eventually, uniform and stable films were formed.

The Langmuir films of the dyads were transferred, by the LB emersion, onto quartz slides. The UV-vis spectroscopy study of these LB films revealed that the Zn porphyrins dissolved in the water subphases were transferred together with the C_{60} adducts. Moreover, the transfer efficiency increased (Fig. 3) in the order: C_{60}py-Zn(TPPS)<C_{60}py-Zn(TMPyP)<C_{60}im-Zn(TPPS)<C_{60}im-Zn(TMPyP), i.e., in accord with the order of increase of the complex stability constants in solution [4].

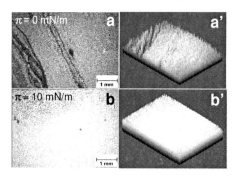
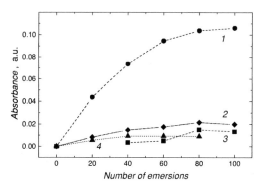

FIGURE 2. BAM images of the C_{60}py film, floating on 0.1 mM Zn(TMPyP), for different values of surface pressure in (a) and (b) 2D as well as (a') and (b') 3D representation.

FIGURE 3. Dependence of the Soret band absorbance at 450 nm in LB films on the number of emersions for C_{60}im-Zn(TMPyP) - (*1*), C_{60}py-Zn(TMPyP) - (*2*), C_{60}py-Zn(TPPS) - (*3*), and C_{60}im-Zn(TPPS) - (*4*).

CONCLUSIONS

Relatively stable, highly oriented molecular structures of non-covalent (ligand-appended C_{60})-(Zn porphyrin) dyads were formed in the "expanded liquid" Langmuir and LB films. Comparison of experimental and calculated values of the area per molecule suggested that, prevailingly, complexes were oriented vertically in the films. The LB film transfer efficiency followed the order: C_{60}py-Zn(TPPS)<C_{60}py-Zn(TMPyP)<C_{60}im-Zn(TPPS)<C_{60}im-Zn(TMPyP), in accord with the respective complex stability constants in solution.

ACKNOWLEDGMENTS

The authors are thankful to the ACS-PRF and NATO for support of this work.

REFERENCES

1. Guldi, D. M. *Chem. Commun.* 321 (2000).
2. Gust, D., Moore, T. A., Moore, L. A. *Acc. Chem. Res.* **26**, 198 (1993).
3. D'Souza, F., Gadde, S., Zandler, M. E., Arkady, K., El-Khouly, M. E., Fujitsuka, M., Ito, O. *J. Phys. Chem.* A **106**, 12393 (2002) and references cited therein.
4. El-Khouly, M. E., Rogers, L. M., Zandler, M. E., Gadde, S., Fujisuka, M., Ito, O., D'Souza, F. *ChemPhysChem* (2003) in press and references cited therein.
5. Maggini, M., Scorrano, G., Prato, M., *J. Am. Chem. Soc.* **115**, 9798 (1993).
6. Noworyta, K., Kutner, W., Deviprasad, G. R., D'Souza, F. *Synth. Metals* **130**, 221 (2002).

Intra-cage Dynamics in Endohedral Fullerenes

K. Vietze* and G. Seifert*

*Institut für Physikalische Chemie, Technische Universität Dresden, D-01062 Dresden

Abstract. Endohedral fullerenes in general and cluster/molecule encapsulated endohedrals in particular exhibit a very complex dynamic behaviour, that has necessarily to be considered when investigating or characterizing these systems by spectrosopic measurements. Molecular dynamics (MD) simulations along with the calculation of dynamic spectra provide a detailed insight into the intra-cage dynamics of these systems and help to clarify its impact on spectroscopic investigations. The effects are shown exemplarily for $Sc_3N@C_{80}$, $Sc_3N@C_{78}$ and $Sc_2C_2@C_{84}$.

INTRODUCTION

The characterization of endohedral fullerenes by spectroscopy is a complicated task, especially for the determination of the cage structure (isomery), the internal structure of the encapsulates and their coupling to the cage. This determination vitally relies on symmetry properties of the system, which are derived from static structure models. But endohedral fullerenes are known for their unique, complex interiour dynamics that lies beyond the harmonic behaviour of standard molecules, in that it allows for very large amplitudes and hence for pronounced anharmonicities, as well as structural fluctuations towards fluxionality. Of course this affects the symmetry properties, and static structure models are insufficient for the description of these systems and consequently for the interpretation of spectroscopic data. So it is essential to utilize dynamic models that take these effects into account.

Computational

The MD calculations have been performed by means of a Density Functional based non-orthogonal Tight Binding scheme (DFTB)[1, 2]. In contrast to the common TB approach all matrix elements are calculated from first principles. The ground state geometries of the endohedrals were globally optimized, including the search for the most stable isomers w.r.t. the cages. The systems were heated and equilibrated over about 30ps in a microcanoical ensemble at different temperatures. (Sc_3N: 575K, $Sc_3N@C_{78}$: 198K/292K/576K, $Sc_2C_2@C_{84}$: 216K/281K/590K). The lengths of the final equilibrium trajectories vary between 70ps and 140ps (up to 1.5ns for $Sc_3N@C_{78}$ at 590 K and $Sc_2C_2@C_{84}$ at 216K). IR spectra and vibrational densities of states were calcuated from the trajectories by means of the Fourier transforms of the dipole moment auto-correlation functions (acf) and the velocity acf's, respectively.

CP685, *Molecular Nanostructures: XVII Int'l. Winterschool/Euroconference on Electronic Properties of Novel Materials,* edited by H. Kuzmany, J. Fink, M. Mehring, and S. Roth
© 2003 American Institute of Physics 0-7354-0154-3/03/$20.00

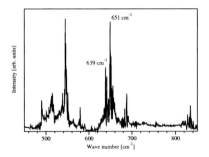 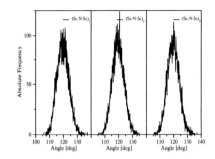

FIGURE 1. $Sc_3N@C_{78}$: Mid-infrared spectrum as calculated from the MD trajectories (left) showing the splitting of Sc_3N-ν_{as} by 12cm^{-1}(experimentally observed: 7cm^{-1}(622+629) [3]), and the distributions of the three Sc-N-Sc angles (right).

RESULTS AND DISCUSSION

A prototype for the importance of the intra-cage dynamics is $Sc_3N@C_{80}$[4]. It has a symmetry of C_{3v} as found by theory and IR/Raman spectroscopy[5], but shows a signature for an icosahedral system when probed with ^{13}C-NMR[6]. The Sc atoms are bonded to pentagons of the carbon cage, but tend to take a position off the center of the ring, closest to one of the five carbon atoms. But due to the icosahedral symmetry of the carbon cage and its spherical shape there is a manifold of energetically degenerate configurations that are separated by rather low barriers. The Sc-C interaction is roughly ten times weaker than the Sc-N interaction, enabling the stiff Sc_3N disc to tumble inside the hollow sphere of the cage. This tumbling consists of a circling motion of the scandiums around the centers (along the carbon atoms) of the respective pentagon within a typical time scale of 4.21×10^{-13}s, and a tilt of the Sc_3N disc where one Sc atom at a time hops from one pentagon to another (typically whitin 1.53×10^{-12}s), lowering the activation barrier for a "rotation" down to 144meV [4]. This, in turn, leads to a fluxional situation, and slow spectroscopic methods like NMR will detect only an average of the scandium positions and the accompanying local fluctuations on the carbon cage. Hence, the icosahedral symmetry is found for the cage although the "real" symmetry of the system is lower (C_{3v}, which is also found with faster spectroscopic methods like IR/Raman, that detect snapshots of the dynamics).

In case of $Sc_3N@C_{78}$, where the encapsulate is the same as in $Sc_3N@C_{80}$, and where the carbon host cage is of almost the same size, the situation is different. The three scandium atoms are now locked to the three pyracelene units in the horizontal mirror plane of C_{78}(5), forming a D_{3h} endohedral fullerene from two D_{3h} subspecies. However, a lower symmetry would be expected from IR/Raman measurements (with a splitting of the anti-symmetric valence vibration of Sc_3N by 7 cm^{-1}(622+629) [3]) and from X-ray diffraction (Sc_3N bond angles of 130°/114°/116°[7]). There are two reasons for these apparent symmetry reductions: One is the slow, large amplitude in-plane rotational mode (R_z) of Sc_3N at 64 cm^{-1}which couples to the Sc_3N in-plane deformation modes (δ) at 198 cm^{-1}, in fact leading to deviations of the Sc-N-Sc bond angles by up to 10°. But the average is 120°, which is also found in ^{13}C-NMR (as D_{3h} symmetry) [7].

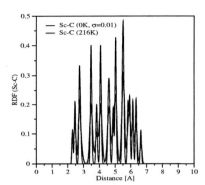

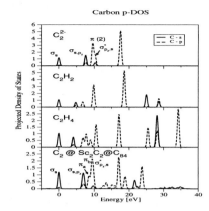

Carbon p-DOS

FIGURE 2. Radial distribution function for Sc – C (cage) in $Sc_2C_2@C_{84}$ (left), and partial carbon DOS for C_2 (right). The C_2 p-DOS is given for the free anion, acetylene, ethylene and C_2 in $Sc_2C_2@C_{84}$ (top to bottom). The resemblence between the latter and a free C_2^{2-} is obvious, the cancellation of the π degeneracy is due to the acetylid bond.

The second reason is a strong dynamic coupling between the internal modes (mainly v_{as}) and the low frequency external modes R_{xy} and T_z, which shift and tilt the Sc_3N disc out of the horizontal mirror plane with amplitudes of the order of 0.5Å at 590K. Due to additional strain on Sc_3N induced by this shifting and tilting the Sc_3N valence vibrations are distorted. Or, in other words, the D_{3h} symmetry is distorted locally in time. The main consequence is a splitting of the intrinsically two-fold degenerated anti-symmetric valence vibration of Sc_3N (fig. 1), which could be (erroneously) interpreted as a reduced symmetry lower than D_{3h} for the system itself. The same holds for the spectroscopic behaviour in X-ray diffraction.

A completely different situation is found in $Sc_2C_2@C_{84}$. The scandium atoms are trapped in two diametrically opposed pyracelene pockets, forming (together with the carbon cage) a very ridig host for a highly mobile central C_2 (fig. 2). In contrast to the tightly bound Sc_3N the Sc_2C_2 forms an acetylid complex (fig. 2, right), that shows a strong interaction along the Sc-Sc axis ($\equiv$ main C_2 axis of C_{84}) via a Sc(d)-$C_2(\pi)$ bond, whose strength is of the same size as that of the Sc_2-cage interaction. In the perpendicular directions the interaction is much weaker, allowing for a high flexibility of the C_2 anion in the central plane. Consequently, there is a strong mixing of intra-molecular with the external vibrational modes (for instance, Sc_2C_2 T_z with v_s and v_{as}, where Sc_2C_2-v_{as} is an almost pure C_2-T_z), and there are 2×2 almost atomic Sc lateral cage modes. Obviously a proper distiction between intra- and inter-molecular modes in terms of a symmetry analysis based description is not longer possible. Moreover, the underlying harmonic approximation itself is insufficient for the description of the system (fig. 3). A better characterization of the system would be $C_2@Sc_2C_{84}$, where there is a highly mobile central C_2 anion, and two laterally independent Sc atoms locked to the carbon cage.

9

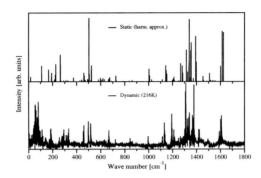

FIGURE 3. Comparison of $Sc_2C_2@C_{84}$ IR spectra: static (T=0, harmonic approximation, upper) vs dynamic (lower). The deviations esp. in the FIR region (C_2 modes, Sc lateral cage modes) are evident.

SUMMARY

The results found in all of our MD calculations (also, e.g., $Eu@C_{74}$[8]) can be generalized as follows. The specific dynamics and its effects on the spectrosopic behaviour strongly depend on the encapsulate or rather on the balance between the intra- and intermolecular interactions within the encapsulated entity and with the host cage, respectively. But even very similar systems can exhibit different dynamics, depending on the respective specific nature of the bonding to the cage (ring vs bond, topology of the bonding site, local curvature, etc), and also the effects on spectroscopic features are different.

In addition, the impact on spectroscopic investigations generally depends on the respective methods (apparent symmetry reductions due to snapshots for fast methods (IR/Raman, X-ray), and apparent symmetry increases due to averaging for slow methods (NMR)), so symmetry assignments, for instance, have always to refer to the spectroscopic method used in particular. Due to the size of vibrational amplitudes found in all endohedrals and the tendencies towards fluxionality it is not longer feasible to map spectroscopic data onto static structure models. Likewise, approaches or models involving or basing on harmonic approximations are beyond their scope of application in many cases.

REFERENCES

1. Seifert, G., Eschrig, H., and Bieger, W., *Z. phys. Chemie*, **267**, 529 (1986).
2. Porezag, D. *et. al.*, *Phys. Rev. B*, **51**, 12947 (1995).
3. Krause, M., private comm.
4. Vietze, K., and Seifert, G., in *Electronic Properties of Novel Materials. AIP conference proceedings 633*, Eds. H. Kuzmany, J. Fink, M. Mehring, S. Roth, AIP, New York, 2002, pp. 39–42.
5. Krause, M., Pichler, T., Kuzmany, H., Georgi, P., Dunsch, L., Vietze, K., and Seifert, G., *J. Chem. Phys.*, **115**, 6596–6605 (2001).
6. Stevenson, S. *et. al.*, *Nature*, **401**, 55–57 (1999).
7. Olmstead, M. *et. al.*, *j-AC*, **113**, 1263–1265 (2001).
8. Vietze, K., Seifert, G., and Fowler, P., in *Electronic Properties of Novel Materiels. AIP conference proceedings 544*, Eds. H. Kuzmany, J. Fink, M. Mehring, S. Roth, AIP, New York, 2000, pp. 131–134.

Scanning probe microscopy and spectroscopy of C$_{60}$ nanorods

M. Mannsberger[1], A. Kukovecz[1,2], V. Georgakilas[3], J. Rechthaler[4], G. Allmeier[4], M. Prato[3], H. Kuzmany[1]

[1] Universität Wien, Institut für Materialphysik, Strudlhofgasse 4, A-1090 Vienna, Austria
[2] Dept. of Appl. and Environ. Chemisty, University of Szeged, Rerrich B. ter 1, H-6720 Szeged, Hungary
[3] Dipart. di Scienze Farmaceutiche, University of Trieste, Piazzale Europa 1., 34127 Trieste, Italy;
[4] Universität Wien, Institut für Analytische Chemie, Strudlhofgasse 4, A-1090 Vienna, Austria

Abstract. We present an AFM/STM analysis of an ionic C$_{60}$ derivative, complemented by Raman measurements, light microscopy and mass spectrometry. It was previously demonstrated that this molecule self-organizes into rod-like structures. Our SPM investigations confirmed the formation of rods and revealed in detail a chiral fashion of the latter. The chirality is suggested to originate from the angle between the apolar C$_{60}$ sphere and it's charged functional group. On the micrometer scale superstructures could be observed which apparently originate from a dendritic growth process. The superstructures withstand heat treatment at 200 °C, but the functional groups are lost, and the remaining C$_{60}$ undergoes a three dimensional polymerization.

INTRODUCTION

Self-organization is the key to transfer the promising results of nanotechnology into manufacturable real-world products. Fullerenes and their derivatives are expected to perform well in this field due to their ability to form nanophases of controlled shape and size [1, 2]. Our goal in this communication is to present a comprehensive probe microscopy and vibrational spectroscopy characterization report on fullerene based self-assembled entities formed from the ionic C$_{60}$ derivative 8-(N-methyl-fullero-pyrrolidinum-1-yl-iodide)-3,6-dioxaoctan-1-ammonium-choride (MFPDAC) (see Figure 1 for formula and model). The self-organizing capability of this molecule has been recently demonstrated by TEM observations [3]. In aqueous solution the MFPDAC molecules form rodlike objects which are tens of nanometers in diameter and up to several micrometers long. We refer to them as "nanorods" from now on.

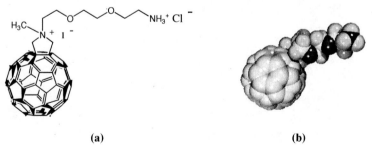

(a)   (b)

FIGURE 1. Chemical structure (a) and space filling optimized geometry model (b) of the nanorod precursor molecule MFPDAC.

CP685, *Molecular Nanostructures: XVII Int'l. Winterschool/Euroconference on Electronic Properties of Novel Materials,* edited by H. Kuzmany, J. Fink, M. Mehring, and S. Roth
© 2003 American Institute of Physics 0-7354-0154-3/03/$20.00

EXPERIMENTAL

Samples were prepared by the repeated sonication of an aqueous solution of MFPDAC as detailed previously [3]. Several droplets of the resulting reddish brown solution were used to coat either a highly ordered pyrolitic graphite plane, a gold plane or a silicon plane kept at 313 K. Finally, the sample was put into a vacuum better than 10^{-4} Pa at room temperature (RT) overnight to remove leftover solvent and volatile contaminants. Heat treatment was also done in vacuum for 12 hours. All AFM measurements were performed at RT in air in tapping mode on a TopoMetrix Explorer instrument using high resonance frequency tips from Veeco Instruments GmbH.

Raman spectra were recorded on the same sample which was prepared by drop-coating a gold mirror. Raman spectra were measured in Stokes mode at 488 nm laser excitation at 80 K in 180 ° backscattering geometry on a triple grating Dilor xy spectrometer equipped with a back thinned CCD detector.

Positive ion laser desorption ionization (LDI) mass spectra were acquired on the high resolution curved field reflectron time-of-flight (TOF) mass spectrometer Axima CFR (Shimadzu Biotech-Kratos Analytical, Manchester, UK) applying delay extraction (delay time: 152 ns and 216 ns). The instrument was equipped with a nitrogen laser (337 nm, 3 ns pulse width, maximum pulse rate: 10 Hz). The acceleration voltage was set to 20 kV and the reflectron voltage to 24 kV.

RESULTS

Clear evidences of MFPDAC self-organization were found by light microscopy and AFM. The molecule self-assembles into rodlike objects which are 10-30 nm wide and up to a few microns long (Figure 2a). Increasing the MFPDAC concentration in the solution results in complex superstructures resembling dendritic growth (Figure 2b). Zooming in on the AFM response of the nanorods reveals more details of the topology.

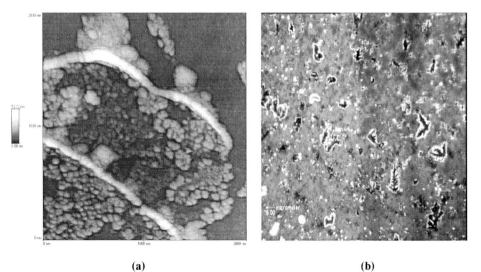

(a)   (b)

FIGURE 2. AFM image showing a typical self-assembled nanorod (a) and a light microscope image (b) where a complex dendritic superstucture is visible.

The rods appear segmented with a worm-like structure. This structure is assumed to be a result of a chiral self-assembly process due to the angle between the apolar C_{60} sphere and it's charged functional group.

The observable Raman spectrum of the nanorods is determined by the individual MFPDAC molecules since the region of intermolecular interactions is too close to the laserline and therefore, it can not be detected. In Figure 3. we present the characteristic low frequency window of the spectrum.

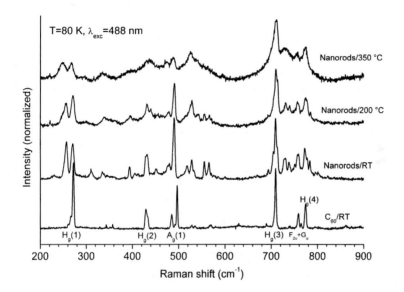

FIGURE 3. Raman spectrum of a C_{60}, as-synthesized nanorods and heat-treated nanorods.

The original I_h symmetry of C_{60} changes to C_1 upon functionalization. This results in the splitting of Raman peaks belonging to the degenerate H representation. The effect is particularly well observable on the $H_g(1)$, $H_g(3)$ and $H_g(4)$ modes in the "Nanorods/RT" spectrum.

A series of heating experiments was performed with the aim to remove the functional groups and thus getting back the original fullerene spectrum. The chances to obtain such results was indeed promising since mass spectrometry clearly evidences that all sidechains are detached at 200 °C [4]. In spite of this it was surprising to see that the Raman spectra "Nanorods/200 °C" and "Nanorods/350 °C" exhibit almost the same features as those of the unheated nanorods. The high symmetry C_{60} spectrum is not retained after heating. This observation indicates that another symmetry breaking process becomes active. This process can be a three-dimensional polymerization of the C_{60} cores which is initiated by the heating of the sample. This reaction is thought to take place simultaneously with the removal of the functional groups. Strong evidence supporting this hypothesis is supplied by AFM observation on the heated sample shown in Figure 4a. The image indicates that the self-assembled nanorods are not destroyed upon heating, but rather retain their rodlike structure. This suggests that the

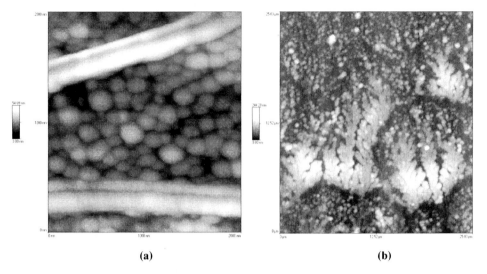

(a)   **(b)**

FIGURE 4. AFM images showing nanorods (a) and dendrites (b) after heating to 200 °C overnight.

fullerenes are interconnected by covalent bonds, which in turn explains the C_1 symmetry of the heated samples. Similarly to Figure 4a also the dendritic superstructures are retained after heating to 200 °C as demonstrated in Figure 4b.

CONCLUSIONS

We presented AFM evidence for the self-assembling nature of the ionic fullerene derivative MFPDAC. The molecule loses its sidegroup and simultaneously undergoes polymerization upon heating which stabilizes the rodlike structure with covalent bonds. We believe that this scheme could serve as a model for preparing stable self-assembled nanoarchitectures.

ACKNOWLEDGMENTS

Financial support from the EU RTN FUNCARS (HPRN-CT-1999-00011) and the Hungarian OTKA F038249 grant is acknowledged.

REFERENCES

1. S. Zhou et al. *Science* **269** (2001) 1944.
2. A.M. Cassell et al. *Angew. Chem.* **111** (1999) 2565.
3. V. Georgakilas et al. *Proc. Nat. Acad. Sci.* **99** (2002) 5075.
4. M. Mannsberger et al. in preparation

Structure and Bonding of polymeric, anionic fullerides:
The case of $[Sr(NH_3)_8]_3(C_{70})_2 \cdot nNH_3$ (n = 20-22)

Martin Panthöfer, Holger Brumm, Ulrich Wedig, and Martin Jansen

Max Planck Institute for Solid State Research, Heisenbergstraße 1, D-70569 Stuttgart, Germany

Abstract. Reduction of C_{70} with strontium dissolved in liquid ammonia results in metal fulleride solvates $[Sr(NH_3)_8]_3(C_{70})_2 \cdot nNH_3$ (n = 20-22) containing linear polymeric, anionic chains $_\infty^1[C_{70}^{3-}]$. The compound was characterised by single crystal structure determination. The accurate crystal structure, determined at atomic resolution, allowed a comparison with results of quantum chemical calculations. The detailed analysis of the bonding character reveals the preservation of the aromatic character of the phenylene-type belt at the equator of C_{70} to be the driving force of this unexpected type of polymerisation.

INTRODUCTION

After the discovery of the first polymeric fullerene phases AC_{60} (A = alkali) [1], a large number of compounds containing covalently linked fullerenes has been synthesised especially via photochemical [2] and pressure induced reactions [3]. A detailed chemical interpretation of the large variety of experimental results obtained to date is restricted by the lack of reliable structural information. Until now a complete determination of the structures of polymeric fulleride species, based on the refinement of all atomic positions without applying any geometrical constraints are only reported for the compounds $AC_{70} \cdot nNH_3$ (A = Sr, Ba), which contain linear polymeric, anionic chains $_\infty^1[C_{70}^{2-}]$ [4,5]

EXPERIMENTAL AND RESULTS

Reducing C_{70} with strontium in liquid ammonia and crystallisation in sealed tubes at autogenous pressure and ambient temperature leads to the compound $[Sr(NH_3)_8]_3(C_{70})_2 \cdot nNH_3$ (n = 20-22) (**1**). The metal to fullerene ratio of 3:2 fits either to the presence of an open shell species C_{70}^{3-} or to the simultaneous presence of C_{70}^{2-} and C_{70}^{4-}. The results presented here point to the existence of the former. (**1**) was

CP685, *Molecular Nanostructures: XVII Int'l. Winterschool/Euroconference on Electronic Properties of Novel Materials,* edited by H. Kuzmany, J. Fink, M. Mehring, and S. Roth
© 2003 American Institute of Physics 0-7354-0154-3/03/$20.00

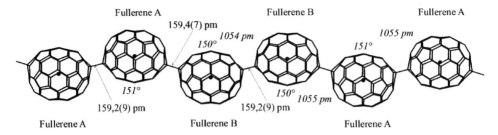

Figure 1. Molecular structure of the polymeric $_\infty^1[C_{70}{}^{3-}]$ anion. The figure includes distances, and angles between centres of gravity, of the C_{70} fullerene units.

characterised by means of single-crystal X-ray structure analysis and quantum chemical methods [6]. The crystal structure consists of linear polymeric $_\infty^1[C_{70}]$ chains (see fig. 1) arranged as a distorted, primitive packing of rods (see fig. 2).

In contrast to high-pressure fullerene polymers and simple binary fullerides, the C_{70} fulleride-ions within these polymer chains are not linked via a two-bond cyclobutane like arrangement, but by solely one carbon-carbon single bond located at the pentagon of each C_{70} cap, close to the former fivefold axis of the neutral monomer. These linear polymeric chains $_\infty^1[C_{70}]$ realise two different, diastereomeric schemes (A and B, see fig. 3) of intercage connection. The arrangement of the two diastereomeric cages follows a syndiotactic scheme AABB. Both cages exhibit almost identical bond lengths, pointing to identical charge states, i.e. trianions $_\infty^1[C_{70}{}^{3-}]$. The lengths of the bridging bonds are 159.2(9) and 159.4(7) pm.

Figure 2. Crystal structure of **1** viewed along the $_\infty^1[C_{70}{}^{3-}]$ chains.

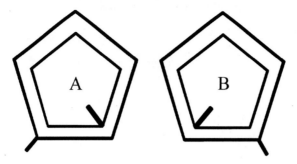

Figure 3. Intercage linkage pattern of the two diastereomeric $[C_{70}^{3-}]$ units with the polymer chains.

Due to the one-dimensional linkage the point group symmetry of the C_{70} unit is reduced from D_{5h} to C_2 (positional symmetry: C_1). Within the limits of error the variation of bond lengths does not break the point group symmetry C_2.

COMPARISON OF $^1_\infty[C_{70}^{3-}]$, $^1_\infty[C_{70}^{2-}]$ AND THE MODEL SYSTEM $[C_{70}(CH_3)_2]^{3-}$

Quantum chemical calculations on model systems $[C_{70}(CH_3)_2]^{n-}$ (n = 2,3) give insight into the bonding properties of the polymeric, anionic fullerides. Already the Hartree-Fock split-valence (SV) basis set level proved to be a good compromise between accuracy and computational expense for the dianion [4] as well as for the trianion. In both cases the structure optimisations in point group C_1 (no structural restrictions) do not deviate more than 2 pm from the corresponding experimental values in the polymers. Further, the optimised structure of the still unknown $^1_\infty[C_{70}^{4-}]$ does not fit to the experimental structure. This is another hint, that the polymer chains in $[Sr(NH_3)_8]_3(C_{70})_2 \cdot nNH_3$ (n = 20-22) consist of trianions $^1_\infty[C_{70}^{3-}]$.

The structural changes due to reduction and polymerisation of C_{70} are significant on the basis of the theoretical as well as the experimental results. In case of the $^1_\infty[C_{70}^{3-}]$-polymer, the bonds to the bridging carbon atoms (connected to an adjacent C_{70} unit) are elongated strongly (> 152 pm), these atoms are of sp^3-type, bonded to four other carbon atoms. Further distortions of the C_{70}-cage are rather localised. They are particularly low in the phenylene-type ring at the equator of the C_{70} cage.

Due to threefold negative charge the experimental bond lengths in $^1_\infty[C_{70}^{3-}]$ are lengthened only by an average value of $\Delta = 0.4(9)$ pm compared to $^1_\infty[C_{70}^{2-}]$. Nevertheless, some local trends are observed for the experimental as well as for the quantum chemical data. Short bonds in the caps of the cage and bonds in the phenylene-type ring are slightly enlarged. Bonds in the five membered rings perpendicular to the equator of the-

cage tend to be larger. This is in agreement with the shape of the HOMO.

The ELF of $[C_{70}(CH_3)_2]^{2-}$ and its topological analysis has been discussed in detail elsewhere [4]. The features of the ELF of the trianion are virtually the same, showing the various bonding types in the carbon cage: single bonds at the bridging atoms with integrated electron numbers of less than 2,03 and a monosynaptic attractor at the neighbouring atom, conjugated bonds in the caps and an aromatic character in the phenylene-type ring.

CONCLUSIONS

Crystal structure analysis reveal $[Sr(NH_3)_8]_3(C_{70})_2 \cdot nNH_3$ (n = 20-22) to consist of polymeric anionic chains $^1_\infty[C_{70}^{3-}]$. Quantum chemical calculations show a hypothetical **monomeric** anion to exhibit its strongest distortions in the phenylene-type belt at the molecule equator [4], this molecular region maintains its structural and bonding character in the **polymeric** anion.

Concentration of all structural distortions into the molecule caps enables the preservation of the aromatic character of the phenylene-type belt at the molecule equator. This is the driving force for the structural response of C_{70} upon reduction, i.e. the formation of polymeric, anionic chains.

REFERENCES

1. Stephens, C. A. *et al.*, Nature **370**, 636 (1994).
2. Rao, A., Zhou, A. and Hanger, G., *Science* **259**, 955 (1993).
3. Iwasa, Y., and Arima, T., *Science* **264**, 1570 (1994).
4. Wedig, U., Brumm, H., and Jansen., M., *Chemistry-A European Journal* **8**, 2769 (2002).
5. Brumm, H., Peters E., and Jansen., M., *Angewandte Chemie Int. Ed.* **40**, 2069 (2001).
6. Panthöfer M., Brumm, H., Wedig, U., and Jansen., M., *Solid State Sciences,* in preparation.

Solid State ^{13}C and ^{1}H NMR Investigations on $C_{60} \cdot 2$ ferrocene

J. Rozen*, R. Céolin†, J. L. Tamarit**, H. Szwarc‡ and F. Masin*

*Matière Condensée et Résonance Magnétique, Faculté des Sciences, Université libre de Bruxelles, Boulevard du Triomphe, CP 232, 1050 Brussels, Belgium
†Faculté de Pharmacie, Laboratoire de Chimie Physique, 4 avenue de l'Observatoire, 75270 Paris Cedex 06, France
**Departament de Fisica i Enginyeria Nuclear, ETSEIB, Universitat Politècnica de Catalunya, Diagonal 647, 08028 Barcelona, Catalonia, Spain
‡Laboratoire de Chimie Physique, UMR 8000, Université Paris Sud, CNRS, bâtiment 490, 91405 Orsay Cedex, France

Abstract. Previous X−Ray experiments[1] have revealed the structure of the crystal thanks to the apparent immobility of the molecules. NMR studies demonstrate that the C_{60} molecules and the ferrocene molecules are in fact moving fast enough to average the magnetic interactions. This is shown by the temperature evolution of the resonance curve and by the shape of the CP−MAS spectrum at room temperature. Further investigations allow to assert that the C_{60} molecules are undergoing $3D$ rotations and that Cp cycles of ferrocene are subject to planar rotations. Finally, analysis of the spectra at low MAS frequency leads to the conclusion that there are two kinds of Cp cycles.

INTRODUCTION

$C_{60} \cdot 2$ ferrocene is one of the only crystals containing fullerene molecules whose structure has been fully identified by X−Ray diffraction. This is due to their apparent immobility. At room temperature, the unit cell is triclinic and space group is $P\bar{1}$. Each molecule of ferrocene has a Cp cycle parallel to a pentagonal face of the neighboring C_{60}. The inter−plane distance is 3.3 Å, which is typical when π−interactions occur between aromatic molecules. In a molecule of ferrocene, the two cycles are almost parallel, the $Cp − Fe − Cp$ angle is 177.8°, and they are slipped sideways by 0.8 Å.

Previous NMR experiments[2] have concluded that the frequencies observed in the solvate are similar to those of the pure crystals: 143.9 ppm for the molecules of C_{60} and 69.5 ppm for those of ferrocene. This excludes any charge transfer between the two kinds of molecules and any conductive property.

In this study, the absence of these electronic correlations between the molecules is confirmed and new features about molecular motions and magnetic properties is reported.

CP685, Molecular Nanostructures: XVII Int'l. Winterschool/Euroconference on
Electronic Properties of Novel Materials, edited by H. Kuzmany, J. Fink, M. Mehring, and S. Roth

EXPERIMENTAL

The NMR spectrometer used is a *Bruker MSL* 300. The magnet generates a field of $7.05\,T$.

Two series of experiences are reported. The first is done at room temperature on ^1H using CP−MAS and decoupling in the probe *HP WB 73A MAS* 4. The 90°pulse has a length of $5\,\mu s$ and the contact time is $5\,ms$. The second is done on ^{13}C between $54\,K$ and $294\,K$ in the probe *HP LP* 50, the temperature being regulated by a *Oxford ITC4*.

The MAS spectra obtained at room temperature have been treated with the Bruker *WIN − MAS* software using the method of Herzfeld and Berger[3] in order to calculate the chemical shift tensor.

RESULTS AND DISCUSSION

Fullerene Dynamics

A MAS frequency of $184\,Hz$ has been used at room temperature to obtain the ^{13}C spectrum of fullerene. The central peak is observed at $143.7\,ppm$, cf. FIGURE 1. This is slightly inferior to the frequency observed for the pure crystal at room temperature ($143.9\,ppm$). This shift is very small and confirms that no charge transfer occurs in the solvate.

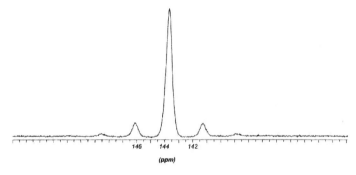

FIGURE 1. ^{13}C CP−MAS spectrum of fullerene molecules in $C_{60} \cdot 2\,ferrocene$ at room temperature under a MAS frequency of $184\,Hz$.

The components of the chemical shift tensor, reported in TABLE 1, are almost identical ($\delta \ll d$). This leads to the conclusion that, at room temperature, the molecules of C_{60} are undergoing fast isotropic rotations. Indeed, the correlation time has to be much smaller than the length of the signal ($FID \simeq 0.03\,s$) in order to average the magnetic interactions.

This result seems to contradict X−Ray experiments which let to think there is no mobility in the crystal. NMR is sensitive to individual ^{13}C on a fullerene molecule whereas X−Rays observe all the carbon atoms without distinction. Nuclear Magnetic Resonance is therefore affected by discrete rotations leading to symmetry−equivalent orientations while X−Rays are not as only an average orientation is observed.

TABLE 1. Components of the chemical shift tensors at room temperature in the solvate. Cp 1 is noted for the left set of C_5H_5 peaks and Cp 2 for the right set. (All values in *ppm*)

	d_1	d_2	d_3	$\bar{d}$	δ	η
C_{60}	144 ± 0.8	145.9 ± 0.7	141.2 ± 0.7	143.7	2.5 ± 1.4	0.76 ± 0.59
Cp 1	96	96.3	23.8	72	48.2	0
Cp 2	97.2	97.2	17.3	70.6	53.3	0

Ferrocene Dynamics and Magnetic Organization

The ferrocene ^{13}C spectrum observed at a MAS frequency of $1196\,Hz$ is presented on FIGURE 2 b. It is unexpected since it contains information that had not been reported previously (cf. below); there are two sets of peaks. Once again the deviation of the central frequency is very small ($69.5\,ppm$ in a ferrocene crystal). This information is in tune with the equivalent finding for fullerene molecules which excludes charge transfer. As seen in the lower part of TABLE 1, the components of the chemical shift tensors can be separated in $d_{\parallel}$ and $d_{\perp}$ ($\eta \simeq 0\,ppm$), which means cycles are undergoing fast uniaxial isotropic rotations compared with the observation time ($FID \simeq 0.035\,s$). This is confirmed by the evolution of the ^1H peak width between $54\,K$ and room temperature reported on the left part of FIGURE 3. The second order momentum, M_2, is reduced upon heating which is typical when molecular movements occur.

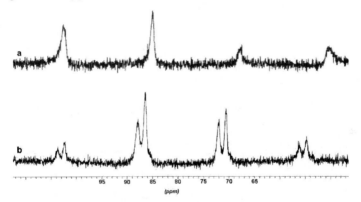

FIGURE 2. Ferrocene ^{13}C CP–MAS spectra at room temperature (a) in the pure crystal at a MAS frequency of $1206\,Hz$, (b) in the $C_{60} \cdot 2\,ferrocene$ solvate at a MAS frequency of $1196\,Hz$.

That two sets of peaks separated by less than $1.5\,ppm$ are present is particular of the $C_{60} \cdot 2\,ferrocene$ solvate. It has been checked that it is not the case in a ferrocene crystal under similar conditions, cf. FIGURE 2 a. In order to get a qualitative idea of the magnetic organization, the areas which are proportional to the number of concerned atoms have been compared. This reveals that 50% of the carbon atoms present in ferrocene contribute to each set. The unit cell contains two molecules of ferrocene and therefore, four cristallographically inequivalent C_5H_5 cycles. It can be concluded that there are two kinds of C_5H_5 cycles magnetically inequivalent in the solvate.

Shannon et al.[4] have shown that a crystal of ferrocene has a particular behaviour when the CP−MAS technique is used. For a given decoupling field, it seems that the width of the peak, Δ, increases when the MAS frequency is risen. From further experiments on deutered ferrocene, it is concluded that MAS frequencies have negative consequences on the spin−spin decoupling. The right part of FIGURE 3 shows that this effect is observed in the $C_{60} \cdot 2\,ferrocene$ solvate where the two peaks are superposed as the MAS frequency is increased. The first NMR spectrum of the solvate was obtained at $4000\,Hz$ under a smaller decoupling field, which could explain why the two peaks were not then distinguished.

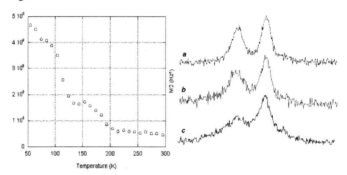

FIGURE 3. ≀ *On the left* ≀ Variation under temperature change of the ^1H peak second order momentum. ≀ *On the right* ≀ ^{13}C CP−MAS spectra at room temperature of ferrocene in the solvate. (a) $v_{MAS} = 1196\,Hz$, (b) $v_{MAS} = 2000\,Hz$, (c) $v_{MAS} = 4001\,Hz$.

CONCLUSION

Comparing the spectra with those of pure crystals, this work has confirmed the absence of conductivity properties in the solvate. Secondly, it has been shown for the first time that, in the solvate, there are two magnetically inequivalent kinds of Cp cycles in ferrocene since two sets of peaks show up in the CP−MAS spectrum. But the main contribution of the experiments is the observation of fast reorientational motions within the solvate lattice, motions that X−Rays had been unable to detect.

ACKNOWLEDGMENTS

Acknowledgements go to W. Stone for reading the manuscript and to P. Pirotte for his useful technical help and support.
This work was supported by the BNB (Banque Nationale de Belgique).

REFERENCES

1. J. D. Crane, P. B. Hitchcock, H. W. Kroto and D. R. M. Walton, *J. Chem. Soc.*, 1764 (1992).
2. E. Shabanova, K. Schaumburg and F. S. Kamounah, *Can. J. Anal. Sci. Spectrosc.*, **43** (2): 53 (1998).
3. J. Herzfeld and A. E. Berger, *J. Chem. Phys.*, **73** (12): 6021 (1980).
4. I. J. Shannon, K. D. M. Harris and S. Arumugam, *Chem. Phys. Lett.*, **6**: 588 (1992).

Comparative study of hydrofullerides $C_{60}H_x$ synthesized by direct and catalytic hydrogenation

A.V.Talyzin, A.Jacob

IMRA-Europe S.A., B.P.213, 06904 Sophia Antipolis CEDEX, France

Abstract. Hydrofullerides with hydrogen content up to 5 Wt.% were obtained by direct and catalytic reaction with H_2 gas. Hydrogen content was monitored *in situ* using gravimetric system and verified by chemical analysis *ex situ*. It was found that pure C_{60} reacts rapidly when exposed to H_2 gas at 673 K and 50-100 Bar. Gravimetric study of this reaction showed that hydrogenation is saturated at about 5 at.% of hydrogen. Mass of the sample goes through a maximum and with a longer reaction time its weight start to decrease. This proves that hydrofullerides with high hydrogen content are not stable and strong hydrogenation results in collapse of C_{60} molecules. XRD study showed that samples prepared by direct hydrogenation without catalyst retain an original FCC structure with increase of cell parameter up to $a=15.1$Å. Catalytic hydrogenation of C_{60} with H_2 gas results in decrease of the reaction temperature and formation of hydrofullerides with new crystal structure (suggestively sc).

1. INTRODUCTION

It is known that C_{60} can be hydrogenated by different methods. Direct reaction of fullerene with hydrogen gas requires high temperature (~673 K) and high hydrogen pressure of 50-100 Bar. As a result, mixture of different hydrofullerides with average composition of $C_{60}H_x$ ($2 < X < 60$) can be obtained [1-3]. It shall be noted that hydrofullerides with high hydrogen content ($X > 36$) remain to be hypothetical or poorly characterized. Relatively pure hydrofullerides with true composition of $C_{60}H_{36}$ and $C_{60}H_{18}$ have been obtained by several other chemical reactions such as Birch reduction, transfer hydrogenation using 9,10-dihydroanthracene et cet. [4-6]. Characterisation of hydrofullerides has been performed by many different methods, including for example Raman and IR spectroscopies, NMR, XRD, XPS [7-9]. Nevertheless, identification of different hydrofullerides by Raman and IR spectroscopy remain to be difficult due to the big number of possible isomers for each hydrofulleride phase, see for example Ref. [10]. Crystal structures of different hydrofulleride phases have been also studied rather insufficient. It is known, for example, that direct reaction of hydrogen (or deuterium) gas with C_{60} at high temperature conditions results in expansion of original fcc structure of C_{60}. The FCC cell parameter has been reported to depend linearly from the number of hydrogen atoms X in the $C_{60}H_x$. It shall be noted that this dependence was based only on three points in the medium hydrogen content $10 < x < 24$ [7]. Few reports are also available for $C_{60}H_{36}$ which exhibited fcc structure with cell parameter of about $a=15.0$Å [11,12], but no systematic data have been reported in the literature for full range of hydrofullerides with different hydrogen content. It is very interesting that structure of the $C_{60}H_{36}$ has been found BCC when some other synthesis methods were used [13].

CP685, *Molecular Nanostructures: XVII Int'l. Winterschool/Euroconference on Electronic Properties of Novel Materials,* edited by H. Kuzmany, J. Fink, M. Mehring, and S. Roth
© 2003 American Institute of Physics 0-7354-0154-3/03/$20.00

In this work we present systematic study of structural modifications in hydrofullerides obtained by direct and catalytic reaction of C_{60} with hydrogen gas at temperatures 573-700 K.

2. EXPERIMENTAL

Most of the experiments were performed using 99.5% and 99.9% C_{60} purchased from MER corporation. Hydrogenation was studied using gravimetric method using Rubotherm balance modified for high temperature experiments. Accuracy of the weight measurement was ± 0.02 mg. Typically, hydrogenation was performed at 50-120 Bar pressure of hydrogen gas and temperature 573-700K.

After hydrogenation the samples were analysed by powder XRD, Raman and IR spectroscopies. Gravimetric data were used only for qualitative analysis and approximate evaluation of hydrogen uptake. Final composition of the samples after experiments was determined by elemental analysis in the commercial laboratory (MIKRO KEMI AB, Uppsala, Sweden). They used a flush combustion gas chromatography method where the samples are completely oxidized into gaseous CO_2 and H_2O. The gases are analyzed by gas chromatography and their concentration determined by a thermal conductivity detector (TCD). The samples were dried in vacuum prior analyses.

X-ray diffraction powder data were collected with a Siemens 5000 diffractometer using CuKα radiation. Raman spectra were obtained by a Renishaw Raman 2000 spectrometer using a 785 nm excitation wavelength with a resolution of 2 cm^{-1}. Only low laser power was used to avoid degradation of sample and photopolymerisation.

For studies of catalytic hydrogenation Ni supported on Al_2O_3/SiO_2 (60% of Ni), Pd supported on Al_2O_3 were used (10% of Pd). Mixing of catalyst with fullerene powder was achieved by ball milling powders for 30-60 min. XRD and Raman analyses showed that no chemical transformations of C_{60} occur during this soft milling.

3. RESULTS AND DISCUSSION
3.1 Non Catalytic Hydrogenation of C_{60}.

In agreement with previous data it was found that reaction of C_{60} with hydrogen proceeds rapidly at 50-100 Bar H_2 pressure and 670-690K. Gravimetric curve goes through the maximum at approximately 1200 min (for pressure 100 Bar) and after that the weight start to decrease. The samples taken in different points of this curve were studied by XRD and elemental analysis. It was found that maximum of the weight increase correspond to hydrofulleride with about 4.9-5.1Wt% of hydrogen. The samples taken after prolonged hydrogenation (from the points on gravimetric curve where weight decreases) exhibited the same elemental composition but become more and more amorphous. The XRD patterns shown in Fig.1 prove that hydrogenated samples maintain their original FCC structure, but the cell parameter is increased up to $a=15.1$Å. Strong broadening of peaks and appearance of amorphous halo is typical for highly hydrogenated samples. Finally if the hydrogenation proceeds even longer the samples transform into a new phase which is liquid at the temperature of hydrogenation. This liquid started to boil and to overflow from the cell. After cooling this phase is black (red in transmission light under microscope) glass-like material and almost amorphous on XRD (see sample 3 in the Fig.1).

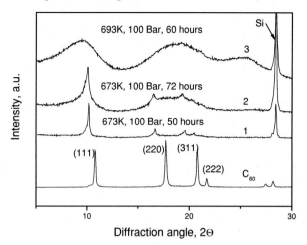

FIGURE 1 XRD patterns recorded from the samples 1,2,3 and compared to the pattern from original C_{60}. Samples 1 and two were obtained in one experiment; sample 3 was obtained in separate experiment. Higher rate of hydrogenation in the Sample 3 is explained by milling of the sample prior hydrogenation and by higher temperature.

Analysis of the gas remained in the reaction chamber showed that decrease of weight during prolonged hydrogenation is due to partial collapse of C_{60} with formation of variety of hydrocarbons (methane, benzene et cet.). The data allow unambiguous interpretation: the maximal hydrogen content in hydrofullerides obtained by direct reaction with hydrogen gas is about 5.1Wt% (correspond to ~ $C_{60}H_{39}$). Hydrofullerides with higher hydrogen content are extremely unstable and induce collapse of C_{60}.

3.2 Catalytic Hydrogenation of C_{60}.

Several transition metal catalysts were tested in order to decrease the temperature of hydrogenation. Pure C_{60} does not react with hydrogen at 573 K even when hydrogen pressure is increased to 200 Bar (although it was reported that the hydrogen released from metal hydride batteries does react even at this temperature). The use of transition metal catalysts allowed to obtain hydrofulleride phases at 573 K. Remarkably, only hydrofulleride phases with hydrogen content below 1-2% exhibited FCC structure. New phase appears as a minor component starting from approximately 2% of hydrogen and with 3.5% (~ $C_{60}H_{24}$) this phase is observed in pure form (See Fig.2 a). The phase was identified as simple cubic with cell parameter a=14.9-15.1 Å (suggestively Pa3 space group). When the hydrogenation of C_{60}/ catalyst mixture was performed at 673 K, the result was again different compared to non catalytic hydrogenation. As it is shown in Fig.2 b, hydrogenation promoted by Ni catalyst resulted in formation of mostly BCC phase with minor FCC part. Body centered cubic phase was reported before in literature for $C_{60}D_{36}$ obtained by catalytic hydrogenation in the presence of iodine and by high pressure hydrogenation [13,14].

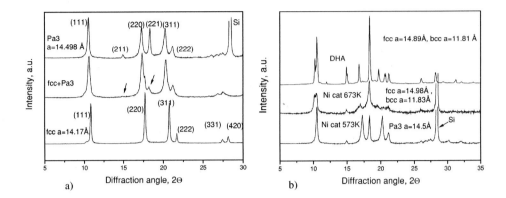

FIGURE 2 a) XRD patterns of the C_{60} (bottom), new Pa3 hydrofullerite phase obtained using Ni catalyst at 573K (top) and patterns of the sample hydrogenated at the same conditions with small amount of catalyst which consist of both fcc and Pa3 phases. Additional peaks due to Pa3 phase are marked with arrows.
b) XRD patterns recorded from samples obtained by transfer hydrogenation reaction (top) and by catalytic reaction with hydrogen gas

It can be suggested that appearance of new structural modifications of hydrofullerite phases is explained by more selective hydrogenation compared to non catalytic reaction. This suggestion is confirmed by comparison of XRD from Ni catalyzed hydrofulleride sample and $C_{60}H_{36}$ sample obtained by transfer hydrogenation method using 9,10-Dihydroanthracene (see Fig.2b). Both of these samples exhibited similar XRD patterns which can be indexed by mixture of two phases one of which with FCC and another with BCC structure. It is known that transfer hydrogenation reaction produce selectively $C_{60}H_{36}$ hydrofulleride represented by only 2-3 isomers. Raman spectra recorded from our sample had confirmed that our sample is, in fact, $C_{60}H_{36}$ in a good agreement with literature data. Catalyst most probably helps to produce hydrofullerides with selected composition $C_{60}H_x$ with only few possible isomers, while direct hydrogenation results in a broad range of X in the mixture of different hydrofullerides. General trend in the structure modifications of hydrofullerides depending on the hydrogen content is shown in Fig.3. Points in the left part of the plot correspond to hydrogenation performed at 573 K, while high hydrogen content was achieved at 670-690 K. Both FCC and BCC samples are included into the plot and only single-phase samples are used. Unlike to previous data, the dependence of cell parameter from hydrogen content is clearly non linear. Expansion of the cell becomes progressively stronger with increase of hydrogen content. At around 5 wt% this curve reach saturation point and prolonged hydrogenation results only in increase of cell parameter. Longer hydrogenation results in collapse of C_{60}

molecules (see Fig.1). Better view on the structural modifications in hydrofulleride phases can be obtained if volumes per molecule of C_{60} are compared.

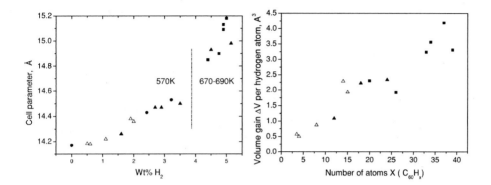

FIGURE 3 a) Cell parameter a calculated from XRD (cubic structure) versus hydrogen percent. Black circles-literature data, filled triangles-Ni catalyzed samples, squares-pure C_{60} samples, open triangles-Pd catalized. Temperature of hydrogenation 573-690K
b) The figure shows dependence of ΔV versus X. Volume gain per hydrogen atom in $C_{60}H_x$ calculated per molecule of C_{60} depending on total amount of hydrogen atoms X. $\Delta V = (V_{C60Hx} - V_{C60})/X$, V_{C60} is taken as volume per molecule in FCC structure of pristine C_{60} (711Å3).

The Fig 3 b) shows dependence of gain volume per molecule normalized to the total number of hydrogen atoms in the $C_{60}H_x$. This dependence is approximately linear. Of course, volumes per molecule calculated this way do not equal directly to the volume of every hydrogen atom in the structure due to rotation of molecules. Instead it should be considered as a "dynamic" volume. Nevertheless, the plot shown in Fig.3 b) allows to see very important trend: the more hydrogen we attach to C_{60} molecule, the bigger volume they occupy. This effect shall be explained by repulsion of hydrogen atoms from each other and weaker C-H bonds in hydrofullerides with higher hydrogen content. It is clear, that elongation of C-H bonds and strain due to repulsion of hydrogen atoms must result in deformation of the shape of C_{60} molecule itself as it was observed for example in $C_{60}H_{18}$ [9]. Finally, at about 5Wt% of hydrogen content this deformation become so strong that C_{60} molecule collapses.   The data shown in Fig.3 are in agreement with calculations which showed that C-H bonds energy must decrease for higher number of hydrogen atoms in the hydrofulleride structure [15].

REFERENCES

1.Loufty R.O., Wexler E.M., Proceedings of the 2001 DOE Hydrogen Program Review. NREL/CP-570-30535

2. Tarasov B.P. ,Shul'ga Yu.M.,Fokina V.N.,Vasilets V.N., Shul'ga N.Yu, Schur D.V., Yartys V.A. , J.Alloys and Compounds, **314**, 296-300, (2001)

3. Shul'ga Yu.M, Tarasov B.P, Fokin V.N., Martynenko V.M., Schur D.V., Volkov G.A., Rubtsov V.I., Krasochka G.A., Chapusheva N.V., Shevchenko W., Carbon, **41**, 1365-1368 , (2003).

4. Darwish A.D., Abdul-Sada A.K., Langley G.J., Kroto H.W., Taylor R, Walton D.R.M., Synth.Metals, **77**, 303-307, (1996).

5. Ruchardt C., Gerst M., Ebenhoch, J., Beckhaus H..D., Campbell E. E. B., Tellgmann R., Schwarz H., Weiske T., Pitter S., Angew. Chem. **105**, 609-611, (1993).

6. Kolesnikov A.I., Antonov V.E., Bashkin I.O., Grosse G., Moravsky A.P., Muzychka A.Yu., Ponyatovsky E.G., Wagner F.E., J. Phys. : Condens. Matter, **9**, 2831-2834, (1997).

7. Yu M Shul'ga, B P Tarasov, V N Fokin, N Yu Shul'ga, V N Vasilets Fiz. Tv. Tela , **41**, 1520-1523, (1999).

8. Okotrub, A. V. ; Bulusheva, L. G. ; Acing, I. P. ; Lobach, A. S.; Shulga, Y. M. J. Phys. Chem., **103**, 716-720, (1999).

9. Darwish, A. D.;Avent, A.G., Taylor, R.; Walton, D. R. M. /J. Chem. Soc., Perkin Trans., **2** , 2051-2054, (1996) .

10. Balasubramanian K., Chem.Phys.Lett., **182**, 257-262, (1991).

11. Bezmelnitsyn V. N., GIazkov V.P., Zhukov V.P., Somenkov V.A., Shilstein S.Sh, in The 4th Biennial International Workshop m Russia Clusters' (Abstracts of Reports). St. Petersburg, 1999 p. 71

12. Kockelmann W., ISIS experimental report, Rhutherford Appleton Laboratory, (2000)

13. Hall L.E., McKenzie D.R., Attalla M.I., Vassallo A.M., Davis R.L., Dunlop J.B., Cockayne D.J.H., J. Phys. Chem. **97**, 5741-5744, (1993).

14. Antonov V.E., Bashkin I.O., Khasanov S.S., Moravsky A.P., Morozov Yu.G., Shulga Yu.M., Ossipyan YU.A., Ponyatovsky E.G., Joumal of Alloys and Compounds **330-332**, 365-368, (2002)

15. Rathna A., Chandrasekhar J., Chem. Phys.Lett., **206**, 217-224 (1993).

Time-scale for Jahn-Teller pseudo-rotations in the organic ferromagnet TDAE-C_{60}

Robert Blinc*, Peter Jeglič*, Tomaž Apih*, Aleš Omerzu* and Denis Arčon*

*Institute Jožef Stefan, Jamova 39, 1000 Ljubljana, Slovenia

Abstract. The correlation times for the Jahn-Teller dynamics of the C_{60}^- ions in the organic ferromagnet TDAE-C_{60} have been extracted from a comparison of experimental and simulated ^{13}C NMR lineshapes. A strong correlation between spin ordering and orientational ordering has been found.

INTRODUCTION

The charge-transfer compound TDAE$^+$-C_{60}^- has the highest transition temperature T_C=16 K of all purely organic non-polymeric ferromagnets [1]. The origin of the ferromagnetic state seems to be strongly related to the orientational ordering of the C_{60}^- ions [2]. This relation is hard to understand if the C_{60}^- ions would be spherical and the unpaired spin density uniformly distributed over the C_{60}^- sphere. A theoretical study [3] indeed yielded for undistorted C_{60}^- ions a T_C which is three orders of magnitude lower than the one observed. Recent ^{13}C NMR data [4, 5] provided quantitative evidence for the occurrence of a Jahn-Teller transition in TDAE$^+$-C_{60}^- resulting in the appearance of a belt-like unpaired electron spin density distribution on the C_{60}^- ions. The dynamics of the Jahn-Teller distortions has however not yet been determined.

It is the purpose of this contribution to determine the time-scale of the Jahn-Teller pseudo-rotations between the three mutually perpendicular distortion directions in TDAE-C_{60} (Fig. 1).

EXPERIMENTAL DETAILS

The ^{13}C NMR spectra of a 40 % ^{13}C enriched TDAE-C_{60} have been measured at a Larmor frequency 95.6 MHz. The ^{13}C enrichment was performed at the C_{60}^- ions and not at the TDAE$^+$ ions.

RESULTS AND DISCUSSION

Two types of motions are considered to affect the ^{13}C NMR lineshape in TDAE-C_{60}. The first type of the motion are physical rotations of the C_{60}^- ion [6, 7]. This type of

CP685, *Molecular Nanostructures: XVII Int'l. Winterschool/Euroconference on Electronic Properties of Novel Materials,* edited by H. Kuzmany, J. Fink, M. Mehring, and S. Roth
© 2003 American Institute of Physics 0-7354-0154-3/03/$20.00

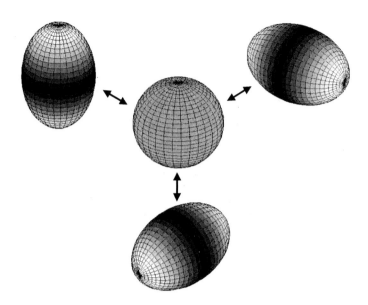

FIGURE 1. Jahn-Teller distortions of the spherical C_{60}^- ion along three mutually perpendicular directions. The shaded areas on the distorted C_{60}^- ions designate areas of larger unpaired electron spin density.

motion has been studied in detail in pure C_{60}. It involves both uniaxial rotations (characterized by a characteristic time τ_R) around the molecular 3-fold axes accompanied by a flipping of these axes (characterized by a characteristic time τ_F) between the symmetry allowed orientations. Below 90 K the C_{60} molecules freeze into two nearly degenerate orientations. Although the orientational potential in TDAE-C_{60} will be rather different from the one in pure C_{60} we assume that the same types of motions are taking place here here too. We assume that the physical motion of the C_{60}^- ions in TDAE-C_{60} is a combination of the uniaxial rotations and flipping of the axes of the rotations.

The second type of motion is the pseudo-rotation of the axes of the Jahn-Teller distortion of the C_{60}^- ion. It is well known that adding electrons to the lowest unoccupied 3-fold degenerate t1u C_{60} molecular orbital (LUMO) should lead to a Jahn-Teller distortion of the molecule [8]. Three different nearly degenerate structures of Jahn-Teller distorted C_{60}^- ions have been identified having D_{5d}, D_{3d} and D_{2h} symmetry [9]. In TDAE-C_{60} we will assume the existence of the D_{2h} structure as it is compatible with the crystal symmetry. It results in a splitting of the three-fold degerate t_{1u} levels into three nondegenerate states [10]. One of them, assigned as LUMO$_z$ corresponds to the belt-like charge density distribution around the elongated z-axis of the C_{60}^- ion and has a lower energy than the other two orthogonal states assigned as LUMO$_x$ and LUMO$_y$. It is therefore the ground state. Further we'll assume that the molecular z-axes can be in general aligned along the three crystal directions a, b and c. The molecular z-axes can jump between these three directions with a characteristic time τ_{JT} (Fig. 1). It is this type of motion, which can dramatically affect the ^{13}C NMR lineshape in view of the electron-nuclear hyperfine coupling as discussed below.

The ^{13}C NMR signal is quite generally given by the relaxation function as

$$G(t) = \exp[i\omega_0 t] \left\langle \exp\left(i\int_0^t \omega\prime(t\prime)dt\prime\right)\right\rangle \qquad (1)$$

where ω_0 is the unperturbed (time independent) ^{13}C NMR frequency and $\omega\prime(t\prime)$ is the fluctuating part of the ^{13}C NMR frequency. The ^{13}C NMR spectra are then obtained by the Fourier transform of the relaxation function.

The time dependent ^{13}C NMR frequency can be expressed as

$$\omega\prime(\vartheta_B, \varphi_B, \vartheta, \varphi) = \omega_L \left[K_\perp + \left(K_\parallel - K_\perp\right)\cos^2\Theta(t)\right] \qquad (2)$$

Here $\Theta(t)$ is the angle between the largest principal axis of the shift tensor **K** at a given nuclear site and the direction of the external magnetic field. ϑ_B, φ_B describe the orientation of the magnetic field with respect to the crystal lattice whereas ϑ, φ describe the position of a given carbon atom on the C_{60}^- ion at time t.

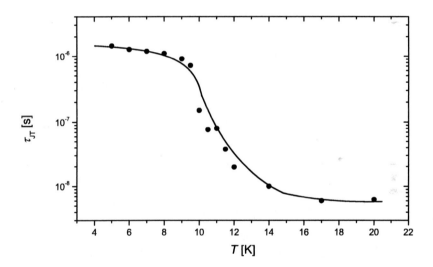

FIGURE 2. The temperature dependence of the correlation time τ_{JT} obtained from fits of the low-temperature ^{13}C NMR spectra. The solid line is a guide for the eye.

In case of spherical symmetry all ^{13}C nuclei on the C_{60}^- ion are equivalent and will experience the same contact hyperfine shift. A Jahn-Teller distortion will make the charge distribution belt-like and different ^{13}C atoms will experience different Fermi-contact shifts. We may assume that both the parallel and the perpendicular components of the anisotropic shift tensor **K** of the carbon atom at ϑ, φ are modified by the same amount $\delta K_{\perp,\parallel} = Bp(\vartheta, \varphi, t)$. Here $p(\vartheta, \varphi, t)$ is the deviation of the unpaired electron spin density distribution function from its spherical form.

A comparison between experimentally measured and simulated ^{13}C NMR spectra at lower temperatures where the physical rotations of the C_{60}^- ions are frozen out allow for the determination of the correlation times τ_{JT} for the Jahn-Teller pseudo-rotations. The obtained results are presented in Fig. 2.

The orientational ordering [11] of the Jahn-Teller distorted C_{60}^- ions maximizes the ferromagnetic coupling and is responsible for the occurrence of the ferromagnetic phase in TDAE-C_{60}.

REFERENCES

1. P.-M. Allemand, K.C. Khemani, A. Koch, F. Wudl, K. Holczer, S. Donovan, G. Gruner, J.D. Thompson *Science* **253**, 301 (1991).
2. D. Mihailovič, D. Arčon, P. Venturini, R. Blinc, A. Omerzu, P. Cevc *Science* **268**, 400 (1995)
3. T. Sato, T. Saito, T. Yamabe, K. Tanaka, H. Kobayashi *Phys. Rev.* **B55**, 11052 (1997).
4. D. Arčon, J. Dolinšek, and R. Blinc, *Phys. Rev.* **B53**, 9137 (1996); R. Blinc, T. Apih, P. Jeglič, D. Arčon, J. Dolinšek, D. Mihailovič, and A. Omerzu *Europhys. Lett.* **57**, 80 (2002)
5. R. Blinc, P. Jeglič, T. Apih, J. Seliger, Arčon, J. Dolinsek, and A. Omerzu *Phys. Rev. Lett.* **8808**, 6402 (2002).
6. R. Blinc, J. Seliger, J. Dolinšek, and Arčon, *Phys. Rev.* **B49**, 4993 (1994).
7. R. Tycko, G. Dabbagh, R.M. Fleming, R.C. Haddon, A.V. Makhija, and S.M. Zahurak, *Phys. Rev. Lett.* **67**, 1886 (1991).
8. M. C.M. O'Brien, *Phys. Rev.* **B53**, 3775 (1996).
9. N. Koga and K. Morokuma, *Chem. Phys. Lett.* **196**, 191 (1992).
10. T. Kawamoto, *Solid State Commun.* **101**, 231 (1997).
11. B. Narymbetov, A. Omerzu, V.V. Kabanov, M. Tokumoto, H. Kobayashi, and D. Mihailovič, *Nature* **407**, 883 (2000).

Mg_4C_{60}:
A New Two-dimensional Fulleride Polymer

F. Borondics, G. Faigel, G. Oszlányi, S. Pekker

Research Institute for Solid State Physics and Optics
H-1525 POB 49. Budapest, Hungary

Abstract. Here we present preliminary results on Mg_4C_{60}, a new fulleride polymer. A series of Mg_xC_{60} compositions were prepared by solid state synthesis. While most samples were multiphase, the nominal composition Mg_4C_{60} provided a single phase material. X-ray powder diffraction data revealed that the structure is rhombohedral based on polymeric sheets of C_{60} molecules. Charge transfer from Mg to C_{60} was estimated by Raman microscopy.

INTRODUCTION

Polymerization in alkali fulleride salts have received extended attention over the last decade. In some alkali fulleride salts the negatively charged fullerene ions spontaneously polymerize and the polymerization-depolymerization cycle is reversible. Intermolecular bonding can take place through [2+2] cycloaddition as in neutral C_{60} but negatively charged C_{60} ions can also be connected by single carbon-carbon bonds. Quantum chemical calculations suggest that polymerization is a general behaviour with charge transfer and single carbon-carbon intermolecular bonds become dominant at high charge states [1]. In spite of prior intensive research, it is remarkable that up to now there has been no report of polymerization in the fulleride salts of alkaline earths or other multiply charged cations. While Ca_xC_{60} and Ba_xC_{60} systems were thoroughly searched for superconductivity, a reinvestigation considering polymerization would be justified. The present work focuses on the yet unexplored and therefore more promising Mg_xC_{60} system.

EXPERIMENTAL

We have prepared 200 mg batches of various Mg_xC_{60} compositions ($0<x<6$) by solid state reaction of C_{60} powder (99.5%) and Mg grains (99.9%). Details of the synthesis is described in Ref. [2]. A typical synthesis lasted over a month so it was advantageous to perform all heat treatments and intermediate grindings inside a glovebox. To activate the surface of Mg, the reaction mixtures were first heated to 480°C for 10 minutes. Then a series of one week heat treatments followed starting at 380°C and ending at 450°C. The reaction took place at C_{60}/Mg interfaces controlled by the solid state diffusion of magnesium. The reaction was followed by x-ray powder diffraction monitoring the amount of unreacted Mg. After four subsequent heat

CP685, *Molecular Nanostructures: XVII Int'l. Winterschool/Euroconference on*
*Electronic Properties of Novel Materials,*edited by H. Kuzmany, J. Fink, M. Mehring, and S. Roth
© 2003 American Institute of Physics 0-7354-0154-3/03/$20.00

treatments shiny dark grey powder formed without significant traces of unreacted magnesium.

X-ray data were measured with highly monochromatic $CuK\alpha_1$ radiation using a HUBER G670 Guinier image plate camera in transmission mode. Samples were held in 0.7 mm diameter glass capillaries which allowed subsequent Raman microscopy to be made on the same samples with a Renishaw System 1000B spectrometer. The laser intensity at 488 nm was carefully limited to avoid sample degradation.

RESULTS

In contrast to their uniform appearance, most samples were multiphase and their powder diffractograms could not be indexed. However, a recurring phase was always present in the x=4-6 composition range. Most clearly it is identified in the nominal Mg_4C_{60} sample. Figure 1. shows the raw diffractogram of this sample which can be indexed as rhombohedral with lattice parameters a=b=9.22 Å, c=25.25 Å, γ=120°.

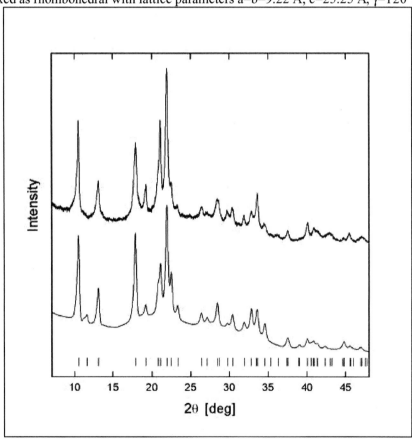

FIGURE 1. Powder diffractogram of Mg_4C_{60}. From the top: (a) raw data, (b) model calculation, (c) allowed reflection positions

Figure 1. also shows a calculated pattern based on the following simplified structural model. The spacegroup is R–3m with symmetry equivalent C_{60} molecules at the fractional coordinates $(0,0,0)$, $(^2/_3, ^1/_3, ^1/_3)$ and $(^1/_3, ^2/_3, ^2/_3)$. The short interfullerene distance along the **a** and **b** directions is an evidence for the formation of polymeric sheets in the z=0, $^1/_3$ and $^2/_3$ planes. The present data did not allow atomic resolution structure refinement so the scattering of C_{60} is approximated by the charge distribution of a spherical shell. Note that the unit cell is similar to that of the high pressure rhombohedral polymer but here the space between the polymeric planes is occupied by magnesium ions and thus the c lattice parameter is slightly larger. In the structure there are two inequivalent magnesium positions at fractional coordinates $(0,0,0.23)$ and $(0,0,0.43)$, both with multiplicity 6. Accordingly, the overall composition is Mg_4C_{60}. This is a chemically reasonable structural model where each C_{60} molecule is surrounded by six magnesium neighbours.

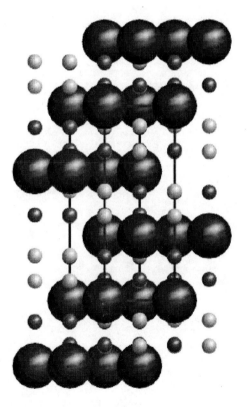

FIGURE 2. Mg_4C_{60} structural model described in the text. Large spheres: C_{60} molecules, small spheres: Mg ions.

The charge state of C_{60} was measured by Raman spectroscopy. Based on the empirical linear shift of the $A_g(2)$ mode [3] a charge state of -10 ± 1 can be deduced. Assuming full charge transfer from magnesium to C_{60} gives an alkaline earth content $x=5\pm0.5$. During the sample preparation multiple grindings may have caused a loss of fine powder while unreacted magnesium particles were always retained. Therefore a $\Delta x=0.5$ shift of the nominal composition can not be excluded.

CONCLUSION

This work is a first step in exploring polymerization of alkaline earth fullerides. Mg_xC_{60} is a largely unexplored system, there is only one paper in the literature on photoelectron spectroscopy of Mg_xC_{60} thin films [4]. In the phase diagram we have identified the composition Mg_4C_{60} and have shown that it is a two dimensional polymer. X-ray powder diffraction revealed the main aspects of the structure and Raman spectroscopy proved that the charge transfer is near to what is expected for the nominal composition. The rhombohedral structure is based on polymeric planes and it is related to the high pressure rhombohedral polymer of pure C_{60}. However, there are important properties of Mg_4C_{60} are yet unresolved like the bonding geometry, the exact charge transfer and the electronic properties. The speed of the synthesis is also a key issue which must be improved to allow complete exploration of the Mg_xC_{60} phase diagram.

ACKNOWLEDGMENTS

We thank Katalin Kamarás and Tamás Pusztai for expert advice on Raman measurements and sample preparation. This research was supported by OTKA grants T029931, T034198 and T032613.

REFERENCES

1. S. Pekker, G. Oszlányi and G. Faigel, *Chem. Phys. Lett.* **282,** 435-441 (1998).
2. F. Borondics, G. Oszlányi, G. Faigel and S.Pekker, *Solid State. Commun.,* in print (2003).
3. T. Wagberg and B. Sundqvist, *Phys. Rev B* **65**, 155421-1-7 (2002)
4. Y. Chen, F. Stepniak, J.H. Weaver, L.P.F. Chibante and R.E. Smalley, *Phys. Rev B* **45**, 8845-8848 (1992).

Structure Analysis of Alkaline Earth Endohedral Fullerenes
M@C_{74}·Co(OEP)·2C_6H_6 (M = Sr, Ba)

D. Flot[1], K. Friese[2], O. Haufe[2], H. Modrow[3], M. Panthöfer[2], A. Reich[2],
M. Rieger[2], G. Wu[4], and M. Jansen[2]

[1]*ESRF, B.P. 220, F-38043 Grenoble Cedex, FRANCE*
[2]*Max Planck Institute for Solid State Research, Heisenbergstr. 1, D-70569 Stuttgart, GERMANY*
[3]*Institute of Physics of the University of Bonn, Nussallee 12, D-53115 Bonn, GERMANY*
[4]*Brookhaven National Laboratory, 75 Brookhaven Ave, 11973 Upton, USA*

Abstract. Structure analysis of alkaline earth endohedral fullerenes M@C_{74} (M = Sr, Ba) has been performed by means of XANES and micro crystal synchrotron diffraction. The experimental results from XANES and simulations based on different exo- and endohedral model structures from ab-initio calculations confirm the endohedral character of these compounds. The crystal structures of M@C_{74}·Co(OEP)·2C_6H_6 (M = Sr, Ba) consist of (M@C_{74})[Co(OEP)]$_2$(M@C_{74}) units arranged in a distorted primitive hexagonal packing. The molecular structure is ordered and exhibits a high level of localization of the endohedral metal atom.

INTRODUCTION

Due to its low band gap, C_{74} has exclusively been reported in the state of the exo- or endohedrally reduced dianion C_{74}^{2-} [1-5]. Although different compounds M@C_{74} have been intensively characterized by means of spectroscopic methods [2-5], there is still a lack of structural data and of a final proof of the endohedral character of this compounds. Therefore, we present XANES and micro crystal synchrotron diffraction studies on alkaline earth endohedral fullerenes, M@C_{74} and M@C_{74}·Co(OEP)·2C_6H_6 (M = Sr, Ba), respectively.

EXPERIMENTAL

Synthesis of alkaline earth endohedral fullerenes by means of the radio-frequency method and HPLC-isolation of the small cage species M@C_{74} (M = Ca, Sr, Ba) was performed according to [4, 5]. Sample preparation for XANES was performed by drop-coating a CS_2 solution of Ba@C_{74} onto a quartz substrate, evaporation of the solvent and finally sealing the thin film of the sample with Kapton foils keeping inert gas conditions in all steps of manipulation. Ba L_{III}-edge XANES spectra were measured in fluorescence mode at beamline BN2 in the ELSA synchrotron radiation laboratory (Bonn, Germany, further details [6]). The accelerator was operated in storage mode at

*CP685, Molecular Nanostructures: XVII Int'l. Winterschool/Euroconference on
Electronic Properties of Novel Materials,*edited by H. Kuzmany, J. Fink, M. Mehring, and S. Roth
© 2003 American Institute of Physics 0-7354-0154-3/03/$20.00

2.3 GeV electron energy with an average current of 50 mA. Energy calibration was performed relative to the Ti K-edge spectrum of a titanium foil measured in transmission mode (E_{Ph} = 4966 eV). Spectra were recorded with a step width of 0.25 eV at an integration time of 2 seconds per point. Due to the small amount of material available, the data quality is limited. To improve statistics, three spectra have been accumulated. A linear pre-edge contribution was subtracted and Fourier-filtering was performed. Micro crystals of $M@C_{74}\cdot Co(OEP)\cdot 2C_6H_6$ (M = Sr, Ba) were grown by slow interdiffusion of solutions of $M@C_{74}$ in benzene and octaethylporphine-cobalt, Co(OEP), in chloroform, respectively. Diffraction intensities were collected from micro crystals (~50 µm) at BNL (X3A1, λ = 64.3 pm) and ESRF (ID13, λ = 73 pm) at 100 K.

The structure solution via direct methods (SIR97, [7]) in a primitive triclinic setting yielded the heavy atom positions. Subsequent refinements and difference Fourier maps (Jana2000, [8]) allowed to locate all C- and N-atoms. Their positions were refined applying a common isotropic displacement parameter. At this stage of the refinement the atomic parameters clearly pointed towards a higher symmetry and the structure was transformed to $C2/m$. Both, C_{74} and Co(OEP) were located on the mirror plane ($4i$). Additional maxima in the residual electron density maps close to the centers of all hexagons and some pentagons, respectively, pointed to a second orientation. However, the electron density map was too diffuse to construct a full second C_{74} cage. C_{74} in the first orientation was therefore replaced by a rigid body, for which the quantum chemically optimized geometrical structure of C_{74}^{2-} was assumed. It has to be pointed out, that the experimental structure of C_{74} as found in the difference Fourier synthesis and the model molecule were in good agreement. Further refinement showed a deviation of the model molecule from the mirror plane, leading to two molecular orientations. This was described either as orientational disorder in space group $C2/m$ and, alternatively, as an ordered structure in space group $C2$ with an additional inversion center as twin operation. Librational motions were described by using the TLS approach. Refinement converged at R_1(obs) = 16.6 % and R_w(all) = 19.1 % for the Ba compound. Quantum chemical calculations of C_{74}, C_{74}^{2-} and $Ba@C_{74}$ were performed in density functional approximation (RI-DFT, LDA, SV and ECP-SV(P)) using the TURBOMOLE 5.3 program package [9].

RESULTS AND DISCUSSION

Quantum chemical calculations on C_{74} and C_{74}^{2-} indicate the structural response of C_{74} upon reduction to be highly localized (Δ_{mean} = 0.2(9) pm) in the belt across the horizontal mirror plane. Maximum bond lengthening (Δ = 1.4 pm) is found in the central bond of the three pyracylene units. According to charge bond length relation, the extra charges are expected to be localized in this bonds. However, the limited accuracy of the population analysis methods does not allow to confirm this assumption. Compared to the equilibrium structure of the dianion, further structural changes in $Ba@C_{74}$, due to the coordination of the inner molecular surface by the metal cation, are small (Δ_{mean} = 0.0 (4) pm) and highly localized. The lengthening of the central bond in the coordinating pyracylene unit is compensated by minor bond-length shortening of the non coordinating pyracylene unit. Therefore, the structure of C_{74}^{2-} is a good approximation of the molecular structure of the C_{74} cage in Ba@C_{74}.

XANES spectra of Ba@C_{74} (Fig. 1) exhibit a well pronounced double maximum structure about 5270 eV. Simulated XANES spectra (FEFF8, [10]) based on different exo- and endohedral equilibrium structures from quantum chemical calculations reproduce these shape resonances exclusively in the case of the off center endohedral structures. This finding might be explained by the higher coordination of the Ba-atom inside the cage compared to the exohedral case. The latter enables strong multiple scattering paths, that significantly contribute to the spectrum. Especially the double maximum structure might hint towards the off center endohedral case. Whereas the shape resonances, i.e. the part of the spectra corresponding to the geometrical environment, fit well to the experimental results, the splitting between white-line and shape resonances is underestimated. This is likely correlated to an underestimation of the effect of the inner shell hole which is created in the photoabsorption process. Full neglect of this effect results in an even smaller splitting. On the other hand, the use of a Z+1 model, which emphasizes this effect, resolves the discrepancy completely.

Assuming the ordered model, M@C_{74}·Co(OEP)·2C_6H_6 (M = Sr, Ba) crystallize monoclinic in space group $C2$. The structure consists of Co(OEP) complexes and M@C_{74} molecules (Fig. 2). Co(OEP) is present in the commonly found *all-syn* conformation which allows the formation of Co(OEP)-dimer. Within these dimer the individual molecules are shifted against each other, thus the Co-cation of one complex is coordinated by one N-atom of the opposite complex, resulting in a quadratic-pyramidal coordination of Co (194 ≤ d(Co-N)/pm ≤ 203 in plane and 297 pm apical, d(Co-Co) = 348 pm). The concave surface formed by the *all-syn* conformation of the ethylgroups of each Co(OEP)-complex coordinates one M@C_{74} molecule. All in all the crystal structure of M@C_{74}·Co(OEP)·2C_6H_6 may be described as a distorted, primitive hexagonal packing of these (M@C_{74})[Co(OEP)]$_2$(M@C_{74}) units. Voids between those dimer are filled by benzene molecules of crystallization.

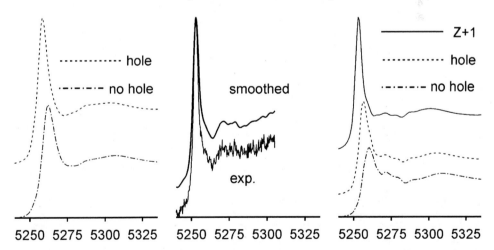

FIGURE 1. Experimental XANES spectra of Ba@C_{74} (mid) and simulated spectra for the exo- (left) and endohedral case (right; abscissa: photonenergy in eV, ordinate: rel. intensity for all graphs).

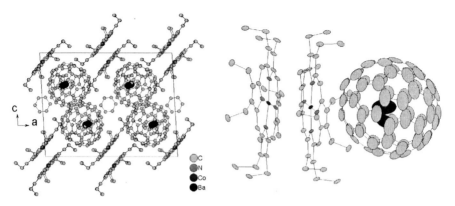

FIGURE 2. Crystal structure of M@C$_{74}$·Co(OEP)·2C$_6$H$_6$ (left; view in [010]) and coordination of M@C$_{74}$ by the Co(OEP) dimer (right; ellipsoids equivalent to 20% probability density).

CONCLUSION

Unlike most endohedral fullerene species, the M@C$_{74}$ compounds (M = Sr, Ba) analysed here do not suffer from a pronounced positional disorder of the endohedrally encaged metal. The endohedral metal atom is displaced off center towards one of the pockets of the cloverleaf shaped C$_{74}$ molecule, as also found from the quantum chemical investigations.

The high degree of order of the endohedral metal atom as well as of the fullerene cage itself, are due to the strong intermolecular interactions of the dipolar M@C$_{74}$ molecule and the Co(OEP)-complex.

REFERENCES

1. Diener, M.D., and Alford, J.M., *Nature*, **393**, 668 (1998).
2. Wan, T.S.M., Zhang, H.-W., Nakane, T., Xu, Z., Inakuma, M., Shinohara, H., Kobayashi, K., Nagane, S., *J. Am. Chem. Soc.*, **120**, 6806 (1998).
3. Kuran, P., Krause, M., Bartl, A., Dunsch, L., *Chem. Phys. Lett.*, **292**, 580 (1998).
4. Haufe, O., Reich, A., Möschel, C., Jansen, M., *Z. anorg. allg. Chem.*, **627**, 23 (2001).
5. Grupp, A., Haufe, O., Jansen, M., Mehring, M., Panthöfer, M., Rahmer, J., Reich, A., Rieger, M., Wie, X.-W., "Synthesis and Isolation of New Alkaline Earth Endohedral Fullerenes M@C$_n$ (M=Ca, Sr; n=74, 76" in *Structural and Electronic Properties of Molecular Nanostructures*, edited by H. Kuzmany et al., , AIP Conference Proceedings 633, New York: American Institute of Physics, 2002, pp. 31-34.
6. Floriano, P.N., Schlieben, O., Doomes, E.E., Klein, I., Janssen, J., Hormes, J., Poliakoff, E.D., McCarley, R.L., *Chem Phys Lett.*, **321**, 175 (2000).
7. Altomare, A., Burla, M.C., Camalli, M., Cascarano, G.L., Giacovazzo, C., Guagliardi, A., Moliterni, A.G.G., Polidori, G., Spagna, R., *J. Appl. Cryst.*, **32**, 115 (1999).
8. Petricek, V., Dusek, M., Jana2000: The crystallographic computing system. Institute of Physics, Praha, Czech Republic (2003).
9. Ahlrichs, R., Turbomole Version 5.3, Quantum Chemistry Group, University of Karlsruhe, Germany (2000).
10. Ankudinov, A., Ravel, B., Rehr, J.J., Conradson, S., *Phys. Rev. B*, **58**, 7565 (1998).

TUMBLING CERIUM ATOMS INSIDE THE FULLERENE CAGE: iCe$_2$C$_{80}$

M. Kanai,* K. Porfyrakis,† A. N. Khlobystov,† H. Shinohara‡, and T. J. S. Dennis*

*Department of Chemistry, Queen Mary, University of London, Mile End Road, London E1 4NS
† Department of Materials, Parks Road, Oxford OX1 3PH
‡ Department of Chemistry & Institute for Advanced Research, Nagoya University, Nagoya 464-8062, Japan.

Abstract. We report the spectroscopic work of two Ce-containing *incar*-fullerenes, iCeC$_{82}$ and iCe$_2$C$_{80}$. UV/Vis, IR, Raman, TOF-MALDI and ^{13}C NMR were employed to investigate the structural and electronic information of these two major isomers of Ce *incar*-fullerenes. Tumbling motion of two Ce atoms inside the I$_h$-C$_{80}$ cage was confirmed and analysed by temperature-dependant ^{13}C NMR.

INTRODUCTION

Fullerenes have nanometer-scale voids capable to encapsulate atoms or even molecules. [82]fullerene-*incar*-metal, iMC$_{82}$ (*e.g.* M = Sc, La, Y, Gd) (endohedral metallofullerene with the C$_{82}$ cage) [1] have been purified and well studied by several groups [2] [3]. This is mainly due to the fact that iMC$_{82}$ is one of the major mono-metal *incar*-fullerenes found in the solvent-extract from arc-generated soot. Similarly, iM$_2$C$_{80}$ as a di-metal *incar*-fullerene has been produced and isolated [2]. iLa$_2$C$_{80}$ is reported to have the C$_{80}$ cage with I$_h$ symmetry by ^{13}C NMR/^{140}La NMR [4] and synchrotron X ray diffraction [5]. Calculation by Nagase *et al* showed that the C$_{80}$\{I$_h$\} becomes stable when the C$_{80}$ cage receives 6 electrons from the incarcerated metal atoms [6]. Ce-containing *incar*-fullerenes have yield relatively higher (yield of iCeC$_{82}$ is *ca.* 6% of all the solvent-extracted fullerenes from arc-generated soot) compared with that of other *incar*-fullerenes such as La and Gd-containing *incar*-fullerenes. Therefore, from the view of mass-productive use of fullerene-based materials, we focused on Ce *incar*-fullerenes and investigated the properties of iCeC$_{82}$ and iCe$_2$C$_{80}$.

CP685, *Molecular Nanostructures: XVII Int'l. Winterschool/Euroconference on Electronic Properties of Novel Materials*, edited by H. Kuzmany, J. Fink, M. Mehring, and S. Roth
© 2003 American Institute of Physics 0-7354-0154-3/03/$20.00

EXPERIMENT

Soot containing Ce *incar*-fullerenes was arc-generated by the so-called reversed arc technique developed by the Nagoya group as described previously [7]. The soot was collected under the anaerobic conditions, and fullerenes were soxhlet-extracted with o-xylene. The residue was then refluxed either in pyridine or in DMF to obtain an *incar*-fullerene enriched solution. The sample was passed through HPLC (JAI, LC-908) for the *incar*-fullerene purification process with toluene eluent. TOF-MALDI mass spectrometry was used to check the purity of each HPLC-obtained fraction. The UV/Vis spectrum of pure iCeC$_{82}$ (FIGURE 1(a)) looks similar to that obtained for iLaC$_{82}$ and iYC$_{82}$. In those cases La and V are supposed to be in the +3 oxidation state. Furthermore, the spectrum looks quite different from that obtained for iScC$_{82}$, where Sc is +2 [8]. IR, Raman and XAFS spectrum of iCeC$_{82}$ also supported this observation that Ce is charged to +3 inside the C$_{80}$ cage. The onset of UV/Vis spectrum of iCe$_2$C$_{80}$ (FIGURE 1(b)) looks quite different from that obtained for empty C$_{80}$ isomers I and II, which were isolated by Wang *et al* [9], suggesting that the C$_{80}$ cage of iCe$_2$C$_{80}$ has a different symmetry from that of empty C$_{80}$.

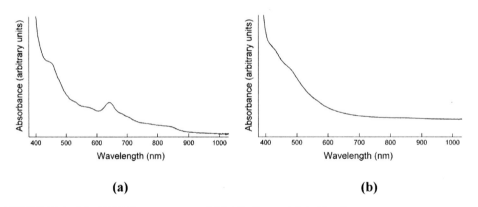

(a)   **(b)**

FIGURE 1. The UV/Vis spectrum of (a) iCeC$_{82}$ and (b) iCe$_2$C$_{80}$.

RESULTS

The ^{13}C NMR spectrum of iCe$_2$C$_{80}$, FIGURE 2, consists of only two peaks - occurring at 148.12ppm and 123.85 ppm with 3:1 intensity ratio. This is consistent with I$_h$ symmetry assuming two Ce atoms average to spherical symmetry by tumbling rapidly within the carbon cage at 300K... Temperature-dependant ^{13}C NMR measurement was also performed on iCe$_2$C$_{80}$ in CS$_2$ solution with d-acetone lock between 200K and 300K.

No signal was observed below 270K. The temperature evolution of the more intense $i\text{Ce}_2\text{C}_{80}$ peak, obtained between 270K and 300K, is shown in FIGURE 3.

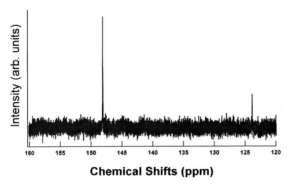

FIGURE 2. The ^{13}C NMR spectrum of $i\text{Ce}_2\text{C}_{80}$ at 300K (in CS_2/d-acetone, 64000 scans, Bruker AX600).

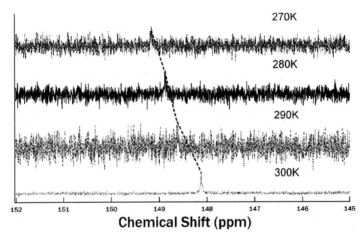

FIGURE 3. The ^{13}C NMR spectra of $i\text{Ce}_2\text{C}_{80}$ measured between 270K to 300K (in CS_2/d-acetone, 10000 scans each except one measured at 300K (64000 scans), Bruker AX600).

DISCUSSION

The intense ^{13}C NMR peak of $i\text{Ce}_2\text{C}_{80}$ (148.12ppm at 300K) shifts almost linearly downfield as the temperature is lowered. This suggests that the diamagnetic current produced by the rotation of two Ce atoms inside the cage has diminished. As the Ce atoms would stop rotating at low temperatures. No change in line-width of the main peak was observed below 270K. No signal observation below 270K suggests that the symmetry of the C_{80} cage has changed below 270K, consistent with the explanation that Ce atoms stop rotating below 270K (which reasonably agrees with the phase transition temperature of C_{60}, 260K).

Assuming that 1) this temperature-dependant shift was mainly owing to the Ce rotation and other terms, such as the T_1 and viscosity change depending on temperature, are smaller; 2) Ce atoms stop rotating at 270K; and 3) the observed shift is linear to the temperature, we can estimate the energy of the diamagnetic current produced by the Ce rotation.

The two Ce atoms produce the diamagnetic current H_D in the centre of the $C_{80}\{I_h\}$ cage,

$$H_D = \frac{1}{c}\int d\tau \frac{r \times j}{r^3} \tag{1}$$

Thus, the magnetic field H which carbons on the C_{80} cage feels will be,

$$H = H_0 - H_D \tag{2}$$

Eq. (2) explains the experimental result that the peak shifted to lower magnetic field at lower temperature, as H_D becomes smaller when the rotation of Ce atoms becomes slower and eventually stop.

For $\Delta T = 30$ K (300 K – 270 K), we obtain H

$\Delta\delta = 1.06$ ppm $= (H$ (Hz) - TMS shift (Hz)) / (Applied magnetic field for ^{13}C)

$$\therefore \quad H = 159 \text{ (Hz)} \tag{3}$$

Therefore, the energy by the diamagnetic current produced by the Ce rotation per one carbon atom,

$$H/(\Delta T \times 80) = 159 \text{ Hz} / (30 \text{K} \times 80) = 2.64 \text{ e}^{-14} \text{ KJ mol}^{-1} \text{ T}^{-1} = 2.20 \text{ e}^{-12} \text{ cm}^{-1} \text{ T}^{-1}$$

is obtained.

A further study on the mechanism of tumbling motion of two Ce atoms inside the C_{80} cage is currently underway, which includes the spin-lattice relaxation time (T_1) measurement of the iCe$_2$C$_{80}$ system.

ACKNOWLEDGMENTS

TJSD is grateful to the Royal Society and EPSRC for the financial support of this work. MK also acknowledge to Queen Mary College for the studentship.

REFERENCES

1. IUPAC recommended nomenclature is used here.
2. H. Shinohara, *Rep. Pro. Phys.* **63**, 843-892 (2000).
3. S. Knorr, A. Grupp, M. Mehring, U. Kirbach, A. Bartl, and L. Dunsch, *Appl. Phy. A* **66**, 257 (1997).
4. T. Akasaka, S. Nagase, K. Kobayashi, M. Waelchli, K. Yamamoto, H. Funasaka, T. Takahashi, T. Hoshino, and T. Erata, *Angew. Chem. Int. Ed. Engl.,* **36**, 1643 (1997).
5. E.Nishibori, M.Takata, M.Sakata, A.Taninaka and H.Shinohara, *Angew. Chem. Int.Ed. Engl.* 40, 2998-2999 (2001).
6. K. Kobayashi, S. Nagase, and T. Akasaka, *Chem. Phys. Lett.,* **245**, 230 (1995).
7. M. Kanai, H. Shinohara, and T. J. S. Dennis, *XVI International Winterschool on Electronic Properties of Novel Materials,* p. 35, The AIP Proceedings Series (2002).
8. M. Inakuma, M. Ohno, and H. Shinohara, *Recent Advances in the Chemistry and Physics of Fullerenes and Related Materials* Vol.2, p. 330-42, Academic Press, New York (1995).
9. C. R.Wang, T. Sugai, T. Kai, T. Tomiyama, and H. Shinohara, *Chem.Commun.,* 557 (2000).

HPLC Separation of Soluble $(C_{60})_n$ Oligomers From Fullerene Photopolymer

Éva Kováts and Sándor Pekker

Research Institute for Solid State Physics and Optics
H-1525 Budapest, POB. 49, Hungary

Abstract. Applying a recent photopolymerization method, we produced polymerized C_{60} in a gram scale and extracted different composition mixtures of its soluble components. We separated the oligomers by high-performance liquid chromatography (HPLC). The major component of the soluble fractions is the (2+2) cycloadduct dimer, C_{120}. Besides the dimer, we detected 3 different trimers and several higher oligomers including tetramers. The variation of the relative amounts of the trimers with the experimental conditions of polymerization is discussed in terms of the rate of formation and thermodynamic stability.

INTRODUCTION

Since the first production of C_{60} photopolymer [1] intensive experimental and theoretical studies [2] revealed that this material is a mixture of various (2+2) cycloadduct oligomers. Powder X-ray diffraction of bulk photopolymer established that linear or planar oligomers formed in the (100) and (111) planes of C_{60} retaining the fcc structure with somewhat contracted cell parameter [3]. Detailed studies of the individual oligomers were prevented by the small amounts of samples. Recently, we worked out a simple monomer recycling method for gram scale production of C_{60} photopolymer [4]. The available large amount of photopolymer made possible the extraction of its soluble components: a pure dimer phase and various mixtures of oligomers were isolated and characterized [5]. Previously, the dimer and some oligomers of C_{60} were also prepared by a mechanochemical method [6]. The trimer components were separated by high performance liquid chromatography (HPLC) and five different geometry isomers were detected by STM [7]. In the photopolymer the expected number of trimers is less because of the more strict topochemical conditions of the single phase polymerization in the fcc lattice. Here we present preliminary results on the separation of soluble photooligomers by HPLC.

EXPERIMENTAL

Photopolymerization of C_{60} was carried out by a large-scale method described previously [4]. A mixture of 800 mg C_{60} and its saturated solution in 10 mL toluene were sealed in a pyrex tube and illuminated by luminescent light sources for a few weeks at a controlled temperature. The raw polymer was subsequently extracted with

CP685, *Molecular Nanostructures: XVII Int'l. Winterschool/Euroconference on Electronic Properties of Novel Materials,* edited by H. Kuzmany, J. Fink, M. Mehring, and S. Roth
© 2003 American Institute of Physics 0-7354-0154-3/03/$20.00

hexane, toluene and 1-methylnaphthalene to obtain unreacted C_{60}, dimer-rich and oligomer-rich phases, respectively. The extraction of the less soluble higher oligomers was assisted by intensive sonication.

The extracted mixtures were separated by a Jasco LC-1500 HPLC system equipped with a UV/Visible detector, an analytical and a semipreparative Cosmosil "Buckyprep" coloumn and an Advantec SF-2100W fraction collector.

RESULTS AND DISCUSSION

According to the HPLC analysis, the first toluene extracts proved to be the mixtures of C_{60} and its dimer. The significant difference of the retention times (RT~8 min for C_{60} and RT~20 min for C_{120}) allowed their preparative separation in toluene, yealding 50 mg dimer of 99.3% purity. The purified dimer was assigned by its IR spectrum [4].

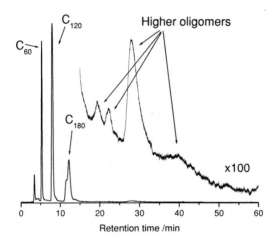

FIGURE 1. HPLC analysis of a fullerene photooligomer mixture. The RT~12 min peaks are mixtures of trimers, the RT~27 min peak is a tetramer.

Fig. 1. shows the chromatogram of a typical methylnaphtalene extract. We used here 20% methylnaphthalene/toluene eluent to prevent the broadening of the peaks of the higher oligomers and to decrease the analysis time. Besides the peaks of C_{60} and the dimer, an unresolved group of peaks can be observed at RT~12 min and a few low intensity peaks at higher retention times. According to their decomposition products, we assigned the RT~12 min and the RT~27 min peaks as trimers and tetramers, respectively: The oligomers of C_{60} decompose readily in solution at ambient temperature and illumination conditions [8]. We separated the oligomers by preparative HPLC then analyzed the collected products. The chromatogram of the RT~12 min material contained the peaks of the monomer and the dimer and only traces of the parent peak indicating that it was a trimer. Similarly, that of the RT~27 min material contained only the three preceeding C_{60}, C_{120} and C_{180} peaks and no other decomposition products. The other low intensity peaks in the range of 15-60 min are probably also tetramers but their amounts are too small for separation.

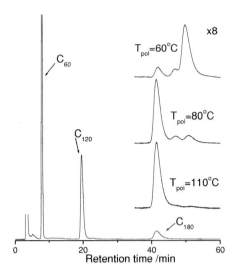

FIGURE 2. HPLC analysis of fullerene photooligomer mixtures. Full chromatogram: polymerization temperature: 110 °C. Insets: blow up of trimer peaks polymerized at different temperatures.

The resolution of the trimer peaks can be improved in toluene eluent due to the increased retention times. Fig. 2. shows chromatograms taken in toluene. The blow up details the peaks of trimers polymerized at different temperatures. Three trimers can be observed at RT= 41 min, 47 min and 50 min, respectively. At lower temperatures the RT=50 min trimer formed in largest amount while at higher temperatures the RT=41 min trimer became predominant. At 110 °C this was the only significant trimer. The purification and the collection of the major trimer components for structural determination is in progress.

Prior to detailed structural studies there are some indirect evidences indicating the possible structures of the observed trimers. Since the crystal structure of the C_{60} photopolymer is fcc [3] the molecular geometry of the trimers should be corresponded to the local symmetry of this structure. In the fcc lattice only four trimers can form with interfullerene angles of 60°, 90°, 120° and 180°. The first one corresponds to a closed triangle while the others are open structures. If the formation of the trimers is statistical, the expected relative abundancy of the above srtuctures is 2:2:4:1, respectively. Such a distribution is expected at low polymerization temperatures where no decomposition of the formed structures has to be taken into account. In these conditions the 120º V-shaped trimer is predominant. Considering the contraction along the cycloadduct bonds, the expected formation of the linear trimer is even less probable because of the increased distance between the reactants. At elevated temperatures the overall yield of the photopolymer decreases [9] while the relative amounts of the dimer and other soluble oligomers increases [4] indicating a dynamic equlibrium between the formation and decomposition of cycloadduct bonds. In these conditions the thermodynamically most stable components can accumulate in the photopolymer. Quantum chemical calculations predict the triangle as the most stable trimer [7] since it has higher number of cycloadduct bonds than the other isomers. The above arguments support that the three observed trimer peaks can be assigned as the 60º, 90º and 120º isomers in the order of their retention times.

CONCLUSIONS

Based on a recently worked out photopolymerization method, we produced various mixtures of soluble oligomers of C_{60} and separated them by HPLC. We isolated the C_{120} cycloadduct dimer in high purity form. For the first time we have shown that the fullerene photopolymer contains three different trimers and a few higher oligomers including at least one tetramer. We assigned the trimers and the tetramer according to their stepwise decomposition: the decomposition products of the higher polymerization degree oligomers consisted only of the lower ones. We detected three different trimers by analytical HPLC at 41 min, 47 min and 50 min retention times in toluene. The relative amounts of these isomers depended strongly on the polymerization temperature: the highest RT component formed predominantly at lower temperatures while the lowest RT one at higher temperatures. Preliminary considerations on the rate of formation and the thermodynamic stability indicate that the trimer peaks of increasing retention times correspond to the 60º, 90º and 120º interfullerene angle isomers, respectively. The preparative HPLC collection of the oligomers and structural studies are in progress.

ACKNOWLEDGMENTS

This work was supported by the grants OTKA: T032613, T029931 and T034198.

REFERENCES

1. Rao, A. M., Zhou, P., Wang, K.-A., Hager, G. T., Holden, J. M., Wang, Y., Lee, W.-T., Bi, X.-X., Eklund, P. C., Cornett, D. S., Duncan, M. A. and Amster, I. J., *Science* **259**, 955 (1993).
2. For a review see: Eklund, P. C., Rao, A. M., *Fullerene Polymers and Fullerene Polymer Composites*, Berlin: Springer, 2000.
3. Pusztai, T., Oszlányi, G., Faigel, G., Kamarás, K., Gránásy, L., and Pekker, S., *Solid State Commun.* **111**, 595 (1999).
4. Pekker, S., Kamarás, K., Kováts, É., Pusztai, T., and Oszlányi, G., *Synthetic Metals* **121**, 1109 (2001).
5. Pekker, S., Kováts, É., Kamarás, K., Pusztai, T., and Oszlányi, G., *Synthetic Metals* **133-134**, 685 (2003).
6. Komatsu, K., Wang, G.-W., Murata, Y., Tanaka, T., Fujiwara, K., Yamamoto, K., and Saunders, M., *J. Org. Chem.* **63**, 9358 (1998).
7. Kunitake, M., Uemura, S., Ito, O., Fujiwara, K., Murata, Y.,and Komatsu, K., *Angew. Chem. int. Ed.* **41**, 969 (2002).
8. Komatsu, K., Fujiwara, K., Tanaka, T., and Murata, Y.,*Carbon* **38**, 1529 (2000)
9. Burger, B., Winter, J., and Kuzmany, H., *Z. Phys. B* **101**, 227 (1996).

Isolation and FTIR analysis of endohedral $L_{3-x}M_xN@C_{80}$ ($0 \leq x \leq 3$) fullerenes

M. Krause[1], J. Noack[1], R. Marczak[1,2], P. Georgi[1], L. Dunsch[1]

[1] *Leibniz-Institute for Solid State and Materials Research Dresden, D-01171 Dresden, Germany*
[2] *Polish Academy of Sciences, Institute of Physical Chemistry, 01-224 Warsaw, Poland*

Abstract. $L_{3-x}M_xN@C_{80}$ ($0 \leq x \leq 3$; M = Sc, Y, Tb, Ho, Er) fullerenes have been isolated and studied by FTIR spectroscopy. Ionic radii and cage size are shown to affect the structure of the encaged trimetal nitride cluster. A correlation between cluster size, metal-nitrogen bond strength and stability of $M_3N@C_{80}$ fullerenes is proposed.

INTRODUCTION

Trimetal nitride (M_3N) fullerenes play a special role among the endohedral fullerenes due to a high abundance in the soot extract and the ability to stabilize non IPR fullerenes [1, 2]. The recent analysis of endohedral $Sc_3N@C_{80}$ fullerene revealed a crucial role of the Sc_3N cluster for the overall stability [3]. This work was now extended to other $M_3N@C_{80}$ structures with uniform and mixed metal clusters, where M = Y, Tb, Ho, and Er. FTIR spectra confirmed the preference of trimetal nitrides to the I_h-C_{80} isomer. Moreover our data point to a strong influence of the metal on the structure of the endohedral clusters and the stability of the clusterfullerenes as a whole.

EXPERIMENTAL

$L_xM_{3-x}N@C_{80}$ ($0 \leq x \leq 3$) clusterfullerenes were prepared by a modified Krätschmer-Huffman method. A reactive arc atmosphere was applied to improve the ratio of trimetal nitride fullerenes to empty fullerenes [4]. Non-fullerene products were removed with common organic solvents. The pre-purified soot was soxhlet extracted by CS_2 for 20 h. High performance liquid chromatography (HPLC) was used for fullerene separation. $M_3N@C_{80}$ (M = Sc, Y, Tb, Ho, Er) were isolated by a single separation step on a 4.6 x 250 mm BuckyPrep column (Nacalai Tesque) in purities > 95 %. For HPLC isolation of $Er_xSc_{3-x}N@C_{80}$ (x = 1, 2) three separation steps on a linear combination of three analytical BuckyPrep colums were used. $La_2@C_{80}$ was isolated by multi stage HPLC using BuckyPrep as well as Buckyclutcher columns. For spectroscopic measurements solutions of 20 - 100 µg fullerene in toluene were used to dropcoat KBr single crystal disks. Residing solvent was removed by heating the polycrystalline films in a vacuum of 10^{-6} mbar at 250 °C for 4 h. Room temperature FTIR spectra were recorded on a IFS 66v spectrometer (Bruker, Germany).

CP685, *Molecular Nanostructures: XVII Int'l. Winterschool/Euroconference on Electronic Properties of Novel Materials*, edited by H. Kuzmany, J. Fink, M. Mehring, and S. Roth
© 2003 American Institute of Physics 0-7354-0154-3/03/$20.00

RESULTS AND DISCUSSION

Tb₃N@C₈₀ was isolated for the first time. The chromatogram of the soot extract in Fig. 1 is dominated by the C_{60}, C_{70} and Tb₃N@C₈₀ peaks. The fullerenes are superimposed to a nonresolved background at shorter retention times due to non fullerene compounds. Quantitative analysis of 25 extracts lead to the following fullerene distribution: 17 % C_{60}, 27 % C_{70}, 45 % Tb₃N@C₈₀ and 11 % other fullerenes. The abundance of Tb₃N@C₈₀ was smaller than for Sc₃N@C₈₀, Ho₃N@C₈₀ and Er₃N@C₈₀ under the same experimental conditions [4, 5]. The mass spectrum in Figure 1 shows the Tb₃N@C₈₀ molecule peak and the expected isotopic pattern. A small peak of C_{60} is also visible. However its intensity is strongly enhanced in the negative ion scan and does not reflect the actual fullerene distribution.

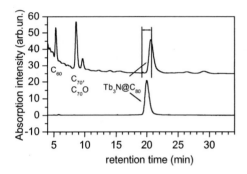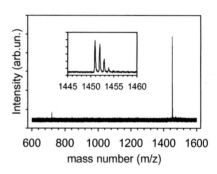

FIGURE 1. Left: Chromatogram of the extract with collection limits for Tb₃N@C₈₀ (upper trace) and of purified Tb₃N@C₈₀ (lower trace); 4.6 x 250 mm BuckyPrep column, 1.6 mlmin⁻¹ flow rate, 100 µl injection volume, eluent toluene; Right: negative ion scan MALDI-TOF mass spectrum of Tb₃N@C₈₀.

In Figure 2 the mid infrared spectra of 6 trimetal nitride fullerenes and La₂@C₈₀ are shown in comparison to a calculated spectrum of I_h-C_{80}^{6-} [6]. The tangential cage mode range is dominated by a very strong, split line group around 1380 cm⁻¹. Three medium intense line groups appeared around 1515, 1450, and 1200 cm⁻¹. Due to the small number of infrared lines and the general agreement of the vibrational structure Tb₃N@C₈₀, Ho₃N@C₈₀, ErSc₂N@C₈₀, Er₂ScN@C₈₀, and Er₃N@C₈₀ are assigned to the I_h-C_{80} cage isomer as it was previously established for La₂@C₈₀ and Sc₃N@C₈₀ [1, 8].

The most prominent feature at lower energies is a very strong, metal dependent line of the trimetal nitride fullerenes, which is split into two components in most cases. This line was not observed for La₂@C₈₀ (Figure 2, left). Explicit spectroscopic data are listed in Table 1. According to the experimental data and calculations for Sc₃N@C₈₀ [3] this line is assigned to the antisymmetric metal-nitrogen stretching vibration - ν_{as}(MN) - of the M₃N cluster.

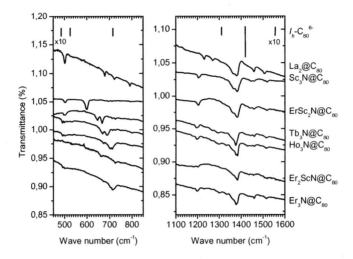

FIGURE 2. Mid infrared spectra of $La_2@C_{80}$ and various $L_xM_{3-x}N@C_{80}$; 500 accumulations, 2.0 cm^{-1} resolution; spectra were shifted along intensity axis for representation; a calculated FTIR spectrum of $C_{80}{}^{6-}$ taken from ref. [6] is included in form of vertical bars.

For a trigonal planar four atomic molecule this mode is doubly degenerated. A splitting into two components must be related to the loss of threefold symmetry. In Table 2 the ionic radii of the metals from our study are given. Tentatively our data suggest the following order for the quantities of influence on the size of the v_{as}(MN) line splitting:

$$\Delta \tilde{v}_{as,\,MN} \; (mixed\;metal\;cluster) \geq \Delta \tilde{v}_{as,\,MN} (ionic\;radius) \geq \Delta \tilde{v}_{as,\,MN} \; (fullerene\;cage)$$

$$(1)$$

TABLE 1. Spectroscopic data of the antisymmetric metal-nitrogen vibration in $L_xM_{3-x}N@C_{80}$ ($0 \leq x \leq 3$); intensity labels as usual

Sample	Wave numbers (cm^{-1})	Splitting energies (cm^{-1})
$Sc_3N@C_{80}$	599 vs	0
$Sc_3N@C_{78}$ (data taken from[7])	622 vs, 629 vs	7
$ErSc_2N@C_{80}$	647 vs, 667 vs	20
$Tb_3N@C_{80}$	669 s, 689 vs	20
$Ho_3N@C_{80}$	703 vs, 711 vs	8
$Er_2ScN@C_{80}$	661 s, 725 s	64
$Er_3N@C_{80}$	704 s, 713 vs	9
$Y_3N@C_{80}$ (data taken from[3])	712 vs, 724 vs	12
$La_2@C_{80}$	-	-

Table 2: Ionic radii of the metals under study (M^{3+} ions, coordination number = 6), after [9]

Sc^{3+}	Y^{3+}	Tb^{3+}	Ho^{3+}	Er^{3+}
74,5 pm	90,0 pm	92,3 pm	90,1 pm	89,0 pm

The frequency of $\nu_{as}(MN)$ allows further conclusions on the structure of the trimetal nitride cluster. For identical, trigonal planar valence force fields the frequency should depend on the reduced mass in the following order:

$$\tilde{\nu}_{as}(Sc_3N) = 1.025 \cdot \tilde{\nu}_{as}(Y_3N) = 1.035 \cdot \tilde{\nu}_{as}(Tb_3N) = 1.036 \cdot \tilde{\nu}_{as}(Ho_3N, Er_3N) \qquad (2)$$

In the experiment the relationship of Eq. (2) is almost fulfilled for $Y_3N@C_{80}$, $Ho_3N@C_{80}$, and $Er_3N@C_{80}$. Compared to these structures $Tb_3N@C_{80}$ and in particular $Sc_3N@C_{80}$ have significantly lower $\nu_{as}(MN)$ frequencies. Hence three force fields have to be taken into account for the 5 uniform $M_3N@C_{80}$ structures under study. This conclusion is strongly supported by the $\nu_{as}(MN)$ splitting energy ordering. The related parameters of the valence force field might be the (MN) valence force constant and the (MNM) as well as (M_3N) bond angles.

Our results emphasize the crucial role of the trimetal nitride cluster for the abundance and stability of trimetal nitride clusterfullerenes. Assuming ideal steric conditions in $Sc_3N@C_{80}$ a larger cluster size and a slight cluster deformation as e.g. in $Er_3N@C_{80}$ can be compensated by an increased metal-nitrogen bond strength without loosing overall stability. If the metal size increases more, the cluster deformation becomes larger, and the metal-nitrogen bond strength decreases. As consequence stability and the abundance of trimetal nitride fullerenes in the soot are reduced as observed for $Tb_3N@C_{80}$.

ACKNOWLEDGMENTS

Cooperation and helpful discussion with K. Vietze and G. Seifert (Dresden), and W. Kutner (Warsaw) are gratefully acknowledged. We cordially acknowledge assistance of H. Zöller, H. Großer, S. Döcke, K. Leger, and F. Ziegs (IFW Dresden).

REFERENCES

1. Stevenson, S., Rice, G., Glass, T., Harich, K., Cromer, F., Jordan, M.R., Craft, J., Hajdu, E., Bible, R., Olmstead, M.M., Maitra, K., Fisher, A.J., Balch, A.L., Dorn, H.C., *Nature* **401**, 55 (1999)
2. Stevenson, S., Fowler, P.W., Heine, T., Durchamp, J.C., Rice, G., Glass, T., Harich, K., Hajdu, E., Bible, R., Dorn, H.C., *Nature* **408**, 427 (2000)
3. Krause, M., Kuzmany, H., Georgi, P., Dunsch, L., Vietze, K., Seifert, G., *J. Chem. Phys.* **115**, 6596 (2001)
4. Dunsch, L., Georgi, P., Ziegs, F., Zöller, H., *Patent claim*, 2002
5. Noack, J., *Diploma thesis*, Dresden, 2002
6. Moriyama, M., Sato, T., Yabe, A., Yamamoto, K., Kobayashi, K., Nagase, S., Wakahara, T., Akasaka, T., *Chem. Lett.* 524 (2000)
7. Georgi, P., unpublished results
8. Akasaka, T., Nagase, S., Kobayashi, K., Wälchli, M., Yamamoto, K., Funasaka, H., Kako, M., Hoshino, T., Erata, T., *Angew. Chem.* **109**, 1716 (1997)
9. Greenwood, N.N., Earnshaw, A., *Chemistry of Elements*, VCH Weinheim, 1990, 1st edition

Electrochemical Response of Metallofullerene Films Casted on Electrodes

Louzhen Fan, Shangfeng Yang, and Shihe Yang[*]

Department of Chemistry, The Hong Kong University of Science and Technology, Clear Water Bay, Kowloon, Hong Kong, China

Abstract. We have studied solution casted films of the major isomer of $Dy@C_{82}$ ($Dy@C_{82}(I)$) by cyclic voltammetry (CV) in acetonitrile. The films are found to display pronounced and stable redox response in the solution. A pair of reversible oxidation and rereduction waves is observed after the reoxidation of a reduced film. The characteristics and the inter-relationship of these waves are uncovered by the CV technique, scanning electron microscopy (SEM), and UV-Vis-NIR spectra.

INTRODUCTION

The unique properties of endohedral metallofullerenes have attracted much interest, especially the electrochemistry of these exotic molecules.[1-3] In 1993, Suzuki et al. first reported the unusual redox properties of dissolved $La@C_{82}$ which significantly different from those of empty fullerenes.[4] In organic solution, $La@C_{82}$ exhibits five reversible reduction and one reversible oxidation processes, and the six reversible couples appear to be arranged in sets of two, unlike those of empty fullerenes.[5-7] So far many lanthanofullerenes $M@C_{82}$ (M= Y, La, Ce, Pr, Nd, Gd, Tb, Dy, Ho, Er, Lu, etc.) have been electrochemically studied.[8-10] Their electrochemical properties are similar. One common feature is that the difference between the first oxidation and the first reduction potentials is very small due to the open-shell electronic structure. As such, the metallofullerenes are strong electron donors as well as strong electron acceptors compared to empty fullerenes.[2-4,8-10] It has also been shown that reduction and oxidation of $La@C_{82}$ take place on the carbon cage, leading to a closed-shell electronic structure, so $La@C_{82}(-)$ and $La@C_{82}(+)$ are stable toward air and water.[11] In addition, two isomers have been extracted and isolated for $La@C_{82}$,[12-14] $Sc@C_{82}$,[15] $Y@C_{82}$,[15] and $Pr@C_{82}$,[16] etc. It was reported that the first oxidation potential of minor isomer shifts negatively by 143 mV relative to that of the major isomer.[14-16]

Electrochemical behavior of thin films of empty fullerenes such as C_{60} has been extensively studied,[17-19] and found to be very different from that of the dissolved species. The main feature of the electrochemical properties of C_{60} film is a large splitting in potential between the reduction and reoxidation waves for the first two electron-transfer reactions, which was interpreted with appreciable reorganization of the film during the reaction. Although interesting electric and magnetic properties are

CP685, *Molecular Nanostructures: XVII Int'l. Winterschool/Euroconference on Electronic Properties of Novel Materials,* edited by H. Kuzmany, J. Fink, M. Mehring, and S. Roth
© 2003 American Institute of Physics 0-7354-0154-3/03/$20.00

expected from doped charge-transfer metallofullerene solids,[4] to our knowledge, electron-transfer reactions of pure metallofullerene films on electrode surfaces have not been reported so far. Most recently, the electrochemistry of La@C_{82} on an artificial lipid film-modified electrode has been studied in water,[20] showing redox responses analogous to those of the solution phase. It was found that without the artificial lipid film, electron transfer of metallofullerene films on electrodes in aqueous solutions is difficult. Here we report the first study on the electron transfer of Dy@C_{82}(I) film on electrode surfaces in an organic solution, e.g., acetonitrile. The electrochemistry of the metallofullerene film is quite different from that of the dissolved Dy@C_{82}(I). Its reduction behavior is similar to that of empty fullerene film,[17-19] i.e., there are large splittings between the first two pairs of reduction and reoxidation waves. However, a pair of reversible oxidation/rereduction waves has been observed with a splitting of only 50 mV.

EXPERIMENTAL SECTION

High-purity Dy@C_{82} (99.5% as estimated with mass spectrometry) is prepared in our laboratory by a combination of standard DC arc-discharge method and isolation method described previously.[21-24] The raw soot from the arc-discharge is subjected to Soxlet extraction using N,N-dimethylformamide (DMF) as the solvent, followed by HPLC separation using a 5PYE column with toluene as the mobile phase. The identity and purity of the sample was verified by methane DCI negative ion mass spectrometry. Acetonitrile (Labscan Asia Co., Ltd) of a high-performance liquid chromatography (HPLC) grade was used as received. Electrochemical-grade tetra-*n*-butylammonium hexafluorophosphate (TBAPF$_6$), and tetra-*n*-butylammonium tetrafluoroborate (TBABF$_4$), and potassium hexafluorophosphate (KPF$_6$) from Aldrich Chemical Co., lithium perchlorate (LiClO$_4$), and sodium perchlorate (NaClO$_4$) from Acros Organics Co. were dried in vacuum at 60°C overnight before use.

Cyclic voltammetry experiments were performed using model 600 Electrochemical Analyzer from CH instruments Inc. U. S. A. in a conventional three-electrode electrochemical cell at ambient temperature. A Pt wire electrode served as the auxiliary electrode, A Ag/AgCl electrode was used as the reference. All electrochemical experiments were performed in a high purity N_2 atmosphere at ambient temperature (22±1°C). Scanning electron microscope (SEM) images were obtained with a field emission microscope (FSEM, JEOL JSM-6300F, Peabody, MA). Dy@C_{82}(I) films were prepared as follows by drop coating of a 1.0 µl solution of Dy@C_{82}(I) in toluene (6.07×10^{-2} mol/ml) on a Pt disk, a Pt plate or an ITO electrode. This was followed by drying in a vacuum at ambient temperature for a period of 4 h. Bulk controlled-potential electrolysis was performed in a conventional H-type cell with a model 600 Electrochemical Analyzer from CH instruments Inc. U. S. A.. The working and auxiliary electrode compartments of the cell were separated with a sintered glass frit. Both the working and the auxiliary electrodes were made of platinum gauze. The anions and cations of Dy@C_{82}(I) (Dy@C_{82}(I)(-) and Dy@C_{82}(I)(+)) were obtained in toluene/acetonitrile (4:1 by volume) containing 0.1 M

TBAPF$_6$ by setting the applied potentials ~100 mV more negative or more positive than E$_{1/2}$ for the redox couples of Dy@C$_{82}$(I)/Dy@C$_{82}$$^-$(I) and Dy@C$_{82}$$^+$(I)/Dy@C$_{82}$(I), respectively. The electrogenerated Dy@C$_{82}$(I)(-) and Dy@C$_{82}$(I)(+) were then transferred from the bulk cell to a quartz cuvette in a N$_2$ atmosphere. UV-Vis-NIR measurements were carried out with Perkin-Elmer LAMBDA 900 UV-Vis-NIR spectrometer under N$_2$ atmosphere. For the film electrode, the same procedure was performed with the film on ITO (indium tin oxide) electrodes.

RESULTS AND DISCUSSION

Figure 1a displays a typical cyclic voltammogram (CV) of a Dy@C$_{82}$(I) film on Pt electrode in acetonitrile containing TBAPF$_6$. Several pronounced redox peaks are observed and highly reproducible, suggesting that the metallofullerenes on the electrode are electroactive in the organic solution. This is in contrast to the case of the La@C$_{82}$ film in water,[20] where electron transfer is obstructed. As shown in Figure 1a, the first and second reduction waves appear at –0.8 V and –1.2 V (vs. Ag/AgCl), respectively. These values are to be compared with the corresponding solution reduction waves at –0.2 V and –0.8 V (vs. Ag/AgCl), respectively (see Figure 1b).

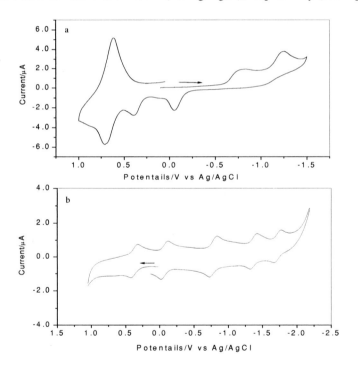

Figure 1. Cyclic voltammograms. (a) A Dy@C$_{82}$(I) film on a Pt electrode (electrolyte: 0.1 M TBAPF$_6$ in MeCN; scan rate: 50 mV/s). (b) A Dy@C$_{82}$(I) solution in toluene/MeCN (4:1, v) (electrolyte: 0.1 M TBAPF$_6$; scan rate: 50 mV/s).

When the potential scan is reversed toward the positive direction, large splittings between the reduction and reoxidation waves for the first and second electron-transfer reactions are observed. This is very similar to redox behavior of C_{60} film,[17-19] but the wave splittings for the $Dy@C_{82}(I)$ film are larger than for C_{60} film (the first reduction wave: 1.2 V vs. 0.5 V; the second reduction wave: 1.2 V vs. 0.2 V). In contrast to the C_{60} film, however, an oxidation wave is observed during the subsequent positive scan and a cathodic wave associated with this oxidation wave appears upon further potential scan reversal to the negative direction. Interestingly, the potential separation between these anodic and cathodic peaks is 50 mV, which is in good agreement with the characteristic value of 59 mV for a reversible one-electron transfer.

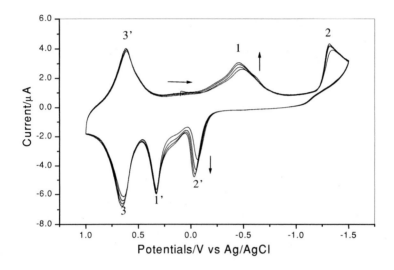

Figure 2. The second and subsequent cyclic voltammograms of a $Dy@C_{82}(I)$ film on a Pt electrode (electrolyte: 0.1 M $TBAPF_6$ in MeCN; scan rate: 50 mV/s).

For the C_{60} film, in addition to the absence of a cathodic wave associated with its oxidation wave, the electroactivity of the film was diminished after cycling over the anodic wave.[17] However, the electroactivity of the $Dy@C_{82}(I)$ film is maintained after many cycles of oxidation and rereduction waves (see Figure 2). From the second scan cycle, the first reduction wave shifts positively to –0.4 V and remains in this position for subsequent scans. The film is stable on Pt or ITO electrodes in the presence of $TBAPF_6$. The first several scans over the first two reduction waves and the first oxidation wave cause only small changes in potential and small increases in peak current. In subsequent cycles, both the potentials and currents of the redox waves are essentially unchanged even after 2 h of continuous scanning at 50 mV/s. However, the film starts to dissolve and the reduction waves become similar to those of the metallofullerene in solution when the potential is scanned beyond the third reduction wave. This is likely due to (1) the elevated negative potential which tends to desorb the negatively charged species, and (2) the increased solubility of the $Dy@C_{82}^{n-}(I)$ species in the polar solvent acetonitrile as n increases.[19]

In order to examine the nature of the redox waves of the metallofullerene films in more detail, potential scans are carried out by setting starting and reversal potentials at different positions and in different directions. When the negative scan reversal is set at −1.1 V, i.e., before the second reduction occurs, only one reoxidation wave at E = +0.37 V (vs. Ag/AgCl) is observed. If the negative potential scan reversal is extended negatively to −1.5 V, a cathodic wave2 appears accompanied by an anodic wave2'. This demonstrates unambiguously that the anodic wave2' and wave1' are associated with the cathodic wave2 and wave1, respectively. On the other hand, if the potential scan is started at 0.1 V in the positive direction, only an anodic wave3' and a cathodic wave3 are observed between 0.1 and 1.0 V. In the subsequent potential scan cycles, the redox couples wave1/wave1' and wave2/wave2' show up again, suggesting that the wave3' is rereduction of the wave3.

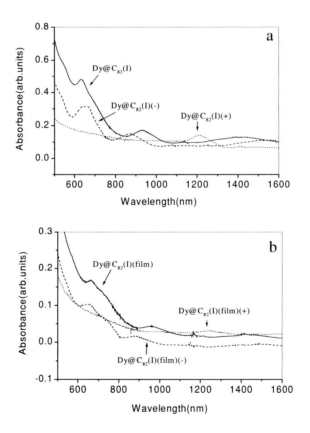

Figure 3. UV-Vis-NIR absorption spectra of $Dy@C_{82}(I)$, $Dy@C_{82}(I)(-)$, and $Dy@C_{82}(I)(+)$. (a) Solution of 0.1 M TBAPF$_6$ in toluene/MeCN (4:1, v). (b) Film on ITO. The $Dy@C_{82}(I)(-)$ and $Dy@C_{82}(I)(+))$ films were obtained by holding the electrode potential, respectively, right after wave1 and wave2 for ~10 s.

Akasaka et al. have shown that the electrogenerated $La@C_{82}^-$ and $La@C_{82}^+$ are very stable in solution due to their closed electronic shell structures.[11,25] Here the electrogenerated metallofullerene monoanion and monocation are also quite stable in the films, and this has allowed us to further confirm the nature of wave1 and wave3 by spectroscopy. For the sake of comparison, we first present the solution phase UV-Vis-NIR absorption spectra of $Dy@C_{82}(I)$, $Dy@C_{82}(I)(-)$, and $Dy@C_{82}(I)(+)$ in Figure 3a. As reported previously,[26] $Dy@C_{82}(I)$ shows characteristic absorption bands at 640, 937, and 1400 nm. For $Dy@C_{82}(I)(-)$, the absorption peaks are shifted to 650, 880, and ~1570 nm, respectively. However, only an obvious absorption peak appears at 1210 nm for $Dy@C_{82}(I)(+)$. Figure 3b shows the UV-Vis-NIR absorption spectra of the corresponding species in the film on ITO. Except for some small spectral shifts, the one-to-one correspondence between the film spectra and the solution spectra is evident. $Dy@C_{82}(I)(film)$ shows three peaks at 695, 950, and 1410 nm. $Dy@C_{82}(I)(-)(film)$ shows three peaks at 655, 875, and ~1520 nm. $Dy@C_{82}(I)(+)(film)$ shows only one peak at 1230 nm. By and large, the absorption peaks of the metallofulerene films are red-shifted from the corresponding spectra in solution due perhaps to the inter-metallofullerene interaction and/or the interaction between the metallofullerenes and the substrate/counter ions.

As we know, the $Dy@C_{82}(I)(-)$ film was obtained by holding the electrode potential right after wave1 for ~10 s. Moreover, the spectrum of this film is similar to that of $Dy@C_{82}(I)(-)$ in solution. These together verify that wave1 is associated with the reduction of $Dy@C_{82}(I)(film)$ to $Dy@C_{82}(I)(-)(film)$. By the same token, it is ascertained that wave3 results from the oxidation of $Dy@C_{82}(I)(film)$ to $Dy@C_{82}(I)(+)(film)$ through the spectral comparison.

The proposed mechanism for the electrode processes is illustrated as follows:

$$(Dy@C_{82}(I))_A + TBA^+ + PF_6^- + e^- \xrightleftharpoons{\text{Reduction Wave}} [\,TBA^+(Dy@C_{82}(I))^-\,]_A + PF_6^-$$

$$(Dy@C_{82}(I))_B + TBA^+ + PF_6^- + e^- \qquad\qquad [\,TBA^+(Dy@C_{82}(I))^-\,]_B + PF_6^-$$

Rereduction wave | Reoxidation Wave

Oxidation Wave

$$[(Dy@C_{82}(I))^+PF_6^-\,]_B + TBA^+ + 2e^- \xrightleftharpoons{} TBA^+ + (Dy@C_{82}(I))_B + e^- + PF_6^-$$

It appears that the structure of the $Dy@C_{82}(I)$ film after reoxidation of the reduced film is responsible for the reversible oxidation of the film. Such a structurally reorganized neutral $Dy@C_{82}(I)$ film is oxidized and rereduced on the electrode accompanied by the incorporation of PF_6^- ions into the film on the positive sweep and the expulsion of the same ions on the subsequent negative sweep. As a result, a very small splitting value between the oxidation and rereduction waves is obtained. As proposed for C_{60} films,[17] we also attribute the large splittings between the first two pairs of the reduction/reoxidation waves of the $Dy@C_{82}(I)$ film to the attendant structural reorganization of the film due to the accommodation of the bulky TBA^+ ions for charge balance. Note that the subscripts A and B represent two different structural forms of the $Dy@C_{82}(I)$ film: B is more stable for the reduced film and A is a more stable form of the neutral film. More specifically, as $Dy@C_{82}(I)$(film) is reduced to $Dy@C_{82}(I)^-$(film), TBA^+ diffuses towards it to form $[TBA^+Dy@C_{82}(I)^-]_A$. Because of the large size of TBA^+, its intercalation requires a structural reorganization of the metallofullerene film. This explains the large splittings of the reduction/reoxidation waves. The reorganized structure B is apparently more uniform and accessible to TBA^+, not mentioning the other ions. Because PF_6^- is much smaller than TBA^+, it can easily get into and out of the reorganized structure B. As such, no further structural reorganization is needed after the oxidation/rereduction of the film, hence a small splitting of the pair is obtained. After the rereduction, the B structure of the film is relaxed to the initial structure A, closing the CV scanning circle.

CONCLUSION

For the first time, we have studied the electrochemical properties of pure metallofullerene films cast from $Dy@C_{82}(I)$ solutions in the acetonitrile solvent. The metallofullerene films are stable against repeated potential cycling. The reduction and reoxidation responses of these films are similar to those of C_{60} films, i.e., there are large splittings between the first two reduction and reoxidation waves. In sharp contrast, however, a reversible pair of oxidation and rereduction waves with a splitting only 50 mV are successfully obtained for the $Dy@C_{82}(I)$ films. It is demonstrated that these reversible oxidation and rereduction waves necessitate the structural reorganizations of the $Dy@C_{82}(I)$ films after reoxidation of the reduced species.

ACKNOWLEDGMENTS

This work was supported by an RGC grant administered by the UGC of Hong Kong. We thank MCPF of HKUST for assistance in sample characterization.

REFERENCES

1. Nagase, S., Kobayashi, K., Akasaka, T., and Wakahara, T., *Fullerenes: Chemistry, Physics and Technology*, edited by K. M. Kadish et al., Electrochemical Society Proceedings, New York: Wiley, 2000, pp. 395-436.
2. Shinohara, H., *Rep. Prog. Phys.* **63**, 843-897 (2000).
3. Yang, S. H., *Trends in Chemical Physics* **9**, 31-43 (2001).
4. Suzuki, T., Maruyama, Y., Kato, T., Kikuchi, K., and Achiba, Y., *J. Am. Chem. Soc.* **115**, 11006-11007 (1993).
5. Li, Q., Wudl, F., Thilgen, C., Whetten, R. L.,and Diederich, F., *J. Am. Chem. Soc.* **114**, 3994-3996 (1992).
6. Dubois, D., Kadish, K. M., Flanagan, S., and Wilson, L., *J. Am. Chem. Soc.* **113**, 7773-7774 (1991).
7. Meier, M. S., Guarr, T. F., Selegue, J. P., and Vance, V. K., *J. Chem. Soc., Chem. Commun.* 63-65 (1993).
8. Suzuki, T., Kikuchi, K., Oguri, F., Nakao, Y., Shinzo, S., Achiba, Y., Yamamoto, K., Funasaka, H., and Takahashi, T., *Tetrahedron* **52**, 4973-4982 (1996).
9. Anderson, M. R., Dorn, H. C., and Stevenson, S. A., *Carbon* **38**, 1663-1670 (2000).
10. Wang, W. L., Ding, J. Q., Yang, S. H., and Li, X. -Y., *Recent Advances in Chemistry and Physics of Fullerenes and Related Materials*, edited by K. M. Kadish et al., Electrochemical Society Proceedings V4, Pennington: Wiley, 1997, pp. 417-428.
11. Akasaka, T., Wakahara, T., Nagase, S., Kobayashi, K., Waelchli, M., Yamamoto, K., Kondo, M., Shirakura, S., Okubo, S., Maeda, Y., Kato, T., Gao, X., Caemelbecke, E. V., and Kadish, K. M., *J. Am. Chem. Soc.* **122**, 9316-9317 (2000).
12. Kikuchi, K., Suzuki, S., Nakao, Y., Nakahara, H., Wakabayashi, T., Shiromaru, H., Saito, K., Ikemoto, I., and Achiba, Y., *Chem. Phys. Lett.* **216**, 67-71 (1993).
13. Yamamoto, K., Funasaka, H., Takahashi, T., and Akasaka, T., *J. Phys. Chem.* **98**, 2008-2011 (1994).
14. Yamamoto, K., Funasaka, H., Takahashi, T., and Akasaka, T., *J. Phys. Chem.* **98**, 12831-12833 (1994).
15. Wakabayashi, T., Okubo, S., Kondo, M., Maeda, Y., Akasaka, T., Waelchli, M., Kako, M., Kobayashi, K., Nagase, S., Kato, T., Yamamoto, K., Gao, X., Caemelbecke, E., and Kadish, K. M., *Chem., Phys. Lett.* **360**, 235-239 (2002).
16. Akasaka, T., Okubo, S., Kondo, M., Maeda, Y., Wakahara, T., Kato, T., Suzuki, T., Yamamoto, K., Kobayashi, K., and Nagase, S., *Chem., Phys. Lett.* **319**, 153-156 (2000).
17. Christophe, J., Bard, A. J., and Wudl, F., *J. Am. Chem. Soc.* **113**, 5456-5457 (1991).
18. Kob, W., Dubis, D., Wloszimier, K., Thomas, M. T., and Kadish, K. M., *J. Phys. Chem.* **97**, 6871-6879 (1993).
19. Chlistunoff, J., Cliffel, D., and Bard, A. J., *Thin Solid Film* **257**, 257, 166-184 (1995).
20. Nakashima, N., Sakai, M., Murakami, H., Sagara, T., Wakahara, T., and Akasaka, T., *J. Phys. Chem.* B **106**, 3523-3525 (2002).
21. Ding, J. Q., and Yang, S. H., *Chem. Mater.* **8**, 2824-2827 (1996).
22. Ding, J. Q., and Yang, S. H., *Angew. Chem. Int. Ed. Engl.* **35**, 2234-2235 (1996).
23. Ding, J. Q., Weng, L. T., and Yang, S. H., *J. Phys. Chem.* **100**, 11120-11121 (1996).
24. Huang, H. J., and Yang, S. H., *J. Phys. Chem.* B **102**, 10196-10200 (1998).
25. Akasaka, T., Wakahara, T., Nagase, S., Kobayashi, K., Waelchli, M., Yamamoto, K., Kondo, M., Shirakura, S., Maeda, Y., Kato, T., Kako, M., Nakadara, Y., Gao, X., Caemelbecke, E., and Kadish, K. M., *J. Phys. Chem.* B **105**, 2971-2974 (2001).
26. Tagmatarchis, N., and Shinohara, H., *Chem. Mater.* **12**, 3222-3226 (2000).

Distortions of C_{60}^{4-} studied by infrared spectroscopy

G. Klupp, F. Borondics, G. Oszlányi and K. Kamarás

Research Institute for Solid State Physics and Optics, Hungarian Academy of Sciences, P. O. Box 49, H-1525 Budapest, Hungary

Abstract. The Jahn-Teller effect plays a crucial role in the explanation of the insulating character of A_4C_{60} (A = K, Rb, Cs). To detect possible phase transitions arising from the interplay between the molecular Jahn-Teller distortion and the distorting potential field of the counterions, we measured the mid-IR spectra of A_4C_{60} compounds in the temperature range 90 - 300 K and found significant spectral changes with temperature in all three compounds. We also compare these spectra to that of Na_4C_{60} in its room-temperature polymeric phase, where the distortion is more pronounced and evident from the structure.

INTRODUCTION

The A_4C_{60} (A = K, Rb, Cs) fullerides are insulating, despite the predictions of band structure calculations. As an explanation, the concept of the Mott-Jahn-Teller insulator has been proposed [1]. In the C_{60}^{4-} molecule the coupling of the threefold degenerate t electronic state with the H_g vibrational modes will lead to a Jahn-Teller distortion that can change the molecular symmetry from I_h to either D_{5d} or D_{3d} (uniaxially distorted), or to D_{2h} (biaxially distorted) [2]. In Cs_4C_{60} there is direct evidence for the distortion: Dahlke et al. observed D_{2h} geometry at both 5 and 293 K by neutron diffraction [3].

Na_4C_{60} has a different structure at room temperature from the above three: it forms two dimensional polymer sheets [4]. In this polymer the fullerene balls are distorted to C_i geometry via the sigma bonds connecting the adjacent monomer units.

Here we present midinfrared (mid-IR) spectra of Na_4C_{60}, K_4C_{60}, Rb_4C_{60} and Cs_4C_{60} to compare the distortions and to draw conclusions concerning the Jahn-Teller effect in A_4C_{60}. We find the spectral pattern consistent with C_i in Na_4C_{60}. In the other three compounds, our results indicate much smaller distortion, varying with temperature. The changes suggest a staggered static distortion or a dynamic Jahn-Teller effect in the high-temperature state, and the dominance of the distortion caused by the potential field of the surrounding ions in the low temperature state.

EXPERIMENTAL

Fulleride salts with more than 95 % purity were prepared by reacting stoichiometric amounts of the alkali metal and C_{60} at 350 °C for K, Rb, Cs, and at 200 °C for Na_4C_{60}. Mid-IR transmittance of the fullerides were measured in KBr pellets pressed in dry box.

CP685, *Molecular Nanostructures: XVII Int'l. Winterschool/Euroconference on Electronic Properties of Novel Materials,* edited by H. Kuzmany, J. Fink, M. Mehring, and S. Roth

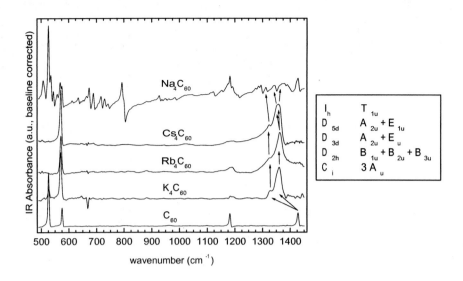

FIGURE 1. Left panel: Room temperature spectra of C_{60}, K_4C_{60}, Rb_4C_{60}, Cs_4C_{60} and Na_4C_{60} with the splitting of the highest frequency T_{1u} mode shown with arrows. Right panel: splitting of T_{1u} modes in different pointgroups.

The IR measurements were performed in dynamic vacuum in a liquid nitrogen cooled flow-through cryostat with a Bruker IFS28 FT-IR spectrometer using 2 cm^{-1} resolution.

RESULTS

The room temperature spectra of the four measured compounds compared to that of C_{60} are shown in Fig.1. In K_4C_{60} and Rb_4C_{60} the two higher-frequency T_{1u} modes of C_{60} show a twofold splitting according to fits with Lorentzians (Fig.2). Assuming a JT distortion twofold splitting occurs when the distortion is D_{3d} or D_{5d} (see right panel of Fig. 1). In Cs_4C_{60} the splitting is threefold (see fits in Fig.2), indicating D_{2h} geometry, which is in accordance with the result of neutron diffraction [3]. In Na_4C_{60} the T_{1u} modes also show a threefold splitting in accordance with the expected C_i geometry [4].

There are numerous spectral features associated with the chemical intermolecular bonds in Na_4C_{60} and the strongly distorted molecular geometry they cause. In contrast to the other A_4C_{60} fullerides, where splitting of the two lower frequency T_{1u} modes is not observed clearly with the present resolution, in Na_4C_{60} all of the T_{1u} modes are split. In addition, many strong new modes appear in the spectrum, while even Cs_4C_{60} – the A_4C_{60} compound having the lowest symmetry fulleride ion – exhibits only few weak new modes. We regard the strong feature around 800 cm^{-1} as direct evidence for the

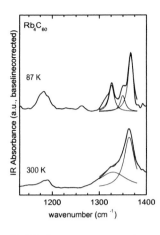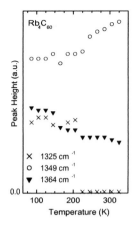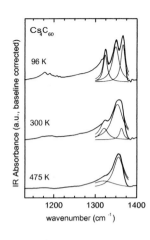

FIGURE 2. Temperature dependence of the spectra of Rb_4C_{60} and Cs_4C_{60}. First panel: the two highest frequency T_{1u} modes of Rb_4C_{60} at 87 K and at 300 K (bold line) with fitted Lorentzian curves (thin line). Second panel: change of peak height of the components of the highest frequency T_{1u} peak in Rb_4C_{60}, indicating a transition between 200 - 220 K. Third panel: the two highest frequency T_{1u} modes of Cs_4C_{60} at 96 K, at 300 K and at 475 K (bold line) with fitted Lorentzian curves (thin line).

single covalent bond, similar to those found in sigma-bonded dimers [5, 6].

The temperature dependence of the IR spectrum of Rb_4C_{60} and Cs_4C_{60} is shown in Fig. 2. The spectra of K_4C_{60} are similar to Rb_4C_{60}, and were shown in Ref. [7]. At low temperature the two highest frequency T_{1u} modes of all three A_4C_{60} salts are split into three, corresponding to a D_{2h} distortion (Fig. 1). On heating these three lines collapse into two, corresponding to a uniaxial distortion. (The latter finding in Cs_4C_{60} is contrary to our earlier statement [8] and reflects the results of the fits shown here.) The change of the spectra with temperature can be best followed by the disappearance of peaks, shown in Fig. 2 and in Ref. [8]. The transitions found occur in the following temperature ranges: K_4C_{60}: 260 - 280 K, Rb_4C_{60}: 200 - 220 K, Cs_4C_{60}: 400 K.

DISCUSSION

In Cs_4C_{60} the low temperature crystal structure is orthorhombic (*Immm*) [3], thus the D_{2h} fulleride ions are orientationally ordered. On heating the crystal structure changes to tetragonal (*I4/mmm*) [3]. By analogy, a similar structural phase transition in K_4C_{60} and Rb_4C_{60} is also likely: NMR relaxation time measurements [9] point in this direction and this was the explanation put forward in our earlier publications [7, 8]. However, a gradual freezing of molecular motion below the IR time scale would also result in similar spectral changes. In this case, the anions could still show orientational disorder occupying two standard orientations in a tetragonal structure. The importance of time scales is also reflected in the fact that the line splitting in Cs_4C_{60} occurs at a higher temperature than in NMR measurements, where a change has been detected between

300 and 350 K [10].

As the potential energy surface of these systems has minima at either D_{5d} or D_{3d} symmetry, and saddle points at D_{2h} symmetry [2], a biaxial distortion can only be realized when the potential field of the surrounding ions lock the molecules in a D_{2h} geometry. As both the tetragonal and orthorhombic arrangements lead to such a potential field, the structure cannot be deducted from the spectra only. Further diffraction studies are needed to clarify this point.

In the high temperature A_4C_{60} phases the crystal structure is bct [11], but the spectra are only consistent with D_{3d} or D_{5d} molecular geometry. As the main axis in these point groups is neither two- nor fourfold, the distortion of the anion obviously is not determined by the crystal field of the solid alone. There are two possible scenarios for the structure of this phase: a staggered static distortion [1, 8] where the molecule's main axis does not coincide with the c axis of the crystal, or a dynamic Jahn-Teller effect where pseudorotation occurs between uniaxial geometries with a characteristic frequency lower than that of the infrared. The fact that the transition temperatures do not scale with cation size is another indication that more than one process is involved. Sorting out the contributions from molecular dynamics and crystal field is a complicated procedure, but exactly this perspective makes these materials interesting.

ACKNOWLEDGMENTS

This work was supported by OTKA grants T 034198 and T 029931.

REFERENCES

1. Fabrizio, M., and Tosatti, E., *Phys. Rev. B* **55**, 13465 (1997).
2. Chancey, C. C., and O'Brien, M. C. M., *The Jahn-Teller effect in C_{60} and Other Icosahedral Complexes* (Princeton University Press, Princeton, 1997).
3. Dahlke, P., and Rosseinsky, M. J., *Chem. Mater.* **14**, 1285 (2002).
4. Oszlányi, G., Baumgartner, G., Faigel, G., and Forró, L., *Phys. Rev. Lett.* **78**, 4438 (1997).
5. Kamarás, K., Tanner, D. B., and Forró, L., *Fullerene Sci. Tech.* **5**, 465 (1997).
6. Kürti, J., Borondics, F., and Klupp, Gy., *AIP Conference Proceedings* **591**, 25 (2001).
7. Kamarás, K., Klupp, G., Tanner, D. B., Hebard, A. F., Nemes, N. M., and Fischer, J. E., *Phys. Rev. B* **65**, 052103 (2002).
8. Kamarás, K., Klupp, G., Borondics, F., Gránásy, L., and Oszlányi, G., *AIP Conference Proceedings* **633**, 55 (2002).
9. Brouet, V., Alloul, H., Garaj, S. and Forró, L., *Phys. Rev. B* **66**, 155122 (2002).
10. Goze, C., Rachdi, F. and Mehring, M., *Phys. Rev. B* **54**, 5164 (1996).
11. Fleming, R. M., Rosseinsky, M. J., Ramirez, A. P., Murphy, D. W., Tully, J. C., Haddon, R. C., Siegrist, T., Tycko, R., Glarum, S. H., Marsh, P., Dabbagh, G., Zahurak, S. M., Makhija, A. V., and Hampton, C., *Nature* **352**, 701 (1991).

Field-doping of C_{60} crystals: A view from theory

Erik Koch[*], Olle Gunnarsson[*], Samuel Wehrli[†] and Manfred Sigrist[†]

[*]Max-Planck-Institut für Festkörperforschung, Heisenbergstraße 1, D-70569 Stuttgart
[†]Theoretische Physik, ETH-Hönggerberg, CH-8093 Zürich

Abstract. The proposal of using the field-effect for doping organic crystals has raised enormous interest. To assess the feasibility of such an approach, we investigate the effect of a strong electric field on the electronic structure of C_{60} crystals. Calculating the polarization of the molecules and the splittings of the molecular levels as a function of the external field, we determine up to what field-strengths the electronic structure of C_{60} stays essentially unchanged, so that one can speak of field-effect doping, in the sense of putting charge carriers into otherwise unchanged states. Beyond these field strengths, the electronic structure changes so much, that on can no longer speak of a doped system. In addition, we address the question of a metal-insulator transition at integer dopings and briefly review proposed mechanisms for explaining an increase of the superconducting transition temperature in field-doped C_{60} that is intercalated with haloform molecules.

The doped fullerenes are materials with very interesting properties. Alkali doped C_{60} with three alkalis per molecule has, e.g., turned out to be metallic, though close to a Mott transition, and superconducting. A problem is, however, that different doping levels can only be realized by preparing separate crystals. Moreover, because of the strong electronegativity of C_{60}, no hole-doping has been achieved. In this context the proposal of using a field-effect transistor for doping pristine C_{60} crystals has raised much interest, in particular since such an approach should allow us to continuously change the doping by simply changing the voltage applied to the gate electrode. Sadly the reports of such field-doping and of spectacular values for the superconducting transition temperatures in such devices [1, 2, 3] have been withdrawn [4] after an investigation showed that the publications were based on fraudulent data [5]. Nevertheless, it is an open question whether field-doping of C_{60} crystals could be achieved in principle. In the following we address several aspects of this question.

EFFECT OF A STRONG ELECTRIC FIELD

Reaching substantial charging (of the order of n electrons per C_{60} molecule) in a field-effect device requires enormous electric fields. As the induced charge is basically restricted to one monolayer of C_{60} [6], a rough estimate can be obtained from basic electrostatics: For neutrality the charge on the gate must equal that on the monolayer, hence the field originating from the gate electrode is given by $E_{\text{gate}} = 2\pi n / A_{\text{mol}}$, where A_{mol} is the area per C_{60} molecule in the monolayer and n is the number of induced electron charges per molecule. Thus the external field is about 1 V/Å per induced charge, corresponding to a voltage drop of about 10 eV across the molecule. In this strong

CP685, *Molecular Nanostructures: XVII Int'l. Winterschool/Euroconference on Electronic Properties of Novel Materials,* edited by H. Kuzmany, J. Fink, M. Mehring, and S. Roth
© 2003 American Institute of Physics 0-7354-0154-3/03/$20.00

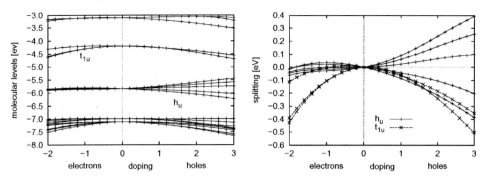

FIGURE 1. Splitting of the molecular levels in the self-consistent multipole field ($l \le 2$) for a (001) monolayer (square lattice) of C_{60} molecules oriented such that one of their two-fold axes points in the direction of the external electric field (perpendicular to the layer).

external field the C_{60} molecules are strongly polarized. Nevertheless, we find that their response is still in the linear regime. Furthermore, in the charged monolayer, the field experienced by a molecule is screened by the polarization of the neighboring molecules. Taking this into account, we find that the field is reduced by about a factor of two. Calculating the splitting of the molecular levels in this screened homogeneous field, we find a quadratic Stark effect, with the splitting of both, the t_{1u} and the h_u level, becoming of the order of the band-width for a field corresponding to an induced charge of three to four charges per molecule. This seems to be consistent with the typical doping levels that had been reported.

For a more realistic description of the electrostatics in the field-effect device, we have, however, to go beyond considering only a homogeneous field. Given the spherical shape of C_{60}, the natural approach is via a multipole expansion [7]: We choose an external field and the corresponding induced charge per molecule. We determine the multipole expansion of the field generated by all other molecules about the molecule centered at the origin. Using the linear response of a C_{60} molecule to multipole fields (calculated *ab initio*), we determine the new charge distribution on the molecules and repeat the procedure until self-consistency is reached.

Figure 1 shows the splitting of the molecular levels in the self-consistent multipole field for different doping levels. While for an external homogeneous field ($l = 1$ multipole) the splitting is independent of the direction of the field, including the effect of the induced charge on the neighboring molecules breaks this symmetry. Surprisingly, the asymmetry in the splitting is quite strong, even though the fields that break the symmetry are fairly weak compared to the homogeneous field. This is because the multipole potentials with even l give rise to a *linear* Stark effect, which changes sign with the external potential and which gives rise to a strong splitting even for weak fields. In addition it turns out [7] that the splitting due to the $l = 1$ and $l = 2$ potentials add or subtract, depending on the sign of the external field: When inducing electrons they add for the t_{1u} level and almost cancel for the h_u, while when inducing holes the situation is reversed. I.e., when a molecular level is filled, the splitting is substantially enhanced. It reaches the order of the band width when inducing about two electrons, or somewhat more than

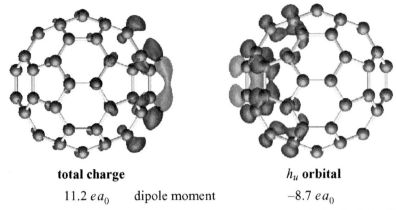

total charge		h_u orbital
11.2 $e a_0$	dipole moment	$-8.7\ e a_0$

FIGURE 2. Polarization of a C_{60} molecule in a homogeneous external electric field of 0.02 a.u. ($\approx$ 1 V/Å). The change in charge density compared to the field-free case is indicated by the $\Delta\rho$-isosurface at $0.0020/a_0^3$. It turns out that the dipole moment for the h_u (HOMO) charge density is of the same order of magnitude as that of the total charge — *but of the opposite sign.*

two holes per molecule. Beyond these fillings the effect of the Stark splitting on the electronic structure of the C_{60} monolayer that carries the induced charge will clearly be very large, and one can definitely no longer speak of doping.

HOMO-ANTISCREENING

As we have seen, the charge density of the C_{60} molecule is strongly polarized in an electric field, and one might expect that the main contribution comes from the polarization of the highest molecular orbital (HOMO). Calculating the change in the HOMO charge density, we find, however, that the dipole moment of the HOMO charge density is of the same order of magnitude as that of the total charge – *but of the opposite sign* (see Fig. 2). This surprising result can be understood, e.g., in terms of perturbation theory: Expanding the wave function to first order in the external field $V = E_z z$ and calculating the dipole moment $p = ez$, we find that the leading term is given by a sum over the matrix element, squared, with all unperturbed molecular orbitals of different parity divided by the characteristic energy denominator. Hence the main contribution comes from energetically close levels, and the sign of their contribution is determined by whether they are energetically above or below the level under consideration. For the HOMO in a molecule with a large HOMO-LUMO gap this means that the contributions mainly come from the molecular levels below – implying antiscreening. We thus see that HOMO-antiscreening should be quite general for molecules with large HOMO-LUMO gap, and, in fact, it can also be found, e.g., in the series of polyacenes: benzene, naphthalene, anthracene, tetracene, and pentacene.

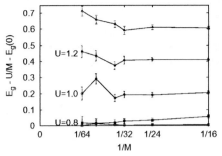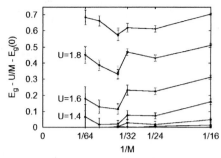

FIGURE 3. Gap $E_g = E(N-1) - 2E(N) + E(N+1)$ with finite-size correction $E_g - E_g(U=0) - U/M$ as calculated by quantum Monte Carlo for the t_{1u} and the h_u band in the (111)-plane of the Pa$\bar{3}$ structure. For the half-filled t_{1u}-band the gap opens between $U_c = 0.8\ldots1.2$ eV, for the half-filled h_u-band between $U_c = 1.2\ldots1.6$ eV.

MOTT TRANSITION

Since the bands in the fullerenes are narrow, while the Coulomb repulsion between two electrons on the same molecule is sizable, the doped fullerenes show effects of strong correlation. It is, e.g., only due to orbital degeneracy that A_3C_{60} is metallic and not a Mott insulator [8]. In field-doped fullerenes the electrons are restricted to a monolayer [6]. Hence the number of nearest neighbors to which an electron can hop is reduced and the bands are even more narrow. It is therefore expected that the Mott transition occurs at critical values of the Coulomb interaction U below those found in the bulk. To determine the transition point, we have performed quantum Monte Carlo calculations [9] for a doped (111)-layer (without Stark splitting) and find that the Mott transition occurs between $U_c = 0.8 - 1.2$ eV for doping with three electrons, and $U_c = 1.2 - 1.6$ eV for the half-filled h_u band (see Fig. 3). For integer dopings other than half-filling the transition is expected to occur for even smaller values of U [9]. Furthermore, the splitting of the molecular levels in the electric field should weaken the effect of the degeneracy on the Mott transition and lead to still smaller values of U_c [10]. In particular for the h_u orbital, the strong electron-phonon coupling might lead to a further reduction of U_c [11]. One has, however, to keep in mind that the Coulomb interaction U depends on the environment of the molecule. Screening due to the polarization of the neighboring molecules is, e.g., responsible for a large reduction of U in the crystal as compared to the value for an isolated molecule [12]. Likewise, it is to be expected that for a molecule in the monolayer next to the gate dielectric U might be substantially changed from the bulk value. This effect is, however, hard to quantify without knowing the microscopic structure of the oxide-C_{60} interface.

ENHANCEMENT OF TRANSITION TEMPERATURE

In A_3C_{60} the superconducting transition temperature T_c increases with increasing lattice constant, i.e., with increasing density of states at the Fermi level [13]. It is therefore natural to try the same for field-doped C_{60}. The simplest way to increase the distance of the molecules in the conducting monolayer is to apply uniaxial stress [14]. An alternative approach is the intercalation of the crystal with inert molecules. In fact, for field-doped C_{60} intercalated with haloform molecules spectacularly increased transition temperatures have been reported [3, 4]. A subsequent analysis of the lattice structure of these crystals revealed, however, that the lattice is mainly expanded perpendicular to the conducting layer, and that the density of states in the doped layer shows no correlation with the reported T_c [15]. Therefore, the additional coupling to the vibrations of the haloform molecules has been proposed as an alternative explanation of the enhancement of the transition temperature [16]. It has, however, turned out that such a coupling is very small and, for the two-fold degenerate modes, is even excluded by symmetry [17].

REFERENCES

1. J.H. Schön, Ch. Kloc, R.C. Haddon, and B. Batlogg, *Science* **288**, 656 (2000).
2. J.H. Schön, Ch. Kloc, and B. Batlogg, *Nature* **408**, 549 (2000).
3. J.H. Schön, Ch. Kloc, and B. Batlogg, *Science* **293**, 2432 (2001).
4. Retractions: *Science* **298**, 961 (2002); *Nature* **422**, 92 (2003).
5. R.F. Service, *Science* **298**, 30 (2002); G. Brumfiel, *Nature* **419**, 419 (2002).
6. S. Wehrli, D. Poilblanc, and T.M. Rice, *Eur. Phys. J. B* **23**, 345 (2001).
7. S. Wehrli, E. Koch, and M. Sigrist (in preparation).
8. O. Gunnarsson, E. Koch, and R.M. Martin, *Phys. Rev. B* **54**, 11026(R) (1996).
9. E. Koch, O. Gunnarsson, and R.M. Martin, *Phys. Rev. B* **60**, 15714 (1999).
10. N. Manini, G.E. Santoro, A. Dal Corso, and E. Tosatti, *Phys. Rev. B* **66**, 115107 (2002).
11. J.E. Han, E. Koch, and O. Gunnarsson, *Phys. Rev. Lett.* **84**, 1276 (2000).
12. V.P. Antropov, O. Gunnarsson, and O. Jepsen, *Phys. Rev. B* **46**, 13647(R) (1992).
13. O. Gunnarsson, *Rev. Mod. Phys.* **69**, 575 (1997).
14. E. Koch, *Phys. Rev. B* **66**, 081401(R) (2002).
15. R.E. Dinnebier, O. Gunnarsson, H. Brumm, E. Koch, P.W. Stephens, A. Huq, and M. Jansen, *Science* **296**, 109 (2002).
16. A. Bill and V.Z. Kresin, *Eur. Phys. J. B* **26**, 3 (2002).
17. E. Koch and O. Gunnarsson, *Phys. Rev. B* **67**, 1614XX(R) (2003).

CARBON NANOSTRUCTURE SYNTHESIS
AND PURIFICATION

Influence of Nickel and Cerium on Carbon Arc and Formation of Carbon Nanostructures

A. Huczko[*], H. Lange[*], M. Bystrzejewski[*], M.Sioda[*], M. Pacheco[†], and M. Razafinimanana[†]

[*]Department of Chemistry, Warsaw University, 1 Pasteur,02-093 Warsaw,Poland
[†]Centre de Physique des Plasmas et Applications de Toulouse, Universite Paul Sabatier, 118 route Narbonne, 31062 Toulouse, France

Abstract. Ni- and Ce-doped graphite anodes were DC arced in helium and water to produce 1D nanocarbons. The products were analyzed by HR SEM and TEM techniques. The synergism of the binary catalysts used has been confirmed. Emission spectroscopy was performed to determine the temperature and C_2 radical distributions in the arc.

INTRODUCTION

Carbon nanotubes (CNTs) have accumulated an impressive list of real and potential applications this due to the unique properties. Despite the hundreds of papers published on the synthesis of CNTs, there are still conflicting reports on both the influence of the operational parameters on the process yield and the mechanism of formation. Regarding a mechanism for nucleation of CNTs the single-walled nanotubes (SWCNTs) were found to grow in the presence of a metal catalyst [1] while the multi-walled nanotubes (MWCNTs) grow without a catalyst [2]. Rao et al. [3] discussed the role of the size of the catalyst nanoparticles in the formation of nanocarbons during the pyrolysis of hydrocarbons. The influence of different kinds of co-evaporated metal catalysts on the various nanostructure formation was also studied [4].

In this paper we report on the formation of CNTs in the carbon arc depending on the anode filling. Arc sublimation of anodes was carried out either at medium pressure in helium or at atmospheric pressure in distilled water. Emission spectroscopy was applied to inter-relate the characteristics of catalyst-seeded plasma and the morphology of the products obtained.

EXPERIMENTAL

The experiments were performed either in a reaction chamber under a static pressure of helium [5] or in an open reaction vessel in water [6]. 6 mm graphite electrodes were used. A hole of 3 mm diameter was drilled in the anode and filled with mixtures of nickel and cerium (as CeO_2) and graphite powders (Table 1). The catalyst content is a fraction of total vaporized material. The obtained products were analyses by high-resolution SEM and TEM. C_{60} content in the resulting soot was evaluated [7], too.

CP685, *Molecular Nanostructures: XVII Int'l. Winterschool/Euroconference on Electronic Properties of Novel Materials,*edited by H. Kuzmany, J. Fink, M. Mehring, and S. Roth
© 2003 American Institute of Physics 0-7354-0154-3/03/$20.00

Plasma diagnostics (temperature and C_2 content evaluation) was carried out following the procedure outlined elsewhere [8].

TABLE 1. Operational Parameters of Experiments Carried out with Different Content of Catalysts in the Anode in Helium

Test No.	Metal content [at. %]		Static pressure [mbar]	Arc current [A]	Arc voltage [V]	Average erosion rate [mg/s]	C_{60} (wt. %) content in soot
	Ni	Ce					
1	0	0	800	55	29	2,1	3,20
2	0	0,2	800	55	35	3,9	0,40
3	0,8*	0	800	55	29	3,0	1,95
4	0,6	0,2	800	55	29	3,5	1,06
5	0,6*	0,2	800	55	29	3,3	1,64
6	1,1*	0,4	800	55	29	2,9	0,10
7	1,9*	0,6	800	53	28	1,3	0,00
8	0,42*	1,4	800	55	29	1,2	0,00

* ultra-fine Ni

RESULTS AND DISSCUSION

Product morphology. Despite the relatively low content of catalysts in carbon gas and fixed U/I parameters in different runs (Table 1) the erosion rate of anodes differed drastically. Also, the content of C_{60} drops dramatically with the increasing concentration of Ni/Ce. Thus, the catalysts influence somehow the quenching phenomena and the plasma itself. This was evidently reflected by a completely different structure of the products (Table 2).

TABLE 2. Comparison of the macroscopic characteristics of products

No.	Metal content (at. %)		Cathode		Soot	Webs	CNTs
	Ni	Ce	Collaret	Core			
1	0	0	Small spongy	Small spongy	Crumbly	None	None
2	0	0,2	Med. Spongy	Ring-like	Crumbly	V. few	None
3	0,8*	0	Small spongy	Small spongy	Crumbly	None	very few(soot)
4	0,6	0,2	Large spongy	Ring-like	Rubbery	Some	soot, web, core
5	0,6*	0,2	Large spongy	Ring-like	Rubbery	Many	As in test 4
6	1,1*	0,4	Large spongy	None	Rubbery	Many	More than in test 1 (soot, web)
7	1,9*	0,6	None	None	Rubbery	None	None
8	0,4*	1,4	None	None	Crumbly	None	Less than test 1 (soot)

* ultra-fine Ni

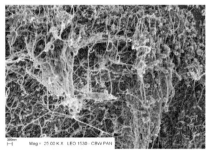
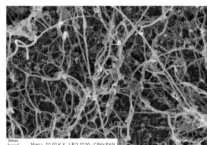

FIGURE 1. SEM images: test 6, web.

Even a slight change in the catalysts ratio and concentration is strongly reflected by the morphology of the product. Carbon nanotubes appear in a soot, cathode deposit and, most of all, in a web. The highest content of CNTs was found in the case of a binary catalyst mixture and the optimal molar ratio of Ni/Ce (run No. 6). Since there has been no quantitative method for CNTs yield evaluation so far, our findings are based only on SEM analyses. Figs 1 and 2 present the examples of the HR SEM images of respected materials.

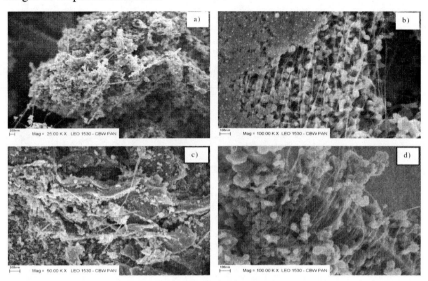

FIGURE 2. SEM images (helium plasma): a) test No 5 , cathode deposit; b) test No 5 , web; c) test No 4 , cathode deposit; d) test No 4 , web.4

HR TEM observations revealed the high abundance of SWCNTs (mostly as ropes) in a web (Fig. 3). Surprisingly, the diameter of many nanotubes was found to be much lower than 1 nm.

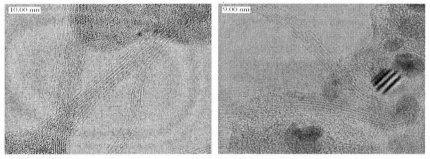

FIGURE 3. HR TEM images of the nanotubes found in the web (test No. 4).

Contrary to the experiments when homogeneous anodes with Gd and Y were used [6] the arc with drilled anodes in water was unstable and the catalyst filling was found as spalls at the bottom of the reaction vessel. Accordingly, the observation of products did not revealed any unusual nanocarbons.

Emission spectroscopy. The arc plasma features were monitored by optical emission. The radiation is mainly associated with the Swan band system $d \; {}^3P_g \rightarrow a \; {}^3P_u$ emitted by C_2. The band 0-0 (516.5 nm) was used for temperature assessment and contribution of this radical in the carbon plasma. The C_2 contents was determined using the effect of self-absorption [8]. The obtained temperatures and C_2 content (in column density units) are illustrated in Fig. 4. The plasma temperature is between 4000-5000 K. The C_2 content depends on anode composition. Generally, the erosion rate of as prepared electrodes, i.e. non-homogenous, strongly fluctuates during arcing resulting in a variation of carbon vapor pressure. However, the high content of C_2 was always observed whenever the plasma contained Ce atoms and ions - curve (b) and (c) in Fig. 4. This effect is related to the pulsed effluent of anode filling caused by drastically different melting and sublimation temperatures of the catalyst and carbon.

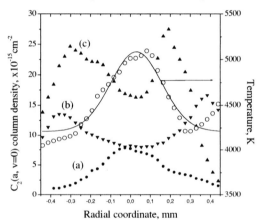

FIGURE 4. Radial temperature (open circles) and column density of C_2 distributions (full markers). Electrode composition: (a) – test 1, (b) – test 5, (c) – test 2.

ACKNOWLEDGMENTS

The work was supported by the Committee for Scientific Research (KBN) through the Department of Chemistry, Warsaw University, under Grant No. 7 T09A 020 20 and BW-1522/22/L (Polonium2002).

REFERENCES

1. Iijima, S. and Ichihashi, T., *Nature,* **363, 603 (1993)**.
2. Iijima, S., *Nature,* **354, 56 (1991)**.
3. Rao, C.N.R., Kulkarni, G.U., Govindaraj, A., Satishkumar, B.C. and Thomas, P.J., *Pure. Appl. Chem.,* **72, 21 (2000)**.
4. Maser, W. K., Bernier, P., Lambert, J.M., Stephan, O., Ajayan, P.M., Colliex, C., Brotons, V., Planeix, J.M., Coq, B., Molinie, P. and Lefrant, S., *Synth. Met.,* **81, 243 (1996)**.
5. Lange H., Baranowski P., Byszewski P., Huczko A., *Rev. Sci. Instr.,* **68, 3723 (1997)**.
6. Lange H., Sioda M., Huczko A., Zhu Y. Q., Kroto H. W., Walton D. R. M., *Carbon,* in press **(2003)**.
7. Huczko A., Lange H., Byszewski P., Popławska M., Starski A., *J. Phys. Chem. A,* **101, 1267 (1997)**.
8. Lange H., Saidane H., Razafinimanana M., Gleizes A., *J. Phys. D: Appl. Phys.,* **32, 1024 (1999)**.

Synthesis of organics/SWNT compounds

T. Takenobu[1,2], T. Takano[1], Y. Iwasa[1,2], M. Shiraishi[3] and M Ata[3]

[1]Institute for Materials Research, Tohoku University, Sendai, 980-8577, Japan.
[2]CREST, Japan Science and Technology Corporation, Kawaguchi, 332-0012, Japan.
[3]Materials Laboratories, SONY Corporation, Yokohama 240-0036, Japan.

Abstract. Here we report synthesis and properties of SWNTs doped with organic molecules, such as TDAE and TCNQ. These organic molecules are well known by their strong ionization energy or electronic affinity and have been used for organic charge transfer compounds. The X-ray profile of all samples strongly suggests encapsulation of molecules inside nanotubes. In optical absorption measurements, the reduction of absorption peak of semiconductive SWNTs was observed and this is understood by filling of the first DOS peak of the conducting band, providing direct evidence of charge (hole or electron) transfer between organic molecules and SWNTs.

INTRODUCTION

Single walled carbon nanotube (SWNT) is the most promising material for molecular electronics, because of its unique structural and electronic properties [1]. Molecular electronics is an emerging area with a goal of using molecular materials as core device components. An advantage is the small size of molecules, surpassing structures attainable by top-down lithography, and could therefore be essential for miniaturization. For further evolution of molecular electronics, the ability to obtain both p- and n-type FETs is important to construct complementary electronics that is known to be superior in performance (for example, low power) to devices consisting of unipolar p- or n-type transistors. Key points to p- and n-type SWNTs are air stability, tuning of doping level, and mass scale production. Several doping methods have been developed for nanotubes, including exposure to gaseous molecules [2], annealing in vacuum [3] or in inert gas [4], intercalation of inorganic materials (alkali metal and halogen) [5], electrochemical experiments [6] and adsorption of organic molecules [7]. To date, the treatment with gas or in vacuum is the only way to obtain n-type SWNTs, however doped materials are slowly de-doped under ambient conditions, and fine tuning of carrier concentration is still very difficult. On the other hand, due to their size and geometry, SWNTs also provide a unique opportunity for nanoscale engineering of novel one-dimensional systems, created by self-assembly of atoms or molecules inside

CP685, *Molecular Nanostructures: XVII Int'l. Winterschool/Euroconference on Electronic Properties of Novel Materials,*edited by H. Kuzmany, J. Fink, M. Mehring, and S. Roth

the SWNT's hollow core. It has been experimentally shown that fullerenes [8] and/or endohedral metallofullerenes [9] can be inserted into SWNTs, forming a peapod-like structure. In this paper, we report synthesis and characterization of organics/SWNT compounds, and demonstrate that the charge transfer between SWNT and organic molecules are controlled by the ionization energy or the electron affinity of guest organic molecules.

EXPERIMENTS

SWNTs were synthesized by a laser ablation method using a Ni/Co catalyst. They were purified by H_2O_2 and aqueous NaOH solution to remove amorphous carbon nano-particles. The detailed procedure and the effect of this purification process were summarized in the literature [10]. Since the H_2O_2 treatment also burns out the caps of SWNTs and the HCl treatment increases defects on the wall, the purified SWNTs already have a sufficient amount of entrance holes for molecules. SWNTs were reacted with vapor of organic molecules in a similar manner to the case of C_{60}-peapod. Organic molecules, summarized in Table 1 [11], were purified by sublimation before reaction with SWNT. Purified and decapped SWNTs were loaded in a glass tube with excess amount of organic powder samples to obtain saturation doping. This tube was evacuated to 2×10^{-6} Torr, and SWNT was degassed by heating with a torch. Organic molecules were placed in a separate position so that molecules stay at room temperature in this process. Then, the ampoule were sealed and annealed for twelve hours in a furnace just above the sublimation temperature of molecules. At this temperature, molecules are mobile and enter SWNTs. We prepared two kinds of samples, *i. e.* as-purified bulk samples and thin films. SWNT films were prepared by spraying SWNT dispersed ethanol on Si or quartz plates using an airbrush. All procedures for measurements were carried out in an atmosphere-controlled Ar glove box, inside glass tubes, or an indium sealed optical cell to avoid the doping effect of oxygen.

TABLE 1. Electron affinity and ionization energy of each molecule.

Molecule	Electron Affinity (eV)	Ionization energy (eV)
Tetrakis(dimethylamino)ethylene (TDAE)		5.36
Tetramethyl-tetraselenafulvalene (TMTSF)		6.27
Tetrathiafulvalene (TTF)		6.4
Pentacene	1.392	6.58
Anthracene	0.56	7.36
3,5-Dinitrobenzonitrile (DNBN)	2.16	
C_{60}	2.65	
Tetracyano-*p*-quinodimethane (TCNQ)	2.8	9.5
Tetrafluorotetracyano-*p*-quinodimethane (F_4TCNQ)	3.38	

For a structural characterization, synchrotron x-ray powder diffraction data were collected at room temperature on the beam line BL02B2 at Super Photon Ring (SPring-8), Japan. For this purpose, bulk organics/SWNT compounds were sealed in thin glass capillaries of 0.5 mm in outer diameter. Doping properties have been investigated by means of Raman and optical absorption spectroscopy on film samples.

RESULTS AND DISCUSSION

Figure 1 (a) shows typical diffraction profiles for pristine and organic molecules doped SWNT materials. The most obvious difference between doped and undoped SWNTs is the strong reduction of peak intensity at about $Q \sim 0.4$ (Å^{-1}), which is indexed as (10) reflection. Such a behavior provides evidence for encapsulation of organic molecules inside SWNT, as encountered in several peapod materials [8] and gas adsorbed SWNTs [12]. Comparative resistivity measurements on pristine SWNT and TCNQ/SWNTs have been carried out with a four-probe method (Figure 2 (b)). To avoid the effect of oxygen, particularly on pristine SWNTs, experiments were made in He atmosphere and samples were degassed at 400 K with assist of vacuum pumping. Resistance of TCNQ doped SWNT was smaller than that of the pristine SWNT sample approximately by a factor of two at room temperature.

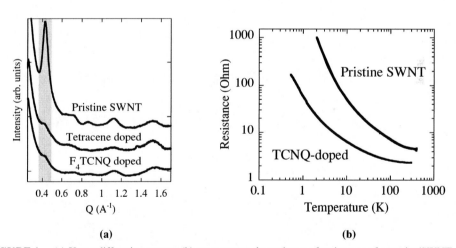

FIGURE 1. (a) X-ray diffraction pattern (b) temperature dependence of resistance of organics/SWNT.

Figure 2 displays Raman spectra of pristine and organic molecule reacted SWNTs. The intensity of RBM in organics/SWNTs samples was strongly dependent on organic molecules. In anthracene or tetracene doped SWNT, the RBM did not show any notable change, while a significant reduction of RBM intensity was observed in TCNQ and TDAE doped SWNT. The latter intensity reduction resembles to that found in K-doped SWNT, being suggestive of occurrence of charge transfer [13]. Evidence for charge transfer is also found in the higher wavenumber region. The two broad components in the G-band located at 1540 cm^{-1} and 1560 cm^{-1} exhibit a slight intensity reduction, which is qualitatively similar to that in K-doped SWNT [13].

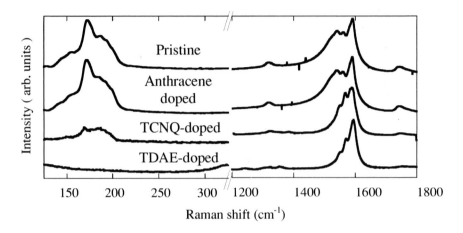

FIGURE 2. Raman spectra of pristine SWNT, and anthracene-, TCNQ-, and TDAE-doped SWNTs.

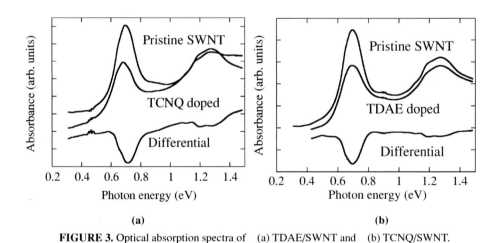

(a)   **(b)**

FIGURE 3. Optical absorption spectra of   (a) TDAE/SWNT and   (b) TCNQ/SWNT.

The Fig. 3 displays optical absorption spectra for TCNQ and TDAE doped SWNTs. The intensity of the $v_s^1 \rightarrow c_s^1$ absorption band at 0.68 eV decreases upon doping. This spectral change was also recorded in SWNTs doped with Cl$_2$TCNQ, TTF, TMTSF, while no appreciable change was detected in SWNTs doped with C$_{60}$, 3,5-Dinitrobenzonitrile, anthracene, tetracene, and pentacene. According to the spectrum in K-doped SWNT, the reduction of peak intensity at 0.68 eV provides direct evidence of carrier injection from organic molecules [5]. The reduction of the interband absorption is strongly dependent on the ionization energy and the electron affinity of molecules, and there is a critical threshold for the charge transfer.

CONCLUSION

In summary, we have synthesized SWNTs encapsulating organic molecules, in which amphoteric carriers are injected to SWNTs. The carrier doping is controlled by ionization energy and electronic affinity of reacted molecules. The described materials exhibit three superior features; air stability, fine tunability of Fermi level, and simple synthesis. These organics/SWNT compound promises to push the performance limit of SWNT based devices for molecular electronics.

REFERENCES

1. Dresselhaus, M. S., Dresselhaus, G. and Avouris, P. (eds.) *Carbon Nanotubes* (Spring, Berlin, 2001).
2. Collins, P., Bradley, K., Ishigami, M. and Zettl., A., *Science* **287**, 1801-1804 (2000).
3. Derycke, V., Martel, R., Appenzeller, J. and Avouris, P., *Appl. Phys. Lett.* **80**, 2773-2774 (2002).
4. Javey, A. *et al.*, *Nature Materials* **1**, 241-246 (2002).
5. Kazaoui, S., Minami, N., Jacquemin, R., Kataura, H. and Achiba, Y., *Phys. Rev. B* **60**, 13339-13342 (1999).
6. Kazaoui, S., Minami, N., Matsuda, N., Kataura, H. & Achiba, Y., *Appl. Phys. Lett.* **78**, 3433-3435 (2001).
7. Kong, J. and Dai, H., *J. Phys. Chem. B* **105**, 2890-2893 (2001).
8. Kataura, H. *et al.*, *Appl. Phys. A* **74**, 1-6 (2002).
9. Lee, J. *et al.*, *Nature* **415**, 1005-1008 (2002).
10. Shiraishi, M., Takenobu, T., Yamada, A., Ata, M. and Kataura, *Chem. Phys. Lett.* **358**, 213-218 (2002).
11. Seki, N and Sato, N., http://www.ossc.kuchem.kyoto-u.ac.jp/database/2ipea/IeEaList.pdf
12. Maniwa, Y. *et al.*, *Jpn. J. Appl. Phys.* **38**, L668-L670 (1999).
13. Iwasa, Y. *et al.*, *NEW DIAM. FRONT. C. TEC.* **12**, 325-330 (2002).

Nucleation as Self-assembling Step of Carbon Deposit Formation on Metal Catalysts

V.L. Kuznetsov[1], A.N. Usoltseva[1], Yu.V. Butenko[1], A.L. Chuvilin[1], M.Yu. Alekseev[2], L.V. Lutsev[2]

[1] Boreskov Institute of Catalysis, Pr. Akademika Lavrentieva 5, 630090 Novosibirsk, Russia
[2] Research Institute "Ferrite-Domen", Chernigovskaya 8, St Petersburg, 196084, Russia

Abstract. A thermodynamic analysis of the carbon nucleation on the metal surface was performed. The master equation for the dependence of critical radius of the carbon nucleus on the reaction parameters, such as the reaction temperature, the catalyst nature, the supersaturation degree of catalyst particle by carbon was obtained. This equation and the phase diagram approach were used for discussion of different scenarios of carbon deposits formation, namely, encapsulated particles, carbon fibers and filaments, multi-wall and single-wall carbon nanotubes. Carbon filament growth via CO disproportionation on cobalt supported catalyst was investigated. Here we demonstrate the possibility to vary the reaction products using the same catalyst just changing the reaction conditions and catalysts pretreatment procedure. For the first time the formation of the carbon filament ropes on Co catalysts was observed.

INTRODUCTION

The processes of self-assembly and self-organization underlying of self-processes, which spontaneously assemble and organize various building blocks into hierarchical structures, have emerged as the most promising techniques for the efficient production of nanostructured materials. In this paper we consider carbon nucleation on metal surfaces as one of the main self-assembling factor, which is important in nanocarbons formation.

For any type of carbon deposits their formation occurs via the common steps including metal-carbon particles formation and carbon nucleation. Carbon deposit nucleation from metal-carbon particles is a crucial stage for the formation of different carbon deposits, namely, catalyst deactivation, carbon filament and nanotubes formation. It can be explained in terms of relatively high thermal stability of nanocarbons due to the low self diffusion of carbon in carbon materials at temperature up to 1500-1700°C. So single wall carbon nanotubes (SWNT) double their diameter only after thermal treatment at the temperature higher then 1500°C [1], while multi-wall nanotube formation from SWNT ropes occurs only at the temperature higher than 2000°C [2]. The formation of stable mosaic closed graphitic nanostructures on the annealed diamond surface at the temperature higher than 1500°C was observed [3]. Thus once formed the nanosize carbon nucleus are stable enough and nucleation step can determine the of carbon deposit types.

CP685, *Molecular Nanostructures: XVII Int'l. Winterschool/Euroconference on Electronic Properties of Novel Materials,* edited by H. Kuzmany, J. Fink, M. Mehring, and S. Roth
© 2003 American Institute of Physics 0-7354-0154-3/03/$20.00

RESULTS AND DISSCUSSION

Theory of Carbon Nucleation on Metal Surface

At present paper we perform the thermodynamic analysis of the carbon nucleation on metal surfaces to estimate the influence of reaction parameters on the nucleation step. We consider the change in the Gibbs free energy using the simplest model of carbon deposit nucleation, where a flat round carbon nucleus with radius r and height h is bonded to the metal surface through its edges. In this case the variation in Gibbs free energy may be written as (for detail please see ref [4]):

$$\Delta G = \frac{\pi r^2 h}{V_M} \cdot RT \ln \frac{x}{x_0} + \pi r^2 \left(\sigma_{nucleus-gas} + \sigma_{nucleus-surface} - \sigma_{surface-gas} \right) + 2\pi r \cdot \frac{\Delta H_{M-C} - \Delta H_{C-C}}{2N_A \cdot r_{C-C}} + 2\pi r \cdot \frac{Q}{4.5h} \quad (1)$$

In this equation the first term presents a free energy change when the graphitic carbon precipitates from the metal-carbon solution and V_M is the molar volume of graphite, R is the universal gas constant, T is a reaction temperature, x/x_0 is a carbon saturation coefficient of the metal-carbon solution. The second term is a free energy change caused by the interaction of the carbon nucleus with the catalyst surface, where σ_i is the corresponding specific surface energies. The third term estimates the energy of chemical bonding between a border atom of the nucleus and the metal surface, and ΔH_i are the enthalpies of formation of a single graphite C-C and metal-carbon bonds, r_{C-C} is the distance between two neighbouring carbon atoms in a zigzag edge of the graphite sheet. The fourth term reflects the strain energy due to the bending of the nucleus graphitic layer during its bonding with the metal surface, where Q is a constant of continuum elasticity formalism [5].

Following the above considerations and proposing $d(\Delta G)/dr = 0$ for critical nucleus we obtain a functional dependence between the critical radius (r_{crit}) of the carbon nucleus and reaction parameters such as reaction temperature, carbon supersaturation degree (x/x_0, where x_0 is equilibrium carbon content in metal), and parameters which characterize the metal catalyst (the enthalpy of the metal-carbon bond formation (ΔH_{M-C}), work of metal adhesion to graphite (W_{ad}):

$$r_{crit} = \left(\frac{\Delta H_{M-C} - \Delta H_{C-C}}{2 N_A \cdot r_{C-C}} + \frac{Q}{4.5h} \right) \cdot \left[\frac{RT \cdot h}{V_M} \ln \frac{x}{x_0} + \left(W_{ad} - 2\sigma_{graphite} \right) \right]^{-1} \quad (2)$$

Figure 1A presents the r_{crit} as a function of the reaction temperature calculated for pure Ni and Fe and catalysts (in solid and liquid state) along with the minimum values of nanotubes radii taken from literature. It could be noted that r_{crit} values are much smaller for liquid catalysts than for the catalysts in solid state due to the difference in the corresponding values of W_{ad}, which is bigger for liquid metal. We also analyzed r_{crit} dependence on other thermodynamic parameters from equation (2) [6]. This analysis has allowed to formulate the requirement for the optimal conditions of single-wall nanotubes (SWNT) growth: a) the temperature increase (with a fixed supersaturation of metal particles by carbon) leads to the formation of smaller nuclei and to the formation of SWNTs, b) SWNT growth occurs preferentially with the participation of liquid metal particles; c) for SWNT synthesis on the solid catalyst particles the high degree of the supersaturation of metal particles by carbon is required; d) the metals with a higher metal-carbon bond energy provide formation of nanotubes with smaller diameters. Figure 1B illustrates our conclusions. The optimal conditions required for SWNT growth, which can be realized by arc discharge and laser ablation techniques, correspond to the line *a'-b'-c'-d'*. In this case carbon nucleation on the liquid catalyst particles (point *a'*) begins after the reaction temperature decrease below the point *b'*. For SWNT growth in the case of isothermal decomposition of hydrocarbon or CO on the surface of solid metal particles (*line a"-b"-c"*) the high degree of carbon supersaturation is required (point *c"*).

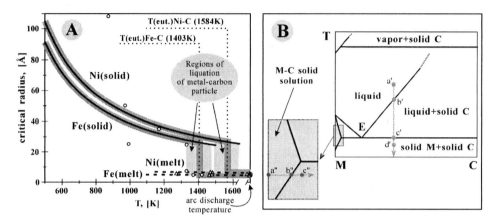

FIGURE 1. **A.** Dependence of r_{crit} on the reaction temperature for Ni and Fe catalysts, along with values extracted from the literature for Ni and Fe catalysts, marked by circles and triangles, respectively [4]. **B.** Changes in the state of metal catalyst according phase diagram consideration. The lines *a'-b'-c'-d'* and *a"-b"-c"* reflect the carbon nucleation from liquid and solid metal particle respectively.

For the last temperature interval SWNT formation is rather complicated because of the necessity to use highly disperse metal particles, which can provide a high carbon supersaturation. One of the most difficult problem is related with sintering of disperse metal particles and carbon diffusion limitations. To overcome these problems chemists use the decomposition of volatile metal compounds in reaction gas mixture containing hydrocarbon or CO to prepare *in situ* disperse metal particles with a high carbon supersaturation to provide the optimal nucleation conditions (see [7] for high pressure CO synthesis of SWNT). Additional improvement can be reached using the promoters, which can decrease the melting point of catalytic particles to provide the high values of W_{ad}. Thus one can not exclude that highly disperse metal particle could be in a liquid-like state during nucleation even at low temperatures because of their formation occurs via the condensation of metal atoms or small clusters.

After the formation of critical nucleus the different scenarios of its evolution can be considered. In the **first scenario**, if x/x_0 value is rather low ($x \approx x_0$, usual for big metal particles), the critical nucleus is relatively large and continues to grow, which results in the formation of graphitic sheets covering a significant part of the surface of the metal particle. After the formation of the first graphitic layer and restoration of required x/x_0 value in subsurface metal region via bulk carbon diffusion the next nucleus can be formed. This alternation of the growth and the following exfoliation of the graphitic sheet produce a variety of filamentous structures with graphitic sheets oriented at an angle or perpendicular to the fiber axis.

In the **second scenario**, a new nuclei form under the primary one, with edges bonded to the metal particle surface. It can be achieved using the disperse metal particles, where the nucleation can proceed only with higher x/x_0 values than in the first case. After the production of a primary nuclei in the case of diffusion limitations the x/x_0 value drops to the ones insufficient for the further formation of nucleus and all carbon atoms diffusing through the metal particle are consumed for the nanotube growth. It should be noted that the internal tube radius can not be smaller than the critical radius of the nucleus, while the external walls have a diameter that is comparable to the size of the metal particle. However, if the x/x_0 value reaches the level required for the nuclei formation nanotube partitions produce. Thus, the oscillations of x/x_0 near the level required for the nuclei formation lead to the bamboo-like nanotube formation.

For the **third scenario** very high x/x_0 values are required. In this case several nuclei precipitate on the surface of the same metal particle and interact to form the tightly packed ropes. If the nucleation frequency is high enough, then a size of the single nanotube tends to a minimal value that corresponds to the critical nucleus radius.

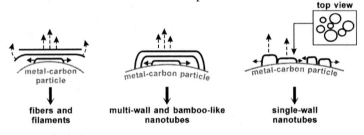

FIGURE 2. The different scenarios of primary nucleus evolution (see text).

Influence of the Catalyst Composition and Reaction Conditions on the Structure of Carbon Deposits

As an example we consider the possibility to vary the reaction products via changing the reaction conditions and the pretreatment procedure of the same Co-based catalyst. The catalyst composition and morphology determine the structure of carbon deposits in accordance with carbon nucleation conditions. Co catalysts were prepared via magnetron sputtering of Co/Si target in Ar atmosphere on Al_2O_3 plate and initially were consisted of Co particles with 2 nm diameter shared SiO_2. Catalytic CO disproportionation was used for the synthesis of carbon nanodeposits. The reaction was performed in the temperature range 500-900°C.

At 500°C catalyst contains uniform metal particles with a size less than 10 nm. As a result bamboo-like carbon nanotubes with narrow diameter distribution and average diameter about 5-10 nm were obtained (figure 3A). At 700°C the formation of the multi-wall carbon nanotubes (MWNT) and bamboo-like nanotubes with a wide diameter distribution (4.5 – 60 nm) and different morphology were observed because of the wide particle size distribution of sintered metal particles (figure 3B). Catalyst pretreatment at 900°C leads to the further sintering with the formation of big metal particles covered with a holed SiO_2 layer. The usage of this catalyst at 500-700°C results in MWNT with a broad diameter distribution (up to 100 nm) along with capsulated metal particles (50 - 200 nm). Furthermore formation of the straight and coiled carbon filaments ropes with a diameter of about 100-500 nm was observed (figure 3C). HR TEM study reveals that these ropes consist of the piles of graphite sheets oriented perpendicular to the filament axis ("pack of card" structure) and have a diameter of about 30-50 nm. Ropes diameters are comparable with the cross-section of catalysts particles. The rope ordering depends on the reaction temperature. The scheme of rope formation is presented in the inset D of figure 3. Thus, the variation of reaction temperature results in the different type of carbon deposits.

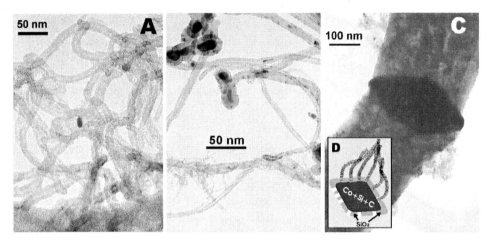

FIGURE 3. Transition electron microscopy (TEM) images of carbon deposits obtained on Co-SiO_2 catalyst after catalyst treatment by CO/H_2 mixture at 500°C (A), 700°C (B), 500-700°C after reduction at 900° C (C) and scheme of carbon filament ropes formation on Co-SiO_2 particle (D).

ACKNOWLEDGMENTS

The work has been supported by INTAS-00-237, RFBR- 02-03-32296, SCOPES No. 7SUPJ062400 and CRDF REC 008

REFERENCES

1. Nikolaev, P.N., Thess, A., Rinzler, A.G., et al., *Chem. Phys. Lett.*, **266** (5-6), 422 (1997)
2. Bougrine, A., Dupont-Pavlovsky, N., Naji, A., et al., *Carbon* **39**, 685 (2001).
3. Kuznetsov, V. L., Chuvilin, A. L., Butenko, Y. V., et al., *Chem. Phys. Lett.*, **289**, 353 (1998).
4. Kuznetsov, V.L., Usoltseva, A.N., Chuvilin, A.L., et al., *Phys. Rev.B.* **64** (23), 5401-5408 (2001).
5. Tomanek, D., Zhong, W., Krastev, E., *Physical Review B*, **48**, 15461 (1993).
6. Kuznetsov, V.L., et al., MRS Symposium Proceedings, v. 706, Z6.22, (2002).
7. Bronikowski, M.J., Willis, P.A., Colbert, D.T., et al., *J. Vac. Sci. Technol. A* **19**(4), 1800 (2001).

Polymerization Of SWCNTs With Di-Nitrens

Michael Holzinger[a], Johannes Steinmetz[a], Damien Samaille[a], Marianne Glerup[a], Patrick Bernier[a], Lothar Ley[b] and Ralf Graupner[b]

[a]GDPC(UMR5581), Université Montpellier II, France,[b]Institute of Technical Physics, Erwin-Rommel-Straße 1, Germany

Abstract. The question of possible technological applications quickly aroused after the discovery of the unique physical properties of single-walled carbon nanotubes (SWCNTs). In this context, a wide range of initial approaches towards organic nanotube chemistry have been developed within the last few years.[1] This also includes the covalent functionalization of the SWCNT sidewalls with nitrenes, carbenes or radicals, to increase solubility in organic solvents.[2] With the use of bis-alkoxycarbony-lazides to create *in-situ* di-nitrenes, it should be possible to link individual SWCNTs and small bundles. The purpose of these experiments is to give the possibility for increasing the understanding of the chemical behaviour of SWCNTs and to provide further proofs for the occourance of this type of reaction. This can also be seen as a new way for forming nanotube materials for exploiting their unique mechanical properties. The results obtained are discussed in this presentation. SEM, XPS, XRD and Raman spectroscopy are used as characterization tools.

INTRODUCTION

It has been shown that nitrenes can be used as a reliable reactant for the functionalization of SWCNT sidewalls.[2] The use of azidocarbonates as precursors is necessary for this type of reaction. Azidocabonates yield thermally induced N_2 extrusion, generating singlet and triplet nitrenes. The singlet nitrenes are transformed into triplet nitrenes by inter-system crossing (ISC) and can also attack the nanotube sidewall in a [2+1] cycloaddition. A triplet state nitrene reacts with the π–system of the nanotube sidewall as a bi-radical, forming an aziridine ring by passing a bi-radicalic transition state.

Azidoalkyls or carboxyls for example would add in a [3+2] cycloaddition followed by the N_2 extrusion. Additionally, azidocarboxyls tend to undergo intramolecular insertions forming isocyanates. All our experiments using azides as 1,3-dipoles in order to form a [3+2] adduct were unsuccessful.

RESULTS

For the addition of nitrenes, the pristine SWNTs, generated by standard arc discharge method, were dispersed in 1,1,2,2-tetrachloroethane (TCE) in an ultrasonic bath. We used as grown material without purification in order to have as few defects as possible on the nanotube sidewall. The saturated solution was heated to 160°C. At this temperature a 20-fold excess (mass) of polyethyleneglycol-(600)-diazidocarbonate (PG600), diluted in TCE, was added drop-wise. After thermally-induced N_2-extrusion, nitrene addition resulted in the formation of alkoxycarbonylaziridino-SWCNTs. Beside the linkage of the nanotubes the addition of the di-functional molecule can also occur along the same sidewall to form tethers (Scheme 1). After cooling down to room temperature by filtration of the modified SWCNTs using a PTFE filter with 5 μm pore size. The paper was washed with TCE and ethanol. After drying at 100°C a highly flexible and stable sheet of cross-linked SWCNTs was obtained. The resulting product is insoluble in water and all common organic solvents.

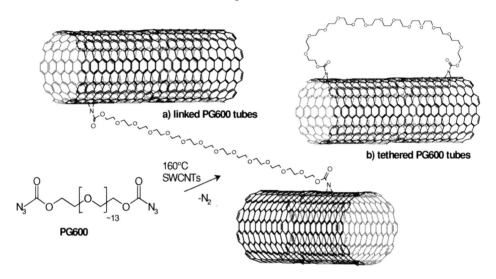

SCHEME 1. possible products after reaction with the bisazidocarbonate PG600. a) linked SWCNTs and b) tethered SWCNTs.

For the characterisation, we prepared a reference by dispersing nanotubes in TCE an adding the same amount of the precursor of the linker. This mixture was filtered without heating. The resulting buckypaper was then used as standard for comparison. The formation of alkoxyaziridino tubes were characterized by various methods such as X-ray diffraction (XRD), Raman spectroscopy and X-ray photoelectron spectroscopy (XPS). XPS is a very promising characterization method for this type of material. It allows not only the concentration of atomic species but also their chemical bonding and the electronic structure in the solid to be determined.[3] The advantage of this method is very high sensitivity: high resolution of the bands of all elements in the sample. The XPS spectrum (Figure 1, black line) of the standard shows a small amount of oxygen (~6%) which was assigned to defects in the tubes, saturated with oxygen. This spectrum also shows that the precursor can easily be washed out when it is not brought to reaction. A contamination with the precursor would lead to a nitrogen signal.

After the reaction, the XPS spectrum of PG600 tubes, (Figure 1, grey line) shows an increase of the amount of oxygen in the sample and a signal from nitrogen is observed. We can calculate the amount of added linker molecules to be in the range of 0.7% from the amount of carbon. However, this value stands for a tendency and not for an absolute value since this method is limited to surface investigations. Fluorine and chlorine are observed in the sample and assigned to be impurities arising from the Teflon filter and residues of the solvent.

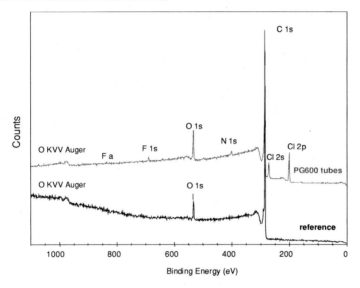

FIGURE 1: XPS spectrum of the standard and the PG600 tubes.

Scanning electron microscopy was used to investigate the topological differences between the standard and the PG600 tubes. By imaging the surfaces of the respective buckypapers, the differences are not easily observed. Figure 2 displays the SEM images of the reference and the PG600 tube sample, recorded at torn edges of the papers. The image of the reference shows densely ordered nanotubes and a quantity of impurities. The dispersion in TCE and the precursor of the linker seem to have no influence on the size of the bundle in the reference compound. The SEM image of the PG600 tube sample show a linked structure which cannot result from a random order. We assign this to a successful linkage and a subsequent splitting of the bundles. In these images, we only observed bundles and no individual nanotubes. Considering that the linker is a polyethylene chain, it seems to be unlikely that this single chain can hold together two bundles without tearing. A stable linkage would be when a sufficient amount of linkers are attached between two nanotubes in a bundle. The nitrenes attack most likely - but probably not exclusively - the defects on the nanotube sidewall. The reaction leads to further sp^3-defects and cause a more localised π-system in the environment. These parts are again preferred for a new attack. A collapse of the nanotube framework is expected when too many addends are attached to a nanotube.

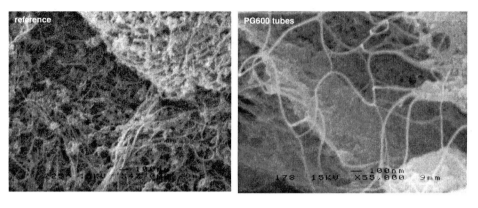

FIGURE 2: SEM images of the reference and the PG600 tubes.

CONCLUSION

Beside the XPS spectroscopy and SEM imaging, X-ray diffractograms and Raman spectra were recorded for characterization which are not discussed in detail in this proceeding. The X-ray diffractograms shows an increase of the distance of nanotubes in the bundles. The characteristic peak for SWCNT bundles at 6.0 (2Θ) changed the position to 5.3 (2Θ). Using Raman spectroscopy, the interpretation of the RBM mode shift of the Raman spectrum leads to similar results. Additionally, the different ratio of the G- and D-line indicates an increase of defects on the nanotube sidewalls.
With these characterization methods, we have given evidence for the formation of akloxyaziridino-SWCNTs but not for the linkage of tubes in different bundles. This only can be proven by other microscopic methods. Here, not only the interconnection can be seen but also the splitting of bundles would be observed, which is an already known side effect of this reaction.[2] The absolute proof of the linkage cannot be given at the moment. Other linkers, more suitable for verifying the linkage, have to be used for making a clear characterisation possible. Also the process has to be optimised for linking of more nanotubes.

REFERENCES

1. J. Chen, M. A. Hamon, H. Hu, Y. Chen, A. M. Rao, P. C. Eklund, R. C. Haddon, *Science (Washington, D. C.)*, *282*, **95**, (1998).
2. M. Holzinger, O. Vostrowsky, A. Hirsch, F. Hennrich, M. Kappes, R. Weiss, F. Jellen, *Angew. Chem., Int. Ed.*, *40*, 4002, (2001).
3. H. Kuzmany, *Solid State Spectroscopy, an Introduction*, Springer-Verlag, Heidelberg, 1998.

The first stages of growth of a C-SWNT

H. Amara[1], J.-P. Gaspard[2], C. Bichara[3], T. Cours[2], G. Hug[1], A. Loiseau[1], and F. Ducastelle[1]

[1] LEM, Onera-Cnrs, BP 72, 92322 Châtillon Cedex, France
[2] Physique de la Matière Condensée, Université de Liège, B5, 4000 Sart-Tilman, Belgium
[3] CRMC2, Cnrs, Case 913, 13288 Marseille Cedex 9, France

Abstract. The nucleation of SWNTs is enhanced by early transition metals or rare earth catalysts. A mechanism is proposed on the basis of tight-binding calculations in order to understand the growth mechanism of a nanotube on a metallic substrate. When electrons are transferred from the metal catalyst to the carbon atoms, the graphene sheet becomes unstable with respect to the formation of odd rings pentagon-heptagon pairs in particular. Six such pairs auto-assemble to build up the cap of a nanotube.

INTRODUCTION

The similarities between the different techniques, observed in transmission electron microscopy, suggest a common growth mechanism [1, 4]. The proposed scenario is the following: first, carbon atoms condense in a low density amorphous state. Then metallic catalyst particles condense in a liquid state which can dissolve significant amounts of carbon. Then, at lower temperatures (1500-2000K) the solubility of carbon decreases. Since the surface energy of a graphene sheet is much lower than that of transition elements, carbon is expected to segregate at the surface.

From the experimental point of view, it is also recognized that in the high temperature route, the SWNT production rate is drastically favoured by the addition of a small amount of rare earth (RE) element such as Y to the transition metal (Ni). The role of RE elements has been studied by coupling different electron microscopy techniques in the case of the Ni-RE (RE = Y, Ce or La) catalyst [5]. It is concluded from this study that the RE modifies the surface catalytic activity of Ni particles by coprecipitating with C at their surface in the form of thin carbide platelets which are clearly visible in the HRTEM images of figure 1.

Theoretical model

Graphene sheets around the metallic particles are indeed frequently observed, but the reason why nanotube growth can be favoured is not clear. To study it, there are two main steps to consider: nucleation and growth. Recent calculations [4] suggest that once a capped nanotube is nucleated, growth can occur through diffusion of surface carbon atoms coming from the root. The study of the nucleation process is much more difficult. The main difficulty is to find the physical effect at the origin of the surface

CP685, Molecular Nanostructures: XVII Int'l. Winterschool/Euroconference on
Electronic Properties of Novel Materials, edited by H. Kuzmany, J. Fink, M. Mehring, and S. Roth
© 2003 American Institute of Physics 0-7354-0154-3/03/$20.00

instability leading to a critical nucleus rather than a graphene sheet. In this paper, we show that an electron donating metal catalyst enhances the nucleation rate.

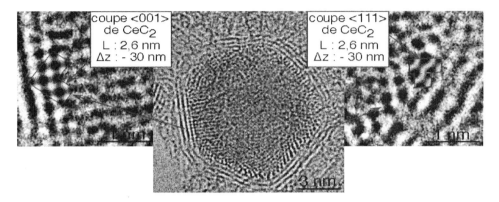

FIGURE 1. HREM image (Jeol 4000EX, Cs=1mm) of a particle detected with Ce catalyst showing a pure FCC Ni core and a CeC_2 platelet at the surface. HREM simulation of the carbide platelet is superimposed and the conditions of simulations are indicated in the inset (L is the projected thickness estimated from the shape of the particle and Δz is the objective defocus).

Since the catalytic metal is necessary to obtain SWNTs, it should necessarily be involved in the nucleation process. In the particular case of Ni-RE-C, the formation of thin carbide platelets provide the nucleation sites of the SWNTs. In this paper, the nucleation control is discussed in terms of charge transfer from the RE to C whose effect would be to stabilize pentagons and heptagons necessary to form a SWNT nucleus (Fig. 2). Since we know that early transition metals (Y, Ti, ...) are much less electronegative than carbon, electrons are transferred towards the carbon atoms. *Ab initio* simulations using the WIEN2K code (Fig. 3) confirm this charge transfer mechanism.

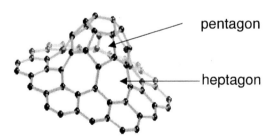

FIGURE 2. SWNT nucleus with pentagons (positive curvature) and heptagons (negative curvature).

To go beyond this approach, we have developed a simple tight-binding model using the moment's method [6]. In this method, the density of states is given by the imaginary part of the diagonal elements of the Green function, whose expansion in terms of continued fractions is well-known. The aim is to build, from the atomic orbital basis set, a new basis in which the tight-binding Hamiltonian is tridiagonal.

Then we denote by a and √b the diagonal and off-diagonal elements of H in the new basis. The coefficients of the continued fraction (a,b) are closely related to the moments of the density of states and are obtained within the recursion method. In our case, we calculate exactly the coefficients up to the (a_3, b_3). The parameters of Xu et al. [8] have been used and adapted to the moment's method limited to the order 6, which allows to calculate efficiently the local energies. Then a repulsive potential is added and the Monte Carlo relaxations of structures are performed.

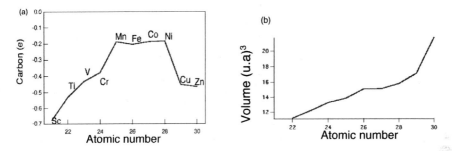

FIGURE 3. Extra charge on a C atom in a FCC structure occupying the octaedric site. Two tendencies are observed along the series: the charge transfer decreases (a) (negative values mean more electrons) and the size effect increases (b) along transition series.

Results

We present different calculations which show the destabilization of the graphene sheet by a charge transfer with respect to structures containing odd rings. In molecules, naphtalene is unstable versus azulene when electrons are added to the molecule (Fig. 4). Similarly, total energy calculations including s and p electrons indicate that a sufficient charge transfer stabilizes 5-7 with respect to hexagonal tilings (Fig .5).

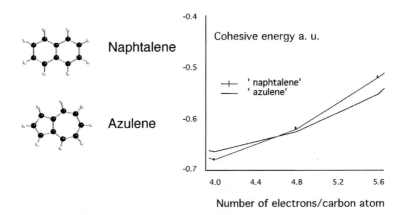

FIGURE 4. Comparison of the π electron Hückel energies of azulene and naphtalene.

93

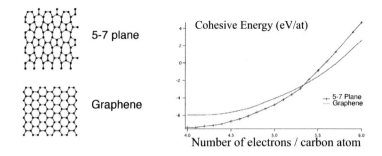

FIGURE 5. Comparison between the total energies of the graphene sheet and of a 5-7 sheet containing only pentagons and heptagons. (tight-binding to the 6th moment + repulsive potential)

Alternatively if charge is transferred to a graphene sheet, a buckling of the sheet is clearly observed (Fig. 6).

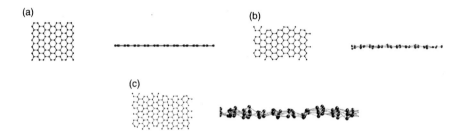

FIGURE 6. Deformation of a graphene sheet by charge transfer. Monte Carlo simulation at 1000K; top view (left) and side view (right). (a) 4 electrons per atom. (b) 4.5 electrons per atom (c) 5 electrons per atom.

CONCLUSION

In the Ni – RE system, charge transfer from the catalyst to C atoms can explain the nucleation of SWNT by a double mechanism : it favours the occurrence of odd-membered rings (pentagons and heptagons) and it strongly modifies the C-C-C bond angles.

REFERENCES

1. Y. Saito et al, Jpn. J. Appl. Phys., **33** (1994) 526
2. H. Kataura et al, Carbon, **38** (2000) 1691
3. J. Gavillet et al, Phys. Rev. Lett., **87** (2001) 275504
4. J. Gavillet et al, Carbon, (2002)
5. J. Gavillet et al, Proceeding Kirchberg (2002)
6. R. Haydock et al, J. Phys. C., **5** (1972) 2845, Solid State Phys., **35** (1980) 216
7. C.H. Xu et al, J. Phys, Condens. Matter, **4** (1992) 6047

CVD-Growth of In-Situ Contacted Single-Walled Carbon Nanotubes

Robert Seidel*, Maik Liebau*, Eugen Unger*, Andrew P. Graham*,
Georg S. Düsberg*, Franz Kreupl*, Wolfgang Hönlein*,
and Wolfgang Pompe[+]

* *Infineon Technologies AG, Corporate Research, 81730 Munich, Germany*
[+] *Institut für Werkstoffwissenschaft, Technische Universität Dresden, 01062 Dresden, Germany*

Abstract. We report on the CVD-growth of single-walled carbon nanotubes (SWCNTs) and their electrical characterization. Growth between catalyst covered Mo-pads yields in-situ contacted SWCNTs. With this approach, only one level of conventional photolithography is required to obtain SWCNTs ready for field-effect measurements. The SWCNT yield depends strongly on the catalyst composition and growth conditions, which have been investigated over a wide range. The growth of isolated SWCNTs in lithographically defined holes is also demonstrated. This is the first step towards the realization of surrounding gate vertical SWCNT field-effect transistors (VCNTFET), which are expected to outperform future silicon-based devices.

INTRODUCTION

Thermal CVD is a simple method to grow SWCNTs directly on a substrate without difficult separation and cleaning procedures [1,2]. For electronic characterization of the as-grown SWCNTs it is further crucial to develop fabrication methods that allow patterned growth and easy contacting. One of the first methods for in-situ contacted growth of SWCNTs was reported by Javey et al. [3] using an alumina supported iron catalyst on predefined Mo-catalyst pads. They used two lithographic steps (Mo-pads, catalyst islands) and spin-on of the catalyst. We have developed a simpler method requiring only one optical lithographic step and PVD deposition of a metal-multilayer system. Our approach further allows to control the yield of the in-situ contacted SWCNTs.

RESULTS AND DISCUSSION

Delzeit et al. [4] observed the enhanced growth of SWCNTs using thin Fe or Fe/Mo catalyst layers on top of 10-20nm thick aluminum layers deposited on oxidized silicon. We have extended the metal-multilayer system by introducing a conducting material beneath the Al-layer. The principal multilayer system consists of the substrate (oxidized silicon), an adhesion layer (e.g. Ti), the electrode metal, a 5-10nm thick Al-layer, and finally the catalyst layer. The metal-layers were deposited in an ion-beam

CP685, *Molecular Nanostructures: XVII Int'l. Winterschool/Euroconference on Electronic Properties of Novel Materials,* edited by H. Kuzmany, J. Fink, M. Mehring, and S. Roth
© 2003 American Institute of Physics 0-7354-0154-3/03/$20.00

deposition system and a thin Fe/Mo catalyst bilayer was used. The Al-layer is essential for the growth of SWCNTs on conducting substrates. SWCNTs could only be grown on top of Mo, Ta, TiN, and TaN if the catalyst was separated from the electrodes by a thin Al-layer. SWCNT growth was not observed on W-electrodes even with an Al-layer.

The CVD growth of the SWCNTs was performed in a 4in. quartz-tube furnace at 900°C. The samples were always inserted into the preheated oven, causing the oxidation of the catalyst-layer and at least a partial oxidation of the Al-layer. The catalyst was reduced in H_2 ambient prior initiating the growth by purging the furnace with a methane/hydrogen mixture. Therefore, the Al in contact with the catalyst is very likely transformed into alumina. This alumina-layer prevents the catalyst from diffusing into the adjacent electrodes, becoming contaminated with the electrode material, or from coalescing into larger particles. The alumina further supports the formation of catalytically active, spherical catalyst particles due to poor wetting characteristics. The thickness of the catalyst layer can be varied to control the yield of SWCNTs. Fig. 1 shows the effect of slightly increasing the thickness of the Fe/Mo-bilayer. The on-resistance of semiconducting SWCNTs was found to be in the MΩ-range using the substrate as a back-gate. A statistical evaluation of a large number of contacted SWCNTs showed a 2/3 abundance of semiconducting SWCNTs, which is expected if there is no preference for certain chiralities during growth.

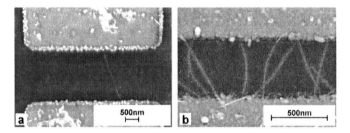

FIGURE 1. SEM images of SWCNTs grown between two Mo-contacts. a) an individual SWCNT (multilayer: 10nm Ti/50nm Mo/10nm Al/0.2nm Mo/0.8nm Fe) b) several SWCNTs after increasing the catalyst layer thickness (multilayer: 10nm Ti/50nm Mo/10nm Al/0.3nm Mo/1nm Fe).

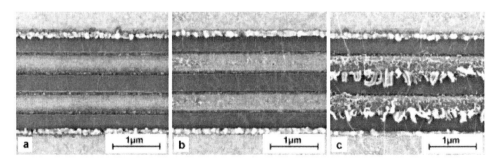

FIGURE 2. SEM images of SWCNTs grown between Mo-electrodes depicting the dependence on the methane partial pressure. All samples were grown at 900°C in a hydrogen/methane mixture with fixed hydrogen pressure of 15kPa. a) sparse SWCNT growth with $p(CH_4) = 5kPa$, b) good SWCNT growth with $p(CH_4) = 10kPa$, and c) good growth but amorphous carbon deposits with $p(CH_4) = 30kPa$.

The partial pressure of the carbon supplying methane is important for the growth of SWCNTs and the amount of co-deposited amorphous carbon. Keeping the temperature (900°C) and H_2 pressure (15kPa) constant we found that a minimum CH_4 pressure of about 5kPa is necessary to observe any SWCNT growth (Fig. 2a). If the CH_4 pressure becomes to high (= 30kPa) an increasing amount of amorphous carbon (Fig. 2c) is obtained. For intermediate partial pressures good SWCNT-growth with no observable amorphous carbon deposits were found (Fig. 2b).

A large number of the in-situ contacted semiconducting SWCNTs prepared in this way showed an ambipolar behavior in back-gate transistor measurements in air. The I_D-V_G curve of an as grown device is shown in Fig. 3a. Fig. 3b gives the I_D-V_G curve of a different device with improved contact resistance after Ti has been deposited on the contacts. The observed hysteresis has been related to adsorbed water molecules [5] or traps in the oxide. There are two possible explanations for the ambipolar behavior:

- Large diameter semiconducting SWCNTs, with a small bandgap [6].
- Double-walled carbon nanotubes (DWCNT) with the outer tube p-type (adsorbed oxygen) and an undoped inner tube, which is n-type.

The first explanation is possible since the diameter of the catalyst is rather poorly defined. Therefore, large diameter particles can form, promoting the growth of large diameter SWCNTs with a correspondingly small bandgap. The tunability of ambipolar devices is also usually less than 3 orders of magnitude further supporting the idea of SWCNTs with a small bandgap. The growth of DWCNTs is likely to happen during thermal CVD but it is difficult to verify whether a particular CNT between two contacts is a DWCNT. The difference between the two minima in each of the plots is related to strong hysteresis effects. Both plots show a rather low off-resistance, which means that the tube cannot be fully turned off at a bias of 100mV. Such behavior is expected for both a SWCNT with a small bandgap and a double-walled tube in which p-type and n-type behavior overlap.

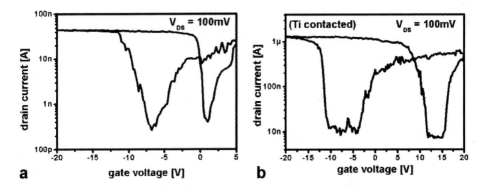

FIGURE 3. Drain current vs. gate voltage curves of SWCNTs showing ambipolar behavior. The bias is 100mV in both cases. The contact resistance of the SWCNT in b) was improved by a Ti-layer on top. The hysteresis is probably caused by adsorbed water.

To date, all attempts to build SWCNTs based field-effect transistors were in a planar fashion, i.e. the SWCNT lies parallel to the substrate. However those approaches do not fully exploit the integration advantages of SWCNTs compared to silicon based technology. In a vertical arrangement a much higher integration density can be realized [7]. We have made the first steps towards the realization of vertical carbon nanotubes field effect transistors (VCNTFET, Fig. 4b). Fig. 4a shows a CNT with a diameter well below 5nm protruding from a 20-30nm diameter hole. Those vertical CNTs were obtained after depositing a metal/oxide multilayer and etching 20-30nm diameter holes by ion milling down to the catalyst. The CNTs were grown according to the growth procedure described above.

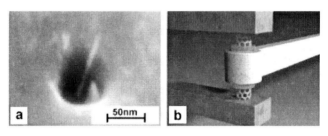

FIGURE 4. a) SEM image of a SWCNT protruding from an about 30nm diameter hole. b) Schematic of a vertical carbon nanotube field effect transistor with a surrounding gate in the middle.

CONCLUSION

In-situ contacted SWCNTs were obtained using a PVD deposited metal-multilayer system. This method allows the easy characterization of as-grown SWCNTs and the study of their electronic behavior under various conditions. The metal multilayer system might further be a suitable approach towards the realization of a vertical carbon nanotube transistor, with the potential for ultimate integration. As the first step we have realized the first SWCNTs grown vertically out of 20-30nm diameter holes.

ACKNOWLEDGMENTS

We are grateful to Z. Gabric for technical support and to W. Pamler for the artwork.

REFERENCES

1. Kong, J., Soh, H., Cassell, A., Quate, C. F., Dai, H.: *Nature,* **395**, 878, (1998)
2. Cassell, A. M., Raymakers, A. J., Kong, J., Dai, H. J.: *Phys. Chem. B.,* **103**, 6484, (1999)
3. Javey, A., Wang, Q., Ural, A., Li, Y., Dai, H.: *Nano Lett.,* **2**, 929, (2002)
4. Delzeit, L., Chen, B., Cassell, A., Stevens, R., Nguyen, C., Meyyappan, M.: *Chem. Phys. Lett.,* **348**, 368, (2001)
5. Kim, W., Javey, A., Vermesh, O., Wang, Q., Li, Y., Dai, H.: *Nanoletters,* **3**, 193, (2003)
6. Javey, A., Shim, M., Dai, H.: *Appl. Phys. Lett.,* **80**, 1064, (2002)
7. Kreupl, F. et al.: "Physics, Chemistry and Application of Nanostructures", *World Scientific,* (2003)

Nanotube Alignment And Dispersion In Epoxy Resin

C. P. Ewels, G. Désarmot, A. Foutel-Richard, F. Martin

DMSC, ONERA, 29 Rue de la Division Leclerc, Châtillon, Paris

Abstract. To date theoretically predicted enhancements of Youngs' Modulus, failure strength, and other mechanical properties of nanotube-composite materials have not been experimentally realised. We discuss the reasons for this and using short fibre composite theory suggest that nanotube straightness is a crucial factor. We examine single walled nanotube – epoxy resin (LY556) composites using microindentation, 3-point bending, electrical resistivity and various microscopy techniques. We test extrusion as a method for aligning the nanotubes, and the use of gum arabic for tube dispersion, straightening and separation.

INTRODUCTION

In polymer matrices, carbon nanotubes have compressive strengths in the range 1.0-1.4TPa, more than 2 orders of magnitude higher than any other fibre [1]. Nanotube composite materials have a bright future, provided various hurdles can be overcome. Good matrix wetting and adhesion is essential if the mechanical properties of the nanotubes are to be fully effectively transferred to the matrix. In addition the straightness and alignment of the nanotubes is also crucial, as discussed below.

In this study we use arc-electric produced single walled nanotubes (SWNTs) from Nanoledge, unpurified. These are mixed with a high strength Araldite epoxy resin, CIBA LY556 (diglycidyl ether of bisphenol A) containing 21.75wt% Araldite HT972 hardener. The gum arabic was Winsor and Newton solution (28.6wt% solid).

Effect Of Nanotube Tangling

TEM images of single walled carbon nanotubes (SWNTs) typically show spaghetti-like entangled networks. It is clear that only short sections of any given tube will lie parallel or near parallel to the direction of an applied stress.

If one considers the short fibre model of Cox [2], a fibre of radius r_f and length l with a Youngs Modulus E_f, encased axially in a cylinder of matrix material radius R with shear modulus G_m undergoes elastic extension ε:

$$\frac{\varepsilon}{\varepsilon_\infty} = 1 - \frac{\cosh\beta(l/2-x)}{\cosh\beta l/2}, \beta = \sqrt{\frac{2G_m}{r_f E_f \ln R/r_f}}. \quad (1)$$

where ε_∞ is the deformation of the fibre when l=∞, the host matrix and the fibre strains being equal ($\varepsilon_\infty = \varepsilon$). In this case there is 100% strain transfer from the matrix

CP685, Molecular Nanostructures: XVII Int'l. Winterschool/Euroconference on Electronic Properties of Novel Materials, edited by H. Kuzmany, J. Fink, M. Mehring, and S. Roth
© 2003 American Institute of Physics 0-7354-0154-3/03/$20.00

to the fibre. If we define a characteristic fibre length l_0, such that $l_0 = 2 / \beta$, we can plot l/l_0 against $\varepsilon/\varepsilon_\infty$:

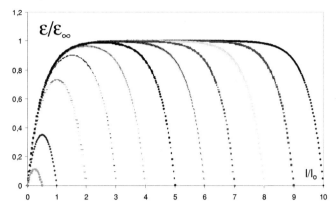

FIGURE 1. Strain transfer from matrix to fibre as a function of fibre length, from short fibre theory.

From Figure 1 it is clear that for long fibre lengths, $l \gg l_0$ there is 100% strain transfer from matrix to fibre except at the fibre ends, but for shorter fibres this is no longer the case. For example for single walled nanotube bundles of radius 15nm and Youngs Modulus 1000GPa, the critical length l_0 is 500nm. We can define γ, the fractional load deformation transfer, to be the average value of $\varepsilon/\varepsilon_\infty$ along the fibre. If the bundle lengths are 10μm then $\gamma=0.95$, *i.e.* there is a 95% deformation transfer from matrix to fibre. However for $l=500$nm, γ is reduced to 52% and for $l=100$nm it drops to only 1.3%. For tangled randomly oriented nanotubes it is clear that although in principle the bundles may be several microns long, there may only be tube sections of order hundreds of nanometres aligned with the applied stress.

The Young's Modulus of a composite, E_c, for unidirectional long fibres is given by the volume average of the fibre (E_f) and matrix (E_m) moduli, *i.e.* $E_c = V_m E_m + V_f E_f$. For a resin with $E_m=3$GPa, containing 15vol% SWNTs, this gives $E_c=153$GPa. However for randomly aligned fibres, $E_c = V_m E_m + {}^1/_6\, V_f E_f$, giving $E_c=28$GPa. Allowing for short fibres as discussed above, $E_c = V_m E_m + {}^\gamma/_6\, V_f E_{f,}$. For $l=500$nm that gives $E_c=8.6$GPa, and for $l=100$nm $E_c=3.1$GPa, practically unchanged from E_m. This is consistent with experimental results which find only small enhancements of Young's Modulus on addition of SWNTs to the host matrix.

Thus random orientation of nanotubes restricts modulus enhancement of the composite, however this effect will be significantly worse for tangled nanotubes. It is clear that we need to develop methods both to align and *straighten* the nanotubes.

Epoxy-Nanotube Composite Viscosity and Extrusion

SWNTs greatly increase resin viscosity (~3 orders of magnitude for 10wt% SWNT). The viscosity is highly temperature dependant, varying by an order of magnitude for a temperature increase of 10°C.

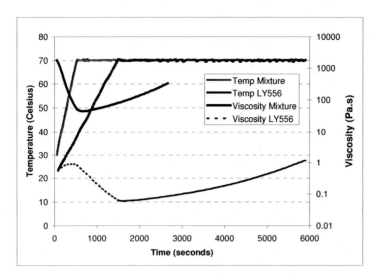

FIGURE 2. Viscosity of pure CIBA LY556 and LY556-9wt% SWNT composite vs time ('Mixture').

1g SWNTs are sonicated in CH_2Cl_2 and mixed with 10g of epoxy (LY556). After solvent evacuation the mixture is pre-cured at 70°C for 45 minutes then transferred to the extruder, which is pre-heated to 50°C to avoid resin solidification in the apparatus. The composite is extruded at ~2.5kbar pressure through a rectangular 15×1mm slot. Extrusion is sensitive to all parameters listed, with only a relatively small range in which extrusion is possible at reasonable pressures, yet the resultant composite strip is not too liquid. The extruded sample is cut at liquid nitrogen temperatures into 48mm strips which are stacked in a mould, out-gassed under vacuum, then cured at 140°C under pressure for 2 hours.

Surface electrical resistivity of the samples is anisotropic, 6-7 times higher perpendicular than parallel to the direction of extrusion, suggesting nanotube alignment. However three point bending tests give a modulus of 3.05 GPa, slightly lower than pure LY556 and suggesting no significant alignment or indeed composite enhancement. If misalignment occurs this could be due to our extrusion conditions, or nanotube redistribution during curing. Microscopy of pre-annealed, post-extrusion samples will be necessary to determine this. We hope to explore further the role of parameters such as extrusion temperature, nozzle width, applied pressure and pre-curing time.

Coating nanotubes with Gum Arabic

Bandyopadhyaya *et al* [3] made aqueous solutions of gum arabic (GA) and nanotubes, with well dispersed nanotubes and break up of bundles due to the gum coating. Their TEM images also suggest that GA adds rigidity to the tubes, since they are all either straight or have only slight curvature. In light of this we investigated the use of GA with SWNTs for epoxy resins composites.

We were able to produce stable well dispersed aqueous solutions of GA-SWNTs via 30 minute sonication, with weight ratios from 1:1 to 6:1. These remained well-dispersed for over three weeks and resembled thick ink. SEM studies suggest however that the GA is coating nanotube bundles rather than individual tubes (Figure 3a), although we note that there may also be individual tubes which are not imaged. AFM shows bundles and isolated tubes (e.g. Figure 3b), although they are hard to distinguish since the tube diameter includes an unknown thickness of GA. Figure 3b also suggests coated tubes are indeed straightened.

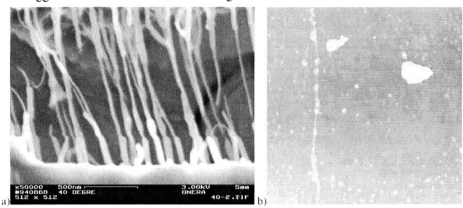

a) b)

FIGURE 3. (a) SEM of gum arabic coated nanotubes (wt ratio 6:1, scale 500nm). (b) AFM image of gum arabic coated nanotube on SiO_2 surface. Image width 2.35µm (courtesy of J. Meyer, Stuttgart).

Unfortunately GA is only soluble in water, and to a lesser extent, in DMSO. Through forming separate aqueous GA-SWNT and ethanol-LY556 solutions and then mixing while sonicating, we formed a reasonably homogenous mixture (1.66wt% SWNT, 8.33wt% GA, 90wt% LY556), which was then filtered to remove large rubber precipitates. Microhardness indentation tests on these non-extruded samples gave an average hardness of 215MPa compared to 195MPa for a similar sample without GA. Note that composition quoted here is pre-filtering; filtering removes lots of GA and SWNTs. This enhanced hardness suggests GA may help separate and straighten nanotube bundles, however further samples and testing is required to confirm this.

We also had problems with porosity during curing, due to reactions between the hardener and GA. Pre-heating the GA-SWNT at 200°C before mixing to burn off the GA led to slightly improved mixing.

ACKNOWLEDGMENTS

Funding for CPE came from the EU TMR network COMELCAN. We thank J Meyer for AFM imaging of gum arabic coated samples.

REFERENCES

1. O. Lourie, D. M. Cox, H. D. Wagner, *Phys. Rev. Lett.* **81** 8 1638 (1998).
2. H. L. Cox, *Brit. J. Appl. Phys.* **3**, 72-79 (1952).
3. R. Bandyopadhyaya, E. Nativ-Roth, O. Regev, R. Yerushalmi-Rozen, *Nanoletters* **2** (1), 25-28 (2002).

Growth Of Patterned SWNTs From Catalysts Fabricated By Direct Photolithography

Shaoming Huang, Jie Liu[*]

Department of Chemistry, Duke University, Durham, NC 27708, USA

Abstract. A novel feasible method to pattern SWNTs with different resolutions on surface in large scale by direct photolithography has been demonstrated. It involves the direct photolithographical patterning of catalysts on substrate by incorporating catalyst precursor into photoresist and then growing SWNT using carbon monoxide chemical vapor deposition (CO-CVD) method. SWNT patterns with micro-scale resolution can be generated using high-resolution photomasks. This direct photolithography method could be applied to other catalyst systems for patterning SWNTs and could be useful for various device applications of SWNTs.

1. Introduction

Single-walled carbon nanotubes (SWNTs) have shown many unique structural, mechanical and electrical properties.[1] They are generally considered as ideal systems for investigating fundamental electronic properties in low-dimension materials and as building blocks for future nanoscale electronics.[1, 2] Many nanoscale devices have been demonstrated using SWNTs as the active components, including quantum wires,[3, 4] logic gates,[5-7] field emitters,[2] field-effect transistors,[8] diodes [9] and inverters[7] etc. Some of the applications, e.g. field emitters in flat panel displays, require the controlled growth and patterning of SWNTs and various technologies such as soft-lithography, photolithography have been used to fabricate SWNT patterns.[2, 10-12] Among these technologies, photolithography is a powerful patterning techniques, which is widely used in the integrated circuit industry. For example, Dai et al.[11] reported the patterned growth of SWNTs on 4 inch SiO$_2$/Si wafers by photolithographically patterning polyethylmethacrylate (PMMA) followed by casting a layer of Al$_2$O$_3$-supported Fe/Mo precursors in methanol suspension. However, to our best knowledge, direct photolithography of catalysts for production of patterned SWNTs has not been demonstrated. In this paper, we report a simple direct photolithography method to produce patterned SWNTs. It involves the formation of catalyst patterns by incorporating catalyst precursors into photoresist followed by oxidation and reduction processes. The growth of patterned SWNTs was achieved by using carbon monoxide chemical vapor deposition (CO-CVD) method. The resolution of the SWNT patterns can be down to micron-scale. Large scale SWNT pattern with 30 microns resolution can be quickly and inexpensively fabricated by using black and white film as photomask. This practical method could be very useful for SWNT field emission device fabrication and applications.

---

[*] To whom correspondence should be addressed. E-mail: j.liu@duke.edu

CP685, *Molecular Nanostructures: XVII Int'l. Winterschool/Euroconference on Electronic Properties of Novel Materials,* edited by H. Kuzmany, J. Fink, M. Mehring, and S. Roth
© 2003 American Institute of Physics 0-7354-0154-3/03/$20.00

2. Experimental

2.1 Catalysts Pattern Formation

In a typical experiment, the silica wafer with 600 nm thick silicon dioxide layer was cleaned in the Piranha solution (a mixture of 98% H_2SO_4 and 30% H_2O_2 at 7:3 v/v) for ca. 30 min. A thin layer (~0.2 μm) of photoresist/catalyst mixture, produced by dissolving 5g dazonaphthoquinone (DNQ)-modified cresol novolak resin and 0.01g~0.05g iron(III) acetylacetonate ($Fe(acac)_3$) in 20 mL of ethoxyethyl acetate, was prepared by spin-casting onto the silica wafer. The wafer was dried in an oven at 80°C for 10 min. Upon UV irradiation through a photomask, the DNQ-novolak photoresist film in the exposed regions was rendered soluble in an aqueous solution of tetramethylammonium hydroxide. The sample was then rinsed with distilled water and baked at 80 °C for 10 min. At this point, a catalyst precursor-containing polymer pattern was fabricated.

2.2 Growth of SWNT

The growth of SWNTs on the surface was carried out using a conventional two-furnace setup with a CO/H_2 mixture as feeding gas as described previously.[13] The first furnace was used for the pretreatment of CO gas (700 °C), and the second one was used for nanotube growth. Before growing nanotubes, the remaining photoresist on the substrate was removed by heating at 600 °C in air for 2 hours. Patterns of catalyst nanoparticles were produced after such treatment. The substrate was then reduced in H_2/Ar mixture at 700 °C for 10 minutes. The growth of nanotube was carried out at 900 °C using CO/H_2 mixture (800 sccm/200 sccm) as feeding gas for 10 minutes.

2.3 Characterization

The samples were characterized by Scanning Electronic Microscopy (SEM, XL-30 FEG, Philips, at 1 KV), Transmission Electronic Microscopy (TEM, Philips) and Atomic force microscopy (AFM, Nanoscope IIIA system with a multi-mode). TEM images were taken on Si_3N_4 TEM window which was used as substrate for in situ growth of SWNT.

3. Results and Discussions

To demonstrate that the mixture of photoresist and Fe(acac)$_3$ can be used directly for SWNT growth on surfaces, we first use TEM grid as the mask for catalysts patterning. The TEM grid was placed on the substrate and the mixture of photoresist and catalyst precursors was spin-coated. After removing the TEM grid, the substrate was used for growth of nanotubes as described in the experimental section. Figure 1a gives the low magnification of as-produced nanotube patterns (white areas are nanotube patterns).

Figure 1. SEM images of (a) low magnification of SWNT patterns using TEM grid as physical mask (b) high magnification of the SWNT patterns showing the random SWNTs.

From higher magnification SEM image (Figure 1b), it was found that dense nanotubes formed in the area where the photoresist/catalyst was deposited. However there are still some nanotubes in the blank regions. This is probably because a small amount of photoresist diffused into these areas during the spin coating process. (Note: SEM images were taken at 1KV, under this condition the size of nanotubes observed under SEM is not the real size of the nanotubes). AFM height measurements (Figure 2a) of produced iron oxide nanoparticles after oxidation process show that the iron oxide nanoparticles have diameters of 2~4 nm which was also confirmed by TEM observation (inset 2a) when the concentration of Fe(acac)$_3$ is 0.2% (Wt%). The nanotubes have the diameters ranging from 1.0 to 3.0 nm based on AFM measurements (Figure 2b). TEM observation gives further evidence of SWNTs in the as-produced nanotubes (inset 2b).

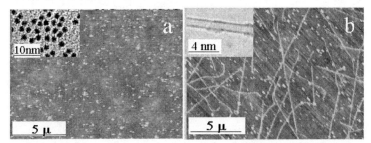

Figure 2. (a) AFM images of iron oxide nanoparticles on surface after removing the polymer by the oxidation process. Inset is the TEM image showing the uniform nanoparticles with 2~4 nm in diameter. (b) Typical AFM image of as-grown SWNTs on surface. Inset is the TEM image of individual nanotubes with diameter of ca. 1.5 nm.

These experimental results indicate that the mixture of photoresist/catalyst precursor could be used to produce catalyst nanoparticles for the growth of SWNTs. We further studied the direct photolithographic patterning of catalysts using such mixtures to achieve high-resolution SWNT patterns. The catalyst-containing photoresist in the regions exposed to UV light was developed because of the photochemical reaction of the photoresist by photogeneration of the hydrophilic indenecarboxylic acid groups from the hydrophobic DNQ[14]. Figure 3a is the low magnification SEM image of nanotube patterns with features of 3μ-resolution lines generated using high resolution photomask. From higher magnification SEM image, there are almost no nanotubes in the resist-free area. The nanotubes have lengthes of several micrometers after 10 minutes reaction.

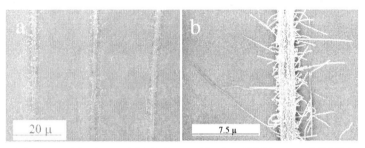

Figure 3. (a) Low magnification and (b) high magnification SEM images of SWNT patterns with 3μ resolution parallel lines.

From certain applications like flat panel displays, we need to grow SWNTs over a large area. We have used conventional black-white films as photomask to fabricate nanotube patterns with resolution of about 30μ. The black-white films were prepared by a printing and photographic reduction method. The patterns designed in Chemdraw software were first printed by a laser printer with resolution 600 dpi onto a plain paper. Then, the patterns were reduced photographically onto 35 mm film using a 35 mm conventional camera to produce a black-white photographic film with resolution of 30 μ. This method can be used to fabricate SWNT pattern quickly and inexpensively over large areas. As an example, regular dots of nanotube patterns with 30μ diameters were produced using such method. Figure 4a and 4b show the pattern of SWNT in large scale and nanotubes on an individual dot.

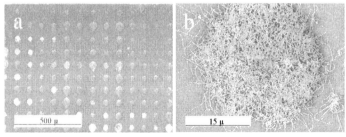

Figure 4 (a) SEM image of large scale SWNT patterns with 30 μ resolution by using film-type photomask and (b) high magnification SEM image of the individual SWNT pattern.

The formation of the nanosized particles was found to be the key issue for SWNT growth in the described method. The photoresist acted as the dispersion media for the catalyst precursor. Precursor molecules dispersed uniformly within the photoresist during spin casting process. These precursor molecules decomposed at 200 °C to form iron oxide nanoparticles within the photoresist. The photoresist was still stable at 200 °C and prevented the aggregation of the nanoparticles. So the concentration of the catalyst precursor in photoresist played an important role in determining the size of nanoparticles. It was found when the concentration of precursor was high than 2% (wt%) only multi-walled CNTs rather than SWNTs were produced because the higher concentration of precursors resulted in larger catalyst nanoparticles which was not favorable for SWNT growth. According to our results, 0.1~0.5 Wt% is the optimal concentration of precursors for SWNT growth by this direct photolithographic method.

4 Conclusion

In summary, we have demonstrated a novel method to grow patterned SWNTs by direct patterning of catalyst precursors incorporated in commercial photoresist resin. The resolution of the SWNT patterns can be as high as a few microns when high-resolution photomask is used. Using a simple high-contract black-white films as photomask, SWNT patterns with 30μ resolution over large area can be quickly and inexpensively fabricated. The concentration of the catalyst precursor, $Fe(acac)_3$, plays an important role in the formation of nanosized catalyst particles for SWNT growth. This highly practical, direct photolithography method could be applied to other catalyst systems for patterned growth of SWNTs and could become useful for various device fabrication techniques.

Acknowledgments

This project is supported by a grant from NASA (NAG-1-01061), Carbon Nanotechnology Inc. and the 2002 Young Professor Award from Dupont.

References

1   M. S. Dresselhaus, G. Dresselhaus, and P. Avouris, *Carbon Nanotubes Synthesis, Structure, Properties, and Applications* (Springer, 2001).
2   O. Zhou, H. Shimoda, B. Gao, et al., Accounts of Chemical Research **35**, 1045 (2002).
3   C. Dekker, Physics Today **52**, 22 (1999).
4   S. J. Tans, M. H. Devoret, H. J. Dai, et al., Nature **386**, 474 (1997).
5   Y. Huang, X. F. Duan, Y. Cui, et al., Science **294**, 1313 (2001).
6   A. Bachtold, P. Hadley, T. Nakanishi, et al., Science **294**, 1317 (2001).
7   V. Derycke, R. Martel, J. Appenzeller, et al., Nano Letters **1**, 453 (2001).
8   S. J. Tans, A. R. M. Verschueren, and C. Dekker, Nature **393**, 49 (1998).
9   M. S. Fuhrer, J. Nygard, L. Shih, et al., Science **288**, 494 (2000).
10 G. Gu, G. Philipp, X. C. Wu, et al., Advanced Functional Materials **11**, 295 (2001).
11 N. R. Franklin, Y. M. Li, R. J. Chen, et al., Applied Physics Letters **79**, 4571 (2001).
12 J. Kong, H. T. Soh, A. M. Cassell, et al., Nature **395**, 878 (1998).
13 B. Zheng, C. Lu, G. Gu, et al., Nano Letters **2**, 895 (2002).
14 J. March, *Advanced Organic Chemistry: Reactions Mechanisms Structure* (John Wiley, New York, 1992).

Purification Effect on the Electronic State of Carbon in HiPco Nanotubes

L.G. Bulusheva[1], A.V. Okotrub[1], A.V. Gusel'nikov[1],
U. Dettlaff-Weglikowska[2] and S. Roth[2]

[1]Nikolaev Institute of Inorganic Chemistry SB RAS, pr. Ak. Lavrentieva 3, 630090 Novosibirsk, Russia
[2]Max-Plank-Institute for Solid State Research, Heisenbergstr. 1, 70569 Stuttgart, Germany

Abstract. Electronic state of carbon in HiPco nanotubes before and after purification was examined using X-ray fluorescent spectroscopy. The $CK\alpha$-spectrum of as prepared HiPco nanotubes was shown to be similar to the spectrum of graphite. The purification of carbon nanotubes by thermal and acid treatment was found to result in an enhancement of high-energy maximum in the $CK\alpha$-spectrum. Quantum-chemical calculation revealed such enhancement can be attributed to the sp-hybridized carbon atoms, which are likely to compose the open tube ends and boundaries of pores developed in the tube walls during the purification.

INTRODUCTION

Utilization of carbon nanotubes for different application usually requires a purification of as prepared material. The purification process is usually a complicated process consisted of several stages. Recently suggested high pressure CO disproportionation (HiPco) seems to be an effective method for continuous production of single-wall carbon nanotubes (SWNTs) of high purity [1]. Besides the SWNTs with average diameter of 1.1 nm the raw HiPco material contains amorphous carbon and Fe particles coated by thin carbon layers. The unwanted products can be selectively removed by multistage procedure included the thermal oxidation in a wet Ar/O_2 environment and washing with concentrated HCl [2]. These stages were shown to increase the total surface area and micropore volume of HiPco material due to probably partial opening of nanotubes [3]. The structural change of SWNTs should effect on the electronic state of carbon.

The aim of the present work is examination of electronic state of carbon in the HiPco material before and after purification using an X-ray fluorescent spectroscopy. X-ray fluorescent spectrum measures a partial density of occupied states in a solid and, therefore, can be directly compared with the result of quantum-chemical calculation. Previously, the calculation on various tube models has been used to interpret an enhancement of high-energy maximum in the spectrum of multiwall carbon nanotubes produced by acetylene decomposition over Fe/Co catalyst [4]. The density of high-energy states is almost not affected by inserting the pentagon-heptagon defects into tube wall. The more noticeable increase of the spectral intensity was shown to be for the tubes with intershell linkages [5].

CP685, *Molecular Nanostructures: XVII Int'l. Winterschool/Euroconference on Electronic Properties of Novel Materials,*edited by H. Kuzmany, J. Fink, M. Mehring, and S. Roth
© 2003 American Institute of Physics 0-7354-0154-3/03/$20.00

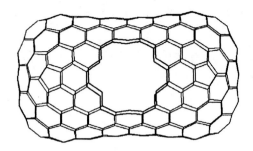

FIGURE 1. AM1 optimized structure of closed (8,8) tube with two hexagonal holes in the side wall.

EXPERIMENTAL DETAILS

SWNTs were produced by high pressure CO disproportionation over Fe catalyst (HiPco technique) at CNI, Houston, USA. The raw sample was purified by controlled thermal oxidation in air and HCl treatment [2]. As prepared HiPco nanotubes (0.5 g) were sonicated in conc. HCl for 15 min. After filtration and washing with deionized water the remaining solid was heated at 225 °C for 18 h. The metal impurities were removed by HCl extraction. The thermal treatment of the sample has been repeated at 325 °C for 1.5 h and at 350 °C for 1 h followed by sonication in HCl. As estimated by inductive coupled plasma spectroscopy the metal impurities decreased to 0.05 wt. % in the final product. TEM observations of the purified material show neither amorphous carbon nor graphitic structures [6].

CKα-spectra of the pristine and purified HiPco nanotubes were recorded with a laboratory X-ray spectrometer «Stearat» using ammonium biphthalate (NH$_4$AP) single crystal as a crystal-analyzer. Near the K-edge of carbon absorption, the reflection efficiency of this crystal has complex dependence from the radiation energy that was corrected by the procedure described elsewhere [7]. The samples were deposited on copper supports and cooled down to liquid nitrogen temperature in the vacuum chamber of the X-ray tube with copper anode (U=6 kV, I=0.5 A). Determination accuracy of X-ray band energy was ±0.15 eV with spectral resolution of 0.5 eV. The spectra were normalized at the maximal intensity.

COMPUTATIONS

The quantum-chemical calculations of perfect and holed (8,8) tubes were carried out using semiempirical AM1 method [8] within GAMESS package [9]. Tube ends were closed by the caps. The holed structure consisted of 264 carbon atoms was obtained by removing two central hexagons from the perfect tube. The geometry of tube models was relaxed by standard BFGS procedure to the gradient value of 10^{-4} Ha/Bohr. Optimized structure of holed (8,8) tube is shown in fig. 1. The total energy of the perfect (8,8) tube is 0.175 eV/atom lower than that of the holed tube.

X-ray transition intensity was calculated by summing the squared coefficients with which carbon 2p atomic orbitals (AOs) involved in the concrete occupied molecular orbital (MO). The energy location of intensity corresponded to the MO eigenvalue. Calculated intensities were normalized by maximal value and broadened by convolution of Lorenzian functions with half width at half maximum of 0.6 eV.

RESULTS AND DISCUSSION

The CKα-spectra of HiPco nanotubes before and after purification are compared in fig. 2a. The spectra show two main features: the intense maximum B around 275.8 eV and high-energy maximum A at 280.4 eV. The relative intensity and position of the maxima in the raw HiPco spectrum is similar to those characterized for the highly oriented pyrolytic graphite [4]. The theoretical CKα-spectrum calculated for all carbon atoms of the closed (8,8) tube (fig. 2b(1)) well agrees with the raw HiPco spectrum. Hence, we can conclude the carbon formed in HiPco process mainly comprises the graphitic-like structures. Analysis of the closed (8,8) tube calculation indicates the maxima A and B correspond to the π- and σ-like state, while the intensity around 278 eV arises due to X-ray transitions from both type orbitals.

The multistage thermal and acid treatment of the HiPco material results in increasing of the spectral maximum A by about 15 % in the intensity. Obviously, the purified sample contains structural defects, which imply different electronic state of carbon atoms comparing to the ideal graphitic structure. One of such state could be sp-hybridized carbon atoms composed the open tube ends or the boundary of holes developed with the purification.

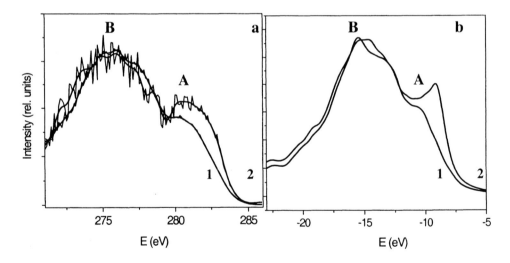

FIGURE 2. Experimental CKα-spectra (a) of pristine (1) and purified (2) HiPco material and theoretical spectra (b) calculated for perfect closed (8,8) carbon tube (1) and that with vacancies (2).

The theoretical spectrum plotted for the boundary carbon atoms of the holed (8,8) tube (fig. 1) is compared to the perfect tube spectrum in fig. 2b. One can see the tendency in the calculated spectra is similar to the experimental data. The spectrum of boundary atoms exhibits a broadening of the maximum A and enhancement of its relative intensity. The dangling bonds of two-fold coordinated atoms composed the boundary of holes was shown by examination of the calculation to provide the maximum enhancement. The high total density observed in the high-energy region of the holed (8,8) tube is caused by accidental near degeneracy in energy of the localized π-states and electron states of the dangling bonds. The orbitals corresponding to the dangling bonds have the energy lower than the energy of highest occupied molecular orbital by about 0.4 eV. This fact can explain the relative chemical inertness of the dangling bonds within the holes.

Geometry optimization of the tube with the smaller size holes, for example the holes obtained by removing of one or two atoms from the carbon network, leads to a hole rearrangement with formation of five-membered rings. Resulting decrease of the number of two-fold coordinated carbon atoms will reduce the maximum A intensity and thus deteriorate the agreement between theory and experiment. We suggest the pores developed with purification of HiPco material should be similar to the six-atomic hexagonal hole (fig. 1) and even have the larger size. Furthermore, the *sp*-hybridized carbon atoms can be occurred at the open tube ends or in an amorphous material saved within the tube cavities.

ACKNOWLEDGMENTS

We thank the INTAS (projects 00-237, 01-254) and the RFBR (Nos.03-03-32286a, 03-03-32336a) for financial support.

REFERENCES

1. Nikolaev, P.; Bronikowski, M.J.; Bradley, R.K.; Rohmund, F.; Colbert, D.T.; Smith, K.A., and Smalley, R.E., *Chem. Phys. Letters* **313**, 91 (1999).
2. Chiang, I.W.; Brinson, B.E.; Huang, A.Y.; Willis, P.A.; Bronikowski, M.J.; Margrave, J.L.; Smalley, R.E., and Hauge, R.H., *J. Phys. Chem. B* **105**, 8297-8301 (2001).
3. Yang, Ch.-M.; Kaneko, K.; Yudasaka, M., and Iijima, S., *Nanoletters* **2**, 385-388 (2002).
4. Bulusheva, L.G.; Okotrub, A.V.; Fonseca, A., and Nagy, J.B., *J. Phys. Chem. B* **105**, 4853-4859 (2001).
5. Bulusheva, L.G.; Okotrub, A.V.; Fonseca, A., and Nagy, J.B., *Synthet. Metals* **121**, 1207-1208 (2001).
6. Dettlaff-Weglikowska, U.; Benoit, J.-M.; Chiu, P.-W.; Graupner, R.; Lebedkin, S., and Roth, S., *Current Applied Physics* **2**, 497-501 (2002).
7. Yumatov, V.D.; Okotrub, A.V., and Mazalov, L.N., *Zh. Structur. Khim.* **26**, 59 (1985).
8. Dewar, M.J.S.; Zoebisch, E.S.; Healy, E.F., and Stewart, J.J.P., *J. Am. Chem. Soc.* **107**, 3902 (1985).
9. Schmidt, M.W.; Baldridge, K.K.; Boatz, J.A.; Elbert, S.T.; Gordon, M.S.; Jensen, J.H.; Kosoki, S.; Matsunaga, N.; Nguyen, K.A.; Su, S.J.; Windus, T.L.; Dupuis, M., and Montgomery, J.A., *J. Comput. Chem.* **14**, 1347 (1993).

Purification of Single-Walled Carbon Nanotubes Studied by STM and STS

A. Vencelová , R. Graupner , L. Ley , J. Abraham[†], M. Holzinger , P. Whelan[†], A. Hirsch[†], F. Hennrich[‡] and M. Kappes[‡]

Universität Erlangen, Institut für Technische Physik II, Erwin-Rommel-Str.1, D-91058 Erlangen
[†]Universität Erlangen, Institut für Organische Chemie, Henkestr.42, D-91054 Erlangen
Université Montpellier II, Groupe Dynamique des Phases Condensées, F-34095 Montpellier
[‡]Forschungszentrum Karlsruhe, Institut für Nanotechnologie, Postfach 3640, D-76021 Karlsruhe

Abstract.
Different chemical purification methods (oxidation in air followed by HCl etch and oxidation by HNO_3) were investigated by scanning tunneling microscopy (STM) and scanning tunneling spectroscopy (STS). These recipes were compared with respect to cleanliness, structural defects, and their effect on the electronic properties. In particular, both p- and n-type doping was observed in individual bundles after the HNO_3 treatment.
Our results imply that the quality of the material in terms of phase purity and structural integrity as well as the electronic properties depend critically on the purification technique.

INTRODUCTION

The physical and chemical properties of single-walled carbon nanotubes (SWCNTs) are more and more attractive for a large number of applications. Since the fabrication of SWCNT material by any method also leads to unwanted by-products (amorphous carbon, transition metals used as catalyst), effective purification techniques are necessary to obtain pure SWCNTs. Annealing in air removes amorphous carbon which is more susceptible to oxidation than CNT material. Oxidizing attack using mineral acids also removes amorphous carbon and in addition leads to the dissolution of metallic remnants. However, several of these recipes are known to modify both the electronic character [1] and the structure of SWCNTs [2]. Scanning tunneling microscopy (STM) and scanning tunneling spectroscopy (STS) allowed us to investigate these processes on an atomic scale. We therefore compared two common purification procedures, i.e. a combined oxidation in air followed by a HCl-etch and the oxidation in HNO_3, in terms of cleanliness, structural defects, and their influence on the electronic properties of the SWCNTs.

EXPERIMENTAL

The SWCNTs were synthesized by laser ablation using a 1 : 1 Co/Ni mixture as catalyst [3]. The raw material was cleaned in different ways: **Sample A** was oxidized in air at 300 C for 1h followed by an etch in 37% HCl, **Sample B** was purified in 10M HNO_3 for

CP685, *Molecular Nanostructures: XVII Int'l. Winterschool/Euroconference on Electronic Properties of Novel Materials,*edited by H. Kuzmany, J. Fink, M. Mehring, and S. Roth
© 2003 American Institute of Physics 0-7354-0154-3/03/$20.00

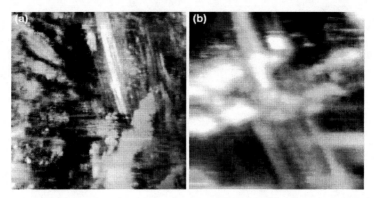

FIGURE 1. STM images taken **(a)** on the raw material (image size 80nm×80nm); **(b)** on sample A (image size 20nm×20nm) showing a decorating of the SWCNTs by amorphous carbon.

1h, and **Sample C** in 3M HNO_3 for 48h under reflux conditions. The purified material was dissolved in dimethylformamide (DMF) and deposited on graphite (HOPG) or Au substrates. Routine investigations of sample quality were carried out by atomic force microscopy (AFM) in air while the STM and STS measurements were performed in ultrahigh vacuum ($<10^{-9}$ mbar). Typical imaging parameters were tunneling currents of 0.4-3 nA at a maximum tip voltage of 300 mV.

RESULTS AND DISCUSSION

In general, atomic resolution could not be achieved by STM on the untreated raw material (Fig. 1a). Furthermore, the density of states measured by STS (not shown) was characteristic for amorphous carbon. We attribute these findings to a decoration of the tubes by amorphous carbon. Moreover, no metallic particles were detected, in accordance with the results of photoelectron spectroscopy [1]. If SWCNTs were identified in the raw material, they seemed to be intact with closed caps and were aligned in bundles with a typical diameter of 10 nm or less.

STM image of the purified **sample A** (Fig. 1b) shows that amorphous particles were still present. STS revealed that most of the investigated SWCNTs were metallic. A typical spectrum of a metallic SWCNT is shown in Fig. 2a. However, on some of the metallic nanotubes additional features were observed in the density of states (Fig. 2b). We attribute this to the formation of defects introduced by the purification process. The structural damage seemed to appear in the adjacent nanotubes within a bundle simultaneously.

After preparation in 10M HNO_3 for 1h (**sample B**) atomic resolution could be routinely achieved (Fig. 3a), which shows that the amorphous carbon and other impurities were effectively removed. The SWCNTs again formed bundles and seemed to be defect-

---

[1] See the article by R. Graupner et al. in this volume

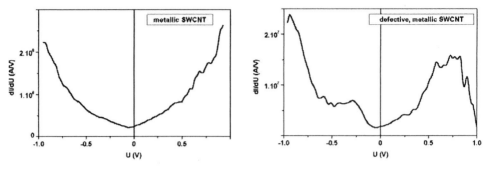

FIGURE 2. STS spectra of metallic SWCNTs in sample A: **(a)** unaffected nanotube; **(b)** defective nanotube

free. Consequently, the STS spectra recorded on different nanotubes were essentially structureless.

The cleanest sample in terms of the amount of by-products was **sample C**. Fig. 3b shows an STM image of a bundle of SWCNTs. In tunneling spectroscopy on semiconducting SWCNTs shifts of the Fermi level to lower energies (Fig. 4a bottom curve) as well as to higher energies (Fig. 4a top curve) relative to mid-gap were observed. Therefore, both p- and n-type doping was found in the very same sample. Interestingly, within an individual bundles the same doping effect (either p- or n-type) was detected. The values of the Fermi-level shifts lie within narrow energy ranges as shown in Fig. 4b. A p-type doping was recently observed in HNO_3 treated bucky-papers of SWCNTs samples using absorption spectroscopy [4]. This doping effect is explained by the intercalation of HNO_3 in-between the tubes [1]. Intercalation can also explain the fact that the p-type doping was observed on all tubes within a bundle. The reason for the simultaneously observed n-type doping is not clear at present, additional experiments are underway to clarify this point.

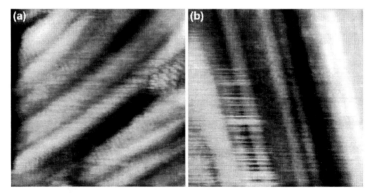

FIGURE 3. Atomically resolved STM images: **(a)** for sample B (image size 10 nm×10 nm); **(b)** for sample C (image size 20nm×20nm).

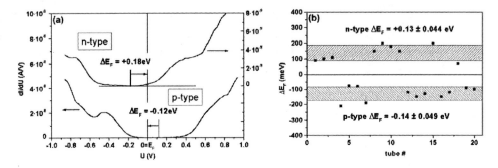

FIGURE 4. (a) Tunneling spectra taken on sample C showing n-type (top) and p-type (bottom) doped nanotubes; (b) Values of the Fermi-level shift observed on different SWCNTs in different bundles; the zero corresponds to a Fermi level position in mid-gap.

CONCLUSIONS

HNO_3 treatment of as prepared SWCNTs leads to an effective removal of unwanted by-products. Strong acid ($10M\ HNO_3$) requires a short purification period (1h). Longer treatment (48h) in $3M\ HNO_3$ generates either p-type or n-type doping of semiconducting SWCNTs within a bundle. The p-type doping for nanotubes within a bundle points to an intercalation process.

Air oxidation (300 C, 1h) followed by an etch in HCl does not eliminate the impurities completely. An altered electronic structure of some metallic SWCNTs may be caused by defects but not all metallic nanotubes are attacked. The structural damage seems to appear in the adjacent SWCNTs of an individual bundle simultaneously.

ACKNOWLEDGMENTS

This work was supported by the Bayerische Forschungsstiftung under the auspices of the "Bayerischer Forschungsverbund für Werkstoffe auf der Basis von Kohlenstoff, FORCARBON".

REFERENCES

1. Bower, C., Kleinhammes, A., Wu, Y., and Zhou, O., *Chem. Phys. Lett.*, **288**, 481–486 (1998).
2. Monthioux, M., Smith, B. W., Burteaux, B., Claye, A., Fischer, J. E., and Luzzi, D. E., *Carbon*, **39**, 1251–1272 (2001).
3. Hennrich, F., Lebedkin, S., Malik, S., Tracy, J., Barczewski, M., Rösner, H., and Kappes, M., *Phys. Chem. Chem. Phys.*, **4**, 2273–2277 (2002).
4. Hennrich, F., Wellmann, R., Malik, S., Lebedkin, S., and Kappes, M. M., *Phys. Chem. Chem. Phys.*, **5**, 178–183 (2003).

Calibration of Raman-Based Method for Estimation of Carbon Nanotube Purity

S.V. Terekhov*[1], E.D. Obraztsova[1], U. Dettlaff-Weglikowska[2], S. Roth[2]

[1]Natural Sciences Center of General Physics Institute, RAS, 38 Vavilov street, 119991, Moscow, Russia
[2] Max-Plank-Inst. fur Festkorperforschung, Heisenbergstr. 1, D-70569, Stuttgart, Germany

Abstract. The set of samples with a calibrated nanotube content has been prepared by mixing pure HipCO nanotubes and an amorphous carbon. The samples have been studied by a Raman-based method for purity estimation, based on a measurement of the tangential Raman mode shift depending on the laser power density applied. The nanotube content values estimated by the Raman-based method have coincided well with the preset ones. The model of a homogeneous two-component medium was used to describe a thermal conductivity of the mixed samples.

INTRODUCTION

Up to now only a few methods are known for estimation of single-wall carbon nanotube (SWNT) percentage in carbon soot synthesized by various techniques. To increase accuracy of one, based on measurement of the tangential Raman mode position in course of a laser heating of the material [1], a calibration procedure was needed. In this work we have performed the calibration using a specially prepared set of 10 samples with the fixed nanotube percentages. The Raman spectra of these samples in course of their heating by a probing laser beam have been measured, and a percentage of the nanotube fraction for each sample has been estimated. For the thermal conductivity description a model of a homogeneous two-component medium was applied.

EXPERIMENTAL

Raman experiments have been carried out with ISA Jobin-Yvon spectrometer S3000 coupled with an "Olympus" microscope. The Ar+-laser with an overall energy output 5W was used for the scattering excitation. Mainly, the laser irradiation with wavelength λ=514.5 nm was used. The power density onto the sample surface varied in the range 2,5 -100 kW/cm^2.

For sample preparation two carbon fractions have been mixed in different proportions: a "Sigma-Aldrich" activated carbon powder and HipCO nanotubes [2], purified by a standard chemical procedure (several consecutive HCl treatments followed by a low temperature oxidation). The paper-like samples of equal thickness have been deposited by co-filtration of these two fractions suspended in an aqueous solution of SDS. The nanotube content in the samples varied from 0.1 to 100 weight

CP685, Molecular Nanostructures: XVII Int'l. Winterschool/Euroconference on
Electronic Properties of Novel Materials, edited by H. Kuzmany, J. Fink, M. Mehring, and S. Roth

%. The sample with the lowest nanotube content was considered as a reference point, corresponding to a pure amorphous carbon.

THERMAL SHIFT OF THE TANGENTIAL RAMAN MODE – A BASIS FOR A PURITY ESTIMATION

The method of a nanotube purity estimation has been proposed recently [1]. The basis of this method is a different laser heating rate for materials with a different nanotube content. For materials with a high nanotube content the rate is low while for the raw soot it is very high. The probing laser beam, used in Raman measurements, may serve as a heater. A frequency of the tangential Raman mode, shifting linearly with the temperature *(Fig.1)*, may serve as a temperature indicator. The calibration of its thermal shift (–0.012 K/cm^{-1} in vacuum) has been performed in an optical oven in the range 80 - 650 K. An advantage of the laser heating in course of Raman measurements is a heat localization and *in situ* observation.

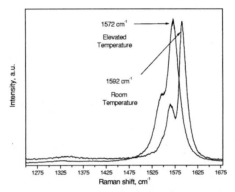

FIGURE 1. The Raman spectra of single-wall carbon nanotubes at room and elevated temperatures.

In case of interaction of a Gaussian beam (with diameter d) with a semi-infinite solid the temperature in a center of the spot on the surface is expressed as [3]:

$$2\Delta T \kappa_{eff} = \sqrt{\pi} d F_0 \qquad (1)$$

where F_0 is the laser beam power density, κ_{eff} – thermal conductivity.

Under the fixed irradiation conditions the material temperature is determined by its thermal conductivity. The nanotube soot includes, usually, several components: nanotubes, amorphous carbon, catalytic metal particles, closed carbon shells. Only a two-component medium approach has been used for consideration of the soot thermal conductivity. The contributions of catalytic particles and closed carbon shells have been neglected due to their nanoscale size and a high degree of isolation.

The thermal conductivity value κ_{eff} for the two-component medium is:

$$\frac{1}{\kappa_{eff}} = \frac{s}{\kappa_{n.}} + \frac{(1-S)}{\kappa_{a.c.}} \quad ; \qquad (2)$$

where S is the nanotube content, k_n and k_{ac} – the thermal conductivity values for the nanotube and amorphous carbon phases, correspondingly.

A MODEL EXPERIMENT WITH THE MIXTURES
CONTAINING A FIXED AMOUNT OF NANOTUBES

To confirm an ability of the laser heating method to estimate quantitatively the nanotube content in different soots a special set of samples with the preset nanotube percentages has been prepared and studied by Raman scattering. The dependencies of the tangential Raman mode shift (from the position 1592 cm^{-1} at room temperature) on a power density of the probing laser beam for the samples with the different nanotube contents are shown in *Fig. 2*. The graph slopes are seen to decrease monotonically while the nanotube content increases.

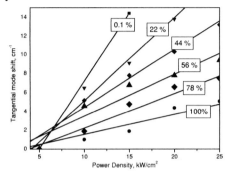

FIGURE 2. A laser-power density dependence of the tangential Raman mode shift (from the position 1592 cm^{-1} at room temperature) for soots with the preset nanotube contents. The content values are indicated on the graphs.

Fig.3 demonstrates a good correlation between the expected (in frames of a two-component medium) and observed values of the graph slope angle *tan* (*a_s*) for soots with the different nanotube contents **S**. A physical meaning of *tan* (*a_s*) is an inverse thermal conductivity. A small deviation of the experimental values of *tan* (*α_s*) from the linear dependence on **S** may arise systematically due to imperfection of the sample preparation. In course of filtration the upper layer of the film may be slightly depleted with the small amorphous carbon particles penetrating easily beneath.

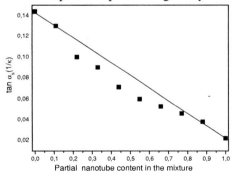

FIGURE 3. The values of *tan* (*α_s*) for mixtures with a preset nanotube content. *Squares* – the experimental points, *line*- a dependence, corresponding to the two-component medium model.

The laser heating method allows also to estimate the thermal conductivity value for mixtures with an arbitrary nanotube content. The tangential Raman mode frequency depends linearly both on temperature and on excitation laser power density. It means that

$$k\Delta T = -\tan \alpha_x P \tag{3}$$

$$\frac{\Delta T_1}{\Delta T_2} = \frac{\tan \alpha_1}{\tan \alpha_2} = \frac{\kappa_2}{\kappa_1} \tag{4}$$

A relative ratio of k_i for pure nanotube and amorphous carbon has been taken from *Fig.2*:

$$\frac{\tan \alpha_n}{\tan \alpha_{a.c.}} = \frac{1}{6.5} = \frac{\kappa_{a.c.}}{\kappa_n} \tag{5}$$

Using a table value of thermal conductivity of amorphous carbon k = 0.016 W/cm*K [4], we have obtained the estimation for HipCO nanotubes:

$$\kappa_{HipCO} = 6.5 \cdot \kappa_{a.c.} = 0.104 \text{ W/cm*K}$$

This value of thermal conductivity coincides well with the data published [5]. It is comparable with the inter-plane thermal conductivity value of pyrolytic graphite [$\kappa_{p.g}$ = 0.057 W/cm*K]. This may mean that the thermal conductivity of SWNT is determined mainly by the inter-tube heat transfer, similar to the inter-plane transport in graphite.

CONCLUSION

The values of the nanotubes content estimated by a Raman-based laser heating method for a set of samples, containing HipCO nanotubes and an amorphous carbon in different proportions, have coincided well with the preset values. The method is based on a measurement of the tangential Raman mode frequency depending on the laser excitation power density. The mode position indicates the material temperature. The rate of laser heating appeared to depend on the material purity. A two-component model of a thermal conductivity of the nanotube-containing soot has been used. The estimations of thermal conductivity values for pure HipCO single-wall carbon nanotubes and different HipCO mixtures with an amorphous carbon have been made.

ACKNOWLEDGMENTS

The work is supported by INTAS-237, RFBR 01-02-17358, 03-02-06266.

REFERENCES

1. Terekhov, S.V., Obraztsova, E.D., Lobach, A.S., Konov, V.I., Appl. Phys. A 74(2002)393-396
2. Hafner, J. H., Bronikowski, M. J., Azamian, B.R., Nikolaev, P., Rinzler, A.G., Colbert, D.T., Smith, K.A., Smalley, R.E, Chem. Phys. Lett. 296, 1998
3. Ready, J.F., "Effect of High-Power laser radiation", Academic Press, New York – London, 1971.
4. Kikoin, I.K. Table of Physical Values, Atomizdat, Moscow, 1976.
5. Hone, J., Whitney, M., Zettl, A., "Thermal conductivity of single-walled nanotubes", Synth. Met., V103, 1999.

The Purification of Single-Walled Carbon Nanotubes studied by X-ray induced Photoelectron Spectroscopy

Ralf Graupner*, Andrea Vencelová*, Lothar Ley*, Jürgen Abraham†,
Andreas Hirsch†, Frank Hennrich** and Manfred M. Kappes**

*Universität Erlangen, Institut für Technische Physik, Erwin-Rommelstr. 1, D-91058 Erlangen
†Universität Erlangen, Institut für Organische Chemie, Henkestr. 42, D-91054 Erlangen
**Forschungszentrum Karlsruhe, Institut für Nanotechnologie, Postfach 3640, D-76021 Karlsruhe,
Germany

Abstract. Using x-ray induced photoelectron spectroscopy (XPS) we compare the purification of Single-Walled Carbon Nanotubes (SWCNTs) through oxidizing treatments, namely either nitric acid (HNO_3) or a combined oxidation in air / hydrochloric acid (HCl)-etch. We find that in the raw material the catalyst metals are covered by amorphous carbon and are exposed by oxidation in air. The catalyst metals are then removed by the HCl etch. After the HNO_3-treatment no catalyst metals are observed, indicating that amorphous carbon and catalyst metals are removed in a one step process. Both acid treatments lead to p-type doped material. In both cases p-doping doping is removed by mild annealing ($\approx 330°C$). For the nitric acid treatment part of the sample is doped n-type, which is stable upon annealing.

INTRODUCTION

Common production processes of SWCNTs, e.g. laser ablation, arc discharge, or the HiPco process do not lead to pure SWCNTs; instead a mixture of SWCNTs, amorphous carbon, and transition metals used as catalyst constitute the raw material. Reliable and controllable purification procedures are therefore a necessary requirement for most application purposes. Among these processes, treatments which rely upon oxidizing attack, such as annealing in air or the preparation using oxidizing acids (HNO_3 or H_2SO_4) are widely used and can be easily scaled up to large quantities. Futhermore, an oxidizing treatment may be the first step for a functionalization of SWCNTs [1]. We therefore present a comparative study of two different purification procedures by oxidizing treatments in air or HNO_3 using x-ray induced photoelectron spectroscopy (XPS). This technique allows the determination of atomic compositions as well as the identification of chemical environments of selected elements by the characteristic chemical shifts of their core lines. Besides, changes of the Fermi-level position caused by a doping of the SWCNTs can be investigated by energy shifts common to all spectral features.

CP685, Molecular Nanostructures: XVII Int'l. Winterschool/Euroconference on
Electronic Properties of Novel Materials, edited by H. Kuzmany, J. Fink, M. Mehring, and S. Roth
© 2003 American Institute of Physics 0-7354-0154-3/03/$20.00

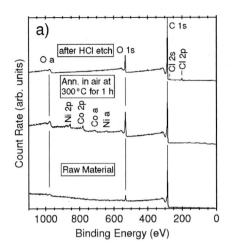 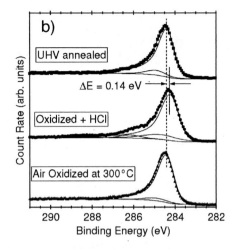

FIGURE 1. a) XPS survey spectra of the raw material (bottom), after anealing in air (middle), and after the subsequent HCl-etch (top); b) C 1s core level spectra after annealing in air (bottom), after subsequent HCl-etch (middle), and after additional annealing in UHV at 330°C (top). The measured spectra (points) have been deconvoluted into up to four component lines using a least squares fit.

EXPERIMENT, RESULTS, AND DISCUSSION

The samples we investigated were produced by laser ablation. Details of the production process are found elsewhere [2]. The raw material was purified by two different processes: process A comprises of oxidation in air at 300°C for 1 h, followed by washing with HCl, whereas in process B the raw material was subjected to 3M HNO$_3$ for 48 h at 150°C under reflux conditions. After both treatments transmission electron microscopy (TEM) revealed the presence of SWCNTs with no or only small residual amounts of amorphous carbon. XPS measurements were performed on the samples in the form of bucky-papers without further preparation in ultrahigh vacuum (UHV). Either monochromized Al K$_\alpha$-radiation ($\hbar\omega = 1486.6$ eV) or synchrotron radiation (PGM-2 at undulator UE 56-2, BESSY, Berlin, $\hbar\omega = 350$ eV) were used as light sources. The overall energy resolution of the C 1s spectra is 0.7 eV for the measurements using the Al K$_\alpha$-line and ≈ 0.3 eV for the synchrotron measurements.

Figure 1a shows the XPS survey spectra for the raw material (bottom), after anealing in air (middle), and after the subsequent HCl-etch (top). For the raw material, only the C 1s and the O 1s core levels, together with the O KVV auger line can be detected. After oxidation in air the characteristic Co and Ni core levels of the catalyst material appear which indicates that these transition metals are covered by amorphous carbon in the raw material. This carbon layer has to be thicker than the escape depth of the photoelectrons in XPS (≈ 10 Å), preventing Co and Ni to be detected in the raw material. The HCl-etch removes the transition metals, however, 2s and 2p core levels of Cl now appear in the spectrum (Fig. 1a, top curve). The C 1s core level of the raw material oxidized in air (Fig. 1b, bottom curve) is located at a binding energy of 284.45 eV, close to the position measured for graphite (284.41 eV, vertical dashed line in Fig. 1b). After HCl-etch, the

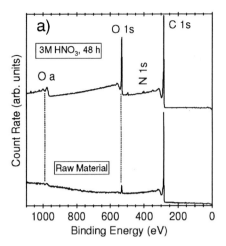
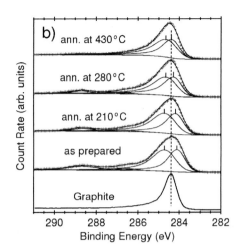

FIGURE 2. a) XPS survey spectra of the raw material (bottom) and after preparation in HNO₃ (process B, top); b) C 1s core level spectra after preparation in HNO₃ and after the indicated annealing temperatures in UHV compared to the C 1s spectrum of graphite (HOPG, bottom).

C 1s position is at 0.14 eV lower binding energy (Fig. 1b, middle). We attribute this shift to a p-type doping of the SWCNTs upon intercalation of HCl / Cl⁻. Annealing this sample should remove the intercalated HCl / Cl⁻, and indeed the C 1s core level shifts back to its intrinsic position after annealing at 330°C in UHV (Fig. 1b, top).

In figure 2a, the XPS survey spectrum of the raw material (bottom) is shown together with the one taken after process B (top). Since in this process the amorphous carbon and the transition metals are removed in a one-step process, no Co and Ni atoms are detected in the survey spectrum. However, in this case the N 1s core level appears after the purification process.

The C 1 s core level spectrum of the SWCNT material taken after process B (Fig. 2b, 2nd from bottom) is considerably broader than the one taken after process A. High resolution C 1s core level spectra using synchrotron radiation shown in Fig. 3 reveal that the core level spectrum of HNO₃-treated SWCNTs consists of two components in contrast to the one-component C 1s line after process A. Accordingly, the spectra in Fig.

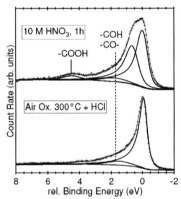

FIGURE 3. High resolution C 1s photoelectron spectra ($\hbar\omega = 350$ eV) after process A (bottom) and after preparation in HNO₃ (top).

2b are fitted using two asymmetric lines of constant width and height ratio for the main component, as well as two additional lines which are caused by the chemically shifted components due to -CO (286.5 eV) and -COO ($\approx$ 288.7 eV) binding configurations. It is interesting to note that the two main components are located at higher and lower binding energy side with respect to the C 1s position of graphite (lowest spectrum in Fig. 2b). Upon annealing, the main component on the low binding energy side shifts towards higher binding energy, whereas the component on the high binding energy side does change its position only by a small amount. We also performed scanning tunneling microscopy and scanning tunneling spectroscopy on the very same sample[1]. In these experiments individual bundles exhibit either p-type or n-type doping. We therefore attribute the two peaks in the C 1s spectrum of the HNO_3 treated material as due to the p- and the n-type doped regions, respectively. The p-type doping is attributed to the intercalation of HNO_3 / NO_3^- [3, 4], similar to the HCl / Cl^- case. Therefore, the shift of the low binding energy component towards the intrinsic position of graphite is consistent with a removal of the intercalated HNO_3 / NO_3^-. The fact that the component on the high binding energy side hardly changes its energetic position points to a dopant which is not merely intercalated. Instead, we assume that covalently bonded N atoms act as donors and are thus responsible for the observed n-type doping.

CONCLUSIONS

Our work shows that both purification methods lead to clean SWCNT material. However, care has to be taken in electronic applications of the material so obtained, as in both cases doping of the SWCNTs is observed. For the HCl treatment, the p-type doping can easily be removed by mild annealing, whereas for the HNO_3 treated material a mixture of both p- and n-type doped NTs are obtained. While the p-type doping is again removed upon annealing at 400°C, the n-type doping is found to be stable up to this temperature. This points to different doping mechanisms: intercalation for p-type and the incorporation of donors for n-type.

ACKNOWLEDGMENTS

This work was supported by the Bayerische Forschungsstiftung under the auspices of the "Bayerischer Forschungsverbund für Werkstoffe auf der Basis von Kohlenstoff, FORCARBON".

REFERENCES

1. Hirsch, A., *Angew. Chem. Int. Ed.*, **41**, 1853–1859 (2002).
2. Lebedkin, S., Schweiss, P., Renker, B., Malik, S., Hennrich, F., Neumaier, M., Stoermer, C., and Kappes, M. M., *Carbon*, **40**, 417–423 (2002).
3. Duclaux, L., *Carbon*, **40**, 1751–1764 (2002).
4. Hennrich, F., Wellmann, R., Malik, S., Lebedkin, S., and Kappes, M. M., *Phys. Chem. Chem. Phys.*, **5**, 178–183 (2003).

---

[1] See the article by A. Vencelová et al. in this volume

PROPERTIES OF SINGLE-WALL
CARBON NANOTUBES

Optical properties of intercalated single-wall carbon nanotubes

X. Liu, T. Pichler, M. Knupfer, and J. Fink

Leibniz- Institute for Solid State and Materials Research, P.O.Box 270016, D-01171 Dresden, Germany

Abstract. The optical properties of alkali-metal intercalated single-wall carbon nanotubes have been investigated using high-resolution electron energy-loss spectroscopy in transmission. The modulation of the filling of the van Hove singularities in the electronic density of states is reflected by the intensity variations of the optical excitations and the appearance of a charge carrier plasmon. The low-energy loss function can be described using a Drude-Lorentz model. Within this model, the optical conductivity at zero frequency σ_0 of Na, K, Rb, and Cs intercalated SWCNT samples were determined and give similar values for maximum intercalation. This model has also been used to derive the dielectric background and the effective (optical) charge carrier mass of the intercalated SWCNTs which is found to be two to three times bigger than in the corresponding graphite intercalation compounds.

INTRODUCTION

The investigation of single-wall carbon nanotubes (SWCNTs) is of great interest, since they consist of only one layer of a bent graphite sheet and can be either semiconducting or metallic depending only on their geometrical structure defined by their chiralities [1]. Therefore, SWCNTs represent ideal building blocks for nanoengineering as a result of their special electronic and mechanical properties. As far as the future application of SWCNTs is concerned, the controlled modification of the electronic properties of SWCNTs becomes very important. Therefore, the possibility to change the electronic properties by doping with electron donors/accpetors, such as for instance intercalation with akali-metal, $FeCl_3$, I_2, Br_2, and HNO_3 etc [2-6], has been studied extensively. Moreover, with the help of the different doping routes, the electronic properties of SWCNTs could be modified in a controlled way.

In this contribution, we present detailed studies of the optical response of intercalated SWCNT with Na, K, Rb, Cs. Their optical properties were characterized by electron energy-loss spectroscopy (EELS) in transmission and analyzed within a Drude-Lorentz model.

EXPERIMENTAL

The nanotubes were prepared by laser ablation, purified and filtrated into mats of a bucky paper as described previously [7,8]. Thin films of SWCNTs with an effective thickness about 100 nm were prepared by dropping an acetone suspension of SWCNTs on a KBr crystal. Subsequently, the films were floated off in distilled water, and mounted onto TEM grids for the EELS measurements. The intercalation was carried

CP685, *Molecular Nanostructures: XVII Int'l. Winterschool/Euroconference on Electronic Properties of Novel Materials,* edited by H. Kuzmany, J. Fink, M. Mehring, and S. Roth
© 2003 American Institute of Physics 0-7354-0154-3/03/$20.00

out in ultra-high vacuum by evaporating alkali metals from a SAES getter source. During intercalation, the sample was kept at 130°C and annealed about 20 min after evaporation to reach homogeneous intercalation. EELS measurements were carried out using a purpose-built 170 keV spectrometer [9] with the energy and momentum resolution being 180 meV and 0.03 Å^{-1}. All measurements are performed at room temperature.

RESULTS AND DISCUSSION

In the electron diffraction experiments, the diffraction peak arising from the bundle structure of SWCNTs was used to determine the intertube distance. When SWCNTs are intercalated with alkali metals the bundle lattice is expanded by the intercalation [4,6]. In other words, the positions of bundle related peaks are shifted to lower q. The bigger the size of the dopant is, the stronger is the lattice expansion. The estimated values are about 0, 1.1, 1.7, and 2.7 Å for fully Na, K, Rb, and Cs intercalated SWCNTs with a mean diameter of 1.37 nm, respectively.

Since the alkali metals are electron donors, the charge transfer occurs from the dopant to the SWCNTs. As a result, the lower conduction bands of SWCNTs are filled which leads to a Fermi level shift to higher energies and results in a quenching of the optically allowed interband transitions. When we turn to the loss function measured in EELS experiments, the low-energy peaks can be assigned to collective excitations caused by these optically allowed transitions [10]. Figure 1(a) shows the change of the loss function with increasing K doping. The K/C ratios were determined from the relative intensities of the C1s and K2p edges [4]. The spectra clearly show that the

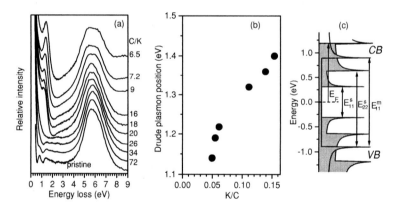

FIGURE 1. (a) The evolution of the loss function with different K-doped levels. (b) The relation between the Drude plasmon energy and the relative atom ratio K/C. (c) The sketch of possible filling of the conduction bands of SWCNTs.

interband transitions disappear for the compound KC_{26}. Then, one new peak appears which can be assigned to the free charge carrier plasmon. The energy position of the plasmon depends on the doping level. These changes are observed in all cases of Na,

128

K, Rb, Cs doping. In the rigid band model, the shift of Femi level depends on the donor concentration. For saturation doping this leads to a Fermi level shift of about 1.2 eV (see sketch in Fig. 1(c)).

In order to obtain more detailed information, the measured loss function was analyzed using a Drude-Lorentz model which can be successfully applied to describe graphite or SWCNT intercalated compounds [4,6]. The dielectric function of the intercalated SWCNTs can be expressed with one Drude plasmon ($E_{p,0}$, and damping Γ_0), and one π and $\pi+\sigma$ plasmon with the oscillator strengths $E_{p,\pi}$, $E_{p,\pi+\sigma}$, energies $E_{T,\pi}$, $E_{T,\pi+\sigma}$, and damping Γ_π, $\Gamma_{\pi+\sigma}$, respectively, i.e.,

$$\varepsilon(E) = 1 - \frac{E_{p,0}^2}{E^2 + i\Gamma_0 E} + \frac{E_{p,\pi}^2}{E_{T,\pi}^2 - E^2 - i\Gamma_\pi E} + \frac{E_{p,\pi+\sigma}^2}{E_{T,\pi+\sigma}^2 - E^2 - i\Gamma_{\pi+\sigma}E}$$

Since the π and $\pi+\sigma$ plasmon are high enough in energy they lead only to an effective screening of the Drude plasmon resulting in a shift to lower energies. Their contribution at low energies can be expressed using a constant dielectric background, derived from the real part of the dielectric function of the π and $\pi+\sigma$ response at zero frequency $\varepsilon_r(0) = \varepsilon_\infty$. Then, the Drude plasmon was fitted with this constant dielectric background ε_∞. The results are shown in Fig. 2 . The unscreened plasma energy, $E_{p,0}$ is

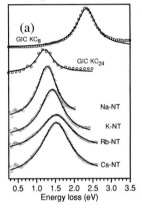

(b)

	Drude		ϵ_∞	$m^\star/m_0$	σ_0 (S cm^{-1})
	$E_{p,0}$	Γ_0			
GIC KC$_8$	5.8	0.2	6.7	0.36	22625
GIC KC$_{24}$	3.48	0.5	6.4	0.33	3277
Cs-NT	3.55	0.9	4	0.83	1890
Rb-NT	3.8	0.98	4.7	0.74	1980
K-NT	3.87	0.7	6.1	0.82	2940
Na-NT	3.38	0.62	5.45	1.11	2470

FIGURE 2. (a) Analysis of the Drude plasmon in the loss function of fully Na, K, Rb, and Cs intercalated SWCNTs compared with GIC (graphite intercalation compounds) KC$_8$ and GIC KC$_{24}$. The solid lines are the fitted curves, the open circles represent the experimental data. (b) The parameters used for the Drude fitting.

about 3.4 - 3.8 eV almost independent of the alkali metals, which suggests that the lattice expansion and the different ionization potentials of the alkali metals affect the plasma energy only weakly. The effective mass of the charge carriers is about 0.74 - 1.11 times that of the free electron mass for the fully intercalated SWCNTs. The scatter between the different intercalants might be related to the uncertainties concerning the exact intercalation level which is assumed to be the same for saturated intercalation independent of the intercalant.

Next, from the Drude-Lorentz model analysis of the intercalated SWCNTs, the optical conductivity can be extracted. Experimental and theoretical studies have shown that intercalation increases the conductivity of SWCNT mats significantly [2]. The optical conductivity at zero energy can be obtained from the expression, $\sigma_0 = E_{p,0}^2 \varepsilon_0 /(\Gamma_0 \hbar)$, as shown in Fig. 2(b). The results show that the optical conductivity increases with higher intercalation level and its value is intercalant dependent. From the combination of the fitting parameters of the Drude plasmon and the π plasmon, the optical conductivity can be calculated, as shown in Fig. 3. These results can be used to be compared with optical measurements. In particular, the larger intercalants, Rb and Cs, result in a lower conductivity which is directly related to the larger width of the charge carrier plasmon. This most probably is caused by a bigger

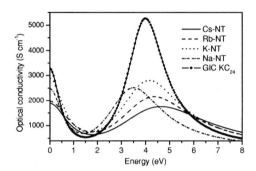

FIGURE 3. The calculated optical conductivity from the fitting parameters in a range including Drude and π plasmons for fully Na (dash-dotted line), K (dashed line), Rb (dotted line), Cs (solid line)-intercalated SWCNTs compared with GIC KC_{24} ($\bullet$).

structural disorder within the nanotube bundles as a consequence of the introduction of big ions and the corresponding required expansion of the bundle lattice.

In summary, we have presented studies of the electronic properties of alkali-metal intercalated SWCNT using EELS. The in-situ measurements show the optical response from the variation of the Fermi level position due to charge transfer, indicating the possibility of tuning the Fermi level to conduction bands upon different doping levels.

ACKNOWLEDGMENTS

We acknowledge financial support from the DFG and the EU SATURN projects.

REFERENCES

1. R. Saito et al., *Physical Properties of Carbon Nanotubes* (Imperial College Press, 1998).
2. R. S. Lee et al., *Nature* **338**, 255 (1997); A. M. Rao et al., *Nature* **338**, 257 (1997).
3. C. Bower et al., *Chem. Phys. Lett.* **288**, 481 (1998).
4. T. Pichler et al., *Solid State Commun.* **109**, 721 (1999) and references therein.
5. L. Duclaux et al., *J .Phys. Chem. Solids* **64**, 571 (2003).
6. X. Liu et al., *Phys. Rev. B* **67**, 125403 (2003).
7. H. Kataura et al., *Synth. Met.* **103**, 2555 (1999).
8. O. Jost et al., *Appl. Phys. Lett.* **75**, 2217 (1999).
9. J. Fink, *Adv. Electr. Electron Phys.* **75**, 121 (1989).
10. T. Pichler et al., *Phys. Rev. Lett.* **80**, 4729 (1998).

^{13}C NMR investigations of the metallic state of Li intercalated carbon nanotubes

M. Schmid[1,2,3], C. Goze-Bac[1], M. Mehring[2],
S. Roth[3], P. Bernier[1]

[1]GDPC, Univ. Montpellier II, F-34095 Montpellier cedex5, France
[2]2. Physikal. Inst., Universität Stuttgart, Pfaffenwaldring 57, D-70550 Stuttgart, Germany
[3]Max-Planck-Institut für Festkörperforschung, Heisenbergstr. 1, D-70569 Stuttgart, Germany

Abstract. ^{13}C Nuclear Magnetic Resonance measurements were performed on pristine and lithium intercalated single wall carbon nanotubes (SWNT). We investigated the NMR signatures by means of static and high resolution Magic Angle Spinning experiments. This allows us to measure in detail the modifications of the lineshape with the Li concentration. Our results can be explained in terms of charge transfer and changes of the metallic state with an increasing density of states at the Fermi level compared to the pristine SWNT.

INTRODUCTION

The utilization of a carbon host to store Li ions in a rechargeable negative electrode has developed to commercial Li-ion cells. The demand for more light and powerful battery cells has led to an intensive investigation of new kinds of carbon species for the use as anode materials. Carbon nanotubes might be a very good candidate for application in batteries as they have shown to satisfy the conditions of high reversibility of energy storage capacities at high discharge rates [1, 2]. In the past, NMR has proven to be a powerful tool for studying the electronic state of such kind of electrode materials. We report here on the metallic state of pristine and Li intercalated SWNT using ^{13}C NMR at room temperature.

SAMPLE PREPARATION

As grown electric arc SWNT were Li intercalated in solution based on aromatic hydro-carbons of a given redox potential and tetrahydrofuran (THF, C_4H_8O). The stoichiometries of LiC_6, LiC_7 and LiC_{10} were obtained using naphthalene, benzophenone and fluorenone, respectively. The experimental details are described elsewhere [3]. Similar to Li intercalation in graphite, we assume THF molecules being cointercalated with Li in the carbon nanotube host and ternary compounds with stoichiometries $Li_x(THF)_yC$ are expected [4].

CP685, *Molecular Nanostructures: XVII Int'l. Winterschool/Euroconference on Electronic Properties of Novel Materials,* edited by H. Kuzmany, J. Fink, M. Mehring, and S. Roth
© 2003 American Institute of Physics 0-7354-0154-3/03/$20.00

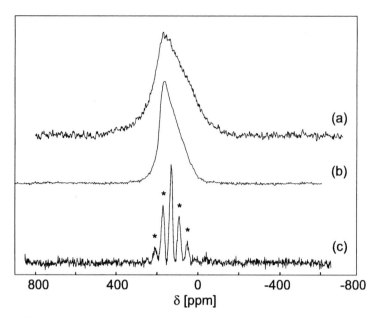

FIGURE 1. ^{13}C NMR at 4.7 T, (a) static spectrum of pristine SWNT; (b) static spectrum of Li intercalated SWNT, LiC$_6$; (c) MAS on LiC$_6$ SWNT at 2 kHz.

EXPERIMENTAL

The ^{13}C NMR experiments were performed using a Bruker ASX200 pulsed NMR spectrometer at a magnetic field of 4.7 T, which corresponds to a ^{13}C Larmor frequency of 50.3 MHz. The high resolution MAS experiments were performed at spinning rate frequencies up to 5 kHz. All spectra were taken by a Fourier transform of the free induction decay (FID).

Fig. 1 (a,b) shows the static ^{13}C NMR spectra of pristine nanotubes and LiC$_6$ respectively and fig. 1 (c) the ^{13}C NMR MAS spectrum of LiC$_6$ at a spinning frequency of 2 kHz. The latter two spectra are typical for all the ^{13}C NMR spectra we measured on Li intercalated SWNT. They are classical for powder pattern spectra of our sp^2 carbon host [5]. Fig. 2 presents on the left the three tensor components for the pristine and the intercalated carbon nanotubes with x=10,7,6 and on the right the corresponding isotropic line positions. All the components were obtained using the Hertzfeld-Berger analysis with low spinning rate MAS experiments (spectra not shown here). A 10 ppm paramagnetic shift is observed of the isotropic line from 126 ppm in pristine SWNT up to 136 ppm in LiC$_6$. In addition, a gradual reducing of the tensor anisotropy appears from LiC$_{10}$ to LiC$_6$. In the following we will discuss the origin of the isotropic line shift as well as the anisotropy reduction upon intercalation.

In pristine SWNT, ^{13}C NMR spectrum and spin-lattice relaxation exhibit two contributions which can be assigned to metallic and semiconducting nanotubes [5]. However, after Li intercalation and charge transfer only one component is observed which sug-

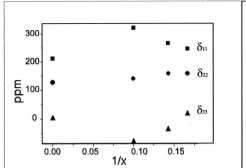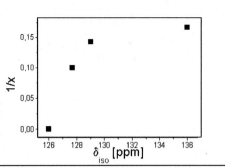

FIGURE 2. (Left) anisotropy components of the CSA tensor for Li intercalated SWNT with different Li concentrations. $1/x = 0$: pristine; $1/x = 0.10$: LiC_{10} , $1/x = 0.14$: LiC_7 and $1/x = 0.16$: LiC_6. (Right) isotropic Knight shifts for the same stoichiometries of Li intercalated SWNT.

gests a strong modification of the electronic band structure of the SWNT in the bundles. The intrinsic electronic state of each nanotube is no more observed suggesting that inter-tube interactions contribute to the appearance of an unique electronic state. Its average density of state at the Fermi level $N(E_F)$ is closely related to the new electronic band structure of the intercalated SWNT bundles. The isotropic Knight shift is a measure for the metallicity of the system. It is proportional to the probability density of the Fermi conduction electrons at the ^{13}C nuclei sites according to

$$K_{iso} \sim |\Psi(0)|^2 N(E_F). \qquad (1)$$

In case of a charge transfer, a paramagnetic isotropic Knight shift is expected to occur as $N(E_F)$ is increased. Since we observe a Knight shift up to 10 ppm in the Li intercalated SWNT, this clearly confirms the assumption of a charge transfer from the alkali metal to the SWNT. The Knight shift and so $N(E_F)$ increase with the intercalation level. No abrupt shift is observed from the expected van Hove singularities which suggests again that the individual electronic properties of the SWNT have been modified through the redistribution of the electronic states.

A complete charge transfer for alkali metals in graphite intercalation compounds (GIC) results to an anisotropic Knight shift of ~ -1800 ppm/electron which corresponds to an anisotropy reduction of ~ -300 ppm in AC_6 (A=alkali metal) [5]. Following the concept of the charge transfer model in alkali GIC, we have to conclude in a limited efficiency of the charge transfer in our case. A limited charge transfer from the lithium metal to the carbon nanotube host could be explained for example by the THF cointercalated in the sample. This changes the crystallographic structure of the alkali-SWNT system and leads to a different electronic band structure. Another explanation for a limited charge transfer could be the ability of Li to produce a Li-C covalency. This is known from graphite where a Li-C covalency can be formed with the graphite layer if Li is placed between 2 C-hexagons at a height of 1.8 Å[6]. In such a case the electron cloud of the alkali metal is distributed on the chemical bond and only a limited charge transfer to the carbon host is observed.

SUMMARY

We recorded static and MAS ^{13}C NMR spectra of pristine and Li intercalated SWNT. The isotropic line is 10 ppm paramagnetic shifted with increasing Li concentration. We suggest that an electronic charge transfer occurs and the electrons are redistributed in the electronic band structure of the SWNT bundle with an increased $N(E_F)$. The observed ^{13}C NMR anisotropy reduction is explained in terms of a pure metallic system with a limited charge transfer from the alkali metal to the SWNT. The reason for the limited charge transfer could be due to change of the crystallographic structure by THF cointercalation or the formation of Li-C covalencies.

In the future, our investigations will be completed by ^{13}C NMR spin-lattice relaxation measurements in order to quantitatively calculate the density of states at the Fermi level.

ACKNOWLEDGMENTS

We thank the COMELCAN (No. HPRN-CT-2000-00128) and FUNCARS (No. HPRN-CT-1999-00011) European Project for funding and support.

REFERENCES

1. Endo, M., Kim, C., Nishimura, K., Fujino, T., and Miyashita, K., *Carbon*, **38**, 183–197 (2000).
2. Wu, Y. P., Rahm, E., Holze, and R., *J. Power Sources*, **114**, 228–236 (2003).
3. Petit, P., Jouguelet, E., and Mathis, C., *Chem. Phys. Lett.*, **318**, 561–564 (2000).
4. Facchini, L., Quinton, M., and Legrand, A., *Physica B*, **99**, 525–530 (1980).
5. Goze-Bac, C., Latil, S., Lauginie, P., Jourdain, V., Conard, J., Duclaux, L., Rubio, A., and Bernier, P., *Carbon*, **40**, 1825–1842 (2002).
6. Conard, J., and Lauginie, P., "Lithium NMR in lithium-carbon solid state compounds" in *New Trends in Intercalation Compounds for Energy Storage*, edited by C. Julien, J. Pereira-Ramos, and A. Momchilov, Kluwer Academic Publishers, 2002, pp. 77–93.

Raman Measurements on Electrochemically Doped Single-Walled Carbon Nanotubes

P. M. Rafailov, M. Stoll, J. Maultzsch and C. Thomsen

Institut für Festkörperphysik der Technischen Universität Berlin, Sekr. PN 5-4, Hardenbergstr. 36, 10623 Berlin, Germany

Abstract. We performed voltammetry experiments and studied the Raman response of electrochemically doped single-walled carbon nanotubes (SWNT) using different salt solutions. The frequency shift of the radial breathing mode (RBM) and the high-energy mode (HEM) were examined as a function of the doping grade. The results are discussed in the frame of a double-resonant model for Raman scattering and a strain estimate from the the high-energy mode shift is proposed.

Introduction. Single-wall carbon nanotubes (SWNT) are novel one-dimensional nanostructures with promising application perspectives. Along with their remarkable electronic properties they combine large surface area, good chemical stability and significant elastic properties[1, 2]. The SWNTs exhibit sharp optical transitions between van Hove singularities in their electronic density of states. Consequently, any charge-transfer doping should lead to significant changes in the electronic properties. A high-degree charge transfer can be achieved by intercalation as the guest species move into the interstitial channels of the SWNT ropes, however, much finer tuning of the added charge is achieved by electrochemical doping[3]. In the latter case the electrolyte ions do not seem to penetrate the SWNT ropes and seems to form a charged double layer only with their external surface. Our results support the evidence that SWNT ropes participate only with their external surface in the double-layer charging and we propose an electrochemical way to determine the effective surface area of SWNT electrodes. We also relate the electrochemically induced shift of the high-energy mode (HEM) frequency to the dimensional changes of the SWNTs and discuss the Raman response in terms of the double-resonant scattering model[4].

Experimental. A stripe of SWNT paper with surface density of $\approx 5 \cdot 10^{-5}$ g/mm^2[5] was prepared as a working electrode in a three-electrode cell equipped with quartz windows. A Metrohm - Potentiostat was employed for cyclic voltammetry measurements. A platinum wire and Ag/AgCl/3 M KCl served as auxiliary and reference electrode, respectively. The working electrode was only partly dived into the solution and was electrically contacted at its dry end.

Several different aqueous solutions (concentration 1 M) were applied in the voltammetric measurements: LiCl, NaCl and KCl. Each solution was purged with N$_2$ gas prior to measurements to remove excessive oxygen. The Raman spectra were recorded with a DILOR triple grating spectrometer equipped with a CCD detector. The 514.5 nm line of an Ar$^+$/Kr$^+$ laser was used and the spectral resolution was 4 cm^{-1}.

CP685, *Molecular Nanostructures: XVII Int' l. Winterschool/Euroconference on Electronic Properties of Novel Materials,*edited by H. Kuzmany, J. Fink, M. Mehring, and S. Roth
© 2003 American Institute of Physics 0-7354-0154-3/03/$20.00

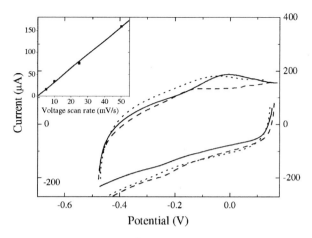

FIGURE 1. Cyclic voltammograms of a 1 mm² SWNT-mat in 1 M aqueous solutions of LiCl (full line), NaCl (dashed line) and KCl (dotted line) at a scan rate of 50 mV/s; all potentials are vs Ag/AgCl. *Inset:* Saddle-point current of the voltammogram for KCl as a function of the voltage scan rate.

Results and Discussion. Figure 1 shows typical voltammograms of a SWNT electrode. We estimated gravimetric capacitances in the range 33 - 38 F/g for the three solutions used which are in line with the ones reported in the literature[2, 3].

We also propose an electrochemical method to determine the effective surface area by means of a glassy carbon rod as a working electrode. A glassy carbon electrode has a smooth shiny surface without any pores which equals its effective surface area in a voltammetric experiment. By putting the measured currents in proportion for both glassy carbon and SWNT electrodes one can therefore determine the effective surface area of the SWNT sample. In this way we obtained similar values of ≈230 m²/g for all three solutions that are slightly smaller than those measured by the conventional N_2 absorption method (285 m²/g)[2]. The ratio between our measured values and theoretical estimates for individual nanotubes (1315 m²/g) is ≈ 0.175, which implies that the nanotube bundles are involved in the double-layer charging only with their external area and that the majority of these bundles contain about 90 nanotubes. The latter value is in good agreement with TEM results[6], which revealed bundles made of 100 nanotubes in average.

We further recorded Raman spectra of the main vibrational modes from the SWNT working electrode at different constant potentials in all three chloride solutions. The recorded spectra (KCl solution) and the phonon frequencies versus applied voltage are shown in Fig. 2 for the RBM and in Fig. 3 for the two most intense peaks of the HEM. It is apparent that up to ≈ 1 V the frequency shift scales linearly with the applied voltage. At higher voltages the frequency shift increases exponentially, which is accompanied by a similar increase in current. We measured the RBM frequency of the original SWNT mat, and of the SWNT working electrode at open circuit before cycling and immediately after taking the Raman spectra at $U = \pm 1$ V. The obtained values coincided to within 0.2 cm⁻¹ with the original RBM frequency. This along with

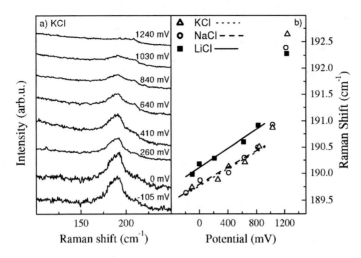

FIGURE 2. *a)* Raman spectra of the radial breathing mode (RBM) at several potentials applied to the SWNT-mat (1 M KCl). *b):* The frequency of the RBM as a function of the applied potential.

the small shift of the RBM below 1 V gives us confidence that the ionic species do not significantly penetrate into the SWNT bundles and the effective surface area remains constant. Such a penetration would alter the van-der-Waals interaction strength, which provides an essential contribution to the RBM frequency[7].

From a linear fit to the frequency points (up to 0.8 V, KCl solution) we estimate a doping-induced shift of 160 cm^{-1}/e$^-$/C-atom for the RBM, 270 cm^{-1}/e$^-$/C-atom for the main HEM peak at 1593 cm^{-1} (HEM1) and 320 cm^{-1}/e$^-$/C-atom for the second-strongest HEM peak at 1567 cm^{-1} (HEM2) in qualitative agreement to previous results[3]. As can be seen in Fig. 3, the frequency slope of HEM2 is slightly but systematically steeper than that of HEM1. In the framework of the double-resonant model for Raman scattering in SWNT[4] such a slope difference may be accounted for. Shifting the phonon branches to higher frequencies upon doping necessarily decreases the wave vector of the phonon to be emitted. The overall hardening of HEM1 is thus weakened while that of HEM2 is enhanced due to the different dispersion of the two phonon branches originating from the graphitic E_{2g} mode.

The actuation strain ε_{zz} of the SWNTs can be estimated from the average HEM frequency shift $\Delta\omega$ utilizing its linear dependence on the doping grade Δf. Assuming equal deformations along the nanotube axis and in circumferential direction and recalling the analogy to the Grüneisen parameter γ for pressure-induced strain[8], we use here a similar phonon deformation potential of order unity and obtain

$$\frac{\Delta\omega(\Delta f)}{\omega_0} = 2\gamma\,\varepsilon_{zz}. \tag{1}$$

Then at ≈ 1 V applied potential the doping grade amounts to $\Delta f \approx 0.005$ e$^-$/C-atom and the strain ε_{zz} to $\approx 0.05\%$.

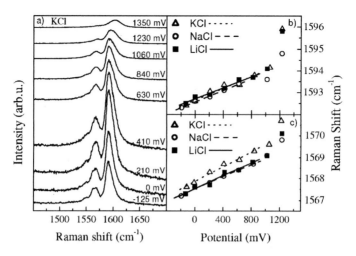

FIGURE 3. *a)* Raman spectra of the high-energy mode (HEM) at several potentials applied to the SWNT-mat (1 M KCl). *b)* The frequency of HEM1 as a function of the applied potential for three different solutions: LiCl (squares), NaCl (circles) and KCl (triangles). *c)* Same as *b)* for HEM2.

Conclusions.   We measured the voltammetric response of a SWNT-paper as a working electrode in an electrochemical experiment, determined its gravimetric capacitance and used a comparative method to estimate its effective surface area with the same experimental setup. We also examined the Raman response of the nanotube vibrational modes upon electrochemical doping in the framework of the double-resonance scattering model. We showed that the high-energy stretching SWNT mode can be used to assess the actuation properties of a carbon nanotube.

Acknowledgments.   We gratefully acknowledge discussions with S. Reich.

REFERENCES

1.   C. Thomsen, S. Reich, H. Jantoljak, I. Loa, K. Syassen, M. Burghard, G. S. Duesberg, and S. Roth, *Appl. Phys. A: Mater. Sci. Process.* , **69A**, 309 (1999).
2.   R. H. Baughman, Ch. Cui, A. A. Zakhidov, Z. Iqbal, J. N. Barisci, G. M. Spinks, G. G. Wallace, A. Mazzoldi, D. De Rossi, A. G. Rinzler, O. Jaschinski, S. Roth and M. Kertesz, *Science* **284**, 1340 (1999).
3.   L. Kavan, P. Rapta, L. Dunsch, M. Bronikowski, P. Willis and R. Smalley, *J. Phys. Chem. B*, **105**, 10764 (2001).
4.   J. Maultzsch, S. Reich and C. Thomsen, *Phys. Rev. B* **65**, 233402 (2002).
5.   prepared by S. Roth and U. Dettlaff, MPI - Stuttgart
6.   A. Thess *et al.*, Science **273**, 483 (1997).
7.   C. Thomsen, S. Reich, A. R. Goñi, H. Jantoljak, P. M. Rafailov, I. Loa, K. Syassen, C. Journet, and P. Bernier, *phys. stat. sol. (b)*, **215**, 435 (1999).
8.   S. Reich, H. Jantoljak, and C. Thomsen, *Phys. Rev. B*, **61**, R13389 (2000).

Valence-Band Photoemission Study
of Single-Wall Carbon Nanotubes

H. Shiozawa[1], H. Ishii[1], H. Kataura[1], H. Yoshioka[2], H. Otsubo[1],
Y. Takayama[1], T. Miyahara[1], S. Suzuki[1], Y. Achiba[1], T. Kodama[1],
M. Nakatake[3], T. Narimura[4], M. Higashiguchi[4], K. Shimada[4],
H. Namatame[4], and M. Taniguchi[4]

[1]*Graduate School of Science, Tokyo Metropolitan University, Tokyo, 192-0397, Japan*
[2]*Department of Physics, Nara Women's University, Nara 630-8507, Japan*
[3]*Photon Factory, High Energy Accelerator Research Organization, Tsukuba 305-0801, Japan*
[4]*Hiroshima University, Higashi-Hiroshima, 739-8526, Japan*

Abstract. We have performed the valence-band photoemission spectroscopy on single-wall carbon nanotubes. Characteristic peak structures originated from one-dimensional van Hove singularities have been successfully observed. The peak structures are well reproduced by the calculation based on the tight-binding model taking into account a diameter distribution of nanotubes.

INTRODUCTION

Single-wall carbon nanotubes (SWNTs) have peak structures in their electronic density of π states due to one-dimensional van Hove singularities and the peak positions strongly depend on their diameters and chiralities. Indeed, diameter dependent peak structures have been observed by optical absorption spectroscopy [1,2] and by scanning tunneling spectroscopy (STS) [3]. However, these experiments have some ambiguities in characterizing the density of states (DOS) of SWNTs [2]. The valence-band photoemission spectroscopy has a great advantage to get direct information about the density of π states. However, no one has succeeded in observing the van Hove singularities by the photoemission spectroscopy to date probably due to less purity of the sample. In this paper, we first report the successful observation of peak structures in the valence-band photoemission spectra due to the van Hove singularity by using extremely high purity SWNT sample. In addition, the photoemission spectrum of SWNT encapsulating the C_{60} fullerene, the so-called C_{60} peapod, was also measured to see the interactions between SWNTs and fullerenes through observing the change in the spectrum by the encapsulations.

CP685, *Molecular Nanostructures: XVII Int'l. Winterschool/Euroconference on
Electronic Properties of Novel Materials,*edited by H. Kuzmany, J. Fink, M. Mehring, and S. Roth
© 2003 American Institute of Physics 0-7354-0154-3/03/$20.00

EXPERIMENTAL

SWNT samples were prepared by laser ablation method and purified by H_2O_2 treatment [1]. Two SWNT samples NT-1 and NT-2 with different tube diameters were prepared. C_{60}-peapod was synthesized using NT-1. The photoemission experiments were performed at the beamline BL-11D of Photon Factory, High Energy Accelerator Research Institute (KEK-PF) and at the beamline BL-1 of Hiroshima Synchrotron Radiation Center (HiSOR), Hiroshima University, using an angle-integrated hemispherical electron energy analyzer. The spectral resolutions were 50 meV for KEK-PF and 15 meV for HiSOR at the excitation energy of 65 eV. The energy position and the resolution were calibrated by the Fermi edge (E_F) of an evaporated Au film. The surface cleaning is the most important process for the success of this experiment, since the valence-band photoemission measurement is extremely surface sensitive. We heated the sample at 200 °C in ultra high vacuum for several hours to get the clean surface. The cleanliness was confirmed by the absence of O $2p$ peak located at a binding energy of 6 eV.

RESULTS AND DISCUSSION

The valence-band photoemission spectra of NT-1 and C_{60} peapod are shown in Fig. 1(a). The spectrum of NT-2 was similar to that of NT-1. The spectrum of the SWNT is essentially composed of sp^2 hybridized σ and π state whose spectral shapes are similar to those of graphite. On the other hand, the spectrum of C_{60} peapod has additional peak structures indicated by vertical bars in the figure, a peak at about 5.5 eV and a shoulder at about 2.2 eV. These additional structures are considered to be originated from the electronic structure of the C_{60} fullerenes in the SWNTs. We now focus on fine structures observed in the binding energies below 1.5 eV, in which the peak structures corresponding to the van Hove singularities are expected. Fig. 1(b) shows the detailed spectra of NT-1, NT-2 and C_{60} peapod. Three characteristic peak structures can be seen in the spectra for all the samples. Comparing with the tight-binding calculation [4], we can identify the first and second peak in order from E_F as the contribution from semiconductive nanotubes, and the third peak as the contribution from metallic nanotubes. We couldn't observe any difference between the peak structure of NT-1 and of C_{60} peapod. This suggests that the charge transfer between SWNT and C_{60} fullerenes are very small. On the other hand, each peak in the spectrum of NT-2 shifts toward the higher binding energy, and the higher peak shows the larger shift compared with those of NT-1. The difference in the peak position can be understood as a consequence of the difference in the mean diameter, as will be mentioned later.

Fig. 2 shows the spectrum calculated by the tight-binding model with taking into

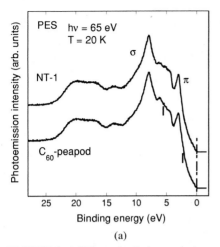 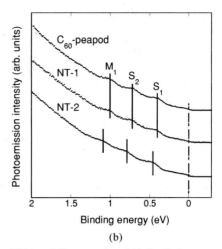

(a)	(b)

FIGURE 1. a) Valence-band photoemission spectra for NT-1 and C_{60} peapod, and b) detailed spectra near the Fermi level for NT-1, NT-2 and C_{60} peapod.

account a diameter distribution of SWNTs, where the nearest-neighbor transfer integral was set to 2.9 eV [4]. In the total DOS calculation of the sample, we assumed that the diameters of the sample are distributed according to a Gaussian function. Then, we convoluted the total DOS by a Lorenzian function with the full width of half maximum of 130 meV to reproduce the spectrum. This broadening is probably due to a finite lifetime of the final state in a photoemission process or diameter and chirality dependences of the work function of SWNTs. We also introduced an energy shift of E_F by 0.07 eV, may be due to the charge transfer

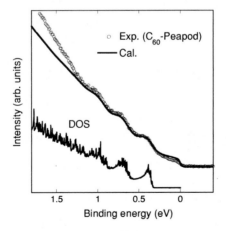

FIGURE 2. Calculated photoemission spectrum and density of states (DOS) using the tight-binding model with taking into account the diameter distribution. Open circles indicate the photoemission spectrum for C_{60} peapod.

between the substrate and the sample. The calculated spectrum well reproduces the experimental result of the C_{60} peapod indicated by open circles in the figure. The deviation of the experimental results from the calculated spectrum at the high binding energies is due to the overlap of the σ band. The difference near E_F may be attributed to one-dimensionality of SWNTs, which would be clarified by high resolution measurement.

The estimated mean diameter and the standard deviation of the diameter distribution for the sample NT-1 or C_{60} peapod were 1.38 nm and 0.10 nm, respectively. The mean diameter of 1.38 nm is adequate for the encapsulation of the C_{60} fullerene with the diameter of 0.71 nm. Those for the sample NT-2 were estimated to be 1.25 nm and 0.12 nm, respectively. The estimated mean diameters were close to the values estimated by Raman spectroscopy and X-ray diffraction measurement [5].

In summary, we have performed the valence-band photoemission spectroscopy on SWNTs and observed the characteristic peak structures originated from the one-dimensional van Hove singularities. The spectra were well reproduced by the tight-binding calculation and the estimated mean diameters are consistent with those of Raman spectroscopy and X-ray diffraction measurement. The results stimulate further investigations on the one-dimensional electronic structure of SWNTs.

ACKNOWLEDGMENTS

This work was performed at the Photon Factory under the approval of the Photon Factory Advisory Committee (2002G185) and at the Hiroshima Synchrotron Radiation Center under the Cooperative Research Program of HiSOR (Accept02-A-12).

REFERENCES

1. Kataura, H. et al., *Synthetic Metals* **103**, 2555-2558 (1999).
2. Ichida, M. et al., *J. Phys. Soc. Jpn.* **68**, 3131-3133 (1999).
3. Wildöer, J. W. G. et al., *Nature* **391**, 59-62 (1998).
4. Maruama, S., private communication.
5. Maniwa, Y. et al., private communication.

Effect Of Gamma-Irradiation on Single-Wall Carbon Nanotube Paper

V. Skákalová[1], M. Hulman[2], P. Fedorko[3], P. Lukáč[3] and S. Roth[1]

[1]Max Planck Institute of Solid State Research, Heisenbergstr. 1, D-70569 Stuttgart, Germany
[2]Institute of Material Physics of University Vienna, Strudlhofgasse 4, A-1090 Vienna, Austria
[3]Department of Chemical Physics, Faculty of Chemical and Food Technology, Slovak University of Technology, Radlinského 9, 81237 Bratislava, Slovakia

Abstract. The mechanical and electrical properties of a bulk material made of single wall carbon nanotubes (SWNT) are, due to weak intermolecular interaction, several orders of magnitude lower than those of the individual molecules themselves.
We studied the effect of gamma-irradiation on SWNT paper in air and under vacuum. For samples irradiated in air, changes in Young modulus and electrical conductivity were observed with maximum value for a dose of 170 kGy. Under vacuum there was only a small effect of irradiation. Raman studies of irradiated samples showed defects formation. Same experiments done with graphite showed similar results.
A likely explanation of the results is that cross-links between nanotubes were induced by irradiation in air.

INTRODUCTION

Single wall carbon nanotubes (SWNT) have been subjected to intensive investigation for their exceptional properties. The Young modulus of an individual molecule reaches values of terapascals [1-3]. The efforts to use the remarkable mechanical properties of SWNT are appearing mostly in the field of polymer composites, mixing nanotubes into a polymer matrix [4-6]. In fibers made of polymer-SWNT composites, the addition of SWNT increases the Young modulus by two orders of magnitude [7-10].

On the other hand, weak van der Waals interactions between nanotubes cause that the mechanical properties of bulk material made of SWNT are rather poor. One way to solve the problem has been through binding various linker-molecules covalently with nanotubes in the hope they would tie them together [11].

In our work we have investigated the effect of gamma-irradiation on the properties of paper made of SWNT, when irradiated either in air or under vacuum. Electrical conductivity and Young modulus were tested and compared for different doses of gamma-irradiation. Structural changes were characterized by Raman spectroscopy and compared with the spectra of gamma-irradiated graphite.

The effect of electron beam irradiation on the stability of SWNT has been known from electron microscopy observations. The mechanism of electron interaction with

CP685, *Molecular Nanostructures: XVII Int'l. Winterschool/Euroconference on Electronic Properties of Novel Materials,* edited by H. Kuzmany, J. Fink, M. Mehring, and S. Roth
© 2003 American Institute of Physics 0-7354-0154-3/03/$20.00

nanotubes was suggested in the work of Smith et al. [12]. To our knowledge, until now there are no studies on gamma-irradiation effect on SWNT.

EXPERIMENTAL

Purified single wall carbon nanotubes commercially prepared by the HiPCO method were obtained from CNI (Texas). The nanotube-paper (NT-paper) was made from a suspension of nanotubes in a 1% solution of SDS in water. The paper was divided into several strips; each of them was sealed in a glass tube either in air or under vacuum. Four glass tubes with powder graphite from Merck were sealed in air as well. The samples were then exposed to gamma-irradiation of Co_{60} used as source, with a power rate 0.37 kGy/hour and a photon energy 1.3 MeV. The samples were taken away from the source after different times of exposition.

The four-probe method was used to determine the electrical conductivity of the irradiated strips of SWNT paper.

The Young modulus was determined from the elastic part of the stress vs. strain curve measured by a force transducer up to 20 N (Hottinger Baldwin Messtechnik, typ 52).

Raman spectra were taken by a FT-Raman spectrometer FRA 106/S (Bruker) with a laser excitation wavelength of 1064 nm. The acquisition times for the spectra vary from several minutes for nanotubes to several hours for graphite. The spectral resolution was 4 cm^{-1}. Graphite samples were made of a powder mixed and pressed with a KBr powder.

RESULTS

Study Of Mechanical And Electrical Properties

Figure 1A shows the stress vs. strain characteristics of the NT-paper irradiated in air with doses of 0, 53, 168 and 505 kGy. The curve of the pristine sample (0 kGy) bends as soon as the strain reaches 5 %. At first, with increasing doses of irradiation, the slope of the curves gets steeper, with purely elastic behaviour. Then, for the highest dose of 505 kGy, the slope of the elastic part decreases and breaks suddenly into a long creeping part. The Young modulus calculated from the elastic parts of the curves is plotted in fig. 2A, with a clear maximum at the dose of 168 kGy.

A similar maximum is observed in the dependence of electrical conductivity on the dose of irradiation (Fig. 2B).

Irradiation under vacuum did not affect the mechanical nor the electrical properties of the NT-paper in the investigated range of doses.

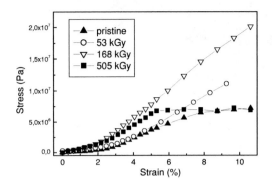

FIGURE 1. Stress vs. strain curves of gamma-irradiated NT-paper

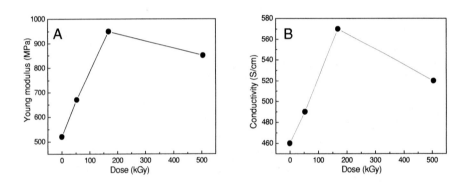

FIGURE 2. A. Young modulus, B. Electrical conductivity vs. dose of gamma-irradiation

Raman Spectroscopy Study

The structural changes were studied by Raman spectroscopy. Figure 3A shows Raman spectra of nanotubes and graphite irradiated in air. The spectra are normalized to the D*-mode (2640 cm^{-1}). Here we focused on the changes of D-mode, which is associated with the concentration of defects in the structure. On the other hand, the intensity of the D* band is assumed to be independent from the defect concentration. Irradiation causes overall changes in the intensity and shape of the spectra. In general, the intensity of the D-mode increases with the time of irradiation as compared to other prominent peaks.

There was only a small effect in Raman spectra found when the nanotubes were irradiated under vacuum.

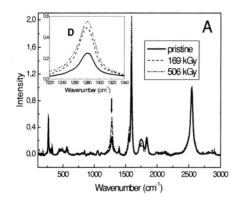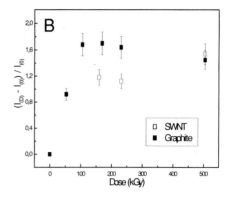

FIGURE 3A. Raman spectra of in-air gamma-irradiated NT-paper

FIGURE 3B. Ratio of [I(d)-I(0)] / I(0) for D-mode vs. dose d of irradiation for NT-paper and graphite

DISCUSSION

An important observation follows from Figure 1. The stress vs. strain curve of the pristine sample has a shape typical for the case where a permanent deformation also takes place. In the pristine sample, the nanotubes slip to new positions when exposed to higher stress, thus causing permanent (plastic) deformation. After gamma irradiation (doses 53 and 168 kGy), the stress vs. strain characteristics contain only a linear part for elastic deformation. It is likely that the irradiation causes the formation of some kind of bonds between nanotubes, fixing their mutual positions. Surprisingly, at the higher dose of 505 kGy, there is a sharp break of the elastic part into an inelastic creeping part. The concentration of defects at this dose is probably very high and the "interconnected" system of nanotubes collapses.

Our simple interpretation is supported by the dependences of Young modulus and electrical conductivity on the doses of irradiation (Fig. 2A and 2B). Beyond the maximum, observed for both these quantities at 168 kGy, the decrease which takes place is consistent with the idea that possible bonds between the nanotubes were broken when the dose reached 505 kGy.

Since the observed effect of gamma-irradiation took place only for irradiation in air, this suggests that air molecules play a crutial role in the structural changes responsible for the improvement of the mechanical and electrical properties of the NT-paper.

In order to visualize changes in Raman spectra caused by irradiation, normalized integrated Raman intensities I(d) of the D band with respect to the D* band were calculated for different doses d. Then the ratio [I(d)-I(0)]/I(0), representing the change of the normalized integrated Raman intensities I(d) relative to the "unirradiated" value I(0), was plotted (Fig.3B). As can be seen, the results for graphite and nanotubes follow the same trend. The integrated intensity of the D band increases considerably at the early stages of irradiation and then saturates when the dose exceeds ~100 kGy. The relative change is also quantitatively very similar for both materials. Such a behavior may be explained by the effect of two competing processes. First, the C-C nanotube bonds are destroyed by energetic gamma–photons generating defects. Once a critical defect concentration is reached, it may be energetically favourable to close the defects again. This process compensates for further defect generation and leads to the constant defect concentration even though irradiation doses increase. In the Raman study of boron doped SWNT done by Maultzsch et al. [13], a similar behaviour of the ratio D/D* was observed when the boron concentration (in other words, the defect concentration) increases.

SUMMARY

Gamma-irradiation of SWNT paper caused changes in the Young modulus and the electrical conductivity with a maximum at the dose of 168 kGy, and changes in Raman spectra as well.

The effect of irradiation was much stronger for samples irradiated in air in comparison to those in vacuum.

The defect formation observed in D-mode of Raman spectra saturated at about 100 kGy. A similar effect was found for in-air irradiated graphite.

ACKNOWLEDGMENTS

This work was supported by the EU Project CARDECOM, NANOTEMP and a grant of Slovak Ministry of Education VEGA 1/0055/03.

REFERENCES

1. Treacy, M. M. J., Ebbesen, T. W. and Gibson, T. M., *Nature* **381**, 680-687 (1996).
2. Wong, E. W., Sheehan, P. E. and Lieber, C. M., *Science* **277**, 1971-1975 (1997).
3. Yu, M. F., Files, B. P., Arepalli, S. and Ruoff, R. S., *Phys. Rev. Lett.* **84**, 5552-5555 (2000).
4. Thostenson, E. T., Ren, Z. and Chou, T. W., *Compos. Sci. Technol.* **61**, 1899-1912 (2001).
5. Mamedov, A. A., Kotov, N. A., Prato, M., Guldi, D. M., Wicksted, J. P. and Hirsch, A., *Nature Materials* **1**, www.nature.com/naturematerials (2002).
6. Zhan, G.-D., Kuntz, J. D., Prato, M., Wan, J., and Mukherjee, A. K., *Nature Materials* **2**, www.nature.com/naturematerials (2003).
7. Carneiro, O. S., Covas, J. A., Bernado, C. A., Caldeira, G., Van Huttum, F. W. J., Ting, J. M., Alig, R. L. and Lake, M. L., ., *Compos. Sci. Technol.* **58**, 401-407 (1998).
8. Lozano, K. and Barrera, E. V., *J. Appl. Polym. Sci.* **79**, 125-133 (2001).
9. Kuriger, R. J., Alam, M. K., Anderson, D. P., Jakobsen, R. L., ., *Composites, Part A* **33**, 53-62 (2002).
10. Sander, J., Werner, P., Shaffer, M. S. P., Demchuk, V., Altstädt, V. and Windle A. H., *Composites, Part A* **33**, 1033-1039 (2002).
11. Dettlaff-Weglikowska, U., Benoit J.-M., Chiu, P.-W., Graupner, R., Lebedkin, S. and Roth, S., Current *Appl. Phys.* **2**, 497-501 (2002).
12. Smith, B. W. and Luzzi D. E., *J. Appl. Phys.* **90**, 3509-3515 (2001).
13. Maultzsch, J., Reich, S., Thomsen, C., Webster, S., Czerw, R., Carroll, D. L., Vieira, S. M. C., Birkett, P. R. and Rego, C. A., *Appl. Phys. Lett.* **81**, 2647-2649 (2002).

Near-Infrared Photoluminescence of Single-Walled Carbon Nanotubes Obtained by the Pulsed Laser Vaporization Method

Sergei Lebedkin [a], Frank Hennrich [a], Tatyana Skipa [b], Ralph Krupke [a], and Manfred M. Kappes [a,c]

[a] *Forschungszentrum Karlsruhe, Institut für Nanotechnologie and* [b]*Institut für Festkörperphysik, D-76021 Karlsruhe, Germany*
[c] *Institut für Physikalische Chemie, Universität Karlsruhe, D-76128 Karlsruhe, Germany*

Abstract. Single-walled carbon nanotubes (SWNTs) with diameters between ~1.1 and ~1.5 nm prepared by the pulsed laser vaporization, dispersed and surfactant-stabilized in D_2O (transparent up to ~1800 nm) show photo- luminescence (PL) from ~1300 up to >1750 nm corresponding to the lowest electronic interband transitions of semiconducting tubes. The characteristic features of this PL such as sharp emission and excitation peaks corresponding to specific (n,m) tubes are similar to those reported recently for SWNTs with smaller diameters around 1 nm [1]. The experimental data are compared with tight-binding model calculations. For the structural assignment of SWNTs we used the empirical relations derived by Bachilo et al. [2].

INTRODUCTION

Single-walled carbon nanotubes (SWNTs) belong to quasi-one-dimensional (1D) systems. Their electronic structure is characterized by van Hove singularities (vHs) in the electronic density of states (Fig. 1) [3], observed, for instance, in optical absorption [4] and resonant Raman spectra of SWNTs [5].

Recently, O'Connell et al [1] discovered a band gap photoluminescence (PL) of SWNTs, which correspond to the lowest transitions $E_{11}{}^S$ between vHs in the conductance and valence zones of semiconducting nanotubes. This relatively bright near-infrared PL

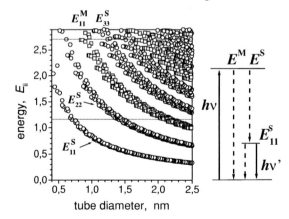

FIGURE 1. Energy separations Eii between pairs of vHs in the conductance and valence zones of semiconducting (dots) and metallic (squares) SWNTs calculated with the tight-binding model ($\gamma_0 = 2.9$ eV). In semiconducting tubes, a radiative transition from the lowest vHs in the conductance zone to the upper vHs in the valence zone ($E_{11}{}^S$) gives rise to the NIR luminescence (hν′).

CP685, *Molecular Nanostructures: XVII Int'l. Winterschool/Euroconference on Electronic Properties of Novel Materials,* edited by H. Kuzmany, J. Fink, M. Mehring, and S. Roth
© 2003 American Institute of Physics 0-7354-0154-3/03/$20.00

has been observed between ~900 and ~1400 nm from raw SWNTs with diameters between ~0.7 and ~1.1 nm prepared by high-pressure catalytic CO decomposition (the HiPco method) and ultrasonically dispersed presumably into individual tubes in a D_2O-surfactant mixture. The PL data have not only provided a valuable information about electronic properties of nanotubes (e.g., a direct measurement of the E_{11}^S and E_{22}^S energies), but also made possible a consistent structural assignment (described by the roll-up indices (n,m)) of all luminescent semiconducting tubes [2].

We have found a similar photoluminescence from D_2O-surfactant dispersions of SWNTs prepared by the pulsed laser vaporization (PLV) [6]. These tubes have diameters between ~1.1 and 1.5 nm and emit from ~1400 up to ≥1750 nm (interband transition energies vary roughly inversely with the tube diameter (Fig. 1)). Thus, the band gap photoluminescence appears to be a common property of different SWNTs materials and makes a promise to become a very powerful method for their characterization.

EXPERIMENTAL

PLV SWNTs were produced as described elsewhere [7]. As-prepared SWNTs were added to D_2O containing 0.3 mg/ml of the Tween-80 nonionic surfactant and dispersed with a quartz ultrasonic tip for ~10 min. A dispersion was then centrifuged in a standard lab centrifuge for 1 hour at 20.000 g. A supernatant solution (of neutral pH) was collected and used for luminescence measurements.

Near-infrared PL of SWNTs was measured in the range of 1200–1750 nm using a double-grating monochromator (spectral slits 10 nm) and a liquid nitrogen cooled germanium detector (Edinburgh Instruments). The excitation range of 620–970 nm was covered with a CW dye laser and a Ti-Sapphire laser (Spectra Physics) pumped with an Ar-ion laser. A typical excitation power was ~50–100 mW. PL spectra were corrected for the relative laser power and the wavelength-dependent detector response. The measurements were performed at room temperature using standard 4 mm quartz cuvettes in a 90°-geometry.

RESULTS AND DISCUSSION

Because of a relatively moderate ultrasonic agitation applied to raw SWNTs, the majority of dispersed nanotubes was likely still in (small) bundles. The PL is strongly quenched in bundles [1], therefore the total PL quantum efficiency was only ~10^{-5} [6]. Nevertheless, the PL spectra could be measured in a short time with a high signal-to-noise ratio due to the powerful laser excitation. Fig. 2 shows a contour plot of the photoluminescence of SWNTs excited in the S2 absorption band corresponding to the E_{22}^S transitions (Fig. 1). Each PL maximum can be attributed to a specific (n,m) nanotube with characteristic E_{11}^S (emission) and E_{22}^S (excitation) energies (Fig. 2). Note that the lower part of the PL contour plot is rather featureless (Fig. 2). This corresponds to optical excitation of nonluminescent metallic nanotubes and to excitation of semiconducting tubes, which is off-resonance with vHs.

Our PL data are in a good agreement with the empirical E_{11}^S, E_{22}^S – (n,m) relations derived recently by Bachilo et al. for HiPco nanotubes [1]. The (n,m) assignment of the luminescent PLV nanotubes on the basis of these relations is shown in Fig. 2 and Table 1.

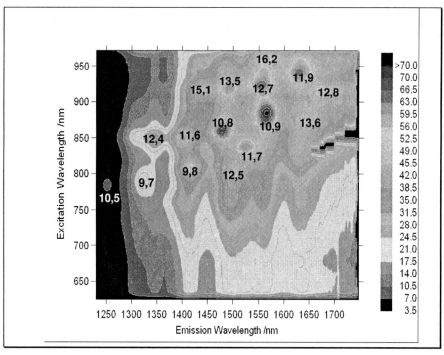

FIGURE 2. Two-dimensional contour plot of luminescence intensity versus emission and excitation wavelengths. The assignment of the luminescence maxima to specific (n,m) nanotubes is shown.

TABLE 1. Energies E_{11}^S and E_{22}^S of SWNTs obtained from the PL spectra and their (n,m) assignment according to the empirical relations of Bachilo et al. [2]. Three last columns indicate the diameter, d, chiral angle, α, and relative PL intensity of the assigned nanotubes.

E_{11}^S, eV exp.(calc.)	E_{22}^S, eV exp.(calc.)	(n,m)	d, nm	α, grad	relative PL intensity
0.74 (0.743)	1.354 (1.345)	(12,8)	1.384	23.4	31
0.756 (0.754)	1.412 (1.411)	(13,6)	1.335	18.0	58
0.759 (0.759)	~1.26 (1.236)	(14,6)	1.411	17.0	~50
0.761 (0.766)	1.308 (1.304)	(11,9)	1.377	26.7	85
0.792 (0.794)	1.395 (1.389)	(10,9)	1.307	28.3	100
0.793 (0.796)	~1.26 (1.242)	(16,2)	1.357	5.8	61
0.798 (0.802)	1.340 (1.322)	(12,7)	1.321	21.4	89
0.813 (0.811)	1.473 (1.475)	(11,7)	1.248	22.7	80
0.826 (0.822)	1.548 (1.556)	(12,5)	1.201	16.6	51
0.832 (0.834)	1.340 (1.330)	(13,5)	1.278	15.6	80
0.839 (0.843)	1.440 (1.419)	(10,8)	1.240	26.3	97
0.860 (0.871)	1.340 (1.330)	(15,1)	1.232	3.2	62
0.875 (0.875)	1.533 (1.526)	(9,8)	1.169	28.1	72
0.884 (0.887)	1.450 (1.433)	(11,6)	1.185	20.4	86
0.921 (0.924)	1.450 (1.436)	(12,4)	1.145	13.9	40
0.935 (0.937)	1.566 (1.555)	(9,7)	1.103	25.9	14
0.989 (0.992)	1.58 (1.562)	(10,5)	1.050	19.1	7

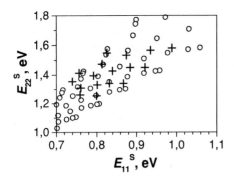

FIGURE 3. Experimental PL data (crosses) compared to the E_{11}^S and E_{22}^S energies calculated in the modified tight-binding model using $\gamma_0^\infty = 2.99$ eV and a constant empirical shift of E_{11}^S values of $+0.135$ eV [8].

According to this assignment, the majority of produced/ dispersed/ emitting semiconducting nanotubes has a relatively large chiral angle, $15° \le \alpha < 30°$. Only two tubes (16,2) and (15,1) with chiral angles smaller than $10°$ and – similar to the HiPco nanotubes [1,5] – no $(n,0)$ 'zigzag' nanotubes with $\alpha = 0°$ were found.

In difference to the simple tight-binding model of SWNTs [3], there is a reasonable agreement between our PL data and the modified model taking into account a curvature of a nanotube surface [8] (Fig. 3). However, an unambiguous (n,m) assignment is still hampered by the large number of different (n,m) tubes in this energy (diameter) interval and by deviations of the calculated values from the observed ones. The assignment is possible only for a few tubes and is consistent with that based on the empirical relations.

ACKNOWLEDGEMENTS

The support of this work by the Deutsche Forschungsgemeinschaft under SFB 551 and by the BMBF is gratefully acknowledged.

REFERENCES

1. O'Connell, M. J., Bachilo, S. M., Huffman, C. B., et al., *Science* **297**, 593 (2002).
2. Bachilo, S. M., Strano, M. S., Kittrell, C., et al., *Science,* **298**, 2361 (2002).
3. Saito, R., Dresselhaus, G., Dresselhaus, M. S., *Physical Properties of Carbon Nanotubes*; Imperial College Press: London, 1998.
4. Kataura, H., Kumazawa, Y., Maniwa, Y., Umezu, I., Suzuki, S., Ohtsuka, Y., Achiba, Y., *Synth. Met.* **103**, 2555 (1999); Jost, O., Gorbunov, A. A., Pompe, W., Pichler, T., Friedlein, R., Knupfer, M., Reibold, M., Bauer, H.-D., Dunsch, L., Golden, M. S., Fink, J., *Appl. Phys. Lett.* **75**, 2217 (1999).
5. Jorio, A., Saito, R., Hafner, J. H., Lieber, C. M., Hunter, M., McClure, T., Dresselhaus, G., Dresselhaus, M. S., *Phys. Rev. Lett.* **86**, 1118 (2001).
6. Lebedkin, S., Hennrich, F, Skipa, T., Kappes, M. M., *J. Phys. Chem. B***107**, 1949 (2003).
7. Lebedkin, S., Schweiss, P., Renker, B., Malik, S., Hennrich, F., Neumaier, M., Stoermer, C., Kappes, M. M., *Carbon* **40**, 417 (2002).
8. Ding, J. W., Yan, X. H., Cao, J. X., *Phys. Rev.* 2002, *B66*, 073401; Hagen, A., Hertel, T., *Nanolett.* **3**, 383 (2003).

Wetting of Single-Wall Carbon Nanotube Ropes and Graphite

Hendrik Ulbricht, Gunnar Moos, and Tobias Hertel

*Fritz-Haber-Institut der Max-Planck- Gesellschaft, Faradayweg 4-6,
D – 14195 Berlin, Germany*

Abstract. We have studied the interaction of a variety of adsorbates, ranging from inert gases to polar molecules, solvents and aromatic compounds with single-wall carbon nanotube and graphite surfaces using thermal desorption spectroscopy (TDS). These studies allow to investigate the wetting properties of such surfaces on the microscopic scale without complications due to microscopic surface roughness arising with some other techniques. The overwhelming majority of adsorbates studied here is found to wet graphite and nanotube surfaces completely while both are found to be wet only partially by water.

INTRODUCTION

The interaction of carbon nanotubes and in particular the interaction of the high-surface area single-wall carbon nanotubes (SWNTs) with adsorbates is crucial for a variety of applications, for example in electronic devices, fuel cells, supercapacitors or in gas sensors. The gas-surface binding energy for example is decisive for the wetting properties of SWNT surfaces and thus also plays an important role for solvation properties. Here, we present first results of a systematic investigation on the interaction of various adsorbates with graphite and SWNT surfaces as probed by thermal desorption spectroscopy (TDS).

RESULTS AND DISCUSSION

Thermal desorption experiments were carried out under ultra-high vacuum conditions. Samples were cleaned in-situ by repeated heating to 1200 K before gas deposition. Details about the experimental setup and sample preparation can be found elsewhere [1,2,3]. In short, gas is admitted to the sample at low temperatures. Subsequent sample heating at a rate of 0.5-1.0 K/s leads to the release of gas from the sample surface and desorbing species are detected using a quadrupole mass-spectrometer. Thermal desorption (TD) spectra are obtained for both, highly oriented pyrolytic graphite (HOPG) and single-wall carbon nanotube samples. The latter are known to consist of a tightly woven mat of nanotube ropes each having a diameter on the order of a few tens of nanometers. The specific SWNT sample surface area of about 100 m^2/g as obtained using TD traces from a Xenon saturated sample is consistent with adsorption on the

CP685, *Molecular Nanostructures: XVII Int'l. Winterschool/Euroconference on Electronic Properties of Novel Materials,* edited by H. Kuzmany, J. Fink, M. Mehring, and S. Roth
© 2003 American Institute of Physics 0-7354-0154-3/03/$20.00

external rope surfaces which is estimated to have a specific surface area on the order of 140 m^2/g.

TD spectra from graphite and SWNT samples taken after exposure to increasing amounts of Xenon are compared in Fig. 1. Most apparent is the shift of the SWNT spectra to higher desorption temperatures as well as a pronounced broadening of the SWNT desorption feature. Note, that the local coverage on nanotube samples is much smaller than on graphite due to the porous nature of these samples where approximately 60% of the sample volume consists of empty space in the form of pores and voids (this does not include the empty endohedral volume inside of SWNTs themselves). The broadening of TD features may be partially related to inhomogeneities and is partially due to the diffusion process that facilitate gas transport from the bulk of the sample into the vacuum [3]. The absence of multilayer desorption features even after prolonged gas exposure is taken as further evidence of the small local coverage on the SWNT surfaces in these experiments. On HOPG multilayer features appear only after complete saturation of the first monolayer. The shift of the low-coverage SWNT desorption feature to higher temperature – if compared to graphite – is most likely related to the presence of adsorption sites with higher local coordination. Such an increase of the local coordination may be due to adsorption in so called grooves on the external rope surfaces [2,3]. Impurities such as lattice defects and vacancies are not expected to play a role for the adsorption of non-polar adsorbates.

The wetting properties are here discussed based on differences in the monolayer and multilayer binding energies. Roughly, we can distinguish complete- from partial- or non-wetting using the difference of the mono- and second layer desorption temperatures in TD spectra. For Xe on HOPG the two are clearly separated which implies that the Xe binding energy to the graphite surface V_{SA} is higher than the Xe binding energy to a xenon surface V_{AA}. Thermodynamically this is expected to favor the completion of the first monolayer before the second layer starts to grow. Such behavior ($V_{SA}>V_{AA}$) is associated with complete wetting of the surface [4]. Partial wetting on the other hand is expected if $V_{SA}=V_{AA}$ i.e. if the temperatures of mono- and multilayer desorption features coincide. Non-wetting with $V_{SA}<V_{AA}$ cannot be discriminated from

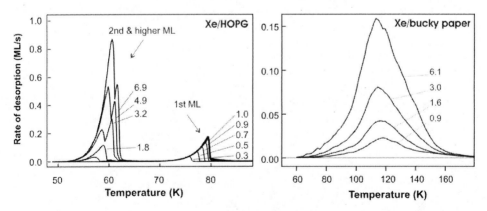

FIGURE 1. Xenon coverage series from graphite and SWNT samples. The numbers on the right of TD spectra denote the exposure prior to desorption in units of close-packed monolayers.

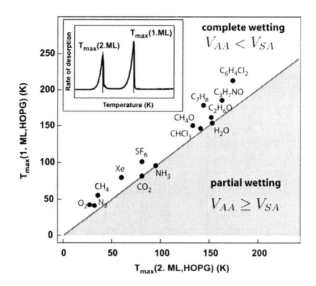

FIGURE 2. Temperature of the desorption peak maximum for the 1. monolayer (1. ML) plotted versus the temperature of the 2. ML for various adsorbates. The TD spectrum in the inset illustrates the two temperatures used for this graph. Adsorbates on the T_{max}(1. ML, HOPG)= T_{max}(2. ML, HOPG) line wet the surface only partially or not at all.

partial wetting by TD spectroscopy because low coverage desorption features cannot appear at lower temperatures than the corresponding multilayer desorption features.

In Fig. 2. we compare temperatures of the desorption peak maxima of monolayer and second-layer features – used here as a measure of mono- and multilayer binding energies. Following the above argument we thus expect a particular adsorbate to completely wet the surface $T_{max}(1.\,ML) > T_{max}(2.\,ML)$. This is the case for all adsorbates in the upper part of the graph. The moderately higher monolayer desorption temperature is in all cases strongly indicative of weak gas-surface interactions and we assume that all adsorbates studied here interact with graphite via van der Waals forces. Carbondioxide, ammonia and water, whose temperature values are located on the grey line with $T_{max}(1.\,ML) \leq T_{max}(2.\,ML)$ are found to wet the graphite surface only partially. The same is true for C_{60} which is not shown in Fig. 3 [5]. The partial wetting of graphite by water is in agreement with the hydrophobic nature of graphite. Dichlorobenzene which is known to be a good fullerene and SWNT solvent has the highest desorption temperature of all adsorbates studied here.

Temperatures characteristic of desorption from SWNT samples are summarized in a similar manner in Fig. 3. Due to the considerable width of desorption features on the SWNT samples we here plot the range of desorption temperatures instead of temperatures at desorption peak maxima. Evidently, one finds that adsorbates with polar groups such as dimethylformamide, ammonia, methanol and ethanol exhibit unusually broad TD features with peak temperatures reaching clearly beyond what is expected for the inert gases due to higher local coordination. We suggest that this is due to the interaction of polar groups with vacancies in the graphene lattice and unsaturated carbon bonds. Hydroxyl or other functional groups are not expected to be present on

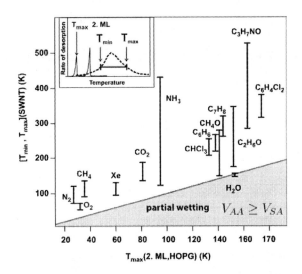

FIGURE 3. Same as Fig. 2 except that desorption features on the SWNT surface are characterized by the a desorption range with T_{min}-T_{max} corresponding to the FWHM of TD features. The insert again illustrates how these temperatures are obtained (dashed line: SWNT spectrum, solid line: HOPG spectrum).

these surfaces due to the high annealing temperatures used for sample cleaning prior to TD experiments. At higher exposure, i.e. with increasing local coverage we find that these TD features shift to lower temperatures supporting the notion that very broad desorption features at low coverage are due to a sampling of minority site functionalities.

The overall increase of desorption temperatures on the SWNT surfaces suggests that these samples are wet by all adsorbates shown in Fig. 3 except for water which is found wet SWNT rope surfaces only partially. The apparently better wetting of SWNT samples as compared to graphite may be attributed to higher surface 'roughness' at the nanoscale and the resulting local increase of van der Waals binding energies.

ACKNOWLEDGMENTS

We would like to thank Nesibe Cindir, Anne-Isabelle Henry and Renju Zacharia for assistance with some of the TDS experiments and Gerhard Ertl for continuing support of this work.

REFERENCES

1. T. Hertel, R. Fasel, and G. Moos, Appl. Phys. A **75**, 449 (2002).
2. H. Ulbricht, G. Moos, and T. Hertel, Phys. Rev. B **66**, 075404 (2002).
3. H. Ulbricht, J. Kriebel, G. Moos, and T. Hertel, Chem. Phys. Lett. **363**, 252-260 (2002).
4. P.G. de Gennes, Rev. Mod. Phys. **57**, 827-863 (1985).
5. H. Ulbricht, and T. Hertel, Phys. Rev. Lett. **90**, 095501 (2003).

Temperature Dependence of Photoconductivity of Single-Wall Carbon Nanotubes

A. Fujiwara[a, b], Y. Matsuoka[a], N. Ogawa[c], K. Miyano[d],
H. Kataura[e], Y. Maniwa[b, e], S. Suzuki[f], Y. Achiba[f]

[a] School of Materials Science, Japan Advanced Institute of Science and Technology
1-1 Asahidai, Tatsunokuchi, Ishikawa 923-1292, Japan
[b] Core Research for Evaluational Science and Technology (CREST)
Japan Science and Technology Corporation (JST)
[c] Department of Applied Physics, School of Engineering, University of Tokyo
7-3-1 Hongo, Bunkyo-ku, Tokyo 113-8656, Japan
[d] Research Center for Advanced Science and Technology, University of Tokyo
4-6-1 Komaba, Meguro-ku, Tokyo 153-8904, Japan
[e] Department of Physics, School of Science, Tokyo Metropolitan University
1-1 Minami-osawa, Hachi-oji, Tokyo 192-0397, Japan
[f] Department of Chemistry, School of Science, Tokyo Metropolitan University
1-1 Minami-osawa, Hachi-oji, Tokyo 192-0397, Japan

Abstract. Temperature dependence of photoconductivity has been investigated for single-wall carbon nanotube (SWNT) films at 0.7eV in the temperature range from 10 K to room temperature (R.T.). Although photoresponse monotonously increases with a decrease in temperature, the behavior depends on the sample. This sample dependence can be attributed to the variation of the transport properties of semiconducting-SWNT circuits. We also relate the potential of photoconductivity measurements for evaluating NTs-devices.

INTRODUCTION

Carbon nanotubes (NTs) [1] have attracted great attention as potential electronic materials because of the one-dimensional tubular network structure on a nanometer scale. The variety of band structures of the NTs, being either semiconducting or metallic depending on the chirality and the diameter of the tube [2,3], is also a novel feature. For applying NTs to electronic devices, more detailed transport properties of semiconducting phase have to be individually clarified. To date, however, selective growth of a single phase of NTs has not been achieved: both the metallic and semiconducting phases can coexist even in a single-wall carbon nanotube (SWNT) bundle [4]. Although Collins *et al.* [5] demonstrated a method for leaving only the semiconducting SWNTs from the mixture of two phases by burning the metallic NTs, this technique is quite difficult. As an easier approach to evaluate transport properties of semiconducting SWNTs in the mixture sample of two phases, we proposed photoconductivity measurements [6-9]: the photoconductivity excitation spectrum shows two peaks around 0.7 and 1.2 eV, which correspond to the first and second peaks of the optical adsorption spectrum in semiconducting NTs, E_{11}^s and E_{22}^s [10,11].

CP685, *Molecular Nanostructures: XVII Int'l. Winterschool/Euroconference on
Electronic Properties of Novel Materials,* edited by H. Kuzmany, J. Fink, M. Mehring, and S. Roth
© 2003 American Institute of Physics 0-7354-0154-3/03/$20.00

In the present study, we report on the detailed temperature dependence of photoconductivity of semiconducting SWNTs by focusing on E_{11}^s. We discuss variation of the temperature dependence of photoconductivity in terms of transport properties of semiconducting-SWNT circuits.

EXPERIMENTAL

The samples of SWNT bundles were synthesized by evaporation of composite rods of nickel (Ni), yttrium (Y) and graphite in helium atmosphere by arc discharge [11,12]. The diameter of the SWNTs used here is determined to be about 1.4 ± 0.2 nm by the Raman frequency of a breathing mode and TEM observation. The typical length of SWNT bundles estimated by scanning electron microscopy (SEM) is a few micrometers. To prepare film samples, soot-containing SWNTs was dispersed in methyl alcohol by ultrasonic vibrator, and suspension of SWNTs was dropped on glass substrate. The typical film sample size is about $100 \times 100 \ \mu m^2$ and the thickness of the film is between 300 and 500 nm. The samples were annealed in vacuum at 10^{-6} Torr and 673 K for 2 hours to remove the absorbed gasses and methyl alcohol from samples. A pair of gold electrodes separated by a 10 μm gap was evaporated in vacuum on to the surface of the film samples and connected to a dc regulated power supply (100.00mV). The samples were mounted in a continuous-flow cryostat and cooled by flowing the vapor of liquid He in the temperature range from 10 K to room temperature (R.T.). As a light source, an optical parametric oscillator (OPO) excited by a pulsed Nd:YAG laser with the duration time of 5 ns was used. The photon energy was set to 0.7 eV which corresponds to the E_{11}^s for SWNTs used in the present work. The temporal profiles of the laser pulse and the photocurrent were monitored with a digitizing oscilloscope. In order to reduce spurious ringing in the fast pulse detection, we were obliged to use the input impedance of the oscilloscope (50 Ω) as the reference resistor despite the obvious disadvantage of lower sensitivity.

RESULTS AND DISCUSSION

Figure 1 shows the temporal evolution of photocurrent for various incident light intensities for two samples at around 10 K and R.T. The current properly follow the laser pulse shape: signal profile with a 5 ns width can be observed and increases with an increase in incident light intensity at R.T. The oscillatory structure for time $t > 10$ ns is considered to be due to ringing in the circuitry. Data at 13.2 K for the sample #A show almost the same behavior to those at R.T., except for the enhancement in the intensity. On the other hand, they change drastically for the sample #B at 10.2 K, as shown in Fig. 1(d). The current shows a pronounced tail superposing on the oscillatory structure after the first peak observed in Figs. 1(a)-(c). This tendency becomes remarkable at low temperatures. Here, we extract the photoresponse from the ratio of the photocurrent to the incident light intensity [6,9]: the photocurrent is defined by fitting *the first peak* in the signal profile with the Gauss function. Temperature dependence of photoresponse normalized by the data at R.T. for samples #A and #B are shown in Fig. 2. Although photoresponse increases monotonously with a decrease

157

in temperature in both samples, the degree of enhancement at low temperatures are different between these two samples. Considering the tail of signal profiles and rather small enhancement of the first-peak intensity [13], it is plausible that trapping of photocarrier affect the photoconductive properties of the sample #B at low temperatures. This result suggests that quality of semiconducting SWNTs is various even in the sample produced by the same method. Although this fact has not been pointed out clearly so far, it is very important information from the viewpoint of application of semiconducting SWNTs to electronic devices. In addition, our results show photoconductive properties are very sensitive to the quality of semiconducting-SWNT circuits. Therefore we can propose that photoconductivity measurements are very effective and powerful technique for evaluating electronic circuit and devices consisting of semiconducting SWNTs.

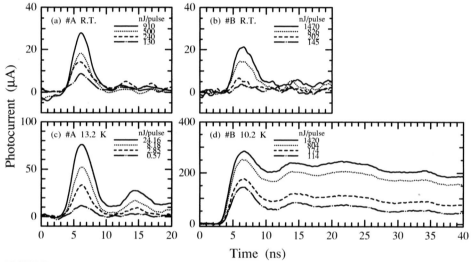

FIGURE 1. Temporal evolution of photocurrent for various incident light intensities for (a) sample #A at R.T., (b) sample #B at R.T., (c) sample #A at 13.2 K, (d) sample #B at 10.2 K.

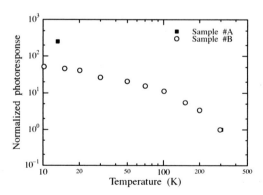

FIGURE 2. Temperature dependence of the photoresponse at 0.7 eV normalized by the data at R.T.

SUMMARY AND CONCLUSIONS

We have investigated detailed temperature dependence of photoconductivity for SWNT bundles. Temporal evolution of profile and its intensity strongly depend on the sample and temperature. The origin of this variation is considered to be due to the difference in trapping effect of the circuits consisting of semiconducting SWNTs. This result suggests that quality of semiconducting SWNTs is various even in the sample produced by the same method, and that photoconductive properties are very sensitive to it. We can propose that photoconductivity measurements are very effective and powerful technique for evaluating electronic circuit and devices consisting of semiconducting SWNTs.

ACKNOWLEDGMENTS

This work was supported in part by the JSPS "Future Program" (RFTF96P00104), Industrial Technology Research Grant in 2001 from New Energy and Industrial Technology Development Organization (NEDO) of Japan, and the Grant-in-Aid for Scientific Research (A) (13304026) from the Ministry of Education, Culture, Sports, Science and Technology (MEXT) of Japan. One of authors (A.F.) was supported by a Grant-in-Aid for Encouragement of Young Scientists (12740202) from the MEXT of Japan.

REFERENCES

1. Iijima, S., *Nature* (London) **354**, 56-58 (1991).
2. Saito, R., Fujita, M., Dresselhaus, G., Dresselhaus, M.S., *Appl. Phys. Lett.* **60**, 2204-2206 (1992).
3. Hamada, N., Sawada, S., and Oshiyama, A., *Phys. Rev. Lett.* **68**, 1579-151581 (1992).
4. Fujiwara, A., Iijima, R., Ishii, K., Suematsu, H., Kataura, H., Maniwa, Y., Suzuki, S., Achiba, Y., *Appl. Phys. Lett.* **80**, 1993-1995 (2001): appears also in *Virtual Journal of Nanoscale Science & Technology* **5**(12) (2002). (http://www.vjnano.org/)
5. Collins, P.G., Arnold, M.S., Avouris, P., *Science* **292**, 706-709 (2001).
6. Fujiwara, A., Matsuoka, Y., Suematsu, H., Ogawa, N., Miyano, K., Kataura, H., Maniwa, Y., Suzuki, S., Achiba, Y., *Jpn. J. Appl. Phys.* **40**, L1229-L1231 (2001).
7. Fujiwara, A., Matsuoka, Y., Suematsu, H., Ogawa, N., Miyano, K., Kataura, H., Maniwa, Y., Suzuki, S., Achiba, Y., *AIP Conference Proceedings* **590**, 189-192 (2001).
8. Fujiwara, A., Matsuoka, Y., Iijima, R., Suematsu, H., Ogawa, N., Miyano, K., Kataura, H., Maniwa, Y., Suzuki, S., Achiba, Y., *AIP Conference Proceedings* **633**, 247-250 (2002).
9. Matsuoka, Y., Fujiwara, A., Ogawa, N., Miyano, K., Kataura, H., Maniwa, Y., Suzuki, S., Achiba, Y., *Sci. Technol. Adv. Mater.*, in press.
10. Ichida, M., Mizuno, S., Tani, Y., Saito, Y., Nakamura, A., *J. Phys. Soc. Jpn.* **68**, 3131-3133 (1999).
11. Kataura, H., Kumazawa, Y., Maniwa, Y., Umezu, I., Suzuki, S., Ohtsuka, Y., and Achiba, Y., *Synth. Met.* **103**, 2555-2558 (1999).
12. Journet, C., Maser, W.K., Bernier, P., Loiseau, A., Lamy de la Chapelle, M., Lefrant, S., Deniard, P., Lee, R., and Fischer, J.E., *Nature* (London) **388**, 756-758 (1997).
13. As discussed in Ref. 9, this small enhancement of the first-peak intensity is also due to the underestimation of the intensity because photocurrent does not respond linearly to incident light intensity but tend to saturate at high light intensity region.

Conduction in Carbon Nanotube Networks

A. B. Kaiser and S. A. Rogers

*MacDiarmid Institute for Advanced Materials and Nanotechnology, SCPS,
Victoria University of Wellington, P O Box 600, Wellington, New Zealand*

Abstract. Recent measurements of the resistivity of single-wall carbon nanotube (SWNT) networks are consistent with our model of metallic conduction interrupted by barriers. We extend our model of thermopower nonlinearities due to peaks in the density of electronic states and apply it to recent thermopower data for carbon nanotube networks.

RESISTIVITY OF SWNT NETWORKS

While individual carbon nanotubes may exhibit ballistic conduction, we suggested [1,2] that the resistance of networks consisting of ropes of single-wall carbon nanotubes was dominated by tunneling through small conduction barriers, such as inter-rope and inter-tube contacts or defects along individual tubes.

We show in Fig. 1 more recent measurements [3-6] of the resistivity of SWNT networks, which support our model of metallic conduction interrupted by barriers. As pointed out by Shiraishi and Ata [3] for their data, the temperature dependence of the resistivity in each case is consistent with the same generic expression that we used for conducting polymers [2]

$$\rho(T) = Q \exp\left(-\frac{T_m}{T}\right) + B \exp\left(\frac{T_t}{T_s + T}\right) \tag{1}$$

where the first term is for quasi-one-dimensional metallic conduction [7] (involving suppression of resistivity at lower temperatures where carriers are not easily backscattered), and the second term represents fluctuation-assisted tunneling through barriers [8]. The interplay between the metallic and nonmetallic terms accounts for the resistivity minimum seen in three of the samples; the barrier sizes deduced from the values of the tunneling parameter T_t are 0.5 - 4 meV. In the other samples showing no minimum, the nonmetallic contribution to the resistivity temperature dependence remains dominant.

These new data tend to be more consistent with a faster-than-linear quasi-1D metallic term in the resistivity expression, rather than a conventional linear increase of metallic resistivity near room temperature that appeared more appropriate for earlier SWNT network samples, especially an individual SWNT rope [1,2]. This makes the new data more similar to the quasi-1D metallic behavior inferred in highly-conducting polymers.

CP685, *Molecular Nanostructures: XVII Int'l. Winterschool/Euroconference on
Electronic Properties of Novel Materials,* edited by H. Kuzmany, J. Fink, M. Mehring, and S. Roth

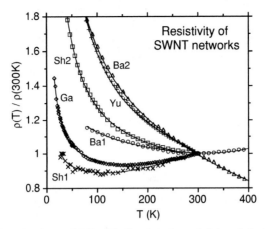

FIGURE 1. Fits of our barrier model Eq. (1) (lines) to the resistivity of single-wall carbon nanotube networks (normalized to its value at room temperature) measured by Shiraishi and Ata (Sh1 a thick mat, Sh2 a thin mat) [3], Bae et al. (Ba1 parallel to the alignment direction of SWNTs in the mat as made, Ba2 after degassing) [4], Gáal et al. (Ga, buckypaper as made) [5], and Yu et al. (Yu, a mat) [6].

THERMOPOWER DUE TO DENSITY OF STATES PEAKS

Thermopower is a property less affected by electrical barriers than conductivity, and so often reveals more intrinsic behavior, as in highly-conducting polymers [2]. We proposed earlier [9] that a density of states (DOS) peak could account for a faster-than-linear increase in thermoelectric power as a function of temperature, with a decrease in gradient at higher temperatures, as sometimes observed in carbon nanotube samples [10-12] (Figure 2(b)). We give here a more complete calculation of the effect of the DOS peak on thermopower and compare the results with our earlier approximation and with experimental data on carbon nanotube networks.

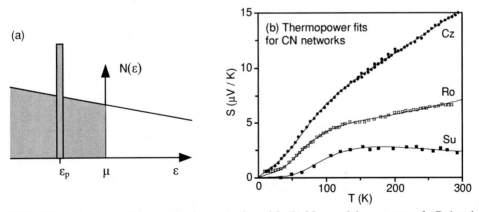

FIGURE 2. (a) Sketch of the density of states peak model. (b) Measured thermopower of a B-doped multi-wall carbon nanotube (MWNT) mat from Czerw et al. (Cz) [10], an O_2-doped SWNT thin film from Romero et al. (Ro) [11], and a degassed SWNT film covered in C_6H_6 from Sumanasekera et al. (Su) [12]. The full lines show fits of the data to our DOS peak model Eq. (4).

A general expression for diffusion thermopower [13] in a uniform material is

$$S_d = \frac{1}{eT} \int (\varepsilon - \mu) \sigma(\varepsilon) \left(-\frac{\partial f}{\partial \varepsilon} \right) d\varepsilon \bigg/ \int \sigma(\varepsilon) \left(-\frac{\partial f}{\partial \varepsilon} \right) d\varepsilon \qquad (2)$$

where e is the electronic charge, μ the chemical potential, f the Fermi function and $\sigma(\varepsilon)$ a partial conductivity function at electronic energy ε. Our model for the density of states is sketched in Fig. 2(a): a sharp DOS peak at an energy ε_p (approximated by a delta function initially) gives rise to a corresponding peak in the partial conductivity

$$\sigma(\varepsilon) = \sigma(\mu)[1 + a(\varepsilon - \mu) + b\delta(\varepsilon - \varepsilon_p)] \qquad (3)$$

where the first two terms represent the usual linear wide-band metallic behavior, and the last term the DOS peak contribution.

We then derive the diffusion thermopower as

$$S_d = \left\{ \beta T + \frac{bT_p \exp(T_p/T)}{eT^2 \left[\exp(T_p/T) + 1 \right]^2} \right\} \bigg/ \left\{ 1 + \frac{b \exp(T_p/T)}{kT \left[\exp(T_p/T) + 1 \right]^2} \right\} \qquad (4)$$

where the term βT is the usual linear Mott diffusion thermopower and $kT_p = (\varepsilon_p - \mu)$. This expression gives rise to the family of curves shown in Fig. 3(a), which show a broad thermopower peak around 0.4 T_p. If we neglect the effect of the DOS peak on conductivity and take the denominator in Eq. (4) as unity, as we did earlier [9], we obtain the dashed lines in Fig. 3(a); for thermopower contributions from the DOS peak of greater than 10 μV/K, this approximation becomes increasingly inaccurate.

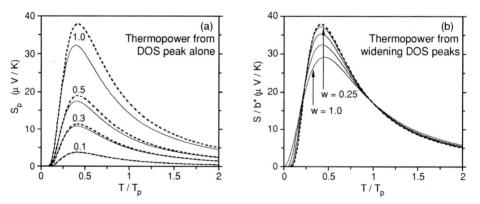

FIGURE 3. (a) Diffusion thermopower arising from a delta-function DOS peak alone calculated using Eq. (4) (full lines), and neglecting the effect of the DOS peak on conductivity (dashed lines); the value of $b^* = b/kT_p$ (which indicates the size of the DOS peak) is given beside each pair of curves. (b) Effect of widening the peak (without the conductivity correction) for halfwidth parameters w = 0.25, 0.5, 0.75 and 1.0 in Eq. (5) (full lines), compared to the corresponding delta-function result (dashed line).

162

Our model gives a good account of thermopower data for carbon nanotube networks that show a fast increase at low temperatures, as shown by the fits in Fig. 2(b). Czerw et al. [10] used our earlier approximate expression [9] to account for their thermopower data in terms of a DOS peak arising from B doping of their multi-wall carbon nanotubes – our new expression gives very similar results owing to the relatively small size of the thermopower nonlinearity.

We have also investigated the accuracy of using a delta function to approximate the DOS peak by making numerical calculations for an inverted parabolic narrow band of total width $2w$:

$$\sigma(\varepsilon) = \sigma(\mu)[1 + (3b/4w)(1 - (\varepsilon - \varepsilon_p)^2 / w^2)] \qquad -w < (\varepsilon - \varepsilon_p) < w \qquad (5)$$

The calculations shown in Fig. 3(b) illustrate the effect of widening of the DOS peak on thermopower. The effect is not dramatic as the thermopower peak is already broad: it broadens further but retains the same overall behavior until the narrow band overlaps the Fermi level ($w > 1$) and the exponential-like thermopower term disappears. We conclude that our delta-function DOS peak expressions give a reasonable description of the thermopower effect for $w < 0.5$, i.e. for a narrow band whose halfwidth w is less than the energy difference between its nearest edge and the Fermi level.

ACKNOWLEDGMENTS

We thank Masashi Shiraishi and David Carroll for helpful discussions, and the Marsden Fund administered by the Royal Society of New Zealand for support.

REFERENCES

1. Kaiser, A.B., Düsberg, G., and Roth, S., *Phys. Rev. B* **57**, 1418-1421 (1998).
2. Kaiser, A B., *Rep. Prog. Phys.* **64**, 1-49 (2001).
3. Shiraishi, M., and Ata, M., *Synth. Met.***128**, 235-239 (2002).
4. Bae, D.J., Kim, K.S., Park, Y.S., Suh, E.K., An, K.H., Moon, J.-M., Lim, Park, S.H., Jeong, Y.H., and Lee, Y.H., *Phys. Rev. B* **64**, 233401 (2001).
5. Gaál, R., Salvetat, J.-P., and Forró, L., *Phys. Rev. B* **61**, 7320-7323 (2000).
6. Yu, H.Y., Jhang, S.H., Park, Y.W., Bittar, A., Trodahl, H.J., and Kaiser, A.B., *Synth. Met.***128**, 235-236 (2002).
7. L. Pietronero, Synth. Met. **8**, 225-231 (1983).
8. P. Sheng, Phys. Rev. B **21**, 2180-2195 (1980) 2180.
9. Kaiser, A.B., Park, Y.W., Kim, G.T., Choi, E.S., Düsberg, G., and Roth, S., *Synth. Met.* **103**, 2547-2550 (1999).
10.Czerw, R., Chiu, P.-W., Choi, Y.-M., Lee, D.-S., Carroll, D.L., Roth, S., and Park, Y.W., *Structural and Electronic Properties of Molecular Nanostructures*, edited by H. Kuzmany et al., AIP Conference Proceedings 633, Melville, New York, 2002, pp 86-91.
11.Romero, H.E., Sumanasekera, G.U., Mahan, G.D., and Eklund, P.C., *Phys. Rev. B* **65**, 205410 (2002).
12.Sumanasekera, G.U., Pradhan, B.K., Romero, H.E., Adu, K.W., and Eklund, P.C., Phys. Rev. Lett. **89**, 166801 (2002).
13.Dugdale, J.S., *The Electrical Properties of Metals and Alloys*, Edward Arnold, London, 1977, p. 107.

Hexagonal diamond from single-walled carbon nanotubes

S. Reich*, P. Ordejón*, R. Wirth†, J. Maultzsch**, B. Wunder†, H.-J. Müller†, C. Lathe†, F. Schilling†, U. Dettlaff-Weglikowska‡, S. Roth‡ and C. Thomsen**

*Institut de Ciència de Materials de Barcelona (CSIC), Campus de la U.A.B., 08913 Bellaterra, Barcelona, Spain
†Geoforschungszentrum Potsdam, Div. 4, Telegrafenberg, 14473 Potsdam, Germany
**Institut für Festkörperphysik, Technische Universität Berlin, Hardenbergstr. 36, 10623 Berlin, Germany
‡Max-Planck-Institut für Festkörperforschung, Heisenbergstr. 1, 70569 Stuttgart, Germany

Abstract. We studied the transformation of single-walled carbon nanotubes into diamond by *ab-initio* calculations and high-pressure and temperature experiments. From the calculations we predict the formation of hexagonal diamond as a high-pressure nanotube phase. High-resolution TEM pictures of single-walled carbon nanotubes subjected to 9 GPa and 700°C clearly indicate the formation of hexagonal diamond grains.

The high-pressure phase diagram of carbon is dominated by diamond, which is energetically the most stable structure at pressures above ≈ 2 GPa. At ambient pressure graphite is the phase with the lowest total energy; however, the difference in total energy between the pure sp^2 and sp^3 phases is only some tens meV per carbon atom.[1] In contrast to the small difference in total energy, the energy barriers between graphite and diamond are well above 300 meV per carbon atom at ambient pressure.[1] They thus prevent a transformation of diamond into graphite even at very long time scales. To overcome these barriers high pressures and high temperature conditions have to be applied. It was shown that graphite and C_{60} transform into diamond at pressures ≈ 15 GPa and temperatures above 1000 K.[2, 3]

In this paper we report on the transformation of single walled carbon nanotubes into diamond. From *ab-initio* calculations we find that single-walled carbon nanotubes form a polymerized high pressure phase above ≈ 7 GPa. This phase transforms into hexagonal diamond by a succesive interlinking of the flattened nanotubes walls. We present high-resolution TEM pictures of single-walled carbon nanotubes which we subjected to high-pressures and temperatures. The interlayer distance of ≈ 4.1 Å observed after the high-pressure treatment clearly indicates the transition into the hexagonal diamond phase.

The basic building blocks of diamond are buckled hexagonal carbon planes. These planes can be arranged in an ABC stacking yielding cubic diamond, which is the phase normally simply called diamond [see Fig. 1(a)]. Similar to the wurzite structure in polar semiconductors, the buckled carbon planes can also be arranged in an ABAB stacking. This phase is called hexagonal diamond; it is shown in Fig. 1b). Hexagonal diamond was

CP685, *Molecular Nanostructures: XVII Int'l. Winterschool/Euroconference on Electronic Properties of Novel Materials,* edited by H. Kuzmany, J. Fink, M. Mehring, and S. Roth

discovered in graphite samples after shock-wave and high-pressure experiments.[4, 5, 6] High-pressure and temperature experiments on graphite sometimes yield cubic, sometimes hexagonal, or a mixture of the two diamond phases. The end product depended on the starting material and the high-pressures equipment in a not-fully understood way.[7, 8, 9, 10] Likewise, twinned cubic-hexagonal crystals were found in *ab-initio* molecular-dynamics calculations of the graphite to diamond transition.[11] It was suggested by Tateyama *et al.*[1] that the graphite to cubic diamond transitions requires a sliding of the graphene planes. In contrast hexagonal diamond results if a sliding of the layers is forbidden.

Single-walled carbon nanotubes are another potential source of artificial diamond. To study the high-pressure phases of single-walled carbon nanotubes and a possible transformation into diamond we performed *ab-initio* total energy calculations of (6,6) nanotube crystals. We used the SIESTA[12, 13] *ab-initio* package within the local density approximation[14] to relax the atomic coordinates of (6,6) nanotube crystals under applied hydrostatic pressure. The relaxations were done for pressures between 0 and 60 GPa by a conjugate gradient minimization under the constrained of an hydrostatic stress tensor. The minimization was considered converged if the forces on the atoms were below 0.04 eV/Å and every component of the stress tensor within $5 - 10\%$ of the desired value. For the *ab-initio* calculation we replaced the core electrons by non-local norm-conserving pseudopotentials;[15] the valence electrons were described by a double-ζ polarized basis set.[13] The basis set cutoffs were 5.12 a.u. for the s orbital and 6.25 a.u. for the p and the polarizing d orbitals. In real space we used a grid energy cutoff of 250 Ry; in reciprocal space we included 64 k points.

At low pressures (0 – 6 GPa) the response of the nanotube crystal to an externally applied pressure is mainly governed by the weak van-der-Waals interaction between the tubes. We found a bulk modulus of 37 GPa in excellent agreement with high-pressure x-ray experiments on single-walled carbon nanotubes.[16] At a pressure of 7 GPa the circular tubes collapsed into a phase with an elliptical cross section. Although this phase requires a large strain energy in parts of the nanotube walls, it is favourable at high pressure, because the empty volume inside the tubes is reduced by 5 % compared to the circular phase at the same pressure. For larger diameter tubes the relative strain energy is much smaller. Thus this transition is expected to occur at lower pressures as suggested by piston-cylinder experiments.[17] At the point of strongest curvature of the nanotube

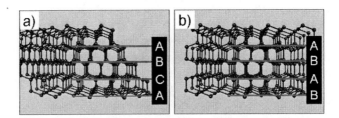

FIGURE 1. a) Cubic diamond with an ABC stacking of the carbon planes. This stacking is obtained by a translation within the carbon planes. b) Hexagonal diamond with an ABAB stacking. This stacking results from a rotation of every second plane by 180° around the axis perpendicular to the planes.

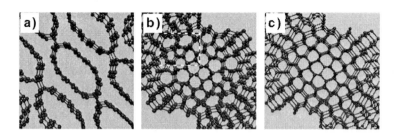

FIGURE 2.   a) (6,6) nanotube crystal at 7.5 GPa pressure. The tubes have an elliptical cross section; diamond-like carbon-carbon bonds connect the tubes. b) (6,6) crystal at 40 GPa. At this pressure the tubes are transformed into a pure sp^3 phase. The dashed rectangular highlights the pentagon-heptagon defects. c) Hexagonal diamond structure obtained from b) by the rotation of one carbon bond per nanotube unit cell. Compare the picture to the ideal hexagonal diamond show in Fig. 1b).

wall the carbon-carbon bonds are of strongly mixed sp^2 and sp^3 character. The structure with an elliptical cross section was unstable when we further increased the pressure. Figure 2a) shows the relaxed structure at a pressure of 7.5 GPa. The flattened nanotubes are connected by carbon-carbon bonds which are almost purely of sp^3 character. In our calculation, the interlinked structure was stable when we released the pressure; evidence for the formation of this phase under pressure was also found in high-pressure and temperature experiments.[18]

Under a pressure of 40 GPa the flattened nanotube walls buckle and subsequently carbon-carbon bonds are formed between the nanotube walls. Finally, the tubes are transformed into the sp^3 structure shown in Fig. 2b). The difference between the high-pressure nanotube phase and the hexagonal diamond structure is the presence of pentagon-heptagon defects which are indicated by the dashed rectangular in Fig. 2b). If we rotate one carbon bond per nanotube unit cell we finally obtain a defect-free hexagonal-diamond crystal presented in Fig. 2c). We calculated the total energy at various pressures for the ambient pressure nanotubes, the pentagon-heptagon defect hexagonal diamond, and the defect-free diamond phase. At ambient pressure the hexagonal diamond with a 5-7 defect is by 80 meV per carbon atom lower in total energy than the crystal of circular nanotubes. The total energy is further reduced by 200 meV in the defect-free hexagonal-diamond phase. Once these phases are formed under high pressure, they are therefore stable under the release of pressure. We also estimated the transition barrier between the interlinked nanotube phase and hexagonal diamond by following the transition path obtained at 40 GPa at various lower pressure. This approach gives an upper bound for the energy barrier, since the saddle point at low pressure might be at a different point in the multi-dimensional phase space. At ambient pressure we found an energy barrier below 150 meV, which is by more than a factor of two lower than for the graphite to diamond transition (350 meV for graphite to cubic diamond and 415 meV for graphite to hexagonal diamond)[1]. The much lower transition barrier suggests that comparatively low pressures are required for transforming nanotubes into diamond at high temperatures. To test this prediction we performed high-pressure and temperature experiments on single-walled carbon nanotubes.

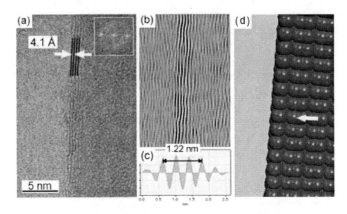

FIGURE 3. a) High-resolution TEM picture of the nanotube sample after high-pressure and high-temperature experiments. The lattice plains running from the top to the bottom of the picture indicate the presence of an hexagonal diamond grain (lattice spacing 4.1 Å). b) Inverse FFT of the lattice fringes in figure a); c) height profile of b). d) Schematic picture of an hexagonal diamond grain. The c axis with a lattice constant of 4.12 Å is indicated by the arrow.

The purified nanotubes were prepared by the laser-ablation method using Ni/Co catalysts. We subjected the tubes to a pressure of 9.0 GPa and 700°C in a MAX80 multi-anvil cell. First, the pressure was increased at room temperature up to 9.0 GPa followed by a heating of the sample to 700°C. Afterwards the sample was rapidly quenched to room temperature and, finally, the pressure was released. We characterized the samples before and after the high-pressure treatment with x-ray and Raman scattering, electron-energy loss spectroscopy and high-resolution transmission electron microscopy. A detailed account of the experiments will be published elsewhere; here we concentrate on the results of the TEM measurements, which indicate the formation of small grains of hexagonal diamond. The high-resolution energy filtered (zero loss) TEM images were obtained with a PHILLIPS CM200 transmission electron microscope equipped with a Lab6 electron source.

After the high-pressure run the samples mostly consisted of carbon nanotubes and some graphitic carbon with a lattice spacing of 3.4 Å. However, we found small grains of an sp^3 carbon phase, which were absent in the starting material or in samples subjected to lower pressures (up to 5 GPa). In Figure 3a) we show a TEM picture of such a spot. The lattice planes running from the top to the bottom in the picture (highlighted by the black lines) have an interlayer spacing of 4.1 Å. The planes are better visible in Fig. 3b) which shows an inverse Fourier transform of the FFT lattice fringes; part c) of the figure shows the height profile of the inverse FFT. Hexagonal diamond is the only carbon phase with sp^3 bonding and a lattice constant of 4.17 Å (along the c axis). Electron-energy loss spectra taken on the grains showed two separate peaks at 292 and 298 eV typical for the $p\sigma^*$ transitions in sp^3 bonded carbon.

The hexagonal-diamond grains we observed in the high-pressure sample typically consisted of only a few carbon layers along the c axis, but were several nm long (> 20 nm). Figure 3d) shows schematically shows an hexagonal-diamond crystal with the same orientation as the grain in Fig. 3a). The large ratio between the length and the width of the grain suggests that after the transformation of the nanotubes the c axis of the hexagonal-diamond crystal is perpendicular to the z axis of the original nanotubes. Precisely this is the transformation path that we predicted from the *ab-initio* calculations.

In summary, form both our calculations and experimentally we found that single-walled carbon nanotubes transform into hexagonal as opposed to cubic diamond. Hexagonal diamond is favoured when using carbon nanotubes as the starting material, because of the polymerized nanotube phase, which forms at intermediate pressure. The carbon-carbon bonds between the tubes prevent a sliding of the flattended nanotubes walls. Further high-pressure experiments aiming at larger, more macroscopic diamond crystals are under way.

We were supported by the Ministerio de Ciéncia y Technología (Spain) and the DAAD (Germany) within the Acciones Intergradas Hispano-Alemanas. St.R. acknowledges a fellowship by the Akademie der Wissenschaften Berlin-Brandenburg. P.O. acknowledges the Fundación Ramon Areces, EU Project SATURN, and a Spain-DGI Project. J.M. was supported by the Deutsche Forschungsgemeinschaft (DFG). High-pressure experiments were performed at the HASYLAB Hamburg. Parts of the calculations were carried out at the CESCA and CEPBA supercomputing facilities.

REFERENCES

1. Tateyama, Y., Ogitsu, T., Kusakabe, K., and Tsuneyuki, S., *Phys. Rev. B*, **54**, 14994–15001 (1996).
2. Clarke, R., and Uher, C., *Adv. Phys.*, **33**, 469–566 (1984).
3. Blank, V. D., Buga, S. G., Dubitsky, G. A., Serebryanaya, N. R., Popov, M. Y., and Sundqvist, S., *Carbon*, **36**, 319–343 (1998).
4. Hannemann, R. E., Strong, H. M., and Bundy, F. P., *Science*, **155**, 995–997 (1967).
5. Bundy, F. P., and Kasper, J. S., *J. Chem. Phys.*, **46**, 3437–3446 (1967).
6. DeCarli, P. S., and Jamieson, J. C., *Science*, **133**, 1961 (1821-1822).
7. Endo, S., Idani, N., Oshima, R., Takano, K. J., and Wakatsuki, M., *Phys. Rev. B*, **49**, 22 (1994).
8. Zhao, Y. X., and Spain, I. L., *Phys. Rev. B*, **40**, 993 (1989).
9. Aust, and Drickames, *Science*, **140**, 3568 (1963).
10. Yagi, T., Utsumi, W., Yamakata, M.-A., Kikegawa, T., and Shimomura, O., *Phys. Rev. B*, **46**, 6031–6039 (1992).
11. Scandolo, S., Bernasconi, M., Chiarotti, G. L., Focher, P., and Tosatti, E., *Phys. Rev. Lett.*, **74**, 4015–4018 (1995).
12. Soler, J. M., et al., *J. Phys.: Condens. Matter*, **14**, 2745–2779 (2002).
13. Artacho, E., Sánchez-Portal, D., Ordejón, P., García, A., and Soler, J., *phys. stat. sol. (b)*, **215**, 809 (1999).
14. Perdew, J. P., and Zunger, A., *Phys. Rev. B*, **23**, 5048 (1981).
15. Troullier, N., and Martins, J. L., *Phys. Rev. B*, **43**, 1993 (1991).
16. Reich, S., Thomsen, C., and Ordejón, P., *Phys. Rev. B*, **65**, 153407 (2002).
17. Chesnokov, S. A., Nalimova, V. A., Rinzler, A. G., Smalley, R. E., and Fischer, J. E., *Phys. Rev. Lett.*, **82**, 343–346 (1999).
18. Khabashesku, V. N., et al., *J. Phys. Chem. B*, **106**, 11155–11162 (2002).

The Electrical Magnetochiral Effect
In Carbon Nanotubes

V. Krstić[1], S. Roth[2], M. Burghard[2], K.Kern[2], G.L.J.A. Rikken[1,3]

[1]*Grenoble High Magnetic Field Laboratory, MPI-FKF/CNRS, F-38042 Grenoble, France.*
[2]*Max-Planck-Institut für Festkörperforschung, D-70569 Stuttgart, Germany.*
[3]*Laboratoire Nationale des Champs Magnétiques Pulsés, CNRS/INSA/UPS, F-31432 Toulouse, France.*

Abstract. Most carbon nanotubes consist of a helical arrangement of carbon atoms. Consequently, tubes of left- and right-handed configuration should exist. Their chiral character is recovered in the magnetoresistance as an odd power contribution in the magnetic field and the current, i.e., depending on the relative orientation of these two quantities. Electrical transport measurements on single-walled carbon nanotubes being subject to an external magnetic field show this dependence. An analytical quantum-mechanical model based on the picture of a free electron confined to a helix is provided in order to yield indications of the microscopic origin of the observed electrical magnetochiral effect.

INTRODUCTION

Carbon nanotubes (CNTs), principally, can be thought of as a graphene sheet [1] of some microns in length and a few nanometer in diameter that has been rolled up seamlessly. Thus they exhibit a large aspect ratio and therefore have a one-dimensional character. The different possible wrapping angles determine the diameter and whether a CNT is metallic or semiconducting [2]. This leads to a variety of CNTs [2], classified by the pair of indices (n,m). The wrapping can occur in two possible directions, leading to right- (D) and left- (L) handed chiral nanotubes. Among the various classes of nanotubes, there are only two special ones which are non-chiral, the so-called armchair (n=m) and zigzag (m=0) tubes [2]. Raman scattering and scanning tunneling microscopy has been used to address the chiral structure of nanotubes but either no information on the handedness is obtained [3] or the results are not consistent [4].

In investigations on charge transport effects so far, e.g. Coulomb-blockade [5], signatures of Luttinger-liquid behavior [6], and ballistic transport [7,8], the CNT's chiral aspects were never reported, although there have been calculations that suggest that the chirality should affect transport properties [9].

Recently, Rikken et al. [10,11] have observed a new magneto-optical phenomenon, called magneto-chiral anisotropy (MChA), in the luminescence/absorption spectra of chiral media, which depends on the magnetic field **B** and the wavevector **k** of the light. The essential features of MChA are (i) its linear dependence on **k·B**, (ii) its dependence on the handedness of the chiral medium (enantioselectivity) and, (iii) its independence of the polarization state of the light. Consequently, the question comes

*CP685, Molecular Nanostructures: XVII Int'l. Winterschool/Euroconference on
Electronic Properties of Novel Materials,* edited by H. Kuzmany, J. Fink, M. Mehring, and S. Roth
© 2003 American Institute of Physics 0-7354-0154-3/03/$20.00

to mind if the to the optical case analogous effect exists for electronic transport in chiral conductors, which could be used to study chirality in CNTs. Similar to the optical case, the existence of MChA can be deduced from symmetry arguments.

In ballistic transport, time- and parity-reversal symmetry arguments yield for the electrical resistance

$$R^{D/L}(\mathbf{I}, \mathbf{B}) = R_0\{1 + \beta \mathbf{B}^2 + \zeta^{D/L} \mathbf{I} \cdot \mathbf{B}\} \qquad (1)$$

where $\mathbf{I}$ is the current and $\mathbf{B}$ the external magnetic field. The parameters $\zeta^D = -\zeta^L$ denote the handedness of the chiral conductor. The parameter β describes the normal longitudinal magnetoresistance. The last term on the right hand side of (1) is named electrical magneto-chiral anisotropy (eMChA). Note that eMChA is outside the linear response regime, and that the Buttiker-Landauer formalism does not necessarily apply a priori. In case of diffusive transport, the Onsager relation has to be used, stating that under time-reversal the conductivity tensor elements σ_{kl} must obey

$$\sigma_{kl}(\mathbf{j}, \mathbf{B}) = \sigma_{kl}(-\mathbf{j}, -\mathbf{B}) \qquad (2)$$

with $\mathbf{j}$ the current density and $\mathbf{B}$ the magnetic field. Similarly, this holds for the elements of the resistivity tensor $\rho(\mathbf{j}, \mathbf{B}) = (\sigma(\mathbf{j}, \mathbf{B}))^{-1}$. Expanding $\rho_{kl}(\mathbf{j}, \mathbf{B})$ under consideration of Onsager's relation and imposing parity symmetry yields (up to second order terms)

$$\rho^{D/L}(\mathbf{I}, \mathbf{B}) = \rho_0\{1 + \alpha \mathbf{j}^2 + \beta \mathbf{B}^2 + \zeta^{D/L} \mathbf{j} \cdot \mathbf{B}\} \qquad (3)$$

which is equivalent to (1) as far as eMChA is concerned. Therefore for both cases, symmetry considerations show that eMChA is an allowed effect in chiral conductors. Note that higher order odd-power terms of the form $\mathbf{j}^{2n+1} \cdot \mathbf{B}^{2m+1}$ (n, m integer) were omitted for simplicity. Recent experiments on macroscopic chiral conductors [12] confirmed the validity of (3).

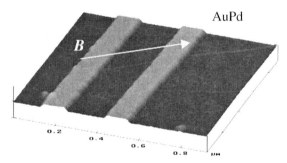

FIGURE 1. Shown is an AFM image of a typical nanotube sample. The external magnetic field B is applied along the SWNT symmetry axis to within a few degrees. The AuPd electrodes are 16 nm in height, 125 nm in width and 195 nm apart. The height of the thin nanotube bundle is about 1.8 nm. For sample preparation see [8].

EXPERIMENTAL RESULTS

The measurements of $\zeta^{D/L}$ have been carried out on two-terminal contacted SWNT bundles with B (magnetic field) along the bundle axis (Fig. 1). Only SWNTs with room-temperature resistances smaller than 13 kΩ, have been investigated implying both a small contact resistance, and a metallic character of the SWNT (resistance values did not change by more than 10% upon cooling to 4.2 K). The resistance anisotropy $\delta R(B,I) \equiv R(B,I) - R(B,-I)$ is determined by standard lock-in techniques. Due to unintentional, small differences between the two metal-tube contacts, $\delta R(B,I)$ can have a contact resistance anisotropy contribution. Such a contribution can however only have an even magnetic field dependence. For elimination, we determine the resistance anisotropy difference $\Delta R(B,I) \equiv \delta R(B,I) - \delta R(-B,I)$. In Fig. 2 the $\Delta R(B,I)$ of the tube in Fig. 1 is plotted as a function of $|B|$ (inset $\delta R(B,I)$ vs. B).

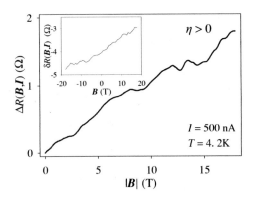

FIGURE 2. $\Delta R(B,I)$ vs. $|B|$ of the SWNT sample in Fig. 1 at $I = 500$ nA and $T = 4.2$ K . In the inset $\delta R(B,I)$ vs. B is shown. The linear term in the eMChA determines the shape of the curve and only small contributions from higher order terms are present.

The term linear in B is clearly dominant and only small contributions from the terms $\propto B^2$ are apparent, which is ascribed to contact artefacts. The absolute value of $\zeta^{D/L}$ can be estimated to be about 500 kΩ/AT. As eMChA slope $\eta \equiv \lim_{B \to 0} \partial \Delta R(B,I)/\partial B$ can be defined, yielding here $\eta > 0$. Samples with $\eta < 0$ were also observed and the abundance of opposite signs of η was found to be equal. The two signs of η are assigned to the two handednesses of the SWNTs [13]. At this time, it is not possible to determine which sign corresponds to which handedness.

THEORETICAL CONSIDERATIONS

In order to obtain some indication for the microscopic origin of the eMChA observed in CNTs, the simple model of a free electron on a helix [14] was extended in the sense that an externally applied magnetic field oriented parallel to the longitudinal axis of a CNT was introduced [15]. In Fig. 3 the mapping of a CNT to a helix is shown.

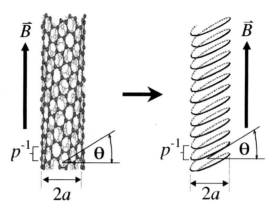

FIGURE 3. Mapping of a CNT to the helix model with $p^{-1} = 2\pi b$.

The Hamiltonian of this system is given by

$$H = \frac{\hbar^2}{2m(a^2 + b^2)}\frac{\partial^2}{\partial\varphi^2} - i\hbar\frac{qa^2}{2m(a^2 + b^2)}B\frac{\partial}{\partial\varphi} + \frac{q^2a^2}{8m}B^2 \qquad (4)$$

and can be solved analytically. First consider the case of diffusive charge transport: using the constant relaxation time approach to calculate the conductivity $\sigma(j_z,B)$, with j_z the current traversing the helix. Taking into account that the components of the average velocity obey $<v_z> = (b/a)<v_\varphi>$ (z: longitudinal component, φ: cyclic component) one finds for the energy

$$E_n = \frac{\hbar^2 n^2}{2mN^2(a^2 + b^2)}\frac{\partial^2}{\partial\varphi^2} + \frac{\pi N a^2}{2}j_{z,n}\cdot B + \frac{q^2a^2}{8m}B^2 \qquad (5)$$

where $j_{z,n}$ is just the standard quantum mechanically calculated current density [15]. The first term on the right-hand side of (5) is the zero-field energy $E_{0,n}$, the third the diamagnetic contribution and the second term describes the eMChA energy E_{eMChA}. On that basis the relative eMChA contribution can be defined by

$$\xi_{dif} = \frac{\sigma(j_{z,n},B) - \sigma(-j_{z,n},B)}{\sigma(j_{z,n},B) + \sigma(-j_{z,n},B)} \approx \frac{\pi a^2 N j_{z,n}\cdot B}{4E_{0,n}} \qquad (6)$$

For a ballistic conductor the contacts have to be taken explicitly into account. The conductance $G = (2e^2/h)\,T_{c\rightarrow h}\,T_{h\rightarrow c}$ has to be determined with $T_{c\rightarrow h}$ ($T_{h\rightarrow c}$) being the transmission probabilities from contact to helix (helix to contact). After straightforward analysis, the relative eMChA contribution ξ_{bal} is found to be proportional to E_{eMChA}/E_F (E_F: Fermi energy of cantact). Evaluation of ξ_{bal}/B and ξ_{dif}/B based on our experimental set-up, yields approximately $3\cdot10^{-7}$ and $6\cdot10^{-5}$ ($\xi_{experiment}/B \approx 10^{-4}$), respectively, i.e. $\xi_{bal} << \xi_{dif}$ [15]. This theoretical result based on the presented quantum-mechanical model is supported by the experimental observation that the eMChA is increasing with decreasing temperature [13].

SUMMARY

The main features of the eMChA could be observed in magnetotransport measurements on SWNTs. The clear observation of eMChA proves that electrical transport through CNTs is sensitive to chirality, and therefore must have a three-dimensional character, implying a cyclic current component. Furthermore, the results suggest a non-enantioselective production process of the CNTs.

Finally, the model of a free electron on a helix was applied to model the eMChA in SWNTs on a microscopic level. The experimental observations and theoretical findings agree qualitatively well.

REFERENCES

1. S. Iijima, *Nature* **56**, 354 (1991).
2. R. Saito, G. Dresselhaus, M.S. Dresselhaus, in *Physical Properties of Carbon Nanotubes*, (Imperial College Press, London, 1998).
3. A. Joria, R. Saito, C.M. Lieber, M. Hunter, T. McClure, G. Dresselhaus, and M.S. Dresselhaus, *Phys. Rev. Lett.* **86**, 1118-1121 (2001).
4. J.W.G Wildoer, L.C. Venema, A.G.Rinzler, R.E. Smalley, and C. Dekker, *Nature* **391**, 59 (1998). W. Clauss, M. Freitag, D.J. Bergeron, and A.T. Johnson, in *Electronic Properties of Novel Materials - Science and Technology of Molecular nanostructures*, XIII International Winterschool, Kirchberg, 1999, AIP Conference Proceedings 486, Eds: H. Kuzmany, J. Fink, M. Mehring, S.Roth, p. 308-312.
5. D.H. Cobden, M. Bockrath, P.L. McEuen, A.G. Rinzler, and R.E. Smalley, *Phys Rev. Lett.* **81**, 681 (1996).
6. M. Bockrath, D.H. Cobden, J. Lu, A.G. Rinzler, R.E. Smalley, L. Balents, and P.L. McEuen, *Nature* **397**, 598 (1999).
7. S. Frank, P. Poncharal, Z.L. Wang, and W. de Heer, *Science* **280**, 1744 (1998).
8. V. Krstić, S. Roth, and M. Burghard, *Phys. Rev. B* **62**, R16353 (2000).
9. Y. Miyamoto, S.G. Louie, and M.L. Cohen, *Phys. Rev. Lett.* **76**, 2121-2124 (1996).
10. G.L.J.A. Rikken, E. and Raupach, *Nature* **390**, 493-494 (1997).
11. G.L.J.A. Rikken, E. and Raupach, *Phys. Rev. E* **58**, 5081-5084 (1998).
12. G.L.J.A. Rikken, J. Fölling, and P. Wyder, *Phys. Rev. Lett.* **87**, 236602 (2001).
13. V. Krstić, S. Roth, M. Burghard, K. Kern, G.L.J.A. Rikken, *J. Chem Phys.* **117**, 11315 (2002).
14. I. Tinoco, and R.W. Woody, *J. Chem Phys.* **40**, 160 (1964).
15. V. Krstić, and G.L.J.A. Rikken, *Chem. Phys. Lett.* **364**, 51-56 (2002).

CHARACTERIZATION OF CARBON NANOTUBES

Dispersive Bands in Graphite and Carbon Nanotubes

A. Jorio*, M. A. Pimenta*, M. Souza*, C. Fantini*, Ge. G. Samsonidze†, G. Dresselhaus†, M. S. Dresselhaus†, A. G. Souza Filho** and R. Saito‡

*Universidade Federal de Minas Gerais, Belo Horizonte-MG, 30123-970 Brazil
†Massachusetts Institute of Technology, Cambridge-MA, 02139-4307 USA
**Universidade Federal do Ceará, Fortaleza-CE, 60455-760 Brazil
‡Tohoku University and CREST JST, Sendai 980-8578, Japan

Abstract. Dispersive features observed in graphite, carbon nanotube bundles and isolated single-wall carbon nanotubes are discussed. Special attention is given to the effects of the van Hove singularities in 1D nanotubes.

INTRODUCTION

The so-called disorder-induced (D) band was the first dispersive feature in the Raman spectra of graphite-like materials to receive serious attention from the physics and chemistry community.[1] The D-band dispersive behavior is now well established as arising from a double resonance mechanism [2, 3] that enhances the probability of Raman scattering events by phonons with wavevector $q \sim 2k$ (k and q are, respectively, the wavevectors for the electron and phonon participating in the Raman scattering process measured from the nearest K point in the 2D Brillouin zone (BZ)). This mechanism was extended to all phonons in the Brillouin zone of graphite in Ref. [4], showing that under special resonance conditions Raman spectroscopy can be used to probe the entire Brillouin zone. Application of this extended double resonance scattering mechanism for special resonance conditions has been used to explain unusual effects, such as Stokes/anti-Stokes frequency asymmetries, [5, 6] previously observed in graphite-like materials.

This paper discusses the behavior of several dispersive features observed in graphite, single-wall carbon nanotube (SWNT) bundles and isolated SWNTs, special attention given to the effect of van Hove singularities in 1D nanotubes. The selection rules for the resonance Raman scattering in carbon nanotubes are considered.

RESULTS AND DISCUSSIONS

The discussion presented here is based on experimental Raman spectra acquired under normal conditions of temperature and pressure, using different Raman lasers (Ti:Sapphire, Argon, Krypton, Helium-Neon). We measured single-wall carbon nanotube (SWNT) bundles grown by the arc method, and isolated SWNTs grown on a

CP685, Molecular Nanostructures: XVII Int'l. Winterschool/Euroconference on
Electronic Properties of Novel Materials, edited by H. Kuzmany, J. Fink, M. Mehring, and S. Roth

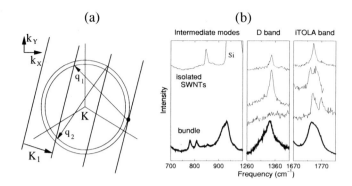

FIGURE 1. (a) Schematic picture for the phonon wavevectors q satisfying multiple-resonance conditions for the 1D Brillouin zone of a nanotube displayed in the context of the 2D Brillouin zone. In a 2D material, any q vector connecting the inner and outer circles would satisfy the double resonance. The largest contribution to the Raman spectra comes from phonons connected to van Hove singularity positions, indicated by the bullet. (b) Raman spectra from the intermediate frequency range, the D band and the combination mode iTO+LA for SWNT bundles and for different isolated SWNTs (we have observed intermediate modes only once for isolated SWNTs) obtained with the same $E_{laser} = 2.41$ eV.

Si/SiO$_2$ substrate by the CVD method.

Effect of van Hove singularities

The dispersive Raman modes in graphite-like materials originate from the E_{laser} dependent phonons that satisfy momentum and energy conservation for a multiple-resonance scattering event.[2, 3, 4] The effect of lowering the dimensionality from a 2D graphene sheet to a 1D nanotube represents, in principle, a confinement of the allowed electron and phonon states in one of the dimensions [see cutting lines in Fig. 1(a)]. However, this confinement effect has even larger consequences, when we consider that the Raman signal comes mostly from resonance states close to the 1D van Hove singularities (vHSs) in the density of electronic scates (DOS), where a 1D cutting line touches an equi-energy contour in the 2D Brillouin zone [see bullet Fig. 1(a)].

Figure 1(b) shows the Raman spectra from dispersive bands in carbon nanotubes (both isolated SWNTs and SWNT bundles). While SWNT bundles exhibit broad features that mostly reproduce the spectra also observed in graphite, the isolated SWNT Raman spectra show a clear and strong dependence on nanotube structure (n,m), that is a result of the multiple resonance process associated with the van Hove singularity positions in the BZ varying with (n,m) indices.

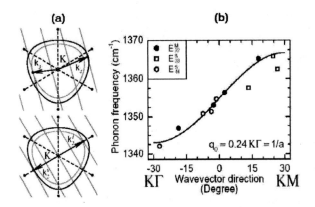

FIGURE 2. (a) Cutting lines for the semiconducting $(17,6)$ and the metallic $(27,0)$ SWNTs. Note that the wavevectors for van Hove singularities depend on both the chiral angle and the specific resonant electronic transition energy E_{ii}. (b) Dependence of phonon frequency on the q direction for a given wavevector magnitude q_0. The frequency difference between phonons for q along KM and $K\Gamma$ provide a measure of the trigonal warping effect for phonons in 2D graphite $\Delta\omega_{ph} = 24\,\mathrm{cm}^{-1}$.[7]

Confinement along the phonon wavevector direction in 1D systems

The momentum conservation requirement for achieving double resonance in a 2D system is responsible for the E_{laser} dependent selection of the magnitude of q (where q is the phonon wavevector in the 2D BZ) for the phonons involved in the scattering process. Therefore in 2D graphite, for a given E_{laser}, the *magnitude* of q is uniquely selected but there is no resonance requirement related to the *direction* of the q vector in the Brillouin zone.

The 1D structure of carbon nanotubes, however, leads to the quantum con£inement of electrons and phonons, thus making the double resonance Raman process selective not only of the *magnitude* of q, but also of the *direction* of q.[7] For example, nanotubes with similar diameters and different chiral angles could exhibit a double resonance mechanism that is equivalent to wave vectors along $K\Gamma$ or KM (and also along intermediate directions) in the 2D unfolded Brillouin zone, depending on the nanotube chirality (n, m) and on the resonant singularity E_{ii} associated with each (n, m) nanotube. Therefore, Raman measurements on isolated SWNTs with similar diameters but different chiral angles allow us to determine the *trigonal warping* of *phonons* in 2D graphite (see Fig. 2).

Selection rules for the Raman scattering

The selection rules for the first-order single-resonance Raman scattering process for a chiral SWNT (C_N symmetry) with Z as the SWNT axis direction and Y as the photon propagation direction can be simply represented by the following relations:[8]

$$E_\mu^v \xrightarrow{\parallel} E_\mu^c \xrightarrow{A(ZZ)} E_\mu^c \xrightarrow{\parallel} E_\mu^v$$

$$E_\mu^v \xrightarrow{\perp} E_{\mu\pm1}^c \xrightarrow{A(XX)} E_{\mu\pm1}^c \xrightarrow{\perp} E_\mu^v$$

$$E_\mu^v \xrightarrow{\parallel} E_\mu^c \xrightarrow{E_1(ZX)} E_{\mu\pm1}^c \xrightarrow{\perp} E_\mu^v$$

$$E_\mu^v \xrightarrow{\perp} E_{\mu\pm1}^c \xrightarrow{E_1(XZ)} E_\mu^c \xrightarrow{\parallel} E_\mu^v$$

$$E_\mu^v \xrightarrow{\perp} E_{\mu\pm1}^c \xrightarrow{E_2(XX)} E_{\mu\mp1}^c \xrightarrow{\perp} E_\mu^v$$

For resonance with VHSs, these selection rules imply that: (1) A modes can be observed for the (ZZ) scattering geometry for resonance with $E_{\mu\mu}$, and for the (XX) scattering geometry for resonance with $E_{\mu,\mu\pm1}$ (the letters between parenthesis denote, respectively, the polarization direction for the incident and scattered photons); (2) E_1 modes can be observed for (ZX), resonance of the incident photon with $E_{\mu\mu}$, or resonance of the scattered photon with $E_{\mu,\mu\pm1}$, and for (XZ), resonance of the incident photon with $E_{\mu,\mu\pm1}$, or resonance of the scattered photon with $E_{\mu\mu}$; (3) E_2 modes can only be observed for (XX), resonance with $E_{\mu,\mu\pm1}$.

For the double resonance process, the phonon wavevectors $\mathbf{q_i}$ are related to the electron wavevectors $\mathbf{k_i}$ by $\mathbf{q_i} = -2\mathbf{k_i} + \mu\mathbf{K_1}$, where i is the vHS index, $\mathbf{K_1}$ is the separation between two adjacent cutting lines (see Fig.1(a)) in the unfolded 2D Brillouin zone, and $\mu = 0$ for ZZ, $\mu = 0, \pm1, \pm2$ for ZX and XZ, and $\mu = 0, \pm2$ for XX polarizations of the incident and scattered photons.[7] A study of the experimental evidence for the polarization dependence of the double resonance features in carbon nanotubes is now under way.

ACKNOWLEDGMENTS

This work has been supported by the Nanoscience Institute, Brazil in Belo Horizonte. The MIT authors acknowledge support under NSF Grants DMR 01-16042 and INT 00-00408. R.S. acknowledges a Grant-in-Aid (No. 13440091) from the Ministry of Education, Japan.

REFERENCES

1. M. S. Dresselhaus, G. Dresselhaus, and P. C. Eklund, *Science of Fullerenes and Carbon Nanotubes* (Academic Press, New York, NY, San Diego, CA, 1996).
2. A. V. Baranov et al., Opt. Spectrosk. **62**, 1036 (1987). [Opt. Spectrosc. **62**, 612 (1987)].
3. C. Thomsen and S. Reich, Phys. Rev. Lett. **85**, 5214 (2000).
4. R. Saito et al., Phys. Rev. Lett. **88**, 027401 (2002).
5. L. G. Cançado et al., Phys. Rev. B **66**, 035415 (2002).
6. Tan et al., Phys. Rev. B **66**, 245410 (2002)
7. Ge. G. Samsonidze et al., Phys. Rev. Lett. **90**, 027403 (2003).
8. A. Jorio et al., Phys. Rev. Lett. **90**, 107403 (2003).

First Observation Of The Raman Spectrum Of Isolated Single-Wall Carbon Nanotubes By Near-Field Optical Raman Spectroscopy

J. Schreiber*[1], F. Demoisson[2], B. Humbert[2], G. Louarn[1], O. Chauvet[1], S. Lefrant[1]

1 IMN-LPC, Institut des Matériaux Jean Rouxel, CNRS/Nantes University, 443222 Nantes, France
2 LCPME,UMR 7564 CNRS - Université Henri Poincaré, Nancy I, France
Joachim.Schreiber@cnrs-imn.fr

Abstract. In this paper, we report the first observation of the Raman spectrum of isolated single-wall carbon nanotubes (SWNTs) by near field optical Raman spectroscopy. Raman spectra have been obtained by using an aperture-based scanning near field optical microscope (SNOM) which is coupled to a T64000 Jobin Yvon spectrometer in collection mode. Surface Enhanced Raman Scattering (SERS) effect is needed in order to amplify the scattered intensity.

1. INTRODUCTION

Since their discovery in 1993, single-wall carbon SWNTs have been the focus of intense interest [1]. Raman scattering has proven to be very useful tool to characterize these materials. Nevertheless, this technique has a major drawback to analyze nanometric materials like SWNTS. The reason is that, nowadays, Raman spectrometers are working with common optical microscopes whose spatial resolution is limited by diffraction to roughly ~ 500 nm. To overcome this limit a lot of efforts are put into methods to disperse and observe individual nanotubes [2,3]. In this paper, another approach is shown with the first Raman spectra of individual SWNTs taken with an aperture SNOM-Raman (Scanning Nearfield Optical Microscope) spectrometer in the near field.

2. DESCRIPTION OF THE SNOM-RAMAN SPECTROMETER

The SNOM spectrometer is composed of two parts, the SNOM microscope and the spectrometer itself (T64000 Jobin Yvon). E. Synge proposed the principle of SNOM [4] in 1928 which was built for the first time by Pohl et al. [5] in 1984. It can be compared to another near field microscope, the Atomic Force Microscope (AFM). The SNOM takes images scanning line per line of a sample while a computer shows the global information (the surface structure and the optical information). In difference to an AFM, this SNOM uses a metallized fiber-optical taper as a probe and a "shear-force mechanism" to detect the near field. The home-made near field microscope is

CP685, *Molecular Nanostructures: XVII Int'l. Winterschool/Euroconference on Electronic Properties of Novel Materials,* edited by H. Kuzmany, J. Fink, M. Mehring, and S. Roth

described elsewhere [6]. The Raman excitation source is a Spectra-Physics laser Stabilite 2017 ($\lambda = 514.5$ nm). The spectrometer used in this study is a Jobin-Yvon Raman T64000, with a liquid nitrogen cooled CCD multichannel detector. An holographic NOTCH filter or a double subtractive stage with gratings with 1800 grooves mm^{-1} is used to remove the Rayleigh scattering. A single dispersion stage with a 1800 grooves mm^{-1} grating is placed in front of the detector. The spectral resolution, with an excitation at 514.5 nm, is about 2.7 cm^{-1}. The confocal Raman microprobe is constituted by an Olympus microscope, equipped with a motorized XY stage with a step of 80 nm.

To couple the SNOM with the spectrometer, the Olympus microscope is focused on the cleaved end of the optical fiber used as probe. In order to obtain a good signal intensity, a x50 objective is used which matches the numerical aperture of the monomode fiber. The metallized optical fiber probe is provided by Veeco.

3. SAMPLE PREPARATION

SNOM-Raman experiments have already been successfully performed, either on fluorescing samples like Rhodamin, on silica which has a very strong Raman cross-section [7] or very recently by A. Hartschuh on SWNTs but with a detection in the farfield [8]. The cross-section of SWNTs is strong too but not on the same order of magnitude. To amplify the Raman signal optimized SERS surfaces are used [9].

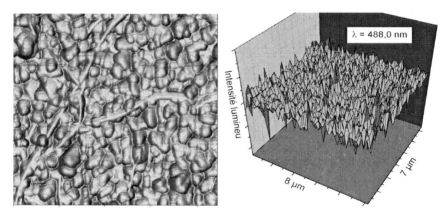

FIGURE 1. On the left: Sample of SWNTs deposited on a silver SERS surface, characterized by AFM. On the right: Optical image of a typical SERS surface taken with the SNOM microscope. The "hot spots", in dark, are the regions with very high electromagnetic field amplification due to surface plasmons.

Purified SWNTs from Tubes@rice are dispersed and sonicated in an ethanol solution. A drop of this solution is put on a SERS substrate. To accelerate the evaporation process and to limit the re-aggregation of SWNTs, the substrate is put in a vacuum chamber. It was found that this was the only way to maintain SWNTs stuck sufficiently strong to the surface, which are then not pulled by the tip, scanning the sample. In figure 1, several bundles of SWNTs can be observed on such a sample, which is characterized by AFM. On the right side of figure 1, we show the optical signal obtained by SNOM of a SERS surface without SWNTs, for purpose of comparison.

In all experiments described here, the sample is illuminated from the far field and the optical signal is detected by the fiber probe in the near field. It can be observed that the local field enhancement due to surface plasmons is very localized in "hot spots", the peaks being much smaller than the optical resolution of the SNOM.

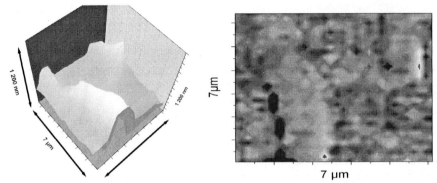

FIGURE 2. On the left: sample of SWNTs deposited on a silver SERS surface, characterized by SNOM; on the right: the corresponding optical image.

4. RESULTS AND DISCUSSION

In order to detect individual nanotubes, the idea is to follow a "big" bundle until small bundles and individual SWNTs are located. This is the only way to proceed because the resolution of the SNOM is not high enough to detect individual SWNTs.

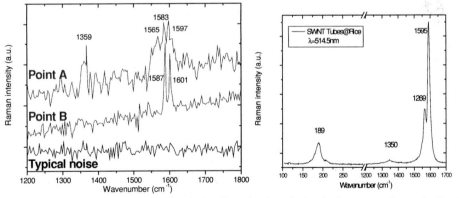

FIGURE 3. On the left: Raman spectra of a bundle, point A; an isolated tube or two tubes, point B and the SERS surface λ = 514,5 nm, point C. On the right: Raman spectra of the same SWNTs in buckypaper at λ = 514,5 nm, recorded in standard conditions.

Then Raman spectra are recorded on the big bundle and next to it. In figure 2, two thick SWNTs bundles can be observed. On the left side, the topographical image is shown and on the right hand side the corresponding optical image. It is detected by a photomultuplier and saved at the same time as the topographical image is taken. Therefore correlations can be made. This advantage is used for the experiments. The optical intensity is observed during the scan. Once the intensity is high and the tip is on or near a bundle, the scan of the sample is stopped and a Raman spectrum is taken. In figure 3, different Raman spectra can be observed. The spectrum "Point A" has been taken on a big bundle with a 300 groove grating. This seems to be very similar to the spectra obtained from the far field. The spectrum in "Point B" has been taken with the SNOM tip close to the big bundle. In this case, we have used the high resolution. This spectrum shows distinct sharp lines with a linewidth of about 4 cm^{-1}. We propose that this Raman spectrum is issued from either one or two individual nanotubes. The typical G mode is upshifted by 5 cm^{-1} but this can be explained by the use of the SERS surface. It is known that these surfaces induce an upshift of all the Raman bands of well dispersed SWNTs. On the other hand, one can not be sure that these bands have the same origin as those observed on SWNTs in the far field. It has to be taken into account that the probe is only 10 nm away from the sample and that therefore, the usually applied selection rules have to be reconsidered. It could be that these bands are infrared bands which became active under these experimental conditions.

Another important thing could be shown with these experiments. It has been possible to record Raman spectra in the near field from SWNTs only when they coincide with "hot spots" with high Raman amplification. This effect has also been observed by Azouly et al. [3].

5. CONCLUSION

The first Raman spectrum of SWNTs taken by an aperture SNOM-Raman has been shown, when "hot spots" coincide with SWNTs, as a way to have a sufficiently strong signal. Nevertheless, further investigations are necessary to clarify the origin of these bands observed in SNOM Raman spectroscopy.

6. REFERENCES

1. Iijima S. and Ichihashi T., *Nature*, **363**, 603 (1993)
2. G.S. Duesberg, I.Loa, M. Burghard, and S. Roth, *Phys. Rev. Lett.*, **85**(25), 5436-5439 (2000)
3. J. Azouly, A. Débarre, A. Richard, P. Tchénio, *Chem. Phys. Lett.*, 331, 347--353 (2000).
4. E. Synge, *Phil. Mag.* 6, 356 (1928)
5. Pohl et al., *Appl. Phys. Lett.* 44, 651, 1984
6. J.Grausem, B. Humbert, A. Burneau, *Appl. Phys. Lett.* 70, 1671 (1997)
7. Zeisel D., Deckert V.,R. Zenobi, Vo-Dinh T., *Chemical Physics Letters* 283, pp 381-385 (1998)
8. A. Hartschuh, E. J. Sanchez, X. Sunney Xie, and L. Novotny, **90**(5), 095503-1 (2003)
9. J. Schreiber, Ph. D thesis, University of Nantes, 2002, unpublished.

Controlled Oxidation of Single-Wall Carbon Nanotubes: A Raman Study

Ferenc Simon, Ákos Kukovecz, Hans Kuzmany

Universität Wien, Institute für Materialphysik, Strudlhofgasse 4, 1090, Wien, Austria

Abstract. Oxidation of single wall carbon nanotubes using H_2O_2 is a common purification and tube opening procedure. We studied the effect of oxidation in well controlled conditions on SWNT samples with different mean diameters, prepared by laser-ablation, CVD and the HiPco process. Detailed multifrequency Raman spectroscopy evidences that oxidation damage depends strongly on the mean tube diameter, damage occuring to small tubes first and successively for larger diameter tubes. We use the peapod filling of the above samples as a control to what extent the tubes are open to the C_{60} molecules.

INTRODUCTION

Related to the synthesis method, SWCNT samples contain inevitable catalytic particles as well as of non-desired carbon compounds. SWCNTs are known to be resistant to oxidation better than any other carbon modifications or metallic particles. This inspires the purification of SWCNTs with oxidating treatment such as refluxing in H_2O_2 or heat treatment in air. Naturally, a reasonable balance have to be achieved when non-wanted side products or catalytic particles are nominally removed and the desired SWCNT sample remains yet in sufficient abundance. Another important aspect of the oxidation is the opening up of the SWCNTs for the environment thus making the encapsulation with materials such as alkali halides or C_{60} possible[1]. Here, we report a systematic gravimetric and Raman study of oxidation of different SWCNT samples using H_2O_2. Raman spectroscopy and in particular multifrequency Raman measurements have been proven to be crucial in the characterization of several properties of SWCNTs[2]. The level of oxidation is controlled by the dilution of the H_2O_2 aqueous solution. We found a strong dependence of oxidation on the nominal tube diameter of the samples, the small tubes being oxidized first, most probably from the tube ends.

EXPERIMENTAL

We studied 3 different SWCNT samples from commercial sources: laser ablation, LA, (Tubes@Rice, Rice University, Houston, Texas), CVD (Nanocyl, Namur, Belgium) and HiPco (Carbon Nanotechnolgies, Houston, USA). The purities as provided by the manufacturers are summarized in Table 1. Oxidation was studied by 2 hours refluxing of the SWCNT in diluted H_2O_2 aqueous solutions. After refluxing, the

CP685, *Molecular Nanostructures: XVII Int'l. Winterschool/Euroconference on Electronic Properties of Novel Materials,* edited by H. Kuzmany, J. Fink, M. Mehring, and S. Roth
© 2003 American Institute of Physics 0-7354-0154-3/03/$20.00

material was filtered and the resulting bucky paper was dried at 140 °C in air. Comparability and reproducibility was assured by the use of the same amount of starting SWCNT material, 15 mg, and the same volume for the aqueous solution, 30 ml, for all treatments. The unit of treatment is defined as 1mg SWCNT in 1 ml of 30 % aqueous H_2O_2 and 1 ml of distilled water. Numbers above unity are achieved by repeating the refluxing, filtering and drying steps. Peapod filling was done following the step in Ref. 3. The peapod concentration in the resulting materials was determined from the Raman signal of the C_{60} $A_g(2)$ following Ref. 4.

TABLE 1. Properties of the studied SWCNT materials

Material	Purity (%)	Mean diam. (nm)	Peapod concentration (%)
HiPco	70	0.98	0
CVD	70	1.20	10
LA	15	1.34	30

RESULTS AND DISCUSSION

Figure 1. shows the weight loss of the SWCNT materials after the described oxidation procedure. Weight loss is generally considered as a poorly controllable quantity since SWCNT materials are known to absorbe an ill defined and different amount of solvents. This originates in the different morphology of the SWCNT bundles. In order to reduce systematic errors related to this, we performed a control experiment, where no H_2O_2 was added to the distilled water. As shown in Fig. 1., the final and starting masses are in a close agreement, which supports the validity of the current gravimetric studies. A surprising observation in Fig. 1. is the complete dissapearance of the HiPco material at treatment unit 1, whereas the LA and CVD compounds seem to tend to a constant value of the final to initial mass ratio. This may originate in two facts: i.) the overally smaller diameters of the HiPco material may be more susceptible for the oxidation treatment than the larger diameter tubes present in the LA and CVD material (see. Table 1.); ii.) it is also possible that a different morphology related to the catalytic particles and the SWCNT may give rise to the difference. This later proposal is based on the fact the for the LA and CVD grown

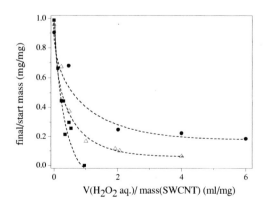

FIGURE 1. Mass loss of SWCNT samples after H_2O_2 treatment (;: HiPco, ●: laser ablation, ': CVD). Dashed lines are guides to the eye.

materials, the SWCNTs are known to grow out of larger catalytic particles, whereas the HiPco based SWCNT are known to be attached to smaller catalytic particles as well as to encapsulate the catalyst itself. The close presence of catalyst particles may catalyze the decay of H_2O_2 into the nascent oxygen reaction agent thus helping the oxidation of nanotubes. In what follows we focus our attention on the HiPco material in order to clearify the full dissappearance of the material with treatment. The peapod concentration attests to what level the SWCNT samples are opened, and whether the tube diameters are large enough to accommodate C_{60}. Our results for the studied samples are summarized in Table 1. Clearly, even when significant number of tube openings are present, the HiPco tubes are too small for a detectable level of C_{60} encapsulation.

Figure 2. shows the RBM mode region of the Raman spectra of the HiPco sample with the oxidation steps defined on Figure 3a. at $\lambda = 488$ nm. Changes associated with the treatment are observed: RBM lines at higher Raman shifts vanish with increasing treatment. SWCNT with larger Raman shifted RBM lines correspond to the thinner tubes as $\nu_{RBM} \vartheta 1/d$, where d is the tube diameter. The evident disappearance of thin SWCNT in the HiPco sample is quantitatively described when the mean tube diameter, d, and its variance, σ, is calculated assuming a monomodal distribution of the tube diameters following Ref. 5. Figure 3b. summarizes the data at 488 and 647 nm: the mean tube diameter gradually decreases that is accompanied by the narrowing of the diameter distribution function. This is understood as oxidation happening to the smallest nanotubes first. A simple calculation has shown that the shifting of d to higher values and the narrowing of the distribution can account for the ~80 % weight loss observed for the '6' treatment.

The Raman D and G modes also hold significant information about the oxidation damage. In Figure 4a. and b. we show the Ramand D and G modes at $\lambda = 488$ nm. The

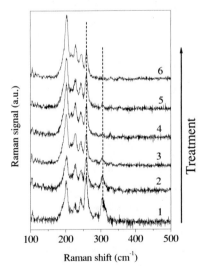

FIGURE 2. Raman spectrum of the oxidized HiPco sample at $\lambda = 488$ nm. Dashed lines indicate Raman lines that vanish with increasing treatment.

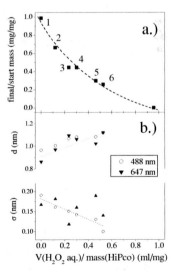

FIGURE 3a. Enlarged view and labelling of the oxidation of the HiPco sample. 3b. Mean and width of the assumed monomodal diameter distribution of the HiPco sample at different laser freqeuncies. Dashed lines are guide to the eye.

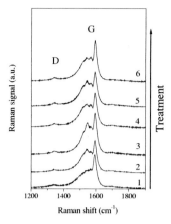

Figure 4a. The Raman D and G modes of the HiPco samples at λ=488 nm.

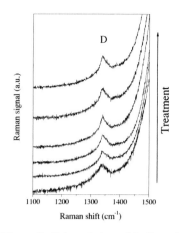

Figure 4b. Enlarged view of the D mode.

disappearance of the small, metallic tubes is evident as the Fano contribution to the G mode gradually dissappears with increasing treatment. The lineshape has been analyzed in detail previously[5]. However, no significant changes are observed for the defect sensitive D mode as shown in Fig. 4b. This evidences that oxidation has no effect on the tube side walls and only tube ends are damaged, no matter how substantial the damage is.

CONCLUSIONS

The oxidation of LA, CVD and HiPco SWCNT samples was studied. We found that oxidation happens to small nanotubes first, starting from the tube ends. This results in a very rapid dissappearance of the HiPco material with oxidation as it consists of nominally small diameter tubes.

ACKNOWLEDGMENTS

This work was supported by the FWF I Austria, Project No. 14983 and by the EU RTN FUNCARS (HPRN-CT-1999-00011). The authors acknowledge the help of Ch. Kramberger in the analysis of SWCNT diameter distribution.

REFERENCES

1. M. Monthioux, *Carbon*, **40** 1809 (2002).
2. H. Kuzmany *et al.*, *Eur. Phys. J B*, **22** 307 (2001).
3. H. Kataura *et al.*, *Synth. Met.*, **121**, 1195 (2001).
4. R. Pfeiffer *et al.*, *Appl. Phys. A*, **76** 1 (2003).
5. Á. Kukovecz *et al.*, *Eur. Phys. J B*, **28** 223 (2002).

Vibrational Spectroscopy Studies of Single-Walled Carbon Nanotubes Subjected to Different Levels of Nitric Acid Oxidation Treatment

Paul Whelan[1], Jürgen Abraham[1], Anderas Hirsch[1], Ralph Graubner[2],
Adrea Dizakova[2], Frank Hennrich[3], Manfred Kappes[3], Jeffery Forsyth[4]

[1]Institute of Organic Chemistry, Henkestraße 42, Erlangen, Germany
[2]Institute of Technical Physics, Erwin Rommel Straße 1, Erlangen, Germany
[3]Institute of Physical Chemistry, Kaiserstraße 12, Karlsruhe, Germany
[4]Materials Research Institute, Howard Street 1, Sheffield, England

Abstract. SWCNTs produced via a laser ablation process were oxidized using different nitric acid strengths and various treatment times. The oxidized samples were characterized using Raman spectroscopy to investigate the level of defects introduced into the material for the different treatment types. FTIR was employed to identify the various functional groups in/on the tube material after the treatment. A larger increase in the defects was observed for samples for a period of 24 to 72 hours using 3M HNO_3 than for samples treaded for 1 to 2 hours with 10M acid. Bands due to $-NO_2$ and -COOH groups were observed in the FTIR spectra.

INTRODUCTION

In crude material the open ends of SWCNTs tubes are often closed by catalytic particles and other impurities. These catalytic particles can be removed by oxidative treatment with, for example, HNO_3, which results in ends largely decorated with carboxyl groups. Side-wall defects can also be introduced under such conditions. Defects in SWCNTs are important in the covalent chemistry of the tubes because they can either serve as anchor groups for further functionalization, or be created by the covalent attachment of further groups [1]. Defects are therefore a promising starting point for the development of SWCNTs functionalization. However, a limited number of defects may be introduced before a sample loses its special electronic and mechanical properties. In the present work SWCNTs were produced by laser ablation using Ni and Co metal catalysts. The SWCNT material was subjected to various HNO_3 treatments and the levels of defects introduced were monitored by vibrational spectroscopy. Raman spectroscopy is used to identify the introduction of defects into the nanotube material and FTIR is used to identify if the defects are due to the introduction of $-NO_2$/carboxyl groups and/or the production of amorphous carbon from the destruction of the SWCNTs, due to the oxidation treatment.

RESULTS AND DISCUSSION

The as prepared material was subjected to different types of HNO_3 treatments, as shown in table 1. These particular treatments were chosen to compare the effects of treating the SWCNTs with a strong

CP685, *Molecular Nanostructures: XVII Int'l. Winterschool/Euroconference on
Electronic Properties of Novel Materials,*edited by H. Kuzmany, J. Fink, M. Mehring, and S. Roth
© 2003 American Institute of Physics 0-7354-0154-3/03/$20.00

acid for a short period of time (samples 2,3 and 4 in table 1) with the effects of treatments with a weaker acid for a longer period of time (samples 5,6, and 7 in table 1).

TABLE 1. SWCNT Nitric Acid Oxidation Treatments.

Sample	Treatment
1	Starting Material (as prepared laser ablasion SWCNTs)
2	1h 10M HNO_3 100°C
3	1h 10M HNO_3 200°C
4	2h 10M HNO_3 200°C
5	24h 3M HNO_3 100°C
6	48h 3M HNO_3 100°C
7	72h 3M HNO_3 100°C

A minor increase in the Raman D band intensity was observed for samples 2,3 and 4 (spectra not shown), indicating a low increase in the defects due to the 10M acid treatments for 1 to 2 hours. However, a large increase in the Raman D band was observed for samples 5,6 and 7 (as shown in figure 1) treated with 3M acid for 24 to 72 hours. The G band positions, bandwidths and I_D/I_G ratios for these samples are compared to the starting material in table 2. There was a liner relationship between the increase in the I_D/I_G ratios and treatment time as shown in the plot in figure 2.

TABLE 2. G band positions, widths and I_D/I_G ratios for the Raman spectra of nitric acid oxidized SWCNT samples

Sample	G band position (cm^{-1})	G bandwidth (cm^{-1})	I_D/I_G Ratio
1	1594	17.3	0.029
5	1606	31.7	0.651
6	1603	39.9	0.915
7	1603	40.3	1.500

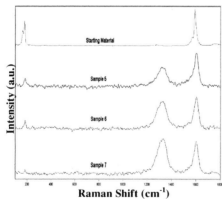

FIGURE 1. Raman spectra of the starting material and nitric acid oxidized SWCNT samples 5, 6 and 7 using 1064nm excitation energy

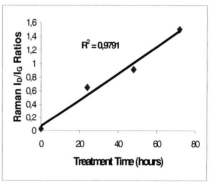

FIGURE 2. Raman I_D/IG Ratios vs 10M HNO_3 Treatment Time

A continual decrease of the RBM intensity was observed with an increase in sample treatment time. This indicated that in addition to introducing defects into the tube material that the treatment may also have destroyed the structure of the tubes and possibly produced amorphous carbon, which would have also contributed to the increase in the Raman D band. Of the two bands in the RBM region the intensity of the band at the higher position is more markedly reduced. This indicates that the smaller diameter tubes were more reactive, which is in agreement with the literature[2,3]. To determine if the reduction of the RBM mode is due to the destruction of tubes or a change in the electronic structure of the tubes due to the intercalation of the nitric acid groups it is necessary to anneal the treated samples and investigate if the RBM signal is recovered. This will be performed in our future work.

Fourier transform infrared (FTIR) spectra of the treated samples were taken to identify the introduction of –COOH and –NO$_2$ functional groups into the SWCNTs. The diffuse reflectance infrared Fourier transform (DRIFTS) and attenuated total reflectance (ATR) spectra are show in figures 3 and 4 respectively. In the DRIFTS spectrum of sample 5 the band at ~1640cm^{-1} is due to –NO$_2$ groups. In the ATR spectra the bands at ~2350cm^{-1} suspected to be due to atmospheric carbon dioxide. The band at ~3,300cm^{-1} could be due to –OH groups present from water present in the samples. However, as this band is larger more prevalent for sample 2 than sample 7 it is also possible that the shorter treatment time with the stronger acid has resulted in the introduction of more –COOH groups into the SWCNTs. Nevertheless, further investigation with both ATR and DRIFTS, of a wider range of samples, is required to obtain a clearer picture of the changes which have occurred.

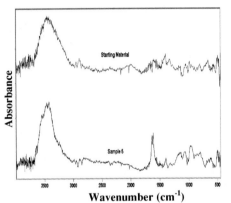

FIGURE 3. DRIFTS spectra of the starting material and sample 5

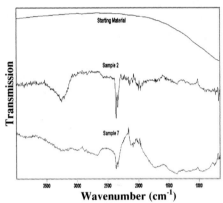

FIGURE 4. ATR spectra of the starting material and samples 2 and 7

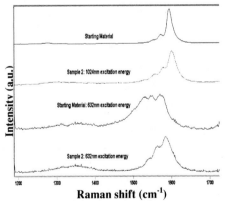

FIGURE 5. Raman spectra of the starting material and sample 5 using both 1064nm and 632nm laser excitation energy

In figure 5 the Raman spectra of the starting material and sample 2 are compared using both 1064 and 632nm excitation energies. 1064nm excitation is known to predominantly probe semi-conducting nanotubes and 632nm is known to probe predominantly metallic tubes [4]. A larger increase in the D band and a and a more dramatic change in TM of the metallic tubes, indicating that the metallic tubes were more reactive than the semi-conducting tubes.

CONCLUSIONS

The Raman spectra show that the HNO_3 treatments introduced additional defects into the nanotube material. Longer treatment times with lower concentration acid introduced greater defects than shorter treatment times with a stronger concentration acid.

The increase in the band at ca. $3300cm^{-1}$ for the ATR spectra indicated more there were possibly carboxyl groups present in the samples subjected to the shorter treatment time. This suggested that the large D band for the Raman spectra of samples 6,7, and 8 might be contributed to by amorphous carbon produced from the destruction of tubes.

The peak at ca. $1640cm^{-1}$ in the DRIFTS spectrum of sample 7 shows there were some NO_2 groups introduced into to the SWCNT material. No carbonyl band was observed. However, further work using DRIFTS is necessary to determine the level of defects introduced for different the different treatments. The Raman spectra of sample 5, using different laser excitation energies, shows that the acid treatment resulted in a larger increase in the D band and a more dramatic change in TM of the metallic tubes, indicating that the metallic tubes were more reactive than the semi-conducting tubes.

ACKNOWLEDGMENTS

We would like to acknowledge the support from the EU 5th Framework project entitled the Chemical Functionalization of Carbon Nanotubes (FUNCARS)

REFERENCES

1. Hirsch, A., *Angew. Chem. Int. Ed.* **41**, 1853-1858 (2002).
2 Zhou, W.; Ooi, Y. H.; Russo, R.; Papanek, P.; Luzzi, D. E.; Fischer, J. E.; Bronikowski, M. J.; Willis, P. A.; and Smalley, R. E. *Chem. Phy. Letters*, **350**, 6-14 (2001)
3. Zhonghua, Y and Brus, L, *J. Phys. Chem B*, **105**, 1123-1134 (2001).
3. Kukovecz, A.; Kramberger, C.; Holzinger, M.; Kuzmany, H.; Schalko, J.; Mannsberger, M. and Hirsch, A. *J. Phys. Chem. B*, **106**, 6374-6380 (2002)

Quantitative Analysis of Optical Spectra from Individual Single-Wall Carbon Nanotubes

Axel Hagen and Tobias Hertel

*Fritz-Haber-Institut der Max-Planck-Gesellschaft, Faradayweg 4-6,
D-14195 Berlin, Germany*

Abstract. We discuss how tight-binding band-structure calculations with a chirality- and diameter-dependent nearest-neighbor hopping integral and an energy independent offset parameter may be used to correlate the geometric structure of single-wall carbon nanotubes (SWNTs) with well resolved features in the UV-VIS-NIR absorption spectra . The assignment of (n,m) indices to interband transitions in specific tube types can support a quantitative analysis of absorption spectra which may eventually be used for rapid screening and optimization of sample composition during SWNT synthesis.

INTRODUCTION

The discovery of band gap fluorescence from semiconducting SWNT's [1] has recently attracted considerable attention because the associated absorption spectra exhibit a series of well-resoled absorption features that facilitate a structural assignment of spectral features to transitions in specific tube types [2,3]. Here, we discuss how quantitative information on the composition of SWNT's ensembles can be obtained from well resolved absorption spectra and show that an (n,m) assignment of absorption peaks is feasible without the availability of fluorescence data [3]. On the basis of tight-binding (TB) calculations we developed a two-parameter model which gives satisfactory agreement between theoretical and experimentally observed gap energies. Furthermore this model also partially accounts for the so-called "ratio problem" which is currently discussed in the literature [4].

EXPERIMENT

UV-VIS-NIR spectra were obtained from tubes synthesized by the high pressure CO decomposition technique (HiPCO-material). This synthesis technique yields SWNTs that are agglomerated in quasi-crystalline SWNT ropes of a few to tens of nanometers in diameter. Tubes in these samples have a mean diameter of 1.0 nm [5,6]. Individual SWNTs were isolated from HiPCO raw material in micelles by the technique described in ref. [1]. At wavelengths below 1350 nm the well resolved spectrum from micelles in H_2O was extended by results from measurements in D_2O.

CP685, *Molecular Nanostructures: XVII Int'l. Winterschool/Euroconference on
Electronic Properties of Novel Materials,*edited by H. Kuzmany, J. Fink, M. Mehring, and S. Roth
© 2003 American Institute of Physics 0-7354-0154-3/03/$20.00

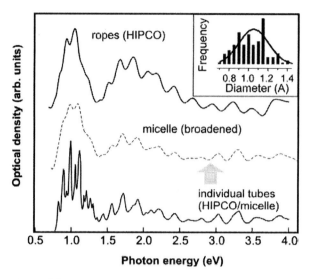

FIGURE 1. Comparison of background-corrected spectra from HiPCO rope material with that of individual tubes in micelles. The numerically broadened spectrum from individual tubes (dashed line) reveals a 50 meV blueshift with respect to the rope spectrum. The diameter distribution in the inset is thus characteristic of both materials and was obtained from refs. [5] (histogram) and [6] (solid line).

RESULTS AND DISCUSSION

For the calculation of optical properties we use a modified tight-binding band-structure that accounts for the misalignment of carbon p_z orbitals on the curved graphene surface of SWNTs [7]. This leads to the introduction of three geometrical factors depending on n and m into the TB band-structure calculation and γ_0^∞ denoting the nearest neighbor hopping integral of a tube with infinite radius .

In addition we included an n and m independent empirical offset parameter Ω, to account for the possibility that gap energies may be shifted to higher or lower energies due to tube-solvent interactions or many-particle effects. Especially the latter are known to play an important role in systems of reduced dimensionality and may be responsible for the so-called "ratio problem" [4]. This refers to the observation that theoretical gap ratios v_{22}^S / v_{11}^S based on one-electron theory as in the TB calculations should asymptotically approach 2 in the limit of large tubes radius whereas the experimentally observed ratios approach an asymptotic value closer to 1.75 [1,2]. If one assumes, as zero order approximation, that many-particle contributions are independent of subband index and tube type, one finds that the theoretical ratios ($v_{22}^S-\Omega$)/($v_{11}^S-\Omega$) can be mapped to the experimental ones by the appropriate choice of the offset parameter Ω.

We have previously devised a scheme which allows to target values for γ_0^∞ and Ω by optimizing the agreement between experimentally observed gap energies v_{11}^{expt} and

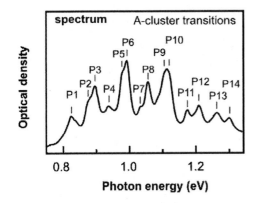

	E_{II}^{S}	Scenario I	Scenario II
γ_0^{∞} (eV)		2.998	2.634
Ω (γ_0^{∞})		-0.045	-0.072
P1	0.825	(13,3)	(10,6)
P2	0.875	(9,8),(14,3)	(9,7)
P3	0.896	(10,6)	(8,7)
P4	0.938	(9,7)	(10,5)
P5	0.977	(8,7)	(8,6)
P6	0.993	(9,5)	(11,3)
P7	1.032	(8,6)	(12,1)
P8	1.056	(12,1)	(9,4)
P9	1.105	(9,2)	(7,5)
P10	1.120	(9,4)	(10,2)
P11	1.174	(10,2)	(11,0)
P12	1.209	(11,0)	(8,3)
P13	1.262	(7,3)	(6,4)
P14	1.301	(8,3)	(9,1)

FIGURE 2. A-cluster region. The 14 gap energies E_{II}^{S} used for determination of the nearest-neighbour hopping integral γ_0^{∞} and energy offset parameter Ω are marked P1-P14.

Table 1. Experimental gap energies and peak assignment for the different scenarios discussed in the text.

calculated gap energies $\nu_{II}^{TB}(n,m)$ for tubes within a relevant diameter range (here, 0.55 nm to 1.35 nm) [3]. A crucial criterion for this analysis is that observed features should be derived by zone folding from the same graphene band-structure, *i.e.* all features should all be accounted for by using one and the same γ_0^{∞} and Ω. For details see ref [3]. This analysis helped to identify two to three target regions in the parameter space within which all peak positions were reasonably well accounted for by the above TB band-structure.

In addition to this analysis, we also performed a nonlinear least squares fit to the absorption spectrum, where the concentrations of all 73 tubes with diameters between 0.65 nm and 1.35 nm were allowed to vary freely.

The combination of these two methods finally identifies two possible scenarios for the assignment of electronic transitions where scenario II in table 1. is clearly favored due to better agreement with the experimental diameter distribution. The resulting peak assignments are summarized in table 1 together with the corresponding values for the two parameters γ_0^{∞} and Ω.

In Fig. 3a we compare experimental with theoretical gap energies calculated using the one- (γ_0^{∞}) and two-parameter (γ_0^{∞} and Ω) TB-model. In order to obtain reasonable agreement of the gap energies predicted by the one-parameter model with experimental values, we would need to increase γ_0^{∞} to 3.4 eV! Fig 3b also illustrates that deviations of experimental and theoretical gap energies derived from the two models are not only larger for the one-parameter model but also show an increasing deviation from experimental values when going from larger to smaller diameters.

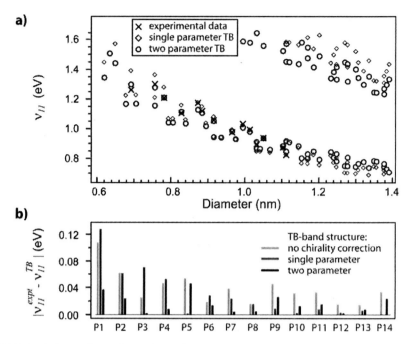

FIGURE 3. a) Comparison of experimental- (scenario II) with calculated gap-energies in the A-cluster region as a function of tube diameter for single- and two-parameter tight-binding (TB) band-structure calculations. b) Deviation of calculated- from the 14 experimental gap-energies marked P1-P14. Best overall agreement is obtained for the two-parameter band-structure including an empirical band offset Ω. The introduction of this offset parameter also leads to improved agreement with the experimentally observed average gap-energy ratio v_{22}/v_{11} of 1.75 [1].

ACKNOWLEDGMENT

We gratefully acknowledge stimulating discussions with L. Brus and G. Dukovic. We also thank them for sharing experimental data prior to publication. It is our pleasure to acknowledge continuing support by G. Ertl.

REFERENCES

1 M.J. O'Connell, S.M. Bachilo, C.B. Huffman, V.C. Moore, M.S. Strano, E.H. Haroz, K.L. Rialon, P.J. Boul, W.H. Noon, C. Kittrell, J. Ma, R. H. Hauge, R.B. Weisman, and R.E. Smalley, Science **279**, 593 (2002).
2 S.M. Bachilo, M.S. Strano, C.Kittrell, R.H. Hauge, R.E. Smalley, R. B. Weisman, Science **298**, 2361 (2002).
3 A. Hagen, T. Hertel, Nano Letters **3**, 383 (2003).
4 C.L. Kane, E.J. Mele, cond-mat/0303528 v1.
5 E. Gregan, S.M. Keogh, T.G. Hedderman, B. McCarthy, G. Farrell, G. Chambers, J.H. Byrne, AIP Conference Proceedings **633**, 294 (2002).
6 A. Kukovecz, Ch. Kramberger, M. Georgakilas, M. Prato, H. Kuzmany, Eur. Phys. J. B **28**, 223 (2002).
7 J.W. Ding, X.H. Yan, J.X. Cao, Phys. Rev. B **66**, 073401 (2002).

Infrared Analysis of Amine Treated Single-Walled Carbon Nanotubes Produced by Decomposition of CO

Frank Hennrich[1], Manfred M. Kappes[1,2], Michael S. Strano[3],
Robert H. Hauge[3], Richard E. Smalley[3]

[1]Forschungszentrum Karlsruhe, Institut für Nanotechnologie, Postfach 3640,
76021 Karlsruhe, Germany
[2]Universität Karlsruhe, Institut für Physikalische Chemie, Kaiserstr. 12, 76128 Karlsruhe, Germany
[3]Rice University, Center for Nanoscience and Technology, MS 100, 6100 Main,
Houston, TX 77005, USA

Abstract. Single walled carbon nanotubes produced by gas-phase decomposition of CO (HiPco process) were treated with nitric acid to generate carboxylic functions which were exploited for three different coupling approaches to functionalize the tubes with an amine to form an amide. Thin films of the reaction products were characterized with infrared spectroscopy.

INTRODUCTION

It is already established that treating single walled carbon nanotubes (SWNTs), which were produced by the Laser ablation method, with oxidizing agents in gas or liquid phase results in the formation of all kinds of oxidic groups at defect sites in nanotubes [1,2,3,4], including carboxylic acid groups. In particular these have been used for functionalization SWNTs [5,6,7,8,9,10,11,12,13,14,15]. IR spectra of oxidized SWNTs treated with organic amines have been published by different research groups [6,7,8,9,10,11,12,13,14,15].

The aim of this paper is to show that treating HiPco SWNTs with nitric acid results in forming oxidic groups which can be used for functionalization similar to what is already shown for Laser SWNTs. Three different approaches were used to couple an amine to carboxylic acid groups. By using transparent, free-standing, thin films with thicknesses in the range of a few hundred nanometers of nanotubes and derivatives the quality of IR-spectra could be noticeably improved which allows a new and detailed interpretation of IR-features.

EXPERIMENTAL

SWNTs were produced by gas-phase decomposition of CO (HiPco process) [16] and will be denoted as SWNTs (I).

CP685, *Molecular Nanostructures: XVII Int'l. Winterschool/Euroconference on
Electronic Properties of Novel Materials,*edited by H. Kuzmany, J. Fink, M. Mehring, and S. Roth
© 2003 American Institute of Physics 0-7354-0154-3/03/$20.00

For oxidizing SWNTs (I), 200 mg HiPco tubes were refluxed in 100 ml 2M nitric acid for 24 hrs. After cooling down, the reaction mixture was centrifuged with 20,000 g for 30 min. and the acid was decanted. The solid was resuspended in deionized water, again centrifuged and the water was decanted. After three Suspension/Centrifugation/Decantation-cycles the remaining solid was suspended in water (~ 0.5 mg/ml) with 1% of the surfactant dodecyl sulfate sodium salt (SDS). The nitric acid treated SWNTs will be denoted as SWNTs (II).

For derivatizing SWNTs (II) with the amine 100 mg of acid treated dried tubes were mixed with 270 mg Octadecylamine (ODA) and the coupling agent Dicyclohexylcarbodiimide (DCC) in dry Dimethylformamide (DMF) and stirred at room temperature for 24 hrs similar to the procedure described in ref.15. After the reaction the reaction mixture was filtered, washed with ethanol and acetone and dried in vacuum (~ 10^{-3} mbar). The ODA-functionalized SWNTs will be denoted as SWNTs (III).

The ODA-coupling via the acidchloride as intermediate was done similar to the procedure described in ref. 6. 100 mg of acid treated, dried tubes were refluxed with 20 ml of $SOCl_2$ and 5 ml of DMF for 2 hrs. After centrifugation, washing with anhydrous DMF and drying, the solid was mixed with 2 g of ODA and heated to 100°C under Argon for 4 hrs. After reaction the reaction mixture was washed several times with DMF and dried in vacuum. The ODA-SWNTs functionalized via $SOCl_2$ will be denoted as SWNTs (IV).

Thin films of all samples were produced by vacuum filtration similar to the procedure described in ref. 15. Before removing the films from the filter using an adhesive tape (Scotch) with an approx. 1/3 cm^2 circular opening they were thoroughly washed with several portions of Acetone and dried in vacuum (~ 10^{-3} mbar) at 40°C for 1h. The mechanical stability of the nanotube films allowed a hole and tear-free removal of the corresponding free-standing area by careful peeling of. The free-standing films were affixed to a metal frame and further dried in vacuum (~ 10^{-3} mbar) for 12 hrs at room temperature.

RESULTS AND DISCUSSION

Fig. 1 shows a typical absorption measurement of a thin film from HiPco SWNTs (I) and of a thin film from Laser SWNTs (I) for comparison (comprising IR and UV-vis-NIR spectra). It demonstrates the advantage of producing thin films because of having the ability of measuring not only UV-vis-NIR-spectra but also IR-spectra in transmission on one film. The range from UV to the NIR will not be discussed in this paper.

The IR-spectrum of SWNTs (I) is featureless except for features at 2852, 2868, 2922 and 2954 cm^{-1} which probably derive from residual DMF that could not be removed completely.

The IR-spectrum of a thin film of SWNTs (II) is very similar to spectra measured with SWNTs which were synthesized by the Laser ablation method and which were also treated with nitric acid. A detailed study and discussion was done in ref. 4. The spectrum shows mainly features at 1130 cm^{-1} (0.14), 1244 (0.28), 1402 (0.17) which could be assigned to C-O- or O-H- stretch motions of alcoholic functions generated during the nitric acid treatment, a broad feature around 1600 cm^{-1} which comprises at least three features at 1598 (0.85), 1618 (0.85) and 1656 (0.61) cm^{-1} corresponding to intercalated HNO_3 and/or NO_2-groups covalently bound to the tubes and a feature at 1754 cm^{-1} (1.00) which corresponds to the C=O-stretch mode. The features in the CH-stretch region at 2850 (0.45), 2921 (0.75) and 2952 cm^{-1} (0.29) probably derive from residual DMF which could not be removed completely.

The IR-spectrum of thin films of SWNTs (III) shows that there is no dramatic difference to the spectrum for the SWNTs (II) accept for additional peaks deriving from ODA at 2953 (0.25), 2924 (1.00), 2852 (0.66), 1468 (0.23) and 723 (0.05) cm^{-1} which are broader than for pure ODA. A huge C=O-band is still visible at 1751 cm^{-1} (0.49) but the broad band assigned to HNO_3/NO_2 in the case of the SWNTs (II) thin film is now shifted to one feature at 1581 cm^{-1} (0.67).

Fig. 2 contains also an IR-spectrum of the same SWNTs (III) film which has been heated to 275°C in vacuum (~ 10^{-3} mbar) for 4 hrs. As result, the feature at 1751 cm^{-1} from the IR-spectrum of the untreated SWNTs (III) film becomes a double peak with a maximum at 1720 (0.55) and 1759 cm^{-1} (0.73). There is only a slight change of intensities for the C=O-vibration region. The sum of the areas of the two peaks at 1720 and 1759 cm^{-1} resembles the area of the peak in the untreated SWNTs (III) spectrum.

The IR-spectrum of a thin film of SWNTs (IV) also shows a double peak in the C=O-region as for the heated SWNTs (III) film with a maximum at 1721 (0.10) and 1755 cm^{-1} (0.15) and huge and broadened peaks in the CH-stretch region at 2953 (0.27), 2924 (1.00) and 2852 cm^{-1} (0.61) coming from the alkyl chains of the ODA.

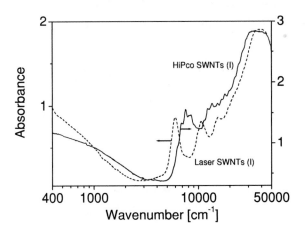

FIGURE 1. Absorption spectra of thin films of HiPco and Laser SWNTs (I).

199

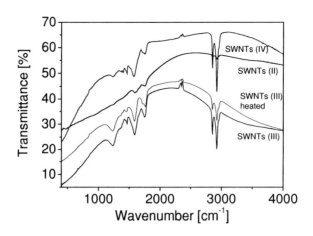

FIGURE 2. FTIR transmission measurements of thin films of SWNTs (I), SWNT (II), SWNTs (III) and SWNTs (IV).

The IR-measurements shown in this work don't contain any of the bands assigned to amide vibrations as claimed by other authors but contains a feature at 1720 resp. 1721 cm^{-1} for samples derivatized by simply heating a mixture of ODA and acid-treated SWNTs as well as samples derivatized via the acidchloride intermediate.

None of the researchers mentioned above discussed that the positions of the amide bands in IR-spectra are heavily dependent on the environment of the amide bond. This is already known from numberless IR-studies on peptides and proteins [17]. The experimental frequencies for amide vibrations of peptides and proteins can roughly lie between 1620 and 1740 cm^{-1} depending on if the samples were measured in polar or unpolar solvents (between 1620 and 1700 cm^{-1}) or if they were measured in gas-phase or isolated in a matrix (between 1700 and 1740 cm^{-1}). Therefore one should include under which conditions the derivatized SWNT-samples were measured.

It is unclear at the moment, how the amide bonds are embedded in nanotube thin films and how they are surrounded by residual solvent molecules such like water or DMF and how they are affected by them in or by the nanotubes itself but the found vibrational frequency of 1720 cm^{-1} for the amide bond implies that the amide bond can vibrate undisturbed in the nanotube film.

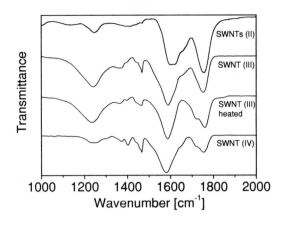

FIGURE 3. 1000 to 2000 cm⁻¹ spectral range of FTIR transmission measurements of thin films of SWNTs (I), SWNT (II), SWNTs (III) and SWNTs (IV) after background substraction.

REFERENCES

1. H. Hu, P. Bhowmik, B. Zhao, M. A. Hamon, M. E. Itkis and R. C. Haddon, *Chem. Phys. Lett.*, 2001, **345**, 25.
2. D. B. Mawhinney, V. Naumenko, A. Kuznetsova, J. T. Yates Jr., J. Liu, R. E. Smalley, *Chem. Phys. Lett.*, 2000, **324**, 213.
3. C. Goze Bac, P. Bernier, S. Latil, V. Jourdain, A. Rubio, S. H. Jhang, S. W. Lee, Y. W. Park, M. Holzinger, A. Hirsch, *Current Appl. Phys.*, 2001, **1**, 149.
4. F. Hennrich, R. Wellmann, S. Malik, S. Lebedkin, M. M. Kappes, *Phys. Chem. Chem. Phys.*, 2003, **5**, 178.
5. J. Liu, A. G. Rinzler, H. Dai, J. H. Hafner, R. K. Bradley, P. J. Boul, A. Lu, T. Iverson, K. Shelimov, C. B. Huffman, F. Rodriguez-Macias, Y.- S. Shon, T. R. Lee, D. T. Colbert, R. E. Smalley, *Science*, 1998, **280**, 1253.
6. J. Chen, M. A. Hamon, H. Hu, Y. Chen, A. M. Rao, P. C. Eklund, R. C. Haddon, *Science*, 1998, **282**, 95.
7. M. A. Hamon, J. Chen, H. Hu, Y. Chen, M. E. Itkis, A. M. Rao, P. C. Eklund, R. C. Haddon, *Adv. Mater.*, 1999, **11**, 834.
8. M. A. Hamon, H. Hu, P. Bhowmik, S. Niyogi, B. Zhao, M. E. Itkis, R. C. Haddon, *Chem. Phys. Lett.*, 2001, **347**, 8.
9. J. Chen, A. M. Rao, S. Lyuksyutov, M. E. Itkis, M. A. Hamon, H. Hu, R. W. Cohn, P. C. Eklund, D. T. Colbert, R. E. Smalley, R. C. Haddon, *J. Phys. Chem. B*, 2001, **105**, 2525.
10. M. A. Hamon, H. Hui, P. Bhowmik, H. M. E. Itkis, R. C. Haddon, *Appl. Phys. A*, 2002, **74**, 333.
11. F. Pompeo, D. E. Resasco, *Nano. Lett.*, 2002, **2**, 369.
12. P. Diao, Z. Liu, B. Wu, X. Nan, J. Zhang, Z. Wei, *Chem. Phys. Chem.*, 2002, **10**, 898.
13. B. Li, Z. Shi, Y. Lian, Z. Gu, *Chem. Lett.*, 2001, 598.
14. E. V. Basiuk, V. A. Basiuk, J.-G. Banuelos, J.-M. Saniger-Blesa, V. A. Pokrovskiy, T. Y. Gromovoy, A. V. Mischanchuk, B. G. Mischanchuk, *J. Phys. Chem. B*, 2002, **106**, 1588.
15. F. Hennrich, S. Lebedkin, S. Malik, J. Tracy, M. Barczewski, H. Rösner, M. Kappes, *Phys. Chem. Chem. Phys.*, 2002, **4**, 2273.
16. P. Nikolaev, M. J. Bronikowski, R. K. Bradley, F. Rohmund, D. T. Colbert. K. A. Smith, R. E. Smalley, *Chem. Phys. Lett.*, 1999, 313, 91.
17. J. Kubelka, T. A. Keiderling, *J. Phys. Chem. A*, 2001, **105**, 10922.

Spectroscopy Study of SWNT in Aqueous Solution With Different Surfactants

V.A. Karachevtsev[1], A.Yu.Glamazda[1], U. Dettlaff-Weglikowska[2], V.S. Leontiev[1], A.M. Plokhotnichenko[1] and S. Roth[2]

[1]Institute for Low Temperature Physics and Engineering, NASU, 61103, Kharkov, Ukraine,
[2]Max-Planck Institute for Solid State Research, Heisenberg Str. 1, 70569 Stuttgart, Germany,

Abstract. Aqueous solutions of HiPCO SWNT with different surfactants (anionic (SDS), cationic (CTAB) and non-ionic (Triton X-100)) have been studied by Raman and Near-infra-red (NIR) absorption spectroscopy. The nanotube interaction with surfactant leads to the spectral shift of lines and its intensity redistribution, compared with the spectrum of SWNT in KBr pellet. The most essential spectral changes are observed for nanotube aqueous solution with the surfactant containing a charge group. Two possible models of micelle formation around the nanotube are discussed.

INTRODUCTION

The potential biological application of single-walled carbon nanotubes (SWNT) is associated with solubility of this nanomaterial in water [1,2]. Unfortunately, nanotubes are poorly soluble in the most of organic solvents and are insoluble in water. This is because SWNTs aggregate in bundles as a result of substantial van der Waals attractions between tubes. To dissolve nanotubes in water without any covalent functionalization, a surfactant would be added into aqueous solution with tubes and then this mixture is suspended by sonication. It is supposed that the sound wave splits bundles in solution. Surfactant in suspension adsorbed on the nanotube surface precludes aggregation of nanotubes in bundles. Sodium dodecyl sulfate, SDS, and Triton X-100 are the most widely used surfactants [1-3]. If the surfactant concentration exceeded some critical point, micelles appeared in solution. The input surfactant concentration for nanotube research exceeded usually critical micelle concentrations (CMC). In spite of wide usage of surfactants as material favored for nanotube water solubility, the SWNT behavior in surfactant micelle surrounding has not been yet investigated properly. In this report we reproduce the results of Raman and NIR absorption spectroscopy study of SWNT in aqueous solution with different surfactants.

EXPERIMENTAL

SWNTs (HiPCO, with 99 % purity) were suspended in aqueous solution with different types of surfactants: a) anionic (SDS) ($CH_3(CH_2)_{11}(SO_4)^- Na^+$), b) cationic (hexadecyltrimethylammonium bromide (CTAB) ($CH_3(CH_2)_{15}N^+(CH_3)_3Br^-$) and c)

CP685, *Molecular Nanostructures: XVII Int'l. Winterschool/Euroconference on Electronic Properties of Novel Materials*, edited by H. Kuzmany, J. Fink, M. Mehring, and S. Roth
© 2003 American Institute of Physics 0-7354-0154-3/03/$20.00

and non-ionic (Triton X-100) (C_8H_{17} C_6H_4 (OCH_2CH_2)$_n$OH, n~10). 0.05 mg/mL nanotube dispersion with surfactant was mixed and then the suspension was sonicated for 40 minutes (20 W 44 kHz). A concentration of surfactants in water solution was 1%. Water was prepared by distillation and then passed through Multi-Q system. The deionized water has resistivity 18 MΩ.

The nanotubes suspension in toluene (0.1 mg/mL) after short sonication was deposited on a quartz substrate to form a thin film for Raman measurements. SWNTs in KBr pellets obtained after pressing were used for NIR absorption study.

Raman experiments were performed in a backscattering configuration. He-Ne laser with energy 1.96 eV (λ_{exc}=632.8 nm) was used for excitation. The spectra were recorded by means of a Raman double monochromator (reverse dispersion 3 Å/mm).

RESULTS

SWNT interactions with different surfactants in aqueous solution result in changes in Raman spectra, as compared with spectra of a nanotube in films (Fig. 1). In this Figure 2 spectral intervals with most intensive bands are shown: the region of the radial breathing mode (RBM) (150-300 cm^{-1}) and the tangential (G) modes (1400-

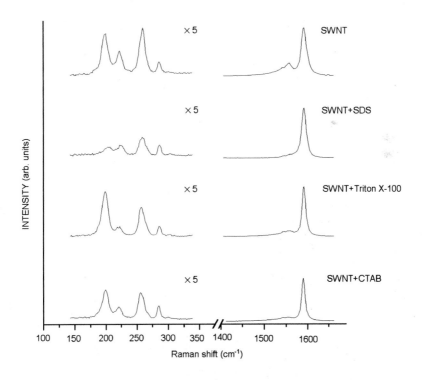

FIGURE 1. Raman spectra of nanotube film (upper) and SWNT in aqueous solution with different surfactants in range of RBM and G mode obtained at λ_{exc}=632.8 nm laser excitation.

1600 cm^{-1}). Each Raman spectra is normalized to the most intense band corresponding to the G-mode. The interaction with different surfactants leads to the intensity decrease and shift of lines attributed to the RBM, compared with the intensive G mode. Intensity of the low-frequency (1500-1580 cm^{-1}) component of the G mode is decreasing also, in comparison with the high-frequency (1580-1600 cm^{-1}) component of this band. It should be noted that the greatest changes are observed for nanotubes suspension with SDS and small ones are revealed for suspension with TX-100.

At He-Ne laser excitation with 1.96 eV, 4 intensive bands (RBM) are observed in Raman spectra of HiPCO nanotubes, corresponding to SWNTs of different diameters and chirality. As was shown [4], 2 lowest-frequency lines correspond to metallic nanotubes and 2 highest-frequency ones to a semiconducting type of nanotubes. The essential changes in spectra are observed for this mode of SWNT in aqueous solutions with SDS. The intensity of the low-frequency band decreases significantly as compared with intensity of other bands and shifts into the high-frequency region by 5 cm^{-1}. Redistribution of band intensities is observed in spectra of the nanotubes in other solutions.

Fig. 2 presents NIR absorption spectra of SWNT in aqueous solutions with different surfactants: SDS, TX-100 and CTAB. For comparison, the absorption spectrum of SWNTs in KBr pellet is shown in this Figure. Light absorption of SWNT in NIR region is caused by electronic transitions between pairs of van Hove singularities in the density of states. So, light absorption in the region 1000-1700 nm can be assigned to electronic transitions E_{11}^S in the semiconducting nanotube (band gap transition), absorption at 700-1000 nm results from the E_{22}^S transitions of these

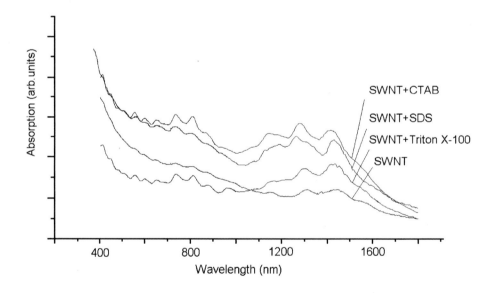

FIGURE 2. Absorption spectra of SWNT in KBr and nanotubes in aqueous solution with different surfactants: SDS, CTAB and TX-100.

types of SWNTs and absorption at 500-700 nm is assigned to the $E_{11}{}^{m}$ transition in the metallic SWNTs [3]. Several bands are observed in the absorption spectrum of HiPCO nanotubes for every electronic transition, corresponding to the tubes with different diameters and chilarity, predominant in the sample. As seen from Figure 2, the bands in the region of the $E_{11}{}^{S}$ transition are shifted in absorption spectra of aqueous solutions SWNTs with various surfactants from each other as well as from the absorption spectrum of SWNT in KBr. Note that spectral shifts are not observed for bands in the region of the $E_{22}{}^{S}$ and $E_{11}{}^{m}$ transitions. As compared with the spectrum of SWNT in KBr, bands corresponding to this transition in nanotubes with CTAB are shifted by about 30-60 nm (16-32 meV) to the higher energy. Small shifts of the same bands are observed for nanotubes solutions with TX-100 but these shifts are smaller by about 10 nm (5-7 meV). The greatest shift of bands (up to 70 meV) can be seen for absorption spectra of nanotubes with SDS.

In nanotubes the aqueous concentration of surfactants was above CMC and micelles were formed around SWNT. As is shown schematically in Fig.3, two types of micelles can be: surfactant molecules surround tubes like a belt (left) [4] or they are distributed uniformly around the tubes (right) [3]. In the first case the alkyl chain and charge group of surfactant molecules lies flat on the graphene SWNT surface along the tube axis. For micelle of the second type the surfactant molecules interact with nanotube by ends of alkyl chain and generally this interaction will be similar for all surfactants.

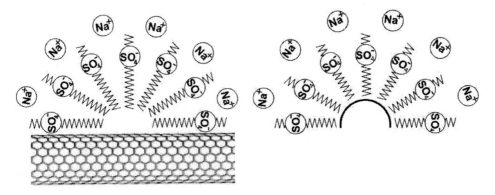

FIGURE 3. Schematic representation of way how surfactant molecule (SDS) may adsorb onto the nanotube: surfactant molecules surround tubes like belt (left) or they are distributed uniformly around tubes (right).

As coupling between the surfactant with a charge group and nanotube leads to more essential transformation of Raman spectra and shift of bands corresponding to $E_{11}{}^{S}$ transition (NIR absorption spectra) in comparison with nonionic surfactant and pristine SWNT, we believe that the role of the charge in forming spectra is very decisive. In the case of the model 2, an effect of the surfactant charge group on nanotube spectra will be lesser due to the rather far location of the charge from the tube. Besides, this effect is weakened due to neutralization by a charge with the opposite sign, present in solution. A possible effect of charge screening of nanotube excited carriers should result in a displacement of density of states (van Hove singularities) that had to affect a shift of energies of E_{11}^S, E_{22}^S and E_{11}^m transitions. In absorption spectra the shift of only the bands corresponding to the E_{11}^S transition is observed. Such a selective influence only on the E_{11}^S transition is possible in the case of doping (or depletion) of the valence zone with a charge [6]. As well, the similar charge effect can occur when the charge group is located near the tube surface and charge transfer takes place (even partially). Thus, on the basis of the above reasoning, it may be supposed that the formation of the first type micelles is preferable.

CONCLUSION

A surfactant interaction with SWNT in aqueous solutions is determined by the surfactant micelle formation around the nanotube. The nanotube interaction with surfactant leads to the spectral shift of lines and its intensity redistribution in Raman and NIR absorption spectra, as compared with the spectrum of SWNT in KBr pellet. The most essential spectral changes are observed for nanotube aqueous solution with the surfactant containing a charge group. Two possible models of micelle formation around the nanotube are discussed. Based on the spectral study, we believe that the model of the micelle formation in which the surfactant molecule lies flat on the graphene SWNT surface along the tube axis is preferable.

ACKNOWLEDGMENTS

This work was supported by INTAS project N 00-237.

REFERENCES

1. M. Shim, N. Kam, R.J. Chen, Y.Li, H. Dai, *Nano Letters* **2**, 285-288 (2002).
2. M.J. O'Connell, P. Boul, L.M. Ericson, C. Huffman, Y.H. Wang, E. Haroz, C. Kuper, J. Tour, K.D. Ausman, R.E. Smalley, *Chem. Phys. Letters* **342** 265-271(2001).
3. O'Connell, M. J.; Bachilo, S. M.; Huffman, C. B., Moore, V. C.; Strano, M. S.; Haroz, E. H.; Rialon, K. L.; Boul, P. J.; Noon, W. H., Kittrell, C.; Ma, J.; Hauge, R. H.; Weisman, R. B.; Smalley, R. E. *Science* **279**, 593-596 (2002).
4. M. F. Islam, E. Rojas, D. M. Bergey, A. T. Johnson, and A. G. Yodh, *Nano Letters* **3**, 269-273 (2003).
5. V.A. Karachevtsev, A.Yu.Glamazda, U. Dettlaff-Weglikowska, V.S. Kurnosov, E.D. Obraztsova, A.V.Peschanskii, V.V. Eremenko and S. Roth. *Carbon* 258 (2003) (in press).
6. M. E. Itkis, S. Niyogi, M. E. Meng, M. A. Hamon, H. Hu, and R. C. Haddon; *Nano Letters* **2**, 155-159 (2002).

Raman Spectroscopy of SWNTs in Zeolite Crystals

M. Hulman[1], L. Li[2], Z.K. Tang[2], O. Dubay[1], G. Kresse[1] and H. Kuzmany[1]

[1]Institut fuer Materialphysik, Universitaet Wien, Austria
[2]Physics Department, Hong Kong University of Science and Technology

Abstract. Single wall carbon nanotubes with diameter 0.4 nm grown in the channels of $AlPO_4$-5 crystals were studied by Raman spectroscopy and *ab initio* density functional calculation. It was found that only two types of nanotubes with different chiralities, (5,0) and (4,2), were responsible for the observed spectra. The frequencies of the radial breathing modes were reliably assigned. A strong response was observed for frequencies around 1250 cm^{-1}. The laser excitation energy of 2.2 eV separates two regions with different line shapes for the G band.

INTRODUCTION

Standard ways of preparation of single wall carbon nanotubes (SWNTs) involve laser ablation, arc discharge, chemical vapour deposition and recently, high pressure CO decomposition. The common feature of all these techniques is that the nanotubes grow in a free space. Moreover, the diameter distribution of SWNTs produced spans a wide range between 0.7 nm and 2.0 nm.

Another possibility is to let the nanotubes grow in templates. This method provides a possibility to produce aligned SWNTs and control the diameter distribution. Crystals of $AlPO_4$ zeolite contain parallel channels of diameter about 0.73 nm. Subtracting twice the van der Waals radius of a carbon atom (0.17 nm), the resulting diameter of nanotubes is around 0.4 nm [1]. Within this diameter range there are nanotubes with three different chiralities: (5,0), (3,3) and (4,2). This means the diameter distribution is very narrow as compared to samples made by other techniques.

The template grown nanotubes were succesfully characterized by Raman spectroscopy [2]. In spite of a very narrow diameter distribution the Raman spectra turned out to be complicated. There are overlapping bands around 550 cm^{-1} where the radial breathing mode is expected. Also the range between 1200 cm^{-1} and 1600 cm^{-1} is crowded with broad features which were recently assigned to various structures in the phonon density of states of 2D graphite or in (5,0), (4,2) and (3,3) nanotubes [1,3].

EXPERIMENTAL

A detailed description of the production of carbon nanotubes in zeolite crystals was reported elsewhere [1]. The typical dimensions of the AFI crystals are about 100 μm in diameter and 300 μm in length.

CP685, *Molecular Nanostructures: XVII Int'l. Winterschool/Euroconference on Electronic Properties of Novel Materials,* edited by H. Kuzmany, J. Fink, M. Mehring, and S. Roth
© 2003 American Institute of Physics 0-7354-0154-3/03/$20.00

Raman spectra were recorded with a triple Dilor monochromator equipped with a CCD detector cooled by liquid nitrogen. Up to 19 lines generated by three different lasers were used for excitation of the spectra. The light was focused on the sample by a 50x objective of the micro Raman system.

RESULTS

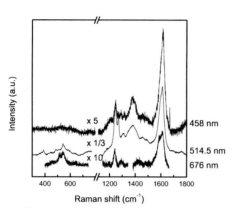

FIGURE 1. Raman spectra of SWNTs embedded in a zeolite crystal for three different laser energies displayed in the figure. The spectra were shifted for clarity and multiplied as indicated.

Fig.1 shows representative spectra of a SWNT/AFI sample for 3 different laser energies. A remarkable characteristic is the strong dependence of the scattering intensity on the laser excitation energy especially in the low energy region around 550 cm^{-1}. The spectra excited with deep blue lasers have a very low intensity there. As the laser wavelength increases a broad band with a peak at 543 cm^{-1} appears. The maximum intensity was found for the laser wavelength of 514 nm. For the longer wavelength of the exciting laser the intensity drops again. The response for excitation with the dye laser (585 nm – 626 nm) gets stronger again. Intensity of spectra excited with red laser lines (wavelengths 676 nm and higher) decreases gradually with increasing laser wavelength (not shown here). The intensity profile can be visualized by evaluating the integrated intensity as demonstrated in Fig.2. The results were corrected for the sensitivity of the Raman system in addition to the laser power normalization. Two dominating peaks are clearly seen at ~ 2.0 eV and ~ 2.4 eV.

The G band between 1550 and 1620 cm^{-1} is the dominating feature in the high energy part of spectra. The recorded line shapes are similar for $E_{laser} > 2.2$ eV (564 nm). For lower excitation energies they change but remain similar for the rest of the laser lines used in the experiments. In contrast to excitation with higher E_{laser}, the lines appear at slightly higher frequencies and the intensity is distributed in a different manner within the bands.

There is an interesting behavior of the spectral part between 1200 cm^{-1} and 1300 cm^{-1}. In the deep blue there is essentially one line centered slightly below 1250 cm^{-1}. For blue lines above 488 nm a new line around 1270 cm^{-1} appears. It remains visible for excitations down to with the yellow lasers and then disappears again. The splitting of the lines is constant with a value of about 20 cm^{-1} for blue excitation energies. In contrast, the lower frequency component changes its position as the laser excitation energy decreases below 2.2 eV. It shifts to lower frequencies as E_{laser} decreases.

DISCUSSION

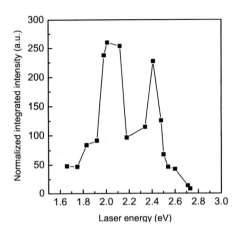

FIGURE 2. Normalized integrated intensity of Raman spectra in the spectral window 450 cm⁻¹ – 650 cm⁻¹ as a function of the laser energy. Squares represent values from the experiments, the solid line is a guide for eye.

The peaks in Fig.2. correspond to resonance enhancement of the Raman spectra. They are closely related to peaks in optical absorption and peaks in the imaginary part of the dielectric function calculated recently [4]. Summarizing all these facts we can safely assign the resonance observed at 2 eV to (4,2) nanotubes and the resonance at 2.4 eV to (5,0) tubes. The last possible candidate, the (3,3) tube, has the resonance energy around 3 eV, beyond the energies accesible with our lasers. The strongest peak at 543 cm⁻¹ for excitation with the 514.5 nm line we ascribe to the radial breathing mode of the (5,0) nanotubes. This agrees very well with the value of 536 cm⁻¹ obtained from density functional theory (DFT) calculations. The latter also gave the value of 538 cm⁻¹ for the RBM of the free (4,2) nanotube. The strongest peak that appeared in the spectrum for E_{laser} < 2.2 eV is located at 514 cm⁻¹. The difference between the experimental and theoretical result for the RBM of the (4,2) nanotubes may be explained by interaction of the tubes with the AFI framework. It is natural to assume that this interaction is stronger for the (4,2) tubes, which have larger

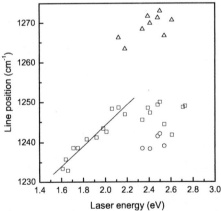

FIGURE 3. The laser energy dependency for the two intense lines in the frequency region from 1200 cm⁻¹ to 1300 cm⁻¹. The higher energy component is displayed by triangles and the two components of the lower energy line are shown by squares and circles for laser energies above 2.2 eV.

diameters than the (5,0) nanotubes.

The G band can be decomposed into several individual lines located between 1550 cm⁻¹ and 1620 cm⁻¹. Frequencies of these modes were calculated recently for free zig-zag nanotubes with diameters 0.55 – 1.5 nm [5,6]. It was found that frequencies of all components of the G band decrease quite strongly as the diameter decreases. For the (5,0) nanotube, the frequencies extend only up to 1563 cm⁻¹ in the present density functional calculations (not shown). Quantitatively, the theoretical value differs considerably from the experimental results. Hardening of the A_{1g} mode may originate from the similarity of both the lattice constants for the AFI crystal and the (5,0) nanotube. The lattice constant of the former is about 1% smaller than the latter. The AFI environment forces nanotubes to fit with its lattice periodicity and squeezes the nanotube lattice. In theoretical calculations, the frequency of the A_{1g} mode shifts to higher frequencies (1600 cm^{-1}), when the nanotube takes on the lattice constant of the AFI crystal.

The dependence of the Raman shift on the laser energy is demonstrated in Fig.3 for the two intense lines in the range from 1200 cm^{-1} to 1300 cm^{-1}. The dispersive behavior found for the lower energy line is characteristic for disorder induced D lines in graphite and SWNTs. The frequency of the lower energy line increases linearly with the laser energy up to 2.2 eV and then stays constant. This behavior is quite different from that observed for conventional SWNTs [7]. According to the DFT calculations for the (5,0) nanotube, there are phonon branches grouped around 1250 and 1280 cm^{-1} at the Γ point of the Brillouin zone (BZ). Most of the branches indeed possess a symmetry that allows them to be Raman active. We assume that the line which is dispersive below $E_{laser} = 2.2$ eV belongs to the (4,2) tube. In contrast to the experiment where we see essentially one line, the DFT calculation yields many phonon branches. Only approximately 1/6 of the total number of the branches is Raman active. Unlike the (5,0) tube, there are branches having the frequency of ~1250 cm^{-1} not only at the Γ point but also at the edge of BZ. This implies that the scenario of the double resonant mechanism is possible, and the line shifts as observed in the experiment. The rate of the line shift is low, 26 ± 3 cm^{-1} eV^{-1}, and reaches only half of the value typical for the larger diameter nanotubes.

ACKNOWLEDGMENTS

M.H. acknowledges a support from the EU project FUNCARS. Prof. Tang and Miss Li acknowledge the support from the RGC of Hong Kong through CERG grant. O.D. acknowledges support by the FWF under grant Nr. P14095.

REFERENCES

1. Z.K. Tang, H.D. Sun, J. Wang, J. Chen, G. Li, Appl. Phys. Lett. **73**, 2287 (1998)
2. M. Hulman, H. Kuzmany, O. Dubay, G. Kresse, Ling Li, Z.K. Tang, J. Chem Phys., submitted
3. A. Jorio, A.G. Souza, G. Dresselhaus, M.S. Dresselhaus, A. Righi, F.M. Matinaga, M.S.S. Dantas, M.A. Pimenta, J. Mendes, Z.M. Li, Z.K. Tang, R. Saito, Chem. Phys. Lett. **351**, 27 (2002)
4. M. Machon, S. Reich, C. Thomsen, D. Sanchez-Portal, P. Ordejon, Phys. Rev. **B 66**, 155410 (2002)
5. O. Dubay, G. Kresse, H. Kuzmany, Phys. Rev. Lett. **88**, 235506 (2002)
6. O. Dubay, G. Kresse, Phys. Rev. **B 67**, 035401 (2003)
7. A. Grueneis, M. Hulman, Ch. Kramberger, H. Peterlik, H. Kuzmany, H. Kataura, Y. Achiba, in Electronic Properties of Molecular Nanostructures, edited by H. Kuzmany, J. Fink, M. Mehring and S. Roth, AIP Conference Proceedings 591, 2001

Artificial neural networks in the analysis of the fine structure of the SWCNT Raman G-band

A. Kukovecz[1,2,*], M. Smolik[1], S.N. Bokova[3], E. Obraztsova[3], H. Kataura[4], Y. Achiba[4], H. Kuzmany[1]

[1] Institut für Materialphysik, University of Vienna, Strudlhofgasse 4., A-1090 Wien, Austria
[2] Dept. of Appl. and Environ. Chemisty, University of Szeged, Rerrich B. ter 1, H-6720 Szeged, Hungary
2 Natural Sciences Center of General Phys. Inst., 38 Vavilov street, 119991, Moscow, Russia
3 Graduate School of Science, Tokyo Metropolitan University, Japan

Abstract. Although the diameter dispersion of the phonons composing the Raman G-band of single wall carbon nanotubes (SWCNTs) is well understood theoretically, systematic experimental studies on the subject are scarce. We investigated 6 different diameter samples between d=1.05-1.57 nm with several excitation lasers and used artificial neural networks (ANN) to explore if there is a connection between the fine structure of the G-band and the sample diameter. An initial screening by a Kohonen self-organizing map revealed that ANN technology is able to identify spectra measured on the same sample. Based on this result several supervised learning algorithms were tested and finally we succeeded in building a resilient propagation ANN with one hidden layer which is able to predict the diameter distribution of a macroscopic SWCNT sample from the structure of its Raman G-band with acceptable accuracy. We suggest that with more extensive calibration this method could be developed into a useful auxiliary technique of SWCNT characterization.

INTRODUCTION

Calculating the diameter distribution of a SWCNT sample from the Raman radial breathing mode (RBM) is a well-established technique. It's main drawbacks are that (i) the RBM shifts considerably due to tube-tube interactions in a bundle which complicates the analysis, and (ii) the RBM is usually 2-10 times weaker than the tangential G-band, therefore, it must be measured longer to obtain the same S/N ratio. In this contribution we show that it is also possible to extract diameter distribution information from the fine shape of the G-band. Although such attempts have already been reported for individual tubes [1], these methods do not appear to be feasible in case of everyday SWCNT samples containing lots of different diameter tubes.

Practically, an Artificial Neural Network (ANN) is an oriented graph. Junction points correspond to neurons and connection vectors to axons and dendrites. Connections have different weights which simulate the different synaptic strengths of a real brain. ANNs learn by gradually modifying the –initially random- connection weights so as to achieve the best match between a given input and the corresponding desired output pattern. Several network architectures (connection schemes) and learning algorithms are available. ANNs are generally useful for handling large

CP685, *Molecular Nanostructures: XVII Int'l. Winterschool/Euroconference on Electronic Properties of Novel Materials,* edited by H. Kuzmany, J. Fink, M. Mehring, and S. Roth

multivariate problems where the analytical relationship between input and output is either unknown or too complex for conventional techniques like peak fitting.

EXPERIMENTAL

Six different SWCNT samples with diameters with mean diameters between 1.05 and 1.56 nm (see Table 1 for details) were measured using 14 different laser lines of an Ar^+-Kr^+ laser between 457-676 nm (sample A was measured at 9 excitation wavelengths only). Spectra were recorded on degassed buckypapers at 80 K in Stokes mode on a Dilor xy triple grating Raman spectrometer.

TABLE (1). Characteristics of the SWCNT samples studied here

| Sample | Gaussian diameter distribution (*) | | Source |
	Center (nm)	Width (nm)	
A	1.05(0)	0.15(0)	HiPco®
B	1.17(4)	0.14(1)	laser abl.
C	1.29(7)	0.14(2)	laser abl.
D	1.39(3)	0.14(8)	laser abl.
E	1.46(7)	0.14(6)	laser abl.
F	1.56(1)	0.15(4)	laser abl.

(*) Determined from the analysis of the Raman RBM

In order to convert the measured spectra into input vectors for the ANN, the following data processing was applied. The G-band region between 1450-1650 cm^{-1} was cut from the spectrum, linearly interpolated to contain 200 data points in each case and converted into a 200 point coefficient vector using the discrete cosine transform (DCT) as shown in Equation 1.

$$DCT[k] = \alpha[k]\sum_{n=0}^{N-1} Raman[k]\cos\left(\frac{\pi(2n+1)k}{2N}\right), \quad \alpha[k] = \begin{cases} \sqrt{\dfrac{1}{N}} & k=0 \\ \sqrt{\dfrac{2}{N}} & k=1,2...N-1 \end{cases} \quad (1)$$

Since such a long input would have been impractical to handle, the coefficient vector was chopped so that in contained only the first 35 coefficients. A series of inverse transformations were performed to prove that the average error introduced into the spectrum by the chopping is less than 0.2 %. Finally, this 35-element vector was normalized to the value of the first element and the first element was replaced with the energy of the excitation laser given in electronvolts.

All neural network simulations were performed using the Stuttgart Neural Network Simulator (SNNS) version 4.2 [2]. Input patterns were fed into the ANN in a random order. The output values were linearly scaled between 0 and 1.

RESULTS

The analysis is based on the fact that nearly all phonons in the nanotube Raman G-band are diameter dispersive, so the diameter of a SWCNT is in principal encoded in the shape of the G-band. Our first goal was to decide if this correlation is really observable. The Kohonen self-organizing map (SOM) is a special ANN which recognizes correlations between input patterns without human supervision [3]. Similar inputs produce output on the same or neighboring neurons which are organized into a square matrix. All SWCNT input vectors were run through the Kohonen learning algorithm until the SOM stabilized itself and the map was analyzed for output clusters originating from input vectors belonging to the same diameter nanotube. The presence of such clusters would prove the assumed correlation while their absence would indicate that there is no extractable relationship between the tube diameter and the shape of the Raman G-band. We have found that the spectra of large diameter tubes (samples D-F) cluster very well, those of sample C from a loose arc on the map while those of A and B do not form tight clusters but are not completely randomly distributed either. Therefore, it appears feasible to establish a more quantitative relationship between the fine structure of the band and the nanotube diameter.

Artificial neural networks do not attempt to find analytical relationships between the input data (G-band) and the output (parameters of the diameter distribution). Rather, they work as a human brain would. In their learning phase they associate known inputs with known outputs. Then in the application phase they take a new input (unknown at the learning stage) and are able to predict the corresponding output with reasonable accuracy on the basis of similarities between the learned input patterns and the new pattern. Several network architectures and learning algorithms were tested until finally we decided in favor of the feed-forward neural network architecture depicted in Figure 1. and the resilient propagation method.

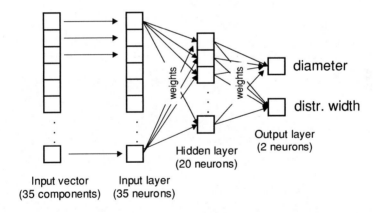

FIGURE 1. Layout of the artificial neural network used in the supervised learning runs.

For the supervised learning runs the 79 measured G-band spectra were randomly divided into three sets: the "Training" set containing 63 spectra was used for the training itself, the "Validation" set containing 8 spectra served prevented the network from overtraining and the 8 spectra of the "Test" set functioned as "unknowns" which could be used to characterize the prediction capacity of the network. The results of five independent runs were compared and averaged to ensure the reproducibility of the ANN training procedure. The final results obtained for the center (d) and width (σ) of the Gaussian diameter distribution from the conventional radial breathing mode (RBM) analysis and the described G-band based method are presented in Table 2.

Table (2)   Comparison of the SWCNT diameter distribution parameters as obtained from the conventional RBM analysis and from the G-band based ANN method described here. ("Test" set)

Sample	d_{RBM} (nm)	$d_{G\text{-band}}$ (nm)	σ_{RBM} (nm)	$\sigma_{G\text{-band}}$ (nm)
A	1.050	1.075	0.150	0.156
B	1.174	1.246	0.141	0.140
C	1.297	1.262	0.142	0.143
C	1.297	1.241	0.142	0.143
D	1.393	1.445	0.148	0.150
D	1.393	1.369	0.148	0.148
E	1.467	1.412	0.146	0.144
F	1.561	1.571	0.154	0.154

It can be seen from the data that the described ANN method was able to predict the diameter distribution parameters of all samples studied with acceptable accuracy. We anticipate that the usefulness of presented G-band based method could be increased (i) if a broader range of known diameter SWCNT samples was available and (ii) if the parameters of the diameter distribution were not derived from the Raman RBM but rather, from an independent technique (e.g. TEM).

CONCLUSIONS

We have shown that it is possible to differentiate between SWCNT samples of different diameters solely on the basis of the fine structure of their Raman G-band. We developed a complete protocol for estimating the diameter distribution of unknown samples from their G-bands. This would not be feasible with conventional peak fitting techniques but is quite possible with artificial neural networks. The method can be made RBM-independent if the training output is obtained from e.g. optical spectroscopy, microscopy or XRD. An advantage of such evaluation is that interaction between the SWCNTs within a bundle would not complicate the analysis anymore.

ACKNOWLEDGMENTS

Financial support from the EU RTN FUNCARS (HPRN-CT-1999-00011) and the Hungarian OTKA F038249 grant is acknowledged.

REFERENCES

[1] A. Jorio et al., *Phys. Rev. B* **65** (2002) 155412.
[2] Stuttgart Neural Network Simulator, User Manual, Version 4.2 (University of Stuttgart, IPVR, 2000)
[3] T. Kohonen, *Biol. Cybern.* 43 (1982) 59-69.
[4] H. Kuzmany et al. *Eur. Phys. J. B* **22** (2001) 307-320.

Raman and HRTEM Monitoring of Thermal Modification of HipCO Nanotubes

E.D. Obraztsova[1], S.N. Bokova[1], V.L. Kuznetsov[2], A.N. Usoltseva[2],
V.I. Zaikovskii[2], U. Dettlaff-Weglikowska[3], S. Roth[3], H. Kuzmany[4]

[1] Natural Sciences Center of General Physics Institute, RAS, 38 Vavilov street, 119991, Moscow,
Russia, elobr@kapella.gpi.ru
[2] Boreskov Institute of Catalysis, RAS, 5 Lavrentieva street, 630090, Novosibirsk, Russia
[3] Max-Plank-Inst. fur Festkorperforscung, Heisenbergstr. 1 D-70569, Stuttgart, Germany
[4] Universitat Wien, Inst. fur Materialphysik, Strudlhofgasse 4, A-1090, Wien, Austria

Abstract. A diameter-dependent oxidative etching of the smallest HipCO nanotubes (diameters 8-10 Å) has been observed by *in situ* Raman measurements both in a laser beam with a variable power density and in an optical oven with a variable temperature. The effect manifested itself as an irreversible disappearance of the Raman breathing modes with frequencies > 220 cm^{-1}. The small tube degradation (up to their complete disappearance) took place at temperatures 150-400°C. Statistical averaging of HRTEM images of HipCO nanotubes, taken after successive temperature steps, has confirmed an obvious shift of the tube diameter distribution maximum toward higher values.

INTRODUCTION

An important advantage of synthesis of single-wall carbon nanotubes (SWNT) by a catalytic decomposition of CO gas under high pressure (HipCO) is a high content of the nanotube fraction (not less than 80%) already in a raw product [1]. After an acid treatment, removing the only contamination- non-capsulated catalytic particles, the material purity becomes close to 100%. A speciality of HipCO nanotubes is a wide distribution over diameter (4-20 Å). A small diameter tail of this distribution may be cut effectively through irradiation of the material by a laser beam with a step-by-step increased power density [2,3]. In our work this phenomenon has been studied by Raman scattering technique. To answer the question, if the laser action is thermal, *in situ* Raman measurements have been performed also for HipCO material heated in an optical oven. The Raman data on diameter distribution of HipCO nanotubes after thermal treatment have been compared with the statistically averaged data of a high resolution transmission electron microscopy (HRTEM).

EXPERIMENTAL

Raman spectra have been measured with *Jobin-Yvon S-3000* spectrometer in micro-configuration. A probing beam of Ar+-ion laser (λ=514.5 nm) with a variable power density (50-600 W/cm^2) was used for spectra excitation and for studying the laser etching effect. An optical oven was supplied with a temperature control in range 25-700°C. HipCO nanotubes, commercially available, have been treated by HCl acid to remove the metal contaminants. A wide distribution of nanotube diameters in a pristine material has been observed by HRTEM investigations *(Fig.1)* and by Raman

CP685, *Molecular Nanostructures: XVII Int'l. Winterschool/Euroconference on
Electronic Properties of Novel Materials,* edited by H. Kuzmany, J. Fink, M. Mehring, and S. Roth
© 2003 American Institute of Physics 0-7354-0154-3/03/$20.00

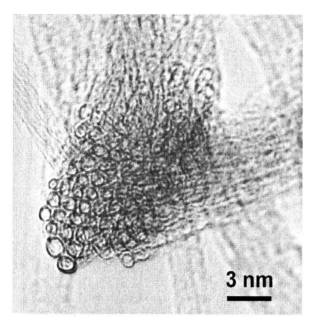

FIGURE 1. HRTEM cross-sectional view of a bundle of HipCO single-wall carbon nanotubes. A variety of the nanotube diameters is evident.

scattering, measured with different excitation wavelengths at room temperature (RT) *(Fig.2).* The Raman breathing mode frequencies (ω, cm^{-1}) have been used to estimate HipCO nanotube diameters *(d, nm)*:

$$\omega = \frac{C_1}{d} + C_2 \quad (1)$$

where C_1=234, C_2=10 cm^{-1} [4]. The Raman modes with frequencies 165 cm^{-1}-306 cm^{-1} *(Fig.2 a-e)* correspond to tubes with diameters 8-15 Å, (according to (1)). HRTEM has revealed also tubes with smaller and bigger diameters *(Fig.5, initial),* but their fractions are small.

RESULTS AND DISCUSSION

The Raman spectra of HipCO nanotubes registered under different power densities of Ar+-laser irradiation (λ=514.5 nm) are shown in *Fig.3.* It is seen that the contribution of the breathing modes with frequencies 250, 266 and 272 cm^{-1} (tube diameter range 8-10 Å) decreases, while the laser power density (P_l) increases. The modes disappear completely at P_l=560 W/cm^2. This transformation is irreversible. The P_l lowering doesn't result in the spectrum recovering *(Fig.3, the upper spectrum).* Due to observation of the effect in air only, it was interpreted as *a selective diameter-dependent oxidative etching of nanotubes.* The laser beam acts as a heater only. This has been confirmed by a reproduction of the selective etching effect in course of HipCO nanotube heating in an optical oven *(Fig.4).* The Raman spectra have been registered at low P_l=50 W/cm^2, excluding the laser heating.

In situ Raman measurements in oven have shown a suppression of a high-frequency wing of the breathing mode contour at T=150-400°C *(Fig.4).* At higher temperatures no Raman modes for tubes with diameter 8-10 Å have been observed. All other tubes have disappeared at 650°C. This value coincides well with thermo-gravimetry data [5]. A linear thermal shift of the tangential Raman mode position (1592 cm^{-1} at RT) has been calibrated in oven. This shift may serve as a temperature indicator in any other experiment. The shift values in air (-0.03 cm^{-1}/grad) and in vacuum (-0,01 cm^{-1}/grad) appeared to be different. The higher value in air may be explained by an increased thermal expansion of defective (after oxidation) nanotubes.

216

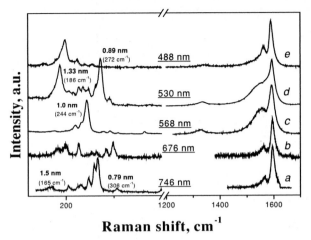

FIGURE 2. The Raman spectra of HipCO single-wall carbon nanotubes registered with different wavelength of laser excitation at room temperature.

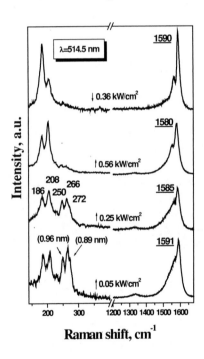

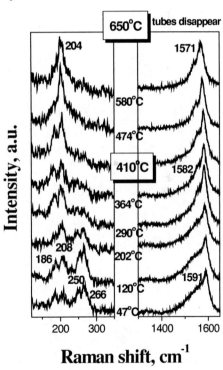

FIGURE 3. The Raman spectra of HipCO nanotubes registered under different power densities of the probing laser beam (in air). *λ= 514.5 nm.*

FIGURE 4. The Raman spectra of HipCO nanotubes registered *in situ* in course of heating of the material in an optical oven (in air). *λ= 514.5 nm*

217

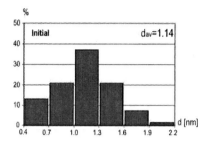

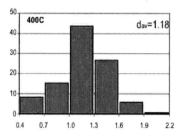

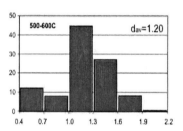

FIGURE 5. HRTEM statistical analysis of redistribution of HipCO nanotube diameters after heating of the material in air under different temperatures. The temperature values and the average nanotube diameter are indicated in Figure.

The temperature threshold values for a complete etching of nanotubes with diameter 8-10 Å have coincided for laser and oven heating processes (~400°C).

HRTEM images of HipCO nanotubes after heating in oven at different temperatures (in air) have been statistically averaged. The diagrams of tube diameter distribution have been built *(Fig. 5)*. They show a step-by-step decrease of small (7-10 Å) nanotube fraction, induced by an increase of the annealing temperature. As a result, the distribution maximum shifts toward higher values. Thus, HRTEM confirms the Raman data on selective oxidative etching of small HipCO nanotubes.

According to HRTEM images a pristine HipCO material includes a remarkable fraction of two-wall nanotubes. The inner shells of these tubes survive at high temperature, protected from oxidation by the outher shells. As a result, a relative content of the nanotubes with diameter less, than 7 Å *(Fig5, bottom)*, increases after annealing.

CONCLUSION

In situ Raman measurements have revealed the effect of a diameter–selective oxidative etching of small HipCO nanotubes under temperatures 150-400°C, taking place in course of laser or oven heating of the material in air. HRTEM data have confirmed the effect.

ACKNOWLEDGEMENTS

The work is supported by projects INTAS-237, RFBR 01-02-17358, RAS Program "Low-dimensional quantum structures".

REFERENCES

1. P. Nikolaev, M.J. Bronikovski, et al., *Chem Phys. Lett., 313(1999) 91.*
2. S.N. Bokova, E.D. Obraztsova, S.V. Terekhov, U. Dettlaff-Veglikowska, S. Roth, *Proc. of Int. Conference IQEC/LAT-2002, Moscow (Russia), June 22-28, 2002, p. 288.*
3. P.Corio, P.S. Santos, M.A. Pimenta, M.S. Dresselhaus, *Chem. Phys. Lett., 360 (2002)557.*
4. H. Kuzmany, W. Plank, H. Hulman, Ch. Kramberger et al., *Eur. Phys. J. B 22 (2000) 307.*
5. I.W. Chiang, B.E. Brinson, R.E. Smalley et al., *J.Phys. Chem. B, 105 (2001) 1157.*

Resonance Raman Scattering in Carbon Nanotubes and Nanographites

M. A. Pimenta[1], A. Jorio[1], M. S. S. Dantas[1], C. Fantini[1], M. de Souza[1], L. G. Cançado[1], Ge. G. Samsonidze[2], G. Dresselhaus[2], M. S. Dresselhaus[2], A. Grüneis[3], R. Saito[3], A. G. Souza Filho[4], Y. Kobayashi[5], K. Takai[5], K. Fukui[5], T. Enoki[5]

[1]*Departamento de Física, Univ. Federal de Minas Gerais, C.P. 702, Belo Horizonte, 30123-970 Brazil,*
[2]*Massachusetts Institute of Technology, Cambridge-MA, 02139-4307 USA,*
[3]*Tohoku University and CREST JST, Sendai 980-8578, Japan,*
[4]*Departamento de Física Universidade Federal do Ceará, Fortaleza-CE, 60455-760 Brazil,*
[5]*Department of Chemistry, Tokyo Institute of Technology, Meguro-ku, Tokyo,152-8551, Japan*

Abstract. In this work, we discuss the resonant Raman process in nanographites and carbon nanotubes, relating the most important Raman features to a first-order (single resonance) or a second-order (double resonance) process. We also show that, in the case of 1D systems, the term "resonance" has a more strict meaning and occurs when the energy of the photon does not simply coincide with the energy of a possible electron-hole pair, but rather matches the separation between van Hove singularities in the valence and conduction bands.

INTRODUCTION

Raman scattering is one of the most common techniques used to study and characterize single walled carbon nanotubes, since the Raman features of SWNTs are very intense, and easily observed even in the case of one isolated nanotube.[1] The anomalous intensity of the Raman spectra of SWNTs is related to the presence of van Hove singularities in the 1D density of electronic states, which concentrate many states in a narrow region of energy, since the signal is proportional to the number of electronic states which participate in the Raman process. Therefore, by changing the energy of the laser used in the experiment, it is possible to get direct information about the 1D electronic structure of the nanotubes from the intensity of the Raman signal.

Since the Raman spectrum of SWNTs is closely related to that of graphite, a number of works have been recently devoted to the study of the Raman spectra of graphite. Besides the first order Raman process, which gives rise to the most intense Raman peak of graphite (the G band), the Raman spectra of graphitic materials also exhibit second-order features, whose frequencies depend on the laser energy. The origin of these dispersive features, including the so-called D-band, is related to a special kind of second-order process with double resonance.[2-4]

In this work, we present the Raman spectra of nanographite on HOPG, and discuss the features associated with the first and second-order (double resonance)

CP685, *Molecular Nanostructures: XVII Int'l. Winterschool/Euroconference on Electronic Properties of Novel Materials,* edited by H. Kuzmany, J. Fink, M. Mehring, and S. Roth
© 2003 American Institute of Physics 0-7354-0154-3/03/$20.00

Raman processes. We will compare these results with similar features observed in carbon nanotubes, and discuss the nature of the Raman scattering process associated with the Raman features of SWNTs. We will show that both first and second-order processes are expected for carbon nanotubes, the first order being the main contribution for the intense RBM and G peaks.

EXPERIMENTAL DETAILS AND RESULTS

The nanographite sample consists of single graphite sheets prepared by a combination of electrophoretic deposition and heat-treatment of diamond nano-particles on a highly oriented pyrolitic graphite (HOPG) substrate. Details of the sample preparation are described in reference [5]. The Raman spectra were acquired on a DILOR XY spectrometer equipped with a CCD detector and coupled to an Olympus microscope (BTH2), which allows a Raman spectra with spatial resolution of about 2μm. The samples were irradiated with the 647.1, 568.2, 514.5, 488.0, 457.9 nm (1.92, 2.18, 2.41, 2.54, 2.71 eV, respectively) lines from an Ar/Kr laser. The power density incident on the sample was smaller than 10 μW/μm^2.

Raman spectra of nanographites

Figure 1 shows the spectra of a sample containing nanographite crystals deposited on a highly oriented pyrolytic graphite (HOPG). The spectrum (a) was taken in a region of the sample without nanographites, and the only feature is the sharp peak around 1580 cm^{-1}, usually called the G peak, which is associated with the zone center optic mode with E$_{2g}$ symmetry. The spectrum (b) of Fig. 1 was taken in a region of the sample containing nanographite crystals. Besides the strong G peak around 1580 cm^{-1}, we can observe two weak features around 1350 and 1620 cm^{-1}, which are usually called D and D* bands, respectively. These bands appear in all kind of disordered carbon materials. The interesting point here is that the width of these two features in Fig. 1(b) is much smaller than those normally observed for disordered carbon materials. Note that the D band is clearly composed by two peaks, as predicted by theory.[6]

The frequency of the D-band depends on the energy of the laser used in the experiment, and this dispersive behavior can be explained by a double resonance Raman scattering mechanism.[2,3] This mechanism involves phonons within the interior of the Brillouin zone of graphite and defects are needed for momentum conservation in the light scattering process. This process involves four steps – electron-hole creation, phonon scattering, defect scattering and electron-hole recombination – and double resonance occurs when two of these steps connect real electronic states. In fact, the double resonance process can occur for phonons belonging to different phonon branches of graphite. There are two kinds of double resonance processes in graphite – the intravalley and the intervalley mechanisms – the first one, involving phonons near the center of the Brillouin zone of graphite, and the second one, involving phonons near the K point.[4] Several dispersive weak features in the Raman spectra of graphitic materials have been explained by the double resonance mechanism, including the D band which is due to an intervalley process.[4] The fact that the D band is much stronger than the other double-resonance features is not yet completely understood. Possibly, this is due to the strength of the coupling of these particular phonons with the electrons involved in the Raman scattering.

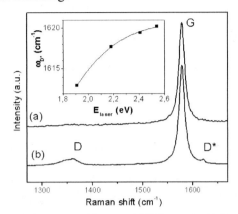

FIGURE 1. Raman spectra of the nanographite on an HOPG sample. (a) Spectrum obtained in a region of the sample without nanographites, showing only the spectrum of HOPG. (b) Spectrum of the nanographites, showing the G peak and the very sharp D and D* peaks. The inset shows the frequency of the D* peak as a function of the laser energy.

In the Raman spectra of disordered graphitic materials, the D* peak, around 1620 cm^{-1}, is very broad and usually appears as a shoulder of the G band. In the case of nanographites, it is significantly narrower (FWHM of 10 cm^{-1}), allowing us to determine precisely the frequency of this peak when we change the laser energy. Also the G band linewidth is narrower, allowing a clear separation of these two features. The inset of Fig. 1 shows the frequency of the D* peak of nanographites as a function of the laser energy. This dispersive behavior shows that this peak is also associated with an intravalley double resonance mechanism.[4] Note that this dependence has a non-linear behavior, reflecting the curvature of the upper optic branch of graphite near the center of the Brillouin zone.

Considering the dispersion of the π and π* electronic states of graphite, which are degenerate at the K point and whose energy separation increases by moving away from this point, the photon from a given laser line is always in resonance with a π-π* optical transition at a given point of the BZ. In this sense, the first-order Raman scattering of graphite, which gives rise to the G band, can be seen as a single resonance process. On the other hand, the D and D* bands are associated with a double resonance process, which involves a phonon and a defect. These peaks are, in general, weaker than the first-order G band, and are practically absent in the spectrum of HOPG (see Fig. 1(a)).

Resonant Raman scattering in carbon nanotubes

A carbon nanotube can be seen as a sheet of graphite rolled up into a cylinder, and as a first approximation we can generate both the electronic and phonon dispersion relations of the nanotubes by folding up the 2D Brillouin zone of graphite.[7] The folding of the π-electron dispersion relations gives rise to a 1D electronic structure characterized by sharp singularities (van Hove singularities) in the density of electronic states (DOS). Considering that the band gap of the semiconducting nanotubes is generally smaller than 1 eV, all photons from a laser beam with energy larger than 1 eV are in resonance with a given optical transition connecting states in the valence and conduction bands. However, only a discrete set of photon energies is in resonance with optical transitions involving electronic states at the van Hove singularities. Therefore, in the case of 1D systems, we have a more strict meaning for the term "resonance". In this sense, we say that a photon is resonant with a given nanotube when its energy not only matches the energy of a possible electron-hole pair, but rather is associated with the separation between the van Hove singularities in the valence and conduction bands.

In a typical spectrum of a carbon nanotube, the most important features are the radial breathing mode (around 200 cm^{-1}), the D band (around 1350 cm^{-1}), and the tangential band (between 1500-1600 cm^{-1}).[8] Similarly to the case of nanographites, both single and double resonance Raman processes can occur in carbon nanotubes. The D band (and its overtone, the G' band) is explained by a double resonance mechanism,[9-12] involving electronic states at the van Hove singularities.[10-12] The phonons involved in this process connect electronic states belonging to different electronic sub-bands. Very weak Raman features are also observed in intermediate frequency regions and they also exhibit a dispersive behavior.[13,14] Similarly to the case of nanographites, the intermediate Raman bands of nanotubes are also associated with a double resonant mechanism in SWNTs.[14]

The resonant behavior of the radial breathing mode is quite different from the dispersive double resonant features. In the case of SWNT bundles, the shape of the RBM Raman band changes completely when the laser energy varies, and this result can be explained by the fact that, in a sample containing a distribution of different SWNTs, each laser line is resonant (in the sense of resonance with van Hove singularities) with a different sub-set of nanotubes present in the bundle.[8] For an isolated SWNT, the RBM peak appears only when the incident or scattered photon is in resonance with the separation between singularities in the DOS, and the E_{laser} dependence of the RBM intensity reflects directly the joint density of electronic states (JDOS).[15] This result is explained by a first-order resonant Raman scattering process, and occurs in the absence of defects in the SWNT structure.

Let us now discuss the possible scattering mechanisms associated with the G band of SWNTs, between 1500 and 1600 cm^{-1}. This band originates from the phonon branch associated with the E_{2g} mode of graphite. Due to the folding of the 2D phonon dispersion of graphite, phonons within the interior of the 2D BZ become zone center modes in the 1D BZ, and those with A_{1g}, E_{1g} and E_{2g} symmetries (A, E_1 and E_2 for non-symmorphic nanotubes) are allowed in a first order Raman scattering.[7] Due to the anisotropy of the force constants, each one of these modes splits into two components related to atomic vibrations parallel and perpendicular to the nanotube axis.

Figure 2(a) shows a single resonance Stokes Raman process, involving a zone-center phonon, which is commonly used to explain the G band in SWNTs.[1,7-8] In this figure, the energy of the incident photon corresponds to the separation of the van Hove singularities, and therefore this is a real resonant process as discussed in the beginning of this section, since there is a huge number of electronic states involved in the scattering mechanism (we are considering a van Hove singularity with an overestimated width of 10 meV, and a phonon associated with the G band with energy of 200 meV). Figure 2(b) shows a double-resonance process within the same electronic sub-band, involving a phonon near the center of the 1D Brillouin zone and a defect. This mechanism has been proposed as an alternative explanation for the G band of SWNTs.[16] Note that, in this case of the double resonance mechanism, the electronic states involved in the Raman process are far from the van Hove singularities. Despite the fact that two steps are resonant (connecting real states), only a small number of electronic states participate in this process. Therefore, in the sense of resonance with van Hove singularities, the double resonant process shown in Fig. 2(b) is, in fact, a non-resonant process.

(a)   (b)

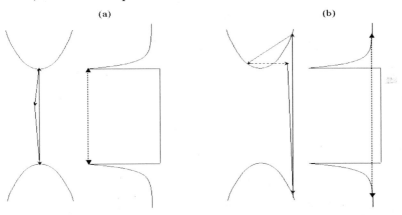

FIGURE 2. Electronic sub-bands and density of electronic states of a carbon nanotube. (a) Single resonance Raman scattering involving the creation of one phonon. The incident photon is in resonance with the van Hove singularities in the valence and conduction bands. (b) Double resonance Raman scattering within the same sub-band, involving a phonon and a defect. In this case, the electron-hole pair energy is much higher than the separation between the van Hove singularities.

In principle, both the single and double resonance processes can contribute to the G band of carbon nanotubes. The single resonance process involves a very large number of electronic states, occurs only for a sharp window of photons, and reflects directly the singularities in the joint density of electronic states.[17] The double resonance process involves a small number of electronic states, and can occur only if there is a defect for momentum conservation.

Similarly to the case of graphite, where the G peak is the most intense feature in the first-order Raman spectrum and is the only peak in the spectrum of a perfect graphite crystal, the G band is also the most intense feature in the Raman spectra of SWNTs. Moreover, the G band of SWNTs is much more intense than the D band, which is known to be the most important double resonance Raman feature for sp^2 carbons. Therefore, the main contribution for the G band of SWNTs comes from the first-order resonant process. On the other hand, we can observe in the Raman spectrum of nanographite (Fig. 1) the weak D* peak of nanographites, which is a double resonant feature associated with phonons near the center of the 2D BZ. We can thus conclude that the G band of SWNTs may also have a small contribution from a double resonance process, much smaller however than the first-order contribution, which is resonant with the van Hove singularities.

ACKNOWLEDGMENTS

This work is supported by CNPq/NSF joint collaboration program (NSF INT 00-00408 and CNPq 910120/99-4) and Instituto de Nanociências, Brazil. A. J. and A.G. S. F. acknowledge financial support from CNPq and CAPES, Brazil. The MIT authors acknowledge support under NSF DMR 01-16042. R. S. acknowledges a Grant-in-Aid (No. 13440091) from the Ministry of Education, Japan.

REFERENCES

1. M. S. Dresselhaus, G. Dresselhaus, A. Jorio, A. G. Souza Filho and R. Saito, Carbon vol. 40, 2043 (2002).
2. A. V. Baranov, A. N. Bekhterev, Y. S, Bobovich, V. I. Petrov, Opt. Spectrosc. 62, 612 (1987).
3. C. Thomsen and S. Reich, Phys. Rev. Letters 85, 5214 (2000).
4. R. Saito, A. Jorio, A. G. Souza Filho, G. Dresselhaus, M.S. Dresselhaus, M.A. Pimenta, Phys. Rev. Letters, 588, 027401 (2002).
5. A. M. Affoune, B. L. V. Prasad, H. Sato, T. Enoki, Y. Kaburagi, and Y. Hishiyama, Chemical Physics Letters, 348, 17 (2001).
6. L. G. Cançado, M. A. Pimenta, R. Saito, A. Jorio, L. O. Ladeira, A. Grüneis, A. G. Souza Filho, G. Dresselhaus, M. S. Dresselhaus, Phys. Rev. B, vol. 66, 035415 (2002).
7. R. Saito, G. Dresselhaus, M.S. Dresselhaus, Physical Properties of Carbon Nanotubes (Imperial College Press, London, 1998).
8. A. M. Rao, E. Richter, S. A. Bandow, B. Chase, P. C. Eklund, K. W. Williams, M. Menon, K. R. Subbaswamy, A. Thess, R. E. Smalley, G. Dresselhaus, M. S. Dresselhaus, Science vol. 275, 187, (1997).
9. J. Maultzsch, S. Reich, C. Thomsen, Physical Review B, vol. 64, 121407 (2001).
10. J. Kurti, V. Zolyomi, A. Grüneis, H. Kuzmany, Phys. Rev. B, vol. 65, 165433 (2002)
11. A. G. Souza Filho, A. Jorio, G. Dresselhaus, M. S. Dresselhaus, R. Saito, A. K. Swan, M. S. Ünlü, B. B. Goldberg, J. H. Hafner, C. M. Lieber, and M. A. Pimenta, Phys. Rev. B, vol. 65, 035404 (2002).
12. Ge.G. Samsonidze, R. Saito, A. Jorio, A.G. Souza Filho, A. Grüneis, M. A. Pimenta, G. Dresselhaus, and M. S. Dresselhaus. Physical Review Letters, vol. 90, 027403 (2003).
13. V. W. Brar, Ge.G. Samsonidze, M. S. Dresselhaus, G. Dresselhaus, R. Saito, A. K. Swan, M. S. Ünlü, B. Goldberg , A.G. Souza Filho, A. Jorio, Physical Review B, vol. 66, 155418 (2002).
14. L. Alvarez, A. Righi, S. Rols, E. Anglaret, and J.L. Sauvajol, Chem. Phys. Lett., vol. 320, 441 (2000)
15. A. Jorio, A. G. Souza Filho, G. Dresselhaus, M. S. Dresselhaus, R. Saito, J. H. Hafner, C. M. Lieber, F. M. Matinaga, M. S. S. Dantas, M. A. Pimenta, Physical Review B, vol. 63, 245416 (2001).
16. J. Maultzsch, S. Reich , C. Thomsen, Physical Review B, vol. 65, 233402 (2002).
17. A. Jorio, M. A. Pimenta, A.G. Souza Filho, Ge.G. Samsonidze, A. K. Swan, M. S. Ünlü, B. Goldberg , R. Saito, G. Dresselhaus, M. S. Dresselhaus, Phys. Rev. Lett., vol. 90, 107403 (2003).

Double-Resonant Raman Scattering in an Individual Carbon Nanotube

C. Thomsen*, J. Maultzsch* and S. Reich[†*]

*Institut für Festkörperphysik, Technische Universität Berlin, Hardenbergstr. 36, 10623 Berlin,
Germany
[†]Institut de Ciència de Materials de Barcelona, Campus de la U.A.B., E-08193 Bellaterra,
Barcelona, Spain

Abstract. Experiments on nearly individual carbon nanotubes were performed showing that double-resonance takes place even on the level of an individual tube. Important consequences for the determination of diameter, chirality and defect concentrations are discussed.

Double-resonant Raman scattering has become a well established process describing a number of features of the Raman spectra of sp^2 bonded carbon compounds such as graphite or carbon nanotubes.[1–4] Characteristic for the double resonance, which involves a photon and a phonon both being resonant in the Raman process, is the excitation-energy dependence of the observed phonon energy. For the D-mode in graphite, bundles of single-walled nanotubes and so-called bucky pearls this shift amounts to 50-60 cm^{-1}/eV excitation energy and is much larger than when observed, e.g., in GaAs quantum wells [5] or Ge [6]. The high-energy mode (HEM) at 1590 cm^{-1},

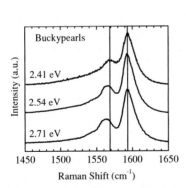
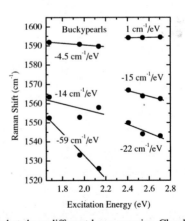

FIGURE 1. *left:* Raman spectra of bucky pearls excited at three different laser energies. Clearly seen is the downshift of the second largest peak for increasing phonon energy. *right:* Peak frequencies in the range of 1.7 to 2.7 eV excitation energy. All peaks have an excitation-energy dependence; it is due to the double-resonance process. The jump in absolute phonon energies and slopes at 2.3 eV is due to a higher electronic band involved in the double resonance as explained in Ref. [2]

CP685, *Molecular Nanostructures: XVII Int'l. Winterschool/Euroconference on Electronic Properties of Novel Materials,*edited by H. Kuzmany, J. Fink, M. Mehring, and S. Roth
© 2003 American Institute of Physics 0-7354-0154-3/03/$20.00

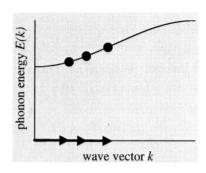

FIGURE 2. *left:* Schematic double-resonant process for a metallic carbon nanotube; different incident phonon energies imply different phonon wave vectors in the double resonance. The defect scattering process and the recombination are not shown for simplicity. *right:* The different scattering vectors correspond to different phonon frequencies.

referred to sometimes as *G*-mode, also displays excitation-energy dependent shifts;[2] they are smaller and sometimes missed in the literature when not evaluated over a large enough energy range.[7] In Fig. 1 we present the Raman frequencies of bucky pearls, which clearly show how the 2nd and 3rd largest peak in the HEM shift to lower energies when excited with increasing photon energies. In contrast to this behavior is that of the *G*-mode in graphite, which is constant in energy for excitations ranging from 1 to 4 eV [8] and thus singly resonant.

In this paper we discuss to which extent the excitation-energy dependence of the Raman spectra in carbon nanotubes is a property of an individual nanotube or whether it is an ensemble property, where specific tubes are selected by the excitation energy for the Raman process. We show with experiments on individual nanotubes that the Raman signal is indeed dominated by the band structure of an individual tube and that the ensemble interpretation may be ruled out as the dominant reason for the excitation-energy dependent shifts in nanotubes. We also discuss the consequences for the interpretation of Raman data as regards the distinction of *metallic* and *semiconducting* spectra, the use of Raman peaks to determine the radius of a nanotube, and its defect concentration.

The double-resonant process is illustrated schematically in Fig. 2. Different wavevectors of phonons corresponding to different incident phonon energies are shown, they scatter the excited electron across the band minimum. For phonons with dispersion, such as the one indicated, the different *k*-vectors correspond to different phonon energies, and hence the excitation-energy dependence of the phonon peak follows naturally. In order to fulfill momentum conservation (both incident and scattered photon momentum are very small) the electron has to be scattered back near to where it was excited. This process is usually ascribed to an elastically scattering defect (*D*-mode and high-energy mode) or a second phonon (2nd order mode scattering). Of course, in a full calculation of the Raman intensity in double resonance the single resonant process (only the photon is resonant) is automatically included; however, its contribution to the total signal is small.[2]

In an alternative attempt to explain the excitation-energy dependence some authors have been focussing on the strength of the van-Hove singularity in the electronic

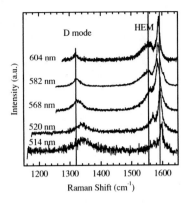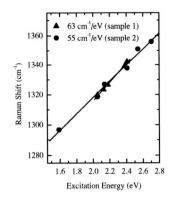

FIGURE 3. *left:* Raman spectra on an isolated or a nearly isolated nanotube; *right:* frequency-dependence of the *D*-mode for two different isolated nanotubes

transition.[3, 9] In this picture the view is taken that a nanotube contributes to the Raman signal significantly only if the optical transition occurs very close to the maximum in the density of states. Varying the excitation energy then selects a different nanotube with a different phonon energy. In this way, an ensemble of tubes with a typical diameter distribution may yield varying Γ-point phonon frequencies.

In summary, in the double-resonant process the excitation-energy dependence reflects the phonon dispersion in an individual nanotube, while in the single-resonant process it resembles a diameter and chirality distribution. Because of the important consequences for the interpretation of the Raman spectra a distinction between these two processes is vital. Obviously, experiment can decide this by measuring the excitation-energy dependence on a single isolated tube, which only in double resonance can have a varying peak frequency in the Raman spectra. In single resonance we would expect only an intensity dependence on excitation energy, but no shift in the Raman frequencies.

For the single-tube experiments we used HiPCo produced nanotubes, which were solution cast onto marked substrates. The density of tubes was ~ 0.5 tubes/μm^2. We used various laser lines of an Ar/Kr lasers and a number of frequencies of a dye laser for excitation. The spectra were dispersed by a Dilor XY triple spectrometer and detected by a CCD detector.

The spectra so obtained are shown in Fig. 3. On the left we show the high-energy region and make the following observations: 1) the *D*-mode, as expected for double resonance, shifts continuously to higher energies with 55-65 cm^{-1}/eV (right of Fig. 3). 2) The high-energy mode frequency varies as well, the highest peak first decreases, then increases; the second largest peak increases monotonically. 3) The lineshape of the high-energy mode changes continuously from a more metallic to a more semiconducting appearance. In the low-energy region (not shown) we find, next to the second-order acoustic peak of Si (~ 300 cm^{-1}), a strong RBM mode (250 cm^{-1}) which slightly shifts to lower energy (245 cm^{-1}) when increasing the phonon energy and then becomes weaker, disappearing for the highest photon energies shown. There is a small second RBM mode at 200 cm^{-1} which also disappears at large excitation energies. There is

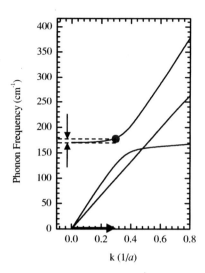

FIGURE 4. Expanded view of the radial breathing mode near the Γ-point of the Brillouin zone of a (10,10) tube. Indicated is the shift of the peak frequency observed in double-resonant Raman scattering compared to the true Γ-point frequency.

thus a single tube or possibly a very thin bundle (two or three tubes) in the experiment shown.

All observations fall naturally into the double-resonance picture and are incompatible with the single resonance interpretation. 1) The D-mode in the individual tube or thin bundle shifts at the same rate as in bulk samples; 2) The high-energy mode resembles the dispersion of a metallic tube. In such tubes the LO-like mode has been shown to soften to about 1550 cm^{-1}.[10] 3) The lineshape change from metallic-looking to semiconducting-looking occurs continuously and is due to scattering near the band minimum ($\sim$ 2.0 eV) and far away from it ($\sim$ 2.4 eV). Detailed calculations showing this lineshape dependence on excitation energy are under way.[11]

We discuss now an important implication for the radial breathing mode, which is frequently used to determine with high accuracy the diameter of a nanotube.[9, 12] While the inverse dependence of its frequency is generally accepted (for deviations at small diameter, see Ref. [13]), the Raman peak frequency, if double resonant, however, does not neccessarily correspond to the Γ-point frequency of this mode. In Fig. 4 we show on an expanded scale the low-energy, low-k region of the phonon dispersion relations of a (10,10) nanotube.[14] For a typical double-resonant phonon wave vector ($\sim$ 0.3 1/a) the dispersion-induced shift corresponds to $\sim$ 10 cm^{-1} in this specific tube. Obviously, the shift depends on the phonon dispersion and on the incident photon energy, but it should be clear that the simple RBM frequency-tube-diameter correspondence does not hold to high accuracy when double resonance is the dominant Raman mechanism.

Finally we discuss briefly the role double resonance has for the determination of defect concentrations in nanotubes. Because the first-order scattering is a defect-induced process, both the D and the G mode intensity are a good measure for the defect concentration. The second-order D^*-mode, on the other hand, is Raman allowed due to the two phonons of equal and opposite momenta involved. For not too large concentrations of defects its intensity may thus be taken as constant and used to normalize the spectra which could be affected, e.g., by a change in absorption due to the defects. In this way we found a useful relative measure of the defect concentrations in nanotubes.[15]

In summary, excitation-energy dependent Raman experiments on nearly isolated tubes have established double-resonant Raman scattering as the dominant mechanism for carbon nanotubes. Diameter and chirality selective scattering, at least for the D and the high-energy mode, can be dismissed. We showed a number of important consequences for the interpretation of the RMB frequency as related to the tube diameter, for the metallic and semiconducting appearance of the high-energy mode spectra and for the determination of defect concentrations in nanotubes.

We thank U. Schlecht and M. Burghard for providing us with samples on marked substrates for these experiments.

REFERENCES

1. C. Thomsen and S. Reich, Phys. Rev. Lett. **85**, 5214 (2000)
2. J. Maultzsch, S. Reich, and C. Thomsen, Phys. Rev. B **65**, 233402 (2002)
3. J. Kürti, V. Zólyomi, A. Grüneis, and H. Kuzmany, Phys. Rev. B **65**, 165433 (2002)
4. R. Saito, A. Jorio, A. G. Souza Filho, G. Dresselhaus, M. S. Dresselhaus, and M. A. Pimenta, Phys. Rev. Lett. **88**, 027401 (2002)
5. V.F. Sapega, M. Cardona, K. Ploog, E.L. Irchenko, and D.N. Mirlin, Phys. Rev. B **45**, 4320 (1990)
6. D.J. Mowbray, H. Fuchs, D.W. Niles, M. Cardona, C. Thomsen, B. Friedl, 20th International Conference on the Physics of Semiconductors, eds. E.M: Anastassakis and J.D. Joannopoulos, Thessaloniki, Greece, (World Scientific, Singapore, 1990), p. 2017
7. A. Jorio, M. A. Pimenta, A. G. Souza Filho, Ge. G. Samsonidze, A. K. Swan, M. S. Ünlü, B. B. Goldberg, R. Saito, G. Dresselhaus, and M. S. Dresselhaus, Phys. Rev. Lett. **90**, 107403 (2003)
8. I. Pócsik, M. Hundhausen, M. Koos, O. Berkese, and L. Ley, in Proceedings of the XVI International Conference on Raman Spectroscopy, ed. A.H. Heyns (Wiley-VCH, Berlin, 1998) p. 64
9. A. Jorio, R. Saito, J.H. Hafner, C.M. Lieber, M. Hunter, T. McClure, G. Dresselhaus, and M. Dressselhaus, Phys. Rev. Lett. **86**, 118 (2001)
10. O. Dubay, G. Kresse, and H. Kuzmany, Phys. Rev. Lett. **88**, 235506 (2002)
11. J. Maultzsch *et al.*, to be published
12. A. Kukovecz, C. Kramberger, V. Georgakilas, M. Prato, and H. Kuzmany, Eur. Phys. J. B **28**, 223 (2002) and R. Pfeiffer, C. Kramberger, C. Schaman, A. Sen, M. Holzweber, H. Kuzmany, H. Kataura, Y. Achiba, this volume
13. J. Kürti and V. Zólyomi, this volume
14. M. Damnjanović, T. Vuković and I. Milošević, J. Phys. A **33**, 6561 (2000) and J. Maultzsch, S. Reich, C. Thomsen, E. Dobardžić, I. Milošević, M. Damnjanović, Solid State Commun. **121**, 471 (2002)
15. J. Maultzsch, S. Reich, C. Thomsen, S. Webster, R. Czerw, D.L. Carroll, S.M.C. Vieira, P.R. Birkett, and C.A. Rego, Appl. Phys. Lett. **81**, (17) 2647-49 (2002)

Photoluminescence of Single-Walled Carbon Nanotubes: Estimate of the Lifetime and FTIR-Luminescence Mapping

Sergei Lebedkin [a], Katharina Arnold [a,c], Frank Hennrich[a], Ralph Krupke [a], Burkhard Renker [b], and Manfred M. Kappes *[,a,c]

[a] Forschungszentrum Karlsruhe, Institut für Nanotechnologie and [b] Institut für Festkörperphysik, D-76021 Karlsruhe, Germany
[c] Institut für Physikalische Chemie, Universität Karlsruhe, D-76128 Karlsruhe, Germany

Abstract. We have applied the efficient FTIR-Luminescence technique to map the near-infrared photoluminescence (PL) of individual micelle-isolated single-walled carbon nanotubes (SWNTs) in a broad UV-visible-NIR excitation range. Monochromatic light of xenon and tungsten halogen lamps was used for PL excitation. Such PL maps provide a rich information about characteristic electronic transitions in SWNTs. The PL intensity shows a nonlinear behavior, when excited with nanosecond laser pulses. This is likely due to quenching interactions between multiple electronic excitations generated and propagating in nanotubes. This effect is similar for different, selectively excited nanotubes and allows us to estimate the PL lifetime (electronic relaxation time) in micelle-isolated nanotubes as ~100 ps.

INTRODUCTION

The recently discovered near-infrared band-gap PL of dispersed, micelle-isolated semiconducting SWNTs [1,2] has opened new perspectives for the study of electronic properties of SWNTs, analysis and structural assignment of nanotubes in various SWNT materials, development of methods for a selective separation of nanotubes with different diameters and chirality (described by the roll-up indices n,m), etc.

The luminescence was first observed for SWNTs produced by high-pressure catalytic decomposition of carbon monoxide (the HiPco method) [1]. These tubes have diameters between ~0.7 and ~1.2 nm and emit in the range of ~900-1400 nm. We have found that water/ surfactant dispersions of SWNTs with larger diameters (~1.1–1.5 nm) produced by pulsed laser vaporization (PLV) of Ni/Co/carbon targets show a similar PL [3]. The luminescence of SWNTs show characteristic sharp excitation–emission maxima corresponding to electronic interband transitions E_{ii}^S between van Hove singularities (vHs) of specific (n,m) nanotubes [1-3]. Interband transition energies vary roughly inversely with the tube diameter [4], therefore the emission of PLV nanotubes (E_{11}^S transitions) is red-shifted to the range from ~1250 up to ≥1700 nm.

We have recently succeeded in a preparation of dispersions with a high fraction of individual PLV nanotubes. This is evidenced by a fine structure of the absorption bands (Fig. 1) and an average quantum yield of PL of ~10^{-3}, i.e. approaching to that of the HiPco nanotubes [1]. Here we show that the PL of such relatively efficiently emitting dispersions of SWNTs can be rapidly (several hours) mapped in the whole UV-visible-NIR excitation range, i.e. from E_{11}^S up to E_{44}^S, E_{55}^S, using a sensitive FTIR-

CP685, Molecular Nanostructures: XVII Int' l. Winterschool/Euroconference on
Electronic Properties of Novel Materials, edited by H. Kuzmany, J. Fink, M. Mehring, and S. Roth
© 2003 American Institute of Physics 0-7354-0154-3/03/$20.00

Luminescence technique for detection and standard monochromatic lamp light sources for excitation. We also estimate the PL lifetime based on the nonlinear behavior of PL under a ns-pulsed laser excitation.

EXPERIMENTAL

The PLV nanotubes were produced as described elsewhere [5]. Raw SWNTs were added to D_2O (transparent up to ~1800 nm) with 1 wt.% of the sodium dodecylsulfate (SDS) surfactant and treated for several hours with a powerful ultrasonic dispergator. After centrifugation for 2 hours at 180.000g, a supernatant dispersion was collected and used for luminescence measurements. A dispersion of HiPco tubes in D_2O/ SDS was obtained from the Rice University.

Photoluminescence spectra of dispersed SWNTs were measured in the range of ~800–1700 nm on a Bruker Equinox 55S/ FRA106 FTIR-Raman spectrometer equipped with a liquid nitrogen cooled germanium detector (Edinburgh Instruments). A monochromatic light of a 400 W xenon lamp and a 150 W tungsten halogen lamp was used for excitation with a spectral width of 8-12 nm. An average quantum yield of the PL of SWNTs (the ratio of photons emitted in the whole PL spectrum to total photons absorbed) was estimated from a comparison with the phosphorescence of singlet oxygen $^1O_2$ photosensitized by C_{70} as described elsewhere [3]. For a ns-pulsed laser excitation of nanotubes in a broad wavelength range, a Coherent Panther OPO-Nd:YAG laser system was applied. Pulsed luminescence was detected with a Hamamatsu R5509 NIR-photomultiplier. All measurements were performed at room temperature using standard 4 x 10 mm quartz cuvettes in a 90°-geometry.

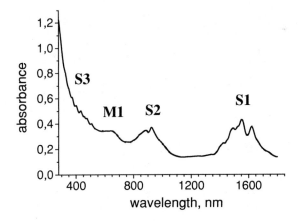

FIGURE 1. Optical absorption spectrum of SWNTs prepared by the pulsed laser vaporization method, dispersed at high ultrasonic power in D_2O/ SDS and centrifuged at 180.000 g. The S1, S2, S3 and M1 absorption bands correspond to the E_{11}^S, E_{22}^S, E_{33}^S and E_{11}^M electronic transitions in semiconducting and metallic SWNTs, respectively.

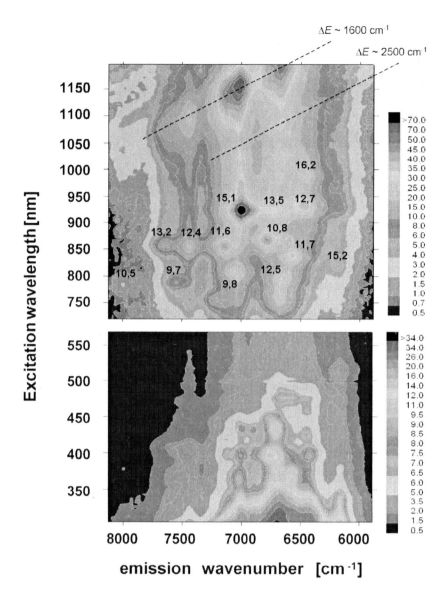

FIGURE 2. FTIR-Luminescence contour map of PLV nanotubes dispersed in $D_2O/$ SDS and excited with a monochromatic light of a xenon lamp (bottom) and a tungsten halogen lamp (top). The (n,m) structural indices were assigned according to Ref. 2. The dashed lines indicate vibronic features (see text). The luminescence intensity is corrected for the excitation intensity, but not corrected for the FTIR spectrometer response.

RESULTS AND DISCUSSION

FTIR-Luminescence mapping of SWNTs. The PL maxima on the FTIR-Lumines-cence contour map for PLV nanotubes (Fig. 2) correspond to the E_{11}^S (emission) – E_{ii}^S (excitation) electronic transitions of specific (n,m) tubes and demonstrate characteristic distribution patterns along the excitation axis similar to those of HiPco nanotubes [2]. The peaks observed in the E_{22}^S excitation region ($S2$ absorption at ~750-1000 nm) agree well with those observed for a moderately sonicated dispersion of PLV nanotubes with a small fraction of individual tubes, but show different relative intensities [3]. This characteristic group of PL peaks was assigned to specific (n,m) nanotubes (Fig. 2) according to empirical relations of Weisman et al. [2].

A simple tight-binding (TB) model of SWNTs [4] provides only a qualitative description of the experimental PL data [1-3]. We have found a much better agreement with a modified model of Ding et al., which takes into account a curvature (chirality) of a nanotube surface [6]. With fitting parameters Hagen and Hertel [7], the modified model reasonably describes the PL peaks pattern in the E_{22}^S excitation region. However, because of the large theoretical number of different semiconducting nanotubes in the PLV material (~30), an unambiguous (n,m) assignment is possible only for a few tubes and for these tubes it is consistent with the relations of Weisman et al.

The PL map shows practically no features in the excitation range of ~500-750 nm corresponding to the E_{11}^M electronic transitions of metallic (non-luminescent) PLV nano-tubes. Below λ_{exc} ~500 nm, a rich pattern of PL peaks corresponding to the E_{33}^S, E_{44}^S and even E_{55}^S excitations is observed. Their assignment with the help of the modified TB model is in progress. Note that E_{44}^S, E_{55}^S transitions practically can not be identified in the absorption spectrum of SWNTs because of the strong plasmon background (Fig. 1).

Another interesting group of features on the PL map are broad and strong vibronic bands shifted relative to E_{11}^S at the characteristic Raman frequencies of SWNTs ($h\omega$ ~1600 and ~2500 cm^{-1}) as marked with the dashed lines in Fig. 2. Their origin is not yet clear. Note that no corresponding bands at $E_{11}^S + h\omega$ could be observed in the absorption spectrum (Fig. 1).

Non-linear effects in the photoluminescence of carbon nanotubes. It has been pointed out that the PL intensity of HiPco nanotubes irradiated with a ns-pulsed N_2-laser is not proportional to the excitation power [1]. We have used an OPO-Nd:YAG laser system to study this effect for different HiPco nanotubes selectively excited at different laser wavelengths in the $S2$ absorption range. A similar behavior was observed for all nanotubes: by increasing the laser intensity I, PL signals are at first proportional to I (like by a steady-state excitation), then they start to deviate from the linear dependence (Fig. 3), and probably approach a saturation at very high laser intensities (not reached in these experiments). Our preliminary results indicate a similar behavior also for PLV nanotubes.

The non-linear behavior can be explained by possible quenching interactions between multiple electronic excitations generated by a strong laser excitation pulse in a single nanotube. If this explanation is correct, it implies that (i) the electronic excitations are rather free to move along the tubes and (ii) their relaxation is not very fast. From a rough estimate of the onset of the PL non-linearity at ~100 photons absorbed per tube (Fig. 3b) within a laser pulse of ~5 ns, we conclude that the PL lifetime (or a characteristic time of

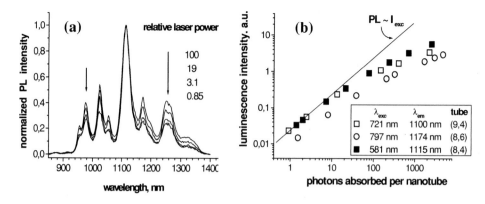

FIGURE 3. Non-linear effects in the luminescence of HiPco nanotubes under a ns-pulsed laser excitation. **a)** Normalized luminescence spectra excited at 581 nm. The luminescence peaks corresponding to different (n,m) tubes show different changes in the PL intensity vs. excitation power. **b)** Dependence of the PL intensity on the average number of photons absorbed per nanotube in each laser pulse. This number was roughly estimated from the concentration of nanotubes (~10 µg/ml) and their average length (~200 nm).

the relaxation of electronic excitations in micelle-isolated HiPco nanotubes) is of the order of ~100 ps. It is much slower than the femtosecond electronic relaxation observed in the solid samples of (bundled) PLV nanotubes [8]. One can expect that electronic relaxation processes depend on the nanotube structure. Indeed, despite a general similarity, relations between the PL intensity and excitation laser power depend also on the nanotube type and excitation wavelength (Fig. 3).

ACKNOWLEDGEMENTS

The support of this work by the Deutsche Forschungsgemeinschaft under SFB 551 and by the BMBF is gratefully acknowledged. The authors thank Prof. R. E. Smalley for a sample of HiPco carbon nanotubes.

REFERENCES

1. O'Connell, M. J., Bachilo, S. M., Huffman, C. B., et al., *Science* **297**, 593 (2002).
2. Bachilo, S. M., Strano, M. S., Kittrell, C., et al., *Science*, **298**, 2361 (2002).
3. Lebedkin, S., Hennrich, F, Skipa, T., Kappes, M. M., *J. Phys. Chem. B***107**, 1949 (2003).
4. Saito, R., Dresselhaus, G., Dresselhaus, M. S., *Physical Properties of Carbon Nanotubes*; Imperial College Press: London, 1998.
5. Lebedkin, S., Schweiss, P., Renker, B., Malik, S., Hennrich, F., Neumaier, M., Stoermer, C., Kappes, M. M., *Carbon* **40**, 417 (2002).
6. Ding, J. W., Yan, X. H., Cao, J. X., *Phys. Rev. B***66**, 073401 (2002).
7. Hagen, A., Hertel, T., *Nanolett.* **3**, 383 (2003).
8. Hertel, T.; Moos, G. *Chem. Phys. Lett.* **320**, 359 (2000).

Multi-spectroscopic investigation of the structure of single-wall carbon nanotubes

Nicolas Izard[1,2], Didier Riehl[1] and Eric Anglaret[2]

1-DGA/DCE/CTA/LOT, 16, bis Avenue Prieur de la Côte d'Or, 94114 Arcueil Cedex, France
2-GDPC, Université Montpellier II, Place E.Bataillon, 34095 Montpellier Cedex 5, France

Abstract. We present a multispectroscopic structural study of various nanotube samples with differents tube diameters. We determine for each sample the mean bundle and tube diameter as well as the tube diameter distribution. The possibility to work on SWNT of various structural characteristics opens new opportunities to correlate the nanotube structure and their physical properties.

INTRODUCTION

Single-Wall Carbon Nanotubes (SWNTs) display unique electronic properties due to their monodimensionality. Their electronic properties as well as their linear and non-linear optical properties directly depend on their structure. In particular, for a better understanding of the optical limiting properties of SWNTs, detailed studies of their structure and electronic properties are required [1].

In these proceedings, we present a multispectroscopic study of SWNT samples produced by different techniques. All samples were studied by scanning and high-resolution transmission electron microscopy, X-ray diffraction, Raman spectroscopy and linear optical absorption. The correlations obtained between these different techniques will be emphasized.

EXPERIMENTAL

SWNTs samples were provided by commercial sources. We studied three kinds of nanotubes : i) SWNTs synthesized by the electric arc process [2] and purified by a multi-step acid treatment, ii) SWNTs produced by pulverisation laser vaporisation, using a double-pulse laser [3], iii) SWNTs synthesized by catalytic decomposition with the HiPCO process [4]. In this paper, these samples will be refered as Electric Arc (EA), Laser and HiPCO, respectively.

Scanning electron microscopy (SEM) images were recorded with a field emission microscope JEOL JSM 6300F. High-resolution transmission electron microscopy (HRTEM) experiments were performed on a 200 kV Philips CM20 microscope. X-ray diffraction data (XRD) were collected using the K_α radiation of a Cu source (λ=1.542 Å) and a curve position sensitive detector (INEL-CPS 120). Raman spectra were measured on a dispersive Jobin-Yvon T64000 spectrometer using the 488, 514.5 and

CP685, *Molecular Nanostructures: XVII Int'l. Winterschool/Euroconference on*
*Electronic Properties of Novel Materials,*edited by H. Kuzmany, J. Fink, M. Mehring, and S. Roth
© 2003 American Institute of Physics 0-7354-0154-3/03/$20.00

647.1 nm lines of an Ar/Kr laser, and on a FT Bruker RFS100 using the fundamental laser line of a Nd:YAG laser at 1064 nm. Optical absorption spectra were recorded with an UV-Visible-NIR Cary 500 spectrophotometer. For these latter measurements, SWNTs were first dispersed in ethanol and then pulverized and dried on an optical glass substrate.

RESULTS

SEM and HRTEM pictures typical of each sample are presented in Fig. 1. From SEM pictures, one remarks that EA and HiPCO SWNTs are of very good purity. The laser sample is also of good purity, but one can observe small amounts of carbon nanoparticles. On HRTEM pictures, one observes that for all samples, tubes assemble into crystalline bundle structures. Note also that at this magnification, nanometer-size catalyst particles are observed on the surface of HiPCO bundles. The mean bundle diameter was estimated from the SEM pictures and the results are reported in Table 1. SWNTs diameter was also estimated from several HRTEM pictures. An FFT-based spatial image filtering was used to measure the periodicity of the SWNTs bundle lattice. The mean-diameter of the nanotubes was estimated from the lattice periodicity, assuming that SWNTs forms a hexagonal close-packed structure.

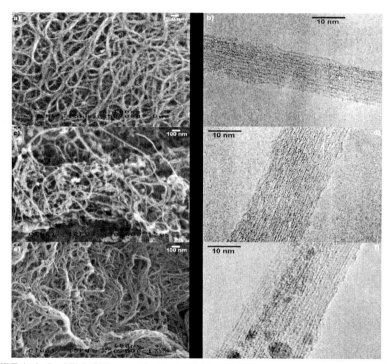

FIGURE 1. Left : SEM (50 k magnification), Right : HRTEM (250 k magnification). From top to bottom : EA, Laser, HiPCO samples.

TABLE 1. Bundle and tube diameters as estimated from different techniques

Bundle diameter	EA	Laser	HiPCO
SEM	20 – 30 nm	15 – 22 nm	7 – 12 nm

Bundle diameter	EA	Laser	HiPCO
Diameter (HRTEM)	1.23 – 1.42 nm	1.34 ± 0.15 nm	0.9 – 1.3 nm
Diameter (XRD)	1.38 ± 0.15 nm	-	0.6 – 1.2 nm
Diameter (Raman)	1.39 ± 0.16 nm	1,34 ± 0.15 nm	0.85 – 1.36 nm
Diameter Range (OA)	1.1 – 1.5 nm	-	0.7 – 1.4 nm

Typical X-Ray diffraction data are presented in Fig. 2. Several points of interest can be noted. First, there is a sharp increase of intensity at very low Q, due to the form factor of individual nanotubes. Second, a series of peaks is observed between 0.4 Å^{-1} and 1.8 Å^{-1} which corresponds to diffraction on the bundle lattice. The most intense peak is the (10) peak, it is measured around 0.42 Å^{-1} for EA and laser samples and around 0.5 Å^{-1} for HiPCO samples. Lastly, one observes the signatures of carbon impurities (graphite and carbon nanoparticles) around 1.8 Å^{-1} for EA and laser samples. Experimental data are fitted using a simple X-Ray diffraction model, which assumes that SWNTs are continuously charged infinite-cylinders, closely packed in a finite-sized bundle, and that each bundle is formed by a set of similar tubes [5]. To calculate the diffraction pattern, we summed the intensity scattered by a set of bundles assuming gaussian distributions of the tube diameter. The distribution which provides the best fit is presented in the insets. For the EA sample, a good fit can be achieved for all the pattern (Fig. 2, left). The data and fit are very close for laser samples (not shown). For HiPCO sample, the fit is more problematic because of the weakness of the signal and of the broadness of the peaks. The distribution of tube diameters is known to be rather large for these samples [4,7]. No good fits could be achieved using a single (monomodal) gaussian distribution, except if the maximum of the gaussian is shifted down to 6 Å (Fig 2, middle). Therefore, we attempted to fit the data using the sum of two gaussian distributions (bimodal distribution). The best fit is presented in the right part of Fig. 2. However, the fit is not selective enough to conclude definitely. Note that a rather poor fit was achieved with the bimodal distribution proposed in ref. [7]. In addition, the broad band centered at 1.5 Å^{-1} is not well fitted in our model. This is due to a not-corrected fluroescence contribution, and maybe also to the limits of our model, which considers monodisperse tubes in each bundle.

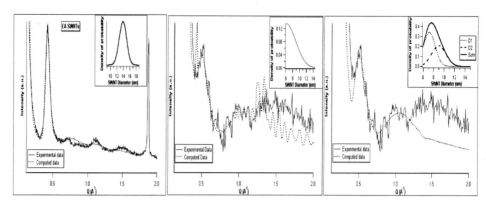

FIGURE 2. XRD data (solid line) and fit (dotted line). Left: EA sample ; Middle: HiPCO sample, single gaussian distribution. Right : same HiPCO sample, sum of two gaussian distributions. The diameter distributions are plotted in the insets.

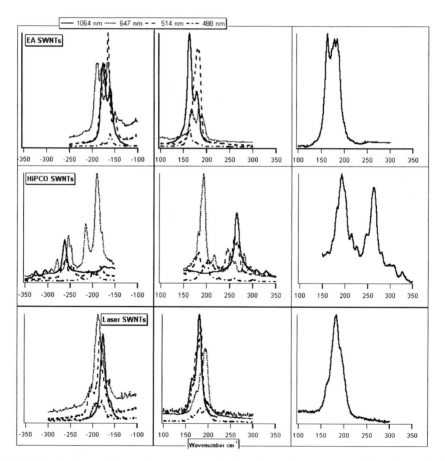

FIGURE 3. Left : Anti-Stokes Raman spectra. Middle : Stokes Raman spectra. Right : Normalized sum of Stokes plus Anti-Stokes Raman spectra.

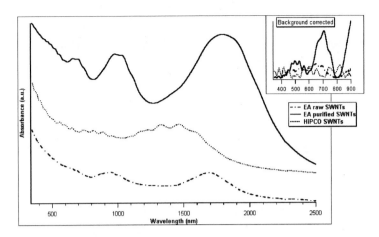

FIGURE 4. Optical absorption spectra of SWNTs. Inset : Background corrected.

Fig. 3 displays Raman spectra for all samples in the frequency range of the radial breathing modes (RBM). Because Raman is a resonant process for SWNTs, different spectra corresponding to different resonant tubes are measured for each exciting laser line [6-9]. By using four wavelengths and measuring Stokes (S) and Anti-Stokes (AS) spectra, we excited the sample at eight different resonant energies. In the right part of Fig.3, we plot the sum of all spectra normalized to the total area of the RBM bands in both S and AS case. In order to determine nanotube diameter from Raman results, we used a phenomenological power law, validated by experiments and calculations [6]:

$$v_{cm^{-1}} = \frac{238}{d_{nm}^{0.93}} \qquad (1)$$

Results are reported in Table 1. Note that there is a minimum of intensity around 230 cm^{-1} in the HiPCO spectra, which corresponds to a small amount of tubes of diameter around 1.05 Å. This "hole" may be due in part to a lack of adapted laser-line to cover this region. Indeed, Kukovecz *et al* measured Raman RBM on HiPCO samples using a laser line at 568.8 nm [7]. However, this can not be ruled out from their data that the distribution of diameters is bimodal, as also suggested by X-Ray data, with a minimum amount of tubes of diameter 1.05 Å.

Fig. 4 displays optical absorption experiments performed on the same samples. The spectra display several major absorption bands in the visible and near IR, due to optical transitions between pairs of van Hove singularities [9]. The allowed optical transitions have been calculated by Kataura *et al* as a function of diameter [9]. Note that the "Kataura plot" is also useful to analyze the resonant Raman results. From the position of the main absorption bands and using the Kataura plot, one can estimate the tube diameter distribution for each sample. Results are reported in Table 1. Note that the main broad bands are the envelopes of series of narrower bands, which are the signature of individual tubes of different chiralities, as expected from the "trigonal wrapping effect" [11]. These bands are better evidenced in the frequency range from

350 to 900 nm after substraction of an exponential background, as displayed in the inset. A detailed analysis of these absorption spectra will be reported elsewhere.

DISCUSSION AND CONCLUSION

Table 1 compares the bundle and tube diameter for each sample, as estimated from SEM, HRTEM, XRD, Raman and optical absorption. There is a good qualitative agreement between all techniques. In EA and laser samples, SWNT present a narrow diameter distribution centered around 1.35 nm. For HiPCO samples, the mean-diameter is shifted towards small values, around 1 nm, and the distribution is much broader but it is not clear from our data whether the distribution is monomodal or bimodal. The size of the bundles is also varying significantly from one sample to another : HiPCO SWNTs assemble into rather small bundles with respect to EA SWNTs. In the future, the possibility of working on different samples of high purity, well-characterized and presenting different structural features opens interesting opportunities to precise the relation between structure and physical properties.

ACKNOWLEDGMENTS

The authors acknowledge A. Loiseau and R. Almairac for fruitful discussions and for their help in HRTEM and XRD measurement.

REFERENCES

1. Vivien. L, Izard, N., Riehl, D., Hache, F., Anglaret, E., these proceedings.
2. Journet, C., Maser, W.K., Bernier, P., Lamy de la chapelle, M., Lefrant, S., Deniard, P., Lee, R., Fisher, J.E., *Nature* **388**, 756-758 (1997).
3. Thess, A., Lee, R., Nikolaev, H., Dai, H., Petit, P., Robert, J., Xu, C., Lee, Y.H., Kim, S.G., Rinzler, A.G., Colbert, D.T., Scuseria, G.E., Tomanek, D, Fischer, J.E., Smalley, R.E., *Sciences* **273**, 483-487 (1996).
4. Nikolaev, P., Bronikowski, M.J., Rohmund, F., Colbert, D.T., Smith, K.A., Smalley, R.E., *Chemical Physics Letters* **313**, 91-97 (1999).
5. Rols, S., Righi, A., Henrard, L., Anglaret, E., Sauvajol, J.L., *The European Physical Journal B* **10**, 263-270 (1999).
6. Rols, S., Righi, A., Alvarez, L., Anglaret, E., Almairac, R., Journet, C., Bernier, P., Sauvajol, J.L., Benito, A.M., Maser, W.K., Munoz, E., Martinez, M.T., de la Fuente, G.F., Girard, A., Ameline, J.C. *The European Physical Jounal B* **18**, 201-205 (2000).
7. A. Kukovecz , Ch. Kramberger , V. Georgakilas , M. Prato , H. Kuzmany, *Euro. Phys. J. B* **28**, 223-230 (2002)
8. Rao A.M., Richter E., Bandow S., Chase B., Eklund P.C., *Science* **275**, 187-191 (1997).
9. Kataura H., Kumazawa Y., Maniwa Y., Umezu I., Suzuki S., Oshtsuka Y., Achiba Y., *Synth. Metals* **103**, 2555-2558 (1999).
10. Alvarez L., Righi A., Rols S., Anglaret E., Sauvajol J.L., *Chem. Phys. Lett.* **320**, 441 (2000).
11. Brown S.D.M., Corrio P., Marucci A., Dresselhaus M.S., Pimenta M.A., Kneipp K., Phys. Rev. B 61, R5137-40 (2000)

(n,m)-Assigned Absorption and Emission Spectra of Single-Walled Carbon Nanotubes

R. Bruce Weisman, Sergei M. Bachilo, Michael S. Strano, Carter Kittrell, Robert H. Hauge, and Richard E. Smalley

Department of Chemistry, Center for Nanoscale Science and Technology, and Center for Biological and Environmental Nanotechnology
Rice University
Houston, Texas 77005 USA

Abstract. Spectroscopic studies of SWNT samples enriched in individual, surfactant-suspended nanotubes have revealed distinct absorption and emission transitions for more than 30 different semiconducting (n,m) species. Through the combined use of pattern analysis, qualitative model calculations, and resonance Raman measurements of radial breathing mode frequencies, (n,m) indices were assigned to the first- and second-van Hove optical transitions of 33 different species. The results show large systematic deviations from a simple dependence of transition wavelength on tube diameter, suggesting the presence of substantial excitonic effects in addition to trigonal warping patterns. Tight-binding models are not successful in simulating the results. An experimental, model-free plot of transition energy vs. nanotube diameter is now available. It is advisable to use this in preference to theoretically constructed Kataura plots, which contain substantial errors. Optical spectroscopy will also prove a valuable tool for selective, sensitive, rapid, and nondestructive analysis of bulk nanotube samples.

INTRODUCTION

One of the most interesting and potentially useful characteristics of single-walled carbon nanotubes (SWNT) is the sensitivity of their electronic properties to physical structure.[1] Each possible SWNT structure can be viewed as a graphene sheet rolled into a seamless cylindrical tube that has a well-defined diameter and chiral wrapping angle. Every such structure can be uniquely indexed by a pair of integers, (n,m), for which m is no greater than n. Because of the nanotubes' very large aspect ratios and resulting quasi-one-dimensionality, their electronic state densities show sharp maxima called van Hove singularities. In the simplest electronic structure models, the energies of van Hove singularities depend inversely on nanotube diameter. In addition, it is predicted that tubes for which $n - m$ is evenly divisible by three show metallic or semi-metallic properties, with a very small or absent band gap at the Fermi level. The remaining tube structures, comprising approximately 2/3 of the total, have significant band gaps and instead behave as semiconductors. As illustrated in Fig. 1, which schematically shows the density of states function expected for a semiconducting structure, the absorption and emission transitions between matching van Hove singularities are predicted to dominate optical spectra of nanotubes.

CP685, *Molecular Nanostructures: XVII Int'l. Winterschool/Euroconference on Electronic Properties of Novel Materials,* edited by H. Kuzmany, J. Fink, M. Mehring, and S. Roth
© 2003 American Institute of Physics 0-7354-0154-3/03/$20.00

Experimental samples of SWNT are currently available only as mixtures containing many different diameters and chiral angles. Although such samples should in principle show superpositions of distinct spectral transitions corresponding to the constituent nanotube structures, observed spectra are instead normally quite diffuse. Much of this spectral shift and broadening is thought to arise from tube-tube interactions within bundles bound by van der Waals forces. Substantially sharper absorption spectra have been obtained through a sample preparation process involving dispersion in aqueous SDS surfactant, followed by ultrasonication to disaggregate tubes from bundles, and centrifugation to separate individually suspended tubes from residual bundles and impurities.[2] Samples of HiPco nanotubes processed in this way show absorption spectra with resolvable maxima within each van Hove region. Most importantly, these samples also display structured emission spectra at near-infrared wavelengths corresponding to transitions between first-van Hove peaks.[2] As we illustrate below, this band gap luminescence enables a more powerful form of optical spectroscopy to be applied to SWNT samples, providing important new insights into nanotube electronic structure.[3]

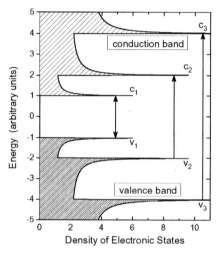

FIGURE 1. Schematic density of states pattern for a semiconducting carbon nanotube.

METHODS AND RESULTS

We used a Spex Fluorolog 3 spectrofluorimeter equipped with a cooled InGaAs detector to measure the intensity of near-infrared $c_1 \rightarrow v_1$ emission resulting from excitation in the range of $c_2 \leftarrow v_2$ or $c_3 \leftarrow v_3$ transitions. The sample contained HiPco nanotubes in an SDS / D_2O suspension, processed as described above for enrichment in individual tubes.[2] D_2O was chosen over H_2O for its improved near-IR transparency. Fig. 2 shows a surface plot of measured (corrected) emission intensity as a function of excitation and emission wavelengths. The group of peaks with excitation wavelengths near 700 nm are attributed to $c_2 \leftarrow v_2$ absorptions of distinct semiconducting (n,m) species, while the those with excitation wavelengths below

400 nm arise from $c_3 \leftarrow v_3$ excitations of the same species. The "valley" of low photoluminescence between these groups corresponds to the lowest energy absorptions of metallic (nonemissive) nanotubes in the sample.

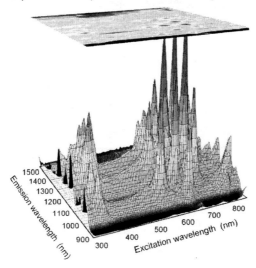

FIGURE 2. Measured emission intensity as a function of excitation and emission wavelengths for a sample of SWNT in D_2O / SDS suspension at room temperature.

DATA ANALYSIS

Our spectral analysis was focused on the set of more than 30 features having excitation values clustered near 700 nm. The coordinates of each peak give the $c_1 \rightarrow v_1$ and $c_2 \leftarrow v_2$ wavelengths for one (n,m) structure. When the ratio of these wavelengths was plotted as a function of emission wavelength, patterns emerged. By comparing these spectral patterns to those seen in simulated data computed using an extended tight-binding model,[4] we identified sets of nanotubes that share the same value of $n - m$. We call these sets "families". The numerical relation between (n,m) values of neighboring spectral points from different families was also deduced, giving the entire pattern of index connectivity among the observed spectral features. Note that our analysis was not based on any quantitative findings from the extended tight-binding model; it used only qualitative patterns from the model's results.

Once the connectivity pattern was established, it was still necessary to choose from among a small number of plausible assignments of peaks to specific (n,m) values. This decision was based on data from resonance Raman spectra measured with a variety of laser wavelengths that allowed resonance with ten different $c_2 \leftarrow v_2$ transitions. The resulting set of radial breathing mode frequencies was most linearly related to deduced inverse tube diameter in only one candidate assignment, which we therefore concluded to be the correct one. This assignment is illustrated in the left panel of Fig. 3 by the (n,m) labels next to the corresponding spectral peak positions.[3]

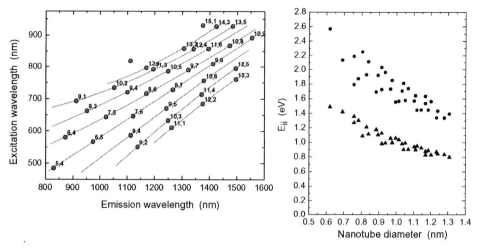

FIGURE 3. Left panel: Positions of measured emission maxima (points), labeled with deduced (n,m) values. Lines connect points belonging to the same family, in which $n - m$ is constant. Right panel: Experimental "Kataura" plot of optical transition energies vs. nanotube diameter. Triangles and circles show first- and second-van Hove transitions, respectively.

With (n,m) indices assigned to each set of first- and second-van Hove transitions, it is simple to construct an experimental Kataura plot of optical transition energies vs. nanotube diameter. This graph is shown in the right panel of Fig. 3. This empirical, model-independent plot differs substantially from the theoretically based Kataura plots that are in widespread use.[5] We strongly recommend that carbon nanotube researchers switch to using the experimental version in order to avoid serious errors in their work.

In the left panel of Fig. 4, data points show the ratio of second-van Hove to first-van Hove transition frequencies vs. the deduced nanotube chiral angle. Points within the same n-m family are connected by solid lines and labeled with the value of n-m. It can be seen that the families having $\mathrm{mod}(n$-$m, 3) = 1$ show higher ratios than families having $\mathrm{mod}(n$-$m, 3) = 2$. Also, in the chiral angle limit of $30°$ (armchair structure), the

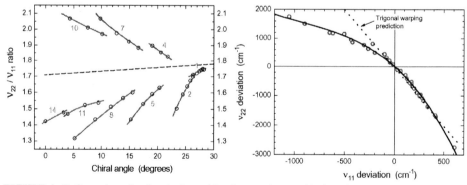

FIGURE 4. Left panel: ratio of optical transition frequencies vs. chiral angle. Numbers label families of points with same $(n$-$m)$ value; lines connect family members. Right panel: correlation between deviations of measured second- and first-van Hove optical transition frequencies from a simple inverse dependence on tube diameter. The dashed line shows the prediction from trigonal warping effects and the solid line is a polynomial fit to the data.

244

ratios appear to converge to a value between 1.75 and 1.80, as illustrated by the dashed line, rather than to 2.0, as predicted by one-electron electronic structure theories. This discrepancy has been termed the "ratio problem" and presumably reflects excitonic effects[6] that may be elucidated by further theoretical analysis.

The optical data for each branch (either first- or second-van Hove) can also be graphed to show transition wavelength vs. nanotube diameter. Each such plot reveals two groups of experimental points separated by a linear boundary region. One group corresponds to tubes having $\mod(n\text{-}m, 3) = 1$; the other, to tubes having $\mod(n\text{-}m, 3) = 2$. We estimated the linear boundary between these groups and compiled the deviations (in frequency units) of each measured point from the boundary line. The right frame of Fig. 4 plots one point for each tube structure, showing the deviation in its second-van Hove transition vs. the deviation in its first-van Hove transition. Clearly, there is a strong correlation pattern. Near the origin, it resembles the straight line of slope -4 predicted from trigonal warping theory,[7] but other effects lead to substantial curvature and a slope approaching -1 for $\mod(n\text{-}m, 3) = 1$ tubes of small diameter and small chiral angle. This curvature exposes subtle aspects of SWNT electronic structure that should be interpretable through appropriate theoretical simulation.

ACKNOWLEDGMENTS

This research has been supported by the NSF (CHE-9900417), the NSF Focused Research Group on Fullerene Nanotube Chemistry (DMR-0073046), the NSF Center for Biological and Environmental Nanotechnology (EEC-0118007), and the Robert A. Welch Foundation (C-0807 and C-0689). Support from NASA (NCC 9-77) for development of the HiPco method is also gratefully acknowledged. We are pleased to thank M. O'Connell and V. Moore for assistance in sample preparation, S. Doorn for Raman spectroscopy, and the University of Illinois Materials Research Laboratory for use of facilities. We are also very grateful to S. Reich for sharing computer codes and extended tight binding results prior to publication, and to E. Mele for helpful and stimulating discussions.

REFERENCES

1. Saito, R., Dresselhaus, G., and Dresselhaus, M. S., *Physical Properties of Carbon Nanotubes*, London: Imperial College Press, 1998.
2. O'Connell, M. J., Bachilo, S. M., Huffman, C. B., Moore, V. C., Strano, M. S., Haroz, E. H., Rialon, K. L., Boul, P. J., Noon, W. H., Kittrell, C., Ma, J., Hauge, R. H., Weisman, R. B., and Smalley, R. E. *Science* **297**, 593-596 (2002).
3. Bachilo, S. M., Strano, M. S., Kittrell, C., Hauge, R. H., Smalley, R. E., and Weisman, R. B., *Science* **298**, 2361-2366 (2002).
4. Reich, S., Maultzsch, J., Thomsen, C., and Ordejon, P., *Phys. Rev. B* **66**, 035412 (2002).
5. Kataura, H., Kumazawa, Y., Maniwa, Y., Umezu, I., Suzuki, S., Ohtsuka, Y., and Achiba, Y., *Synth. Metals.* **103**, 2555-2558 (1999).
6. Ando, T., *J. Phys. Soc. Jpn.* **66**, 1066-1073 (1997).
7. Saito, R., Dresselhaus, G., and Dresselhaus, M. S., *Phys. Rev. B* **61**, 2981-2990 (2000).

Assignment of (n,m) Raman and Absorption Spectral Features of Metallic Single-Walled Carbon Nanotubes

Michael S. Strano, Erik H. Haroz, Carter Kittrell, Robert H. Hauge, and Richard E. Smalley

Department of Chemistry, Center for Nanoscale Science and Technology, and Center for Biological and Environmental Nanotechnology
Rice University
Houston, Texas 77005 USA

Abstract. The (n,m) spectral features for isolated metallic single walled carbon nanotubes were deduced by examining Raman excitation profiles. Correlation of the radial breathing mode frequency with diameter identifies the (n,m) index of the metallic tube. Observation of the energy of Raman intensity maximum provides experimental values for the optical transitions directly, and allows for model independent estimation of peak splitting and diameter scaling. The results were extrapolated to all metallic nanotubes using a parameterized functional form deduced from the framework of the Tight Binding approximation.

INTRODUCTION

Identification of spectroscopic features and correlation with nanotube geometric structure will aid attempts to purify, separate and sort nanotubes based on their electronic structure. Carbon nanotubes are labeled by two integers *n and m* that define the circumferential length, πd_t and chiral angle α of the vector connecting two periodic atom locations as a conceptual graphene sheet is "rolled" into the tube [1]. Advances in the solution phase dispersion and processing of carbon nanotubes [2], and spectrofluormetric detection have lead to a definitive (n,m) assignment of semiconducting features by exploiting their band-gap fluorescence [3].

Alternatively, metallic nanotubes do not fluoresce, but still posses absorption maxima corresponding to inter-band electronic transitions. Features in the Raman spectrum which are distinct for particular *(n,m)* nanotubes couple to these transitions as the probing laser energy becomes commensurate with the transition energy [4]. In this way, by looking at the intensity profile of the Raman spectrum through a range of closely spaced excitation energies, these features can be correlated and assigned to (n,m) nanotubes.

Nanotubes produced by CO disproportionation over Fe generally have smaller diameters (0.6 to 1.2 nm) compared to those produced by laser ablation, CVD and electric arc methods and this is ideal for spectroscopic assignment purposes [5-7]. As the

CP685, Molecular Nanostructures: XVII Int'l. Winterschool/Euroconference on Electronic Properties of Novel Materials, edited by H. Kuzmany, J. Fink, M. Mehring, and S. Roth

nanotube diameter increases, the norm of the electronic wave vector becomes small and chirality differences in electronic structure are minimized [8]. Also, the radial breathing mode frequency (RBM) in the Raman spectrum is inversely proportional to nanotube diameter [9], meaning that larger diameter distributions may provide unfavorable resolution between discrete nanotube frequencies. Recent solution phase dispersion methods have demonstrated the ability to produce isolated, individual nanotubes for spectroscopic analysis [2]. In this work, the optical transitions of metallic nanotubes are measured directly using Raman excitation profiles in the range from 627 to 565 nm and also between 458 to 502 nm laser excitation. Correlation of the RBM frequency with diameter allows for an unambiguous (n,m) assignment.

Experimental

Raman spectra were obtained using sodium dodecyl sulfate suspended nanotubes in aqueous solution prepared as reported previously [2]. The pH of the sample was adjusted to 10 by the addition of 1 N NaOH to fully restore optical transitions that become diminished at lower pH [10]. Raman spectroscopy was performed by lasing a Coherent 599 dye laser employing a Rhodamine 590 dye pumped using the 488 nm line of a Lexel 95 water cooled Ar^+ laser. A liquid N_2 cooled CCD array detector was used for detection. Discrete Ar^+ lines at 457.5, 476.7, 488, 496.9, 502 and 514.5 nm were also employed. Wavelength and intensity were calibrated using Ne and black body emission lamps respectively.

Radial Breathing Mode/Diameter Relationship

The graphene plane almost uniquely quantizes nanotube diameters. The frequency of the Raman radial breathing mode is a monotonic function of the diameter and modeled as linear in inverse diameter with an offset [3, 9].

$$\omega_{RBM} = \frac{c_1}{d_t} + c_2 = \frac{\pi c_1}{a_{C-C}\sqrt{3(n^2 + nm + m^2)}} + c_2 \qquad (1)$$

Here, ω_{RBM} is the mode frequency, $a_{c-c} = 0.144$ nm, d_t is the nanotube diameter (nm) and c_1 and c_2 are empirically derived parameters. Previously, our work on the assignment of the semiconducting nanotubes [3] has yielded constants $c_1 = 223.5$ (nm cm) and $c_2 = 12.5$ (cm^{-1}.) Metallic nanotubes for which $|n-m| = 3q$ where q is an integer are separated in diameter to greater extent than semi-conducting nanotubes. Hence, probing the region between 457 and 627 nm where the first inter-band transitions of the metallic nanotubes are dominant for HiPco produced material should produce only those discrete mode frequencies predicted by these constants. For example, the (12,0) and (8,5) appear sequentially in a list of metals sorted by diameter and they are separated by 0.051 nm (12 cm^{-1}.) When these frequencies are plotted versus $1/d_t$ as in Fig 1, statistically identical parameters c_1 and c_2 are obtained. This provides an additional confirmation of this correlation and a useful mapping of ω_{RBM} and (n,m.) For example, with few exceptions Eq. 1, can be used to uniquely assign (n,m) indices of metallic RBM features in the Raman spectrum.

Figure 1: Observed Raman radial breathing mode frequencies assigned to closest (n,m) species versus inverse diameter. The linear relationship is statistically identical to that produced in [3] and agrees with independent diffraction measurements.

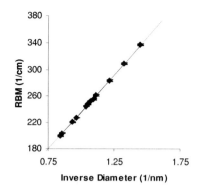

Correlation of Optical Transitions

Chiral metallic nanotubes (n≠m) are predicted to demonstrate peak splitting of the v1→c1 transition with positive and negative deviations from an armchair (n=m) curve [1, 8, 11] due to trigonal warping. We represent the form of this curve (for metallic v1→c1) transitions as a truncated asymptotic expansion in d_t:

$$\lambda_{11}^{Armchair} = \frac{hc}{A}\left(\frac{d_t}{6a_{c-c}} + \frac{a_{c-c}B}{4d_t}\right) \tag{2}$$

Here, A and B are adjustable parameters. In the simplest single electron model, A is related to the C-C interaction energy and equal to the conventional γ_o parameter with B = 1. The split bands of zig-zag metallic nanotubes can be scaled as positive and negative deviations from eq. 2 such that:

$$\frac{1}{\lambda_{11}^{+}} = \frac{1}{\lambda_{11}^{Armchair}} + \frac{2\beta}{hc}\left(-\frac{3}{2}\left(\frac{a_{c-c}}{d_t}\right)^2 + 3\delta\left(\frac{a_{c-c}}{d_t}\right)^3\right)Cos[3\alpha] \tag{3}$$

and

$$\frac{1}{\lambda_{11}^{-}} = \frac{1}{\lambda_{11}^{Armchair}} + \frac{2\beta}{hc}\left(\frac{3}{2}\left(\frac{a_{c-c}}{d_t}\right)^2 + 3\delta\left(\frac{a_{c-c}}{d_t}\right)^3\right)(Cos[3\alpha])^2 \tag{4}$$

Here, α is the nanotube chiral angle. The latter two parameters β and δ reflect the extent of band splitting and in the simplest example of a single parameter tight binding model are equal to the conventional γ_o and 1 respectively. In this case A and B are also γ_o and 1 respectively. Note that in the limit as the norm of the **k** vector approaches zero, eq. 3 and 4 approach the familiar limit of $\lambda_{11} = \frac{hcd_t}{6Aa_{c-c}}$ with A = γ_o. The $Cos[3\alpha]^n$ factor (n = 1 for λ^{+} and 2 for λ^{-}) scales the armchair deviation for chiral nanotubes.

We note that the same approach can be used to describe transitions of the semiconducting nanotubes for validation. The armchair curves are scaled by factors of 1/3 and 2/3 for the v1→c1 and v2→c2 semiconductor transitions respectively:

$$\lambda_{11}^{Armchair} = \frac{hc}{A}\left(\frac{d_t}{2a_{c-c}} + \frac{a_{c-c}B}{4d_t}\right) \text{ (for v1→c1)}, \quad \lambda_{11}^{Armchair} = \frac{hc}{A}\left(\frac{d_t}{4a_{c-c}} + \frac{a_{c-c}B}{4d_t}\right) \text{ (for v2→c2) (5,6)}$$

For both sets of transitions:

$$\frac{1}{\lambda_{11}^{\pm}} = \frac{1}{\lambda_{11}^{Armchair}} + 2\beta\left(\pm\frac{3}{2}\left(\frac{a_{c-c}}{d_t}\right)^2 + 3\delta\left(\frac{a_{c-c}}{d_t}\right)^3\right)Cos[3\alpha] \tag{7}$$

However, the negative root represents the species for which mod 3 of (n-m) = 1 with the positive corresponding to mod 3 of (n-m) = 2 for v1→c1. For the v2→c2, the signs are exactly reversed. For comparison, the fitted values for semiconducting transitions are: A = 3.73 eV, B = 1.63, β = 1.07 eV and δ = -1.62 for v1→c1 and A = 3.33 eV, B = 3.28, β = 2.77 eV and δ = -0.52 for v2→c2.

Metallic Transitions from Raman Excitation Profiles

In the Rhodamine 6G range, four resolvable profiles corresponding to metallic nanotubes are observed and assigned as the (11,2) with λ_{11} = 569 nm, (11,5) with λ_{11} = 589 nm and (12,0) with λ_{11} = 575 nm and (8,8) at λ_{11} = 589 nm. The (11,2) assignment was selected over the (7,7) despite their identical diameters because of the close proximity of the (12,0) transition. The framework outlined above indicates that the closeness of the transitions suggests similar chiral angles for these two species. Alternatively, the higher energy Ar$^+$ laser range monitors primarily two smaller diameter metallic nanotubes (12,0) and (8,5) which demonstrate transitions near 502 nm.

The optimal values of (A, B, β, and δ) for metallic nanotubes were found by nonlinear regression of these observed transitions mapped using the dye laser experiment described above and the two prominent maxima obtained from discrete Ar+ laser energies. These optimal values are: A = 3.0 eV, B = 4.47, β = 2.23 eV and δ = -0.78. The value of A falls within the range of γ_o parameters reported for optical calculations [1, 4, 11] and suggests that excitonic effects, if operative for metallic transitions, are minor compared to semiconducting nanotubes. Figure 2 compiles the (n,m) assignment and extrapolated transition wavelengths with the observed (n, m) species highlighted. We note that the extrapolated model is able to predict the appearance of weaker metallic RBM features from both scanning experiments. The (7,4) nanotube at 304 cm^{-1} appears in only the 457 and 476 nm Ar+ spectra and we note that it has two predicted transitions near these energies. The (8,2) feature at 316.5 cm^{-1} is apparent in the 502 nm spectrum with a predicted transition at 491 nm.

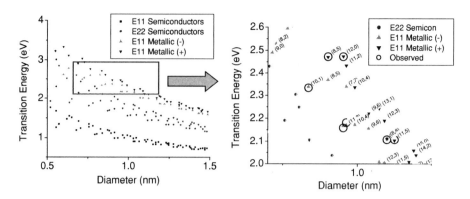

Figure 2: *(left)* A revised Katura plot from equations (2) through (7) for electronic transitions of single walled carbon nanotubes probed by optical methods. *(right)* Observed and extrapolated metallic carbon nanotube assignment for the 1st inter-band

Acknowledgements

This research has been supported by the NSF (CHE-9900417), the NSF Focused Research Group on Fullerene Nanotube Chemistry (DMR-0073046), the NSF Center for Biological and Environmental Nanotechnology (EEC-0118007), and the Robert A Welch Foundation (C-0689 and C-0807). Support from NASA (NCC 9-77) for development of the HiPco method is also gratefully acknowledged.

References

1. Saito, R., G. Dresselhaus, and M.S. Dresselhaus, *Physical Properties of Carbon Nanotubes.* 1998, London: Imperial College Press.
2. Strano, M.S., *et al., The role of surfactant adsorption during ultra-sonication in the dispersion of single walled carbon nanotubes.* J. Nanosci. and Nanotech., 2003. (in press.).
3. Bachilo, S.M., *et al., Structure-Assigned Optical Spectra of Single-Walled Carbon Nanotubes.* Science, 2002.
4. Dresselhaus, M.S., *et al., Raman spectroscopy on isolated single wall carbon nanotubes.* Carbon, 2002. **40**(12): p. 2043-2061.
5. Thess, A., *et al., Crystalline ropes of metallic carbon nanotubes.* Science, 1996. **273**(5274): p. 483-487.
6. Reich, S., C. Thomsen, and P. Ordejon, *Electronic band structure of isolated and bundled carbon nanotubes.* Physical Review B, 2002. **65**(15): p. 155411.
7. Bronikowski, M.J., *et al., Gas-phase production of carbon single walled nanotubes from carbon monoxide via the HiPco process: A parametric study.* Journal of Vacuum Science & Technology, 2001.
8. Saito, R., G. Dresselhaus, and M.S. Dresselhaus, *Trigonal warping effect of carbon nanotubes.* Physical Review B, 2000. **61**(4): p. 2981-2990.
9. Sauvajol, J.L., *et al., Phonons in single wall carbon nanotube bundles.* Carbon, 2002. **40**: p. 1697-1714.
10. Strano, M.S., *et al., Reversible, Band-Gap Selective Protonation of Single-Walled Carbon Nanotubes.* Journal of Physical Chemistry B, 2002.
11. Reich, S. and C. Thomsen, *Chirality dependence of the density-of-states singularities in carbon nanotubes.* Physical Review B, 2000. **62**(7): p. 4273-4276.

FUNCTIONALIZATION OF
CARBON NANOTUBES

Synthesis Procedures for Production of Carbon Nanotube Junctions

K. Niesz[1], Z. Kónya[1], A. A. Koós[2], L. P. Biró[2], Á. Kukovecz[1], I. Kiricsi[1]

[1]*Applied and Environmental Chemistry Department, University of Szeged, Rerrich Béla tér 1, H-6720 Szeged, Hungary*
[2]*Research Institute for Technical Physics and Materials Science, H-1525 Budapest, P.O.Box 49, Hungary*

Abstract. Two different procedures of the preparation of carbon nanotube junctions were achieved. In the first method interconnection of the tubes is generated over chemical groups built from the reaction between modified nanotubes and diamines. The resulted carbon nanotube junctions have been investigated by TEM and STM. In the second method catalyst particles have been deposited on the outer surface of carbon nanotubes and branches of nanotubes were produced at this contact point by catalytic chemical vapor deposition (CCVD) of acetylene. The product has been characterized by TEM.

INTRODUCTION

Effort directed towards tuning and modifying synthesis procedures to obtain carbon nanotube junctions of T, Y, H or X shapes lead to inappropriate results concerning the industrial or large scale production1. However, the importance and the demand for these junctions are quite large, since these may be the secondary building units of carbon nanotubes based chips or even more complex nanoelectronic devices.

Recently, some novel solutions of their preparation have been published. A Taiwanese group described a method to prepare multi-junctioned carbon nanotubes on a mechanically pretreated silicon surface applying chemical vapor deposition (CVD) technology[2] using the decomposition of methane at 1373 K. The nanotubes were nucleated following the lines prepared by scratching the surface with 600-grit sand paper. Chemical reactions can also be used for preparation of carbon nanotube junctions. P.W. Chiu et al. reported interconnecting reactions between functionalized carbon nanotubes[3]. By the described method, the carboxyl groups on the wall of single wall carbon nanotubes are converted to carbonyl chloride groups by reaction with $SOCl_2$ at room temperature. The formed COCl groups are very reactive on the outer surface and can be reacted easily with various amines, particularly diamines resulting in the formation of acidic amide bonding. When two functionalized carbon nanotubes react with such an amine molecule interconnected tubes are generated. The resulting carbon nanotube junctions were investigated by STM.

In this paper, we report on the results obtained on the preparation of carbon nanotube junctions applying two different procedures. The first method is similar to Chiu's one which was mentioned above, i.e. we used functionalized multiwall carbon

CP685, *Molecular Nanostructures: XVII Int'l. Winterschool/Euroconference on Electronic Properties of Novel Materials,* edited by H. Kuzmany, J. Fink, M. Mehring, and S. Roth
© 2003 American Institute of Physics 0-7354-0154-3/03/$20.00

nanotubes and successfully interconnected them by propylene-diamine as proved by TEM and STM. The second method demonstrates a novel principle: catalyst material has been deposited on the outer surface of carbon nanotubes and branches of nanotubes were produced at this contact point by catalytic chemical vapor deposition (CCVD) of acetylene. The product has been characterized by TEM.

EXPERIMENTAL

Synthesis and Purification of Carbon Nanotubes

MWNTs were used in the experiments. The production was performed by catalytic chemical vapor deposition of carbon from acetylene in an acetylene-nitrogene mixture. The catalyst for this procedure was cobalt and iron supported alumina. The acetylene:nitrogene flow ratio was 1:10. The reaction temperature was 973 K. The freshly prepared catalyst was placed on a quartz boat in thin layer and put into the preheated zone of the reaction tube. After drying and removal of water the preheated sample was pushed into the heated part of the reactor and the flushing nitrogen was switched to the reaction mixture. Generally the reaction was complete in 60 minutes. Then the stream was changed for nitrogen, and the reactor was cooled to ambient temperature, and the raw material was collected.

Functionalization of Tubes for Producing Junctions

We first built COOH groups onto the outermost shell of the carbon nanotubes by stirring them in the mixture of concentrated sulfuric and nitric acid. When the carbon nanotubes possess carboxyl groups at their ends and/or their outer shell they may take part in various chemical reactions. They can be converted into carbonyl chloride groups simply by reacting them with $SOCl_2$. The portion of nanotubes containing – COCl groups was reacted with propylene-diamine. This reaction gives a very complex product mixture since there are several possible reaction paths. The reaction scheme is depicted in Figure 1.

The successful interconnection of the tubes by propylene diamine has been proven by TEM and STM. The STM investigation of nanotubes was carried out on highly oriented pyrolytic graphite surface (HOPG). The STM images were acquired using commercial Pt/Ir tips with setpoint currents of 200 pA at a bias of 1V. The nanotubes were ultrasonicated in isopropanol and droplets of the suspension were drop dried on the HOPG surface.

The second way to produce junctions on carbon nanotubes is a further CCVD of acetylene carried out using a catalyst prepared by deposition of iron and cobalt salts as catalyst precursor on the outer surface of pre-prepared carbon nanotubes. This technique, to the best of our knowledge, is unique in the field of carbon nanotube chemistry. In this case the product obviously has the catalytic metal particle in the interior of the side branch of the junction. It is assumed that upon acid treatment these particles may be dissolved from the tubes.

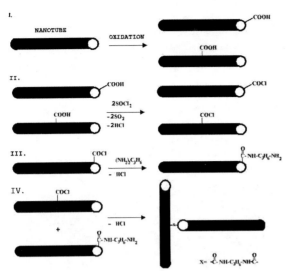

I.

II.

III.

IV.

$X= \overset{O}{\underset{\|}{C}}-NH-C_3H_6-NH-\overset{O}{\underset{\|}{C}}-$

FIGURE 1 Formation mechanism of carbon nanotube junctions

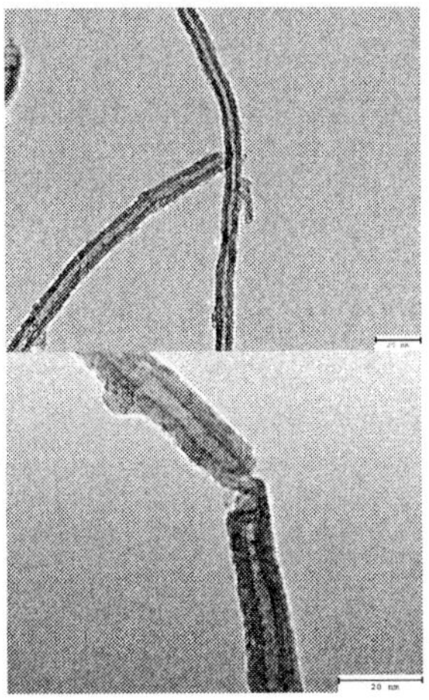

FIGURE 2. TEM images of double junctions with different shapes

RESULTS AND DISCUSSION

The evidences accumulated for characterizing the MWNT junctions revealed that using functionalized MWNTs and propylene-diamine as coupling molecule carbon nanotube junctions can be obtained. In Figure 2 the TEM images show some examples for the double junctions with various shapes. The shape of the junction strongly depends on the location of the functional groups. Besides this technique we investigated the structure of junctions by STM as well (Fig. 3). Figure 4 shows two consecutive STM images taken in the same region. The group of nanotubes on the upper part of the images (point A) is shifted without the alteration of its global arrangement about 200 nm relative to the cleavage step on the HOPG (point B), while the nanotube on the lower part of the images is rotated with 30 degrees. The nanotubes on the upper part of the images move together and their relative position remains approximately unchanged. This indicates that the binding force between the nanotubes is larger than the van der Waals interaction between the nanotubes and the substrate; consequently the tubes are strongly interconnected chemically by functional groups.

It is evident from the STM investigations that the individual tubes are not only laying on each other in the junctions, but they are strongly connected by chemical forces over functional groups.

Figure 5 shows TEM images about junctions prepared by the second CVD of acetylene on nanotube supported transition metal catalyst. However, when the interaction between the metal and the surface is weak, the metal particle embedded in the tip of the tube is moving away from the surface. Beside the catalytic action the metal particle may have a second role in the production of carbon nanotube junctions.

It may act as sealing material between the tubes. Since it is metal, it may enhance the conductivity between the tubes.

CONCLUSION

We described a procedure for preparing multi wall carbon nanotube junctions by a slightly modified method originally known from the very recent literature. We produced different shapes of double junctions (L, Y, T) using chemical coupling agents such as propylene-diamine. We presented a second method to produce multibranched nanotubes in which a novel technology is described using a second CCVD step and a novel catalyst consisting of metal particles deposited on carbon nanotubes. In both cases well graphitised carbon nanotube junction were obtained.

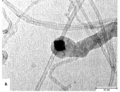

FIGURE 4. Consecutive STM images about moving a chemically connected group from nanotubes

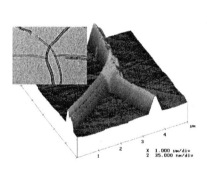

FIGURE 3. STM image of a double junction. The inset shows a TEM picture about two Y junctions

FIGURE 5. TEM image about junctions using nanotube supported iron (A) and cobalt (B) catalysts.

ACKNOWLEDGMENTS

Authors thank the financial help to the European Commission and the National Science Foundation of Hungary (RTN Program, NANOCOMP network, RTN1-1999-00013, OTKA T037952). KZ acknowledges support from the Bolyai fellowship and the Hungarian Ministry of Education (FKFP 216/2001, OTKA F038249).

REFERENCES

1. D. Zhou and S. Seraphin, *Chem. Phys. Letters* **238,** 286-289 (1995)
2. Jyh-Ming Ting and Chi-Chi Chang, *Appl. Phys. Letters* **80,** 324-325 (2002)
3. P.W. Chiu, G.S. Duesberg, U. Dettlaf-Weglikowska and S. Roth, *Appl. Phys. Letters* **80,** 3811-3813 (2002)

Chemical Modification of SWNTs

Yutaka Maeda,[1] Tadashi Hasegawa,[1] Takatsugu Wakahara,[2] Takeshi Akasaka,[2] Nami Choi,[3] Hiroshi Tokumoto,[3] Said Kazaoui,[3] Nobutoshi Minami[3]

[1]Department of Chemistry, Tokyo Gakugei University, Koganei, Tokyo 184-8501, Japan

[2]Center for Tsukuba Advanced Research Alliance (TARA Center), University of Tsukuba, Tsukuba, Ibaraki 305-8577, Japan

[3]Nanotechnology Research Institute, National Institute of Advanced Industrial Science and Technology, Tsukuba, Ibaraki 305-8562, Japan

Abstract. One step synthesis of highly pure soluble single-walled carbon nanotubes (s-SWNTs) under mild condition is reported. SWNTs were functionalized long-hydrophobic groups by oxidation and amidation. Raman and near infrared (NIR) absorption spectral data conformed that s-SWNTs maintain the inherent properties of SWNTs.

Since their discovery in 1991,[1] the unique physical and electrical properties of carbon nanotubes (CNTs) have attracted much attention to this area. Many potential applications of carbon nanotubes including hydrogen storage,[2] scanning probe microscopy tip,[3] field emission,[4,5] electronic device[6,7] and so forth have been investigated. For many of these usages, however, solution phase handling would be promisingly useful. Two methods for solubilization of single walled carbon nanotubes (SWNTs) have been investigated to lead to new material science; i) SWNTs-composites[8-11] with surfactant and polymer and ii) chemical functionalization of SWNTs.[12-16] Many SWNTs-composite have been reported since these methods are effective to change not only the solubility but also the property such as the optical one. Haddon et al. reported the soluble SWNTs terminated with octadecylamine at the open-end and/or defects site of SWNTs. To separate the SWNTs and carbonaceous impurities, they purified these soluble SWNTs with HPLC technique.[17] Here we simplify their method to one step synthesis of highly pure soluble SWNTs under mild condition. Especially, we have designed the functionalization of the open-end SWNTs for further modification so that the resulting SWNTs can keep the inherent properties. Because in the case of the sidewall functionalization of SWNTs, the characteristic absorptions corresponding to the band gap of metallic and semiconductor nanotubes disappeared.[18] Then term band gap in metallic CNT is applied to the energy separation of the first pair of singularities in the electronic density of states (DOS) of metallic CNT.

Two different types of SWNTs were used as starting materials in this study, which were purchased from Tubes@Rice and CarboLex. SWNTs of Tubes@Rice (SWNTs-L)

CP685, Molecular Nanostructures: XVII Int'l. Winterschool/Euroconference on Electronic Properties of Novel Materials, edited by H. Kuzmany, J. Fink, M. Mehring, and S. Roth
© 2003 American Institute of Physics 0-7354-0154-3/03/$20.00

are produced by the pulsed laser vaporization of a metal/carbon target in a furnace at 1100°C, and purified at Rice University by a previously reported procedure.[19] The resulting material consists of 90 vol% SWNTs-L. SWNTs-L were centrifuged, filtered and washed with methanol to remove the metal catalyst and surfactant. SWNTs of CarboLex (SWNTs-A) are grown by a modified arc-discharge method developed at the University of Kentucky. The pristine SWNTs-A was purified by the combining two-step processes of acid treatment and thermal annealing in air. To remove metal catalysts, SWNTs were treated by HCl solution. The modified thermal annealing, i.e., air oxidation method, was carried out to remove the amorphous carbon in air by using a rotary system. This methodology is very effective for the removal of amorphous carbon, because the sample exposed to the surface evenly.[20,21]

Fullerene pipes were prepared by treating SWNTs with concentrated H_2SO_4/HNO_3 according to a method developed by Liu et al.[22] The as-prepared tubes were further treated with a 4:1 mixture of concentrated H_2SO_4 and 30% aqueous H_2O_2 to form the oxidized short SWNTs (SWNTs-COOH).

SWNTs-COOH was dispersed in a dry N,N-dimethylformamide (DMF) solution containing 1% aniline.[23] Excess amount of n-octadecylamine and dicyclohexyldiimide (DCC) was added to DMF and then the resulting solution was kept at 120°C for 60 h while magnetically stirring. DCC is widely used for an effective coupling reagent for the direct condensation of carboxylic acids with amine in peptide synthesis.[24,25] After filtering and washing with a PTFE filter the product of s-SWNTs was obtained. (Scheme)

Figure 1 shows the absorption spectra of s-SWNTs-A and s-SWNTs-L from visible to near infrared region in THF solution. Both s-SWNTs are also soluble in chloroform.[12] The spectrum of s-SWNTs-A displays three major peaks at 1846 nm, 1016 nm, and 710 nm. These peaks are usually observed in a SWNTs sample, the first two of which are assigned to the first and second lowest energy gaps of semiconducting SWNTs and the last to that of metallic SWNTs.[12,26] In the case of the sidewall functionalized SWNTs, the absorption spectra show no characteristic feature in a NIR region. In this context, the absorption feature of s-SWNTs-A suggests that the number of the defects and the substituents on the sidewall is very small. Three major

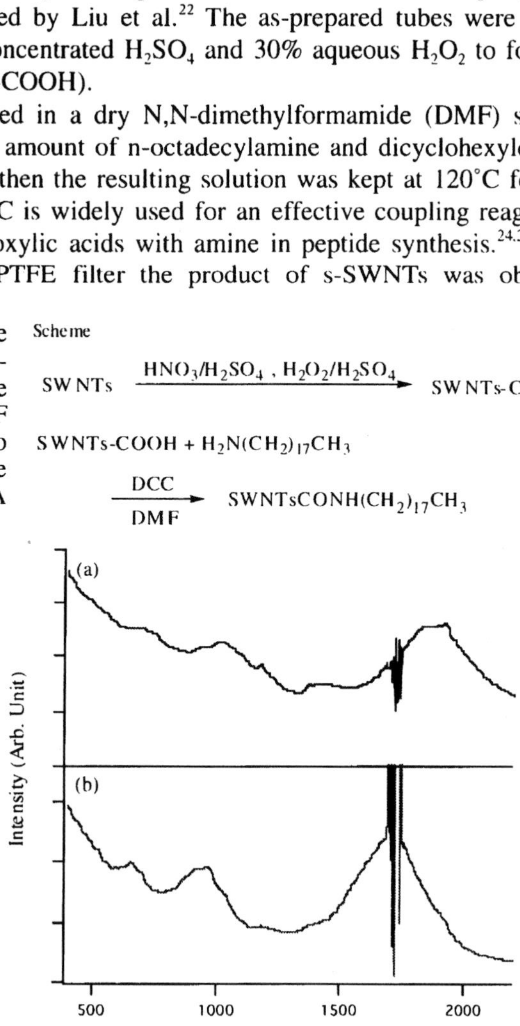

Scheme

$$SWNTs \xrightarrow{HNO_3/H_2SO_4 \ , \ H_2O_2/H_2SO_4} SWNTs\text{-}COOH$$

$$SWNTs\text{-}COOH + H_2N(CH_2)_{17}CH_3$$

$$\xrightarrow[DMF]{DCC} SWNTsCONH(CH_2)_{17}CH_3$$

FIGURE 1. NIR Spectra of s-SWNTs. (a) CarboLex (b) Tubes@Rice.

peaks at 1668 nm, 960 nm, and 650 nm in the absorption spectrum of s-SWNTs-L is shifted in comparison with those of s-SWNTs-A. Chen et al. also reported this type of peak-shift in the absorption spectra of two different SWNTs according to their diameters.[27]

Raman spectroscopy of SWNTs was performed at room temperature by using the 515 nm excitation line from an Ar ion laser. The Raman absorption peaks of s-SWNTs-A at 166 and 1591 cm^{-1} are similar to those of AP-SWNTs-A at 161 and 1583 cm^{-1}, and SWNTsCOOH-A at 160 and 1594 cm^{-1}. No obvious change occurs in the Raman absorption at 182 and 1587 cm^{-1} of SWNTs-L upon treatment for solubilization. The radial breathing modes (RBM) of both s-SWNTs-A and -L around at 160 cm^{-1} are in good agreement with those of the starting materials, indicating that the structures of SWNTs-A and -L are not significantly changed. The difference in RBM position between s-SWNTs-A and s-SWNTs-L may depend on their diameter distributions and their band-gaps affording the different absorption spectra.

The s-SWNTs were investigated using atomic force microscopy (AFM) in the tapping mode of operation with a single crystal conventional Si tip. AFM samples were prepared by casting THF solutions of s-SWNTs on mica substrates.

By AFM observation, the average length of s-SWNTs-L is estimated 1 μm. On the other hand, the average length of s-SWNTs-A is 0.8 μm. The diameter of s-SWNTs suggests the bundle formation of s-SWNTs. The DMF solution with 1% amine is known to be effective to protect SWNTs from the bundle formation,[16] but may not be good for s-SWNTs. The AFM images of s-SWNTs without repeated washing by organic solvent shows large amount of the residual reactant. Solubility of the non-purified s-SWNTs is much higher than that of the purified s-SWNTs, which suggests that residues act as surfactant and polymer for increasing solubility. These residues are not active for raman spectra and have no absorption in NIR region; this means that AFM analysis is very important for the characterization of s-SWNTs.

In conclusion, we have described one step synthesis of highly pure soluble SWNTs under mild condition. Raman and NIR spectra suggest that s-SWNTs maintain the inherent properties of SWNTs. The characteristic feature of s-SWNTs produced from arc discharge and pulsed laser vaporization methods is quite different. These s-SWNTs may surely be a good precursor for further modification, since s-SWNTs could react in a homogeneous system.

ACKNOWLEDGMENTS

This work was supported in part by Industrial Technology Research Grant Program'02 from New Energy and Industrial Technology Development Organization (NEDO) of Japan and by a Grant-in-Aid and the 21st Century COE Program "Promotion of Creative Interdisciplinary Materials Science" from the Ministry of Education, Culture, Sports, Science and Technology of Japan.

REFERENCES

1. Iijima, S. Nature 1991, 354, 56.

2. Dillon, A. C.; Jones, K. M.; Bekkedahl, T. A.; Kiang, C. H.; Bethune, D. S.; Haben, M. J. Nature 1997, 386, 377.

3. Dai, H.; Hafner, J. H.; Rinzler, A. G.; Colbert, D. T.; Smalley, R. E. Nature 1996, 384, 147.

4. Rinzler, A. G.; Hafner, J. H.; Nikolaev, P.; Lou, L.; Kim, S. G.; Tománek, D.; Nordlander, P.; Colbert, D. T.; Smalley, R. E. Science, 1995, 269, 1550.

5. De Herr, W. A.; Châtelain, A.; Ugarte, D. Science, 1995, 270, 1179.

6. Bockrath, M.; Cobden, D. H.; McEuen, P. L.; Chopra, N. G.; Zettl, A.; Thess, A.; Smalley, R. E. Science 1997, 275, 1922.

7. Tans, S. J.; Devoret, M. H.; Dai, H.; Thess, A.; Smalley, R. E.; Geerligs, L. J.; Dekker, C. Nature 1997, 386, 474.

8. Bandow, S.; Rao, A. M.; Williams, K. A.; Thess, A.; Smalley, R. E.; Eklund, P. C. J. Phys. Chem. 1997, 101, 8839.

9. Holzer, W.; Penzkofer, A.; Gong, S. H.; Bleyer, A.; Bradley, D. D. C. Adv. Mat. 1999, 8, 974.

10. O'Connell, M. J.; Boul, P.; Ericson, L. M.; Huffman, C.; Wang, Y.; Haroz, E.; Kuper, C.; Tour, J.; Ausman, K. D.; Smalley, R. E. Chem. Phys. Lett. 2001, 342, 265.

11. Chen, J.; Liu, H.; Weimer, W. A.; Halls, M. D.; Waldeck, D. H.; Walker, G. C. J. Am. Chem. Soc. 2002, 124, 9034.

12. Chen, J.; Hamon, M. A.; Hu, H.; Chen, Y.; Rao, A. M.; Eklund, P. C.; Haddon, R. C. Science, 1998, 282, 95.

13. Holzinger, M.; Vostrowsky, O.; Hirsch, A.; Hennrich, F.; Kappes, M.; Weiss, R.; Jellen, F. Angew. Chem. Int. Ed. 2001, 40, 2002.

14. Bahr, J. L.; Yang, J.; Kosynkin, D. V.; Bronikowski, M. J.; Smalley, R. E.; Tour, J. M. J. Am. Chem. Soc. 2001, 123, 6536.

15. Georgakilas, V.; Kordatos, K.; Prato, M.; Guldi, D. M.; Holzinger, M.; Hirsch, A. J. Am. Chem. Soc. 2002, 124, 760.

16. Tagmatarchis, N.; Georgakilas, V.; Prato, M.; Shinohara, H. Chem. Commun. 2002, 2010.

17. Niyogi, S.; Hu, H.; Hamon, M. A.; Bhowmik, P.; Zhao, B.; Rozenzhak, S. M.; Chen, J.; Itkis, M. E.; Meier, M. S.; Haddon, R. C. J. Am. Chem. Soc. 2001, 123, 733.

18. Boul, P. J.; Liu, J.; Mickelson, E. T.; Huffman, C. B.; Ericson, L. M.; Chiang, I. W.; Smith, K. A.; Colbert, D. T.; Hauge, R. H.; Margrave, J. L.; Smalley, R. E. Chem. Phys. Lett. 1999, 310, 367.

19. Rinzler, A. G.; Liu, J.; Dai, H.; Nikolaev, P.; Huffman, C. B.; Rodriguez-Macias', F. J.; Boul, P. J.; Lu, A. H.; Heymann', D.; Corbert', D. T.; Lee, R. S.; Fischer, J. E.; Rao, A. M.; Eklund, P. C.; Smalley, R. E. Appl. Phys. A 1998, 67, 29.

20. Ebbesen, T.; Ajayan, P. M.; Hiura, H.; Tanigaki, H. Nature 1994, 367, 519.

21. Moon, J.; Hyeok, K.; Lee, Y. H.; Park, Y. S.; Bae, D. J.; Park, G. J. Phys. Chem. B 2001, 105, 5677.

22. Liu, J.; Rinzler, A. G.; Dai, H.; Hafner, J. H.; Bradley, R. K.; Boul, P. J.; Lu, A.; Iverson, T.; Shelimov, K.; Huffman, C. B.; Rodriguez-Macias', F.; Shon, Y.; Lee, T. R.; Colbert, D. T.; Smalley, R. E. Science 1998, 280, 1253.

23. Choi, N.; Kimura, M.; Kataura, H.; Suzuki, S.; Achiba, Y.; Mizutani, W.; Tokumoto, H. Jpn. J. Appl. Phys. 2002, 41, 6264-6266.

24. Liu, Z.; Shen, Z.; Zhu, T.; Hou, S.; Ying, L.; Shi, Z.; Gu, Z. Langmuir 2000, 16, 3569.

25. Sano, M.; Kamino, A.; Okamura, J.; Shinkai, S. Science 2000, 293, 1299.

26. Kataura, H.; Kumazawa, Y.; Maniwa, Y.; Umezu, I.; Suzuki, S.; Ohtsuka, Y.; Achiba,Y. Synth. Met. 1999, 103, 2555.

27. Chen, J.; Rao, A. M.; Lyuksyutov, S.; Itkis, M. E.; Hamon, M. A.; Hu, H.; Cohn, R. W.; Eklund, P. C.; Colbert, D. T.; Smalley, R. E.; Haddon, R. C. J. Phys. Chem. B 2001, 105, 12525.

Interaction of SWNT with Simple Dye Molecules

T.G. Hedderman, L. O'Neill, S.M. Keogh, E. Gregan, G. Chambers, H.J. Byrne

Facility for Optical Characterisation and Spectroscopy (FOCAS), School of Physics, Dublin Institute of Technology, Kevin Street, Dublin 8, Ireland.

Abstract: Single wall carbon nanotubes are insoluble in most organic solvents such as toluene. Improvements in the solubility of the single wall carbon nanotubes are however seen as a result of specific interactions with dye molecules such as terphenyl and anthracene. Suspensions formed in toluene with these dye molecules and the single wall carbon nanotubes are seen to be stable over prolonged periods. Spectroscopic analysis clearly shows an interaction between the carbon nanotubes and the dye molecules. It is proposed in this study that the use of these more simple molecular systems may provide a route towards selectively processing different types of tubes.

INTRODUCTION

Single wall Carbon Nanotubes (SWNT) are known to be insoluble in most common organic solvents [1]. Several methods for enhancing the solubility of SWNT have been reported [2] the most successful employing the use of conjugated polymer systems such as poly(m-phenylenevinylene-co-2,5-dioctyloxy-p-phenylene) (PmPV) [3] .However the exact nature of the polymer interaction with the SWNT is still unclear. It is proposed in this study that further light may be cast upon the nature and extent of the interaction between large conjugated polymers systems and SWNT by exploring the potential of simple molecular systems such as anthracene and terphenyl to selectively interact and map onto the SWNT lattice. [4]. Here it is proposed that the dye molecules terphenyl and anthracene will interact with a specific SWNT type based upon the size an structure of the dye molecule. The structure of anthracene is similar to the back bone of an armchair SWNT while terphenyl is similar to a zigzag SWNT and is therefore proposed to selectively interact with these specific SWNT types Presented are a number of spectroscopic studies which indicted that the dye molecules can selectively interact with specific SWNT The studies were performed in toluene as this solvent is known to have a poor affinity for the retention of SWNT and hence can act as a good indicator for enhanced solubility of the SWNT due to the presence of the dye.

CP685, *Molecular Nanostructures: XVII Int'l. Winterschool/Euroconference on Electronic Properties of Novel Materials,* edited by H. Kuzmany, J. Fink, M. Mehring, and S. Roth
© 2003 American Institute of Physics 0-7354-0154-3/03/$20.00

EXPERIMENTAL

Various concentrations of the dyes terphenyl and anthracene in toluene were prepared as reported in reference [5]. The dye solutions were characterised using UV-Vis-NIR (Perkin Elmer Lambda 900), fluorescence (Perkin Elmer LS55) and Raman (Instruments SA LabRam 1B) spectroscopy. Arc discharge SWNT's obtained from Rice University (Tubes@rice, Houston, TX) were added to the dye solutions prepared above in a 1:1 ratio by weight (SWNT:oligomers). The suspensions were then sonicated for fifteen minutes at 45^0C and allowed to settle for 24 hours, after which the supernatant was carefully pipetted off from the precipitated SWNT's. The suspensions were then allowed to settle for a further 24 hours before being characterised by Raman, UV-Vis-NIR and fluorescence spectroscopy. For Raman measurements, samples were dropcast onto glass slides from suspension and onto a quartz plate for UV-Vis-NIR.

RESULTS

The first observation was that solubility of the SWNT is enhanced upon addition of dye molecules and after six weeks it was noted that the SWNT's did not appear to aggregate, where as SWNT's simply suspended in toluene aggregated and precipitated within 24 hours. This observation indicates an enhanced stability of the SWNT suspension in the presence of the dyes.

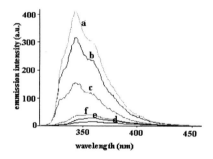

FIGURE 1. Fluorescence of terphenyl before and after the addition of SWNT

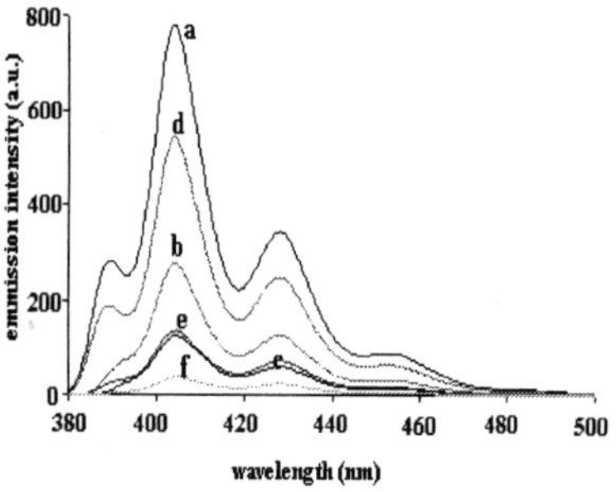

FIGURE 2. Fluorescence of anthracene before and after the addition of SWNT

Increasing concentration of the dyes in solution increased the fluorescence. Fig. 1 shows the fluorescence of terphenyl in toluene before (a), (b), (c) and after (d), (e), (f), the addition of SWNT. Before the addition of SWNT the terphenyl emission is seen to have a peak maxima at 345nm. On addition of SWNT, the fluorescence is quenched dramatically and the peaks red shift by 20nm. The red shifting is a result of increased conjugation and is brought about by the alignment of the three phenyl rings of terphenyl so that mapping to the SWNT can occur. The quenching is most probably due to energy transfer from the dye to the SWNT. This result is a strong indication of mapping between the dye and the SWNT. The fluorescence of anthracene, fig. 2 indicated a similar trend but there was no shift in either the blue or red direction on the addition of SWNT. This is due to rigid structure of the molecule.On examination of figure 3 it is noted that concentrations 0.10×10^{-5}M to 0.25×10^{-5}M of terphenyl, the fluorescence decreases. This indicates that terphenyl is interacting with the SWNT. Concentration 0.25×10^{-5} M is the maximum quenching point and the point at which the weight of terphenyl to SWNT is at approximately a 2:1 ratio. After this point the fluorescence increases indicating that maximum interaction has occurred and the terphenyl that is added after this point is free in solution. At 1.0×10^{-5}M the fluorescence begins to decrease. The trend occurring after 0.25×10^{-5}M has been seen for solution of terphenyl in toluene [6] with the absence of SWNT. It was found that the decrease in fluorescence in the case where there is no SWNT present was as a result of π-π stacking of the free terphenyl and so fluorescence energy is lost between terphenyl molecules. Because the trend for both the terphenyl solution and the SWNT/terphenyl solution are the same one can also say that the π-π stacking is the case for the SWNT/terphenyl. These findings highlight that after 0.25×10^{-5}M there is no additional interaction occurring between the SWNT and the terphenyl. Anthracene shows similar trend features as fig. 3. Some similarities are that for terphenyl and anthracene the maximum quenching is at 0.25×10^{-5}M. The optimum ratio by weight for SWNT: anthracene is 4:1. The fluorescence of anthracene increases up to 2.5×10^{-5}M and is then known to decrease after that point.These results point towards specific interaction at well-defined ratios.

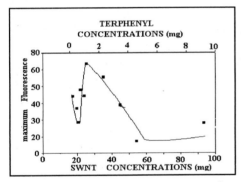

FIGURE 3. Concentrations of terphenyl at 10^{-5}M   **FIGURE 4.** UV/Vis/NIR of SWNT

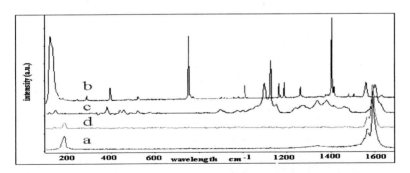

FIGURE 5. The Raman spectra at laser line 514nm of a (a) pristine SWNT, (b) anthracene crystal, (c) composite spectrum, (d) washed composite of SWNT.

UV/Vis/NIR was run on a pristine sample of SWNT and on the dye composites. The purpose of this was to investigate the diameter range that the dye molecules were interacting with when compared to the pristine sample. Fig. 4 suggests that terphenyl is interacting with a narrow range of SWNTof about 1.38nm and anthracene, is interacting with a broad range of diameters of 1.30-1.46nm. Raman, further supports the hypothesis that there is interaction between SWNT and dyes. Fig. 5 shows the Raman spectra of (a) pristine SWNT (b) anthracene crystal, (c) a composite spectrum taken at 514nm When spectra (a) and (b) are compared with (c) one notices that it is not a weighted summation of (a) and (b) but a unique spectrum that has a number of new peaks particularly in the 900-1450cm^{-1} region. The composite was then washed with excess toluene and the spectrum revealed a SWNT (d) which is a similar spectrum to that of (a). This suggests that the dye has indeed interacted with a SWNT.

CONCLUSION

Results show that a stable suspension of SWNT in toluene is formed as a result of the interaction with the dyes. The electronic properties of the dyes are greatly effected by the presence of SWNT and this is clearly shown in the quenching of fluorescence of the dyes upon addition of SWNT. Fluorescence also gives the point of maximum interaction between the dyes and SWNT. The UV/Vis/NIR gives an indication of the diameter range of the SWNT with which the dyes are interacting. The results reported support the idea that the dye is mapping to the SWNT. Evidence of interaction is further enhanced by the composite and washed composite Raman spectra. The results suggest that this can be fine tuned to give selective processing of armchair and zigzag.

ACKNOWLEDGMENTS

FOCAS is funded under the Irish Government Programme for Research in Third Level Institutions.

REFERENCES

1. Bahr J.L., Mickelson E.T, Bronikowski J., Smalley R.E, Tour J.M.: Chem Comm 193-194 (2001).
2. Chen J. , Hamon M.A., Hu H., Chen Y., Rao A.M., Eklund, P.C., Haddon R.C.: Science282 95-98 (1998).
3. Dalton A.B, Stephens C., Coleman J.N., McCarthy B, Ajayan P.M, Lefrant S., Bernier P., Blau W.B., Byrne H.J.: J. Phys. Chem. B **104** 10012-10016 (2000).
4. Keogh S., Hedderman T.G., Gregan E., Farrell G., Dalton A.B., McCarthy B., Chambers G., Byrne H.J.: Proceedings of the international winter school on electrical properties of novel material. (2002) pp 570-573.
5. Hedderman T.G., O'Neill L., Maguire A., Keogh S.M., Gregan E., Mc Carthy B., Dalton A.B., Chambers G., Byrne H.J.: Proceedings of the SPIE. (2002).
6. O'Neill L., Final year BSc. Project, D.I.T. Kevin Street (2002).

The Physical Interactions between HiPCo SWNTs and Semi-Conjugated Polymers

[1]S.M. Keogh, [1]T.G. Hedderman, [1]A. Maguire, [2]M.G. Rüther, [1]E. Gregan, [1]G. Chambers, [1]H.J. Byrne

[1]Facility for Optical Characterisation and Spectroscopy (FOCAS), School of Physics, Dublin Institute of Technology, Kevin Street, Dublin 8, Ireland

[2]Materials Ireland, Physics Department, Trinity College Dublin, Dublin 2, Ireland

Abstract. Hybrid systems of the conjugated organic polymer poly (p-phenylene vinylene-co-2,5-dioctyloxy -m-phenylene vinylene) (PmPV) and HiPco SWNT are explored using spectroscopic and thermal techniques to determine specific interactions. Vibrational spectroscopy indicates a weak interaction and this is further elucidated using Differential-Scanning Calorimetry and Temperature Dependent Raman Spectroscopy. Two distinct transitions in region of -60°C and + 60°C are investigated.

INTRODUCTION

Single walled carbon nanotubes (SWNTs) exhibit many unique physical and chemical properties [1]. Although there are many different methods for SWNT synthesis, none have been successful in producing 100% pure SWNTs in large quantities. HiPco samples contain a high amount (90%) of high purity (99%) SWNTs [2]. The diameter distribution is predominantly 0.7nm-1.2nm (average diameter of 1nm[2]). Specific interactions between arc-discharge SWNT within polymer composites have been documented [3][4]. This proposed system for purification uses the organic semi-conjugated polymer poly (p-phenylene vinylene-co-2,5-dioctyloxy-m-phenylene-vinylene) (PmPV). The chemical structure of PmPV has been shown to allow solubility due to the presence of floppy side chains and conjugation along the polymers backbone, which allows wrapping of the polymer around the nanotubes [3]. Although the exact nature of interaction has not yet been clearly defined, in this proposed system the polymer maps onto the nanotube lattice in a periodic and well-defined manner [4]. In this study hybrid systems of the organic conjugated polymer PmPV (provided by Trinity College, Dublin), and HiPco from toluene solution are explored in order to further elucidate the specific nature of the interaction. Temperature dependent Raman (T.D.Raman) spectroscopy at 514.5nm in conjunction with Differential Scanning Calorimetry (DSC) are employed to obtain thermal measurements of the composite.

CP685, *Molecular Nanostructures: XVII Int'l. Winterschool/Euroconference on Electronic Properties of Novel Materials,*edited by H. Kuzmany, J. Fink, M. Mehring, and S. Roth
© 2003 American Institute of Physics 0-7354-0154-3/03/$20.00

EXPERIMENTAL

Composites of HiPco SWNTs (0.1% by weight) were mixed in 1g/L of PmPV in toluene. The composites were sonicated for 10 hours, left to stand for 48 hours and then decanted. Films were made by drop casting onto glass slides and characterised by the (Instruments S.A Labram 1B) Raman system and a Pyris Diamond DSC from Perkin Elmer.

RESULTS

Differential Scanning Calorimetry (DSC) was used to study the thermal properties of the composites by monitoring the thermodynamics of the system as it passes through phase transitions. Fig 1(a) shows a thermogram of the nanotube composite and pristine polymer from the temperature range of -55°C to -30°C. Fig 1(b) shows thermograms of the nanotube composite and pristine polymer from the temperature range of 30°C to 60°C. These broad features between 30°C and 60°C are due to the melting of the polymer backbone and are present in both the composite and pristine polymer thermograms. The composite shows a new doubly peaked endotherm between -30°C and -50°C, which is not present in the pristine polymer thermogram, and as there are no distinct features in the thermogram of raw tubes this new feature is proposed to be due to an interaction between the polymer and the nanotubes. Therefore there are three distinct regions of interest in the composite. **A**: this is before the composite peaks occur in the thermogram. **B**, this is after the composite peak and before the melting of the polymer backbone, and **C**, this is after the polymer backbone has melted. These three regions were further investigated using temperature dependant Raman spectroscopy at 514.5nm in order to gain a greater understanding of the changes that occur in the thermograms. Figure 2 presents temperature dependant Raman spectrum at 514.5nm for the 0.1% HiPco composite. The spectra correspond to points labeled **A**, **B** and **C** in the thermograms.

Transition 1: Changes in composite only

Transition 2: Broad features occur in composite and Polymer

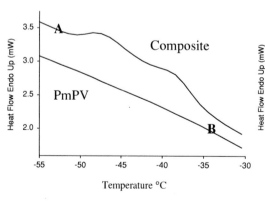

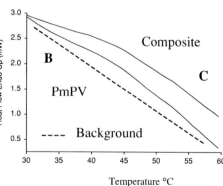

Figure 1(a). Thermogram of the SWNT Composite showing low temperature regions

Figure 1(b). Thermogram of the SWNT Composite showing high temperature region

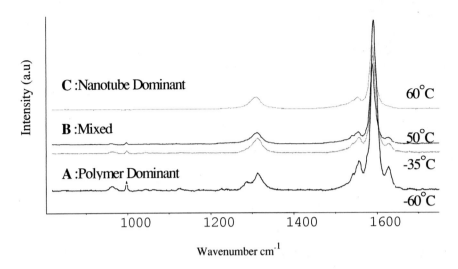

Figure 2. Temperature Dependant Raman spectra at 514.5 nm of the
HiPco SWNTs in polymer composites

Region A: Polymer Dominant Composite Spectrum

Spectrum A (-60°C) corresponds to region A in the thermogram. Two distinct polymer peaks are present at $960cm^{-1}$ and at $1000cm^{-1}$, which are also present in the pristine polymer Raman spectrum. These peaks are more dominant at lower temperatures and are attributed to the vibration of the side chains of the polymer. It is noted that at -35°C the peak at $960cm^{-1}$ has almost vanished, suggesting that the polymer side chains have melted. The D-line region of the composite has a peak at $1312cm^{-1}$ and a shoulder at $1283cm^{-1}$. This doubled peak feature is due to the combination of the polymer's phenyl ring and double bond stretching and occurs also in the pristine polymer Raman spectrum. The shoulder at $1283cm^{-1}$ is no longer present at 50°C. In the G-line region of the composite six peaks are present at $1541cm^{-1}$, $1557cm^{-1}$, $1577cm^{-1}$, $1591cm^{-1}$, $1627cm^{-1}$ and $1653cm^{-1}$. This is similar to the pristine polymer, but the relative intensities have changed. The peaks at $1627cm^{-1}$ and $1577cm^{-1}$ are less intense in the composite than in the pristine polymer.

Region B: Mixed Composite Spectrum

Traditionally presented Raman spectra occur between the temperatures of -35°C and 50°C, labeled B in figure 2. This corresponds to the temperature after which the composite peaks occur in the thermogram. Composite A differs from composite B, as the polymer is no longer dominating the spectrum. Here the G-line region presents peaks at $1627cm^{-1}$, $1589cm^{-1}$ and $1555cm^{-1}$ and shows a mixed polymer/tube contribution. The composite peak at $1627cm^{-1}$ is less intense. The D-line region shows

267

a peak present at 1312cm^{-1}, the shoulder at 1283cm^{-1} is no longer present. Also there is a decrease in intensity of the polymer peak at 1000cm^{-1}.

Region C: Nanotube Dominant Composite Spectrum

Spectrum C corresponds to region C in the thermogram, which is dominated by changes in the pristine polymer. The composite spectrum is dominated by nanotubes, suggesting that the nanotubes tangential modes are no longer restricted due to change in the pristine polymer. The polymer peaks at 960cm^{-1}, 1000cm^{-1}, and at 1283cm^{-1} are no longer present and the D-line is symmetrical. This is similar to a raw nanotube Raman spectrum, however the composite peak at 1627cm^{-1} is still present but less enhanced.

CONCLUSION

From the DSC and Raman results obtained, it is clear that as the temperature of the composite changes so does its interaction properties. It can be concluded that three different composite types occur between the temperature ranges of –60 °C and +60°C. At -60°C, the composite presents a Raman spectrum which is dominated by the polymer. This suggests that the polymer is restricting the nanotubes tangential modes from vibrating. At -35°C, after the first transition in the thermogram, which only appears in the composite, the Raman spectra becomes mixed, suggesting that the polymer has a weaker interaction with the tubes than at -60°C. Raman spectroscopy suggests side chains are involved in this interaction as they are not as intense in the Raman spectrum at -35°C. At 60°C, after the second transition, the composite is influenced by the melting of the polymer. The Raman spectrum of the composite is dominated by nanotubes, suggesting that the polymer backbone is also involved in this interaction. Raman spectroscopy suggests that both the polymer side chains and the polymer backbone are involved in composite formation, but as the temperature is increased the interaction becomes weaker.

ACKNOWLEDGEMENTS

FOCAS is funded through the National Development Plan 2000-2006 with support from the European Development fund. SMK acknowledges DIT Scholarship support, and Stefanie Maier and Anna Drury, Trinity College Dublin for providing PmPV.

REFERENCES

1. Yakobson, B.I., Smalley, R.E., *Am. Sci.* **85**, (1997) 324.
2. Nikolaev, P., Smalley, R.E et al :*Chem Phy Lett* **313** (1999) 91-97
3. Shi, Z.J. Lian, Y.F., Zhou, X.H., Gu Z.N., Zhang, Y.G., Iijima, S., Zhou, L.X., Yue, K.T., Zhang, S.L., *Carbon* **37** 1449 (1999) 1449.
4. Dalton, A.B., Blau, W.J., Chambers, G., Coleman, J.N., Henderson, K., Lefrant, S., McCarthy, B., Stephan, C, Byrne H.J: *Synth. Met.* **121** (2001)1217-1218.

Effect of Nanotube Type on the Enhancement of Mechanical Properties of Free-Standing Polymer/Nanotube Composite Films

M. Cadek[1], J. N. Coleman[1], D. Blond[1], A. Fonseca[2], J. B. Nagy[2], K. Szostak[3], F. Béguin[3] and W. J. Blau[1].

1) Materials Ireland Polymer Research Centre, Department of Physics, Trinity College Dublin, Dublin 2, Ireland.
2) Lab. de Résonance Magnétique Nucléaire, Fac. Univ. N.D. de la Paix, rue de Bruxelles 61, 5000 Namur, Belgium.
3) Centre de Recherche sur la Matière Divisée, CNRS, 1b, rue de la Férollerie, Cedex 02, F-45071 Orléans, France.

Abstract. In this work it will be shown that the mechanical reinforcement of polymer composites using carbon nanotubes is dependant on the total surface area of carbon nanotubes dispersed into the polymer matrix. Furthermore, it will be shown that the total length of carbon nanotubes dispersed in the polymer matrix does not have any influence on the mechanical properties. Four differently produced carbon nanotube samples were introduced into a polyvinyl alcohol matrix and it can be reported that the sample with the smallest diameter, thus the nanotube sample with the highest surface area, increased mechanical properties such as tensile modulus by 120% while adding less than 1wt% of carbon nanotubes. Measurements were accomplished on bulk free standing polymer films by tensile testing.

INTRODUCTION

Carbon nanotubes (CNT) are known since their discovery by Suomi Ijiima [1] in the early nineties as the potential reinforcement material. It is well known and reported that CNT's have outstanding mechanical properties such as Young's modulus of 1-2TPa [2,3] and tensile strength of up to 60GPa [3]. Therefore numerous research groups investigated the reinforcement of polymer composites using CNT's [3-6]. Several problems have to be overcome using CNT's as a reinforcement agent. It is of immense importance to insure a good dispersion of CNT's in the polymer matrix as well as a strong interfacial bonding between CNT's and the polymer chains. This is essential in order to insure that the applied load is carried by the reinforcement agent rather that by the polymer matrix.

In this work we used differently produced carbon nanotubes. Catalytically grown multi walled carbon nanotubes (cMWNT) [7], arc discharged grown multi walled carbon nanotubes (aMWNT)[8] and single walled carbon nanotubes (SWNT) produced by the HipCo method [9]. It will be shown that small diameter multi walled carbon nanotubes (MWNT) with the highest total surface area dispersed into the polymer matrix show the best reinforcement properties. The tensile modulus could be

CP685, *Molecular Nanostructures: XVII Int'l. Winterschool/Euroconference on Electronic Properties of Novel Materials,*edited by H. Kuzmany, J. Fink, M. Mehring, and S. Roth
© 2003 American Institute of Physics 0-7354-0154-3/03/$20.00

improved by a factor of 2 while introducing less than 1wt% of cMWNT. In addition to that we will be able to show that the reinforcement of polymers does not necessarily depend on the length of CNT's.

EXPERIMENTAL METHODS

In order to bypass impurity problems such as catalytically particles or amorphous and turbostratic graphite a method developed by Coleman et al. and optimized by Murphy et al. was used for non destructive purification of the different nanotube samples. As a purification and matrix material, polyvinyl alcohol (PVA) was chosen. PVA is soluble in water and was mixed in a concentration of 30g/L using an ultra high power sonic tip for 1min.

As reinforcing agents two different cMWNT were used. (N-cMWNT's and O-cMWNT's). N-cMWNT's show an average diameter of 14nm and lengths of approximately 50μm. They were pre-purified by oxidation [7]. O-cMWNT were measured to have an average diameter of 16nm and lengths of up to 1-5μm [10]. O-cMWNT's were pre-purified by acid treatment in HCL. For comparison we used aMWNT's having an average diameter of 25nm and an average length of 1μm. The purity of the raw material was 30-40%. HipCo-SWNT's purchased from CNI are known to have an average diameter of 1nm but were arranged in bundles of an average diameter of 7nm. The purity of HipCo-SWNT's was found to be 80% whereas 20wt% catalytically iron particles could be observed.

In order to disperse the nanotube samples 1wt% of N-cMWNT, O-CMWNT and HipCo-SWNT were sonicated into the PVA/H_2O solution with an ultra high power sonic tip for 5min and placed afterwards for 2hrs into a low power sonic bath. This was followed by an additional sonication for 5 more min using the sonic tip.

In case of the aMWNT sample 25wt% of nanotubes were added into the PVA/H_2O solution and sonicated for 1min using the sonic tip. This was followed by a mild sonication for 2hrs in the sonic bath.

All samples were afterwards left undisturbed for 24hrs in order to allow impurities to form a sediment on the bottom of the sample bottle. The resulting solution rich in nanotubes was then decanted while the sediment consisting of impurities remained in the sample bottle.

These solutions were then blended with PVA/H_2O solution to produce a range of various mass fractions. 1ml of each sample was dropped onto a polished Teflon disk (30mm in diameter) and placed into a 60°C heated oven in order to allow the solvent to evaporate and form a polymer film. This procedure was repeated 4 times resulting in a 200-300 micron thick bulk polymer film. These films could be easily peeled of the Teflon disks and were cut into 10x4mm rectangular stripes in order to perform tensile testing.

Tensile testing was carried out on a Zwick Z100 using a 100N load cell and a cross head speed of 0.5mm/min.

In order to analyse the results mass fractions were transferred into volume fraction assuming a density of 2.16g/cm³ for both aMWNT's and cMWNT's [6] and a density of 1.5g/cm³ for HipCo-SWNT's [11].

RESULTS AND DISCUSSION

As seen in figure 1 the tensile modulus of polyvinyl alcohol could be increased by a factor of more than 2 from 1.9GPa to 4.1GPa while adding less than 1wt% of N-cMWNT to the matrix. This corresponds to a volume fraction of 0.6vol%. In case of the a-MWNT the tensile modulus could be increased by a factor of 2 from 1.9 GPa to 3.75 GPa while having 1.8wt% of nanotubes in the matrix. The relative tensile modulus increase confirms earlier results published elsewhere [4]. For the O-cMWNT it can be observed that an increase of 70% from 1.9GPa to 3.11GPa can be seen while adding a volume fraction of 0.6vol% of nanotubes.

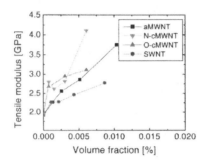

FIGURE 1: Tensile modulus increase plotted versus volume fraction for polyvinyl alcohol reinforced with catalytically grown and arc grown MWNT and SWNT produced by the HipCo process.

The Hipco-SWNT/PVA composite showed only an increase of 50% from 1.9GPa to 2.77GPa. This can be explained by the well known effect of rope formation by SWNT's and will be confirmed in figure 2a.

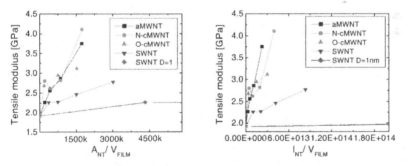

FIGURE 2: a) Tensile modulus increase for polyvinyl alcohol plotted versus total curface area of nanotube per unit volume of polymer film. b) Tensile modulus increase for polyvinyl alcohol plotted versus total length per unit volume of polymer film.

As the polymer/nanotube interaction occurs at the nanotube surface the nanotube surface area is expected to be an important parameter. Total nanotube surface area is related to the volume fraction. In figure 2a volume fraction has been converted to surface area using the measured nanotube diameters and it can be observed that all the tensile moduli for MWNT's fall onto the same line. This is a remarkable result. Using the real diameter value of HipCo-SWNT's of 1nm it can be seen that the results do not agree with the MWNT samples. Using the measured average diameter value of 7nm it can be observed that the tensile modulus increases in line with MWNT's, but not as much as expected. This leads to the assumption that buddle formation leads to lower performance. The interfacial shear strength in a nanotube buddle is known to be lower than between nanotube and a polymer matrix and leads to the assumption that sliding of the tubes within the bundle occurs and therefore a low increase of tensile modulus can be seen. Bundle formation of SWNT's can also be observed in scanning electron microscopy graphs (not shown here).

Figure 2b shows the tensile modulus increase plotted versus the calculated total length of nanotubes induced per volume of polymer film. It can be observed, similarly to figure 2a, that the tensile moduli for MWNT's are falling to the same line whereas the SWNT's results using the real diameter value of 1nm are not agreeing with these results.

The fact that the tensile moduli for all the MWNT samples falling into the same area indicates that the length of nanotube does not play a major role for mechanical reinforcement of composites. As it is well known that catalytically grown MWNT are an order of magnitude longer than arc discharge grown MWNT.

CONCLUSION

It has been shown that small diameter multi walled carbon nanotubes show the highest increase in tensile modulus compared to the other nanotubes. While adding less than 1wt% the tensile modulus could be increased by 120%. In addition when plotting tensile modulus versus surface area the graphs of MWNT fall all on the same line which indicates that reinforcement is critically dependent on the surface area rather than on length or aspect ratio.

REFERENCES

1. S. Iijima, Nature **354**, 56 (1991).
2. O. Lourie and H. D. Wagner, Appl. Phys. Lett. **73**, 3527 (1998); M. M. J. Treacy, T. W. Ebbesen, and J. M. Gibson, Nature **381**, 678 (1996).
3. R. H. Baughman, A. A. Zakhidov, and W. A. d. Heer, Science **279**, 787 (2002).
4. M. Cadek, J. N. Coleman, V. Barron, K. Hedicke, and W. J. Blau, Appl. Phys. Lett. **81**, 5123 (2002).
5. J. Sandler, M. S. P. Shaffer, T. Prasse, W. Bauhofer, K. Schulte, and A. H. Windle, Polymer **40**, 5967 (1999); M. S. P. Shaffer and A. H. Windle, Adv. Mater. **11**, 937 (1999); J. N. Coleman, K. P. Ryan, M. S. Lipson, A. Drury, M. Cadek, M. i. h. Panhuis, R. P. Wool, and W. J. Blau, XVI. International Winterschool/Euroconference, Kirchberg/Tirol, Austria, **633**, 557 (2002).
6. D. Qian, E. C. Dickey, R. Andrews, and T. Rantell, Appl. Phys. Lett. **76**, 2868 (2000).
7. N. Pierard, A. Fonseca, Z. Konya, I. Willems, G. V. Tendeloo, and J. B. Nagy, Chem. Phys. Lett. **335**, 1 (2001); Y. Soneda, L. Duclaux, and F. Beguin, Carbon **40**, 965 (2002).
8. W. Krätschmer, L. D. Lamb, K. Fostiropoulus, and D. R. Huffman, Nature **347**, 354 (1990); T. W. Ebbesen and P. M. Ajayan, Nature **358**, 220 (1992); M. Cadek, R. Murphy, B. McCarthy, A. Drury, B. Lahr, M. i. h. Panhuis, J. N. Coleman, R. Barklie, and W. J. Blau, Carbon **40**, 923 (2002).
9. M. J. Bronikowski, P. A. Willis, D. T. Colbert, K. A. Smith, and Richard E. Smalley, J. Vac. Sci. Tech. **19**, 1800 (2001).
10. F. Bénguin, Personal communication (2003).
11. J. N. Coleman, W. J. Blau, A. B. Dalton, E. Munoz, S. Collins, B. G. Kim, J. Razal, M. Selvidge, G. Vieiro, and R. H. Baughman, Appl. Phys. Lett. **82**, 1682 (2003).

Raman Spectroscopy of Carbon Nanotube-Polyaniline and Functionalized CNT/SOCl$_2$ Films

Núria Ferrer-Anglada[1], Martti Kaempgen[2], Viera Skákalová[2], Ursula Dettlaff-Weglikowska[2], Siegmar Roth[2]

[1] Departament de Física Aplicada, Universitat Politècnica de Catalunya, Campus Nord B4, J. Girona 3-5, 08034 Barcelona, Spain

[2] Max-Planck-Institute für Festkörperforschung, Heisenbergstrasse 1, D-70569 Stuttgart, Germany

Abstract. Different transparent conducting heterostructures as thin films on transparent plates have been obtained: a thin network of Carbon Nanotubes (CNT), 80% to 93% transparent and enough conducting to be used as anode, on which Polyaniline is deposited electrochemically. In order to obtain information about the possible interaction between CNT and the polymer, we analysed the Raman spectra, on these hetero-thin films obtained at different chemical and electrochemical conditions. To compare, we analysed by Raman spectroscopy self-standing CNT-Polyaniline films where polyaniline was electrochemically deposited on a thick network of CNT (bucky-paper). Electrical conductivity and optical transparency are measured: Polyaniline improves the conductivity by a factor from 2 to 3 for films where the transparency is 67%. The best transparency, 93% to 97%, is obtained with SOCl$_2$ functionalized CNT thin networks.

INTRODUCTION

Conducting transparent materials would be interesting to be used as new transparent electrodes, substituting the well known ITO or other Transparent Conducting Oxides. It is known that carbon nanotubes (CNT) are good conductors, the conductivity of bucky-paper could be near 2000 S/cm, and is limited by the inter-tube processes [1]. Polyaniline is one of the best studied conducting polymers, either by transport [2] or spectroscopic properties [3]. Some published results on composites from carbon nanotubes and polyaniline show that polyaniline properties are improved in the composite material, when a small proportion of CNT are present [4-6]. The objective of the present work is to obtain transparent conducting thin films, in a way that depositing a slight amount of polyaniline on it, the interconnection between CNT sould be better.

CP685, *Molecular Nanostructures: XVII Int'l. Winterschool/Euroconference on Electronic Properties of Novel Materials,* edited by H. Kuzmany, J. Fink, M. Mehring, and S. Roth

EXPERIMENTAL

Different heterostructures as thin films on plastic, glass or quartz plates have been obtained: thin networks of Single Walled Carbon Nanotubes (CNT) obtained by HIPCO process, transparent and enough conducting to be used as anode, on which Polyaniline can be deposited electrochemically. We analyse by Raman spectroscopy, optical absortion and electrical conductivity at room temperature, the properties of these networks of nanotubes (CNT) before and after deposition of polyaniline (CNT-PA); and networks using $SOCl_2$ functionalized carbon nanotubes, $CNT/SOCl_2$ without and with polyaniline.

On figure 1 we present the Atomic Force Microscopy image of CNT network with polyaniline, CNT-PA, were we can observe that the polymer grows mostly on the CNT, showing its granular structure.

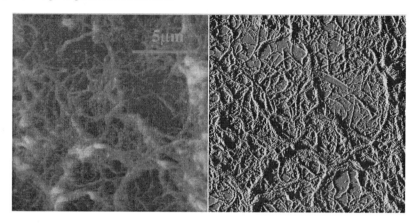

FIGURE 1. AFM image of a transparent thin Single Walled Carbon Nanotube network, CNT-PA, the polymer grows mostly on the carbon nanotubes.

Electrical Resistance and Transparency

TABLE 1. Electrical Resistence and transparency of different representative films.

	CNT network	CNT-PA	CNT/SOCl$_2$	CNT/SOCl$_2$-PA
Transparency, %T	80 – 90%	62 – 90%	93 – 99%	-
Electrical Resistence, R(Ω)	660	245	$1,1 . 10^3$	$7 . 10^3$

With functionalized $CNT/SOCl_2$ the transparency is notably higher, the electrical conductivity of the network is lower than CNT-PA but still quite high (R/square from 1 to 3 kohm). Nevertheless, when polyaniline is electrochemically deposited on these networks, the conductivity decreases dramatically, R/square increases by 5-10 times, even at very low PA content.

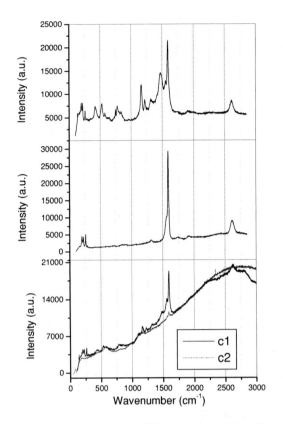

FIGURE 2. Raman spectra with a 633 nm excitation laser line, of transparent thin networks of CNT-Pa composites, a), compared to spectra of thin networks of CNT, b); increasing the electrodeposition time when growing polyaniline some cain of degradation occours: corresponding time for sample c1, t= 35 min., for sample c2 t=115 min., c).

Raman Spectroscopy

We analysed the Raman spectra, of those thin films obtained at different chemical and electrochemical conditions. To identify the PA peaks, we compare these results with those shown on the Raman spectra of self-standing carbon nanotubes with Polyaniline, BP-PA, where polyaniline was electrochemically deposited on a thick network of CNT (bucky-paper) at the same conditions. Raman spectra are very sensitive to polyaniline presence and is characteristic of conducting emeraldine salt form, see figure 2. On fig. 2 c) we can observe that, when the electrochemistry deposition time of polyaniline is too long (near two hours in that case), some degradation occurs, in that case the electrical resistance increases again. Figure 3 shows the Raman spectra of CNT thin networks modified with $SOCl_2$. After depositing PA the CNT lines nearly disappear, indicating degradation.

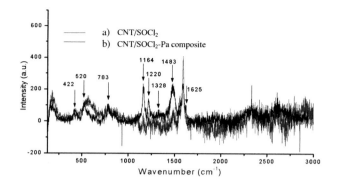

FIGURE 3. Raman spectra of: thin networks of functionalized carbon nanotubes CNT/SOCl$_2$, a); CNT/SOCl$_2$-PA composite, b).

DISCUSSION

Our results show that we could obtain transparent and electrically conductive large area samples. The electrical resistences ($\geq 200\Omega$) are still one order of magnitude greater than in the currently used transparent conductive oxides (TCO) like ITO, in which the resistivity are tipically from 0,1 to 1 mΩ cm, or R= 10–100Ω/square. Some other polymers and functionalization groups should be tested in a future work.

ACKNOWLEDGMENTS

N. Ferrer-Anglada thanks the MECD of the Spanish Government for a mobility grant Ref. PR2002-0050, for a research stage at MPI-FKF., Stuttgart.

REFERENCES

1. Ajayan, P. M., Zhou, Q. Z., *Top. Appl. Phys.*, **80**, 391 (2001).

2. Reghu, M., Cao, Y., Moses, D., Heeger, A. J., *Phys. Rev. B*, **47**, 4, 1758-1764 (1993).

3. Cochet, M., Buisson, J.-P, Wéry, J., Jonusauskas, G., Faulques, E., Lefrant, S., *Synth. Met.,* **119**, 389-390 (2001).

4. Cochet, M., Maser, W. K., Benito, A. M., Callejas, M. A., Martinez, M.T., Benoit, J-M., Schreiber, J., Chauvet, O., *Chem. Commun* **16**, 1450 (2001).

5. Deng, J., Ding, X., Zhang, W., Peng, Y., Wang, J., Long, X., Li, P., Chan, S.C., *European Polym. J*, **38**, 2497 (2002).

6. Zengin, H., Zhou, W., Jin, J., Czerw, R., Smith, D. W., Echegoyen, L., Carroll, D. L., Foulger, S. H., Ballato, J., *Adv. Matter.*, **14**, 20, 1480-1483 (2002).

Interconnection of Chemically Functionalized Single-Wall Carbon Nanotubes via Molecular Linkers

Urszula Dettlaff-Weglikowska[1], Viera Skakalova[1], Ralf Graupner[2], Lothar Ley[2], and Siegmar Roth[1]

[1]Max-Planck-Institut fuer Festkoerperforschung, Heisenbergstr. 1, 70569 Stuttgart, Germany
[2]Institut für Technische Physik, Universitaet Erlangen, Erwin-Rommel Str. 1 91058 Erlangen, Germany

Abstract. Single wall carbon nanotubes produced by high pressure CO decomposition (HiPco process) were oxidized in air at 350 °C and treated by thionyl chloride to generate acylchloride groups. Subsequent coupling reactions were carried out with different amine molecules to form intertube junctions. Tripropylentetramine, 2,4,6-triaminopirimidine and 2,5-bis(4-aminophenyl)-1,3,4-oxadiazole have been chosen as a molecular linker. The reaction products were characterized by XPS and Raman spectroscopy. Also electrical and mechanical properties of bucky papers prepared from chemically modified tubes were studied and are discussed.

INTRODUCTION

Chemical modification of single wall carbon nanotubes (SWNTs) is essential for many applications. Recently, we have shown that end and side wall functionalization allows tubes interconnection via organic molecular linkers [1]. These nanotubes T-junctions have been used for preparing of nanosized all carbon transistors, where both the conducting channel and the gate are made of nanotubes. Interconnection of tubes by chemical bonding could also affect mechanical and electrical properties of entangled nanotube network in the bucky paper.

The aim of present work is to investigate the influence of the linker molecules on the electrical and mechanical properties of the bucky papers prepared from interconnected tubes.

EXPERIMENTAL

SWNTs were produced by high pressure decomposition of CO (HiPco process) at CNI, Huston, USA. For functionalization 50 mg of the tubes were oxidized at 350 °C in air for 10 min and treated with $SOCl_2$ at 45 °C for 24 h. The mixture was diluted with DMF, filtered on a PTFE membrane filter and washed with DMF. For coupling reaction with molecular linker the resulting bucky paper was devided into 3 parts. Each part was resuspended in 25 ml DMF solution of appropriate amine. After sonication in ultrasonic bath for 15 min and 15 h stirring at room temperature the

CP685, *Molecular Nanostructures: XVII Int'l. Winterschool/Euroconference on Electronic Properties of Novel Materials*, edited by H. Kuzmany, J. Fink, M. Mehring, and S. Roth
© 2003 American Institute of Physics 0-7354-0154-3/03/$20.00

suspension was filtered, washed with DMF and dried at 75 °C. Tripropylentetramine (linker I), 2,4,6-triaminopirimidine (linker II) and 2,5-bis(4-aminophenyl)-1,3,4-oxadiazole (linker III) have been used as an organic molecular linker. All amines were purchased from Aldrich.

The four probe method was used to determine the electrical conductivity of the bucky papers. The Young's modulus was determined from the elastic part of the stress vs. strain curve measured by a force transducer up to 20 N (Hottinger Baldwin Messtechnik, typ 52). Raman spectra were measured using an excitation wavelength of 632,8 nm.

RESULTS AND DISCUSSION

The basic step in chemical functionalization is oxidation which results in formation of carboxyl groups SWNT-COOH attached to the defects and open ends of the nanotubes. The $SOCl_2$ treatment converts the carboxyl groups into acidic chlorides SWNT-COCl. These functional groups react with amine groups of the linker molecule connecting the tubes via amide linkage SWNT-CONHR [2].

As a linker we have chosen long chain aliphatic tripropylentetramine (linker I) and two aromatic amines 2,4,6-triaminopirimidine (linker II) and 2,5-bis(4-aminophenyl)-1,3,4-oxadiazole (linker III). Corresponding chemical reactions and formulas of the molecular linkers are illustrated by the scheme 1.

$$\text{SWNT} \xrightarrow[\text{air}]{O_2} \text{SWNT-COOH} \xrightarrow[45\ °\text{C},\ 24\ \text{h}]{SOCl_2} \text{SWNT-COCl}$$

$$\downarrow\ H_2N\text{-R-}NH_2$$

$$\text{SWNT-CO-NH-R-HN-CO-SWNT}$$

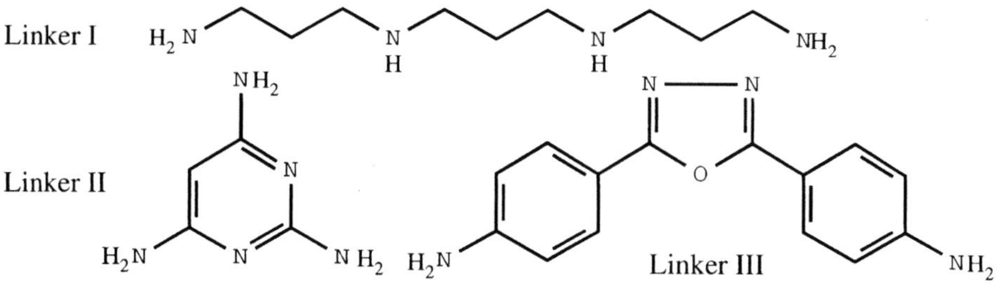

Scheme 1. Coupling reactions with organic linkers leading to tubes interconnection.

We analysed the interconnected tubes using X-ray photoelectron spectroscopy. The carbon C (1s) spectra of SWNTs (not shown here) measured on a thin films prepared from pristine, treated with $SOCl_2$ and modified with aliphatic and aromatic diamine linker show significant changes. We observe that the position of the sp^2 carbon peak changes after each reaction step. Shift to lower binding energy indicates electron acceptor behavior of chlorine atoms attached by $SOCl_2$ treatment while shift to higher binding energy reveals donor behavior of amide groups created by coupling reactions [1].

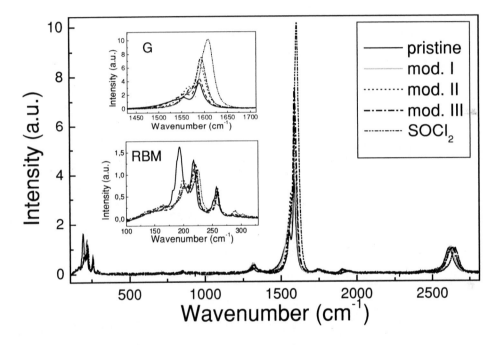

FIGURE 1. Raman spectra of SWNTs chemically modified by three molecular linkers.

Figure 1 shows room temperature Raman spectra for pristine SWNTs, tubes after reaction with $SOCl_2$ and after coupling reaction with each linker-molecule. The insets display the changes in RBM and G mode of the spectra. The peak shift in the tangential vibration mode reveals that the attached chemical functional groups facilitate charge transfer with the nanotube host. Similarly, changes in the RBM-mode indicate a change in the electronic character of the tubes after chemical reaction.

The investigation of mechanical properties of SWNTs modified by molecular linkers is presented in figure 2. Electrical and mechanical properties are summarized in table 1. The results show low toughness value of pristine bucky paper due to the creeping under mechanical stress, often appearing problem in the application of bucky papers. Toughness described mechanical stability of the material, e.g. the energy needed to break it under mechanical stress. Increasing of toughness after chemical functionalization is remarkable. Especially, if a long aliphatic chain is used as a linker, the toughness increases by factor 20 in comparison with usual values of toughness for pristine bucky papers. The possible explanation for this property is the elasticity of the chain molecule forming cross-links through the nanotube network. No significant improvement of the Young's modulus value in comparison with that of pristine bucky paper has been observed. Similarly electrical conductivity was not improved by molecular linkage, in contrary with the bucky paper functionalized with $SOCl_2$ only. In this case, the electrical conductivity was improved by factor 2,5.

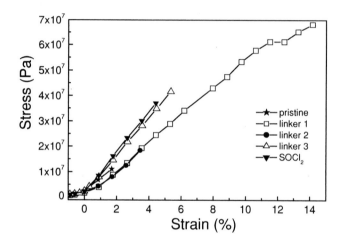

FIGURE 2. Stress as a function of strain measured on the bucky papers prepared from SWNTS modified by molecular linkers.

SAMPLE	Young's modulus (Mpa)	Toughness (kJ m^{-3})	El. Conductivity (S cm^{-1})
Pristine	660	90	605
Linker I	550	4800	530
Linker II	550	310	590
Linker III	770	1130	720
SOCl$_2$	840	830	1420

TABLE 1. Young's modulus, toughness and electrical conductivity determined on bucky papers prepared from SWNTs modified by molecular linkers.

CONCLUSIONS

The experimental results show that the intermolecular SWNTs interconnection are formed by coupling chemically functionalized tubes with organic molecular linkers. No significant improvement of the Young's modulus and electrical conductivity value in comparison with that of the pristine bucky paper has been observed. It has to be emphasised that aliphatic chain linker tripropylenetetramine increases significantly the toughness of the bucky paper.

ACKNOWLEDGMENT

The present work was financially supported by the European Project CARDECOM.

REFERENCES

1. P. W. Chiu, G. S. Duesberg, U. Dettlaff-Weglikowska and S. Roth, *Appl. Phys. Letters* **80**, 3811 (2002).
2. M. A. Hamon, J. Chen, H. Hu, Y. Chen, M. E. Itkis, A. M. Rao, P. C. Eklund and R. C. Haddon, *Adv. Mater.* **11,** 834 (1999).

Organic Functionalization of Carbon Nanotubes

Dimitrios Tasis[a], Nikos Tagmatarchis[a], Vasilios Georgakilas[a], Davide Pantarotto[a,b], Lisa Vaccari[a], Alberto Bianco[b], Dirk M. Guldi[c] and Maurizio Prato[a]*

[a] Dipartimento di Scienze Farmaceutiche, Università di Trieste, Trieste, Italy
[b] Immunologie et Chimie Thérapeutiques, IBMC, UPR 9021 CNRS, Strasbourg, France
[c] Radiation Laboratory, University of Notre Dame, Notre Dame, USA

Abstract. A simple and versatile process to achieve covalent functionalization at the endcaps and sidewalls of carbon nanotubes is presented. The reaction is based on the 1,3-dipolar cycloaddition of azomethine ylides. Various functional groups can be attached and the modified nanotubes have shown interesting applications in biology or materials science.

INTRODUCTION

Carbon nanotubes (CNTs) have attracted a great interest among the scientific community because of their extraordinary properties [1]. However, the inherently difficult handling of CNTs poses serious obstacles to their further development. Essentially, dissolution of CNTs may be necessary for proper physico-chemical examination and use [2-5].

Herein, we describe our method for functionalizing CNTs by 1,3-dipolar cycloaddition of azomethine ylides [6-8]. This approach had been successfully applied to fullerene chemistry [9].

RESULTS

Covalent functionalization. The functionalization methodology is based on the 1,3-dipolar cycloaddition of azomethine ylides, generated by condensation of an α-amino acid and an aldehyde. The CNTs were suspended in DMF, together with an excess of aldehyde and of appropriate modified amino acid (Scheme 1) [6-8]. The heterogeneous reaction mixture was heated at 130 °C for 4 days. After workup, a brown-colored solid was obtained.

The functionalized carbon nanotubes were characterized spectroscopically by ^{1}H-NMR, IR, UV-Vis and transmission electron microscopy (TEM). In general, absorption spectra of functionalized CNTs show a broad band with maximum at 250 nm which decreases almost exponentially by reaching the near-IR region. The characteristic UV-Vis transitions are lost which means that we obtain extensive functionalization on the graphitic network.

CP685, Molecular Nanostructures: XVII Int'l. Winterschool/Euroconference on Electronic Properties of Novel Materials, edited by H. Kuzmany, J. Fink, M. Mehring, and S. Roth

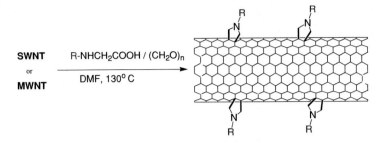

SWNT
or
MWNT

$R\text{-}NHCH_2COOH \,/\, (CH_2O)_n$

DMF, 130° C

(1) R = $-CH_2CH_2OCH_2CH_2OCH_2CH_2OCH_3$
(2) R = $-CH_2CH_2OCH_2CH_2OCH_2CH_2NHCOOC(CH_3)_3$
(3) R = $-CH_2CH_2OCH_2CH_2OCH_2CH_2NHCOFc$ (Fc : ferrocene)
(4) R = $-(CH_2)_7\text{-}CH_3$

Scheme 1 : *Dipolar cycloaddition of azomethine ylides generated by condensation of an α–amino acid and paraformaldehyde.*

The ^{1}H-NMR and IR spectra of modified carbon nanotubes with the ferrocene (**3**) and octyl (**4**) group, are shown in Figure 1 and 2, respectively. In the ^{1}H-NMR spectra, the typical peaks were observed for ferrocene (4-5 ppm) and aliphatic protons (0.8-1.4 ppm), whereas the IR spectra allow the identification of the carbonyl group in the region 1600-1700 cm^{-1}.

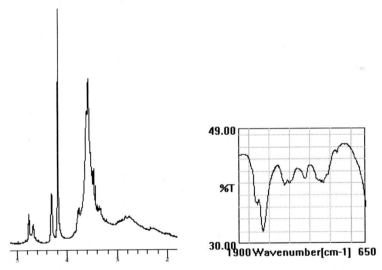

Figure 1 : *^{1}H-NMR and IR spectra of ferrocene-modified carbon nanotubes 3.*

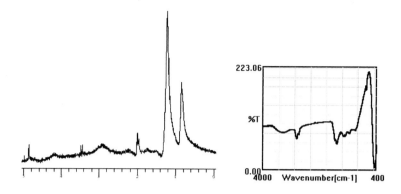

Figure 2 : *¹H-NMR and IR spectra of octyl-modified carbon nanotubes* **4**.

The thermal degradation studies of modified carbon nanotubes have provided further evidence for covalent functionalization. Actually, a certain amount of material is treated thermally under inert atmosphere until the temperature reaches 350 °C. After the treatment, a 25 % weight loss was observed. This is attributed to the total amount of functional groups which are eliminated by heating. This result means that one organic moiety corresponds roughly to one hundred carbon atoms of the graphitic surface.

Applications. The purification of CNTs is a matter of great importance. Samples of raw CNTs include impurities, such as amorphous carbon or metal nanoparticles. We have recently developed a new method for the purification of HiPCo CNTs which consists on the following steps: a) organic functionalization of pristine carbon nanotubes b) purification of the soluble adducts c) recovery of the starting material by thermal detachment of functional groups [10]. Combination of these steps led to highly pure material. The first step is based on chemical modification by 1,3-dipolar cycloaddition onto the surface of nanotubes. This chemical route resulted in solubilization of the functionalized nanotubes leaving the metal particles insoluble in the solvent used (DMF). Functionalized amorphous carbon was still present since it is soluble in the reaction medium. To further purify the functionalized nanotubes a process consisting of dissolution and slow precipitation by using a combination of solvents, was employed. The purified material was treated thermally at temperature reaching 350 °C under inert atmosphere. The resulting solid was annealed at 900 °C and found to be insoluble in any solvent, while TEM images showed that CN were recovered and most importantly were found to be free from impurities (Figure 3).

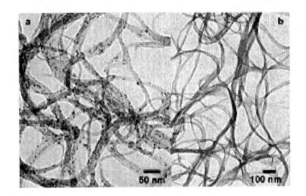

Figure 3 : *TEM images of a) pristine nanotubes (HipCO); b) recovered nanotubes after thermal treatment.*

Potential applications of CNTs in biosciences are not in full blossom yet because, of the difficulty in solubilizing these materials in aqueous media. We described very recently the preparation of water soluble nanotubes for further development as useful bioactive molecules [7]. The chemical approach is shown in Scheme 2.

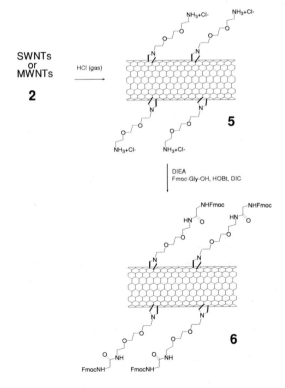

Scheme 2 : *Synthesis of water-soluble CNTs (5) and amino acid derivatized material (6)*

Functionalized CNTs **2** are treated with gaseous HCl affording the ammonium-modified material **5**. The latter was found to possess a remarkably high solubility in water and characterized by homo- and hetero-nuclear NMR, IR and TEM. The free amino groups was further derivatized with a N-protected glycine residue using standard coupling reagents. This is actually the first step towards the synthesis of peptide-carbon nanotube conjugates.

Another application of functionalized CNTs studied in our laboratory is the performance and function of this material in donor-acceptor ensembles [8]. We report for the first time photoinduced intramolecular electron transfer in the ferrocene-modified nanotubes **3**. In this system, the ferrocene chromophore acts as an electron donor whereas the nanotubes as an electron acceptor. To confirm the electron transfer mechanism, various spectroscopies were used, including fluorescence, cyclic voltammetry, bulk electrolysis, laser flash photolysis and pulse radiolysis. This result opens the way to use these kinds of systems as integrative components in solar energy conversion.

ACKNOWLEDGEMENTS

This work was carried out with partial support from the EU (RTN networks "FUNCARS" and "CASSIUSCLAYS"), MIUR (PRIN 2002, prot. 2002032171), the Universities of Trieste and the Office of Basic Energy Sciences of the U.S. Department of Energy. This is document NDRL-4447 from the Notre Dame Radiation Laboratory. We thank Mr. Claudio Gamboz and Prof. Maria Rosa Soranzo (CSPA, Trieste) for kind help with the TEM experiments.

REFERENCES

1. Falvo, M. R.; Clary, G. J.; Taylor, R. M.; Chi, V.; Brooks, F. P.; Washburn, S.; Superfine, R. *Nature* **389**, 582-584 (1997).
2. Hirsch, A. *Angew. Chem. Int. Ed.* **41**, 1853-1859 (2002).
3. Bahr, J. L.; Tour, J. M. *J. Mater. Chem.* **12**, 1952-1958 (2002).
4. Niyogi, S.; Hamon, M. A.; Hu, H.; Zhao, B.; Browmik, P.; Sen, R.; Itkis, M. E.; Haddon, R. C. *Acc. Chem. Res.* **35**, 1105-1113 (2002).
5. Sun, Y.-P.; Fu, K.; Lin, Y.; Huang, W. *Acc. Chem. Res.* **35**, 1096-1104 (2002).
6. Georgakilas, V.; Kordatos, K.; Prato, M.; Guldi, D. M.; Holzinger, M.; Hirsch, A. *J. Am. Chem. Soc.* **124**, 760-761 (2002).
7. Georgakilas, V.; Tagmatarchis, N.; Pantarotto, D.; Bianco, A.; Briand, J.-P.; Prato, M. *Chem. Commun.* 3050-3051 (2002).
8. Guldi, D. M.; Marcaccio, M.; Paolucci, D.; Paolucci, F.; Tagmatarchis, N.; Tasis, D.; Vazquez, E.; Prato, M. *Angew. Chem. Int. Ed.* **2003**, submitted.
9. a) Maggini, M.; Scorrano, G.; Prato, M. *J. Am. Chem. Soc.* **115**, 9798-9799 (1993); b) Prato, M.; Maggini, M. *Acc. Chem. Res.* **31**, 519-526 (1998); c) Tagmatarchis, N.; Prato, M. *Synlett.* in press.
10. Georgakilas, V.; Voulgaris, D.; Vazquez, E.; Prato, M.; Guldi, D. M.; Kukovecz, A.; Kuzmany, H. *J. Am. Chem. Soc.* **124**, 14318-14319 (2002).

SideWall Electrophilic Functionalization of Carbon Nanotubes

Nikos Tagmatarchis[ab], Vasilios Georgakilas[a], Dimitrios Tasis [a],
Maurizio Prato[a]* and Hisanori Shinohara[b]*

[a] Dipartimento di Scienze Farmaceutiche, Università di Trieste, Piazzale Europa 1, 34127, Trieste,
Italy, [b] Department of Chemistry, Nagoya University, Nagoya 464-8602, Japan

Abstract. A method for functionalizing the sidewalls of HiPco SWNTs via electrophilic addition is presented. Further organic transformations were carried out on free functional groups present onto the surface of nanotubes in order to analytically characterize and directly image the new material.

INTRODUCTION

In principle, chemical functionalization of the sidewalls of cabon nanotubes with organic molecules renders soluble the as produced functionalized material [1-5]. As a consequent result, solubility of carbon nanotubes gives the possibility for a systematic study of their novel properties and also enhances the opportunity for fabricating novel nanostructures.

Herein, we describe our method for functionalizing HiPco SWNTs by electrophilic addition of chloroform [6]. The motivation for such kind of functionalization originates from a recent report on electrophilic addition of polychloroalkanes to [60]fullerene with the simultaneous observation and characterization of alkyl-[60]-fullerenyl cation intermediates [7, 8].

RESULTS

Commercially available HiPco SWNTs were ground with an excess of the strong Lewis acid $AlCl_3$ in the absence of oxygen and moisture. Then, dry chloroform was added and the mixture was refluxed for 2 days. Under these conditions, one molecule of $CHCl_3$ (in the forms of $CHCl_2$- and Cl-) is expected to add onto the surface of nanotubes. However, the labile intermediate was not isolated but was further hydrolyzed to the more stable material **1**. In the following step, acylation of the free hydroxyl groups with propionyl chloride was performed in order to generate the corresponding ester functionalized SWNT **2**, as shown in Scheme 1. The purified functionalized SWNTs were analytically characterized by infra-red (IR) and nuclear magnetic resonance (NMR) spectroscopies and visualized by transmission electron microscopy (TEM) at an accelerating voltage of 100 kV.

CP685, *Molecular Nanostructures: XVII Int'l. Winterschool/Euroconference on
Electronic Properties of Novel Materials,*edited by H. Kuzmany, J. Fink, M. Mehring, and S. Roth
© 2003 American Institute of Physics 0-7354-0154-3/03/$20.00

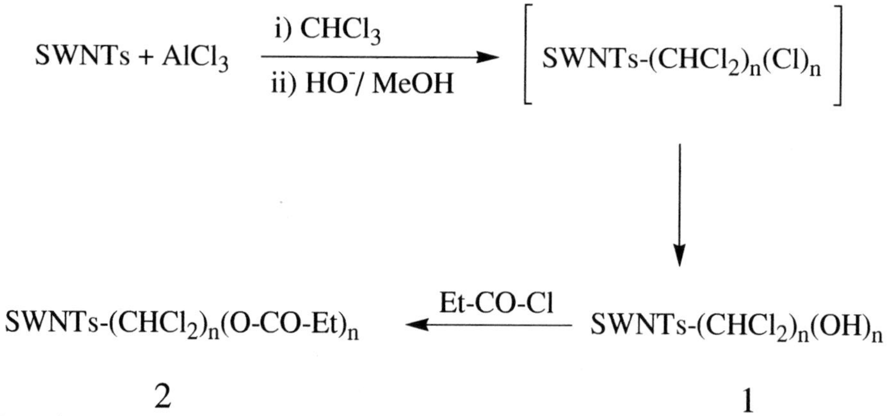

SWNTs + AlCl3
i) CHCl3
ii) HO-/ MeOH
SWNTs-(CHCl2)n(Cl)n
SWNTs-(CHCl2)n(O-CO-Et)n
Et-CO-Cl
SWNTs-(CHCl2)n(OH)n
2
1

Scheme 1. Reaction scheme for the present functionalization of HiPco SWNTs.

DISCUSSION

The solid-state grinding of HiPco SWNTs with $AlCl_3$ is believed to facilitate the electrophilic addition reaction of chloroform most likely by initially producing thinner bundles of nanotubes. When the same batch of HiPco SWNTs used for the current studies heated (but without any mechanical grinding) in the presence of Lewis acid and chloroform, no such reaction took place.

In the following figure 1, the IR spectra of materials **1** and **2** clearly show the characteristic stretching modes of the functional groups attached onto the sidewalls of nanotubes. Thus, in the left spectrum, the presence of free hydroxyl groups in **1** can be easily detected as a broad band around 3370 cm^{-1}, while this band disappears as soon as the hydroxyl groups acylated with propionyl chloride to the corresponding esters in material **2**. In the latter case, the carbonyl band of the ester is clearly seen on the right spectrum at 1725 cm^{-1}.

a)
b)
4000
3500
3000
2500
2000
1500
1000
500
Wavenumbers

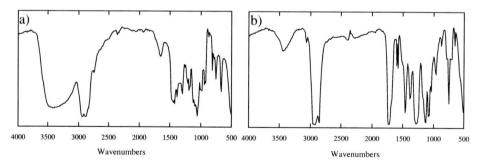

Figure 1. Infra-red spectra of functionalized HiPco SWNTs; a) in **1**, the presence of free HO- groups results in the broad band at 3370 cm^{-1}, b) in **2**, the presence of –C=O shows the characteristic absorption at 1725 cm^{-1}.

The improved solubility of functionalized SWNTs **2** due to the introduction of propyl alkyl chains as substituents of the ester moieties allowed the performance of nuclear magnetic resonance (NMR) spectroscopy measurements in deuterated chloroform. The protons introduced as part of the functional groups were identified, as shown in the following figure 2. Briefly, the methyl (CH$_3$-) and methylene (CH$_2$-) protons of the propyl alkyl chains resonate in the region 2.00-2.80 ppm while the methine (CH-) ones at 4.70-4.90 ppm, respectively. However, due to the statistical distribution of the functional groups onto the nanotube surface and the low symmetries of functionalized SWNTs, these protons give broad signals.

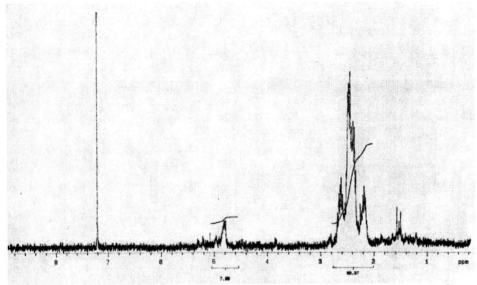

Figure 2. ^{1}H NMR spectrum of functionalized SWNTs **2** in CDCl$_3$.

With transmission electron microscopy (TEM) functionalized nanotubes **1** and **2** were directly imaged. Clearly, the presence of nanotubes can be visualized in the samples monitored, while considerable differences can be also found in the nature of the functionalized nanotubes before and after the acylation reaction. In the following figure 3, the left side image corresponds to **1** while the one on the right corresponds to **2**. The former image shows thick bundles as a result of the poor solubility of **1** in organic solvents, probably due to the large number of hydrophilic hydroxyl groups that favors nanotube aggregation through formation of hydrogen bonds. However, as soon as the esters **2** were formed the solubility was substantially increased due to the introduction of numerous alkyl chains resulting in thinner bundles of nanotubes, as shows the right image of figure 2.

In summary, we have shown that electrophilic addition reactions can be performed onto the sidewalls of carbon nanotubes and produce soluble nanotube-based materials. Moreover, free functional groups that are present on the nanotube skeleton can be further derivatized with common organic reactions producing interesting materials.

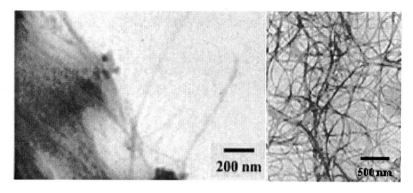

Figure 3. Transmission Electron Microscopy (TEM) images of functionalized HiPco SWNTs **1** (left panel) and **2** (right panel), respectively.

ACKNOWLEDGMENTS

This work in Italy was carried out with partial support from the European Union, Human Potential Network "FUNCARS" (HPRN-CT-1999-00011), MIUR (PRIN 2002, prot. n. 2002032171). The work in Japan was supported by the Japan Society for the Promotion of Science and the Future Program on New Carbon Nano-materials.

REFERENCES

1. Hirsch, A. *Angew. Chem. Int. Ed.* **41**, 1853-1859 (2002).
2. Bahr, J. L.; Tour, J. M. *J. Mater. Chem.* **12**, 1952-1958 (2002).
3. Niyogi, S.; Hamon, M. A.; Hu, H.; Zhao, B.; Browmik, P.; Sen, R.; Itkis, M. E.; Haddon, R. C. *Acc. Chem. Res.* **35**, 1105-1113 (2002).
4. Georgakilas, V.; Kordatos, K.; Prato, M.; Guldi, D. M.; Holzinger, M.; Hirsch, A. *J. Am. Chem. Soc.* **124**, 760-761 (2002)
5. Georgakilas, V.; Tagmatarchis, N.; Pantarotto, D.; Bianco, A.; Briand, J.-P.; Prato, M. *Chem. Commun.* 3050-3051 (2002).
6. Tagmatarchis, N.; Georgakilas, V.; Prato, M.; Shinohara, H. *Chem. Commun.* 2010-2011 (2002).
7. Kitagawa, T.; Sakamoto, H.;. Takeuchi, K. *J. Am. Chem. Soc.* **121**, 4298-4299 (1999).
8. Kitagawa, T.; Takeuchi, K. *Bull. Chem. Soc. Jpn.* **74**, 785-800 (2001).

Covalent Functionalization of Arc Discharge, Laser Ablation and HiPCO Single-Walled Carbon Nanotubes

Juergen Abraham[a], Paul Whelan[a], Andreas Hirsch[a], Frank Hennrich[b], Manfred Kappes[b], Damien Samaille[c], Patrick Bernier[c], Andrea Vencelová[d], Ralf Graupner[d] and Lothar Ley[d]

a) Institute of Organic Chemistry, Henkestrasse 42 ,Erlangen, Germany;
b) Institute of Physical Chemistry, Kaiserstrasse 12, Karlsruhe, Germany;
c) GDPC, University Montpellier II, France;
d) Institute of Technical Physics, Erwin Rommel Strasse 1, Erlangen, Germany.

Abstract. The work we present here is based on the chemical functionalization of single walled carbon nanotubes (SWCNTs) *via* the known nitrene reaction. For this work we used single walled carbon nanotubes, produced by the arc discharge, laser ablation and HiPCO technique. The comparison of the maintained functionalized single walled carbon nanotubes with the SWCNT starting material enabled us to draw conclusions on the reactivity of the nanotubes. Our studies revealed a remarkable difference in the behavior of the variety of studied nanotubes, towards chemical functionalization.

INTRODUCTION

The discovery of single walled carbon nanotubes (SWCNTs) revealed the unique physical properties of this new material. On the way to use these properties for technical applications, it was soon obvious that the tailoring of large ensembles of as pure and as uniform nanotubes as possible will be one of the major challenges. In this context the modification of single walled carbon nanotubes *via* chemical functionalization attracted considerable interests in single walled carbon nanotube research. The work we present here is based on the functionalization of arc discharge, laser ablation and HiPCO single walled carbon nanotubes. The functionalization was achieved by utilizing the known nitrene reaction.[1,2]. The comparison of the synthesized compounds allowed us to determine the behavior of each kind of tubes towards chemical functionalization.

RESULTS AND DISCUSSION

To investigate the influence of the technique used for the production of single walled carbon nanotubes on their behavior towards chemical functionalization, we

CP685, *Molecular Nanostructures: XVII Int'l. Winterschool/Euroconference on Electronic Properties of Novel Materials,* edited by H. Kuzmany, J. Fink, M. Mehring, and S. Roth
© 2003 American Institute of Physics 0-7354-0154-3/03/$20.00

treated a suspension of SWCNTs in o-dichlorobenzene with ethylazidocarbonate to achieve functionalization on the sidewall of the nanotubes (scheme 1).

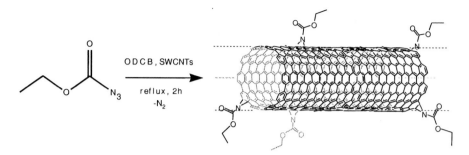

SCHEME 1. Reaction scheme for the functionalization of the SWCNTs with ethylazidocarbonate to give ethoxycarbonylaziridino-SWCNTs.

This reaction was carried out with single walled carbon nanotubes produced by the use of the laser ablation, arc discharge and by HiPCO technique. All three functionalization processes gave compounds (summarized in table 1) of functionalized SWCNTs as black solids.

TABLE 1. Compounds.

Starting SWCNTs	Functionalized SWCNTs
Pristine Laser Ablation-SWCNTs (LaPris)	Ethoxycarbonylaziridino-SWCNTs (LaEth)
Pristine Arc Discharge-SWCNTs (ArcPris)	Ethoxycarbonylaziridino-SWCNTs (ArcEth)
Purified HiPCO-SWCNTs (HiPCOPu)	Ethoxycarbonylaziridino-SWCNTs (HiPCOEth)

AFM microscopy of the functionalized SWCNTs revealed an increase in the purity of the functionalized samples compared to the SWCNTs used as starting material. Further studies with TEM confirmed this, but also show that there are still

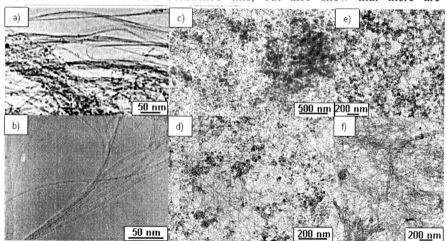

FIGURE 1. TEM images of HiPCOPu (a), HiPCOEth (b), LaPris (c), LaEth (d), ArcPris (e) and ArcEth (f).

292

remains of amorphous carbon in the samples after the functionalization process. The purity of the samples is diverse. The TEM images (figure 1) of the functionalized HiPCO and arc discharge SWCNTs prove the high purity of the samples. However in contrast to this we have to note here that the purity of the laser ablation tubes is significantly lower, that might be due to the higher content of amorphous carbon in this SWCNT starting material. TEM images with a higher

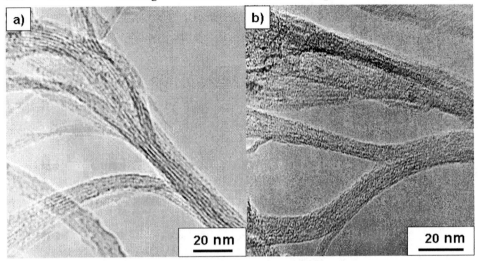

FIGURE 2. TEM images of ArcPris (a) and ArcEth (b) at higher resolution.

resolution reveal only for the arc discharge samples a loss of the lattice structure of the individual nanotubes within the bundles (figure 2) after the functionalization process. This might indicate a less pronounced graphitization of the nanotube sidewall of the arc discharge tubes.

TABLE 2. Carbon to Oxygen to nitrogen ratio as determined by XPS survey spectra.

Compound	C (%)	O (%)	N (%)
LaPris	94	6	0
LaEth	76.6	18.6	4.8
ArcPris	99	1	0
ArcEth	90.5	6.6	2.9
HiPCOPu	93	6	1
HiPCOEth	78	16	6

Later spectroscopic studies via XPS revealed that the reaction succeeded in the way we proposed (figure 3). The comparison of the atomic compositions determined by the XPS survey scans (table 2) show a dramatic increase in the amount of oxygen and nitrogen in all the functionalized compounds. The atomic ratio of the upcoming nitrogen compared to the oxygen fits quit well to our prediction of a 2:1 ratio (oxygen to nitrogen) (compare scheme 1).

Raman spectra recorded with an excitation wavelength of 1064 nm reveal a shift of the RBM-band for all of the functionalized compounds compared to the corresponding SWCNT starting material (figure 4). In contrast to the laser ablation and the arc discharge material, the HiPCO SWCNTs show a shift of the RBM-band to a lower wavenumber within the functionalized material.

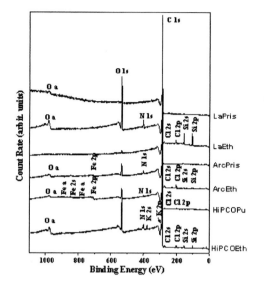
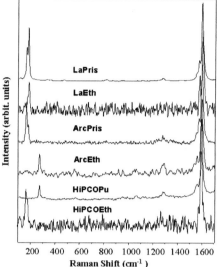

FIGURE 3. XPS survey spectra of SWCNT starting material and of the functionalized SWCNTs.

FIGURE 4. Raman spectra (λex=1064 nm) of SWCNT starting material and of the functionalized SWCNTs.

All of the Raman spectra of the functionalized SWCNTs exhibit a decrease in the intensity of the RBM-band and an increase in the intensity of the D-band compared to the G-band.

CONCLUSION

We were able to show the deviant behavior of laser ablation, arc discharge and HiPCO single walled carbon nanotubes towards chemical functionalization via reactive nitrenes. The functionalization process leads to a remarkable increase in the purity of the functionalized samples compared to the starting material. The comparison of the TEM images illustrates that the tubes remain unharmed, only the ArcEth tubes seem to lose the graphitic lattice structure on the sidewalls. XPS spectra reveal that the HiPCO SWCNTs are the most reactive ones in the studied range. The Raman spectra show a strong shift of the RBM positions of all functionalized samples compared to the starting materials.

REFERENCES

1. M. Holzinger; O. Vostrowsky; A. Hirsch; F. Hennrich; M. Kappes; R. Weiss; F. Jellen; Angew. *Chem. Int. Ed.* **40,** 4002-4005 (2001).
2. M. Holzinger; J. Abraham; P. Whelan; R. Graupner; L. Ley; F. Hennrich; M. Kappes; A. Hirsch; *J. Am. Chem. Soc.;* **in press.**

NANOTUBE FILLING AND
DOUBLE-WALL CARBON NANOTUBES

Defect Free Inner Tubes in DWCNTs

R. Pfeiffer[*], Ch. Kramberger[*], Ch. Schaman[*], A. Sen[*], M. Holzweber[*], H. Kuzmany[*], T. Pichler[†], H. Kataura[**] and Y. Achiba[‡]

[*]Institut für Materialphysik, Universität Wien, Vienna, Austria
[†]Institut für Materialphysik, Universität Wien, Vienna, Austria
and Leibniz Institute for Solid State and Materials Research, Dresden, Germany
[**]Department of Physics, Tokyo Metropolitan University, Tokyo, Japan
[‡]Department of Chemistry, Tokyo Metropolitan University, Tokyo, Japan

Abstract. By annealing fullerene peapods at high temperatures in a dynamic vacuum it is possible to produce double wall carbon nanotubes (DWCNTs). A Raman investigation revealed that the inner SWCNTs are remarkably defect free showing very strong and very narrow radial breathing modes (RBMs). Lorentzian line widths scale down to about $0.35 \, cm^{-1}$ which is almost 10 times smaller than reported for single tubes so far. Also, the scattering intensities for the inner RBMs can be more than 10 times higher than those for the outer RBMs. Additionally, all radial modes of the inner tubes are split into at least two components.

INTRODUCTION

Fullerenes [1, 2] and single wall carbon nanotubes (SWCNTs) [3, 4, 5] have attracted a lot of scientific interest over the last decade due to their unique structural and electronic properties. A few years ago, it was discovered that these new forms of carbon can be combined to form so-called peapods [6], *i.e.*, fullerenes encapsulated in SWCNTs. By annealing such peapods at high temperatures in a dynamic vacuum it became possible to transform the enclosed C_{60} or C_{70} peas into SWCNTs within the outer tubes, producing double wall carbon nanotubes (DWCNTs) [7, 8, 9]. The growth process of the inner tubes is a new route for the formation of SWCNTs in the absence of any additional catalyst. As we will show in the following, this growth under shielded conditions results in highly defect free inner tubes [10]. This is a further indication that the interior of SWCNTs may be used as a "nano cleanroom" for the formation of unique materials. Additionally, the diameters of the inner tubes can become rather small, coming close to SWCNTs grown in zeolites. This provides the possibility to study tubes with a more coarse grained diameter distribution.

EXPERIMENTAL

As starting material for our DWCNTs we used C_{60} peapods (in the form of buckypaper) produced with a previously described method [11]. The outer tubes had a mean diameter of about 1.39 nm as determined from the RBM Raman response [12, 13]. The filling of the tubes large enough for C_{60} to enter was close to 100 % as evaluated from a Raman

CP685, *Molecular Nanostructures: XVII Int'l. Winterschool/Euroconference on Electronic Properties of Novel Materials*, edited by H. Kuzmany, J. Fink, M. Mehring, and S. Roth

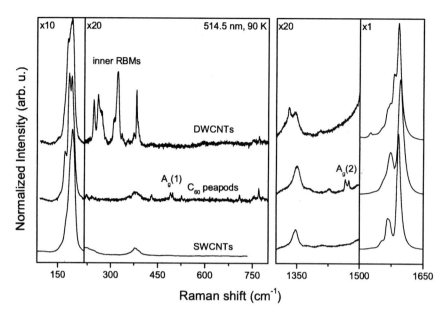

FIGURE 1. Normal resolution Raman spectra of SWCNTs (bottom), C_{60} peapods (middle) and DWC-NTs (top) measured with $514.532\,\mathrm{nm}$ at $90\,\mathrm{K}$. The spectra were normalized such that the intensity of the G-mode was equal for all samples.

[14, 15] and EELS analysis [16]. These peapods were slowly heated up to $1300\,^{\circ}\mathrm{C}$ in a dynamic vacuum for 12 hours and were then slowly cooled down to room temperature. With TEM measurements we checked that in the resulting material the C_{60} peapods had really transformed into DWCNTs.

The Raman spectra were measured with a Dilor xy triple spectrometer using various lines of an Ar/Kr laser, a He/Ne laser and a Ti:sapphire laser. The spectra were recorded at $90\,\mathrm{K}$ in normal resolution mode (NR, $\bar{\nu}_{\mathrm{NR}} = 1.3\,\mathrm{cm}^{-1}$ in the red) and at $20\,\mathrm{K}$ in high resolution mode (HR, $\bar{\nu}_{\mathrm{HR}} = 0.5\,\mathrm{cm}^{-1}$ in the red). In these measurements the samples were glued on a copper cold finger with silver paste.

RESULTS AND DISCUSSION

The effect of the filling with C_{60} peas and the following heat treatment on the Raman spectrum of SWCNTs is demonstrated in Fig. 1. The peapod spectrum is more or less the superposition of the empty SWCNT spectrum with the spectrum of pristine C_{60}. It is worth mentioning that the totally symmetric $A_g(1)$ and $A_g(2)$ modes of C_{60} are split when measured inside a SWCNT environment [14, 17]. The top spectrum in Fig. 1 is obtained from DWCNTs produced from C_{60} peapods. The C_{60} modes have vanished and many lines between 200 and $450\,\mathrm{cm}^{-1}$ have emerged. These latter modes are the RBMs of the inner SWCNTs corresponding to tube diameters between 0.5 and $1.2\,\mathrm{nm}$. The shape of the RBM of the outer tubes was also slightly altered by the treatment.

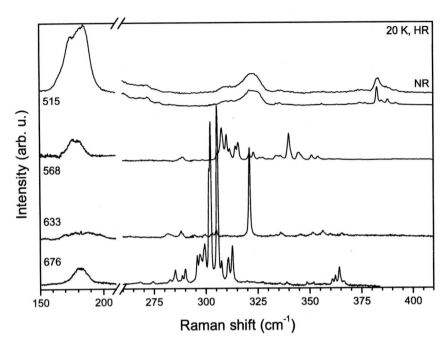

FIGURE 2. High resolution Raman spectra of DWCNTs recorded at 20K for selected excitations indicated in nm. For comparison the 515 nm normal resolution spectrum is also shown.

Additionally, the D- and G-mode regions were also affected by the heat treatment. The D-mode is broadened to lower Raman shifts and shows an additional line (for excitations between 502 and 568 nm) whose origin is unclear at the moment. The G-mode is also more structured and extended to lower wave numbers. It is interesting to note that for all excitations the G-mode of the peapods is observed at the highest Raman frequency compared to the other samples.

The radial part of the high resolution Raman spectra of DWCNTs is shown in more detail in Fig. 2 for different laser lines from green to red. For the green excitation (515 nm) the inner RBMs have only about a fourth of the intensity of the outer RBMs and show no distinct structure. For the yellow excitation (568 nm) the response of the inner tubes is already stronger than that of the outer ones and the inner tube RBMs have a very pronounced structure. Finally, in the two spectra excited with red lasers (633 and 676 nm) the intensity of the inner RBMs is up to 10 times higher than that of the outer tubes. In these last two spectra one can see very sharp and distinct lines. The 633 nm spectrum even shows only one very strong mode at about 320 cm^{-1} while one can observe a lot of strong modes between 280 and 315 cm^{-1} in the 676 nm spectrum. Since the difference in excitation energy is only about 50 meV the inner RBMs have very sharp resonances. A closer inspection of the spectra revealed that there are in total more modes than geometrically allowed tubes such that the inner RBMs are split into at least two components.

In order to determine the intrinsic line widths of the inner RBMs we deconvoluted

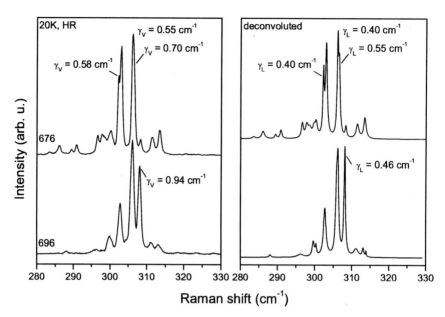

FIGURE 3. High resolution Raman spectra of DWCNTs recorded at 20 K with laser excitations of 676 (top) and 696 nm (bottom). Left: the as measured spectra, right: the corresponding deconvoluted spectra. γ_V and γ_L are the Voigtian and Lorentzian line widths, respectively.

our spectra in the following way: the laser line itself was measured and fitted with a Gaussian and the recorded spectrum was fitted with Voigtians where the Gaussian part was kept constant at the width of the laser line. The FWHM γ_L of the Lorentzian parts of the Voigtians was then considered to be the intrinsic line width.

Fig. 3 compares two spectra recorded with red lasers and their deconvolutions. After deconvolution the line widths are as small as $\gamma_L = 0.4\,\mathrm{cm}^{-1}$ which is almost 10 times as narrow as line widths of SWCNTs reported so far [18].

In summary, we have shown that the RBMs of the inner tubes of DWCNTs can be very narrow and much stronger than the RBMs of the outer tubes. These unusual narrow Raman lines indicate a very long lifetime for the phonons and are evidence for highly shielded and defect free growth conditions inside the tubes. This makes SWCNTs a perfect reaction space in the nano world. The splitting of the RBMs suggests that the interaction between the inner and outer tubes cannot be neglected.

ACKNOWLEDGMENTS

The authors acknowledge financial support from the FWF in Austria, project P14893, and from the EU, project RTN HPRNCT-1999-00011. H. K. acknowledges for a Grant-in-Aid for Scientific Research on the Priority Area "Fullerenes and Nanotubes" by the Ministry of Education, Science, and Culture of Japan.

REFERENCES

1. Kroto, H. W., Heath, J. R., O'Brien, S. C., Curl, R. F., and Smalley, R. E., *Nature*, **318**, 162–163 (1985).
2. Krätschmer, W., Lamb, L. D., Fostiropoulos, K., and Huffman, D. R., *Nature*, **347**, 354–357 (1990).
3. Iijima, S., *Nature*, **354**, 56–58 (1991).
4. Iijima, S., and Ichihashi, T., *Nature*, **363**, 603–605 (1993).
5. Thess, A., Lee, R., Nikolaev, P., Dai, H., Petit, P., Robert, J., Xu, C., Lee, Y. H., Kim, S. G., Rinzler, A. G., Colbert, D. T., Scuseria, G. E., Tománek, D., Fischer, J. E., and Smalley, R. E., *Science*, **273**, 483–487 (1996).
6. Smith, B. W., Monthioux, M., and Luzzi, D. E., *Nature*, **396**, 323–324 (1998).
7. Bandow, S., Takizawa, M., Hirahara, K., Yudasaka, M., and Iijima, S., *Chem. Phys. Lett.*, **337**, 48–54 (2001).
8. Bandow, S., Chen, G., Sumanasekera, G. U., Gupta, R., Yudasaka, M., Iijima, S., and Eklund, P. C., *Phys. Rev. B*, **66**, 075416 (2002).
9. Kataura, H., Abe, M., Fujiwara, A., Kodama, T., Kikuchi, K., Misaki, Y., Suzuki, S., Achiba, Y., and Maniwa, Y., "Fullerene Peapods and Double-Wall Nanotubes: Structures and Optical Properties," in [19], pp. 103–107.
10. Pfeiffer, R., Kuzmany, H., Kramberger, C., Schaman, C., Pichler, T., Kataura, H., Achiba, Y., Kürti, J., and Zolyomi, V., *Phys. Rev. Lett.* (2003), submitted.
11. Kataura, H., Maniwa, Y., Kodama, T., Kikuchi, K., Hirahara, K., Suenaga, K., Iijima, S., Suzuki, S., Achiba, Y., and Krätschmer, W., *Synthetic Met.*, **121**, 1195–1196 (2001).
12. Hulman, M., Plank, W., and Kuzmany, H., *Phys. Rev. B*, **63**, 081406(R) (2001).
13. Kuzmany, H., Plank, W., Hulman, M., Kramberger, C., Grüneis, A., Pichler, T., Peterlik, H., Kataura, H., and Achiba, Y., *Eur. Phys. J. B*, **22**, 307–320 (2001).
14. Pfeiffer, R., Pichler, T., Holzweber, H., Plank, W., Kuzmany, H., Kataura, H., and Luzzi, D. E., "Concentration of C_{60} Molecules in SWCNT," in [19], pp. 108–112.
15. Kuzmany, H., Pfeiffer, R., Kramberger, C., Pichler, T., Liu, X., Knupfer, M., Fink, J., Kataura, H., Achiba, Y., Smith, B. W., and Luzzi, D. E., *Appl. Phys. A*, **76**, 449–455 (2003).
16. Liu, X., Pichler, T., Knupfer, M., Golden, M. S., Fink, J., Kataura, H., Achiba, Y., Hirahara, K., and Iijima, S., *Phys. Rev. B*, **65**, 045419 (2002).
17. Pfeiffer, R., Kuzmany, H., Pichler, T., Kataura, H., Achiba, Y., Melle-Franco, M., and Zerbetto, F., *Phys. Rev. B* (2003), submitted.
18. Jorio, A., Saito, R., Hafner, J. H., Lieber, C. M., Hunter, M., McClure, T., Dresselhaus, G., and Dresselhaus, M. S., *Phys. Rev. Lett.*, **86**, 1118–1121 (2001).
19. Kuzmany, H., Fink, J., Mehring, M., and Roth, S., editors, *XVI International Winterschool on Electronic Properties of Novel Materials*, vol. 633 of *AIP Conference Proceedings*, American Institute of Physics, Melville, New York, 2002.

Tailoring double-wall carbon nanotubes?

Yasunori Sakurabayashi[1], Marc Monthioux[2], Keisuke Kishita[1],
Yoshinao Suzuki[1], Takuya Kondo[1], Mikako Le Lay[3]

[1] Toyota Motor Corp., Material Engineering Division 2, 1,Toyota-cho, Toyota, Aichi, 471-8572 Japan
[2] CEMES, UPR A-8011 CNRS, BP 4347, F-31055 Toulouse cedex 4, France
[3] Toyota Motor Engineering and Manufacturing Europe, Tech. Cent., Hoge Wei 33B, Zaventern, Belgium

Abstract. Double-wall carbon nanotubes (DWNTs) are interesting nano-objects derived from single-wall nanotubes (SWNTs) whose potentiality for some applications (such as composite reinforcement) are even more promising than for SWNTs. Depending on the respective chirality of the inner and outer SWNTs involved, electronic properties may also vary, with subsequent possible applications in nanotechnology. Up to now, DWNTs were prepared either from catalytically-enhanced thermal cracking of gaseous hydrocarbons, electric arc, or thermal annealing of C_{60}@SWNTs (peapods). A drawback of the former as opposed to the latter is that DWNTs hardly concern the whole nanotube production as far as they are always mixed with SWNTs and other MWNTs. Whatever, in any DWNT (or MWNT) from the literature so far, the graphene-graphene distance looks about equal to the d_{002}-spacing in turbostratic, polyaromatic carbon, i.e., ~0.34 nm. For peapod-derived DWNTs prepared from regular 1.35 nm wide SWNTs, for instance, it means that the inner diameter is ~0.7 nm, i.e. that of the former contained C_{60} molecules before coalescence. We report here the unexpected experimental observation that the inter-tube distance and/or the inner tube diameter of DWNTs prepared from peapods by combining both electron irradiation and heating may possibly vary. This is likely to interestingly bring variations to the overall electronic properties of DWNTs. On the other hand, it would allow the preparation of SWNTs (the inner tube) with diameters narrower than 0.7 nm, for which no common route exists yet.

INTRODUCTION

The principle of making double-wall carbon nanotubes (DWNTs) from "peapods" (i.e., single-wall nanotubes – SWNTs - whose the cavity is filled with fullerene molecules [1]) was first proposed by Smith et al. [2] to explain the concomitant occurrence of such DWNTs (whose the inner tube had the diameter of C_{60}) and peapods in purified then annealed SWNT-based samples. In this case, the formation of the inner tubes - thus forming DWNTs - was supposed to be due to the coalescence of the formerly encapsulated C_{60} molecules under the effect of the 1200°C annealing treatment. The latter assumption was then first experimentally confirmed by Smith and Luzzi [3].

On the other hand, electron irradiation was reported [1,2] then investigated [4] to also result in the coalescence of C_{60} molecules when encapsulated in SWNTs. However, such a process does not allow DWNTs to fully develop, due to the competition between damaging effect and coalescence effect. Only distorted elongated capsules form instead for the best, before the tubular morphologies collapse and the overall material is transform into either an amorphous or a regular

CP685, *Molecular Nanostructures: XVII Int'l. Winterschool/Euroconference on
Electronic Properties of Novel Materials,*edited by H. Kuzmany, J. Fink, M. Mehring, and S. Roth

multi-wall, polyaromatic material, with respect to the irradiation conditions (at least in the voltage range investigated (80-400 kV).

This paper reports preliminary results from experiments where both electron irradiation and thermal treatment were applied onto peapod materials in the mean time in an transmission electron microscope (in-situ experiments). Though yet to be completed and confirmed, the results obtained are worth being published because they open the door to the possible control of a basic feature of DWNTs such as the tube-tube distance and/or the inner and outer tube diameters, that are known to be somehow related to their chirality.

EXPERIMENTAL

Peapods were prepared from SWNTs obtained by the arc discharge method (provided by Dr. Kataura from Tokyo Metropolitan University, using Ni and Y catalysts) and 99.9% pure C_{60} molecules (MER Co.). 20-30 mg of both materials were put in a sealed quartz vessel first vacuumed (10^{-3} Pa) then filled with argon and heat-treated at 500°C for 8 hours.

The peapod material obtained was then deposited onto nickel microgrids (from a droplet of a sonicated suspension in ethanol) and introduced into a JEOL 2010 transmission electron microscope (TEM) equipped with a GATAN 628 heating sample holder. Peapods were then subjected to various in-situ treatments combining both electron irradiation (in the range 120-200kV) and temperature (in the range 200-450°C), using a heating ramp of 10°C/min. Treatment durations were determined by the time required for a successful coalescence of the fullerene molecules.

For comparison, 1200°C heat-treatments were also performed in sealed quartz vessel in isothermal conditions (12 hour dwell) with no electron irradiation (ex-situ experiments).

RESULTS AND DISCUSSION

The various treatment conditions used led to various results regarding the extension of C_{60} molecule coalescence, whose Figure 1 provides few examples. Starting from peapods (Figure 1a), low temperature treatment (400°C) barely succeeds in the coalescence of few C_{60} only (Figure 1b), while a 1200°C heat treatment forms genuine DWNTs (Figure 1c), which is consistent with previous experiments [3,4]. On the other hand, a 200kV electron irradiation merely leads to the destruction of the irradiated peapods, starting by inducing many structural defects revealed by an increasing amount of distortions brought to the tube wall (Figure 1d). On the other hand, applying a moderate temperature while peapods are irradiated, the contained fullerene molecules actually coalesce within a SWNT until forming a tube, hence leading to the formation of a DWNT (Figure 1e).

Interestingly, the inner tube diameter of the DWNT formed in Figure 1e (combined heating + irradiation conditions) looks narrower than that of the inner tube from the DWNT obtained from high temperature treatment only (Figure 1c). The outer tube diameters being similar in both case, the difference in the inner tube diameters from both experiments necessarily induces a difference in the tube-tube distance.

303

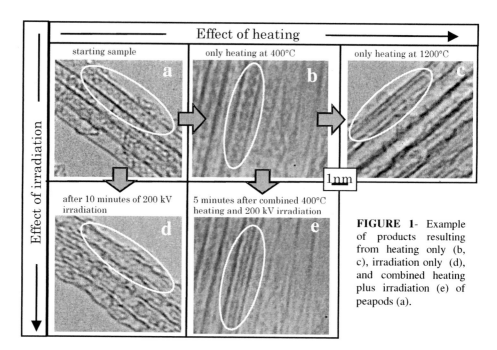

Effect of heating ➤

| starting sample | only heating at 400°C | only heating at 1200°C |
| a | b | c |

| after 10 minutes of 200 kV irradiation | 5 minutes after combined 400°C heating and 200 kV irradiation |
| d | e |

1nm

FIGURE 1- Example of products resulting from heating only (b, c), irradiation only (d), and combined heating plus irradiation (e) of peapods (a).

Investigating more systematically the irradiation-heating experimental conditions for various temperature and voltage values, plots of the inner/outer tube diameter ratios and of the outer-inner tube diameter differences were drawn as Figure 2 and Figure 3 respectively.

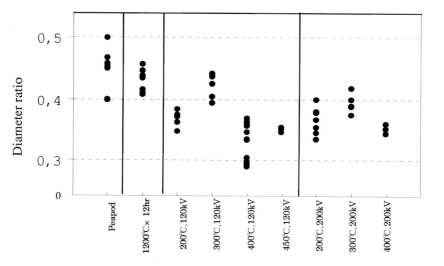

FIGURE 2 – Plot of the inner/outer tube ratio of DWNTs obtained for various combined heating + irradiation conditions ("peapod" = starting material)

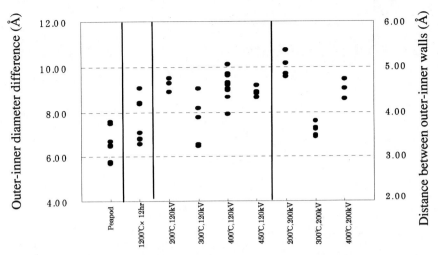

FIGURE 3 - Plot of the outer-inner tube difference (or tube-tube distance) of DWNTs obtained for various combined heating + irradiation conditions ("peapod" = starting material)

Figure 2 shows that some conditions (for instance 400°C at 120 kV) induce significantly lower inner/outer diameter ratios (~0.25-0.35) as opposed to the starting values (~0.4-0.5). On the other hand, Conditions such as 200°C at 200 kV induce a significantly larger inter-tube distance (~0.5-0.55 nm) as opposed to the fullerene-SWNT distance in the starting material (~0.28-0.38 nm), i.e. far beyond the 0.34 nm van der Waals gap regularly encountered in turbostratic, polyaromatic materials, and in the same range as in proto-polyaromatic structures such as di-coronenes in kerogens [5].

Such a variability in DWNT featrures was not anticipated. A systematic investigation taking into account various factors - such as the possible interactions between neighbouring peapods within bundles, or the accurate influence of the starting outer tube diameter - must therefore be carried-out to confirm these results and understand the related mechanisms. Then, it will be interesting to investigate whether tayloring DWNT features using appropriate temperature + irradiation combinations is actually possible, so that to tentatively obtain specific, electronic properties for instance. The latter actually depend not only on chiralities, as for regular SWNTs, but also on radius of curvature [6] and mutual positioning [7].

REFERENCES

1. Smith B.W., Monthioux, M., Luzzi D.E., Nature **396**, 323-324 (1998).
2. Smith B.W., Monthioux, M., Luzzi D.E., Chem. Phys. Lett. **315**, 31-36 (1999).
3. Smith B.W., Luzzi D.E., Chem. Phys. Lett. **321**, 169-174 (2000)
4. Luzzi D.E., Smith B.W., Carbon **38**, 1751-1756 (2000)
5. Durand B., In "Kerogen", Technip, Paris (1980)
6. Lambin P., Philippe L., Charlier J.C., Michenaud J.P., Comput. Mater. Sci. **2**, 350-356 (1994)
7. Charlier A., MacRae E., Heyd R., Charlier M.F., Moretti D., Carbon **37**, 1779-1783 (1999)

Catalyst Free Growth of Single-Wall Carbon Nanotubes (SWCNTs)

M. Holzweber[1], Ch. Kramberger[1], F. Simon[1], A. Kukovecz[1], H. Kuzmany[1], H. Kataura[2]

[1]*Institut für Materialphysik,Universität Wien, Strudlhofgasse 4, A–1090 Wien, Austria*
[2]*Department of Physics, Metropolitan University of Tokyo, Tokyo, Japan*

Abstract. We report on the growth process of single wall carbon nanotubes from a catalyst free carbon source. The carbon for the growth process is provided from C_{60} fullerenes encapsulated into the cage of the tubes (peapods). The growth process is studied by Raman scattering at various stages for the transformation of the C_{60} peas to a new SWCNT inside the primary tube. The growth process was found to start for tubes with the smallest geometrically allowed diameters. At the same time the response from the C_{60} molecules started to disappear. Eventually the signal from the latter was gone while only the thin inner tubes have reached their final concentration.

INTRODUCTION

Since their discovery in 1991, carbon nanotubes (CNT) have raised a growing interest in the scientific community [1,2]. This interest is mainly based on their unique quasi one-dimensional structure and on their high application potential. A disadvantage of the material originates from the rather large concentration of impurities present after the growth process. A great part of these impurities are encapsulated particles of the catalyst needed for the growth process. A few years ago, it became possible to combine fullerenes and single wall carbon nanotubes (SWCNTs) to form so called peapods [3]. Recently, it was discovered that the C_{60} molecules inside the tubes can be transformed into new SWCNTs accommodated inside and concentric to the primary tubes [4]. This transformation was established by annealing the system at high temperatures in dynamic vacuum. The growth process of the inner tubes is unusual as it proceeds without any additional catalysts.

Raman spectroscopy has been used extensively to study SWCNTs. This holds in particular for the tubes grown inside the primary tubes. Three main signatures are important in the Raman spectra. At low frequencies, between 150 and 450 cm^{-1}, one observes the response from the radial breathing mode (RBM), at high frequencies, between 1450 and 1600 cm^{-1} the graphitic lines (G-lines) appear and at medium high frequencies, between 1300 and 1400 cm^{-1}, a defect induced line (D-line) is seen.

For standard diameter tubes the RBM exhibits a broad structure of overlapping lines. The interest of the inner tubes is based on their small diameter in the range 0.5 – 1.0 nm. For inner tubes sharp RBM lines are observed in spectra for all geometrically allowed chiralities [5,6]. The concentric system of the two grown nanotubes is now

*CP685, Molecular Nanostructures: XVII Int'l. Winterschool/Euroconference on
Electronic Properties of Novel Materials,* edited by H. Kuzmany, J. Fink, M. Mehring, and S. Roth

conveniently assigned as double wall carbon nanotubes (DWCNTs). There is also interest in the system since it combines the stiffness of MWCNTs with the small diameters of SCWNTs.

To study the growth process of the inner tubes in detail, we analyzed various components of the transformation process. The following factors are expected to have a significant influence: the annealing temperature, duration of annealing, and the details of the heating and cooling phase. In the following we concentrate on the annealing time. This was done in order to see intermediate states of the transformation process. The Raman response from these states was compared with the response from fully transformed samples.

EXPERIMENTAL

For the heat treatment we used a horizontal tube furnace with a tube diameter of 38 mm. The samples were placed in a quartz tube and pumped to a dynamic vacuum better than 10^{-7} mbar during the heat treatment. For each treatment peapods from the same charge were used as starting material. Two types of experiments were performed.

In a first type of experiment (type 1), the furnace was preheated to 1280 °C before the samples were put in. After 30 minutes of heat treatment the samples were removed from the furnace and then cooled down to room temperature rapidly.

In the second experiment (type 2) the samples were put into the furnace at room temperature. The furnace was then heated slowly up to 1280 °C. After 1 hour at the maximum temperature the furnace cooled down to room temperature in 3 hours.

TABLE 1 Summary of treatments and characterizations for four samples.

Sample	Heat Treatment	Normal Dispersion	High Dispersion
D	Peapods	RBM & G-line 488	
C	1 x Type 1	RBM & G-line 488,568,647,676	RBM 568, 647, 676
B	2 x Type 1	RBM & G-line 488,568,647,676	RBM 568, 647, 676
A	Type 2	RBM & G-line 488,568,647,676	RBM 568, 647, 676

For our Raman measurement we use a Dilor xy triple monochromator spectrometer. Specta were excited at 90 K with 4 different laser lines as litsted in Table 1. The RBM was measured with high dispersion and normal dispersion mode. The G-line was measured with normal dispersion mode with all 4 laser lines. We used the blue laser line, 488 nm, to check the pea signal from the $A_g(2)$ line.

RESULTS AND DISCUSSION

The duration of the annealing process had a significant influence on the resulting Raman spectra. In all experiments where a second tube was grown inside the primary tube very narrow lines were observed in the spectral range of the RBM as demonstrated in Fig.1 (left).

After the first annealing step the signatures of the inner tubes already appeared.

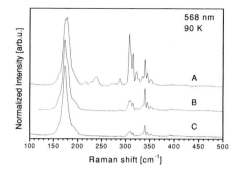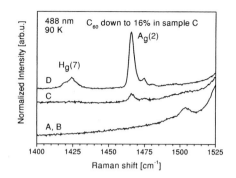

FIGURE 1. Left: RBM of sample A (annealed for 1 hour), B (annealed for 2 x 30 min), and C (annealed for 30 min), all spectra excited with a yellow laser at 568 nm; right: Ag(2) and Hg(7) lines of C_{60} excited with 488 nm. Intensities are normalized to G-Line area, D: peapod starting material

They increased with the second annealing step but do not yet reach the final height as it is observed for the sample which was continuously annealed. Simultaneously with the growth of the inner tubes the signal from the peas inside the tubes located at 1425 and 1465 cm^{-1} had considerably decreased after the first step as demonstrated in Fig. 1(right). This behavior continues with the second step of the annealing process even though a quantitative correlation between the loss of signal from the C_{60} molecules and a gain of signal from the inner tubes is not provided. This can be considered as an indication that the inner tubes are not formed out directly from the C_{60} molecules. We conclude that there exists at least one intermediate state which cannot be observed by Raman with the excitation frequencies used. The line shape of the RBM of the outer tubes changes only marginally during the transformation process.

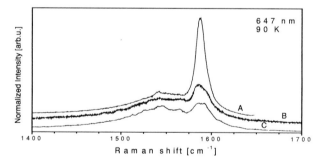

FIGURE 2. G- Line of samples A, B, and C with 647 nm at 90 K
A: treated for 1 hour,
B: treated for 2 x 30 min,
C: treated for 30 min

In the spectra measured with the red laser lines we found a major change in the line shape of the G-line. The broad structure of the G-line has a Fano shape which originates from metallic primary tubes. A sharp line of semiconducting inner NTs appears in the spectra. The narrow structure shows no Fano shape and is not found in the spectra of the starting material. It is assigned to the resonance for an E_{22} transition of semiconducting inner tubes.

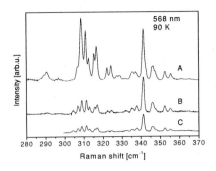 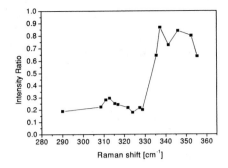

FIGURE 3. High resolution spectra of the RBM of inner tubes excited with 568 nm at 90 K,. A: treated for 1 hour, B: treated for 2 x 30 min, C: treated for 30 min

FIGURE 4. Ratio of the line intensities (relative signal heights between spectrum B and spectrum A) shown in Fig 3.

The extremely narrow lines are demonstrated by the high resolution spectra in Fig. 3. The figure exhibits the response from inner tubes for 3 different samples. Sample A heated for 1 hour at 1280 °C shows the largest signal. Sample B was heated 2 x 30 min. The peaks of this measurement do not reach the same height although the overall treatment time and the temperature are the same. Sample C was heated for 30 min. The signal from this sample was the smallest as we expected. The signals increase as a function of time. Tubes with different diameters form out at different times starting with small diameter tubes. Large diameter tubes appear to need longer treatment to reach the maximum signal height. Fig. 4 shows the ratio of RBM intensities for two differently annealed samples. It is evident that in the stepwise annealed sample B the number of small diameter tubes is comparable to sample A but that the larger inner tubes were not formed or were destroyed during the annealing.

In summary, we have produced single wall tubes from C_{60} peas with different annealing procedures. We have shown that the duration of the thermal treatment plays an important role in the transformation process. The growth process was found to start for tubes with the smallest diameters.

ACKNOWLEDGMENTS

This work was supported by the FWF in Austria, Project 14893 and by the EU RTN FUNCARS (HPRN-CT-1999-00011)

REFERENCES

1. Iijima, S., and Ichihashi, T., *Nature*, **363**, 603–605 (1993).
2. Thess, A. *et al.*, *Science*, **273**, 483–487 (1996).
3. Smith, B. W., Monthioux, M., and Luzzi, D. E., *Nature*, **396**, 323–324 (1998).
4. Bandow, S. *et al.*, *Chem. Phys. Lett.*, **337**, 48-54 (2001).
5. Pfeiffer, R.,. *et al.*, V., submitted.
6. Kramberger, Ch. *et al.*, in this volume.

Assignment of Chiral Vectors in Carbon Nanotubes

Ch. Kramberger[1], R. Pfeiffer[1], Ch. Schaman[1], H. Kuzmany[1], H. Kataura[2]

[1]Institut für Materialphysik, Universität Wien, Strudlhofgasse 4, A - 1090 Vienna, Austria
[2]Department of Physics, Metropolitan University of Tokyo, Tokyo, Japan

Abstract. Double wall carbon nanotubes (DWNTs) were derived from peapods by annealing at high temperature. We report on the Raman response of the RBM of the inner tubes. As a consequence of their small diameter the Raman spectra of the inner components of the DWNTs show well distinguishable RBM lines of the different inner tubes, besides the usual response from the outer tubes. The evaluation of these spectra has led to a full assignment of the chiral vectors to the spectroscopic lines. We found $\nu_{RBM} = 235.5 / d + 13.8 \cdot d$,ν in cm^{-1}, d in nm.

INTRODUCTION

The transformation of C_{60} inside single walled NTs (SWNTs) into another thinner NT was reported several times in recent years [1], and stimulated extensive interest in the scientific community. This so formed double walled NTs (DWNTs) are a novel material and especially the inner NT are of major interest. The latter have much smaller diameters (down to 0.5 nm) than NTs produced in any other common way, such as laser ablation, CVD, arc discharge or the HiPco process. Furthermore they are synthesized in a chemical pure carbon environment. There is no catalyst needed and they are believed to be NTs of perfect purity without any further purification steps. Each of these inner pure NTs has a well defined environment given by the surrounding outer NT. This DWNTs were studied with Raman spectroscopy. The RBM response of this inner NTs is found to exhibit extremely narrow lines down to 0.35 cm^{-1}. Additionally the narrow RBM lines of these very thin NTs are spread over a wide spectral range as the RBM frequency scales as the inverse diameter. Thus DWNTs provide excellent material for the assignment of chiral vectors.

EXPERIMENTAL

Peapods (NTs with encapsulated C_{60}) as well as pristine NTs, both prepared as bucky paper, were used as a starting material. From these peapods we obtained DWNTs by annealing them in an active vacuum better than 5.10^{-7} mbar at 1280 °C for twelve hours and furnace cooled from this temperature. The Raman spectra were recorded by a xy triple monochromator spectrometer from Dilor. The spectrometer can operate in an additive and a subtractive mode for high and normal dispersion. Each spectrum was recorded twice, once in normal dispersion mode and additionally in high

CP685, *Molecular Nanostructures: XVII Int'l. Winterschool/Euroconference on Electronic Properties of Novel Materials,*edited by H. Kuzmany, J. Fink, M. Mehring, and S. Roth

dispersion mode. The samples were mounted inside a target cell with an active vacuum better than 10^{-6} mbar. A copper cooling tip allowed cooling the samples down to either 90 K with liquid N_2 or 20 K with liquid He. All measurements were done in 180° backscattering geometry through the front side window of the cell. An ArKr laser, a HeNe laser and a tunable Ti-Sapphire laser, pumped by an Ar laser, were used for excitation with 18 different laser lines.

RESULTS AND DISCUSSION

The Gaussian diameter distribution of the pristine NTs is centered at 1.39 nm with a standard deviation of 0.12 nm and the filling of the peapods was found to be close to 100% by analysis according to work published previously in [2] and [3]. Figure 1a shows a comparison of Raman spectra of the pristine peapods and the DWNTs. The RBM of the inner NTs is located above 200 cm^{-1} and shows much more distinguishable features than the one from the outer NTs below 200 cm^{-1}. Figure 1b shows that the high dispersion mode and lower temperatures reveal even more structure in the RBM of the inner NTs.

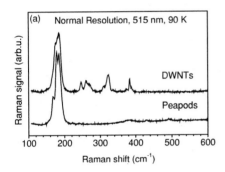 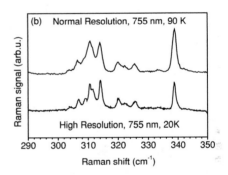

FIGURE 1.
(a) Raman spectra in the spectral range of the RBM recorded for peapods and DWCNTs with 515nm at 90K
(b) Blow up response from the inner tubes for 755 nm at 90K (high resolution) and at 20K (normal resolution)

The narrow RBM lines and their spacing are a unique property of DWNTs. Each line corresponds to one specific inner NT diameter. The Raman cross section of NTs changes dramatically with the excitation energy. In single spectra there is always just a selection of NTs visible. By combining all the spectra and taking into account experimental errors of up to 2 cm^{-1} we obtained a set of 23 main line positions, ranging from 280 cm^{-1} to 440 cm^{-1} [5]. Then we fitted (1) to various full assignments. Full assignments means that we assume to have observed all geometrically allowed chiralities in this spectral range.

$$v_{RBM}[cm^{-1}] = C_1/d + C_2 d \text{ with } d[nm] = 0.0783\sqrt{m^2 + mn + n^2} \quad (1)$$

For the fit we use only one data point for pairs of chiralities with exactly or nearly the same frequency. E. g. assuming a C_1 of 240 nm cm^{-1}, the frequencies of (8,4) and (10,1) have a difference of 1.3 cm^{-1}. Table 1 shows the best possible full assignment. Chiralities that are in the same line are assigned to the same frequency.

TABLE 1. Assignment of Chiral Vectors to Spectroscopic Lines

Chiral Vector(s)	Diameter (nm)	RBM (cm^{-1}) Experiment	RBM (cm^{-1}) 235.5/d+13.8*d
(7,0)&(5,3)	0.548	437.0	437.1
(6,2)	0.564	423.2	424.8
(7,1)	0.591	405.9	406.5
(5,4)	0.611	396.6	393.5
(6,3)	0.621	390.3	387.4
(8,0)	0.626	384.6	384.5
(7,2)	0.641	374.7	376.2
(8,1)	0.669	365.3	361.2
(5,5)	0.678	357.5	356.6
(6,4)	0.683	353.5	354.4
(7,3)	0.696	346.1	347.9
(9,0)	0.705	340.1	343.8
(8,2)	0.718	335.4	338.0
(6,5)&(9,1)	0.747	324.6	325.5
(7,4)	0.755	321.9	322.2
(8,3)	0.771	314.4	316.0
(10,0)	0.783	311.4	311.5
(9,2)	0.795	306.7	307.3
(7,5)&(6,6)	0.817	302.7	299.3
(8,4)&(10,1)	0.829	296.8	295.6
(9,3)	0.847	289.3	289.7
(11,0)	0.861	285.1	285.3
(10,2)	0.872	282.6	282.1

Figure 2a shows the RMS (root mean square) error of different assignments. Assignment shift 0 refers to the assignment shown in table 1. At negative shift each chirality is assigned to the next lower frequency. E. g. at assignment shift –3 (5,4) is no longer assigned to 396.6 cm^{-1} but to 374.7 cm^{-1} as it was proposed in [4]. In the plotted range from –4 to 4 the value of C_1 changes from 200 nm cm^{-1} to 270 nm cm^{-1}. This covers the whole possible range of C_1. Assignment 0 has a RMS of 1.9 cm^{-1} and yields C_1=235.5 nm cm^{-1} and C_2=13.8 nm^{-1}cm^{-1}. No other assignment in this range has a RMS better than 3 cm^{-1}. Figure 2b shows a lineup of the fitted and experimental line positions. It is obvious that the drawn connection lines are the best possible match of the experimental pattern to the fitted pattern. The specific pattern in the fitted points originates from the coarse distribution of the NT diameters. C_1 and C_2 are just shifting and stretching this pattern. It is important to stress that not the frequencies themselves produce significant errors to different assignments. The gaps in between them are important. When the NT diameter becomes larger than 0.9 nm there are no longer any significant gaps in the diameter distribution, and the average step in frequency reaches the observed experimental deviation of line positions in different spectra.

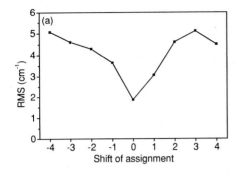 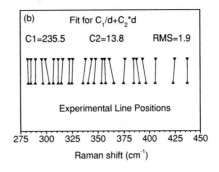

FIGURE 2..
(a.) Root mean square (RMS) error for 9 different assignments. Assignment 0 is listed in table 1. At negative shift each chirality is assigned to the next lower frequency and vice versa with positive shift.
(b.) Lineup of experimental (▲) and fitted (▼) line positions for the assignment 0. Correlated points are connected. This assignment has a significantly low RMS.

Seven lines have been assigned to metallic NTs. This may be surprising but there is definitely no way to assign our data taking only semiconducting NTs into account. To reach the same density of RBM frequencies of semiconducting NTs a value of C_1 below 180 nm cm^{-1} would be required. We conclude that metallic NTs become observable because of either their unusual small diameter or because they are part of a DWNT. This matter is addressed in more detail in [5].

ACKNOWLEDGMENTS

This work was supported by the FWF in Austria, Project 14893 and by the EU RTN FUNCARS (HPRN-CT-1999-00011)

REFERENCES

1. S. Bandow, M. Takizawa, K. Hirahara, et al. *Chem. phys. Lett.* **337**, 48-51 (2001)
2. H. Kuzmany, W. Plank, M. Hulman, et al. *Eur. Phys. J. B* **22**, 307-320 (2001)
3. H. Kuzmany, R. Pfeiffer, Ch. Kramberger, et al. *Appl. Phys. A* **76**, 449-455 (2003)
4. S. Bachilio, M. Strano, C. Kittrel, et al. *Scienc* **298**, 2361-2366 (2002)
5. R. Pfeiffer, H. Kuzmany, Ch. Kramberger, et al. *Phys. Rev. Lett.* (2003), submitted

Listing of Raman Lines from Double-Walled Carbon Nanotubes

A. Sen[1], Ch. Kramberger[1], Ch. Schaman[1], R. Pfeiffer[1], H. Kuzmany[1], and H. Kataura[2]

1 Institut für Materialphysik, Strudlhofgasse 4, A-1090 Vienna
2 Department of Physics, Tokyo Metropolitan University, Tokyo, Japan

Abstract. Double walled carbon nanotubes were derived from C_{60} peapods by annealing at 1280 °C for 2 hours in a dynamic vacuum. Resonance Raman spectroscopy was used to investigate the resulting tube systems. The spectral range of the radial breathing mode of the inner shell nanotubes was found to consist of well separated narrow lines. From the 41 geometrically allowed tubes between 250 and 470 cm^{-1} 36 were identified. The observed number of lines was more than two times the number of allowed tubes. This effect is found to be due to a splitting of all Raman lines into doublets or triplets.

INTRODUCTION

The transformation of a linear C_{60} chain inside single wall carbon nanotubes (SWCNTs) into an extra inner shell tube was reported several times recently [1-3], and stimulated extensive interest in the scientific community. These so formed double wall carbon nanotubes (DWCNTs) are a novel material. Since the inner tubes have smaller diameters than nanotubes produced in any other common way, they are of major interest.

Raman spectroscopy is a key technique in the investigation of SWCNTs. The Raman spectra of DWCNTs show the radial breathing modes (RBMs) of both shells. Since the RBM frequency scales as the inverse nanotube diameter the response of the two shells is clearly separated. The RBMs of the inner nanotubes are located between 250 and 470 cm^{-1}.

In this report we present Raman investigations on DWCNTs synthesized by annealing C_{60} peapods for 2 hours at 1280 °C in a dynamic vacuum better than 10^{-5} mbar. We studied the resonance enhancement of the Raman signal with different excitation energies. Special attention is devoted to the accurate listing of the observed RBM lines. There are 41 geometrically allowed tubes in the spectral range from 250 to 470 cm^{-1}, 36 of them were identified. The 5 missing lines are expected to belong to metallic tubes with large diameters.

EXPERIMENTAL

As starting material we used C_{60} peapods in the form of buckypaper, prepared as previously reported [4]. The outer tubes had a mean diameter of 1.39 nm [5]. These peapods were filled close to 100 % as determined from Raman and EELS analysis [6,7]. The Raman spectra were recorded with a Dilor xy triple spectrometer with a

CP685, Molecular Nanostructures: XVII Int'l. Winterschool/Euroconference on Electronic Properties of Novel Materials, edited by H. Kuzmany, J. Fink, M. Mehring, and S. Roth

liquid N_2 cooled CCD detector. Various laser lines between 488 and 755 nm were used for excitation. The spectra were recorded in high resolution ($\Delta\nu_{HR} = 0.5$ cm^{-1} at 647 nm) as well as in normal resolution ($\Delta\nu_{NR} = 1.5$ cm^{-1} at 647 nm). During the recording the samples were cooled to 20 K or 90 K.

RESULTS AND DISCUSSION

Figure 1 shows a spectrum of DWCNTs for 633 and 746 nm excitation at 90 K in the normal dispersion mode. The broad structures in Fig. 1 in the frequency range between 150 and 220 cm^{-1} represent the RBM response from the outer shell tubes. The sharp lines above 230 cm^{-1} originate from the inner shell tubes. The latter show very sharp resonance by changing the excitations energy.

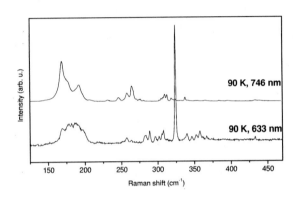

Figure 1. Raman spectra of DWCNTs in the normal dispersion mode at 90 K as excited with 633 nm and 746 nm.

By comparing the spectra measured with excitation wavelengths from 488 to 755 nm we obtained a set of 36 different line positions in the RBM of the inner NT. This set was assigned to distinct species of nanotubes [8]. The assigned line position gave very good agreement to line positions calculated from an $1/d$ law, where d is the diameter of the tubes. To demonstrate this agreement, deviations between measured and calculated line positions are depicted in Fig 2. The root mean square error of the deviation is 2.0 cm^{-1}.

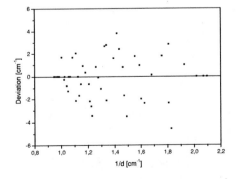

Figure 2. Discrepancy between the calculated and the experimental line position. Since each line was assigned to one specific species of NT, the deviation is plotted versus the inverse diameter.

During this evaluation special attention had to be devoted to the line positions in high resolution mode. In this mode absolute line positions are affected by the absolute central position of the spectra. Successive spectral windows of one spectrum do not fit to each other. They have ambiguous values for the position of common lines. In order to study this systematic error in more detail, we have recorded the incident laser line several times at different spectrometer positions. This shift is interpreted as a systematic error and can therefore be eliminated by rescaling the frequencies in the spectra. Figure 3 shows the Raman response of the inner tube RBMs for a 746.0 nm excitation in high resolution. Spectrum (a) is plotted as measured and line positions between overlapping windows do not match. This error stems from the nonlinearity of the frequency scale. By applying a linear correction to both sides of the central position of each window this error was eliminate, as it is shown in spectrum (b). The spectra in the normal dispersion mode have reliable line positions. Therefore, they can be used to confirm the rescaled frequency axis of the corrected high resolution spectra.

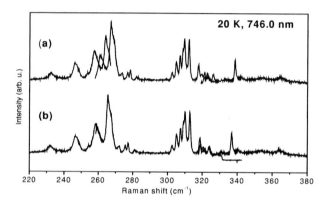

Figure 3. Raman spectra of DWCNT in high resolution at 20K, as recorded (a) and after rescaling the frequency axis (b).

After correcting the Raman spectra in high resolution mode, there were still two to three times more lines than geometrically allowed tubes. A possible explanation for this unexpected large number of lines is provided by assuming a splitting of the inner shell RBM lines in doublets or triplets. The splitting is not uniform, neither in number of split components (singlets, doublets and triplets) nor in the distance in between them. However, the splitting is limited to below 3 cm^{-1}. We suggest, that this behavior is a consequence of the variety of possible DWCNT in the sample. One kind of inner CNT may be found inside one, two or three different outer CNT. Since the shell to shell distance in a DCWNT have a non negligible influence on the RBM frequency, the Raman response looks like a split line.

The mutual alignment of the spectra recorded with different lasers is not straight since the resonance from the individual tube species are very sharp. This means a response from a particular tube appears often just in on Raman spectrum. This suggests to measure spectra with very small differences in excitation energies or, in other words, the use of tunable lasers. An example is given in Fig. 4 where a Ti:Sapphire laser was used to probe the resonance enhancement of the inner CNT between 705 and 754 nm. The figure shows the resonance enhancement for various excitation energies in this range. The energetic distance between successive excitations is as low as 4.5 meV. Comparing the spectrum at the top with the spectrum at the bottom obviously causes difficulties in the mutual correlation. These difficulties are lifted by exciting spectra with 10 intermediate energies. The FWHM of the resonance profile of the response at 319 cm^{-1} is 60 meV. The dashed vertical lines are guides for the eye and demonstrated the error on the frequency scale. As can be seen this error is smaller than $\pm$ 1 cm^{-1}.

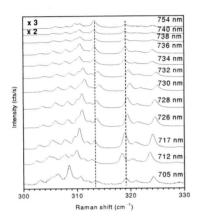

Figure 4. As measured high resolution Raman spectra of the RBM of inner shell carbon nanotubes, excited with various lines from a Ti:Sapphire laser.

Summarizing, we reported on the Raman response of DWCNTs for excitation with a large number of laser lines. Frequency correction and alignment of the spectra led to an identification of 36 RBM Raman lines which can be used for chirality assignment.

ACKNOWLEDGEMENT

Work supported by the Austrian FWF, Project 14893.

REFERENCES

1. S. Bandow et al., Chem. Phys. Lett. **337**, 48 (2001).
2. S. Bandow et al., Phys. Rev. B **66**, 075416 (2002).
3. R. Pfeiffer et al., Phys. Rev. Lett. **90**, 225501 (2003).
4. K. Hirahara et al., Phys. Rev. Lett. **85**, 5384 (2000).
5. H. Kuzmany, W. Plank, M. Hulman, Eur. Phys. J. B **22**, 307 (2001).
6. H. Kuzmany et al., Appl. Phys. A **76**, 449 (2003).
7. T.Pichler et al., Phys. Rev. B **65**, 045419 (2002).
8. Ch. Kramberger et al., in this volume.

Stability of Heated Tube as a Pod and Raman Studies for Pristine and Br$_2$ Doped Double-Wall Carbon Nanotubes

S. Bandow [a], T. Yamaguchi [a], K. Hirahara [a],
G. Chen [c], P. C. Eklund [c,d], and S. Iijima [a,b]

[a] Department of Materials Science and Engineering, Meijo University,
[b] Japan Science and Technology Corporation (SORST-JST),
1-501 Shiogamaguchi, Tenpaku, Nangoya 468-8502, Japan

[c] Department of Physics and [d] Materials Science, The Pennsylvania State University,
University Park, PA 16802, USA

Abstract. A single nanotube was cut by electron beam heating in transmission electron microscope. The lifetimes of the tubes due to electron beam heating became the longest when the tubes were heated in dry air at 500 °C in advance. This behavior can be explained by considering the quantity of contaminated carbons on the tube and the number of defects on the tube-wall. The excitation energy dependence on the Raman spectra showed the decrease of the C-C stretching mode vibration frequency as decreasing the tube diameter. The softening of the frequency becomes conspicuous for the tube with diameter less than ~1 nm. The chemical doping of Br$_2$ to double-wall carbon nanotubes (DWNTs) indicated that most of electron holes would be generated only for the outer tubes and the electron hole doping to the inner tubes was difficult to take place. This fact can be explained when we consider the electronic structures of outer and inner tubes and less interlayer interaction in DWNT.

INTRODUCTION

The nanotube space is considered to be a quasi-one dimensional being able to fill the kinds of molecules. The first transmission electron microscopy (TEM) observation for filling the fullerenes inside the single-wall carbon naotube (SWNT) was achieved by Luzzi *et al.* at the university of Pensylvania in 1998 [1]. The fullerenes in the nanotube space is well arranged and the shape is similar to the peas in the pod. Therefore, such fullerene encapsulating nanotube is frequently called as *peapod*. The method for high yield synthesis of the peapod was established in 2000 independently by Bandow *et al.* [2] and by Kataura *et al* [3]. Essential for their method is based upon to make the open-ended or side-wall-opened SWNTs, from these windows the vaporized fullerenes can intrude. This model is based on the total-energy electronic-structure calculation indicating that the energy of ~0.5 eV will be gained for the peapod [4,5], but the energy gain for the C$_{60}$ molecule physisorbing the outside of the tube is only ~0.09 eV [5].

In order to keep open the window for entering the guest materials into the host tube, the heating in dry air is necessary for the acid treated purified SWNTs or simply the as-prepared SWNTs. Although the yield almost reaches to 100 % [2] when we

CP685, *Molecular Nanostructures: XVII Int' l. Winterschool/Euroconference on Electronic Properties of Novel Materials,* edited by H. Kuzmany, J. Fink, M. Mehring, and S. Roth

use purified SWNTs followed by heating in dry air at appropriate temperature, the yield cannot be so high when using the as-prepared SWNTs followed by heating. This might be cased by attaching the remaining amorphous carbons onto the windows during vaporizing the fullerenes at high temperature.

The purification is based on refluxing the sample in the 70 % HNO_3 at 160 °C. In which process, the impurity carbons are burnt out or to form hydroxide carbons which will float in the liquid. The centrifugal sedimentation of the SWNTs and decanting the liquid above sediments for several times will reduce the floating impurities. Even reducing such floating impurities, they will attach on the side-wall of SWNT when the liquid was evaporated. These carbons would close the windows. Therefore, the heating will be necessary, and it is easy to deduce that the heating temperature closely depends on the rinse condition for washing off such floating impurities.

LIFETIME MEASUREMENTS FOR HEATED NANOTUBES

In the present experiment, we used the HiPco sample for investigating the stability of the tube due to the convergent electron beam heating in TEM (120 kV, ~10 pA/nm^2). In the HiPco sample, some impurity carbons are frequently attached on the tube surface and they are easy to remove by heating. Heating of HiPco sample was carried out in the flow of compressed dry air (~100 ml/min) with the temperatures at 420, 500 and 600 °C. The lifetimes of the tubes were

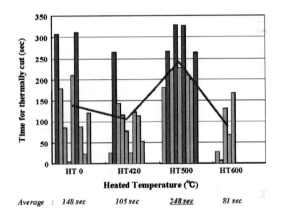

FIGURE 1. Lifetimes for heat treated tubes in dry air

determined by TEM when the single tube was cut off. Measurements were done for several tubes and the results are in Fig. 1, where HT0, HT420, HT500 and HT600 mean, respectively, the samples without heat treatment, heat treated at 420, 500, and 600 °C for 60 min. The tube diameter dependence on the lifetime could not be seen in the present experiments and the tube diameters were ranged from ~0.9 to ~1.4 nm. For HT0, the lifetimes of the tubes are rather scattered; the lifetime for some tubes reached to ~300 sec but very short lifetime tubes (< ~30 sec) also existed. Such scattering feature decreased with increasing the heat-treating temperature. The mean lifetimes against the heat-treated temperatures were 148, 105, 248 and 81 sec, respectively, for HT0, HT420, HT500 and HT600 samples. Of course, the number of the tubes in the sample was decreased upon increasing the heat-treating temperature. The mean lifetime became the longest when heat-treating at 500 °C. Individual single tube TEM observations for HT0, HT420, HT500 and HT600 samples indicated that the quantity of impurity carbons attaching on the tube-wall became the smallest for HT500 sample.   As a tendency, some parts of the tube were very clean for HT0 sample, but other parts of the tube were partially wrapped by impurity carbons. Normally, the impurity carbons are chemically unstabler than the sp^2 bonded carbon nanotubes.   Therefore, such carbon impurities would start to burn at lower temperatures than to burn the nanotubes.   Decrease on the amount of impurity carbons will be realized with increasing heat-treating temperature.   On the other hand, such burning of the impurity carbons would damage the tube wall and would make defects.   Much higher temperature heating, like in the case of HT600, would make larger size of defect-holes by destroying the tubular structure.   Illustrations for above structural models are indicated in Fig. 2 with typical electron micrographs corresponding to individual illustrations.   Here we back to the consideration why did the lifetime of HT500 tube become the longest.   The reason should closely relate the cleanness of the surface and the number of defect on the tube wall.   It is considered that the impurity carbons are also easy to re-arrange the atom positions or to deform by the electron beam heating, whose re-arrangement would damage the tube surface resulting the short lifetime of the tube.   For HT500 tube, most of the impurity carbons are removed from the surface and also expected to have small number of defect, being the longer lifetime tube.   As described above, larger defect-holes or destroyed tubular structures are expected for the HT600 tube, whose tube should be very weak against the heating and easy to destroy.

For the peapod formation, the tubes need to have suitable size of defect which is enough for entering the fullerene molecule [5].   The heating in dry air certainly removes the impurity carbons attaching on the surface of the tube and makes the defect-holes as illustrated in Fig. 2.   However, it was found that the higher temperature heating was not good for making effective defect-holes for fullerene entrance due to breakage of the tubular structure. We now conclude that the suitable heating temperature lies between 420 and 500 °C, and it depends on the tube surface cleanness.

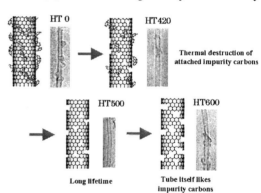

FIGURE 2. Schematic for the defect size on the tube

EXCITATION ENERGY SELECTIVE RAMAN SCATTERING FROM DOBLE-WALL CARBON NANOTUBES

Using the SWNTs (from pulsed YAG laser vaporization using the Fe-Ni catalyst and purified) heated in dry air with above mentioned condition, the fullerene molecules can effectively enter the nanotube space by the vapor phase reaction at 400 °C for 24 hours and make a chain like structure [2].   The yield for the peapods is reached almost 100 %, and the 1200 °C heating in vacuum for 24 hours guarantees the conversion of the inside fullerenes to the inside tube [6].   The Raman scattering study for double-wall carbon nanotubes (DWNTs) was carried out by using the different

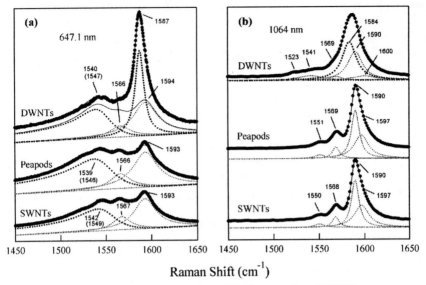

FIGURE 3. Raman spectra from SWNTs, peapods and DWNTs

laser excitation energies at 488 (2.54), 514 (2.41), 647 (1.92) and 1064 nm (1.17 eV). From the radial breathing mode (RBM) vibration frequencies, we can estimate the diameter range for outer and inner tubes, resulting the diameter ranges roughly from ~1.3 to ~1.6 nm for outer tubes and from ~0.6 nm to ~0.9 nm for inner tubes [7].

According to the diameter dependence of the energy separation (E_{ii}) between the sets of mirror image DOS spikes for the nanotubes [7,8], the laser energies are resonantly absorbed to the tubes when the excitation energy matches with the value of E_{ii}. For 488 and 514 nm excitations, the E_{33} or E_{44} of semiconducting outer tubes and the E_{22} of ~0.6-0.7 nm diameter class semiconducting inner tubes will match to the laser energies. For 647 nm excitation, the same kinds of things can be expected in the E_{11} of metallic outer tubes, and in the E_{22} of ~0.8-0.9 nm diameter class semiconducting inner tubes. For 1064 nm excitation, the E_{22} of semiconducting outer tubes and the E_{11} of ~0.7 nm diameter class semiconducting inner

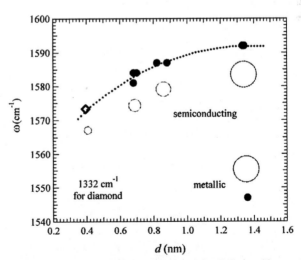

FIGURE 4. Tube diameter dependence on the C-C stretching mode vibration frequency. The data for 0.4 nm tube is from Ref. 9

tubes will be matched. Due to such selectivity on the resonance properties of the tubes, we succeeded to determine the C-C stretching mode vibration frequency for small diameter tubes.

The Raman spectral features recorded with the excitations of 647 nm and 1064 nm are indicated in Fig. 3. The spectra were taken for the samples of SWNTs, peapods and DWNTs, which were prepared by step-by-step. Therefore, we can pick up the Raman signal from the inner tubes by de-convoluting each spectrum. In Fig. 3a, the Raman spectral shapes for SWNTs and peapods are very similar; combination of two lorentzian components (thin dotted lines) and the asymmetric Briet-Wagner-Fano (BWF) component (thick dotted line) well represent the spectra observed. Here the BWF component is well known to associate with the metallic tubes and the lorentzian one from semiconducting tubes. For DWNTs, an intense lorenzian component centered at 1587 cm^{-1} is overlapped with the signal from SWNTs as recognized in Fig. 3a. This newly appeared lorentzian line should associate with the signal from the inner tubes. In addition, when we consider the resonance condition at 647 nm excitation, the tube diameter can be estimated at ~0.8 nm. Therefore, the C-C stretching mode vibration frequency for ~0.8 nm tube is to be 1587 cm^{-1}. By following the same consideration way, we can estimate the C-C stretching vibration frequency for ~0.7 nm tube to be 1584 cm^{-1} from Fig. 3b. Therefore, it can be found that the C-C stretching vibration frequency decreases with decreasing the tube diameter. The tube diameter dependence on the C-C stretching vibration frequency (ω) is indicated in Fig. 4. The closed circles are from the present experiments and the open diamond is from Jorio *et al* [9].

Br$_2$ DOPING TO DOUBLE-WALL CARBON NANOTUBES

The Br$_2$ doping to the DWNTs is also studied by the Raman scattering with two excitation energies, 514 (2.41) and 1064 nm (1.17 eV) [10]. Upon exposing the Br$_2$ vapor to DWNTs in a vacuum sealed glass ampoule at room temperature, the RBM signals from the outer tubes were completely disappeared, while those from inner tubes were rarely affected, suggesting that the hole doping only to the outer tubes occurs but not or less to the inner tubes. In the G-band Raman spectra, those charge transfer features were clearly detected as the shift of the Raman peak as indicated in Fig. 5. In the top pannel of Fig. 5, the G-band signal detecting at ~1591 cm^{-1} is to associate with outer tubes and the signal at ~1581 cm^{-1} can

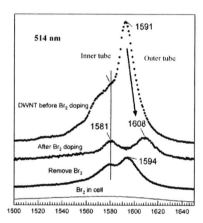

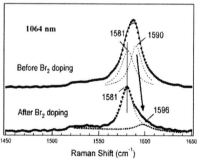

FIGURE 5. Raman spectral changes due to Br$_2$ doping

be assigned to be ~0.7 nm diameter inner tubes (see Fig. 4). It is obvious to see the blue-shift of the outer tube C-C vibration, but not for the inner tubes after Br_2 doping. The Raman intensity from the inner tubes is stronger for 1064 nm excitation than for 514 nm one. This behavior is easy to understand since the resonance condition for ~0.7 nm diameter inner tube is better to satisfy at 1064 nm excitation than at 514 nm one. We therefore can clearly see the inner tube Raman feature at 1064 nm excitation, indicating that almost unchanged Raman spectrum from the inner tubes after Br_2 doping (see bottom panel of Fig. 5). The blue-shift for the 1591 cm^{-1} line by 17 cm^{-1} indicates that the hole doping ratio is to be one hole per 26 carbon atoms [10], and the holes are mostly generated on the outer tube in

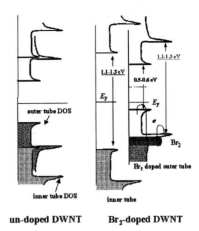

FIGURE 6. Schematic of electronic structures

DWNT. From these Raman results, the schematic of the electronic structures for un-doped and Br_2 doped DWNTs can be drawn as Fig. 6. The electrons in the outer tube band above the 2nd van Hove singularity DOS would be transferred to the Br band, while the electrons in the inner tube band could not be transferred to Br band due to the relative relation of the band structures. Therefore, it concludes that the holes are difficult to generate on the inner tubes and the resonance condition has not been changed after Br_2 exposure.

ACKNOWLEDGEMENTS

The author S.B. thanks to the Japanese Ministry of Education, Culture, Sports, Science and Technology for the foundation of the Grant-in-Aid for Scientific Research (C), No. 14540545. K.H. and S.I. acknowledge the support by the US office of Naval Research (ONR-N000140010762), and the Meijo group is also founded by the 21st century COE program of the Japanese Ministry of Education, Culture, Sports, Science and Technology.

REFERENCES

1. B. W. Smith, M. Monthioux, and D. E. Luzzi, *Nature* **396**, 323 (1998).
2. K. Hirahara, K, Suenaga, S. Bandow *et al.*, *Phys. Rev. Lett.* **85**, 5384 (2000).
3. H. Kataura, Y. Maniwa, T. Kodama *et al.*, *Synthetic Metals* **121,** 1195 (2001).
4. S. Okada, S. Saito, and A. Oshiyama, *Phys. Rev. Lett.* **86**, 3835 (2001).
5. S. Berber, Y.-K. Kwon, and D. Tomanek, *Phys. Rev. Lett.* **88**, 185502 (2002).
6. S. Bandow, M. Takizawa, K. Hirahara *et al.*, *Chem. Phys. Lett.* **337**, 48 (2001).
7. S. Bandow, G. Chen, G. U. Sumanasekera *et al.*, *Phys. Rev.* **B 66**, 075416 (2002).
8. H. Kataura, Y. Kumazawa, Y. Maniwa *et al.*, *Synthetic Metals* **103**, 2555 (1999).
9. A. Jorio, A. G. Souza Filho, G. Dresselhaus *et al.*, *Chem. Phys. Lett.* **351**, 27 (2002).
10. G. Chen, S. Bandow, E. R. Margine *et al.*, *Phys. Rev. Lett.* submitted.

Vibrational properties of double-walled carbon nanotubes

J. Maultzsch*, S. Reich*†, P. Ordejón†, R. R. Bacsa**, W. Bacsa**,
E. Dobardžić‡, M. Damnjanović‡ and C. Thomsen*

*Institut für Festkörperphysik, Hardenbergstr. 36, 10623 Berlin, Germany
†Institut de Ciència de Materials de Barcelona (CSIC), Campus de la Universitat Autònoma de
Barcelona, E-08193 Bellaterra, Barcelona, Spain
**Laboratoire de Physique des Solides, Université Paul Sabatier, 118 route de Narbonne, 31062
Toulouse Cédex 4, France
‡Faculty of Physics, University of Belgrade, P.O. Box 368, 11001 Belgrade, Serbia

Abstract. We study the vibrational properties of double-walled carbon nanotubes by *ab initio*
calculations and Raman scattering. Furthermore, we investigate the stability of double-walled tube
configurations with other interlayer distances than in graphite.

An increasing number of carbon nanotube studies focused very recently on double-walled tubes (DWNT). The interaction between the inner and outer tube, for example, has been investigated in relation to the breathinglike phonon modes [1, 2]. The comparatively small diameter of the inner tube in typical DWNT samples has led to a number of Raman studies which attempted an assignment of the inner tube chirality based on the radial breathing mode (RBM) frequencies [3, 4, 5]. In these investigations, the interaction between the tube walls is often modeled by adding a constant to the dependence of the RBM frequency on the inverse tube diameter. In both theoretical and experimental studies the distance between the tube walls is usually assumed the same as the interlayer distance in graphite, i.e., ≈ 3.4 Å.

In this paper we study the vibrational properties of double-walled carbon nanotubes first by an analytically solvable spring-constant model and second by *ab initio* density-functional calculations for several pairs of armchair tubes. We show that double-walled nanotubes with interwall distances different from the value in graphite may exist. In the DWNT with a smaller wall distance [(4,4)@(8,8)], the phonon modes exhibit a strong mixing; in all other investigated DWNTs each of the constituents vibrates independently. In the (3,3)@(8,8) tube, the change in phonon frequencies due to the wall interaction is larger for the high-energy optical phonon modes than for the RBM. Finally, we compare our results with Raman spectra.

In order to estimate the RBM frequecies in DWNT in a simple approximation, we modeled the tubes by homogeneous cylinders and the wall-wall interaction by an effective spring constant. If the inner tube diameter d_1 is below 10 Å, an analytical expression can be used for the RBM frequencies Ω_1 and Ω_2 of the out-of-phase and the in-phase

CP685, *Molecular Nanostructures: XVII Int' l. Winterschool/Euroconference on
Electronic Properties of Novel Materials,* edited by H. Kuzmany, J. Fink, M. Mehring, and S. Roth

TABLE 1.   Total energy per carbon atom after relaxation of the atomic positions.

	E_{tot} eV/atom	radius Å		E_{tot} eV/atom	radius (inner) Å	radius (outer) Å	energy gain meV/atom
(3,3)	-154.8697	2.102	(3,3)@(8,8)	-155.1776	2.103	5.458	195
(4,4)	-155.0691	2.769	(4,4)@(8,8)	-155.1983	2.725	5.546	-23
(8,8)	-155.2663	5.463	(3,3)@(9,9)	-155.1873	2.106	6.121	96
(9,9)	-155.2803	6.141	(4,4)@(9,9)	-155.2341	2.772	6.123	188

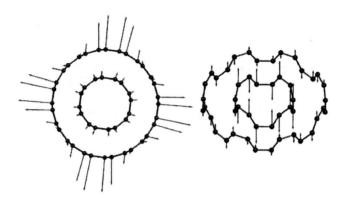

FIGURE 1.   (4,4)@(8,8) tube: in-phase breathing mode at 232 cm^{-1} (left) and out-of-phase LO mode at 1582 cm^{-1} (right).

motion, respectively:

$$\Omega_{1/2} \approx \omega_{1/2} + \frac{\kappa d_1}{2m_{1/2}\,\omega_{1/2}},$$

where $\kappa = 1.159 \cdot 10^5$ amu cm^{-2} Å^{-1} is found from the graphite B_{2g} mode at 127 cm^{-1}, $\omega_{1/2} \propto 1/d_{1/2}$ are the RBM frequencies and $m_{1/2}$ the linear mass densities of the inner/outer tube [6].

To investigate the full vibrational spectrum of DWNT, we performed *ab inito* calculations using the SIESTA code within the local-density approximation [7]. A double-ζ, singly polarized basis set of localized atomic orbitals was used for the valence electrons. In Table 1 we summarize the total energy per carbon atom after the relaxation of the atomic positions in the investigated double-walled tubes. All of the four DWNTs, with the inter-wall distance ranging from 2.8 Å [(4,4)@(8,8)] to 4.0 Å [(3,3)@(9,9)], appear stable. The energy gain per carbon atom from the sum of the two isolated constituents to the double-walled tube is given in the last column of Table 1.

In the (4,4)@(8,8) and the (3,3)@(9,9) tube, the two constituents have a fourfold and threefold rotational axis in common, respectively, i.e., they are commensurate with respect to rotations about the tube axis. In the other two DWNTs, the only symmetry operations besides the translation along the tube axis are the horizontal mirror plane and

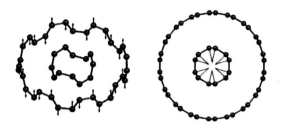

FIGURE 2. Longitudinal optical mode in the (3,3)@(8,8) tube (left) and RBM mode in the (3,3)@(9,9) tube (right). In both DWNT, the inner and outer tube vibrate independently.

the U axis [8, 9]. Thus for the incommensurate DWNTs a very low friction between the walls is expected if the tubes are rotated against each other, but there is interaction between the walls in the commensurate ones [10]. Since the calculation of the acoustic phonon frequencies is not accurate enough, we investigated this prediction by rotating the inner tube in the (4,4)@(8,8) and the (3,3)@(8,8) and comparing the total energy before and after the rotation. We found a change in total energy less than 1 meV/atom for the (3,3)@(8,8) when the (3,3) tube was rotated; in the (4,4)@(8,8) tube the total energy increased by several meV per carbon atom upon rotation of the (4,4) tube, in agreement with the group-theoretical prediction.

We calculated the Γ-point phonon frequencies of the DWNTs, of their isolated constituents and, for comparison, of their constituents with the same (unrelaxed) atomic positions as in the DWNTs. As expected because of their small interwall distance and the commensurability, the RBM frequencies in the (4,4)@(8,8) tube are upshifted by 25-55 cm^{-1} with respect to the RBM in single-wall tubes. The longitudinal optical (LO) phonon mode is upshifted as well by up to 100 cm^{-1}. Moreover, the phonon displacements exhibit a strong mixing in the (4,4)@(8,8) tube. The in-phase and out-of-phase displacements of the two tubes are clearly seen (Fig. 1). Due to the symmetry reduction in the double-walled tube, the amplitude of the RBM vibration does not neccessarily remain constant along the circumference of the tube, e.g., the in-phase RBM amplitude exhibits a modulation by $\pi/2$ along the circumference.

In the (3,3)@(8,8) tube, we found no significant change in the RBM frequencies. Surprisingly, the transversal optical (TO) phonon mode of the (8,8) tube exhibits a downshift by 30 cm^{-1}. Most of the phonon eigenvectors in this DWNT are independent vibrations of their constituents; either the inner or the outer tube is displaced, while the amplitude of the other tube's displacement is zero (see Fig. 2). Also in the (3,3)@(9,9) tube the RBM frequencies are not affected by a wall-wall interaction. This might partly be due to the large interwall distance of 4 Å. On the other hand, from the commensurability of the tubes we would expect some interaction between the walls.

For comparison, we show in Fig. 3 the Raman spectra from two different spots of the same DWNT sample, both of which were typically found. The spectrum to the right is "metallic"-like, however, the downshift of the upper peak to below the graphite

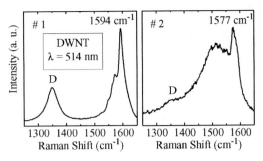

FIGURE 3. Raman spectra of different spots on the same sample of CVD-grown double-walled carbon nanotubes. The laser wavelength is 514 nm.

frequency is even for a metallic tube very unusual. This could be a hint to the downshift of the TO mode in the $(3,3)@(8,8)$ tube predicted by the *ab initio* calculations.

In summary, we have shown that double-walled nanotubes with interwall distances between 2.8 and 4.0 Å may exist. We found from *ab initio* calculations that for some DWNTs the high-energy optical phonon modes are even more affected by the wall-wall interaction than the RBM modes. Since the effect on the RBM frequency of the inner tube is predicted to increase strongly with the tube diameter [1, 2, 6], *ab initio* calculations for larger DWNTs are neccessary. The effect of the symmetry of the tubes and of their interwall distance on the wall-interaction has to be investigated by future experiments and calculations.

ACKNOWLEDGMENTS

P. O. acknowledges support from Fundación Ramón Areces, EU project SATURN, and a Spain-DGI project. S.R. acknowledges a fellowship by the Akademie der Wissenschaften Berlin-Brandenburg. J.M. acknowledges support from the Deutsche Forschungsgemeinschaft under grant number Th 662/8-1.

REFERENCES

1. Popov, V., and Henrard, L., *Phys. Rev. B*, **65**, 235415 (2002).
2. Benoit, J. M., *et al.*, *Phys. Rev. B*, **66**, 073417 (2002).
3. Bandow, S., *et al.*, *Phys. Rev. B*, **66**, 075416 (2002).
4. Bacsa, R. R., *et al.*, *Phys. Rev. B*, **65**, 161404(R) (2002).
5. Kramberger, C. (2003), this volume.
6. Dobardžić, E., Maultzsch, J., Milošević, I., Thomsen, C., and Damnjanović, M. (submitted 2003).
7. Soler, J. M., *et al.*, *J. Phys.: Cond. Mat.*, **14**, 2745 (2002).
8. Božović, N., Božović, I., and Damnjanović, M., *J. Phys. A: Math. Gen.*, **18**, 923 (1985).
9. Damnjanović, M., Milošević, I., Vuković, T., and Sredanović, R., *Phys. Rev. B*, **60**, 2728 (1999).
10. Damnjanović, M., Vuković, T., and Milošević, I., *Eur. Phys. J. B*, **25**, 131 (2002).

Synthesis, Characterization and Chirality Identification of Double-Walled Carbon Nanotubes

Wen-Cai Ren, Feng Li, and Hui-Ming Cheng[a]

Shenyang National Laboratory for Materials Science, Institute of Metal Research,
Chinese Academy of Sciences, 72 Wenhua Road, Shenyang 110016, China

Abstract. Catalytic decomposition of methane with the addition of sulfur was used to synthesize double-walled carbon nanotubes (DWNTs) and their aligned long ropes. High-resolution transmission electron microscopy (HRTEM) and laser Raman resonance spectroscopy were employed to study the structure and diameter distribution of the as-prepared DWNTs. HRTEM observations show that the outer and inner diameter of the DWNTs are in the range of 1.6-3.6 and 0.8-2.8 nm, respectively. Moreover, we found that the interlayer spacing of DWNTs is not a constant, ranging from 0.34 to 0.41 nm. The DWNTs show corresponding peaks in the radial breathing mode (RBM) from resonant Raman spectra, which are considered to be associated with the inner tubes as well as outer tubes of DWNTs. Moreover, we can also distinguish metallic and semiconducting nanotubes based on the resonantly enhanced RBM features of the DWNTs, which provides a basis for their applications in electronic field.

---

[a] Corresponding author and participant; electronic mail: cheng@imr.ac.cn

CP685, *Molecular Nanostructures: XVII Int'l. Winterschool/Euroconference on*
Electronic Properties of Novel Materials, edited by H. Kuzmany, J. Fink, M. Mehring, and S. Roth

INTRODUCTION

The properties of carbon nanotubes (CNTs) closely depend on their structures, so the preparation of CNTs with unique structure is very important for extensive researches. Great achievements have been so far obtained in the synthesis of CNTs, such as large-scale synthesis of single-walled carbon nanotubes (SWNTs) and growth of aligned multi-walled carbon nanotube (MWNT) arrays [1-3]. However, there are still big challenges about controllable synthesis of CNTs, such as chirality, shell number and aligned SWNT array.

Double-walled carbon nanotubes (DWNTs) can be regarded as a critical structure between SWNTs and MWNTs. A DWNT, as a combination of two coaxial SWNTs, has some special applications in nanodevices, such as molecular conductive wire, molecular capacitor in a memory device and nano-bolt-nut, due to its special double walls structure and its different chirality combination [4, 5].

However, the studies of DWNTs were hindered due to the difficulty of the controllable synthesis. Recently, J.L. Hutchison et al. and S. Bandow et al. synthesized DWNTs by different methods [6, 7]. However, the purity of the products and the controllability were poor. Moreover, the rope-like product is not reported yet, though it is very important for applications, such as highly conducting microcables, mechanically robust electrochemical microactuators, sensors, probes and emission gun [8, 9], due to its special aligned rope structure.

In this paper, DWNTs and their aligned long ropes were synthesized by catalytic decomposition of methane with the addition of sulfur. The microstructure of the DWNTs was characterized by scanning electron microscopy (SEM), transmission electron microscopy (TEM), high-resolution TEM (HRTEM) and laser Raman resonance spectroscopy. Moreover, the Raman spectra of the DWNTs were studied in detail and their chirality was identified, based on their resonantly radial breathing mode (RBM).

EXPERIMENTAL

The apparatus for synthesis of DWNTs by catalytic decomposition of methane was similar to that described in a previous article [2]. In our experiment, methane was used as the carbon source, hydrogen as the carrier gas and ferrocene as the catalyst precursor. Methane, with a constant total flow rate, was divided into two parts and one of them was used as carrier gas carrying thiophene, a sulfur-containing compound, to enhance the growth of CNTs. During the preparation, methane, hydrogen, ferrocene and thiophene vapors were mixed and carried into the reactor, maintained at 1373K.

After 10 minutes synthesis, some web-like and rope-like products were obtained.

The microstructure of the products was observed using scanning electron microscopy (SEM), transmission electron microscopy (TEM Philips EM420 100kV), high-resolution TEM (HRTEM JOEL 2010 200kV) and Raman spectroscopy.

Raman spectra of the DWNTs were obtained using three Raman systems: (i) a micro-Raman spectrometer (Renishaw 1000B, objective ×50) equipped with an air cooled charge coupled device and notch filters, (ii) a micro-Raman spectrometer (Jobin Yvon HR800, objective ×50), and (iii) a modular research micro-Raman spectrograph with Kaiser optical system (Hololab 5000R). The 488.0 nm (2.54eV) and 514.5nm (2.41eV) lines from an Ar$^+$ laser in system (i), the 632.8nm (1.96eV) line from a He-Ne laser in system (ii), and the 785nm(1.58eV) line from a Ti:sapphire laser in system (iii) were used to obtain the Raman spectra. The laser power impinging on the samples was controlled to be less than 1mW over laser spot of ~1µm^2.

RESULTS AND DISCUSSION

Structural Characterization of the As-Prepared DWNTs

Morphology of the As-Prepared DWNTs

Under optimum experimental conditions, two types of products (web-like and rope-like) were obtained, and look semi-transparent and very light. Fig.1 (a) is the TEM image of the web-like DWNTs and it is found that the DWNTs mostly form bundles and entangle with each other. Fig. 1 (b) shows a SEM image of an aligned DWNT ropes. It can be found that the rope is composed of thousands of roughly-aligned filaments which were identified to be DWNT bundles.

HRTEM observations provide further structural information of the products obtained. It revealed that more than 70 percent of the product was DWNTs and the rest was mostly catalysts capsulated by graphite layers and amorphous carbon. A few

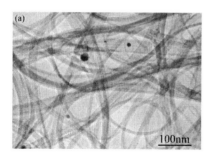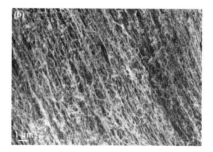

FIGURE 1. (a) A TEM image of as-prepared web-like DWNTs. (b) High magnification SEM image of an as-prepared DWNT rope.

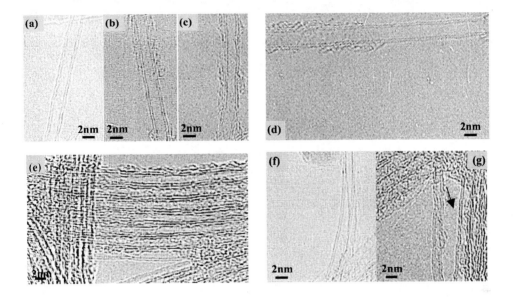

FIGURE 2. Some typical HRTEM images of DWNTs [10], 2002, reprinted with permission from Elsevier.

SWNTs and three-walled carbon nanotubes were occasionally found. Fig. 2 (a) to (d) show typical images of four isolated DWNTs. Most of the DWNTs form bundles due to the Van der Waals attraction between the tubes, and these bundles always consist of many well-aligned DWNTs. Fig. 2 (e) shows a DWNT bundle with a diameter of 15.1 nm in which seven parallel DWNTs with an even diameter (outer diameter about 2.0 nm) aligned one by one.

In addition to the double walls structure, another important structural characteristic of the DWNTs is that their interlayer spacing is not a constant, ranging from 0.34 nm to 0.41 nm [10], from the experimental observations, in consistent with the results obtained by other investigations [6, 7]. Moreover, there is not a one-to-one relation between the outer diameter and the inner one of a DWNT. For example, we found from HRTEM observations that the inner tube diameters are different, even though the outer tube diameters of the DWNTs are identical, such as those shown in Fig. 2 (g) (outer diameter 2.0 nm and inner diameter 1.23 nm) and in Fig. 2 (f) (outer diameter 2.0 nm and inner diameter 1.34nm). A. Charlier et al. considered that this phenomenon was related to the different chirality combinations of coaxial tubes [11]. Moreover, theoretical calculation shows that an interlayer spacing range is possible from the point of energy stability [5].

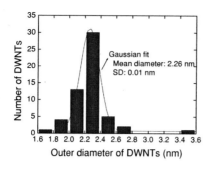 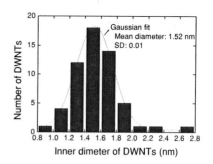

FIGURE 3. Diameter distribution of the as-prepared DWNTs, based on the HRTEM observations. The Gaussian fit to these data gives a mean diameter of 2.26 ± 0.01 nm in outer diameter and 1.52 ± 0.01 nm in inner diameter [10], 2002, reprinted with permission from Elsevier.

Diameter Distribution of the As-Prepared DWNTs

Most of the isolated DWNTs that were observed have outer tube diameters ranging from 1.6 to 2.8nm and inner diameters ranging from 1.0 to 2.0 nm [10]. The diameter distribution result is shown in Fig. 3 based on the HRTEM observations, and the Gaussian mean outer diameter and inner diameter are 2.26 ± 0.01nm and 1.52 ± 0.01 nm, respectively. These results show that the DWNTs synthesized by our methane catalytic decomposition method have smaller diameters and more narrow distribution than those grown by the hydrogen arc discharge method. Moreover, these diameters are mostly within the resonant windows formed by the E_{22}^{S}, E_{33}^{S}, E_{11}^{M} and E_{44}^{S} transitions and the incident photon $E_{laser}=1.96$ eV, according to the theoretical calculations about isolated SWNTs [12]. Therefore, if the interaction between the outer and inner tube is weak enough, we can expect that two peaks in the radial breathing mode (RBM) corresponding to the outer and inner nanotube of a DWNT, respectively, should be observed.

Raman Spectra of the As-Prepared DWNTs

It is well known that resonant Raman scattering measurements associated with the radial breathing mode (RBM) of small diameter CNTs is a particularly valuable tool for detecting SWNTs and measuring their diameter, since the electronic states are highly sensitive to the diameter of the nanotube [12, 13]. Moreover, the corresponding unique chirality of a SWNT can be identified from its diameter, along with one interband energy E_{ii} [12].

An important application of DWNTs is expected as nanoelectronic devices, so it is very important to identify their conductivity. However, the first problem we face is whether the RBM of DWNTs can be detected, similar to that of the SWNTs, which is determined by their diameter and the interaction of two SWNTs constituting a DWNT. In fact, the RBM was detected for those DWNTs derived from the chains of fullerenes in SWNTs [7], while it was not observed for the DWNTs with large diameter synthesized by hydrogen arc discharge method [6]. Carbon nanotubes with large diameter are not expected to exhibit a large (resonantly enhanced) Raman cross-section, their RBM band is therefore much more difficult to be detected. Because of smaller diameter of our DWNTs, their RBM is expected to be detected.

A DWNT can be considered as a combination of two coaxial SWNTs separated radially by the interlayer distance of graphene sheet, about 0.34nm in a geometric shape. Theoretical calculation indicates that the interlayer interaction does not open the energy gap. Thus it is concluded that the outer and inner layer retain the basic electronic properties of each constituent graphene monolayer tubule, even though the detailed energy dispersion relations are affected by the interlayer interaction [4]. We consider that the effects of coupling between adjacent graphene layers on the vibrational and electronic states are not of sufficient strength to significantly affect the physical properties of DWNTs due to the weak van der Waals interaction, as similar to the intra-bundle and inter-tube coupling in a SWNT bundle [10]. Therefore, it is not expected that many new Raman modes actually be observed, and the spectral intensity should still be dominated by the selection rules for the two isolated SWNTs constituting a DWNT [10].

To realize the *in situ* Raman scattering measurements, the as-prepared sample in a grid was used, which was shown to contain more than 70 percents of DWNTs by HRTEM observation. Fig. 4 shows a typical Raman spectrum of our as-prepared

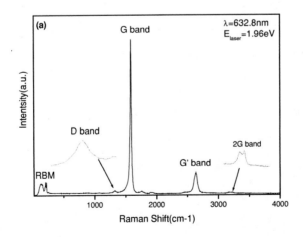

FIGURE 4. A typical Raman spectrum of the as-prepared DWNTs excited by 1.96eV laser [14].

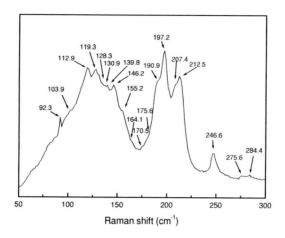

FIGURE 5. A typical RBM of the as-prepared DWNTs excited by 1.92 eV excitation laser energy [10], 2002, reprinted with permission from Elsevier.

DWNTs [14]. A Raman spectra profile similar to that of SWNTs was found for the DWNTs, which has resonantly enhanced RBM, G band and weak D band.

RBM of the As-Prepared DWNTs

Two different characteristics of the RBM band of the DWNTs (Fig. 5) were found, comparing to that of SWNTs. First, it mainly consists of two components at low frequency region forming pair peaks, centered at $112cm^{-1}$ and $197cm^{-1}$. Another characteristic of the RBM band is that each RBM band has some Raman shifts with a small frequency interval, which is thought to be caused by the two-graphene layers in the DWNT and their chirality.

The outer diameters and inner diameters of the as-prepared DWNTs were calculated, according to the Raman shift [15]. Table 1 summarizes the peak positions and their calculated tube diameters. It is found that the diameters corresponding to the lower frequency band obtained by the calculation are almost equal to the outer diameters obtained by the measurements from HRTEM images, and the higher frequency band corresponds to the inner diameter [10]. Moreover, HRTEM observation shows that few SWNTs with diameter about 2.3nm and 1.3nm were found. Therefore, we can conclude that the two mean peaks are the RBM of the two coaxial SWNTs constituting a DWNT, which also supports the theoretical prediction about the weak interlayer interaction of the DWNTs, i.e., the outer and inner layer retain the

334

TABLE 1. Raman Peak Positions and the Calculated Diameters for the As-Prepared DWNTs [10], 2002, reprinted with permission from Elsevier.

Outer Diameter of the As-Prepared DWNTs, $\omega_r (d_t)$ [a] cm^{-1} (nm)		Inner Diameter of the As-Prepared DWNTs, $\omega_r (d_t)$ [a] cm^{-1} (nm)	
92.3	(2.75)	128.3	(1.98)
		130.9	(1.94)
103.9	(2.44)	146.2	(1.74)
		155.2	(1.64)
112.9	(2.25)	164.1	(1.55)
		170.5	(1.49)
		175.6	(1.45)
119.3	(2.13)	190.9	(1.33)
		197.2	(1.29)
128.3	(1.98)	197.2	(1.29)
		207.4	(1.22)
		212.5	(1.20)
130.9	(1.94)	207.4	(1.22)
		212.5	(1.20)
139.8	(1.82)	246.6	(1.03)
146.2	(1.74)	275.6	(0.92)
		284.4	(0.89)

[a] ω_r is the Raman shift, and d_t is the corresponding diameter of a tube.

basic electronic properties of each constituent graphene monolayer tubule.

From Table 1, it was found that an outer diameter incorporates with several possible candidates of inner tubes, which is consistent with our HRTEM observations. And, conversely, an inner diameter may incorporate with several possible candidates of outer tubes. For example, if the inner diameter is 1.20 nm, the corresponding outer diameter may be 1.98 or 1.94 nm. The tube diameters cannot take continuous but discrete values due to the effect of their chirality [11]. Therefore, several secondary inner tubes with different diameters can be grown near an ideal tube diameter with a minimum energy determined only by the van der Waals interaction with parent outer tube, even slightly increasing the total electronic energy of DWNTs [5, 10].

Chirality Identification of the As-Prepared DWNTs

The determination of chirality of a nanotube, i.e., the (n, m) index, by resonance Raman scattering depends primarily on the determination of E_{ii} using the unique

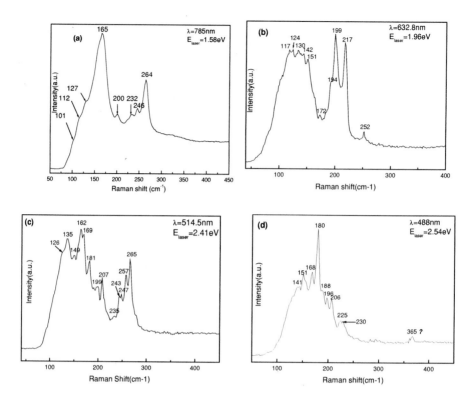

FIGURE 6. RBM spectra of DWNTS with different E_{laser} value of (a) 1.58eV, (b) 1.96eV, (c) 2.41eV, (d) 2.54eV. The Lorentz fitting process obtained the positions of RBM frequencies [14].

relation between E_{ii} and (n, m) in Ref [12], since d_t can be obtained from the Raman shift [12]. Therefore, for obtaining the detailed structural information of the DWNTs, i.e., the chirality of their outer tubes and inner tubes, the various incident laser energy (E_{laser}) values were used to measure the product and the results are shown in Fig. 6. In Fig. 6 (a), (b) and (c), the RBM frequencies can be divided into two well-distinguished groups associated with the inner and outer layers of the DWNTs, similar to Fig. 5. This implies that E_{22}^S and E_{11}^M satisfy the resonance condition for the inner layer, while E_{33}^S and E_{44}^S satisfy the resonance condition for the outer layer [12, 14]. This suggests that the inner and outer layer pairs made by pairing up tentatively assigned DWNTs satisfy the [E_{22}^S and E_{33}^S] or [E_{11}^M and E_{44}^S] resonance conditions, respectively.

However, the ω_{RBM} grouping shown in Figs. 6 (a) and (c) appears to be different, compared to Fig. 6 (b), and the groups at low frequency are downshifted. Since resonance enhanced Raman signal from many large diameter nanotubes could not be detected, the RBM bands could not be assigned to the inner or outer layers clearly. Also, the RBM spectrum in Fig. 6 (d) does not clearly divide into two groups because a large number of tubes with different diameters are in resonance with the same E_{laser} excitation. However, a common characteristic of the four Raman spectra is that the line shape of their RBM spectra is broad, indicating that different groups of inner and outer walls of the nanotubes in our sample are resonant with each E_{laser}. The intensity of the RBM band is much larger, and the spectra are more complex than those in previously reported works on DWNTs [7, 16]. The complexity probably arises from the mixing of many different (n, m; n', m') constituents in our DWNT samples, where the pairs of integers in brackets denote the inner tubes (n, m) and outer tubes (n', m'), respectively.

Subsequently, the spectra in Fig. 6 were tentatively assigned to (n, m) indices, based on their d_t and E_{ii} values [17], except for the two modes observed in Fig. 6 (a) at 200cm^{-1} and Fig. 6 (b) at 252cm^{-1} that could not be assigned, based on tight-binding calculations. Nanotubes with the same ω_{RBM} that are resonant with different laser excitation energies, such as 168cm^{-1} in Fig. 6 (d) and 169cm^{-1} in Fig. 6 (c), were assigned to the same (n, m) indices. On the other hand, within the resonance window of a given excitation energy, more than one set of (n, m) index can be assigned to a given ω_{RBM}. For example, ω_{RBM}=246cm^{-1} in Fig. 6 (a) can be assigned to (8, 7) or (11, 3) since they both are within the resonant windows of the 1.58eV laser excitation. The differences between the experimental ω_{RBM} and the theoretical assignments for these d_t values are less than 4%.

Since most of the diameter determinations based on the RBM frequency in Raman experiments employ tight binding results in their energies, departure from tight binding predictions can lead to deviations in diameter and (n,m) assignments for small diameter tubes. The nanotubes that are resonant with E_{laser} below 1.53eV, between 1.68 and 1.91eV, from 2.01 to 2.36eV and above 2.59eV cannot be observed with our four available laser lines. At the same time, as a result of the notch filter cut-off, we cannot measure Raman shifts less than 100cm^{-1} (d_t>2.5nm). Therefore, a few tube diameters found in HRTEM observations can not be detected from the Raman spectra.

The indices of possible inner and outer layer tubes that can be used to assemble DWNTs are shown in Table 2. For the assembled DWNTs, the inner and outer constituent SWNTs can either be identified by the same E_{laser} excitation, such as (10, 7; 24, 1), where both constituents are in resonance with the 632.8nm laser excitation, or by being resonant with different laser lines. As an example, for the (12, 5; 24, 1) tube, the inner and outer tubes are in resonance with the 488nm and 632.8nm line, respectively. Most of the diameters of the assembled DWNTs can be confirmed by HRTEM observations except for the small diameter DWNTs, such as (8, 1; 11, 9), (8, 1; 15, 4) and (8, 1; 17, 3). From Table 2, we learned that two DWNTs with the same outer diameters could have inner tubes with different diameter. This observation is confirmed by our HRTEM measurements. The tentatively assembled DWNTs indicate that all combinations of diameters and chiralities are possible for the two layers of

TABLE 2. Conductive Type Calculated from Assigned Indices and the Assembled DWNTs [14].

DWNT$_{S-S}$	(8, 1; 11, 9) (8, 1; 15, 4) (8, 1; 17, 3) (12, 1; 18, 7) (12, 1; 20, 4) (8, 7; 18, 7)
	(8, 7; 20, 4) (8, 7; 21, 3) (11, 3; 20, 4) (11, 3; 18, 7) (11, 4; 18, 7) (11, 4; 20, 4)
	(11, 4; 23, 1) (12, 5; 24, 1) (12, 5; 19, 9) (11, 7; 24, 1) (11, 7; 19, 9) (11, 7, 20, 9)
	(16, 0; 19, 9) (16, 0; 20, 9) (10, 9; 20, 9) (10, 9; 19, 12) (10, 9; 19, 12) (12, 7; 20, 9)
	(11, 9; 19, 12) (15, 4; 19, 12) (17, 3; 24, 7) (15, 7; 24, 7) (18, 5; 26, 9) (14, 10; 26, 9)
	(18, 7; 26, 9) (23, 1, 26, 9)
DWNT$_{M-S}$	(11, 2; 18, 5) (11, 2; 14, 10) (11, 2; 18, 7) (11, 2; 20, 4) (10, 4; 18, 7) (10, 4; 20, 4)
	(9, 6; 18, 7) (9, 6; 20, 4) (13, 1; 18, 7) (13, 1; 20, 4) (13, 1, 23, 1) (10, 7; 23, 1)
	(10, 7; 24, 1) (10, 7; 19, 9) (12, 6; 24, 1) (12, 6; 19, 9) (12, 6; 20, 9) (11, 8; 20, 9)
	(17, 2; 19, 12) (14, 8; 24, 7)
DWNT$_{S-M}$	(8, 1; 17, 2) (11, 7; 16,13) (16, 0; 16, 13) (10, 9; 16, 13) (12, 7; 16, 13)
DWNT$_{M-M}$	(10, 7; 16, 13) (12, 6; 16, 13) (11, 8; 16, 13)

nanotubes, provided that the minimum inter-layer separation is maintained.

If the nanotube growth process is random, then 1/3 of the tubes would be expected to be metallic, and 2/3 are semi-conducting [13]. This implies that the ratio of S-S: M-S: S-M: M-M should be 4:2:2:1. However, the assembled DWNTs include 32 S-S, 20 M-S, 5 S-M and 3 M-M DWNTs, as shown in Table 2. The ratio of S-S and M-S or S-M and M-M is close to 2. But the ratio of M-S to S-M pairs is far from 1. This suggests that the outer layers of our DWNTs are predominately semi-conducting.

CONCLUSIONS

1. DWNTs were synthesized by the decomposition of methane in the presence of Fe catalyst, with the addition of sulfur. Their outer diameter and inner diameter, based on HRTEM observations, are in the range of 1.6-3.6 nm and 0.8-2.8 nm, respectively.
2. The interlayer spacing of DWNTs is not a constant, ranging from 0.34 nm to 0.41 nm, which is in agreement with Raman spectra analysis.
3. The as-prepared DWNTs show corresponding radial breathing mode (RBM) Raman shift associated with their inner tubes as well as outer tubes, which supports the independent electronic structure of each constituent graphene monolayer tubule from theoretical predictions. Moreover, the chirality of the DWNTs was identified based on the RBM excited by four different laser excitation energy and the results show that the outer tubes of the DWNTs are predominately semiconducting.

ACKNOWLEDGEMENTS

The authors acknowledge NSFC grants (No.50025204, 50032020 and 90206018) and the Special Fund for Major Basic Research Projects (G2000026403) for support of this work.

REFERENCES

1. Thess, A., Lee, R., Nikolaev, P., Dai, H.J., Petit, P., Pobert, J., Xu, C.H., Lee, Y.H., Kim, S.G., Rinzler, A.G., Colbert, D.T., Scuseria, G.E., Tomanek, D., Fischer, J.E., Smalley, R.E., *Science* **273**, 483-487 (1996).

2. Cheng, H.M., Li, F., Su, G., Pan, H.Y., He, L.L., Sun, X., Dresselhaus, M.S., *Appl. Phys. Letters* **72**, 3282-3284 (1998).

3. Ren, Z.F., Huang, Z.P., Xu, J.W., Wang, J.H., Bush, P., Siegal, M.P., Provencio, P.N., *Science* **282**, 1105-1107 (1998).

4. Saito, R., Dresselhaus, G., Dresselhaus, M.S., *J. Appl. Phys.* **73**, 494-500 (1993).

5. Saito, R., Matsuo, R., kimura, T., Dresselhaus, G., Dresselhaus, M.S., *Chem. Phys. Letters* **348**, 187-193 (2001).

6. Hutchison, J.L., Kiselev, N.A., Krinichnaya, E.P., Krestinin, A.V., Loutfy, R.O., Morawsky, A.P., Muradyan, V.E., Obraztsova, E.D., Sloan, J., Terekhov, S.V., Zakharov, D.N., *Carbon* **39**, 761-770 (2001).

7. Bandow, S., Takizawa, M., Hirahara, K., Yudasaka, M., Iijima, S., *Chem. Phys. Letters* **337**, 48-54 (2001).

8. Zhu, H.W., Xu, C.L., Wu, D.H., Wei, B.Q., Vajtal, R., Ajayan, P.M., *Science* **296**, 884-886 (2002).

9. Baughman, R.H., Zakhidov, A.A., de Heer, W.A., *Science* **297**, 787-792 (2002).

10. Ren, W.C., Li, F., Chen, J., Bai, S., Cheng, H.M., *Chem. Phys. Letters* **359**,196-202 (2002).

11. Charlier, A., Mcrae, E., Heyd, R., Charlier, M.F., Moretti, D., *Carbon* **37**, 1779-1783 (1999).

12. Dresselhaus, M.S., Dresselhaus, G., Jorio, A., Souza Filho, A.G., Saito, R., *Carbon* **40**, 2043-2061 (2002).

13. Dresshaus, M.S., Eklund, P.C., *Adv. In Phys.* **49**, 705-814 (2000).

14. Li, F., Chou, S.G., Ren, W.C., Swan, A.K., Unlu, M.S., Goldberg, B.B., Cheng, H.M., Dresselhaus, M.S., *J Mater. Res.* In press.

15. Cheng, H.M., Li, F., Sun, X., Brown, S.D.M., Pimenta, M.A., Marucci, A., Dresselhaus, G., Dresselhaus, M.S., *Chem. Phys. Letters* **289**, 602-610 (1998).

16. Bacsa, R.R., Laurent, C., Peigney, A., Bacsa, W.S., Vaugien, T., Rousset, A., *Chem. Phys. Letters* **323**, 566-571 (2000).

17. Jorio, A., Saito, R., Hafner, J.H., Lieber, C.M., Hunter, M., McClure, T., Dresselhaus, G., Dresselhaus, M.S., *Phys. Rev. Letters* **86**, 1118-1121 (2001).

Commensurate Double-wall Nanotubes: Symmetry and Phonons

E. Dobardžić, I. Milošević, B. Nikolić, T. Vuković and M. Damnjanović

Faculty of Physics, University of Belgrade, P. O. Box 368, Belgrade 11001, Serbia and Montenegro

Abstract. For translationally periodic double-wall carbon nanotubes stable configurations and full symmetry groups are determined. Using this, the phonon dispersions and eigenvectors are calculated and assigned by the complete set of conserved quantum numbers. Breathing-like modes, friction modes and acoustic modes are studied. The latter are found to be linear in wave vector k.

Recently, double-wall carbon nanotubes (DWCNs) have been generated by coalescence of C_{60} molecules encapsulated into the single-wall CNs [1] and by chemical vapor deposition [2]. Resonant Raman measurements have been used to determine the inner and outer diameter of the DWCNs [3]. Theoretically, stable structures of DWCNs have been calculated [4] and their breathing-like phonon modes have been studied [5]. However, up to the knowledge of the authors, no systematic theoretical study of the DWCNs stable configurations, symmetry and lattice dynamics has been reported yet. In this contribution we find the stable configurations and symmetry groups of the commensurate (i.e. translationally periodic) DWCNs and calculate their phonon dispersions and atomic displacements.

DWCN W@W$'$ is a pair of co-axial single-layer tubes: the inner W$=(n_1,n_2)$ and the outer W$' = (n_1',n_2')$. Such a tube is commensurate (CDWCN) if the ratio a/a' of the layers' periods is rational. We study CDWCNs with inner diameters $2.8\text{Å} \leq D \leq 50\text{Å}$. Assuming that the inter-layer distance is $3.44\pm0.2\text{Å}$, there are 318 of them (roughly 0.5% of all DWCNs with such diameters). Frequently, in 178 cases, the walls have the same or opposite chiral angle, Tab. 1; such tubes we denote as $(\hat{n}_1,\hat{n}_2)n@n'$ and $(\hat{n}_2,\hat{n}_1)n@n'$, respectively, where $(\hat{n}_i = n_i/n = n_i'/n'$, $i = 1,2)$. Particularly, here are 60 zig-zag $ZZ_n = (1,0)n@(n+9)$, and 35 armchair, $AA_n = (1,1)n@(n+5)$ tubes.

Here we sketch the results of the extensive symmetry based research of these CDWCNs. This includes determination of the stable configurations, full symmetry analysis, phonon frequencies and displacements and their assignment by conserved quantum numbers. To this end we use the *POLSym* code [6], recently applied to SWCNs [7].

Symmetry group of CDWCN is the intersection of the groups of its layers [8]. It depends on the relative position of the walls, which is completely determined by the angle of rotation Φ and length of translation Z (around and along the tube axis, respectively) which are to be performed on x- to match x'-axis (here x- and x'-axis pass through the centers of carbon hexagons of W and W$'$; layers z- and z'-axis coincide).

The stable configuration is found numerically, as the minimum of the W-W$'$ interaction potential $V(\Phi,Z) = \frac{1}{2}\sum v(\boldsymbol{r}_\alpha - \boldsymbol{r}'_\alpha)$ (summation over all atoms α of W and α' of W$'$)

CP685, Molecular Nanostructures: XVII Int' l. Winterschool/Euroconference on
Electronic Properties of Novel Materials, edited by H. Kuzmany, J. Fink, M. Mehring, and S. Roth

TABLE 1. Rays of CDWCNs with the diameter $2.8\text{Å} \le D \le 50\text{Å}$. The tubes of the ray given in column 1 are obtained by substituting allowed values of n from column 4 (the values in brackets should be omitted). The corresponding line and isogonal groups are in columns 2 and 3, respectively. In the last column number of found CDWCNs (total 223) is given. Translational periods are given in the units of graphene period $a_0 = 2.46\text{Å}$.

CDWCN ray	Line Group	Isogonal	z values	No.
$z(1,0)@(z+9)(1,0)$	$T(\sqrt{3})D_{1d}$	D_{1d}	$4,5,\ldots,62$ $(6,9,12,\ldots,60)$	40
	$T(\sqrt{3})D_{3d}$	D_{3d}	$6,9,12,\ldots,60$ $(9,18,\ldots54)$	13
	$T(\sqrt{3})D_{9d}$	D_{9d}	$18,27\ldots,63$	6
	$T(\sqrt{3})T_c S_{18}$		9	1
$z(1,1)@(z+5)(1,1)$	$T(1)D_{1d}$	D_{1d}	$2,3,\ldots,36$ $(5,10,15,20,25,30,35)$	29
	$T_c(1)S_{10}$	D_{5d}	5	1
	$T(1)D_{5d}$		$10,15,\ldots,35$	6
$z(3,2)@(2+z)(3,2)$ $z(3,2)@(2+z)(2,3)$	$T^1_2(\sqrt{57})D_1$	D_2	$1,3,\ldots,13$	14
	$T(\sqrt{57})D_2$		$2,4,\ldots,14$	14
$z(4,1)@(2+z)(4,1)$ $z(4,1)@(2+z)(1,4)$	$T^1_2(\sqrt{7})D_1$	D_2	$1,3,\ldots,13$ $1,3,\ldots,11$	7 6
	$T(\sqrt{7})D_2$		$2,4,\ldots,12$ $2,4,\ldots,12\ (6)$	6 5
	$T^{13}_{14}(\sqrt{7})D_1$	D_{14}	13	0
	$T^3_{14}(\sqrt{7})D_2$		6	1
$z(7,3)@(1+z)(7,3)$ $z(7,3)@(1+z)(3,7)$	$T(\sqrt{237})D_1$	D_1	$1,2,\ldots,7$	14
$z(8,1)@(1+z)(8,1)$ $z(8,1)@(1+z)(1,8)$	$T(\sqrt{219})D_1$	D_1	$1,2,\ldots,7$	14

with Lenard-Jones inter-atomic potential [4]. It turns out that in all cases the high symmetry configurations are optimal, and in addition to the roto-translational symmetries, various parities are present. In contrast to Ref. [4], we find the stability of CDWCNs to depend essentially on the chirality difference of the layers and to be scarcely sensitive to variations of the diameter difference of the outer and inner layer (around the value that corresponds to the graphite inter-layer distance).

Whenever at least one layer of CDWCN is chiral, the stable configuration is $(\Phi, Z) = (0,0)$; thus, the coinciding $x = x'$ axis is common U-axis. The only CDWCNs with both achiral walls are the mentioned series ZZ_n and AA_n. For AA_5 (Fig. 1, right) and ZZ_9 the stable configuration is $(\Phi, Z) = (\pi/8n, a/4)$, while in all other cases $(\Phi, Z) = (0, a/4)$. One easily sees that there is common U axis, while the perpendicular plane is glide plane for AA_5 and ZZ_9, and mirror plane in other cases. Therefore, the resulting symmetry groups are the line groups of fifth, tenth or ninth family. For the 223 tubes the groups are specified in Tab. 1. Although symmetry of the CDWCNs is considerably reduced in comparison to the SWCNs symmetry, it is still nontrivial, having various parities (in addition to the linear and angular quasi-momenta k and m) that are manifested as the conserved quantum numbers. The irreducible representations are labeled in the form ${}_k\Gamma^{\Pi}_m$: U-parity is given by $\Pi = \pm$, while Γ is A (or B) for the one-dimensional representations odd/even with respect to vertical mirror plane, and E and G for two-

and four-dimensional representations. This way the degeneracy of the bands is a priori given, as well as the selection rules for the inter-band transitions. For both achiral walls, the isogonal point group is D_{Qd} with odd order principle axis $Q = N$; thus, it contains spatial inversion, excluding simultaneous Raman and infrared activity.

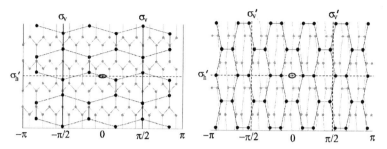

FIGURE 1. Stable configurations of the CDWCNs ZZ_3 and AA_5 in the unfolded picture: ○ denotes the two-fold horizontal axis while solid (dashed) lines denote mirror (glide and roto-reflection) planes.

The dynamical model considered involves force constants used for the SWCNs [9, 7] for the pair of atoms from the same layer and Lenard-Jones potential [4] otherwise. The calculations are performed by the *POLSym* (E) package, fully implementing line group symmetry through the modified group projector technique [10].

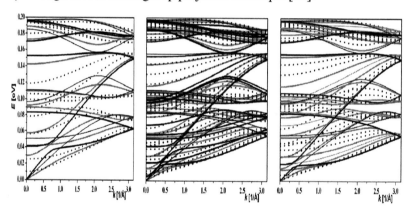

FIGURE 2. Phonon dispersions of (5,5) (left), (10,10) (right) and CDWCN AA_5 (middle).

The inter-layer interaction may be understood as the perturbation of the intra-layer one. Therefore, the CDWCN phonon bands are effectively the perturbed bands of the layers (Fig. 2). The perturbation decrease with frequency, but increase with diameter.

The four acoustic modes of each layer (longitudinal, LA, double degenerate transverzal, TA and circular, or twisting, CA) are mixed to give four CDWCN acoustic modes (LA, TA and CA) and four friction modes (LF, TF and CF): in all these modes walls oscillate like rigid bodies: in phase (for acoustic) or out of phase (for friction). The acoustic branches are linear in k; their slope is close to that of SWCNs, giving the sound velocities: $v_{TA} = 954$km/s, $v_{LA} = 2064$km/s and $v_{TC} = 1518$km/s. Variations between different CDWCNs is within 1%. The friction modes are low energy ones. Particularly,

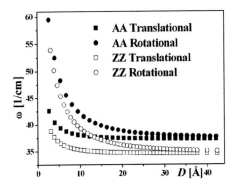

	Chiral wall			Achiral walls		
Q	1	2	≥ 3	1	2	≥ 3
LA, LF	$_0A_0^-$	$_0A_0^-$	$_0A_0^-$	$_0A_0^-$	$_0A_0^-$	$_0A_0^-$
TA$_y^x$, TF$_y^x$	$_0A_0^+$ / $_0A_0^-$	$_0A_1^+$ / $_0A_1^-$	$_0E_1$	$_0B_0^+$ / $_0A_0^-$	$_0B_1^+$ / $_0A_1^-$	$_0E_1^+$
CA, CF	$_0A_0^-$	$_0A_0^-$	$_0A_0^-$	$_0B_0^-$	$_0B_0^-$	$_0B_0^-$

Plot legend: ■ AA Translational, ● AA Rotational, □ ZZ Translational, ○ ZZ Rotational. Axes: ω [1/cm] vs D [Å].

FIGURE 3. Frequencies of LF and CF modes as a functions of tube diameter. Acoustic and friction modes are assigned by the irreducible representations in the table on the right: For each order of the isogonal group principle axis, Q, the corresponding representations are given for the tubes with at least one chiral wall and for those with both walls achiral. Note that for $Q < 3$, the degenerate TA modes' quantum numbers distinguish the $x-$ and $y-$displacements.

frequencies of LF and CF modes (Fig. 3) agree with the previous predictions and experimental verification of the telescope effect [8, 11, 12]. The acoustic and friction like modes are assigned by the same quantum numbers (see table at Fig. 3).

There are two breathing like modes, in-phase and out-of-phase. Both may be derived by perturbative methods, from the radial breathing modes of the walls [5]. They transform according to the identical representation $_0A_0^+$ of the symmetry group. Our results are in agreement with the predictions of Ref. [5].

REFERENCES

1. B. W. Smith and D. E. Luzzi, Chem. Phys. Lett. **321** (2000) 169.
2. R. R. Bacsa, Ch. Laurent, A. Peigney, W. S. Bacsa, Th. Vaugien and A. Rousset, Chem. Phys. Lett. **323**, 566 (2000); W. Z. Li, J. G. Wen, M. Sennett and Z. F. Ren, Chem. Phys. Lett. **368**, 299 (2002).
3. S. Bandow, G. Chen, G. U. Sumanasekera, R. Gupta, M. Yudasaka, S. Iijima, and P. C. Eklund, Phys. Rev. B **66**, (2002) 075416; S. Bandow, M. Takizawa, K. Hirahara, M. Yudasaka and S. Iijima, Chem. Phys. Lett. **337**, 48 (2001); R. R. Bacsa, A. Peigney, Ch. Laurent, P. Puech and W. S. Bacsa, Phys. Rev. B **65**, (2002) 161404.
4. R. Saito, R. Matsuo, T. Kimura, G. Dresselhaus and M. S. Dresselhaus, Chem.Phys.Lett. **348**, (2001) 187.
5. V. N. Popov and L. Henrard, Phys. Rev. B **65**, (2002) 235415.
6. I. Milošević, A. Damjanović and M. Damnjanović, Ch. XIV in *Quantum Mechanical Simulation Methods in Studying Biological Systems* ed. D. Bicout and M. Field (Springer-Verlag, Berlin 1996).
7. E. Dobardžić, I. Milošević, B. Nikolić, T. Vuković and M. Damnjanović, Phys. Rev. B (2003, submited).
8. M. Damnjanović, I. Milošević, T. Vuković and R. Sredanović, Phys. Rev. B **60**,(1999) 2728.
9. R. A. Jishi, L. Venkataraman, M. S. Dresselhaus, G. Dresselhaus, Chem. Phys. Lett. **209**, (1993) 77.
10. M. Damnjanović, T. Vuković and I. Milošević, J. Phys. A **33**, (2000) 6561.
11. M. Damnjanović, T. Vuković and I. Milošević, *Eur. Phys. J. B* **25** (2002) 131; T. Vuković, M. Damnjanović and I. Milošević, *Physica E* **16** (2003) 259.
12. J. Cumings and A. Zettl, Science **289** (2000) 602.

Tuning of Electronic Structure of C_{60}@SWCNT and C_{70}@SWCNT (Peapods): *In-Situ* Raman and Vis-NIR Spectroelectrochemical Study

Ladislav Kavan[1,2], Lothar Dunsch[2] and Hiromichi Kataura[3]

[1] *J. Heyrovský Institute of Physical Chemistry, Academy of Sciences of the Czech Republic, Dolejškova 3, CZ-182 23 Prague 8*
[2] *Institute of Solid State and Materials Research, Helmholtzstr. 20, D - 01069 Dresden*
[3] *Tokyo Metropolitan University, 1-1 Minami-Ohsawa, Hachioji, Tokyo 192-0397, Japan*

Abstract. Charger-transfer on fullerene peapods (C_{60}@SWCNT and C_{70}@SWCNT) was studied by electrochemistry in 0.2 M $LiClO_4$ + acetonitrile and in butylmethylimidazolium tetrafluoroborate (ionic liquid). The latter medium offers broader window of accessible electrochemical potentials, good electrical conductivity and favorable optical properties for spectroelectrochemistry. Similar to empty SWCNT, the electrochemistry of peapods is dominated by capacitive double-layer charging. Vis-NIR spectra evidence reversible doping-induced bleaching of the transitions between Van Hove singularities. The bleaching of optical transitions is mirrored by quenching of resonance Raman scattering in the region of tube-related modes. The Raman modes of intratubular C_{60} exhibit considerable intensity increase upon anodic doping of peapods, but these modes are not enhanced at cathodic charging. The "anodic enhancement" of Raman scattering from intratubular C_{60} is not reproduced in C_{70}@SWCNT. All relevant Raman modes of intratubular C_{70} show symmetric charge-transfer bleaching as the RBM/TM lines. A tentative interpretation follows from two arguments: (i) the LUMO of C_{60} is located unusually close to the Fermi level of SWCNT and (ii) C_{70}@SWCNT (peapods) may change the configuration from lying to standing one. This keeps the LUMO of C_{70} at higher energies even in wide tubes.

INTRODUCTION

The fullerene peapods, C_{60}@SWCNT [1] and C_{70}@SWCNT (SWCNT = single-walled carbon nanotube) present new challenges to electrochemistry of nanocarbons, because both the SWCNT and C_{60}/C_{70} show specific redox responses. Electrochemical tuning of the Fermi level position allows convenient monitoring of electronic transitions between Van Hove singularities, and of the resonance Raman spectra of SWCNT [2] and C_{60}@SWCNT (peapods) [1]. This paper is concerned with the electrochemical doping of peapods, which is, especially for the hole-doping, superior to chemical redox doping. Our previous short communication [1] focussed on electrochemical doping of C_{60}@SWCNT. Here we present new results on a detailed spectroelectrochemical study of C_{60}@SWCNT and C_{70}@SWCNT.

EXPERIMENTAL SECTION

Fullerene peapods (C_{60}@SWCNT, filling ratio 85% and C_{70}@SWCNT, filling ratio 72 %) were prepared and purified as described elsewhere [3,4]. Freshly sonicated

CP685, *Molecular Nanostructures: XVII Int'l. Winterschool/Euroconference on Electronic Properties of Novel Materials,* edited by H. Kuzmany, J. Fink, M. Mehring, and S. Roth

ethanolic slurry of peapods was deposited by evaporation onto Pt or ITO (indium-tin oxide conducting glass) electrodes. The film was treated at 100-150°C in vacuum and further handled under nitrogen atmosphere. Electrochemical experiments were carried out using conventional potentiostatic set-up with Pt auxiliary and Ag-wire pseudo-reference electrodes. The potential of Ag-wire pseudo-reference electrode was calibrated against the redox potential of ferrocene (Fc/Fc^+). Several droplets of ferrocene+acetone solution were added to the electrolyte solution at the end of each set of spectro/electrochemical measurements. The electrolyte solution, 0.2 M $LiClO_4$ + acetonitrile, contained <10 ppm H_2O (Karl Fischer titration). Butylmethylimidazolium tetrafluoroborate (Fluka) was purified by extraction with water, then toluene and acetone, followed by treatment with active carbon and Al_2O_3. Finally, the ionic liquid was outgased and dried at 80°C in vacuum. The ITO-supported film of peapods served for *in-situ* Vis-NIR spectroelectrochemistry (Shimadzu 3100 spectrometer). Due to limited stability of ITO against reductive breakdown, the Vis-NIR spectra were studied at potentials > -0.4 V vs. Fc/Fc^+. Raman spectra were excited by Ar^+ laser at 2.41 eV or 2.54 eV, respectively (Innova 305, Coherent). The Raman spectra were recorded on a T-64000 spectrometer (Instruments, SA) interfaced to an Olympus BH2 microscope (objective 50x, the laser power impinging on the cell window was ca. 1.5 mW).

RESULTS AND DISCUSSION

Cyclic voltammograms of peapods in 0.2 M $LiClO_4$ + acetonitrile or in ionic liquid are dominated by capacitive double-layer charging similar to empty SWCNT [1,2]. The butylmethylimidazolium tetrafluoroborate (ionic liquid):

allows broader electrochemical window (-2.4 to 1.7 V *vs.* Fc/Fc^+). The used ionic liquid has also favorable optical properties for in-situ Vis-NIR and Raman spectroelectrochemistry.

No distinct redox peaks of the reduction of fullerenes [5,6] were detected at the peapods. The capacitance of empty SWCNT was ca. 40 F/g in 0.2 M $LiClO_4$ + acetonitrile [2]. For an ideal double layer capacitor, the change in number of electrons transferred per one carbon atom, Δf equals:

$$\Delta f = M_C C \Delta U / F \qquad (1)$$

where M_C is atomic weight of carbon, ΔU is potential difference and F is Faraday constant. Equation (1) yields $\Delta f = 0.005$ e⁻/C-atom for $\Delta U = 1$ V and $C = 40$ F/g. A one electron reduction of C_{60} represents $\Delta f = 0.017$ e⁻/C-atom (for C_{70}: $\Delta f = 0.014$ e⁻/C-atom). Hence, the redox-process C_{60}/C_{60}^- or C_{70}/C_{70}^- corresponds to ca. 3 times more electrons per C-atom compared to double layer charging of the wall. Cyclic voltammetry evidences that the electroreduction of intra-tubular C_{60}/C_{70} is hampered.

This is in accord with our earlier data on C_{60}@SWCNT [1] as well as with the fact that even a stronger chemical reduction of C_{60}@SWCNT with K-vapor was sluggish, starting with a charge transfer to the nanotube wall [7]. The charge-compensation "through-wall" is not sufficient, due to a limited double-layer capacity (cf. Eq. 1).

Similar to empty SWCNT [2], dry peapods (or in the electrolyte solution at open-circuit potential) showed three optical bands: at 0.7 eV ($v_s^1 \rightarrow c_s^1$), 1.25 eV ($v_s^2 \rightarrow c_s^2$), and 1.8 ($v_m^1 \rightarrow c_m^1$) eV. No distinct bands of C_{60}/C_{70} (expected at ca. 2.3 - 2.6 eV) were found. Anodic polarization shifts the Fermi level, and the singularities are depleted in the sequence: c_s^1, c_s^2, c_m^1. Analogously, cathodic polarization leads to sequential filling of the singularities: v_s^1, v_s^2, v_m^1. In both cases, the optical bands reversibly disappear in the same sequence [1,2]. In ionic liquid, the Vis-NIR spectra can be recorded up to 2.45 V vs. Fc/Fc$^+$. At high anodic potentials, a new optical band appears between 1.1 eV to 1.3 eV. It reminds the doping-induced transitions [8,9] for chemically (Br$_2$) p-doped SWCNT ($v_s^n \rightarrow v_s^1$, $v_s^n \rightarrow v_s^2$; n≥3) [1,2,8,9].

Figures 1 and 2 show *in-situ* Raman spectra of electrochemically charged C_{60}@SWCNT and C_{70}@SWCNT, respectively. Similar spectra were obtained also in ionic liquid. The assignment of C_{60} peaks (cf. arrows in Fig. 1) is as follows: 270 cm^{-1} - $H_g(1)$; 430 cm^{-1} - $H_g(2)$; 494 cm^{-1} - $A_g(1)$; 709 cm^{-1} - $H_g(3)$; 769 cm^{-1} - $H_g(4)$; 1424 cm^{-1} - $H_g(7)$; 1465 cm^{-1} - $A_g(2)$. Both $A_g(1)$ and $A_g(2)$ modes show a satellite line at larger frequency (see also Refs. [2,7]). Anodic charging leads to gradual disclosure of $H_g(8)$ line at ca. 1573 cm^{-1} due to the blue-shift of TM. The $H_g(8)$ mode is normally hidden in tube-related features and cannot be observed in dry peapods.

Arrows in Fig. 2 point to eleven intense C_{70} lines, which can be observed also in bare C_{70}. Their tentative assignment is as follows: [10,11]: 260 cm^{-1} - (A_1, E_2"); 450 cm^{-1} - (A_1, E_1"); 563 cm^{-1} - (A_1, E_1'); 699 cm^{-1} - (A_1, E_1', E_2'); 740 cm^{-1} - (A_1, E_1", E_2'); 1061 cm^{-1} - (A_1, E_2'); ; 1181 cm^{-1} - (A_1, E_1', E_2'); 1226 cm^{-1} - (A_1, E_1", E_2'); 1256 cm^{-1} - (E_1", E_2'); 1443 cm^{-1} - (A_1, E_1', E_2'); 1466 cm^{-1} - (A_1, E_1'). Also in this case, we can see significant splitting of some C_{70} lines: at 260 cm^{-1}, 450 cm^{-1} and 699 cm^{-1}.

Cathodic doping of C_{60}@SWCNT causes overall decrease of Raman intensities (Fig. 1). However, at anodic potentials, we see an interesting enhancement of Raman intensities of intratubular C_{60} (Fig. 1). The "anodic enhancement" is apparent for all C_{60} related modes, except the satellite lines of $A_g(1)$ and $A_g(2)$ modes at 500 and 1474 cm^{-1}, respectively. The $A_g(2)$ line shows no frequency shift upon doping (Fig. 1), compared to its position in pristine peapods (1465 cm^{-1}). This demonstrates, as expected, (i) no transfer of holes from the nanotubes to the C_{60}, and (ii) no dimerization or photodecomposition of C_{60} at the used experimental conditions.

The "anodic enhancement" of intratubular C_{60} does not reproduce in C_{70}@SWCNT (Fig. 2). The relevant Raman modes of C_{70}-peas show the "normal" symmetric potential-dependence (as the RBM/TM lines), although not all the peapod-bands are equally sensitive to charging (cf. lines at 699/705 cm^{-1}, and 450/460 cm^{-1} in Fig. 2). As in C_{60}@SWCNT, doping allows disclosure of two lines (1332 cm^{-1} and 1564 cm^{-1}; the latter being the strongest line in Raman spectrum of bare C_{70}). These bands are normally hidden in peapods by overlapping D- and G-lines of SWCNT.

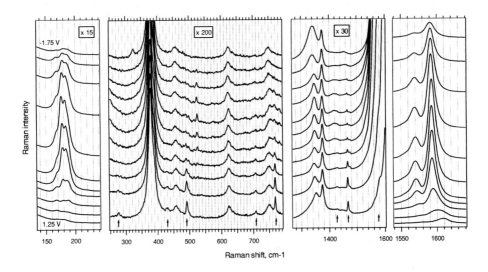

FIGURE 1. Raman spectra of C_{60}@SWCNT on Pt electrode (excited at 2.41 eV) in 0.2 M $LiClO_4$ + acetonitrile. The electrode potential varied by 0.3 V from -1.75 V to 1.25 V vs. Fc/Fc$^+$ for curves from top to bottom. Spectra are offset for clarity, but the intensity scale is identical for the respective window. The intensities are zoomed in the first three windows. Arrows indicate the expected Raman lines of C_{60}. The peaks at 378.5 and 1374.5 cm^{-1} belong to acetonitrile.

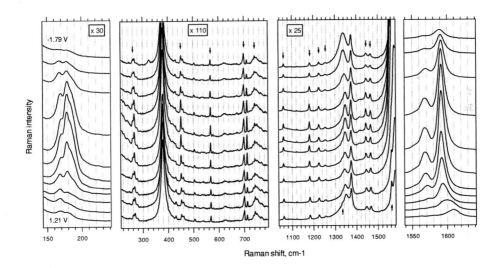

FIGURE 2. Raman spectra of C_{70}@SWCNT on Pt electrode (excited at 2.41 eV) in 0.2 M $LiClO_4$ + acetonitrile. The electrode potential varied by 0.3 V from -1.79 V to 1.21 V vs. Fc/Fc$^+$ for curves from top to bottom. Spectra are offset for clarity, but the intensity scale is identical for the respective window. The intensities are zoomed in the first three windows. Arrows indicate the expected Raman lines of C_{70}. The peaks at 378.5 and 1374.5 cm^{-1} belong to acetonitrile.

The non-symmetric anodic Raman enhancement in C_{60}@SWCNT can be interpreted as follows: The LUMO level of C_{60} (t_{1u}) is unusually close to the Fermi level of SWCNT [12]. This anomalous position is due to the hybridization of nearly free electron state of SWCNT and the π-state of C_{60} [12]. Extra charge interacts primarily with the peapod wall, causing the depletion/filling of states close to the Fermi level. This leads to symmetric anodic/cathodic bleaching of the RBM/TM intensities and the frequency shifts of TM. Cathodic charging of C_{60}@SWCNT causes extra electrons to quench efficiently the HOMO-LUMO transition. However, anodic charging virtually does not influence the electronic structure of C_{60}.

In C_{70}@SWCNT, the LUMO is located above the conduction band of SWCNT. This is the case also for wider tubes, in which the intratubular C_{70} can rearrange from lying to standing configuration [4]. Such a steric compensation is inherently excluded for C_{60} because of its spherical symmetry. The "thick" C_{60} peapods have their LUMO-band very near the Fermi level, while it can even be partly filled and the Raman resonance is partly quenched [12]. Anodic charging of thick C_{60} peapods will deplete the half-filled LUMO, which efficiently enhances the overall Raman intensity.

ACKNOWLEDGEMENTS

This work was supported by IFW Dresden, by the Academy of Sciences of the Czech Republic (contract No. A4040306) and by the Czech Ministry of Education (CZ-JP cooperation grant ME487). We thank Martin Kalbac and Dirk Rohde for experimental support.

REFERENCES

1. Kavan, L., Dunsch, L. and Kataura, H. *Chem. Phys. Lett.* **361**, 79-85 (2002).
2. Kavan, L., Rapta, P., Dunsch, L., Bronikowski, M. J., Willis, P. and Smalley, R. E. *J. Phys. Chem. B* **105**, 10764-10771 (2001).
3. Kataura, H., Maniwa, Y., Kodama, T., Kikuchi, K., Hirahara, K., Suenaga, K., Iijima, S., Suzuki, S., Achiba, Y. and Kratschmer, W. *Synth. Metals* **121**, 1195-1196 (2001).
4. Kataura, H., Maniwa, Y., Abe, M., Fujiwara, A., Kodama, T., Kikuchi, K., Imohori, H., Misaki, Y., Suzuki, S. and Achiba, Y. *Appl. Phys. A* **74**, 349-354 (2002).
5. Jehoulet, C., Obeng, Y. O., Kim, Y. T., Zhou, F. and Bard, A. J. *J. Am. Chem. Soc.* **114**, 4237-4247 (1992).
6. Janda, P., Krieg, T. and Dunsch, L. *Adv. Mater.* **17**, 1434-1438 (1998).
7. Pichler, T., Kuzmany, H., Kataura, H. and Achiba, Y. *Phys. Rev. Lett.* **87**, 7401(2001).
8. Kazaoui, S., Minami, N., Jacquemin, R., Kataura, H. and Achiba, Y. *Phys. Rev. B* **60**, 13339-13342 (1999).
9. Jacquemin, R., Kazaoui, S., Yu, D., Hassanien, A., Minami, N., Kataura, H. and Achiba, Y. *Synth. Metals* **115**, 283-287 (2000).
10. Schettino, V., Pagliai, M. and Cardini, G. *J. Phys. Chem. A* **106**, 1815-1823 (2002).
11. Gallagher, S. H., Armstrong, R. S., Bolskar, R. D., Lay, P. A. and Reed, C. A. *J. Am. Chem. Soc.* **119**, 4263-4271 (2002).
12. Okada, S., Saito, S. and Oshiyama, S. *Phys. Rev. Lett.* **86**, 3835-3838 (2001).

One-dimensional System in Carbon Nanotubes

H. Kataura[1], Y. Maniwa[1], T. Kodama[1], K. Kikuchi[1], S. Suzuki[1],
Y. Achiba[1], K. Sugiura[1], S. Okubo[1,2], K. Tsukagoshi[2]

[1]*Graduate School of Science, Tokyo Metropolitan University, Tokyo 192-0397, Japan*
[2]*RIKEN, Wako, Saitama 351-0198, Japan*

Abstract. An inside space of the single-wall carbon nanotube (SWNT) can be used as a template for one-dimensional molecular crystals simply by filling the space with molecules. Here we show some results of the filling with fullerenes, organic molecules, and DNA. High-density filling was confirmed by X-ray diffraction analysis for the each sample. It was found that the relative intensity change in radial breathing modes in the Raman spectra from the SWNT, not from the molecules inside, can be used as a filling monitor similarly to the (10) peak intensity change in the X-ray diffraction.

INTRODUCTION

Single-wall carbon nanotubes (SWNTs) are regarded as an ideal one-dimensional system although they have cylinder like, two-dimensional structure. Indeed, valence band photoemission spectra of the SWNTs show beautiful oscillations due to one-dimensional van Hove singularities caused by quantization of wave vectors in the circumference direction. [1] On the other hand, an inside space of SWNT is also expected to be an ideal one-dimensional space where we can easily make one-dimensional molecular crystals simply by filling. Actually, we found that 1D-crystals of C_{70} inside SWNTs do not show orientational phase transition due to the one-dimensional fluctuations. [2] Furthermore, it was found that water molecules form ice-nanotubes inside SWNTs at low temperature. [3] We believe that the one-dimensional system inside SWNT is one of the most interesting system not only from the view point of physics but also the future applications. In this paper, we will show some results on filling nanotubes with many kinds of molecules such as fullerenes, metallo fullerenes, organic molecules and DNA.

EXPERIMENTAL

Diameter controlled single-wall carbon nanotubes (SWNTs) were synthesized by the laser ablation method using NiCo alloy catalyst.[4] Although an initial purity of the SWNTs prepared by the laser ablation is high, a further purification is required for the molecule filling. At first, soot containing SWNTs was washed in toluene to remove fullerenes. Amorphous carbons were removed by refluxing in 15% hydrogen peroxide water solution for three hours at 100 °C. After careful filtration of byproducts

CP685, *Molecular Nanostructures: XVII Int'l. Winterschool/Euroconference on*
Electronic Properties of Novel Materials, edited by H. Kuzmany, J. Fink, M. Mehring, and S. Roth
© 2003 American Institute of Physics 0-7354-0154-3/03/$20.00

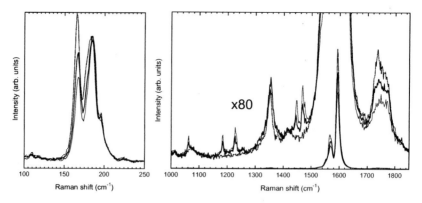

FIGURE 1. Raman spectra of SWNTs, empty, partially filled, and fully filled with C_{70} denoted by dotted, thick solid, and thin solid curves, respectively. The excitation is 488.0 nm.

produced in the purification process, SWNT papers were formed by peeling off the dried SWNTs film on a membrane filter. SWNT papers were heated in high-vacuum to vaporize metal particles. Then SWNT papers were kept at 420 °C in air for one hour to prepare sufficient number of entrance holes on the tips and probably on the wall of SWNTs.

Many kinds of molecules were filled in SWNTs from a vapor phase or from the solution phase. After the filling process, excess molecules outside SWNTs were removed by heating in high-vacuum or by washing in solvent. In general, the macroscopic filling factors were estimated by X-ray diffraction measurements. Where the (10) peak intensity from two-dimensional triangle lattice of SWNTs is very good indicator of molecule filling, since the (10) peak locates very close to the zero point of the structure factor of SWNT and hence the total structure factor is sensitively changed by the additional structure factor of filled molecules.

RESULTS AND DISCUSSION

Fullerene filling

Because of the large binding energy between fullerenes and SWNTs [5], fullerene filling in SWNTs is one of the easiest cases as was first demonstrated by Smith *et al.* [6] X-ray diffraction (XRD) analysis indicates that over 95% filling was achieved for C_{60} and 99% for C_{70}. Raman spectra indicate that all the Raman active modes of C_{60} can be observed at 4.2 K, but at room temperature Ag(2) mode intensity is decreased drastically by blue laser irradiation, which suggests a formation of the one-dimensional photopolymer inside SWNTs. In the case of C_{70}-peapods, we also observe all Raman active modes of C_{70}, but the relative intensity ratio between peaks is somewhat different from three-dimensional C_{70} crystal. In both cases, by a careful treatment, we can estimate the filling factor from the Raman intensity of peculiar modes of C_{60} and C_{70}. [7]

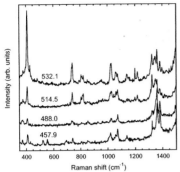 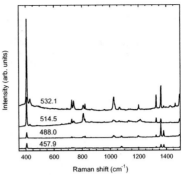

FIGURE 2. Resonance Raman spectra of Pt-porphyrin in SWNT (left figure) and thin film (right figure) . The excitation wavelengths are indicated in the figures in nm.

Figure 1 indicates Raman spectra of SWNTs that are empty, partially filled, and fully filled with C_{70}. Change in filling factor is clearly indicated by change in Raman intensity of the peaks originated from C_{70} molecules. It is clearly seen that radial breathing mode (RBM) peak A at 165 cm^{-1} becomes much weaker by filling C_{70} molecules while the intensity of peak B does not change. Synchronously, the peak intensity of high-frequency mode (G-band) has decreased. The similar tendency is also observed in C_{60}, C_{78}, Dy@C_{82}, La@C_{82}, Gd@C_{82}, and Ca@C_{82} filling.

Organic molecule filling

Filling organic molecules is also easy just like filling fullerenes, if the target molecules have appropriate sizes for inner space of SWNT. However, we have to pay attention to the temperature control, since in general, the organic molecules are thermally unstable compared with the fullerenes. It was confirmed by XRD analysis that many kinds of organic molecules can be encapsulated in SWNTs at high-density. Pt-porphyrin is one of the best cases among them, since the porphyrin is rather tough for high-temperature treatment and shows beautiful resonance Raman spectra instead of the intense photoluminescence that is dominant in the other organic molecules. Figure 2 indicates resonance Raman spectra of Pt-porphyrin encapsulated in SWNTs and of thin film. Observed resonance effect of filled Pt-porphyrin is slightly different from that of the film, which is suggesting that the electronic structure of Pt-porphyrin is slightly modified by the encapsulation. Some splits in Raman peaks suggest a distortion of the molecular structure by the encapsulation. The relative intensity change in RBM peaks by the filling which is observed in fullerene filling is also observed in Pt-porphyrin, Zn-diphenyl porphyrin, rhodamine-6G, and chlorophyll filling. Here, some organic molecules were filled from the solution phase.

DNA filling

DNA filling in SWNTs was done in the solution phase, since the DNA is fragile to the heat treatment and sonication. We used synthesized DNA with length of 20 MER. Although TEM photograph is unclear, XRD analysis indicates high-filling and also we observed RBM intensity change just like fullerene filling and organic molecule filling.

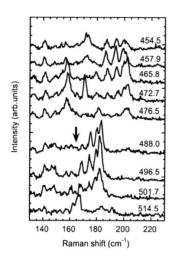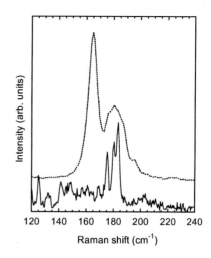

FIGURE 3. Raman spectra of SWNTs synthesized by laser ablation method using RhRu catalyst at 1150 °C. Left figure shows resonance RBM Raman spectra where excitation wavelengths are indicated in the figure. No peak is observed at 165 cm^{-1} for 488.0 nm excitation. In the Right figure, a dotted curve shows Raman spectra of high-purity SWNTs with thick bundles as a comparison for 488.0 nm excitation.

Raman spectra as filling monitor

As discussed above, all the filling systems show common RBM intensity change accompanying an intensity drop in G-band. To explain this phenomenon, we introduce Raman spectra of low purity SWNT sample synthesized by laser ablation method using RhRu catalyst as indicated in Fig.3. Since the synthesis condition is far from the optimum, the Raman spectra indicate very widely spread diameter distribution and low purity. Interestingly, very narrow peaks are observed with very sharp resonance feature. By considering the low purity and the wide diameter distribution, SWNTs are regarded to be well isolated in amorphous carbon. Surprisingly, there is no peak at 165 cm^{-1} that is typical RBM in the high-purity SWNTs when the excitation is 488.0 nm.

Here we want to discuss about the origin of the RBM peak at 165 cm^{-1}. A theoretical calculation indicates that $E_{i,i}$ transition is allowed when the light polarization is parallel to the tube axis while $E_{i,i+1}$ and $E_{i,i-1}$ transitions are allowed for the perpendicular polarization. However, in the case of an individual SWNT, only $E_{i,i}$ transition should be observable because a large depolarization effect suppresses allowed transitions for perpendicular polarized filed, where the depolarization effect is caused by the local field induced by applied filed. [8] On the other hand, purified SWNTs form very thick bundles, typically about 100 to 300 nm in diameter. In this case, local field correction for the SWNTs inside the bundle is modified by the polarizations of surrounding SWNTs. A rough estimation suggests that the local filed in the bundle is weaker than that of isolated SWNTs. Consequently, it leads to the additional absorption band in the bundled SWNTs. If the RBM peak at 165 cm^{-1} is enhanced by the optical transition perpendicular filed to the tube axis, it can not be observed in the isolated SWNTs but is observed in the bundles. In the case of filled SWNTs, the molecules inside SWNT bring additional electrons, and then increase the dielectric function of the system, which increases the local field. As a result, the RBM intensity at 165 cm^{-1} is decreased again by the depolarization effect.

SUMMARY

We have demonstrated high-rate molecule filling inside SWNTs for some kinds of fullerenes, organic molecules, and DNA. For all the cases, we have observed similar intensity change in the RBM Raman spectra by the molecule filling. We claim that this spectral change can be used as filling monitor just like (10) peak intensity change in XRD profile. The origin of the RBM intensity change is not fully understood but probably connected with the difference between the electronic structure of isolated SWNT and the bundle.

RBM at 165 cm^{-1} is not observed in the isolated SWNTs because of the large depolarization effect. After constructing thick bundles, the peak appears since the depolarization is weakened by the local field of surrounding SWNTs. Since molecule filling enhances the dielectric function of the system, the depolarization effect rises again, and then the RBM intensity is weakened. The increase of the dielectric function is confirmed by a small but apparent red shift of optical absorption peaks due to one-dimensional van Hove singularities. [9] We found some other RBM peaks which show similar characteristics on the filling for another excitation wavelength.

ACKNOWLEDGMENTS

Authors thank Y. Misaki and K. Hirahara for TEM observations. This work was supported partly by the Grant-in-Aid for Scientific Research (A) No. 13304026 from the Ministry of Education, Culture, Sports, Science and Technology of Japan, and partly by Industrial Technology Research Grant Program in '02 from New Energy and Industrial Technology Development Organization (NEDO) of Japan.

REFERENCES

1. H. Shiozawa *et al.*, "Valence-Band Photoemission Study of Single-Wall Carbon Nanotubes", in this issue.
2. Y. Maniwa *et al.*, J. Phys. Soc. Jpn. **72**, 45-48 (2003).
3. Y. Maniwa *et al.*, J. Phys. Soc. Jpn. **71**, 2863-2866 (2002).
4. H. Kataura *et al.* Carbon **38**, 1691-1697 (2000).
5. S Okada, S. Saito, and A. Oshiyama, Phys. Rev. Lett. **86**, 3835-3838 (2001).
6. B.W. Smith, M. Monthioux, D.E. Luzzi, Nature **396**, 323-324 (1998).
7. H. Kuzmany, *et al.*, Appl. Phys. A **76**, 449-455 (2003).
8. H. Ajiki and T. Ando, Solid State Commun. **102**, 135-142 (1997).
9. H. Kataura *et al.*, Appl. Phys. A. **74**, 349-354 (2002).

The direct imaging and observed packing behaviour of othro-carborane molecules within single walled carbon nanotubes

D.A. Morgan[1], J. Sloan[1,2] and M. L. H. Green[1]

[1]:*Wolfson Catalysis Centre (Carbon Nanotechnology Group), Inorganic Chemistry Laboratory, University of Oxford, South Parks Road, Oxford, OX1 3QR U.K.*
[2]*: Department of Materials, University of Oxford, Parks Road, Oxford, OX1 3PH U.K.*

Abstract. *Ortho*-carborane molecules have been inserted into single walled carbon nanotubes (SWNTs) and imaged directly by high resolution transmission electron microscopy (HRTEM). Direct imaging revealed that both discrete molecules and zig-zag 1D chains of *o*-carborane molecules were observed to pack into SWNT capillaries. Upon further e-beam irradiation, partial decomposition and rearrangement of clusters of *o*-carborane molecules was observed.

INTRODUCTION

In the same year that the first foreign material was encapsulated within single walled carbon nanotubes (SWNTs) [1], Smith *et al.* found that tubules prepared via pulsed laser vapourisation (PLV) contained close-packed 1D chains consisting of C_{60} and other fullerene molecules (i.e. C_n), termed 'peapods' or, more formally, C_n@SWNTs [2]. Subsequently, quantitative filling of SWNTs with specific fullerenes was achieved [3] followed by incorporation of endofullerenes, resulting in $[La_2@C_{80}]@SWNT^4$ and $[Gd@C_{82}]@SWNT$ [5] composites. Scanning tunnelling microscopy (STM) studies performed on C_{60}@SWNT [6] and $[Gd@C_{82}]@SWNT$ [7] reveal significant band-gap modulations, lending credibility to the argument that similar composites could form components in devices such as a solid-state quantum computer [8] or a quantum cascade laser [9]. We have now extended this filling chemistry to boron-containing molecular species and report here the imaging and packing properties of *o*-carborane molecules inserted into SWNTs.

Ortho-carborane (1,2-dicarbododecaborane) has the molecular formula $(CH)_2(BH)_{10}$ and consists of an icosahedral cluster of C_2B_{10} with hydrogen atoms attached at the vertices, in a structure analogous to borane or $(B_{12}H_{12})^{2-}$ [10]. In the bulk, this structure crystallises in a *fcc* lattice in which the lattice points are occupied by entire molecules as is the case for C_{60} [10,11]. At r.t., these molecules rotate rapidly, as shown by NMR, and may thus be regarded as spheres 0.81 nm in outside diameter (Fig. 1(a)) as compared to an analogous diameter of 0.10 nm for C_{60} molecules (Fig. 1(b)) [10]. HRTEM image simulation studies (Fig. 1(c) **I-IV**) show that the imaging properties of individual *o*-carborane molecules are fundamentally different to those of

CP685, *Molecular Nanostructures: XVII Int'l. Winterschool/Euroconference on Electronic Properties of Novel Materials,*edited by H. Kuzmany, J. Fink, M. Mehring, and S. Roth
© 2003 American Institute of Physics 0-7354-0154-3/03/$20.00

the fullerenes and image as blurred spots corresponding to 12 weakly scattering atoms (i.e. H, C and B) arranged in a tightly packed cluster (Fig. 1(c), top) rather than as dark circles (Fig. 1(c), bottom) corresponding to C atoms arranged in a 0.7 nm diameter sphere [12].

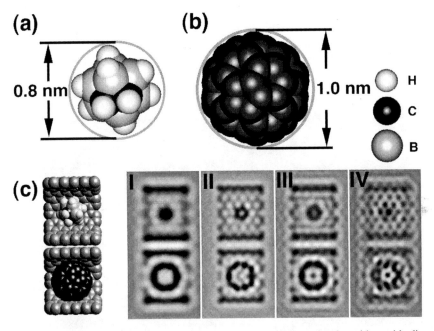

FIGURE 1. (a) and (b) space filling models of *o*-carborane and C_{60} molecules with outside diameters at r.t. indicated (c) Space filling models (left) of o-carborane (top) and C_{60} inside a (10,10) SWNT fragment and corresponding computed focal series (**I** –49 nm; **II** –39 nm; **III** –29 nm; **IV** –19 nm defocus).

EXPERIMENTAL

Samples of SWNTs were prepared by a modified catalytic arc synthesis route [12]. Selected sample were subjected to acid treatments (i.e. shaking in at r.t. 35% HCl followed by washing and oven drying) to assess the impact on filling yield. For each experiment, ~50 mg SWNTs together with ~100 mg of o-carborane were ground together and transferred to a quartz ampoule in a dry box. The ampoule was sealed under vacuum and then heated to 350°C in a tube furnace. For the later runs the SWNTs were heat treated first at a temperature of 420°C under a dynamic flow of 20% O_2/80% Ar for 20 minutes. The products were examined in a 300kV JEOL JEM-3000F field emission gun HRTEM (C_s = 0.6 mm). Images were acquired on a Gatan model with Si [110] lattice spacings. Electron energy loss (EEL) 794 1k CCD camera for which the magnification was calibrated spectra were recorded with a Gatan image filter equipped with a 2k 794IF/20 MegaScan CCD located below the 3000F column.

Image simulations were performed with a standard multi-slice algorithm employing representative parameters for the 3000F microscope.

RESULTS AND DISCUSSION

The presence of *o*-carborane within SWNT bundles and individual SWNTs was confirmed by complimentary EELS studies, and HRTEM imaging and simulation (Fig. 2). It proved difficult to obtain reliable EELS spectra from individual filled SWNTs as these damaged rapidly when exposed to a focused 1 nm electron probe. Spectra were instead recorded from SWNT bundles with a spread beam (Fig. 2(a)) and clearly demonstrated the presence of B in the composite material (Fig. 2(b)). HRTEM images indicated the presence of discrete molecules (Fig. 2(c) and (d)) and short chains of *o*-carborane within SWNTs (Fig. 2(g) and (h)). Where single molecules were observed, it was reasonable to conclude that these are located inside the SWNTs, as indicated in Fig. 2(e), as o-carborane material located on the outside of the SWNTs rapidly vaporized in the electron beam whereas discrete molecules remain 'fixed' inside the tubule upon continued exposure to the electron beam for 1-2 min. The observed imaging behaviour of these molecules was also consistent with computer image simulations (Figs.1(c) and 2(f)). In the case of short chains formed inside moderately wide (i.e. 1.4-1.6 nm diameter) SWNTs (Figs. 2(g) and (h)), these invariably formed 'zig-zag' chains (i.e. rather than linear chains as reported for C_n@SWNTs [2]) as represented schematically in Fig. 2(i). We assume that this packing behaviour results from clusters of molecules being compressed into these staggered arrangements by the van der Waals surface of the encapsulating SWNT.

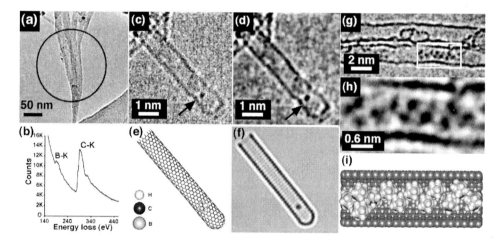

FIGURE 2 (a) Low magnification HRTEM image of a bundle of SWNTs treated with *o*-carborane. (b) EELS spectrum obtained from indicated region in (a). (c) and (d) HRTEM image and noise-filtered image of a discrete *o*-carborane molecule within the tip of a capped 1.2 nm diameter SWNT. (e) and (f) structure model and Scherzer focus simulation a single *o*-carborane molecule within a SWNT tip. (g) and (h) HRTEM image and detail obtained from a short chain of *o*-carborane molecules formed within a ~1.6 nm diameter SWNT. (i) Schematic representation of composite in (h).

The o-carborane molecules within the 'zig-zag' chains apparently packed somewhat closer (i.e. 0.5-0.6 nm) than the 'nearest neighbour' distance of ~0.7 nm predicted from the bulk structure model [10]. Additionally, this packing behaviour differs markedly from that typically observed for fullerene and endofullerene molecules encapsulated within similar diameter SWNTs which typically form linear chains in which individual molecules are separated by slightly less than the predicted van der Waals separation of ~0.34 nm due to van der Waals compression by the SWNT [13]. The shortening of the intermolecular distances can be attributed partially to the packing behaviour as the molecules will stagger in projection with the result that the observed separation will seem less in a two-dimensional image. Additionally, van der Waals compression effects may also contribute to a shortening of the observed spacing by analogy with ref. 13. The observation of 'zig-zag' rather than linear chains can be directly attributed to the smaller size of the o-carborane molecules compared to C_{60} (Fig. 1(a) and (b)). The latter fit relatively snugly inside narrower (i.e 1.2 and 1.6 nm SWNTs) whereas the smaller o-carborane molecules can more readily compress into staggered arrays (i.e. Figs. 2(h) and (i)).

Estimated filling yields for these experiments were quite low, varying from 5-20%, and were somewhat sensitive to the processing of the SWNT sample used. The highest yields came from using acid treated SWNTs although SWNT samples pre-treated, by heating in 20% O_2 at 420ºC for 20 minutes, showed no improvements to filling yields, contrary to reports that this drastically increases the filling yields of fullerene encapsulation for acid treated PLV SWNTs [3]. This could possibly be due to the sensitivity of purification and opening treatments to the type of SWNT material, as the samples were prepared using arc-synthesised SWNTs.

Upon extended beam irradiation in the HRTEM (approximately 30,000 300kV electrons/second/molecule), clusters of encapsulated o-carborane molecules under went apparent dissociation followed by fusion, forming clusters that resemble enlarged fullerenes or higher spheroidal or ovoid carborane molecules (Figs. 3(a) and (b)). This behaviour resembles similar phenomena reported for fullerene molecules formed within SWNTs [2,3,12]. If correct, then this process presumably involves radiolytic fission of the carborane spheres to form fragments, followed by coalescence to form the enlarged carboranes. The latter structures would be stabilised and partially templated by the encapsulating SWNT.

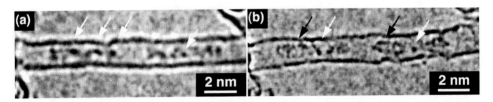

FIGURE 3 (a) HRTEM image obtained from a discrete SWNT containing several o-carborane clusters, some of which are indicated by white arrows. (b) HRTEM image showing the same SWNT after beam irradiation (at *ca.* 30,000 300kV electrons/molecule/sec) for 5 min. Some carborane molecules remain discrete (white arrows) whereas others have apparently coalesced to form enlarged ovoid structures (dark arrows).

It has been shown that successful encapsulation of o-carborane into SWNTs has been achieved, using a sublimation technique, thus performing the first introduction of a boron containing species into carbon nanotubes and the only 'spherical' molecules, other than fullerenes, to be observed packing in them. This encapsulation is confirmed both by HRTEM and EELS observations. There has been one report of the encapsulation of boron carbide with multi-walled graphitic layers [14], but there have been no reported cases of filling carbon SWNTs with boron containing species. The resulting composites represent the first example of non-fullerenic spheroidal molecules forming 'peapod' (where the 'peas' are effectively 'petit-pois' due to their smaller size compared to conventional fullerenes) -type structures.

ACKNOWLEDGEMENTS

We are grateful to Dr. Andrew Hughes of the University of Durham for supplying a sample of o-carborane. We also acknowledge financial support from the Leverhulme Trust, the Petroleum Research Fund, administered by the ACS (Grant No. 33765-AC5) and the EPSRC (Grant Nos. GR/L59238 and GR/L22324). J.S. is indebted to the Royal Society for a University Research Fellowship.

REFERENCES

1. Sloan, J., Hammer, J. Zweifka-Sibley, M., and Green, M.L.H., *J. Chem. Soc., Chem. Commun.*, 1998, 347-348 (1998).
2. Smith, B.W., Monthioux, M., and Luzzi, D.E., *Nature*, **396**, 323-324 (1998).
3. Smith, B.W., and Luzzi, D.E., *Chem. Phys. Lett.*, **321**, 169-174 (2000).
4. Smith, B.W., Luzzi, D.E., and Achiba, Y., *Chem. Phys. Lett.*, **331**, 137-142 (2000).
5. Suenaga, K., Tence, M., Mory, C., Colliex, C., Kato, H., Okazaki, T., Shinohara, H., Hirahara, K., Bandow S., and Iijima, S., *Science*, **290**, 2280-2282 (2000).
6. Hornbaker, D.J., Kahng, S.-J., Misra, S., Smith, B.W., Johnson, A.T., Mele, E.J., Luzzi, D.E. and Yazdani, A., *Science*, **295**, 828-831 (2002).
7. Lee, J., Kim, H., Kahng, S.-J., Kim, G., Son, Y.-W., Ihm, J., Kato, H., Wang, Z.W., Okazaki, T., Shinohara H., and Kuk, Y., *Nature*, **415**, 1005-1008 (2002).
8. Toth, G., and Lent, C.S., *Phys. Rev. A*, **63**, 0521315 (2001).
9. Faist, J., Capasso, F., Sirtori, C., Sivco, D.L., Hutchinson, A.L., Cho, A.Y., *Appl. Phys. Lett.*, **67**, 3057-3062 (1995).
10. Baughman, R.H., *J. Chem. Phys.*, **53**, 3781-3789 (1970).
11. David, W.I.F., Ibberson, R.M., Matthewman, J.C., Prassides, K., Dennis, T.J.S., Hare, J.P., Kroto, H.W., Taylor R., and Walton, D.R.M., *Nature*, **353**, 147-149 (1991).
12. Sloan, J., Dunin-Borkowski, R.E., Hutchison, J.L., Coleman, K.S., Williams, V.C., Claridge, J.B., York, A.P.E., Xu, C., Bailey, S.R., Brown, G., Friedrichs S., and Green, M.L.H., *Chem. Phys. Lett.*, **316**, 191-198 (2000).
13. Hirahara, K. Bandow, S., Suenaga, K., Kato, H., Okazaki, T., Shinohara, H., and Iijima, S. *Phys. Rev. B*, **64**, 115420-1 (2001).
14. D. Zhou, S. Seraphin, and J.C. Withers, *Chem. Phys. Lett.* **234**, 233-239 (1995).

NON-CARBONACEOUS NANOTUBES

Electronic structure and optical properties of boron doped single-wall carbon nanotubes

T. Pichler[1], E. Borowiak-Palen[1], G.G. Fuentes[1], M. Knupfer[1], A.Graff[1], J. Fink[1], L. Wirtz[2], A. Rubio[2]

[1]Leibniz Institute for Solid State and Materials Research Dresden, P.O. Box 270016 D-01171 Dresden, Germany.
[2]Department of Material Physics, University of the Basque Country, centro Mixto CSIC-UPV, and Donostia International Physics Center, Po. Manuel de Lardizabal 4, 20018 Donostia-San Sebastián, Spain.

Abstract: We present a study of the electronic structure and the optical properties of boron doped single walled carbon nanotubes which have been produced by a substitution reaction from nanotube templates. The morphology and crystal structure of the samples have been characterized by transmission electron microscopy and electron energy-loss spectroscopy. Clean boron doped SWCNT with an average boron content of 15 at% have been produced. The B1s and C1s core level spectra reveal that boron is in an sp^2 configuration and that the effective charge transfer is about 0.5 holes per boron to the C-derived states. The boron substitution also leads to new features in the optical absorption spectra which can be attributed to the appearance of an acceptor band about 0.1 eV above the top of the valence band of the SWCNT. These changes in the electronic structure and in the optical properties upon boron substitution are in good agreement with state of the art *ab initio* calculations.

INTRODUCTION

The control of the structural and electronic properties of single wall carbon nanotubes (SWCNT) still represents a big challenge in the nanotube research field since bulk samples usually contain metallic and semiconducting tubes with a finite diameter and chirality distribution. A controlled modification of the electronic properties of SWCNT by substitution of carbon by heteroatoms represents one of the most promising routes to achieve a uniform sample. For instance B doping of SWCNT drives all SWCNT metallic SWCNT, whereas only semiconducting SWCNT are observed for BN substitution or for the formation of BN nanotubes [1,2]. Previously, the substitution of carbon by boron and nitrogen was investigated in detail [3]. We adapted this method to optimize the preparation regarding high purity multiwall BN nanotubes and highly B-doped SWCNT (B content up to 20 at%) [4]. So far, only a few experimental studies have reported on the chemical and structural properties mainly by transmission electron microscopy (TEM) and local electron energy-loss spectroscopy (EELS) in a TEM [3] and on the electronic structure of B-doped MWCNT by scanning tunneling spectroscopy (STS) on individual tubes [5]. In this contribution we report first results on the electronic properties of highly B-doped SWCNT using bulk sensitive optical spectroscopy and EELS as probes. The optical response is then compared with the results from *ab initio* calculations within the local density approximation (LDA)

EXPERIMENTAL

For the preparation of the B-doped SWCNT, raw material produced by the standard laser ablation technique, which has been optimized for high yield (70wt% SWCNT with a mean diameter of 1.23 nm) [6] was used as a template and an adapted substitution reaction with ammonia as carrier gas and at a reaction temperature of

1150°C as described in detail elsewhere was applied [4]. For the characterization by TEM, EELS and optical absorption a well sonicated suspension of the sample in acetone was produced. Then a homogenous thin film was prepared by dropping onto a KBr single crystal. The optical absorption spectra were conducted in a BRUKER 113V/88 spectrometer in the spectral range from the mid infrared to the ultraviolet. The samples for TEM and EELS were prepared directly from the film on KBr, which was floated off in distilled water and mounted onto standard microscopy grids. The EELS measurements were carried out in a purpose built spectrometer with an energy and momentum resolution set to 330 meV and 0.1 Å^{-1}, respectively [7].

RESULTS AND DISCUSSION

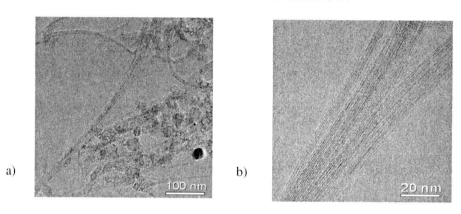

a)

b)

FIGURE 1. Typical TEM micrographs of B-substituted SWCNT with 10-20 at% substitution level: (a) overview and (b) extended scale.

Figure 1 shows a typical TEM micrograph of the prepared B-doped SWCNT sample. Typical bundles of SWCNT as in the raw material are observed. Local TEM-EELS measurements across a SWCNT bundle revealed a B content of 10-20 wt% with no reminiscent traces of nitrogen in the sample within the experimental resolution [4]. Besides such a local probe, bulk sensitive EELS measurements with high energy and momentum resolution were conducted. The results of EELS core level excitation spectra of B-doped SWCNT are depicted in Fig. 2. Hence the average B content on a macroscopic scale levels up to 15 wt% is up to 15 wt% and again with no traces of nitrogen within the experimental limit (0.5at%). Additional information about the charge transfer in the doped SWCNT can be extracted from changes in the fine structure of the excitation edges (see inset of Fig. 2). A comparison of the C1s edges of pristine SWCNT and the doped compound provides evidence for the appearance of additional unoccupied states of carbon character in the electronic structure of the B doped SWCNT. At the C1s threshold of the doped tubes a clear low energy shoulder appears which is roughly 0.7 eV below the corresponding threshold of the pure SWCNT. The additional area under the shoulder is about 6% of the total area, suggesting a hole charge transfer to carbon derived states of about 0.5 holes per boron atom, which is slightly lower than the estimation by Mele et al. for B doped graphite (~0.8 holes per B) [9].

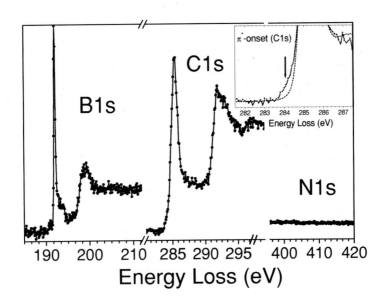

FIGURE 2. High resolution EELS core level excitation spectra of B-doped SWCNT in the range of the B1s, C1s and N1s edges. The inset shows the onset of the C1s excitation edge in SWCNT (dotted line) and B doped SWCNT (solid line) in an extended range. The additional doping induced shoulder is indicated by the arrow.

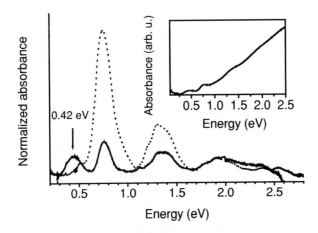

FIGURE 3 Optical absorption spectra of B-doped SWCNT (solid line) and pristine SWCNT (dotted line) after background subtraction. The inset shows the absorption spectra of the B-doped SWCNT before subtraction. The arrow points out the new doping induced absorption feature.

Additional information about the evolution of the electronic properties upon doping can be extracted from optical absorption measurements. In the left panel of Fig. 3 the optical absorption spectra of the SWCNT raw material and the B doped SWCNT are shown in the range of the first two allowed transitions of the semiconducting SWCNT and the first allowed transition of the metallic SWCNT. The general shape of the absorption peaks remains unchanged upon doping. The intensity of the absorption peaks of the semiconducting SWCNT is strongly bleached, whereas the peak of the metallic tubes is hardly affected by the doping. Interestingly, a new absorption feature appears in the B doped SWCNT at about 0.4 eV, as indicated by the arrow.

In order to get more insight into the electronic structure of the semiconducting B-doped SWCNT ab initio band structure calculations have been performed. The *ab initio* calculations were conducted with the code ABINIT [10] using density functional theory in the LDA (for further details see Ref. [8] in this volume).

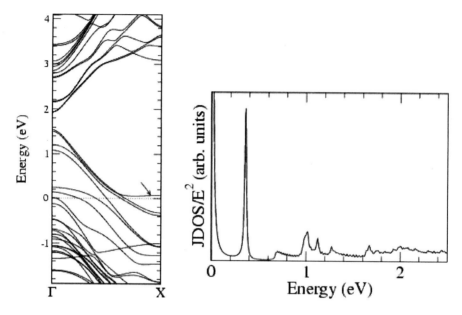

FIGURE 4 Left panel: Band structure of a semiconducting (14,0) SWCNT with 12.5% B substitution. The dotted line denotes the Fermi level. The arrow indicates the boron-like acceptor band around the X-point. Right panel: Joined DOS of the (14,0) tube shown in the left panel.

Fig. 4 shows the example of a (14,0) tube with 12.5 % B content which has about the same mean diameter as our SWCNT (more examples using different substitution levels and different coordination of the B atoms are given in Ref [8] in this volume). In addition to a doping induced shift of the Fermi level there is a formation of additional bands which have strong B character. For this calculation, one of these bands has is flat around the X point about 0.1 eV above the Fermi level. Two other bands with strong B character have a flat region at the X point about 0.3 eV below the Fermi level. This gives rise to a strong peak at about 0.4 eV in the joined density of states (DOS). Hence, the transitions between these bands, which can be at different position of the Brillouin zone for different substitution and ordering [8], are in good agreement with the observed new optical absorption peak at 0.4 eV. The formation of a flat band just above the Fermi level is analog to the formation of an acceptor level in doped semiconductors. However, there are distinct differences since the dopant concentration in our case is orders of magnitude higher than in classical semiconductor physics. Due our high doping level the boron levels hybridize with the carbon levels and form strongly dispersive "acceptor bands". This effective charge transfer also means a finite delocalization of charge around the B site. The band structure of the valence band is strongly distorted upon doping, because the levels of carbon atoms that are substituted by boron move upwards in energy. These results show that in the case of highly B doped SWCNT a simple rigid band model as used in semiconductor physics and as has been applied previously to doped SWCNT is not sufficient to explain the changes in the electronic properties and additional effects like charge localization and doping induced band structure changes play an important role.

To summarize, we have shown that in 15 wt% B doped SWCNT the effective charge transfer from the C derived valence bands is about 0.5 holes per substituted B atom which leads to a shift of the Fermi level of about 0.7 eV. Optical absorption indicated the formation of an acceptor band in the band structure of the semiconducting tubes which is nicely confirmed by *ab initio* calculations of the electronic structure of SWCNT with 12.5% B content.

Acknowledgements: We thank the DFG (PI 440) and the EU (SATURN, COMELCAN) for funding. The authors are also grateful to K. Müller for the technical assistance and O. Jost for delivering of as-produced SWCNT raw material.

REFERENCES

[1] N.G. Chopra et al., *Science*, 269, 966 (1995).
[2] A. Loiseau et al., Phys. Rev. Lett, 76, 4737 (1996).
[3] W. Han et al., *Appl. Phys. Lett.*, 73, 3085 (1998) and *Chem. Phys. Lett.*, 299, 368 (1999) ; D. Golberg et al., *Chem. Phys. Lett.*, 323, 185 (2000) and *Carbon*, 38, 2017 (2000).
[4] E. Borowiak-Palen et al., submitted and in this volume.
[5] D.L. Carroll et al., *Phys. Rev. Lett.* 81, 2332 (1998).
[6] O.Jost et al., *Chem. Phys. Lett.* 339, 297 (2001).
[7] J. Fink, *Adv. Electron. Electron Phys.* 75, 121 (1989) and references therein.
[8] L. Wirtz et al. in this volume.
[9] E.J. Mele, J.J. Ritsko, *Phys. Rev. B* 24, 1000 (1981).
[10] The ABINIT code is a common project of the Universite Chatolique de Louvain, Corning incorporated and other contributors (see also www.abinit.com).

Synthesis, structural and chemical analysis, and electrical property measurements of compound nanotubes in the B-C-N system

Dmitri Golberg, Yoshio Bando, Pavel S. Dorozhkin, Zhen-Chao Dong, Cheng-Chun Tang, Masanori Mitome, and Keiji Kurashima

Advanced Materials and Nanomaterials Laboratory, National Institute for Materials Science, Tsukuba, Ibaraki 305-0044, Japan

Abstract. Homogeneous or heterogeneous multi-walled nanotubes composed of B, C and N atoms were prepared through high-temperature chemical reactions (1700-2000 K) between C (or CN_x) nanotubes, boron oxide and nitrogen. Morphology, helicity, chemical composition and atomic species distribution in nanotubes were studied using field emission 300 kV transmission electron microscopes (TEM) JEOL-3000F equipped with a Gatan 766 Electron energy loss detector and JEOL-3100FEF equipped with an Omega Filter. Transport and field emission properties of nanotubes were studied using a low energy electron point source microscope.

INTRODUCTION

Performance of nanotubes (NTs) in downsized electrical circuits and devices is on the forefront of their practical applications [1]. To date, the research on electrical properties of nanotubular structures has been mainly focused on pure carbon NTs. However NTs of various chemical compositions form in the complex B-C-N system, including N-doped and/or B-doped C NTs, ternary B-C-N and pure BN NTs [2,3]. The synthesis parameters for these compound NTs have not yet been well established leading to limited experimental data. The compound NTs have a whole range of advantages over pure C counterparts. For instance, the environmental stability of a NT increases with increase in BN fraction within C shells, e.g. BN NTs were found to be more chemically and thermally stable than C NTs [4].

The electrical response of pure C NTs may vary from metallic to semiconducting, being a complex function of NT helicity, morphology, number of layers, diameter and topological defects [5]. To date, none of these parameters has been precisely controlled during the syntheses, making performance of pure C NTs in a given electronic device unpredictable. By contrast, B-C-N NTs are suggested to be reliable semiconductors with the band gap being a function of the relative C and BN fractions.

CP685, *Molecular Nanostructures: XVII Int'l. Winterschool/Euroconference on Electronic Properties of Novel Materials*, edited by H. Kuzmany, J. Fink, M. Mehring, and S. Roth

Electrical response of NTs [*I-V* characteristics, field emission, (FE)] is of prime importance in nanotechnology. The unique geometry of nanotubes suggests that they are ideal FE materials for flat panel displays. In fact, such displays based on C nanotubes have already been realised. However, the prospects of B-C-N NTs usage as parts of conducting or FE devices remain poorly established.

In this paper the production, structural and chemical characterisation, and electrical response of well-structured compound B-C-N multi-walled NTs is reported. The electrical properties of individual ropes consisting of dozens of NTs were measured.

EXPERIMENTAL

B-C-N NTs were prepared from CVD-grown pure C or CN_x ($x<0.1$) NTs as described in our recent papers [6-10].

The resultant nanotube powders were imaged using a field emission JEOL-3000F high-resolution transmission electron microscope operated at 300 kV and equipped with a "Gatan 766" electron energy loss spectrometer (EELS) 2D-DigiPEELS, and a high-resolution energy-filtered field emission transmission electron microscope (HRTEM, JEOL-3100F; Omega filter) with the estimated spatial resolution on the energy-filtered images of ~0.5 nm at the optimal conditions. The B, C and N elemental map acquisitions during energy-filtered TEM were performed using a conventional 3 window procedure. The slit width of the Omega filter was adjusted to 20 eV. Exposure times during elemental mapping were set at 8 for B, 10 s for C and 12 s for N map collections.

Electrical testing was performed in a low energy electron point source (LEEPS) microscope [8-10]. Prior to the measurements, a tungsten tip of the microscope was cleaned in vacuum by electron bombardment to obtain an atomically clean surface as confirmed by taking FE patterns.

RESULTS AND DISCUSSION

All the experimental facts are summarized as follows: (i) resultant compound B-C-N NTs frequently assemble in bundles several µm in length and ~50-300 nm in diameter; (ii) a BN/C atomic ratio of NTs increased with increase in synthesis temperature; (iii) island-like crystallisation of BN-rich fragments accompanied with BN-rich/C-rich domain separation across and along of NTs, or growth of homogeneously structured B-C-N tubular layers both took place depending on growth conditions (temperature, vapour pressure, usage of specific promoters and first of all - the type of the starting material); (iv) transport properties of NTs were widely ranged from metals through semiconductors to insulators depending on relative BN/C fractions and spatial distribution of BN-rich and C-rich tubular layers; (v) thermally and chemically stable BN-rich NT ropes displayed excellent electron FE properties with an emission current *per* rope easily reaching several µA.

Figure 1 depicts TEM data taken on a NT bundle viewed edge-on, that reveals BN-rich fragments on the external tube surface and C-rich domains on its internal surface, as additionally confirmed by the B, N and C composition profiles across the line shown. The former two fully correlating profiles have peak separation of ~10 nm, whereas the latter one exhibits narrower separation of ~8 nm. Present open-ended NTs may be suggested to be insulating nanocables as implied by natural conducting/semiconducting behaviour of C-rich B-C-N NTs and insulating properties of BN-rich NTs [6-10]. In fact, such tubes revealed a giant resistance of more than ~1 GΩ when their periphery is touched by the LEEPS tungsten tip, but they displayed perfect FE and thus good conductance of the internal layers, Fig. 1b.

In order to make sure that the measured high-resistance of the internal layers is not solely due to a contact resistance we additionally prepared nanorods of pure BN (no use of C NT templates) of nearly the same external diameter as BN-insulated B-C-N NT bundles. The resistance was ~5 GΩ or more, thus displaying properties very similar to those of external layers of BN-rich insulated NT cables and bulk hexagonal BN. Therefore the prepared B-C-N NTs having the outermost BN-rich domains and the innermost C-rich layers may serve as parts of complex NT networks with electrically-insulated fragments or as nanoscale field-effect transistors with built-in electrical insulation.

By contrast, homogeneous NTs made of uniformly distributed B, C and N atoms, Figure 2, exhibit stable semiconducting properties and high currents during FE (albeit FE is rather unstable). We also envisage a significantly enhanced thermal and chemical stability of such emitters due to high B-N fraction in them.

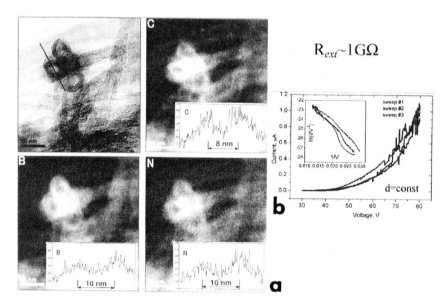

FIGURE 1. (a) Zero-loss TEM image and elemental maps of B-C-N NT bundle displaying BN-rich and C-rich domain separation across individual tubes (viewed edge-on) as confirmed by cross-section elemental profiles shown in the insets. The bundle revealed high-resistant external BN-rich layers (R~1GΩ) and excellent FE (and thus conductance of the internal layers) as confirmed in (b).

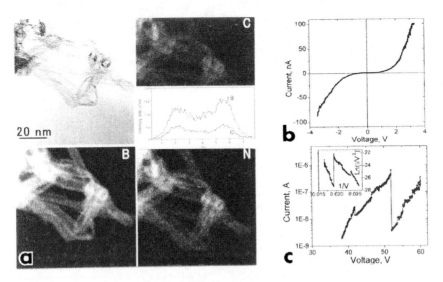

FIGURE 2. (a) Zero-loss TEM image and elemental maps of B-C-N NT bundle displaying homogeneous B-C-N tubular shells, as confirmed by cross-section elemental profiles taken on an representative individual 4-layered NT (inset). The bundle displayed semiconducting *I-V* curve in (b) and high, albeit not stable, emission current in (c).

The presently synthesized heterogeneous and homogeneous B-C-N NTs reflect a decent route to tuning NT electrical properties *via* chemical modification, rather than through a complicated helicity control yet achieved by any research group.

The authors are grateful to M. Terrones, N. Grobert, H. Terrones and M. Reyes-Reyes for supply of CN_x starting material used for B-C-N NT syntheses.

REFERENCES

1. Baughman R.H., Zakhidov A.A., and de Heer W.A., *Science* **297**, 788-790 (2002).
2. Stephan O., Ajayan P.M., Colliex C., Redlich P., Lambert J.M., Bernier P., and Lefin P., *Science* **266**, 1683-1685 (1994).
3. Chopra N.G., Luyken P.J., Cherrey K., Crespi V.H., Cohen M.L., Louie S.G., and Zettl A., *Science* **269**, 966-968 (1995).
4. Golberg D., Bando Y., Kurashima K., and Sato T., *Scripta Mater.* **44**, 1561-1565 (2001).
5. Wildoer J.W.G., Venema L.C., Rinzler A.G., Smalley R.E., and Dekker C., *Nature* **391**, 59-61 (1998).
6. Golberg D., Dorozhkin P., Bando Y., Hasegawa M., and Dong Z.-C., *Chem. Phys. Lett.* **359**, 220-227 (2002).
7. Golberg D., Bando Y., Mitome M., Kurashima K., Grobert N., Reyes-Reyes M., Terrones H., and Terrones M., *Chem. Phys. Lett.* **360**, 1-7 (2002).
8. Golberg D., Dorozhkin P., Bando Y., Dong Z.-C., Tang C.C., Grobert N., Reyes-Reyes M., Terrones, and M. Terrones, *Appl. Phys. A: Mater. Sci.& Proc.* **76**, 499-508 (2002).
9. Dorozhkin P., Golberg D., Bando Y., and Z.-C. Dong, *Appl. Phys. Lett.* **81**, 1083-1085 (2002).
10. Golberg D., Dorozhkin P., Bando Y., Dong Z.-C., Grobert N., Reyes-Reyes M., Terrones H., and Terrones M., *Appl. Phys. Lett.* **82**, 1275-1277 (2003)

Temperature Activated BN Substitution Of SWCNT

F. Hasi, F. Simon, M. Hulman, H. Kuzmany

Institut für Materialphysik, Universität Wien, Strudlhofgasse 4, A-1090 Wien, Austria

Abstract. A promising recipe for the production of single wall boron nitride nanotubes (BNT) is the so-called substitution reaction. We used SWCNT (HiPCo) as a starting material, B_2O_3 and N_2 gas as boron and nitrogen sources respectively. An important complication arises due to the presence of the reacted variants in the final product. We performed optical and multi frequency Raman measurements to identify the phonon modes of BNSWNT and unwanted side-products.

INTRODUCTION

Boron nitride nanotubes, where electronic properties are independent of helicity [1] have been synthesized using almost the same methods as for the production of carbon nanotubes. Arc discharge [2, 3], high preassure laser heating [4] and oven-laser ablation result in low yields with respect to the amount and purity of the final product. Later on, a new method was presented the so-called "synthesis of boron nitride nanotubes (BNT) by a substitution reaction" [5]. Using carbon nanotubes (CNT) as a template for this chemical reaction, one could produce considerable amounts of BN-nanotubes as well as mixed systems like BC or BCN nanotubes [6]. In this work, we tried to investigate the conditions in which the reaction takes place and to identify the unwanted side-products which play a significant role in the optical response of the final material.

EXPERIMENTAL

The substitution of carbon atoms in SWCNTs by B and N atoms can be represented formally in the following way:

$$B_2O_3 + 3C \text{ (nanotubes)} + N_2 \rightarrow 2BN \text{ (nanotubes)} + 3CO$$

In the above reaction B_2O_3 vapor generatet from molten B_2O_3 reacts with CNTs and N_2 gas. Pure single wall carbon nanotubes (SWCNT) produced by Carbon Nanotechnologies, Inc (Houston, USA) by the HiPCo process were heated gradually together with B_2O_3 in a flowing nitrogen atmosphere at 1150-1270 °C, held at this temperature for 2-4 hours and then cooled down to room temperature over 2 h. The heating was carried out in a horizontal tube furnace, in a quartz tube and the B_2O_3 powder was placed on an Al_2O_3 crucible covered with SWCNTs. The N_2 gas flows continuously at ambient preasure. After the heat treatment the reaction product was

CP685, *Molecular Nanostructures: XVII Int'l. Winterschool/Euroconference on Electronic Properties of Novel Materials,* edited by H. Kuzmany, J. Fink, M. Mehring, and S. Roth
© 2003 American Institute of Physics 0-7354-0154-3/03/$20.00

extracted from the crucible and samples in the form of pellets were prepared with KBr powder for IR and multi-frequency Raman spectroscopy.

IR, FT-Raman and visible Raman spectra of the final product were recorded in an BRUKER IFS-66V spectrometer, FT 106 ($\lambda = 1064$, resolution = 4 cm^{-1}) spectrometer and Dilor xy triple monochromator ($\lambda = 514.5$ nm, resolution = 1.5 cm^{-1}) with a liquid cooled CCD detector, respectively.

RESULTS AND DISCUSSION

Substitution reaction below 1200 °C: IR spectra (Fig. 1b,c) taken from samples from the substitution reaction at 1150 °C show differences from initial SWCNT (which have no IR activity). The modes at around 800 cm^{-1} are also shifted by 10-15 cm^{-1} compared to the hexagonal BN, h-BN (Fig. 1a), which is a stable form of boron nitride and a side product of the reaction. Similarly for the high frequency part of the spectra the response from the reacted material is shifted by 15 cm^{-1} as compared to h-BN. The question rises whether this IR activity comes from boron substituted SWCNT or it is due to nano-crystalline BN phase. Nano-crystalline BN may have different IR response from h-BN.

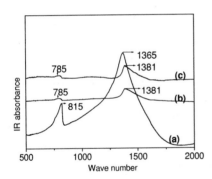

Figure 1. IR spectra of **(a)** h-BN, **(b)** substituted SWCNT at 1150 °C, 2h, **(c)** substituted SWCNT at 1150 °C, 4h

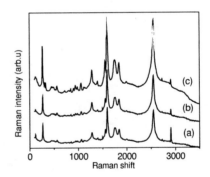

Figure 2. Raman spectra of **(a)** substituted SWCNT at 1150°C, 4h. **(b)** substituted SWCNT at 1150 °C, 2h. **(c)** pure SWCNT (HiPCo)

Fig. 2a, b show FT-Raman spectra of substituted SWCNTs at 1150 °C in 4 and 2 hours, respectively. The low resolution, 4 cm^{-1}, of the experiment did not allow the detection of any differencies in the Raman shift between untreated and substituted SWCNTs sample. Fig. 2c. shows untreated SWCNTs.

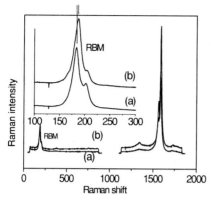

Figure 3. Raman spectra of **(a)** B-substituted SWCNT at 1150°C, 2h. **(b)** pure SWCNT (HiPCo)

The better resolution, 1.5 cm^{-1}, of the multichannel Raman apparatus allowed the detection of differencies in Raman shift Fig. 3 shows the Raman shift at $\lambda = 514.5$ nm for the treated (Fig. 3a.) and non treated (Fig. 3b.) SWCNTs. We observe a change in the Raman shift of the radial breathing modes, RBM, in the order of 3 wave numbers An explanation for that can be found by the breaking of the C≡C symetry from boron substitution.

Substitution above 1200 °C: When the substitution reaction is done above 1200 °C the resulting material has a rich IR activity (Fig. 4a.). Modes detected for materials with this reaction temperature can not be immediately attributed to BN nanotubes. This is evidenced by a control experiment where the same IR active modes were detected when no SWCNT material was used: $B_2O_3 + N_2$, at 1270 °C for 4 hours (Fig. 4b.). The material (b) was therefore investigated by Transmission Electron Microscopy (TEM) and no tubes were detected.

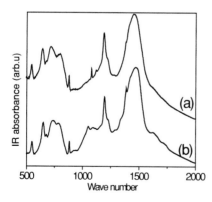

Figure 4. IR spectra of **(a)** the resulting material: SWCNT at 1270 °C, 4h. **(b)** heat treated B_2O_3 at 1270 °C / 4h in N_2

CONCLUSIONS

We have shown that the substitution process is rather complex and the final material is contaminated from side products which have several IR active modes. Below 1150 °C no reaction takes place between the starting materials. Between 1150 °C and 1270 °C reaction temperature, a weak shift in the heat treated tubes indicates some substitution reaction. A better understanding of the process to determine the right temperature in which the reactions take place and trying other sources of boron as well as other carrier gases is the ongoing work. Raman characterization of the substitution level seems to be more useful than infrared as the final material has a complex IR response that strongly depends on the reaction temperature.

ACKNOWLEDGMENTS

F. Hasi acknowledges a support from the FWF project No.14386 TPH. The autors are grateful to Thomas Pichler for providing the TEM measurements.

REFERENCES

1. X. Blase, A. De Vita, J.C. Charlier, and R. Car, *Europhysics Letters*. **28**, 335 (1994)
2. N.G. Chopra, R.J. Luyken, K. Cherrey, V.H. Crespi, M.L. Cohen, S.G. Louie, and A. Zett, Applied *Physics Letters*. **269**, 966 (1995).
3. A. Loiseau, F. Willaime, N. Demoncy, G. Hug, and H. Pascard, *Physical Rev. Letters*.**76**, 4737, (1996).
4. D. Golberg, Y. Bando, M. Eremets, K. Takemura, K. Kurashima, and H. Yusa, *Applied Physics Letters*. **69**, 2045 (1996).
5. W. Han, Y. Bando, K. Kurashima, and T. Sato, *Applied Physics Letters*. **73**, 21 (1998).
6. D. Golberg, P. Dorozhkin, Y. Bando, M. Hasegava, Z.C. Dong, *Chemical Physics Letters*. **359**, 220 (2002).

Production and characterization of MWBNNT and B-doped SWCNT

E. Borowiak-Palen[1,2] , T. Pichler[1], G. G. Fuentes[1], A. Graff[1], R.J. Kalenczuk[2], M. Knupfer[1] and J. Fink[1].

1) Leibniz-Institute for Solid State and Material Research, P. O. Box 270016 Dresden, Helmholtzstr.20, D-01171 Dresden, Germany
2) Institute of Chemical and Environment Engineering, Technical University of Szczecin, Poland

Abstract: We present a study of mutiwalled boron nitride nanotubes (MWBNNT) and boron doped single wall carbon nanotubes (SWCNT) which have been produced by substitution reaction from SWCNT templates. Pure MWBNNT with a 1:1 BN atomic ratio as well as clean SWCNT with B-substitution levels up to 20% have been produced. The morphology and crystal structure of the samples have been characterized by transmission electron microscopy (TEM), electron diffraction and electron energy-loss spectroscopy (EELS).

INTRODUCTION

Single wall carbon nanotubes (SWCNT) exhibit different electronic properties depending on their chirality and diameter. The control of the properties of carbon nanotubes is still one of the big challenges in the nanotube research field. Since as-produced bulk samples of SWCNT usually contain a mixture of metallic and semiconducting nanotubes, the application in nanoelectronic devices is still very difficult. One of the possible methods to achieve SWCNT with only metallic character is a p-type doping by substituting carbon with boron. This also leads to a modification of the chemical composition of the nanotubes. Similarly, only semiconducting SWCNT are observed for BN substitution leading to the formation of BN nanotubes, which have been first produced using the arc discharge technique [1]. Previously, the substitution of carbon by boron and nitrogen was in detail investigated by Golberg et al. [2-5]. We adapted this method to optimize the preparation regarding high purity multiwalled boron nitride nanotubes (MWBNNT) and highly B-doped SWCNT (B content up to 20 at%). In both cases we used single-walled carbon nanotubes produced by laser ablation as a template [6].

EXPERIMENTAL

For the preparation of MWBNNT, the SWCNT raw material (template) produced by the standard laser ablation technique, which has been adapted for high yield (70wt% SWCNT), was mixed with molybdenum oxide and boron oxide in the mass ratio 1:2:5. Nitrogen was introduced to the system as a carrier gas for 30 minutes at 1500°C [7]. B-doped SWCNT were prepared in a similar way, but with ammonia instead of nitrogen as carrier gas and at a reaction temperature of 1150°C [8]. The strong oxidation agent MoO_3 was not used in this experiment.

In both cases the mixture of the starting materials was first put in a ceramic crucible, which was inserted into a high vacuum tube. Then the system was evacuated until a base pressure of 10^{-7}mbar was reached. Before heating the carrier gases (nitrogen or ammonia) were introduced until the partial pressure was 10^{-5} mbar. As a last step of the process the system was heated up to the reaction temperature and kept there for 30 minutes. For both experiments we used the same apparatus schematically shown in Fig.1

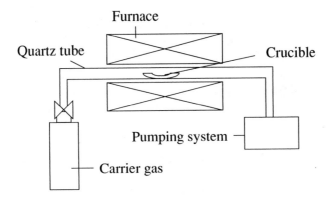

FIGURE 1. Scheme of the system used for preparation of the substituted SWCNT.

For the characterization by optical spectroscopy, a well sonicated suspension of the sample in acetone (B-doped SWCNT) and in CCl_4 (MWBNNT) was produced. Then a homogenous thin film was prepared by dropping onto a KBr single crystal. The optical absorption spectrum was measured from the mid-infrared to the ultraviolet region with the spectral resolution set to 2 cm^{-1} (0.25 meV) using a Bruker IFS 113V/88 spectrometer. The samples for transmission electron microscopy (TEM) and TEM-EELS were prepared directly from the film on KBr, which was floated off in distilled water and mounted onto standard microscopy grids.

RESULTS AND DISCUSSION

A) MWBNNT

The adapted substitution process to obtain boron nitride nanotubes occurs n two steps, first the conversion of single-walled carbon nanotubes into multiwalled carbon nanotubes has taken place and then the complete substitution of the carbon atoms by nitrogen and boron [7]. Fig.2a shows the morphology of the SWCNT starting material used as a template. It is a typical laser ablation product of SWCNT with mean tube diameters of about 1.25nm. The reaction product consists of multiwalled nanotubes with 2-30 walls and a wide range of diameters between 1-

10nm (Fig.2b). The contrast seen in TEM micrographs of MWBNNT is higher than that observed for carbon nanotubes. This observation is in good agreement with a previous contribution [9]. In addition, after using the abovementioned parameters we have found that 95% of the reaction product is MWBNNT with only few percent of amorphous material [10].

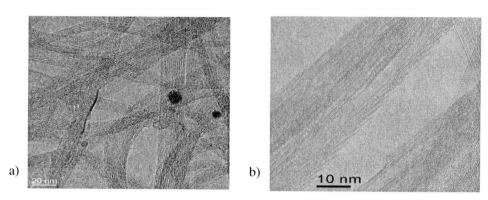

FIGURE 2. TEM micrographs of the raw SWCNT material used as a template (a) and the product of conversion (complete substitution) to MWBNNT (b).

EELS measurements reveal a boron to nitrogen stoichiometry of 1:1 (not shown here). B1s and N1s excitation edges were compared to the edges of h-BN, respectively. This reveals a very similar chemical environment in both cases and thus demonstrates the sp^2 hybridization in the MWBNNT.

The infrared response of MWBNNT is dominated by two characteristic BN-vibrations at 799 cm^{-1} and at 1375 cm^{-1}, which are discussed in detail elsewhere [7]. The first feature at 1.7 $Å^{-1}$ of the electron diffraction pattern of MWBNNT reveals a lattice enlargement perpendicular to the BN sheets of around 6% with respect to the reference sample of h-BN (i.e. 0.33 nm). This implies a reduction of the interaction between the nanotube walls in comparison to the interaction between the h-BN layers.

B) B-doped SWCNT

The TEM micrograph shown in the left panel of Fig 3. reveals the structural properties of the sample. We observe that the high temperature treatment strongly reduces the amount of the amorphous material present in the starting raw material of SWCNT. In order to estimate the boron concentration in the samples we used TEM-EELS on a local scale and bulk sensitive high resolution EELS. A typical example of a TEM-EELS spectrum measured at one of the positions indicated by the arrows in the left panel of Fig. 3 is depicted in the right panel of the Figure. The amount of boron incorporated into the nanotube is up to 20 at% with a mean boron content of 15 at%. The average boron content is in good agreement with the results from the bulk sensitive EELS measurements. Both EELS spectra did not show any reminiscence of

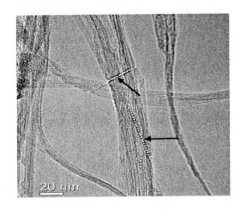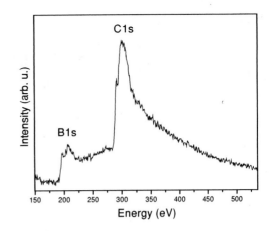

FIGURE 3. Left panel: TEM images of B-doped SWCNT bundles. The arrows and the white circles point to the position on a doped SWCNT bundle where the TEM-EELS spectra were performed to determine the chemical composition of the sample. Right panel: Typical TEM-EELS spectrum taken from the B-doped SWCNT bundles after strapping the background.

nitrogen demonstrating that no BN phase is formed at this temperature. This points to the well known fact, that ammonia is helpful to create defects in the walls [11] giving new active places where the boron can easily incorporate into vacancies. In this case NH_3 is much more active than N_2 and gives higher boron concentration than it was considered before [4,12].

To summarize, we have shown how to prepare B-doped SWCNT and MWBNNT in an efficient way by an adapted substitution method using SWCNT templates. The structural properties and the stoichiometry of the samples have been analyzed. Pure MWBNNT with a 1:1 BN atomic ratio as well as clean SWCNT with B-substitution levels up to 20% have been produced.

Acknowledgements: We thank the DFG (PI 440) for funding. The authors are greateful to K. Müller, E. Kockrick for technical assistance and O. Jost for delivering of the as-produced SWCNT raw material.

REFERENCES

[1] N. G. Chopra et al., *Science*, 269, 966 (1995)
[2] W. Han et al., *Appl. Phys. Lett.*, 73, 3085 (1998)
[3] D. Golberg et al., *Chem. Phys. Lett.*, 323, 185 (2000)
[4] W. Han et al., *Chem. Phys. Lett.*, 299, 368 (1999)
[5] D. Golberg et al., *Carbon*, 38, 2017 (2000)
[6] O.Jost et al., *Chem. Phys. Lett.*, 339, 297 (2001)
[7] E. Borowiak-Palen et al., *Chem. Commun.*, pp. 83 (2003)
[8] E. Borowiak-Palen et al., submitted
[9] A. Loiseau et al., *Phys. Rev. Lett.*, 76, 4737 (1996)
[10] G. G. Fuentes et al., *Phys. Rev. B*, 67, 035429 (2003)
[11] R. Kaltofen et al., *Thin Solid Films*, 290, 112 (1996)
[12] D. Golberg et al., *Chem. Phys. Lett.*, 308, 337 (1999)

Electronic Properties of
Multiwall Boron Nitride Nanotubes

<u>G.G. Fuentes</u>[1*], Ewa Borowiak-Palen[1,2], T. Pichler[1], A. Graff[1], G. Behr[1], R.J. Kalenczuk[2], M. Knupfer[1], J. Fink[1]

[1]*Leibniz Institute for Solid State and Materials Research Dresden, P.O. Box 270016 D-01171 Dresden, Germany.*
[2]*Technical University of Szczecin, Institute of Chemical and Environment Engineering, Szczecin, Poland.*
*Present address: AIN. San Cosme y San Damián s/n. E-31191 Pamplona (Spain).

Abstract. We report on infra-red and electron energy-loss spectroscopy studies of high purity multiwall boron nitride nanotubes synthesised by substitution reactions. The IR pattern of the BN tubes shows the presence of both tangential (~800 cm^{-1}) and longitudinal (~1400 cm^{-1}) modes characteristic of h-BN. We show that the different energy positions and the LO-TO splitting observed for the BN-tubes can be explained by geometrical considerations. The dielectric function ε of the BN-tubes obtained from the electron energy-loss function reveals an intense π-π^* interband transition at 5.4 eV, which is shifted to lower energies by 0.6 eV when compared to hexagonal BN. In addition, the absorption onset of the optically allowed transitions is less abrupt and begins at lower energies than that of h-BN. We ascribe this effect to the presence of tubes with low inner diameter (below 3 nm) in good agreement with recent theoretical band structure calculations.

INTRODUCTION

The existence of the nanotubular allotropic form of hexagonal boron nitride (h-BN) was recently proposed using tight binding calculations by Rubio et al.[1] on the basis of the similarities between lattice structure and bond length of graphite and h-BN. The band structure for this new class of metastable structures was predicted to have a large band gap leading to promising potential uses in blue and violet photoluminescence devices[2]. During the last years BN nanotubes have been fabricated using different methods. Loiseau et al.[3] synthesised bundles of hexagonal-like multiwall BN nanotubes using arc-discharge of HfB$_2$ electrodes in a nitrogen atmosphere. Han et al.[4] proposed a route based on the chemical substitution of carbon by B and N from a template of single wall carbon nanotubes (SWCNT) using B$_2$O$_3$ powder as boron source in a N$_2$ atmosphere. Under the same conditions, Golberg et al.[5,6] found a sharp enhancement of the production yield of multiwall BN nanotubes by adding MoO$_3$ as catalyst. More recently, Zhi et al.[7] characterised the photoluminescence of aligned multiwall BN nanotubes synthesised by CVD and pulsed-arc-discharge.

In this work we present a TEM, infra-red (IR) and electron energy-loss spectroscopy (EELS) study of high purity multiwall BN nanotubes[8]. The dielectric properties and electronic structure of the BN nanotubes are compared with those reported in the literature for bulk h-BN, as well as with recent theoretical and experimental results of single wall and multiwall BN nanotubes.

*CP685, Molecular Nanostructures: XVII Int'l. Winterschool/Euroconference on
Electronic Properties of Novel Materials,* edited by H. Kuzmany, J. Fink, M. Mehring, and S. Roth

EXPERIMENTAL

Multiwall BN nanotubes have been grown by following the route by Golberg et al.[6] based on the substitution of carbon atoms from pristine SWCNT´s by boron and nitrogen. The TEM pictures shown in Figure 1 reveal the presence of multiwalled structures with a B:N atomic ratio 1:1. The tubes exhibited an averaged inner diameter of around 3 nm, although tubes with both larger and smaller diameters were found. The reaction was carried out in a N_2 atmosphere at 1500°C for 30 min using a mixture of 5:2:1 in weight of B_2O_3, MoO_3, and SWCNT, respectively. The BN tubes were deposited on a KBr crystal using a suspension of raw material in CCl_4 until an effective thickness of about 100 nm was reached. The as obtained films were used for the IR characterisation. Subsequently, the films were floated off in distilled water and mounted on standard 1000 mesh copper microscopy grids for EELS. The EELS measurements were carried out using a dedicated 170 keV spectrometer described in detail elsewhere[9]. The energy and momentum resolution for the electron energy-loss spectra was set to 180 meV and 0.03 Å^{-1}. In the case of the core level excitations we used an energy and momentum resolution of 330 meV and 0.1 Å^{-1}, respectively. The probing areas of both IR and EELS provide information representative for a large number of nanotubes.

 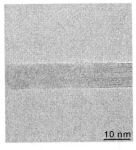

Fig.1 High resolution TEM micrographs of typical multiwalled boron nitride nanotubes after reaction.

RESULTS

A. Infra-red characterisation

In Figure 2 we present the measured IR response of the BN tubes[10] (solid line) and a polycrystalline h-BN reference for comparison (dotted line). The spectrum of the BN tubes is dominated by vibrations with a strong dipole moment centred at 800 cm^{-1} (vibration perpendicular to the tube axis) and at 1372 cm^{-1} with a shoulder at 1540 cm^{-1} (vibration along the tube axis). In the case of polycrystalline h-BN the corresponding modes are observed at 811 cm^{-1} and at 1377 cm^{-1} with a weak shoulder at 1514 cm^{-1}. Apparently, both IR responses show different characteristics. In fact, there is a small up-shift of the high energy TO mode by 5 cm^{-1}, which is characteristic for the BN nanotubes. However, most of the observed differences can be explained by texturing effects. To show this issue we should bring in mind the work by Geik et al.[11] reporting on the IR active modes in pyrolytic h-BN measured with the electric field polarized parallel and perpendicular to the c-axis. In this work it is shown that the LO-TO splitting, (i.e. the separation between the longitudinal and transversal modes) is maximal when the electric field is polarized perpendicular to the c-axis of pyrolitic h-BN. In this scheme, the large LO-TO splitting observed in the case of the BN tubes (much larger than that present in polycrystalline BN) may be compatible with a 2D arrangement, similar to that characteristic for crystalline h-BN, given by a preferential orientation of the tubes parallel to the KBr substrate. The position of the tangential mode in the BN tubes (i.e 800 cm^{-1}) is downwards shifted in energy by 11 cm^{-1} with respect to that of polycrystalline h-BN. This shift is also compatible with the 2D arrangement of the tubes parallel to the substrate, since the spectral weight of such tangential modes in pyrolityc h-BN is minimum in energy in the case of the electric field perpendicular to the c-axis, in good agreement with the explanation given for the large LO-TO splitting of the longitudinal modes.

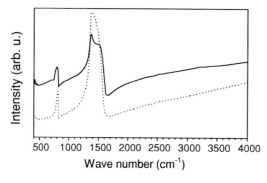

Fig. 2 IR spectrum of multiwalled boron nitride nanotubes (solid line) and h-boron nitride (dotted line). The peak at low frequency corresponds to a vibration between the layers (tubes), the second peak at high frequency to vibration in layer (in the tube).

B. High resolution EELS

Figure 3 (top) represents the B1s excitation spectrum of multiwall BN nanotubes (solid), and a X-ray absorption spectrum of B1s (dashed) from polycrystalline h-BN[12]. The B1s spectrum of the multiwall BN nanotubes exhibits a sharp peak at 192 eV due to the excitonic B1s→π^* transition, which is only allowed for momentum transfers perpendicular to the BN sheets[13,14]. The B1s→σ^* excitations give rise to two peaks at 198.2 eV and 199.5 eV. Figure 3 (bottom) represents the N1s excitation spectrum of the multiwall BN nanotubes and the reference N1s x-ray absorption spectrum from h-BN. The low energy π^* resonance is observed at 401 eV in good agreement with that expected for h-BN. In addition, the broad σ^* resonance around 408 eV also resembles that observed for h-BN. As in the case of the B1s spectra, the intensity of the π^* resonance relative to the σ^* resonance is larger as compared to the π^*/σ^* ratio for the reference material. We attribute this effect to the fact that the nanotubes are oriented preferentially along the surface plane of the KBr crystal and therefore parallel to the surface of the copper grid, so that, for core level measurements, where the momentum transfer is chosen perpendicular to the surface of the copper grid, the contribution of the electronic transitions to final states of π^* character which originate from the $2p_z$ orbital perpendicular to the BN sheets is enhanced. This interpretation is consistent with the texturing effects discussed for the IR results.

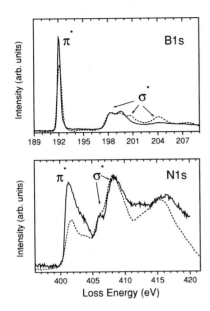

Fig. 3 B1s and N1s excitation spectra of BN nanotubes as measured by EELS (solid). B 1s and N1s excitation spectra of polycrystalline h-BN (dashed) as measured using synchrotron radiation. (taken from Ref. [11]).

Figure 4 shows the loss spectra between 0 and 15 eV of the BN tubes (solid line) and of polycrystalline h-BN (dots+solid line). The low energy loss spectrum of the BN tubes is dominated by a double collective excitation at 6.7 eV and 7.7 eV related to π-π^* electronic interband transitions, which are shifted to lower energies with respect to the same excitation observed for h-BN.

A double peak centred at 11 eV and 12 eV can also be observed in the loss spectrum of the BN tubes. In the case of the h-BN such excitations appears as a single peak. The experimental loss function of the BN tubes exhibits a large band gap of around 4 eV, similar to that observed for h-BN. The absorption onset of the loss function of the BN tubes shows however a more progressive slope than that observed in the case of h-BN.

The low energy-loss spectra can be better understood in terms of the real and imaginary parts of the dielectric functions, which can be obtained by a Kramers Kronig analysis (KKA) of the measured loss spectra. Figure 4 shows the loss function (upper part), ε_1 (dashed line), and ε_2 (solid line, bottom part) obtained from the KKA of the loss function. ε_2 exhibits an intense excitation at 5.4 eV (ω_1) and a weaker feature at 6.8 eV (ω_2), both corresponding to electronic interband transitions of π character leading to collective excitations (at 6.6 eV and 7.8 eV), as evidenced by the two maxima in the loss function. In addition, ε_2 reveals a double feature centred at 11 eV ($\omega_{3,4}$) which is also reproduced in the loss function and a broad oscillator at 14.7 eV (ω_5), the latter being associated to electronic σ-σ^* transitions of h-BN[13].

Interestingly, the first optical transition (ω_1) is downshifted in energy by 0.6 eV with respect to the same excitation in h-BN (cf. Ref. [15], i.e. ω_1=6.1 eV). This energy shift leads the π plasmon in the loss function to peak at lower energies as compared to that of the h-BN, as shown Fig. 4.

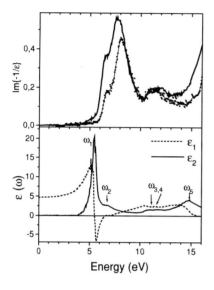

Fig. 4 (top) Electron energy loss function of BN tubes (solid line) and h-BN (dots) between 0 and 16 eV. (bottom) Real and imaginary part of the dielectric function of BN tubes derived from the loss function using a Kramers-Kronig analysis. See text for details.

In order to understand this effect we have to refer to band structure calculations by Rubio et al.[1] where the transport gaps of BN single wall nanotubes were calculated theoretically as a function of their inner diameters. Essentially, the study by Rubio et al.[1] predicted that the transport gap of BN tubes increases as the inner diameter increases, but reaching a saturation value for diameters of ~1 nm for zigzag tubes, and ~3 nm in the case of armchair tubes. As mentioned above, the analysis of the TEM pictures revealed an averaged inner diameter of the tubes of 3 nm that is very close to the limit in which the arm-chair tubes present a reduction in the energy of the first optical excitations. Moreover, the fact that there is not a narrow diameter distribution, i.e. a non negligible number of tubes have diameters below 3 nm, implies that the absorption onset expected for the BN tubes may not be abrupt, as in the case of the h-BN (cf. Fig.4), but more progressive. This assumption is clearly demonstrated by closer inspection of the loss function of the BN tubes around 4 eV, where a progressive increase of the absorption intensity is manifested.

CONCLUSIONS

In summary, we have studied high purity multiwall BN nanotubes, as produced by substitution reactions, by TEM and IR absorption and EELS. IR and core level absorption spectroscopies revealed the similarities between the lattice structure (phonons) and chemical environment (B1s and N1s) of the BN tubes and h-BN. However, orientation effects of the nanotubes can give rise to apparent changes in the shape of the IR and core level absorption spectra.

The optical response function as obtained by KK revealed a shift to lower energies of the first optical transition in the BN tubes by 0.6 eV with respect to that of h-BN. We propose that this fact is related to the presence of tubes with a diameter below 3 nm, in agreement with theoretical predictions in the literature.

ACKNOWLEDGEMENTS

We thank the EU (IST-NID-Project SATURN) and DFG (PI 440) for funding. G.G. Fuentes acknowledges the financial support through the EU´s Marie Curie program. We also thank the help by the IFW technical staff.

REFERENCES

[1] A. Rubio, J. L. Corkill, M. Cohen, Phys. Rev. B 49, 5081 (1994)
[2] X.D. Bai, E.G. Wang, J. Yu, Appl. Phys. Lett. 77, 67 (2000)
[3] A. Loiseau, F. Willaime, N. Demoncy, G. Hug, H, Pascal, Phys. Rev. Lett. 76, 4737 (1996)
[4] W. Han, Y. Bando, K. Kurashima, T. Sato, Appl. Phys. Lett. 73, 3085 (1998)
[5] D. Golberg, Y. Bando, W. Han, K. Kurashima, T. Sato, Chem. Phys. Lett. 308, 337 (1999)
[6] D. Golberg, Y. Band, K. Kurashima, T. Sato, Chem. Phys. Lett. 323, 185 (2000)
[7] C.Y. Zhi, J.D. Guo, X.D. Bai, E.G. Wang, J. Appl. Phys. 91, 5325 (2002)
[8] G.G. Fuentes, E. Borowiak-Palen, T. Pichler, X. Liu, A. Graff, G. Berg, R.J. Kalenczuk, M. Knupfer, J. Fink. Phys Rev. B 67, 035429 (2003)
[9] J. Fink, Adv. Electron. Electron Phys. 75, 121 (1989) and references therein.
[10] E. Borowiak-Palen, T. Pichler, G.G. Fuentes, A. Graff, G. Berg, R.J. Kalenczuk, M. Knupfer, J. Fink. Chem. Comm. 82-83, 1 (2003)
[11] R. Geick, C.H. Perry, G. Rupprecht, Phys Rev. 146, 543 (1966)
[12] R. Gago, I. Jiménez, J.M. Albella, L.J. Terminello, Appl. Phys. Lett. 78, 3430 (2001)
[13] John Robertson, Phys. Rev. B 29, 2131 (1984)
[14] I. Tanaka, H. Araki, Phys. Rev. B 60, 4944 (1999)
[15] Y-N. Xu, W. Y. Ching, Phys. Rev. B 44, 7787 (1991)

Raman Spectroscopy of Single-wall BN Nanotubes

R. Arenal de la Concha[1,2], L. Wirtz[2], J.Y. Mevellec[3], S. Lefrant[3], A. Rubio[2], A. Loiseau[1]

[1] *LEM, Onera-Cnrs, 29 Avenue de la Division Leclerc, BP 72, 92322 Châtillon, France*
[2] *Department of Material Physics, University of the Basque Country, Centro Mixto CSIC-UPV, and Donostia International Physics Center, Po. Manuel de Lardizabal 4, 20018 Donostia - San Sebastián, Spain.*
[3] *Laboratoire de Physique Cristalline, IMN, BP 32229, 44322 Nantes, France*

Abstract. We present results on the vibrational properties of BN-SWNTs together with a study of the synthesis material by transmission electron microscopy. Phonon modes have been investigated by Raman spectroscopy with laser excitation wavelengths in the range from 363.8 to 676.44 nm. The assignment of the modes is guided by *ab-initio* calculations.

INTRODUCTION

Raman spectroscopy has is an efficient and nondestructive technique for characterizing C-NTs, since it provides useful information not only on the geometric, electronic and vibrational structure, but also on physical effects [1,2]. In the case of boron nitride NTs (BN-NTs), the experimental progress on studying their physicals properties has been slowed because the quantity of single wall (SW) NTs was very limited. Recently, a continuous laser vaporization process [3] was reported that produced BN-SWNTs in gram quantities.

In this paper, we present results of the first experimental study of Raman scattering at SW-BNNTs and *ab initio* calculations in support of experimental data.

EXPERIMENT

BN-NTs have been produced by laser ablation of a h-BN target with a continuous CO_2 laser under a nitrogen flow at a pressure 1 bar [3]. Transmission electron microscopy (TEM) was carried out using a Jeol 4000FX (at 400kV) and the chemical compositions were examined using energy electron spectroscopy (EELS) in a VG501 (at 100kV). Raman spectra were collected with the Jobin-Yvon T64000 Raman spectrometer using in the visible range a microscope in a backscattering geometry. The laser excitation wavelengths of 363.8, 457.9, 488, 514.5 and 676.44 nm were used. The laser beam was focused on a spot size of a few μm in diameter which allowed to explore different areas of the sample. The laser power was below 10 mW in order to prevent burning of the sample.

CP685, Molecular Nanostructures: XVII Int'l. Winterschool/Euroconference on Electronic Properties of Novel Materials, edited by H. Kuzmany, J. Fink, M. Mehring, and S. Roth

RESULTS AND DISCUSSION

Description of the samples

The TEM image in Fig. 1 a) shows a general view of the synthesis products. The two dominant structures are nanotubes and closed polyhedral structures. From TEM analysis, we estimated that the yield of BN-NTs could be of the order of 25% of the ablated material and that about 80% of the NTs are single walled, and 15% double walled nanotubes.

The BN-SWNTs appear either as individual tubes or as bundles. Fig. 1 b) is a typical HRTEM image, showing a bundle and a SWNT. The diameter distribution is centered at 1.4 nm (FWHM=0.6 nm) and 1.6 nm (FWHM=0.3 nm) for isolated tubes and bundles, respectively. The tube length is between 100 and 400 nm with a few tubes exceeding 1000 nm.

The bonding state and chemical composition of the individual nanostrucures were determined by EELS. The spectra taken from the NTs show clearly the presence of boron and nitrogen in sp^2 bonding and confirmed the expected BN stoichiometry [4].

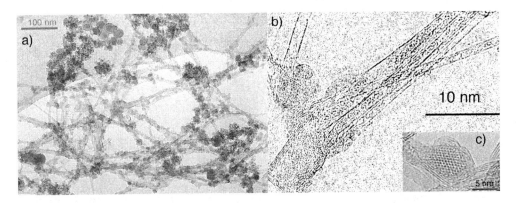

FIGURE 1 a) Low magnification TEM image of a BN sample, showing bundles of SW-BNNTs, cages of h-BN and clusters of nanonaparticles b) HRTEM picture of a bundle of SW-BNNTs and two isolated SW-BNNTs c) The inset shows a HRTEM image of a boron nanocrystal.

Polyhedral nanostructures consists of particles either nested into h-BN shells or located at the tip of NTs. The particles are identified from EELS analysis as having a pure boron core. This core is often crystalline and has a rhombohedral form (Fig. 1 c)). Very often its surface is covered by a thin boron oxide layer [4].

Raman Experiments

The Raman experiments carried out on BN-NT samples turned out to be very delicate for several reasons. First, the samples appeared rather heterogeneous at the scale of the laser beam size. Depending on the spot position, different spectra were recorded either resembling that of pure h-BN [5] or assigned to areas rich in NTs. Second, some regions of the samples were easily burnt even with the lower laser power used. Finally a strong luminescence is always observed whatever the spot position, the temperature and the wavelength.

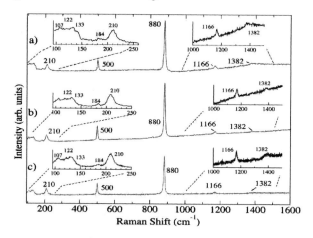

FIGURE 2 Room temperature Raman spectra of unpurified SW-BNNTs at the same area and at different laser energies (a) 457.9; b) 488; c) 514.5 nm). Insets show enlargements of the low and high frequency regions.

Fig. 2 shows the Raman spectra associated with BN-NTs for different wavelengths at room temperature. No dependence on the excitation wavelength is observed. In contrast to C-NTs [2], the absence of resonant modes even at 363.8 nm confirms the insulating character of BN NTs and indicates a gap larger than 3.4 eV. Many peaks can be distinguished in three different regions of the spectra. The low-frequency region displays a set of peaks at 107, 122, 133, 184 and 210 cm^{-1} [5]. In the high-frequency range, two peaks at 1166 and 1382 cm^{-1} can be discerned. Finally, sharp and intense peaks at 500 and 880cm^{-1} are seen in the medium range.

Calculations and modes assignment

We have performed an *ab initio* density-functional study in the local density approximation using the code ABINIT [6]. We have employed Troullier-Martins pseudopotentials and a plane-wave energy cut-off of 80 Ry. A more detailed analysis is presented in [7]. Results are presented in Fig.3 for a selection of tubes together with the variation of the phonon frequencies obtained by the zone-folding (ZF) method as a function of the tube diameter. Except for the breathing mode which cannot be calculated with the ZF method, Fig.3 demonstrates an excellent agreement between *ab initio* values and frequencies derivated from ZF.

We now turn to the comparison between experiments and theory. In the low frequencies range, we can identify the radial breathing modes (RBM). Owing to the diameter distribution of our samples, the calculated RBM and others active low frequency modes are consistent with the peaks observed at 107, 122, 133 and 184 cm^{-1}. The peak at 210 cm^{-1} can be associated with a tangential mode (E_1). The broad and weak band at 1382 cm^{-1} can be easily associated with a signal of tubes (A_g and E_{1g} modes). The weakness intensity of the peaks is partially due a strong luminescence background. The assignment of the most intense peaks at 500 and 880 cm^{-1} is not possible with the frequencies obtained from the calculations. Furthermore, these peaks cannot be attributed to the contaminants detected by TEM and EELS in the samples (amorphous C, Ca, amorphous BN, SiO_2 and B_2O_3). In particular, we do not observe features typical of Boron even though boron particles are mixed with the tubes. We therefore suspect that the peaks are due to another contaminant which has not yet been identified but which is only present in the areas containing the nanotubes.

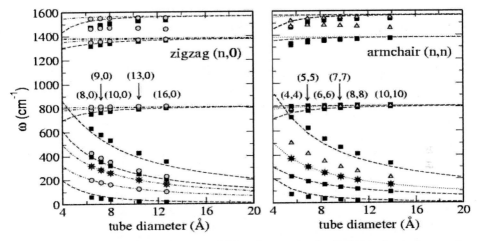

FIGURE 3 Phonon frequencies of Raman and IR actives modes for (infinitely long) zigzag and armchair NTs as a function diameter. Comparison of *ab initio* values (symbols) with ZF method (lines). Squares/dashed lines: only Raman active; triangles/dotted lines: only IR active; circles/dash-dotted lines: Raman and IR active. The RBM is in addition marked by asterisks. For zigzag-tubes, the IR modes are a subset of the Raman ones, while for armchair tubes, the two sets are disjoint.

CONCLUSION

We have presented a combined study of first-principle calculations for phonon modes and Raman experiments on BN-SWNTs synthesized by laser vaporization of h-BN. The non resonance of Raman spectra at excitation wavelengths ranging from 676.44 to 363.8 nm indicates a band gap larger than 3.4 eV. Peaks at low and high frequencies have been identified as a clear signal of the tubes in spite of their weakness. Thin peaks at medium frequencies are not yet explained and are expected to arise from an unknown contaminant.

ACKNOWLEDGMENTS

This work has been supported by the European research and training network COMELCAN (HPRN-CT-2000-00128).

REFERENCES

1   R. Saito, G. Dresselhaus, M. S. Dresselhaus, *Physical properties of C-NTs*, Imperial College Press, London, 1998; M. S. Dresselhaus, G. Dresselhaus, Ph. Avorius (Eds.), *C-NTs: synthesis, structure, properties and applications*, Springer-Verlag, Berlin, 2001.
2   A. M. Rao, E. Richter, S. Bandow, B. Chase, P. C. Eklund, K. A. Williams, S. Fang, K. R. Subbaswamy, M. Menon, A. Thess, R. E. Smalley, G. Dresselhaus, M. S. Dresselhaus, *Science* **275**, 187 (1997).
3   R. S. Lee, J. Gavillet, M. Lamy de la Chapelle, A. Loiseau, J.-L. Cochon, D. Pigache, J. Thibault, F. Willaime, *Phys. Rev. B* **64**, 121405 (2001).
4   R. Arenal de la Concha, A. Vlandas, O. Stephan, A. Loiseau, (to be published).
5   R. Arenal de la Concha, L. Wirtz, J.Y. Mevellec, S. Lefrant, A. Rubio, A. Loiseau, submitted to *Chem. Phys. Lett.* (2003).
6   The ABINIT code is a common project of the Université Catholique de Louvain, Corning Incorporated, and other contributors (URL http://www.abinit.org)
7   L. Wirtz, R. Arenal de la Concha, A. Loiseau, A. Rubio, submitted to *Phys. Rev. B* (2003).

Scanning tunneling microscopy and spectroscopy of boron nitride nanotubes

Masa Ishigami, Shaul Aloni and A. Zettl

Physics Department, University of California, Berkeley, CA 94720 USA
Materials Sciences Division, Lawrence Berkeley National Laboratory, Berkeley CA 94720 USA

Abstract. We have investigated electronic properties of boron nitride nanotubes using a low temperature scanning tunneling microscope (STM) operated at 7K. STM images of the tubes reveal hexagonal lattices or stripe patterns, which can be caused by interlayer coupling or scattering of electronic states of the nanotubes. In addition, scanning tunneling spectroscopy measurements indicate that the tubes have band gaps exceeding 4 eV, and reveal van Hove singularities confirming the one-dimensional nature of electronic states of the nanotubes.

INTRODUCTION

Boron nitride nanotubes[1] (BNNTs) can be described as sheets of hexagonal boron nitride (h-BN) rolled into seamless concentric cylinders. Much like carbon nanotubes, they have a high Young's modulus[2] and have been predicted to be exceptionally good thermal conductors. Unlike their carbon analogue, BNNTs are expected to be semiconductors with a 4 to 5 eV band gap irrespective of their chirality[1]. Although BNNTs were first experimentally synthesized in 1995[3], so far no confirmations of these predicted electronic properties have been reported. In the work reported here, we have used a low temperature high resolution scanning tunneling microscope (STM) to investigate this newer class of nanotubes.

EXPERIMENTAL

BNNTs were synthesized using an arc-discharge technique as described elsewhere[4]. Carefully collected as-grown soot was first ultrasonically suspended in 1,2-dichloroethane and then deposited from the solution onto Au(111) surfaces. The nanotubes in our STM samples are mostly double-walled with 3 nm outer diameters as confirmed by atomic force microscopy and transmission electron microscopy. The samples were outgassed at 623 to 723 K for 3 hours in ultra high vacuum prior to the STM investigations. Scanning tunneling microscopy and spectroscopy experiments were preformed in a homemade ultra-high vacuum low-temperature STM operated at 7 K. Scanning tunneling spectroscopy (STS) was performed using a bias modulation lock-in technique. The bias voltage was modulated at 20-50mV at a frequency of ~2.7kHz.

CP685, *Molecular Nanostructures: XVII Int'l. Winterschool/Euroconference on Electronic Properties of Novel Materials,*edited by H. Kuzmany, J. Fink, M. Mehring, and S. Roth
© 2003 American Institute of Physics 0-7354-0154-3/03/$20.00

RESULTS AND DISCUSSION

Scanning tunneling microscopy

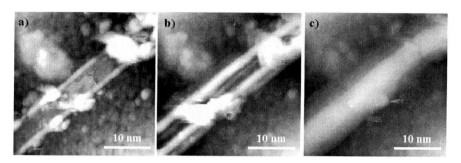

FIGURE 1. Constant current STM images of a boron nitride nanotube taken with –2.0(a), -4.0(b), -7.0(c) volt sample biases respectively with a tunneling current of 0.5 nA.

Fig. 1 shows a typical sample bias dependence of STM images of a BNNT. The nanotube appears as a cylinder when imaged at high positive and negative sample biases. When sample bias magnitude is lowered below threshold of about 4 volts, nanotube-related topographical features diminish in height and eventually reach the substrate corrugation level at a bias magnitude of approximately 1 volt. Below the threshold, the tube appears with a "hollow core", as seen in Fig. 1a and 1b. This structure indicates that no electronic states are available in the nanotube at these energies, suggesting that BNNTs have wide band gaps.

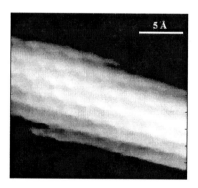

FIGURE 2. A constant current STM image of a boron nitride tube showing a triangular lattice. V_{sample} = -4.8 volts and I_{tunnel} = 1.0 nA.

At high sample bias magnitudes, BNNTs appear as featureless cylinders. When the magnitudes of the sample biases are lowered to values just above the threshold, two types of intramolecular features are resolved. About 10% of the nanotubes display hexagonal patterns 20 pm in height and with lattice constants of 2.5 Å; an example is shown in Fig. 2. The lattice constant is consistent with that of bulk h-BN. The vast majority of the tubes exhibit stripe patterns, as shown in Fig. 3 with periodicities ranging from 2 to 12 Å. In some cases, two different periodicities are observed simultaneously on the same tube as shown in figure 3b.

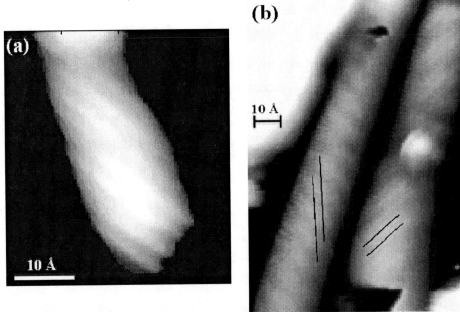

FIGURE 3. a) An STM image of a boron nitride tube showing a stripe pattern. V_{sample} = - 3.2 volts and I_{tunnel} = 0.2 nA. **b)** Two nanotubes displaying two different stripe patterns simultaneously. Two black lines on each tube highlight stripes with smaller periodicities (~ 3 Å). Larger periodicity stripes have periodicities of about 10 Å. V_{sample} = - 4.25 volts and I_{tunnel} = 0.5 nA.

The periodicities of these stripe patterns seem to be unrelated to the h-BN lattice constant. The lack of correlation rules out an adsorbate layer as the cause for the stripes. Furthermore, if the stripes were STM tip-convolutions of the atomic arrangements, the periodicities should depend on chiral angles and would be distributed in a fairly narrow range of values around the h-BN lattice constant[5]. Artifact effects due to tip shapes have also been ruled out because different stripe patterns can be imaged simultaneously with a same tip, as shown in Fig. 3b. The stripes are intrinsic to the nanotubes and reflect spatial variations of local electronic densities of states.

There are two mechanisms which cause similar stripe patterns in graphite or carbon nanotubes. First, as seen on carbon nanotubes[6], the stripe patterns can be interference patterns generated from electrons scattered by tube ends or defects. An example of a defect in a BNNT can be seen as a bright spot on the right tube shown in Fig. 3b. Second, since most of the BNNTs imaged in our experiment are double-walled[4], the stripe patterns may result from interlayer coupling. The observed periodicities would then depend on chiralities, diameters, and relative positions of the outer and the inner tube. We would expect the coupling to create superlattice modulations best described by Moiré patterns. Such modulations from interlayer coupling have been observed in STM of graphite[7] and multi-walled carbon nanotubes[8].

Scanning Tunneling Spectroscopy

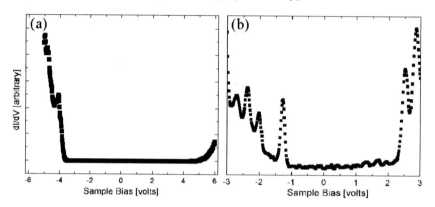

FIGURE 4. A typical scanning tunneling spectra (STS) acquired on a selected BNNts. Both spectra were taken on tubes displaying the stripe patterns.

Fig. 4a shows a tunneling spectrum taken on the majority of the BNNTs. A total lack of features between -3.7V and +5V would naively suggest a band gap of over 8 eV. Such a high band gap value is unexpected for these nanotubes and indeed seems unphysical. Increasing the sensitivity of tunneling spectroscopy by decreasing tip-sample separation did not lead to detection of any additional electronic states in the band gap. In order to understand this unusually large measured band gap, the geometry of our tunnel junction must be considered. In our case, electrons must travel from the tip, through the tube, into the gold substrate. It is likely that bias voltages, applied between the gold substrate and the tip, have not completely fallen in the tip-tube tunneling junction, broadening the apparent gap. Assuming that the tube-substrate contact resistance is low and the tube resistance is comparable to the tunnel junction resistance, the voltage drop in the tunnel junction will be smaller than the applied sample bias. In this case, true band gap of the nanotube can be determined by measuring the exact resistance across the tube. On the other hand, if the tube-gold substrate contact resistance is high, the extraction of the true band gap is more complicated requiring additional considerations of both tip-tube and tube-substrate tunnel junctions. In either case, determination of the band gap requires appropriate theoretical models and additional experiments.

Fig. 4a also shows a sharp peak at -4 volts. Similar sharp features have been observed near the onset of conductance in positive or negative voltages in other spectra. We believe that these are signatures of van Hove singularities, which are manifestation of the one-dimensional electronic structures of the BNNTs.

A few BNNTs revealed spectra which are quite different. As shown in Fig. 4b, in these spectra, van Hove singularities of BNNTs are clearly visible. The first set of the features on this particular spectrum is separated by 3.8 eV. If the locations of these peaks can be interpreted as the onsets of the conduction and the valence bands, the value of the gap is slightly smaller than expected[1]. However, recent theoretical calculations[9] show that an electric field applied across a BNNT can induce a giant Stark effect which reduces the band gap. Field of 1 V/nm is expected to reduce a 4.5 eV band gap of a 3 nm diameter BNNT to 2.8 eV[9]. Therefore, the measured band gap of 3.8 eV maybe reflective of an intrinsic band gap somewhat larger (say 4 to 5 eV) reduced by a tip-induced Stark effect.

REFERENCES

1. Rubio, Angel, Corkill, Jennifer L., and Cohen, Marvin L., *Physical Review B* **49**, 5081-4 (1994).
2. Chopra, N. G., and Zettl, A., *Solid State Communications* **105**, 297-300 (1998).
3. Chopra, N. G., Luyken, R. J., Cherrey, K., Crespi, V. H., Cohen, M. L., Louie, S. G., and Zettl A., *Science* **269**, 966-7 (1995).
4. Cumings, John, and Zettl, A., *Chemical Physics Letters* **316,** 211 (2001)
5. Venema, L. C., Meunier, V., Lambin, Ph., and Dekker, C., *Physical Review B* **61**, 2991-6 (2000).
6. Clauss, W., Bergeron, D. J., and Johnson, A. T., *Physical Review B* **58**, R4266 (1998)
   Kane, C. L., and Mele, E. J., *Physical Review B* **59**, R12759-62 (1999).
7. Xhie, J., Sattler, K., Ge, M., and Venkateswaran N., *Physical Review B* **47**, 15835-41 (1993).
8. Ge M. H. And Sattler K., Science 260 , 515 (1993).
9. Khoo, K. H., Mazzoni, M. S. C., and Louie, S. G., submitted for publication

Computer Modelling of One-Electron Density of States for Carbon and Boron Nitride Nanotubes

A.V. Osadchy, E. D. Obraztsova, S. V. Terekhov, V. Yu. Yurov

Natural Sciences Center of General Physics Institute, RAS, 38 Vavilov str., 119991, Moscow, Russia

Abstract. The computer program has been developed to calculate numerically one-electron density of states (DOS) for single-wall carbon and boron nitride nanotubes. The result has been obtained by numerical integration of dispersion curves of hexagonal graphite and boron nitride. The allowed values of wave vector have been derived using "zone folding" model. The Monte-Carlo numerical integration method was used. The DOS parameters calculated for different nanotubes, including chiral ones, corresponded well the data published. Experimentally, DOS parameters have been estimated for tubes geometries corresponding to dominating "breathing" modes in Raman spectra registered under different laser excitation energies. The calculated and experimentally observed values of resonance energies have coincided well.

One-electron DOS for boron nitride nanotubes of different geometry have been calculated. Due to a big gap value in boron nitride no metallic nanotubes have been revealed.

Single-wall carbon nanotubes (SWNT) are unique due to their one-dimensional structure [1]. A very small diameter is a reason of a specific view of electron density of states (DOS) demonstrating a set of singularities. Their positions for the fixed nanotube depend on its geometry [2]. Experimentally, the information about electronic structure may be obtained by STM, optical absorption and resonance Raman spectroscopy. For interpretation of the experimental data the modelling of SWNT DOS is necessary.

This work is devoted to development of the computer program for numerical calculations of one-electron DOS for single-wall carbon and boron nitride nanotubes [3], applicable to nanotubes of any geometry. An experimental proof of calculations has been done by Raman spectroscopy.

Algorithm of DOS calculation

Single-wall carbon nanotubes

For single-wall carbon nanotubes the DOS calculations have performed by Monte-Carlo numerical integration method. A two-dimensional dispersion relation for graphene sheet has been chosen in form, proposed in [4]:

$$E(\overline{K}) = \pm\gamma_0 \cdot \left\{ 1 + 4 \cdot \cos\left(\frac{\sqrt{3} \cdot K_x \cdot a_0}{2}\right) \cdot \cos\left(\frac{K_y \cdot a_0}{2}\right) + 4 \cdot \cos^2\left(\frac{K_y \cdot a_0}{2}\right) \right\}^{\frac{1}{2}}, \qquad (1)$$

CP685, *Molecular Nanostructures: XVII Int'l. Winterschool/Euroconference on Electronic Properties of Novel Materials,* edited by H. Kuzmany, J. Fink, M. Mehring, and S. Roth
© 2003 American Institute of Physics 0-7354-0154-3/03/$20.00

where K_x, K_y are the X- and Y- components, correspondingly, of a wave vector **K**, $\gamma_0 =$ 2.95 eV, and $a_0 = 0.246$ nm.

A view of the dispersion relation for *2D*-graphite (graphene) is shown in *Fig.1a*. For each carbon nanotube geometry this relation has been differentiated over all allowed wave vector values in the first Brillouin zone. The examples of DOS, calculated for fixed nanotubes, are shown in *Fig.2*.

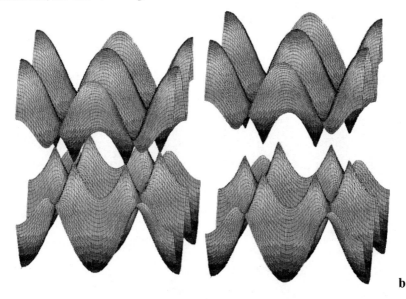

a

b

FIGURE 1. A view of two-dimensional (2D) dispersion relations for graphene (a)
and hexagonal boron nitride (b).

Single-wall boron nitride nanotubes

The calculation of DOS for single-wall boron nitride nanotubes has been performed using the same method as for carbon nanotubes. But *2D*-dispersion for boron nitride is different from the graphene one. The relation has been deduced on the basis of published one-dimensional dispersion relations for few directions in Brillouin zone of hexagonal BN [5]. An inequality of B and N atoms has been taken into account. A view of *2D*-dispersion for boron nitride is show in *Fig.1b*. The corresponding equation is:

$$E(K_x, K_y) = \frac{\tilde{\varepsilon} \pm \sqrt{\tilde{\varepsilon}^2 + 4 \cdot \tilde{\gamma}_0 \cdot \omega(K_x, K_y)}}{2}, \text{ where} \qquad (2)$$

$$\omega(K_x, K_y) = \left(1 + 4 \cdot \cos\left(\frac{\sqrt{3} \cdot K_x \cdot \tilde{a}_0}{2}\right) \cdot \cos\left(\frac{K_y \cdot \tilde{a}_0}{2}\right) + 4 \cdot \cos^2\left(\frac{K_y \cdot \tilde{a}_0}{2}\right)\right)^2$$

and $\tilde{a}_0$, $\tilde{\gamma}_0$, $\tilde{\varepsilon} \equiv \varepsilon_{2p1} - \varepsilon_{2p2}$ - constants

The constant values have been optimized to reach the minimal deviation from the dispersion along known directions [5]. The constant values obtained are:

$$\tilde{a}_0 = 0.249 \text{ nm}, \ \tilde{\gamma}_0 = 2.85 \text{ eV}, \ \tilde{\varepsilon} = 4.3 \text{ eV} \qquad (3)$$

The examples of DOS calculated for BN nanotubes are shown in *Fig. 3*. For all nanotube geometries DOS demonstrates a big (more than 4 eV) first mirror gap (E_{11}). It means that there are no BN single-wall nanotubes with metallic properties.

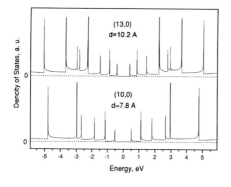

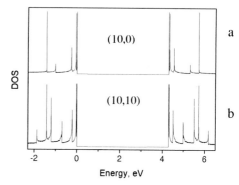

FIGURE 2. The calculated one-electron density of states for single-wall carbon nanotubes of different geometry.

FIGURE 3. The calculated one-electron density of states for BN single-wall nanotubes of different geometry.

Raman proof of DOS calculated for single-wall carbon nanotubes

The resonance Raman technique has been chosen for comparison of the experimental and calculated parameters of one-electron DOS for single-wall carbon nanotubes of fixed geometry. The Raman spectra of carbon SWNT (in a "breathing" mode" spectral range), registered with the different excitation energies, are shown in *Fig. 4*. The nanotubes have been synthesized by laser ablation method. *Fig. 4* shows that different peaks go through the resonance at different excitation wavelength. These peaks correspond to nanotubes of different geometry. Their diameters (*d*) may be estimated from the "breathing" mode frequency (*ω*)[6]:

$$\omega = \frac{C_1}{d} + C_2, \qquad (4)$$

where $C_1 = 234$, and $C_2 = 10$ см$^{-1}$, ω is measured in *cm^{-1}*, *d* is measured in *nm*.

According to (4) the modes dominating under excitation energies 2.4 eV and 2.5 eV correspond to nanotubes with diameters 13.2 Å (186 cm^{-1}) and 12.5 Å (197 cm^{-1}). The DOS for all possible geometrical configurations for nanotubes with these diameters have been calculated. Among them two geometries have been chosen: **(12,7)** – for tubes with diameter 13.2 Å and **(14,3)** – for tubes with diameter 12.5 Å. The calculated values of mirror gaps E_{33} for them have coincided, in fact, with the

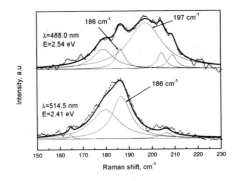

FIGURE 4. Transformation of the "breathing" Raman mode
induced by variation of laser excitation energy.

resonance energies values fixed experimentally by the excitation laser energies. This coincidence is a real proof of a plausibility of DOS calculations.

Conclusion

The computer program has been developed for numerical calculations of one-electron density of states for single-wall carbon and boron nitride nanotubes of any geometry. A semi-empirical expression of two-dimensional dispersion for the hexagonal boron nitride has been obtained using the data published on one-dimensional dispersion relations for different directions in Brillouin zone of hexagonal BN. The results of DOS calculations for carbon SWNT with fixed diameters have been verified experimentally by the resonance Raman spectroscopy.

Acknowledgments

The work is supported by PFBR 01-02-17358, INTAS-00-237, SCOPES 7SUPJ062400. A. V. Osadchy is grateful to ISSEP and Moscow Government for the post-graduate grant in 2002-2003.

References

1. M.S. Dresselhaus, G. Dresselhaus, P.C. Eklund, *in "Science of Fullerenes and Carbon Nanotubes", New York, NY, San Diego, CA: Academic Press, 1996.*
2. J. W. Mintmire, C. T. White, *Phys Rev. Lett., 81, 2506 (1998).*
3. A.V. Osadchy, E.D. Obraztsova, S.V. Terekhov, V.Yu. Yurov, *JETP Letters 77(2003) 480.*
4. M.S. Dresselhaus in *"Carbon Filaments and Nanotubes: Common Origins, Differing Applications?", ed. L.P. Biro et al., NATO Science Series E, 372 (2000) 11.*
5. Y.-N. Xu, W. Y. Ching, *Phys. Rev. B 44 (1991) 7787.*
6. H. Kuzmany, W. Plank, H. Hulman, Ch. Kramberger et al., *Eur. Phys. J. B 22 (2000) 307.*

Density Functional Calculations of Nanotube Bundles from Calcium Disilicide

S. Gemming* and G. Seifert*

*Institut für Physikalische Chemie und Elektrochemie, Technische Universität Dresden,
Mommsenstr. 13, D-01062 Dresden, Germany.

Abstract. Bundles of (6,6) tubes from $CaSi_2$ were studied by density functional calculations. The Ca ions are included both in the intratubular region and on the outer surface of the tube. According to the formation energies $CaSi_2$ tubes are in the experimentally accessible range. Metallic conductivity along the tube interior and a strongly ionic character of the outer tube surface make this tube an ideal nano-wire. The regular arrangement of positive and negative charges on the outer surface assists the aggregation of neighbouring tubes within a bundle of tubes.

INTRODUCTION

In the present publication the results of a density-functional band-structure study on nanotube bundles of $CaSi_2$ trilayers are described. This approach is qualitatively different from the investigations of single, free-standing nanotubes or wires by large supercell models or by non-periodic methods. Beyond the investigation of the single tube it provides an assessment of the inter-tubular interaction which has become of great interest lately as a handle to manipulate patterning or the build-up of regularly structured nanotube arrays. Especially the partially ionic character of the Si-Ca interaction makes $CaSi_2$ nanotubes promising candidates for self-organisation into bundles and further to textured surfaces. Furthermore, a similar investigation on AlB_2 nanotube bundles showed that the interaction between adjacent tubes provides a stabilising effect. [1] If it can be achieved that the metallicity of bulk $CaSi_2$ is preserved also along the tubular structure, $CaSi_2$ tubes can also act as a template for the current-assisted deposition of other nanowires or, later, as nanocontact for the deposited structure. Investigations on the structure and properties of the isoelectronic phosphorous nanotubes [2] and on $CaAlSi_2$ nanotubes [3] indicate that this goal can indeed be reached.

COMPUTATIONAL DETAILS

The structural and electronic properties of bulk and tubular structures of $CaSi_2$ were determined by local density-functional band-structure calculations with the program ABINIT (www.abinit.org). Norm-conserving pseudopotentials for the configurations [Ar] $4s^2$ of Ca and [Ne] $3s^2 3p^2$ of Si and a plane-wave cutoff of 333 eV were employed. The Monckhorst-Pack meshes for the calculation of the Brillouin-zone integrals were chosen to match the density of the 10 x 10 x 10 mesh, which was used for the reference

CP685, *Molecular Nanostructures: XVII Int'l. Winterschool/Euroconference on Electronic Properties of Novel Materials,* edited by H. Kuzmany, J. Fink, M. Mehring, and S. Roth

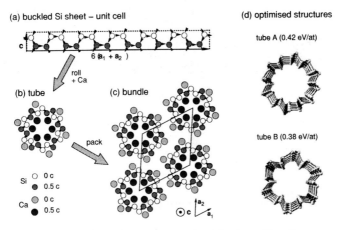

(a) buckled Si sheet – unit cell

$6 (a_1 + a_2)$

roll + Ca

(b) tube

(c) bundle

pack

Si ○ 0 c
   ● 0.5 c

Ca ◎ 0 c
   ● 0.5 c

⊙ c a_2 a_1

(d) optimised structures

tube A (0.42 eV/at)

tube B (0.38 eV/at)

FIGURE 1.   Schematic drawing of $CaSi_2$ (6,6) tube and bundle structure.

calculation of the 1R structure. The curves of the densities of valence electron states (DOS) were broadened by convolution with a Gaussian function of 0.5 eV width.

NANOTUBE STRUCTURES

Fig. 1 depicts the construction of the (6,6) nanotube bundles by rolling up a buckled sheet of Si hexagons (a) along the $6a_1 + 6a_2$ lattice vectors. Among several possible arrangements of the Ca atoms the one depicted in (b) is the most stable one: 6 Ca atoms reside each on the outer tube surface and in the tube interior, with a shift of 0.5 c with respect to each other. The packing of the tubes to hexagonal bundles (c) is achieved by applying trigonally periodic boundary conditions perpendicular to the tube's c axis and optimising the cell size. Two prototypical arrangements are obtained as only every second Ca site on the tube surface is occupied. The most stable arrangement is obtained, if the Ca rows of neighbouring tubes avoid each other as much as possible.

The structure determined in Ref. [4] for a negatively charged pure Si (6,6) nanotube was employed as a starting geometry for all structure optimisations. The formation energies for the corresponding tube bundle amounts to +0.42 eV/at referred to bulk $CaSi_2$ in the 1R structure. A more stable tube of +0.38 eV/at formation energy is obtained upon further optimisation. Both structures are depicted in Fig. 1 (d), for clarity the Ca atoms are omitted.

Tube A experiences a stronger hexagonal distortion from a purely circular shape than tube B, and the Si backbone in tube B has a smaller average diameter than the one of tube A. A closer inspection of the Si backbone reveals the major difference between the two structures. In A the bulk buckling of the hexagonal Si network is preserved exactly as in the structures obtained for (6,6) tubes of Si$^-$ and of P previously by density-functional based methods. [4, 2] This turbine-wheel-like structure is the curved analogue of the planar rhombohedral modification of black phosphorus. In B the buckling is partially

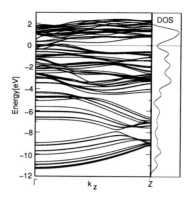

FIGURE 2. Band structure and DOS of CaSi$_2$ (6,6) bundle.

reversed to the atom arrangement of the orthorhombic modification of black phosphorus. The corresponding Si backbone is shaped like a flower and deviates significantly from an ideally circular shape. This particular atom arrangement allows for a reduction of the tube strain: Under the assumption that CaSi$_2$ consisted of Ca^{2+} and Si$^-$ ions less Si lone pairs are compressed into the tube interior in structure B.

ELECTRONIC PROPERTIES

Fig. 2 gives a representation of the electron bands along k$_z$ for tube B bundles along with the density of states. The Fermi level is set to zero. The Si 3s manifold splits between -11.4 eV and -4.1 eV. The dispersion of the Si-derived 3p bands spreads from -4.1 eV up to the Fermi level. The Ca-derived levels are introduced mainly at the lower edge of the conduction band. The mixing of Si- and Ca-based bands occurs in the vicinity of the Γ point, and the gap at Z amounts to 1.1 eV Thus, CaSi$_2$ tubes of type A exhibit channels for metallic conductivity along the wire. For tubes of type B the same behaviour is obtained and will be described in more detail elsewhere. Thus, the gross features of the density of states of the CaSi$_2$ tubes are determined by the Si-derived bands. The tiny, but important details around the Fermi level are modified by the Ca-derived levels. There is also a remakable correspondence to the DOS published earlier for the isoelectronic and structurally related nanotubes of pure phosphorus.

The response of the electronic structure to an electric field applied along the directions of the three basis vectors was additionnally calculated for both the pure bulk crystal and for the bundles of type B. The dielectric constant of bulk CaSi$_2$ is only marginally anisotropic with values of 23 ε_0 along a$_1$ and a$_2$ and of 29 ε_0 along c. In the tube bundles the anisotropy of the dielectric constant is more pronounced and amounts to 11 ε_0 between the tubes, but to 36 ε_0 along the tube. These values confirm the picture of a tube with metallic conductivity along the tube, but with ionic, insulator-type interactions between adjacent tubes.

The Born effective charge tensors were calculated from the response of the wave function to an external electric field. The charges of the outermost Ca and Si ions amount to +1.6 and -1.2, respectively. Both the Ca ions located in the intratubular region and the innermost Si atoms remain virtually neutral. The two other sets of Si ions are located between the inner and outer tube surfaces and bear effective charges of about -0.07 for the inner set and of about -0.7 for the outer set. Hence, there exists a chemical gradient along the tube diameter which allows for a homogeneous transition from the ionic bonding at the outer tube surface to the more metallic bonding at the tube interior.

CONCLUSIONS

Density functional band structure calculations were carried out for bundles of $CaSi_2$ tubes with the prototypical (6,6) structure. The most stable tubular structures contain Ca ions both in the intratubular region and on the outer surface of the tube. For compounds with a low inter-layer interaction the strain energy of bending a single sheet to a tube is often employed as a measure for the experimental accessibility. For ionic compounds like $CaSi_2$ the interlayer interaction is not negligible any longer. Thus the formation energy with respect to the bulk should be discussed, because it includes both the strain energy and the energy for the ablation of a single layer. The formation energies calculated here for the $CaSi_2$ bundles amount to 0.38 eV/at and 0.42 eV/at. These values compare favourably with the strain energies of 0.25 eV/at which had been calculated previously for the same (6,6) structures of MoS_2 or P by the DF-TB method. Hence, it is proposed on the basis of the present calculations that $CaSi_2$ nanotubes could be experimentally accessible.

ACKNOWLEDGMENTS

The authors acknowledge financial support from the German-Israel-Foundation (G.S.) and from the Deutsche Forschungsgemeinschaft through the Graduiertenkollegs "Struktur-Eigenschaftsbeziehungen bei Heterocyclen" and "Akkumulation von Molekülen zu Nanostrukturen" (S.G.).

REFERENCES

1. Quandt, A., Liu, A. Y., and Boustani, I., *Phys. Rev. B*, **67**, 125422 (2001).
2. Seifert, G., and Hernandez, E., *Chem. Phys. Lett.*, **318**, 355 (2000).
3. Shein, I. R., Ivanovskaya, V. V., Medvedeva, N. I., and Ivanovskii, A. L., *JETP Lett.*, **76**, 189 (2002).
4. Seifert, G., Köhler, T., Urbassek, H. M., Hernandez, E., and Frauenheim, T., *Phys. Rev. B*, **67**, 193409 (2001).

Band structure of boron doped carbon nanotubes

Ludger Wirtz* and Angel Rubio*

*Department of Material Physics, University of the Basque Country, Centro Mixto CSIC-UPV, and
Donostia International Physics Center, Po. Manuel de Lardizabal 4,
20018 Donostia-San Sebastián, Spain

Abstract. We present *ab initio* and self-consistent tight-binding calculations on the band structure of single wall semiconducting carbon nanotubes with high degrees (up to 25 %) of boron substitution. Besides a lowering of the Fermi energy into the valence band, a regular, periodic distribution of the p-dopants leads to the formation of a dispersive "acceptor"-like band in the band gap of the undoped tube. This comes from the superposition of acceptor levels at the boron atoms with the delocalized carbon π-orbitals. Irregular (random) boron-doping leads to a high concentration of hybrids of acceptor and unoccupied carbon states above the Fermi edge.

INTRODUCTION

The electronic properties of single wall carbon nanotubes depend sensitively on the diameter and the chirality of the tubes. Therefore, in order to use tubes as elements in nano-electronical devices, a controlled way to produce and separate a large quantity of tubes of specific radius and chirality has to be found. Alternatively, doping of tubes by boron and nitrogen [1] may lead to electronic properties that are more controlled by the chemistry (i.e., the amount of doping) than the specific geometry of the tubes. Indeed, theoretical investigations of BCN nanotube heterojunctions have predicted that the characteristics of these junctions are largely independent of geometrical parameters [2]. It has been predicted that even stochastic doping may lead to useful electronic elements such as chains of random quantum dots and nanoscale diodes in series [3].

The realization of p-type doping of pure carbon nanotubes by a substitution reaction with boron atoms [4, 5, 6] offered the possibility to transform semiconducting tubes into metallic tubes by lowering the Fermi level into the valence band. Indeed, transport measurements [7, 8, 9] have shown a clearly enhanced conductivity of B-doped tubes. The metallic behavior of B-doped multiwall carbon nanotubes was also confirmed by scanning tunneling spectroscopy [10].

In bulk semiconductors, typical doping concentrations of impurity atoms are around 10^{-3} %. At such low concentration, the presence of group III atoms (e.g., B) in a group IV semiconductor leads to the formation of an acceptor state (non-dispersive band) in the band gap at low energy above the valence band edge [11]. For B doping in a C(8,0) tube, Yi and Bernholc [12] have calculated (using a B/C ratio of 1/80) that this acceptor state is located at 0.16 eV above the Fermi energy. However no discussion has been made up-to-date on the evolution of this acceptor-like level with the degree of B-doping. Since in recent experiments [6], Boron substitution up to 15 % has been reported, we investigate in this paper the band structure of strongly B-doped single wall carbon nanotubes.

CP685, *Molecular Nanostructures: XVII Int'l. Winterschool/Euroconference on
Electronic Properties of Novel Materials,* edited by H. Kuzmany, J. Fink, M. Mehring, and S. Roth

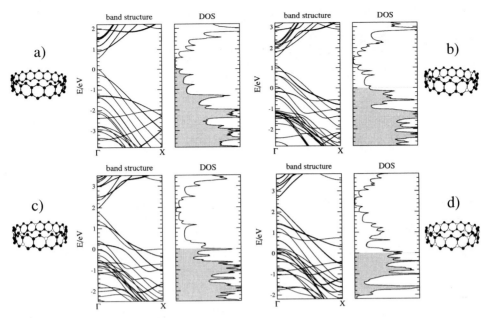

FIGURE 1. Band structure and DOS of a C(16,0) tube with a) 0%, b) 6.25%, c) 12.5%, and d) 25% boron doping. Zero energy denotes the Fermi edge. Filled states are indicated by grey shadowing. Insets display the corresponding unit cells (C atoms black, B atoms white).

COMPUTATIONAL METHOD

In order to demonstrate the effect of B doping on a semiconducting tube, we have chosen a (16,0) tube with a diameter of 12.5 Å which is close to the average radius of commonly produced SWNT samples. Since the distribution of borons in B-doped tubes is still unknown, we have performed two sets of calculations. 1.) For regular periodic distributions that are commensurate with the unit-cell of the undoped tubes, we perform *ab initio* calculations [13] of electronic band structure and density of states (DOS). 2.) In order to test the effect of disordered B doping, we employ a supercell comprising 6 unit-cells (384 atoms) and distribute the boron atoms "randomly" with the only constraint that they cannot occupy nearest neighbor positions on the hexagonal carbon grid. For this large system, we use a self-consistent tight-binding (SCTB) method [3, 15].

RESULTS

Fig. 1 shows the band structure and DOS of a C(16,0) tube without doping and with various concentrations of B-doping in a periodic manner (see insets). For the undoped tube it can be noted, that the DOS is symmetric around the band gap for the 1st and 2nd Van-Hove singularities (VHSs). However, beyond this small energy regime around the band gap, asymmetries arise due to the mixing of p and s orbitals. Doping of the tubes

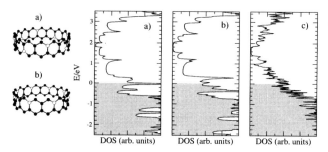

FIGURE 2. DOS of a C(16,0) tube with 12.5% B doping: a) regular doping (as in Fig. 1), b) regular doping (alternative geometry, c) random doping.

leads to a lowering of the Fermi level into the valence band of the undoped tube. Clearly, the stronger the doping, the stronger is also the shift of the Fermi level. Above the highest occupied bands of the undoped tube new bands are formed. These bands correspond to the formation of an acceptor level in semiconductors with very low dopant concentration. Due to the strong concentration in the present case, the boron levels hybridize with the carbon levels and form strongly dispersive "acceptor-like bands". The band structure of the valence band is strongly distorted upon doping, because the levels of carbon atoms that are substituted by boron move upwards in energy. In addition, the lowering of the symmetry by doping leads to many clear avoided crossings between states that perform a real crossing in the case of the undoped tube where they posses a different symmetry. Clearly, in all cases displayed in Fig. 1, the doping with B leads to a metallic character of the tube. However, the density of states at the Fermi level strongly depends on the geometry of the structure and varies non-monotonously with the doping concentration. In fact, in BC_3 tubes with a different geometry than in Fig. 1 d), a gap between π-type orbitals and σ type orbitals opens up at the Fermi level and renders the tubes semiconducting in this particular case [17].

Fig. 2 compares the DOS of a C(16,0) tube with 12.5% boron substitution in three different geometries. The two regular structures both display a pattern with the familiar pronounced Van Hove singularities of a 1-dim. band structure. However, the different ordering of the boron atoms gives rise to a different hybridization of the bands and thereby to a pronounced difference in the peak structure of the DOS. The DOS at the Fermi level strongly depends on the ordering of the B atoms. In the DOS of the randomly doped tube, the pattern of the VHSs is mostly smeared out [18]. Due to the missing translational geometry, the 1-dim. band structure of the regularly doped tubes is transformed into the DOS of a 0-dimensional structure, i.e. of a large molecule or cluster without periodicity. The states between the Fermi level and the original valence band edge of the undoped tubes are hybrids of acceptor levels and unoccupied carbon levels. The electronic excitations into these levels should explain the optical absorption spectra of B-doped carbon tubes [6]: In addition to the pronounced absorption peaks that are commonly affiliated with the transitions E_{ii} ($i = 1, 2$) from the first/second occupied VHS to the first/second unoccupied VHS of the pure semiconducting carbon tubes, the spectra of B-doped tubes display additional absorption at energies lower than E_{11}.

CONCLUSION

The calculations on the band structure of boron doped carbon nanotubes clearly confirm the expectation (and experimental observation) that these tubes are metallic with low resistance. For regularly doped structures, we have observed the formation of a dispersive "acceptor" band while the Fermi level is shifted downwards into the valence band of the undoped tubes. Electronic excitations into these hybrids of acceptor states and unoccupied carbon levels are expected to play an important role in optical absorption spectra. Randomly doped tubes display the same downshift of the Fermi edge but cease to display strongly dispersive bands. We hope that in the near future, spatially resolved TEM/EELS will help to elucidate the exact geometry of the boron dopants and scanning tunneling spectroscopy will probe the exact density of states.

Work supported by COMELCAN (contract number HPRN-CT-2000-00128). We acknowledge stimulating discussion with T. Pichler, J. Fink and G. G. Fuentes.

REFERENCES

1. O. Stéphan et al., Science **266**, 1683 (1994); M. Terrones et al., Chem. Phys. Lett. **257** (1996); Ph. Redlich et al., Chem. Phys. Lett. **260**, 465 (1996); Y. Zhang et al., Chem. Phys. Lett. **279** 264 (1997); R. Sen, A. Govindaraj, and C. N. R. Rao, Chem. Phys. Lett. **287**, 671 (1998); D. Golberg et al., Chem. Phys. Lett. **359**, 220 (2002); D. Golberg et al., Chem. Phys. Lett. **360**, 1 (2002);
2. X. Blase, J.-C. Charlier, A. De Vita, and R. Car, Appl. Phys. Lett. **70**, 197 (1997).
3. P. E. Lammert, V. H. Crespi, and A. Rubio, Phys. Rev. Lett. **87** (2001).
4. W. Han, Y. Bando, K. Kurashima, and T. Sato, Chem. Phys. Lett. **299** 368 (1999).
5. D. Golberg, Y. Bando, W. Han, K. Kurashima, and T. Sato, Chem. Phys. Lett. **308**, 337 (1999).
6. E. Borowiak-Palen et al., this issue; T. Pichler et al., this issue.
7. M. Terrones et al., Appl. Phys. A **66**, 307 (1998).
8. B. Wei, R. Spolenak, P. Kohler-Redlich, M. Rühle, and E. Arzt, Appl. Phys. Lett. **74**, 3149 (1999).
9. K. Liu, Ph. Avouris, R. Martel, and W. K. Hsu, Phys. Rev. B **63**, 161404 (2001).
10. D. L. Carroll et al, Phys. Rev. Lett. **81**, 2334 (1998).
11. see, e.g., N. W. Ashcroft and N. D. Mermin, *Solid State Physics*, Saunders College Publishing, Orlando, 1976.
12. J.-Y. Yi and J. Bernholc, Phys. Rev. B **47**, 1708 (1993).
13. The *ab initio* calculations have been performed with the code ABINIT, using density-functional theory (DFT) in the local density approximation (LDA). We use a plane wave expansion with an energy cutoff at 50 Ry. Core electrons are simulated by Troullier-Martins pseudopotentials. A supercell geometry is employed with a large inter-tube distance (10 a.u.). The density is calculated self-consistently using a sampling of 8 k-points for the (quasi-one dimensional) first Brillouin zone of the tubes. A non-self consistent calculation yields the energies of occupied and unoccupied states on a grid of 50 k-points. Since the tubes have large diameter, we have neglected the effect of local atomic distortion induced by doping which leads to a small outward movement of the boron atoms (< 0.11 Å [12]).
14. The ABINIT code is a common project of the Université Catholique de Louvain, Corning Incorporated, and other contributors (URL http://www.abinit.org).
15. In the SCTB calculation (following Ref. [3]) we use a non-orthogonal tight-binding basis with s and p valence orbitals, fitted to LDA calculations [16]. The Coulomb potential generated by the charge density is self-consistently included as a shift in the on-site potentials. We have checked for the systems displayed in Fig. 1 that the SCTB yields results in very close agreement with the *ab initio* calculations.
16. D. Porezag, Th. Frauenheim, Th. Köhler, G. Seifert, and R. Kaschner, Phys. Rev. B **51**, 12947 (1995).
17. Y. Miyamoto, A. Rubio, S. G. Louie, and M. L. Cohen, Phys. Rev. B **50**, 18360 (1994).
18. The system still possesses a translational symmetry even though with a large unit cell. Therefore, some of the dispersive band structure, including Van-Hove singularities, is still visible.

Optical absorption in small BN and C nanotubes

L. Wirtz*, V. Olevano†, A. G. Marinopoulos†, L. Reining† and A. Rubio*

*Department of Material Physics, University of the Basque Country, Centro Mixto CSIC-UPV, and Donostia International Physics Center, Po. Manuel de Lardizabal 4, 20018 San Sebastián, Spain
†Laboratoire des Solides Irradiés, UMR 7642 CNRS/CEA, Ecole Polytechnique, F-91128 Palaiseau, France

Abstract. We present a theoretical study of the optical absorption spectrum of small boron-nitride and carbon nanotubes using time-dependent density-functional theory and the random phase approximation. Both for C and BN tubes, the absorption of light polarized perpendicular to the tube-axis is strongly suppressed due to local field effects. Since BN-tubes are wide band-gap insulators, they only absorb in the ultra-violet energy regime, independently of chirality and diameter. In comparison with the spectra of the single C and BN-sheets, the tubes display additional fine-structure which stems from the (quasi-) one-dimensionality of the tubes and sensitively depends on the chirality and tube diameter. This fine structure can provide additional information for the assignment of tube indices in high resolution optical absorption spectroscopy.

Just as a carbon nanotube can be thought of as a rolled up graphene sheet, a hexagonal single sheet of BN can be used to construct a BN nanotube. These tubes are isoelectronic to carbon tubes, but carry over some of the characteristic differences of hexagonal BN with respect to graphite. In particular, BN-tubes have a bandgap similar to h-BN, mostly independent of the tube diameter and chirality [1, 2]. Related to this large band gap (the DFT-band gap is 4 eV (see Fig. 1) while the quasi-particle band gap in the GW approximation amounts to 5.5 eV [2]) one expects a high thermal stability and relative chemical inertness for BN-tubes as compared to its carbon counterparts.

After first synthesis of multi-wall BN-tubes was reported in 1995 [3], multi and single wall BN-tubes are now routinely produced in several groups, the latest success being the production of single-wall BN-tubes in gram quantities [4]. The challenge now consists in the spectroscopical characterization of nanotube samples (both C and BN) and, if possible, single isolated nanotubes. The final goal is to find a unique mapping of the measured electronic and vibrational properties onto the tube indices (n, m). One possible spectroscopic method is optical absorption spectroscopy where direct excitation from occupied to unoccupied states leads to photon absorption.

The energy difference E_{ii} between corresponding occupied and unoccupied Van Hove singularities (VHSs) in the 1-dimensional electronic density of states (DOS) of C-nanotubes is approximately inversely proportional to the tube diameter d. In resonant Raman spectroscopy and scanning tunneling spectroscopy, this scaling is employed for the determination of tube diameters. A recent, spectacular example is the fluorescence spectroscopy of single carbon tubes in aqueous solution, where E_{22} is probed through the frequency of the excitation laser and E_{11} is probed simultaneously through the frequency of the emitted fluorescent light [5]. For the distance between the first VHSs in

CP685, *Molecular Nanostructures: XVII Int' l. Winterschool/Euroconference on Electronic Properties of Novel Materials,* edited by H. Kuzmany, J. Fink, M. Mehring, and S. Roth
© 2003 American Institute of Physics 0-7354-0154-3/03/$20.00

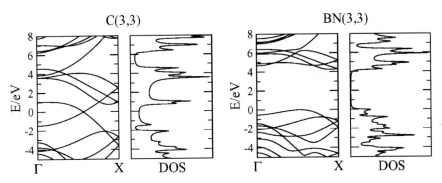

FIGURE 1. Electronic band structure and density of states (DOS) of a C(3,3) nanotube (left side) and a BN(3,3) nanotube (right side). Zero energy denotes the upper edge of the valence band. The calculation has been performed using DFT in the local density approximation.

semiconducting C-tubes, a simple π-electron tight-binding fit yields the relation $E_{11} = E_{22}/2 = 2a_{C-C}\gamma_0/d$, where a_{C-C} is the distance between nearest neighbor carbons. The value for the hopping matrix element γ_0 varies between 2.4 eV and 2.9 eV, depending on the experimental context in which it is used. This fact is a clear indication that the above relation gives only qualitative and not quantitative information on the tube diameter. Furthermore, for small tubes the band structure completely changes with respect to the band structure of large diameter tubes, including a reordering of the VHSs in the density of states and displaying fine structure beyond the first and second VHSs (Fig. 1). This structure sensitively depends on the tube indices and may be probed by optical absorption spectroscopy over a wider energy range (possibly extending into the UV regime).

The scope of this paper is to use *ab initio* techniques to go beyond the tight-binding estimate for excitation energies and to uncover some of these additional features present in the optical spectra of small C and BN nanotubes. In order to develop an intuitive understanding of the optical absorption in tubes, we compare with the absorption in single sheets of graphene and h-BN. In this paper we present results for the very small C(3,3) and BN(3,3) tubes while work for larger tubes is in progress. C(3,3) tubes seem not to exist as free single-wall tubes, but they have been grown and shown to be stable inside the cylindrical channels of a zeolite crystal [6]. The band structure and density of states (DOS) of the two tubes are shown in Fig. 1. The following calculations will show which vertical excitations from unoccupied to occupied states yield the dominant contributions to the absorption spectra.

The cross section $\sigma(\omega)$ for optical absorption at frequency ω is proportional to the imaginary part of the macroscopic dielectric response function of the system [7]: $\sigma(\omega) \propto \text{Im}(\varepsilon_M(\omega))$. We evaluate ε_M using linear response theory [8] within the general framework of time-dependent density-functional theory (TDDFT).

First, we determine the ground state equilibrium geometry of the system using the code ABINIT [9, 10]. From the ground state density we compute the one-particle states $|n, \mathbf{k}\rangle$ and energies $\varepsilon_{n,\mathbf{k}}$ (labeled by Bloch wave vector $\mathbf{k}$ and band index n) for all occupied and a large set of unoccupied bands.

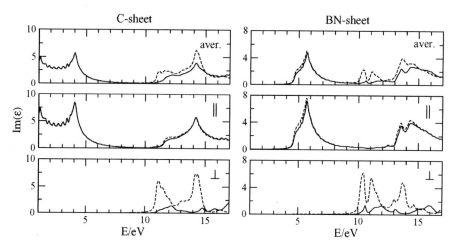

FIGURE 2. Imaginary part of the computed dielectric function (arb. units) for a graphene sheet (left side) and a single hexagonal BN-sheet (right side). Lower panels: light polarization perpendicular to the plane; middle panels: polarization parallel to the plane; upper panels: spatial average. Solid lines: calculation with local field effects; dashed lines: without LFE (see text for details).

The next step is the calculation of the independent particle polarizability χ^0 [11]. It involves a sum over excitations from occupied bands to unoccupied bands [12]:

$$\chi^0_{\mathbf{G},\mathbf{G}'}(\mathbf{q},\omega) =$$

$$2\int \frac{dk^3}{(2\pi)^3} \sum_n^{occ.} \sum_m^{unocc.} \left[\frac{\langle n,\mathbf{k}|e^{-i(\mathbf{q}+\mathbf{G})\mathbf{r}}|m,\mathbf{k}+\mathbf{q}\rangle\langle m,\mathbf{k}+\mathbf{q}|e^{i(\mathbf{q}+\mathbf{G}')\mathbf{r}'}|n,\mathbf{k}\rangle}{\varepsilon_{n,\mathbf{k}} - \varepsilon_{m,\mathbf{k}+\mathbf{q}} - \omega - i\eta} - (m \leftrightarrow n) \right], \quad (1)$$

where $(m \leftrightarrow n)$ means that the indices m and n of the first term are exchanged. The result is checked for convergence with respect to the number of bands [13] and the discrete sampling of k-points within the first Brillouin zone [14]. Using the random phase approximation [8], the "longitudinal" dielectric function is obtained through $\varepsilon_{\mathbf{G},\mathbf{G}'}(\mathbf{q},\omega) = 1 - V_c(\mathbf{q}+\mathbf{G})\chi^0_{\mathbf{G},\mathbf{G}'}(\mathbf{q},\omega)$, where $V_c(\mathbf{q}) = 4\pi/|\mathbf{q}|^2$ is the Coulomb potential in reciprocal space. Finally, the macroscopic dielectric response is given by $\varepsilon_M(\omega) = 1/\varepsilon_{00}^{-1}(\mathbf{q} \to 0, \omega)$. The limit $\mathbf{q} \to 0$ depends on the direction of $\mathbf{q}$, i.e., the polarization of the electric field. The difference between ε_{00} and ε_M is due to the inhomogeneity of the response of the system and is called "local field effects" (LFE). For convergence of ε_M, $\varepsilon_{\mathbf{G},\mathbf{G}'}$ has to be calculated for a sufficient number of $\mathbf{G}$-vectors. The matrix is then inverted and ε_M is obtained from the head of the inverse matrix as $(\varepsilon_{00}^{-1})^{-1}$.

The optical absorption spectra of a single graphene sheet and a sheet of h-BN are displayed in Fig. 2 (extended far into the region of UV light). The spectra are strongly dependent on the polarization of the laser beam. The main difference between C and BN can be seen for the polarization parallel to the plane: While the C-sheet absorbs for all frequencies in the visible light range (the "color" of graphite is black), absorption in BN only sets in above 4 eV, i.e. in the region of UV light (since DFT underestimates the band-gap, we expect a blue-shift of the onset by ≈ 1.5 eV). The high frequency part

408

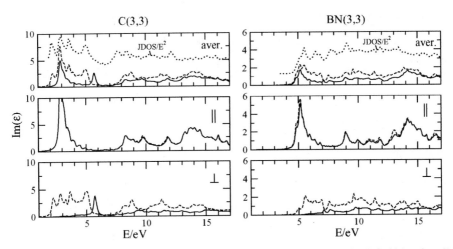

FIGURE 3. Imaginary part of the dielectric function (arb. units) for a C(3,3) tube (left side) and BN(3,3) tube (right side). Lower panels: light polarization perpendicular to the tube; middle panels: polarization parallel to the tube axis; upper panels: spatial average. Solid lines: calculation with local field effects; dashed lines: without LFE. The joint density of states (divided by E^2) is indicated by dotted lines.

of the spectra are quite similar because C and BN are isoelectronic and the high-lying unoccupied states are less sensitive to the difference in the nuclear charges than the states at and below the Fermi energy. Local field effects turn out to be unimportant for parallel polarization. The absorption spectra in perpendicular polarization are remarkably similar for C and BN: up to 9 eV both sheets are completely transparent. Furthermore, in both cases, LFE lead to a strong reduction of absorption at energies higher than 9 eV (redistribution of oscillator strength to even higher energies).

Fig. 3 displays the absorption spectra for the (3,3) tubes of C and BN. For comparison, in the panel of the spatially averaged spectra, we have also included the joint density of states (JDOS), divided by the square of the transition energy. If LFE are neglected, most peaks of the JDOS are visible in the averaged absorption spectra while some peaks are suppressed due to small or vanishing oscillator strength in Eq. (1). Proper inclusion of LFE leads to a smoothing of the spectra and to a redistribution of oscillator strength to higher energies for polarization perpendicular to the tube axis. However, some fine structure survives and may be discernible in high-resolution optical absorption experiments. This fine structure is not an artifact of low k-point sampling but is due to the presence of VHSs in the 1-dim. DOS of the tubes (Fig. 1). In the absorption spectrum polarized along the axis of the C(3,3) tube, the pronounced peak at 3 eV corresponds to the first (dipole-allowed) transition between VHSs in Fig. 1. Even though the tube is metallic, absorption at lower energies is suppressed due to the dipole-selection rules. The position of the first absorption peak in C-tubes not only depends on the tube radius but also (for fixed radius) on the index pair (n, m) and varies up to 2 eV for very small tubes[6, 15, 16, 17]. In contrast, the onset of absorption in BN-tubes corresponds to the threshold in the BN-sheet and is mostly independent of the tube indices (only the peak structure above the onset varies with the tube indices). Between 7 and 12 eV, both C and BN tubes display a similar pattern of peaks which are absent in the corresponding sheets. As in the case of the sheets, LFE are almost negligible for parallel polarization but lead to a strong depolarization in perpendicular direction: The C(3,3) tube is almost transparent up to 5 eV [17] in agreement with the experimental observation [6].

In conclusion, the calculations for optical absorption of small C and BN-nanotubes display a variety of features beyond the excitation between first and second Van-Hove singularities. In C-tubes the position of the first absorption peak strongly varies with the tube indices while in BN-tubes the first peak is determined by the band gap of BN and is therefore mostly independent of (n, m). Some of the fine-structure which distinguishes BN-tubes and C-tubes of different chirality is only visible in the UV region which gives rise to the hope that this energy regime will be probed in the future. Still, features related to electron-hole attraction (excitonic effects) are missing in the calculations. They can play a role in both C and BN tubes, leading to new structure in the band-gap and a redistribution of oscillator strength. They may explain the anomalous E_{11}/E_{22} ratio [5] measured recently for C-tubes [18]. Work along these lines for BN-tubes is in progress.

REFERENCES

1. A. Rubio, J. L. Corkill, and M. L. Cohen, Phys. Rev. B **49**, 5081 (1994).
2. X. Blase et al., Europhys. Lett **28**, 335 (1994); Phys. Rev. B **51**, 6868 (1995).
3. N. G. Chopra et al., Science **269**, 966 (1995).
4. R. S. Lee et al., Phys. Rev. B **64**, 121405(R) (2001).
5. S. M. Bachilo et al., Sciene **298**, 261 (2003).
6. Z. M. Li et al., Phys. Rev. Lett. **87**, 127401 (2001).
7. The dielectric function $\varepsilon_M(\omega)$ is defined in the context of 3-dim. periodic systems. Isolated tubes have only a 1-dimensional translational symmetry. However, the use of a plane-wave expansion requires effectively the use of a three-dimensional super-cell. Therefore, we perform calculations for a periodic array of nanotubes with an intertube distance of 10 a.u. This keeps the interaction between neighboring tubes low and renders nearly isolated tubes. Analogously, for the 2-dim. "single" sheets of C and BN, we perform calculations for an infinite stacking of sheets with an inter-sheet distance of 12 a.u. The absolute value of $\varepsilon_M(\omega)$ scales with the dimension of the employed super-cell, but we are interested only in the relative absorption cross section $\sigma(\omega)$ in arbitrary units.
8. For a recent review article with an extensive list of references see, e.g., G. Onida, L. Reining, and A. Rubio, Rev. Mod. Phys. **74**, 601 (2002).
9. The ABINIT code is a common project of the Université Catholique de Louvain, Corning Incorporated, and other contributors (URL http://www.abinit.org).
10. The cutoff energy of the plane-wave basis is 50 Ry. Core-electrons are simulated by Troullier-Martins pseudopotentials. Calculations are performed in the local-density approximation (LDA).
11. χ^0 is expanded in reciprocal lattice vectors $\mathbf{G}, \mathbf{G}'$ and depends besides on ω on the momentum transfer $\mathbf{q}$. It describes the change of the density ρ in response to the change of the total potential v_{tot} (which in turn is composed of the Hartree, the exchange-correlation and the external potentials): $\delta\rho = \chi^0 \delta v_{tot}$.
12. We use the linear response code DP (URL http://theory.polytechnique.fr/codes/dp)
13. For all systems, we compute bands up to $N = 4n_{occ}$, where n_{occ} is the number of occupied bands.
14. For the 2-dim. Brillouin zone of the C-sheet, we use a sampling of 87×87 k-points, for the BN-sheet 67×67. For the 1-dim. Brillouin zones of the C-tubes, we use of 44 k-points, for the BN-tubes 20.
15. H. J. Liu and C. T. Chan, Phys. Rev. B **66**, 115416 (2002).
16. M. Machón et al., Phys. Rev. B **66**, 155410 (2002).
17. A. G. Marinopoulos et al., submitted (2002).
18. See also the contributions of S. G. Louie and E. Mele in this volume.
19. Work supported by the European research and training networks COMELCAN (HPRN-CT-2000-00128) and NANOPHASE (HPRN-CT-2000-00167), by NABOCO and by MCyT (MAT2001-0946).

The Strongly Correlated 1D Spin State in Li-doped Mo-S Nanotubes

D.Mihailovic[1], Z.Jaglicic[2], R.Dominko[3], and A.Mrzel[1]

1Jozef Stefan Institute, Jamova 39, 1000 Ljubljana, Slovenia and Faculty of Mathematics and Physics, University of Ljubljana, Slovenia
2 Institute for Mathematics, Physics and Mechanics, Jadranska 19, 1000 Ljubljana, Slovenia
3. National Chemistry Institute, Hajdrihova 19, Ljubljana, Slovenia

Abstract. We report on the magnetic properties of Li-doped Mo-S nanotubes (NTs), which appears to be a realisation of a near-ideal 1D strongly correlated state in a bulk system. The material is exceptional because of its inherently extraordinarily weak coupling between 1D subunits, which is an order of magnitude weaker even than in carbon nanotube ropes. In spite of clear evidence for strong electronic correlations from a giant paramagnetic susceptibility, no transition to an ordered state is observed down to very low temperatures. Instead,the system appears to be stabilised in a paramagnetic state by fluctuations characteristic of 1-dimensional systems, with no evidence of a quantum critical point as $T = 0$ is approached.

INTRODUCTION

The magnetic properties of pure one-dimensional systems are largely unexplored because of the practical difficulty of measuring the macroscopic magnetic properties of isolated chains, ladders, wires, tubes - etc. Short of levitation in free space, such measurements are very difficult to achieve. Indeed magnetism is not generally associated with nanotubes or nanowires.

It has been shown recently that MoS_{2-y} NTs are extremely weakly coupled to each other in bundles, and their properties are possibly the closest to 1-dimensional of any solid state system known [1]. The present paper describes the magnetic properties of Li-doped MoS_{2-y} NTs, whose magnetic properties are found to be rather extraordinary and unexpected. The observation of magnetic behaviour is particularly strange, since Mo does not readily form ferromagnetic compounds, while Mo metal itself is diamagnetic.

EXPERIMENTAL METHOD

The synthesis of MoS_{2-y} NTs was performed by the method reported in Ref. [2]. The Li doping was performed electrochemically as reported elsewhere [3,4]. The exact positions of the Li ions are not known at present. However, it is known that if the NTs are strongly dispersed in solvent before Li insertion, the capacity of the material for Li

CP685, *Molecular Nanostructures: XVII Int'l. Winterschool/Euroconference on Electronic Properties of Novel Materials,*edited by H. Kuzmany, J. Fink, M. Mehring, and S. Roth
© 2003 American Institute of Physics 0-7354-0154-3/03/$20.00

storage is drastically reduced, indicating that the Li occupies the interstitial space between the nanotubes (see Fig 1) rather than the space inside the NTs themselves.

The magnetisation measurements were performed in a Quantum Design SQUID magnetometer in Valencia as reported previously [3]. Subtraction of the background was carefully performed and a number of samples from different synthesis batches were measured to ensure that the results were reproducible. Importantly, the undoped NTs have a very small susceptibility, which cannot be reliably distinguished from the background with the small quantities of sample available at the time of writing. Exposing the sample to air results in a gradual decrease of the magnetic signal, and after approximately 24 hours the signal disappears.

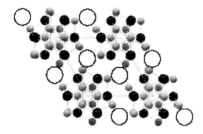

FIGURE 1. The structure of the Mo-S NT system. The Li ions are interstitial in between the NTs (indicated by empty circles).

This observation is important for eliminating possible magnetic impurities which could give rise to a spurious magnetic signal. The reported findings are so unusual, that we had to repeat the measurements and calculations many times to be sure that the results are not spurious. In the end, the fact that the SQUID magnetic measurements and the ESR [5] independently gave the same order of magnitude, gave us confidence in the data.

SUSCEPTIBILITY DATA

The normalised magnetic susceptibility χ of a typical sample at 1kOe is shown in the left panel of Figure 2. It is rather weakly temperature-dependent, with a linear dependence in the range from 300K to about 30K. In some samples, the susceptibility increases slightly at low temperatures with a Curie-like temperature dependence. A fit

$$\chi = \frac{C}{T+\theta} + \chi_0 + \chi_1 T$$

to the data using a two-component susceptibility :

gives $\chi_0 = 1.4 \times 10^{-2}$ emu/mol, $\chi_1 = -6.8 \times 10^{-6}$ emu/mol and $C = 2 \times 10^{-2}$ emu/mol with $\theta < 2K$. The susceptibility of a number of different samples is shown in the right panel of Fig. 2. Very similar behaviour is observed in ESR measurements, with similar magnitude of χ_0 and a similar T-dependence [5].

DISCUSSION

The magnetic state of these 1D objects is difficult to explain in terms of a Fermi liquid description, even with strongly exaggerated Van Hove (VH) bands. Using as an estimate the value of the density of states from density-functional theory (DFT) calculations of $N(E)=10$ states/eV-mol, the Pauli susceptibility $\chi_P = \mu_B^2 N(E) \approx 1.4$ x 10^{-4} emu/mol. This is comparable with the susceptibility of common metals, and is 2

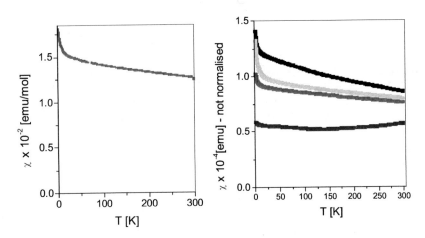

orders of magnitude too small to account for the measured susceptibility.

FIGURE 2. Magnetic susceptibility of Li:Mo-S NTs. The left panel is in normalised units. The right panel shows a number of different samples showing qualitatively similar behaviour.

Supposing the susceptibility were due solely due to a VH band at the Fermi level, which would increase the density of states even more than the DFT estimate, then its bandwidth W would have to be very narrow, and significantly less than 0.1 eV. Such a band would result in *a strongly-temperature-dependent susceptibility*. Such behaviour is well known in some heavy Fermion compounds for example. The absence of such a strongly T-dependent susceptibility appears to effectively exclude a simple VH scenario for explaining the susceptibility, and one has to invoke an unusual form of strongly electronically correlated state to understand these data.

The usual phenomenological approach to such strongly correlated systems is to invoke a Stoner enhancement factor S, such that $\chi = \chi_0 S$, where S $=[1-IN(E)]^{-1}$ and I is the exchange-correlation integral. Stoner's criterion for formation of a FM state is $IN(E) > 1$. From the data in Figure 2 at $H = 1$ kOe, we obtain S = 20 and at $H = 10$kOe, S=3.3. Both of these values are very large, indicating that the product $IN(E)$ is very close to 1, implying a proximity to a ferromagnetic (FM) state.

Considering the weak inter-wire coupling from mechanical measurements, we consider some Luttinger liquid models of strongly correlated electrons in one dimension. A number of predictions have been calculated in the past[6], the most recent being the one of Nelisse et al [7]. The general prediction indeed gives a T-independent susceptibility, but the magnitude is similar to the Pauli susceptibility χ_P

and is given by $\chi_{LL} = \mu_B^2/t$, where t is the hopping integral. Estimating $t = 0.1$, we obtain $\chi_{LL} \approx 10^{-4}$ emu/mol, which is far too small to account for the observed value of χ. We conclude that the present models of strongly interacting electrons in one dimension do not apply to the present system.

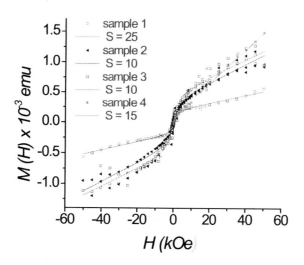

FIGURE 3. Magnetisation of a number of different Li:Mo-S samples. The lines are fits to the data using different values of spin S in the Brillouin function.

Our general understanding of the observed behaviour is as follows. The system apparently exists in a highly correlated spin state close to a ferromagnetic instability. The magnetisation behaviour $M(H)$ indeed confirms this (Figure 3). A fit to the data using a sum of two terms to the $M(H) = \chi^*H + M_0B[T,H,S]$, where $B[T,H,S]$ is the Brillouin function, gives very good agreement. The non-linear behaviour at the origin desecribed by the second term in the magnetisation appears to be clear evidence of the existence of superparamagnetic clusters with S = 10 ~ 25, as expected for a system close to a FM instability. However, the remaining linear term implies a high-field susceptibility $\chi^* = 2 \times 10^{-3}$ emu/mol, which is still very large compared to typical Pauli susceptibility values, and implies that a paramagnetic state exists up to very high fields and over an exceptionally large temperature range. Normal materials never reach such high susceptibilities, but rather prefer to form a spin-ordered state, or even a superconducting one, which is very common in some families of transition metal chalcogenide compounds. The most plausible explanation for the observed behaviour is that the paramagnetic 1D state is stabilised by fluctuations. This effect would be entirely due to the 1D nature of the system, and would not be present in higher dimensions. At high temperature, spin fluctuations, as well as structural ones (including normal phonons) are important. At low temperatures however, these fluctuations would be expected to diminish, and in the absence of quantum fluctuations, the system would be expected to have a quantum critical point at $T = 0$.

The present data measured down to 2K do not indicate the existence of an ordered state down to 2K, suggesting that quantum fluctuations in 1D may prevent the system from ordering even at $T = 0$. Nevertheless, magnetic measurements performed well below 2K might still reveal the existence of a critical behaviour approaching a QCP as $T \rightarrow 0$.

ACKNOWLEDGMENTS

We wish to acknowledge the ESF MOLMAG programme for support and the Valencia group of E. Coronado for their hospitality on their SQUID magnetometer.

REFERENCES

1. A.Kis, D.Mihailovic, A.Mrzel, A.Jesih, I.Piwonoski, A.J.Kulik, W.Benoit, L.Forro. Advanced Materials, to be published (2003)
2. M.Remskar, A.Mrzel, Zora Skraba, Adolf Jesih, Miran Ceh, Jure Demsar, Pierre Stadelman, Francis Levy and D.Mihailovic, Science **292**, 479 (2001)
3. D.Mihailovic, Z.Jaglicic, D.Arcon, A.Mrzel, A. Zorko, M.Remskar and V.V.Kabanov, R.Dominko, and C. J. Gómez-García. J. M. Martínez-Agudo. E. Coronado., Phys. Rev. Lett. **90**, 146401 (2003)
4. R. Doinko, D.Arcon. A.Mrzel, A.Zorko, A., P.Cevc, P.Venturini, M. Gaberscek, M.Remskar, D.Mihailovic. Advanced Materials **14,** 1531-1534 (2002)
5. D.Arcon, A,Zorko, P.Cevc, A.Mrzel, M.Remskar, R.Dominko, M.Gaberscek, D.Mihailovic Phys.Rev.B **67**, 126423 (2003)
6. G.Juttner et al., Nucl.Phys. B **522**, 471 (1998)., F.Mila and K.Penc, Phys. Rev. B **51**, 1997 (1995)
7. H.Nelisse et al., Eur.Phys.J. B **12**, 351 (1999)

Temperature dependent ESR of doped chalcogenide nanotubes

Denis Arčon[*], Andrej Zorko[†], Pavel Cevc[†], Aleš Mrzel[†], Maja Remškar[†], Robert Dominko[**], Miran Gaberšček[**] and Dragan Mihailović[†]

[*]Faculty of Mathematics and Physics, University of Ljubljana, Jadranska 19, 1000 Ljubljana, Slovenia
[†]Institute Jožef Stefan, Jamova 39, 1000 Ljubljana, Slovenia
[**]National Chemistry Institute, Hajdrihova 19, 1000 Ljubljana, Slovenia

Abstract. Recently discovered single-wall subnanometer-diameter molybdenum disulfide tubes (nMoS$_2$) were electrochemically doped with Li and then studied with X-band ESR. While undoped nMoS$_2$ show no X-band ESR signal between room temperature and 4 K we found in heavily doped nMoS$_2$ samples two distinct ESR components: a narrow component with a linewidth of few Gauss and attributed to the formation of small spin clusters and a broad component with a linewidth of more than 800 G. The broad ESR component is characteristic of Mo d-orbital-derived band. The temperature dependence of the ESR spin susceptibility and the linewidth of the broad ESR component can be discussed either in terms of conducting electrons coupled to defects or in terms of random-exchange Mo Heisenberg chain model.

INTRODUCTION

Fullerene-like objects can be prepared from many different inorganic layered materials. Dichalcogenide fullerene-like and tube-like nanoparticles[1, 2, 3, 4] were reported. Very recently the synthesis of self-assembled single-wall subnanometer-diameter molybdenum disulfide tubes (nMoS$_2$) was reported [5]. A structural investigation of bundles of nMoS$_2$ revealed that identical nanotubes are packed into hexagonal array with the nanotube center-to-center distance being 0.96 nm [5]. Individual nMoS$_2$ nanotubes are proposed to consist of sulfur-molybdenum-sulfur cylinders. This material is due to a high degree of order and uniformity of MoS$_2$ nanotubes in each bundle suitable for a study of microscopic properties of nanotubes even with a bulk techniques like electron-paramagnetic resonance (EPR) or nuclear magnetic resonance (NMR).

Here we report on the study of the low-temperature electronic properties of electrochemically doped nMoS$_2$ with Li. We have found that large amounts of Li$^+$ ions can be intercalated into the channels between the individual nMoS$_2$ nanotubes. Intercalation goes mainly through one-dimensional diffusion of Li$^+$ ions into the host structure donating an electron to the nMoS$_2$ nanotubes. It seems that the Li$^+$ ion intercalation leads to a formation of many defects, which effectively decrease the nMoS$_2$ tube length as compared to the electron correlation length. These defects, which can be either topological or paramagnetic or even a result of inhomogeneous doping, determine the low-temperature electronic properties of doped nMoS$_2$ nanotubes and may be responsible for the relatively large irreversible losses during the first charging cycle.

CP685, Molecular Nanostructures: XVII Int'l. Winterschool/Euroconference on
Electronic Properties of Novel Materials, edited by H. Kuzmany, J. Fink, M. Mehring, and S. Roth
© 2003 American Institute of Physics 0-7354-0154-3/03/$20.00

EXPERIMENTAL DETAILS

Single-wall nMoS$_2$ were grown by a standard method published earlier [5]. Electro-chemical lithium insertion into nMoS$_2$ NTs bundles or 2H MoS$_2$ was carried out in a laboratory-made three-electrode cell [6]. The working electrodes were prepared by mixing as-grown nMoS$_2$ nanotube bundles or dispersed nMoS$_2$ nanotube bundles with 10 % wt. of sulphonated polyaniline (for improved electrical contact) [7]. Before assembling, the working electrode and separators (Celgard No.2402) were soaked with electrolyte (1M LiClO$_4$ solution in propylene carbonate).

Typically the charge capacity of as-grown nMoS$_2$ sample is in the first cycle approximately 385 mAh/g, which corresponds to an intercalation of about 2.3 Li per MoS$_2$ [7]. This is significantly higher than in layered MoS$_2$ with the same cell configuration, where we could intercalate up to $\sim$ 1 Li per one MoS$_2$ unit.

After electrochemical doping, the Li$_x$$nMoS_2$ sample was scratched from the substrate in a glove box and sealed into an ESR quartz tube. Continuous wave (cw) X-band ESR measurements were performed on a Varian dual resonator with a reference sample in the second resonator to account for the slight changes in the Q-factor during the measurements. The temperature was stabilized within 0.2 K in a continuous flow cryostat ESR 900. To estimate the ESR spin susceptibility Cu(SO$_4$)$_2$·5H$_2$O has been used as a standard sample.

RESULTS AND DISCUSSION

As prepared nMoS$_2$ samples show no X-band ESR signal at any temperature between 300 K and 4 K. This means that if nMoS$_2$ were metallic then the density of states at the Fermi level should be extremely small. However electrochemical doping of nMoS$_2$ lead to the observation of a strong ESR signal (see the inset to Fig. 1). A typical ESR spectrum of Li$_x$$nMoS_2$ (for $x \sim 2.3$) consists of two quite different components: a narrow component with a g-factor $g_1 = 2.0029$ and with a linewidth of few Gauss and a broad component with a g-factor $g_2 \sim 2.15$ and linewidth of more than 800 G. The narrow component is attributed to the formation of small spin clusters. Its temperature dependence and origin will be discussed in detail elsewhere [8].

We stress here that the relative intensities of the two components vary from sample to sample depending on the sample pretreatment and quality. The fact that we found in doped samples prepared from dispersed nMoS$_2$ only a very weak narrow ESR component strongly suggests that Li$^+$ ions mainly occupy inter-tubular sites and not intra-tubular ones. It also suggests that the broad ESR line is connected with the same intercalation process.

The measured broad ESR component is very similar to the broad line observed in Rb saturated layered MoS$_2$. The line in these samples was assigned to d-band conduction electrons of the host 2H-MoS$_2$ that are donated by the Rb intercalant [9]. A similar line has also been measured in Li doped layered 2H-MoS$_2$ [8]. We also note that it has been theoretically suggested[10] that in nMoS$_2$ the Mo d states especially dominate the upper valence band and the lower conduction band edge. The assignment of our broad

ESR signal to electron spin resonance of Li donated electrons in the Mo d-state derived conducting band of nMoS$_2$ is thus a very likely possibility.

However a temperature dependence of the broad ESR component of Li$_{2.3}n$MoS$_2$ shown on Fig. 1 is rather atypical for a conducting electron spin resonance (CESR) in a one-dimensional system. The intensity (Fig. 1a) of the ESR line gradually increases on cooling. At the same time the ESR line broadens (Fig. 1b). However, as the broadening is very strong it gets nearly impossible to measure the line below 40 K (see for instance the inset to Fig. 1b). Traditionally, it is assumed that the observed CESR is broadened and shifted by the so-called Elliot mechanism. Within this model the CESR linewidth should decrease with decreasing temperature. This is obviously not the case for the broad ESR line in Li$_x n$MoS$_2$. Furthermore, assuming a 1D electronic structure in Li$_x n$MoS$_2$, the spin-orbit scattering should be reduced as the momentum space into which the electrons can scatter is also reduced. One thus expects much narrower lines than they were observed.

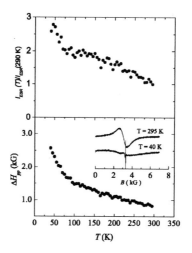

FIGURE 1. Temperature dependence of (a) the relative intensity and (b) the linewidth of the broad ESR component measured in Li$_{2.3}n$MoS$_2$. In the inset we compare the spectra measured at 295 K and 40 K.

We note however that doping of Li can lead to a formation of defects, which "cut" the individual nanotube to short segments (Fig. 2). Furthermore if the doping would be inhomogeneous, the conducting electrons would feel a potential which varies from site to site. In such a case, the electrons may even get localized within individual segments of the nanotube at low temperatures. The measured ESR signal is then a sum of the contributions from nanotube segments of different length

$$I(\omega) = \frac{1}{N} \sum_{k=1}^{N} \chi_k \frac{\Gamma_k}{\pi(\omega^2 + \Gamma_k^2)}. \tag{1}$$

Here χ_k and Γ_k are the susceptibility and the linewidth of the electrons localized at a tube segment k. N is simply a number of different segments. The line is thus expected to become inhomogeneously broadened at low temperatures. At higher temperatures the electrons may tunnel between different segments of the nanotube and the signal should become more and more symmetric as the temperature is increased. Such a behavior has been indeed observed. Within this model the conductivity should exhibit activated type of behavior.

FIGURE 2.   Schematically presented view of the $Li_x nMoS_2$ ground state. Please note that the "defects" (grey circles) "cut" the nanotube to shorter segments.

CONCLUSIONS

In conclusion, temperature dependence of the X-band ESR signal has been measured in $Li_x nMoS_2$. The results suggest intercalation of Li^+ ions into channels between individual nanotubes accompanied by a charge transfer to the host $nMoS_2$. A very inhomogeneous ground state is formed with many defects which could ultimately cut the nanotube into shorter segments and may lead to localization of electrons within individual segment.

ACKNOWLEDGMENTS

AD acknowledges financial support of NATO through SfP 976913 grant.

REFERENCES

1. R. Tenne, L. Margulis, M. Genut, G. Hodes, *Nature* **360**, 444 (1992).
2. L. Margulis, G. Salitra, R. Tenne, M. Talianker, *Nature* **365**, 113 (1993).
3. M. Remškar, Z. Škraba, F. Cleton, R. Sanjines, and F. Lévy, *Appl. Phys. Lett.* **74**, 633 (1999).
4. M. Remškar, Z. Škraba, M. Regula, C. Ballif, R. Sanjinés, F. Lévy, *Adv. Mater.* **10**, 246 (1998).
5. M. Remškar, A. Mrzel, Z. Škraba, A. Jesih, M. Čeh, J. Demšar, P. Stadelmann, F. Lévy, D. Mihailović, *Science* **292**, 479 (2001).
6. M. Gaberscek, M. Bele, R. Dominko, J. Drofenik, S. Pejovnik, *Electrochem. Solid-State Lett.*, **3**, 171 (2000).
7. R. Dominko, D. Arčon, A. Mrzel, A. Zorko, P. Cevc, P. Venturini, M. Gaberšček, M. Remškar, and D. Mihailović, *Adv. Mater.* **14**, 1531 (2002).
8. D. Arčon, A. Zorko, P. Cevc, A. Mrzel, M. Remškar, R. Dominko, M. Gaberšček, and D. Mihailovič, *Phys. Rev.* **B67**, 125423 (2003).
9. S. Bandow, Y. Maruyama, X.X. Bi, R. Ochoa, J.M. Holden, W.T. Lee, and P.C. Eklund, *Mat. Sci. Eng.* **A204**, 222 (1995).
10. G. Seifert, H. Terrones, M. Terrones, G. Jungnickel, and T. Frauenheim, *Phys. Rev. Lett.* **85**, 146 (2000).

THEORY OF NANOSTRUCTURES

Topological coordinates for deformed nanotubes

István László * and André Rassat [†]

*Department of Theoretical Physics, Institute of Physics, Budapest University of Technology and Economics, BUTE Center for Applied Mathematics and Computational Physics, H-1521 Budapest, Hungary
[†]Département de Chimie, Ecole Normale Supérieure, 24 rue Lhomond, 75231 Paris Cedex 05, France

Abstract. Starting from the topological arrangement of carbon atoms an algorithm is given for the construction of nanotube Cartesian coordinates. The final relaxed structures were obtained by a molecular mechanics calculation where the carbon-carbon interactions were supposed only between neighboring atoms of the initial tiling. In a given tiling we obtained toroidal or helical structures depending on the special position of the super cell parallelogram.

INTRODUCTION

The geometric and electronic properties of polyhex carbon nanotubes can be changed significantly by appropriate distribution of pentagonal and heptagonal defects. Thus we can speak toroidal coiled and Y-branched carbon nanotubes[1, 2, 3, 4, 5, 6, 7, 8, 9, 10, 11]. Here we present a method for the construction of the (x, y, z) Cartesian coordinates of atoms in toroidal and helical carbon structures. Starting from the topological coordinate method for fullerenes[12, 13] and tori[14, 15] we can produce these coordinates from only the topological arrangement of the carbon atoms. The final relaxed structures were obtained by a molecular mechanics method, based on the Brenner potential[16]. Here we present only the algorithm of the method. The technical details and the mathematical background can be found in reference [11].

THE METHOD

Let us suppose that the topological arrangement of carbon atoms is described by the $\mathbf{a}_1$, $\mathbf{a}_2$ unit cell vectors and the $\mathbf{b}_1 = n\mathbf{a}_1 + m\mathbf{a}_2$, $\mathbf{b}_2 = p\mathbf{a}_1 + q\mathbf{a}_2$ super cell vectors, where n, m, p and q are integers. The tiling is given by the neighbors of the n_c atoms in the unit cell. The graph G of the nanotube is obtained by identifying two opposite edges of the parallelogram (the super cell), and the identification of each pair of opposite edges yield the graph G of the torus. Using only the n, m, p and q integers and the connectivity structure of the carbon atoms of the unit cell we can construct the (x, y, z) Cartesian coordinates of the carbon atoms of a nanotube or torus in the following way.

From n, m, p and q we construct the $\mathbf{A}$ adjacency matrix of the corresponding torus with $A_{ij} = 1$ if i and j are adjacent and $A_{ij} = 0$ otherwise. We chose four bilobal

CP685, *Molecular Nanostructures: XVII Int'l. Winterschool/Euroconference on Electronic Properties of Novel Materials*,edited by H. Kuzmany, J. Fink, M. Mehring, and S. Roth
© 2003 American Institute of Physics 0-7354-0154-3/03/$20.00

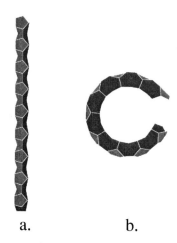

a.   b.

FIGURE 1. Topological coordinates for $(n,m,p,q) = (1,0,0,9)$ in tiling I (a.) and the corresponding relaxed structure (b.)

eigenvectors $\mathbf{c}^{k_1}$, $\mathbf{c}^{k_2}$, $\mathbf{c}^{k_3}$ and $\mathbf{c}^{k_4}$, and the topological coordinates of the nanotube are calculated as

$$x_i = S_3 C_i^{k_3}, \tag{1}$$

$$y_i = S_4 C_i^{k_4}, \tag{2}$$

$$z_i = R \arccos(S_1 C_i^{k_1}/R) \quad if \quad C_i^{k_2} \geq 0 \tag{3}$$

and

$$z_i = R(2\pi - \arccos(S_1 C_i^{k_1}/R)) \quad if \quad C_i^{k_2} < 0 \tag{4}$$

These formula are obtained by a simple transformation from our formula for the torus. This is why $\mathbf{A}$ is the adjacency matrix of the corresponding torus. For the definition of bilobal eigenvectors, the S_i scaling factors and the R radius see [11, 12, 13, 14, 15]. From the topological coordinates the final structure was obtained with the help of molecular mechanics calculation based on the Brenner potential[16] using interactions between first neighbors defined in the tiling.

RESULTS

In the present work tiling *I* and tiling *II* were studied. In tiling *I* there were only pentagonal and heptagonal faces and tiling *II* contained pentagonal, heptagonal and hexagonal faces. Figures 1. and 2. show results of tiling *I* and in Figures 3. and 4. structures obtained from tiling *II* are presented. In each cases Figure a. presents the structures obtained by the topological coordinates, and the corresponding relaxed structures are in Figures b.

FIGURE 2. Topological coordinates for $(n,m,p,q) = (1,-1,5,5)$ in tiling I (a.) and the corresponding relaxed structure (b.)

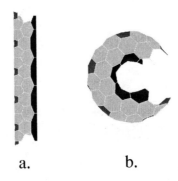

FIGURE 3. Topological coordinates for $(n,m,p,q) = (1,0,0,5)$ in tiling II (a.) and the corresponding relaxed structure (b.)

The relaxed structure depends both the tiling and the super cell. The pentagons are gei-nerating the positive Gaussian curvatures and the heptagons the negative ones. The final structures have toroidal or helical shapes depending on the distribution of pentagons, heptagons and the position of the super cell. The purely polyhex tiling produces only nanotubes independent of the super cell, as there are not pentagons and heptagons that are necessary for positive and negative Gaussian curvatures.

FIGURE 4. Topological coordinates for $(n,m,p,q) = (1,-2,4,2)$ in tiling II (a.) and the corresponding relaxed structure (b.)

ACKNOWLEDGMENTS

I.L. thanks the grants from OTKA (T038191 and T043231).

REFERENCES

1. Macky, A. L., and Terrones, H., *Nature (London)*, **352**, 762 (1991).
2. Dunlap, B. I., *Phys. Rev. B*, **46**, 1933 (1992).
3. Chernozatonskii, L. A., *Phys. Lett. A*, **172**, 173 (1992).
4. Scuseria, G. E., *Chem. Phys. Lett.*, **195**, 534 (1992).
5. Ihara, S., Itoh, S., and Kitakami, J.-I., *Phys. Rev. B*, **47**, 12908 (1993).
6. Ihara, S., Itoh, S., and Kitakami, J.-I., *Phys. Rev. B*, **48**, 5643 (1993).
7. Dunlap, B. I., *Phys. Rev. B*, **49**, 5643 (1994).
8. Terrones, H., Terrones, M., Hernández, E., Grobert, N., Charlier, J.-C., and Ajayan, P., *Phys. Rev. lett*, **84**, 1716 (2000).
9. László, I., and Rassat, A., *Int. J. Quantum Chem.*, **84**, 136 (2001).
10. Bíró, L. P., Márk, G. I., Koós, A. A., Nagy, J. B., and Lambin, P., *Phys. Rev. B*, **66**, 165405 (2002).
11. László, I., and Rassat, A., *J. Chem. Inf. Comput. Sci.*, **43**, 519 (2003).
12. Manolopoulos, D. E., and Fowler, P. W., *J. Chem. Phys.*, **96**, 7603 (1992).
13. Fowler, P. W., and Manolopoulos, D. E., *An atlas of fullerenes*, Clarendon Press, Oxford, 1995.
14. László, I., Rassat, A., Fowler, P. W., and Graovac, A., *Chem. Phys. Letters*, **342**, 369 (2001).
15. László, I., Rassat, A., Fowler, P. W., and Graovac, A., "Topological Coordinates for Carbon Nanostructures," in *Electronic Properies of Molecular Nanostructures*, edited by H. Kuzmany, J. Fink, M. Mehring, and S. Roth, AIP Conference Proceedings 591, American Institute of Physics, Melville, New York, 2001, pp. 438–441.
16. Brenner, D. W., *Phys. Rev. B*, **42**, 9458 (1990).

Ab initio studies of electron-phonon coupling in single-walled nanotubes

M. Machón[*], S. Reich[*†], J. M. Pruneda[**], C. Thomsen[*] and P. Ordejón[†]

[*]*Institut für Festkörperphysik, Technische Universität Berlin, Hardenbergstr. 36, 10623 Berlin, Germany*
[†]*Institut de Ciència de Materials de Barcelona (CSIC), Campus de la U.A.B. E-08193 Bellaterra, Barcelona, Spain*
[**]*Dept. of Earth Sciences, Cambridge University, Downing Street, CB2 3EQ, Cambridge U.K.*

Abstract.
We present *ab initio* calculations of electron-phonon coupling in single-walled nanotubes and graphene. The perturbation of the electronic energies due to the atomic distortion caused by totally symmetric phonons was calculated, yielding the matrix elements of the electron-phonon interaction. For the radial breathing mode (RBM) we obtained a decrease of the electron-phonon interaction with increasing diameter. This is in good agreement with the fact that the equivalent mode for graphene is an out-of-plane translation which cannot affect the electronic system. The matrix elements for the RBM and the optical A_{1g} mode show different behaviours for armchair and zig-zag nanotubes.

Raman spectroscopy is a powerful technique, which yields not only information about the vibrational properties of physical systems, but also about the electronic properties and the interaction between electrons and lattice vibrations. For a single resonant Raman process, the Raman scattering cross section is given by [1]:

$$P \propto \left| \sum_{a,b} \frac{<f|H_{e-r}|b><b|H_{e-ph}|a><a|H_{e-r}|i>}{(E_{laser} - E_{ai} - i\gamma)(E_{laser} - \hbar\omega_{ph} - E_{bi} - i\gamma)} \right|^2. \tag{1}$$

In this work, we are interested in the electron-phonon coupling matrix element $< b|H_{e-ph}|a >$, which scales the intensity of the Raman peaks. We will concentrate on the totally symmetric vibrational modes which give origin to two of the main features of the Raman spectrum of carbon nanotubes [2]. A totally symmetric phonon can only cause transitions between electronic states of the same symmetry, thus, we will concentrate on diagonal matrix elements. Khan *et al.* [3] showed that such a matrix element can be related to the shift of the electronic bands under the atomic displacement pattern of the phonon as follows:

$$< \mathbf{k}, n|H^i_{e-ph}|\mathbf{k}, n > = \sum_a \sqrt{\frac{\hbar}{2MN\omega^i}} \varepsilon^i_a \frac{\partial E_n(\mathbf{k})}{\partial \mathbf{u}_a}, \tag{2}$$

where the sum runs over all atoms in the unit cell. $\mathbf{k}$ and n denote, respectively, the wave vector and band index of the electronic state, i indexes the phonon, M is the atomic mass, N the number of atomic cells taken into account, ε^i is the polarization vector of the phonon, $E_n(\mathbf{k})$ the electronic energy and $\mathbf{u}_a$ the atomic displacement.

CP685, *Molecular Nanostructures: XVII Int'l. Winterschool/Euroconference on Electronic Properties of Novel Materials,* edited by H. Kuzmany, J. Fink, M. Mehring, and S. Roth

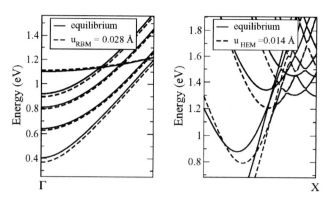

FIGURE 1. **Left:** Solid lines: equilibrium band structure of a (10,0) nanotube. Dotted lines: band structure of a (10,0) nanotube distorted after the pattern of the RBM, with an atomic displacement of 0.028 Å. The first quarter of the Brillouin zone is represented. **Right:** Solid lines: equilibrium band structure of an (8,8) nanotube. Dotted lines: Band structure of an (8,8) nanotube distorted after the pattern of the A_{1g} optical mode, with an atomic displacement of 0.014 Å. The last fourty percent of the Brillouin zone is shown. In both panels the Fermi energy lies at 0 eV.

Following this idea, we studied the electron-phonon coupling for totally symmetric phonons based on *ab initio* band structure calculations for achiral single-walled nanotubes with diameters between 4 and 15 Å. All calculations were performed with the SIESTA code [4, 5] within the local density approximation, as parameterized by Perdew and Zunger.[6] The core electrons were replaced by non-local norm-conserving pseudopotentials.[7] A double-ζ, singly polarized basis set of localized atomic orbitals was used for the valence electrons, with cutoff radii of 5.12 a.u. for the s and 6.25 a.u. for the p and d orbitals as determined from an energy shift of 50 meV by localization. [8, 9]

First, the structures were relaxed minimizing the atomic forces down to 0.04 eV/Å and the phonon spectrum at the Γ-point was calculated by the finite differences method. Then, the electronic band structure was calculated for the relaxed geometry and for different values of the atomic displacement. The shift of the bands showed a linear dependence on the atomic displacement, as expected for small displacements.

As can be seen in Fig. 1 both displacement patterns yield band shifts, but no splitting, as expected for an A_{1g} phonon. The band shift shows a strong dependence on the specific electronic state under study. In this work, we are interested on the effect of the electron-phonon interaction on Raman spectra, thus we focus in the regions of the Brillouin zone in which the absorption of photons takes place, that is the Γ-point for zig-zag nanotubes and the region close to the K-point for armchair nanotubes.

In Table 1 the calculated electron-phonon matrix elements for the RBM at the conduction and valence bands yielding the optical transition of lowest energy are listed. The highest matrix elements correspond to the smallest nanotubes, they become lower for bigger radii. The same RBM yields a smaller change on the bond lengths for bigger diameters, so the effect on the electronic states is weaker. This is in agreement with the fact that the RBM mode corresponds to an acoustic out-of-plane mode in graphene, which

428

TABLE 1. Calculated diameters, frequencies and electron-phonon coupling matrix elements (in eV) for the RBM at the absorbing region of the Brillouin zone for the lowest conduction band (c) and the highest valence band (v).

	(5,0)	(3,3)	(6,0)	(10,0)	(6,6)	(8,8)	(15,0)	(11,11)	(19,0)
d(Å)	4.1	4.2	4.8	7.9	8.2	10.9	11.8	15.0	15.0
ω (cm^{-1})	520	542	446	287	278	209	188	151	149
c	**0.30**	0	**0.07**	**0.23**	0.04	0.05	**0.17**	0.03	**0.16**
v	**0.86**	1.56	**0.03**	**0.22**	0.11	0.09	**0.26**	0.05	**0.16**

TABLE 2. Same as Table 1 for the A_{1g} optical mode.

	(5,0)	(3,3)	(6,0)	(10,0)	(6,6)	(8,8)	(15,0)	(11,11)	(19,0)
ω (cm^{-1})	1598	1531	1587	1661	1626	1623	1567	1611	1635
c	**0**	0.52	**0.36**	**1.00**	0.58	0.38	**1.07**	-	**1.03**
v	**1.5**	0.61	**0.62**	**1.00**	0.43	0.32	**1.06**	-	**1.01**

cannot affect the electronic system. Thus, the electron-phonon coupling must tend to zero in the limit of infinite diameter. This trend becomes clearer when separating zig-zag and armchair nanotubes. The matrix elements of zig-zag nanotubes are a factor of 5 higher than those of armchair nanotubes with similar radii, except for the smallest nanotubes which do not share the general features of the bigger nanotubes due to the high curvature.

In Table 2 the analogous matrix elements are shown for the A_{1g} optical mode. Again, we see different behaviors for armchair and zig-zag nanotubes. For zig-zag nanotubes bigger than 5 Å, the electron-phonon coupling matrix elements are practically constant, while for armchair nanotubes they tend to diminish for increasing radius. This trend is characteristic of the bands closest to the Fermi energy, and is related to the vanishing coupling at the K-point. With increasing radius, the absorbing zone of the band shifts towards the K-point and the matrix element tends to vanish.

Comparing the two modes, is clear that the matrix elements of the A_{1g} optical mode are, in general, significantly higher than those of corresponding to the RBM.

In Fig. 2 the band structure of graphene close to the K-point is shown for equilibrium geometry, and for displaced atoms after the E_{2g} in-plane optical phonon which corresponds to the optical A_{1g} mode in achiral nanotubes. Within the zone folding scheme (only applicable to nanotubes with big diameters), the Γ-point of the lowest absorbing bands of big zig-zag nanotubes correspond to the shown graphene bands close to the K-point. As can be seen, close to the K-point, the band shift is almost constant, which explains the constant matrix elements obtained for zig-zag tubes. However, changes may be expected for bands with higher energies.

Summarizing, we studied the electron-phonon interaction in achiral carbon single-walled nanotubes. We were able to show different behaviors for zigzag and armchair nanotubes, for both the RBM and the A_{1g} optical mode. For the RBM matrix element, a decreasing trend with increasing radius was found, as well as a factor 5 difference

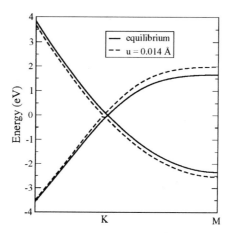

FIGURE 2. Electronic band structure of graphene. Solid line: equilibrium geometry. Dotted line: atoms displaced after the E_{2g} in-plane optical mode analogous to the optical A_{1g} mode in achiral nanotubes.

between both chiralities; the zig-zag nanotubes have the highest matrix elements. The A_{1g} optical mode was shown to have larger matrix elements than the RBM, which are practically constant for the zig-zag nanotubes, and decreasing for increasing radius for the armchair nanotubes, at least for the bands close to the Fermi energy.

ACKNOWLEDGMENTS

We acknowledge the Ministerio de Ciencia y Tecnología (Spain) and the DAAD (Germany) for a Spanish-German Research action (HA 1999-0118). P. O. acknowledges support from Fundación Ramón Areces (Spain), EU project SATURN, and a Spain-DGI project. S.R. acknowledges a fellowship by the Akademie der Wissenschaften Berlin-Brandenburg.

REFERENCES

1. Martin, R. M., and Falicov, L. M., "Resonant Raman Scattering," in *Light Scattering in Solids I: Introductory Concepts*, edited by M. Cardona, Springer-Verlag, Berlin, Heidelberg, New York, 1983, vol. 8 of *Topics of Applied Physics*, p. 79.
2. Maultzsch, J., Reich, S., and Thomsen, C., *Phys. Rev. B*, **65**, 233402 (2002).
3. Khan, F., and Allen, P., *Phys. Rev. B*, **29**, 3341 (1984).
4. Sanchez-Portal, D., Ordejón, P., Artacho, E., and Soler, J. M., *Int. J. Quantum Chem.*, **65**, 453 (1997).
5. Soler, J. M., Artacho, E., Gale, J. D., García, A., Junquera, J., Ordejón, P., and Sánchez-Portal, D., *J. Phys. Condens. Mat.*, **14**, 2745 (2002).
6. Perdew, J. P., and Zunger, A., *Phys. Rev. B*, **23**, 5048 (1981).
7. Troullier, N., and Martins, J., *Phys. Rev. B*, **43**, 1993 (1991).
8. Junquera, J., Paz, O., Sánchez-Portal, D., and Artacho, E., *Phys. Rev. B*, **64**, 235111 (2001).
9. Artacho, E., Sánchez-Portal, D., Ordejón, P., García, A., and Soler, J., *phys. stat. sol. (b)*, **215**, 809 (1999).

Optical and Vibrational Spectra of Narrow Nanotubes: A Symmetry Based Approach

I. Milošević, B. Nikolić, E. Dobardžić and M. Damnjanović

Faculty of Physics, University of Belgrade, P. O. Box 368, Belgrade 11001, Serbia and Montenegro

Abstract. By use of the tight binding method for induced representations (based on the line group symmetry) polarized optical conductivity, radial breathing and high energy vibrational mode frequencies of the narrow single-wall carbon nanotubes are calculated. The absorption spectra features are assigned by the complete set of conserved quantum numbers and the results obtained are discussed in relation to the previously reported calculations and measurements.

Recently, narrow single-wall carbon nanotubes (SWCNTs) have been synthesized in a porous zeolite [1]. By high-resolution transmission electron microscope diameter of these tubes is determined to be 0.42 ± 0.02 nm [2, 3]. The polarized optical absorption spectra of the narrow SWCNTs has been measured [2] and modeled by first principle electronic band structure calculations based on local density function approximation and SWCNTs with chirality (5,0), (3,3) and (4,2) are suggested to be the only possibilities [2, 4, 5]. Apart from the dipole transitions assignment [2, 5], based, however, on the (more or less) incorrectly determined isogonal point groups, symmetry has not been used in the previously reported calculations. As for the vibrational properties, radial breathing mode (RBM) frequency for tubes (5,0), (3,3) and (4,2) has been calculated by H. J. Liu and C. T. Chan [4] and compared to the (by Z. K. Tang and X. D. Xiao) measured Raman spectra.

In this Contribution we evaluate numerically, using the line group theoretical methods [6], parallelly polarized (i.e. along the tube axis) optical conductivity for the nanotubes (5,0), (4,2), (3,3) and (5,1) in the energy region $0 - 4$ eV. Also we carry out calculations of the RBM and high energy mode (HEM) frequencies in these tubes.

The full line group symmetry [7] of the SWCNTs is used and the calculations are carried out by use of the *POLSym* package [6] which is based on the tight binding (TB) method for representations of the induced type [8]. In the calculations of the electronic band structure, the results of the density functional tight binding (DFTB) calculations [9] are taken as input data. By using a basis set consisting of one s and three p orbitals (per carbon atom) hybridization of the graphitic σ, π, σ^* and π^* states is taken into account. Within the dipole approximation, the optical transition matrix elements are calculated out of the completely symmetry adapted Bloch functions [10]. More details on the method of the optical absorption calculations can be found in Ref. [11]. As for the vibrational spectra calculations, we use the force-constant model, starting with the graphite constants [12] and adjusting them (kinematically and dynamically) to the geometry of a SWCNT [13].

The calculated (real part of) optical conductivity for the nanotubes (5,0), (3,3), (4,2)

CP685, *Molecular Nanostructures: XVII Int'l. Winterschool/Euroconference on Electronic Properties of Novel Materials,* edited by H. Kuzmany, J. Fink, M. Mehring, and S. Roth
© 2003 American Institute of Physics 0-7354-0154-3/03/$20.00

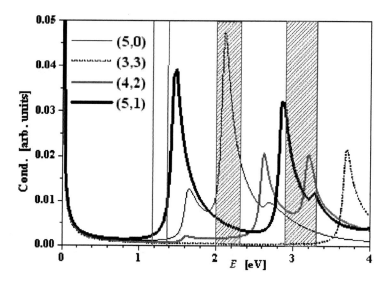

FIGURE 1. Calculated parallelly polarized component of the optical conductivity for the nanotubes (5,0), (3,3), (4,2) and (5,1). Vertical lines indicate the measured absorption spectra features [2]: sharp peak $\sim 1.4\,\text{eV}$ and broad bands (shadowed) $\sim 2.1\,\text{eV}$ and $\sim 3.1\,\text{eV}$.

and (5,1), in a case of the parallel polarization of the incoming field, is given in Fig. 1. First order time-dependent perturbation theory is used and SWCNTs are assumed to be ideally structured, infinite and isolated. (Diameters of the tubes are given in Tab. 1, last row.) The symmetry assignation of the dipole inter-band transitions corresponding to the calculated absorption spectra features is given in Tab. 1. It is found to be quite different from the one presented in Ref. [2]. Namely, symmetry transformations of SWCNTs form non symmorphic line groups [7] which means that for SWCNTs, isogonal point groups are not subgroups of the symmetry groups. In particular, $\mathbf{D}_{10h}$, $\mathbf{D}_{6h}$, $\mathbf{D}_{28}$ and $\mathbf{D}_{62}$ are the isogonal point groups of the nanotubes (5,0), (3,3), (4,2) and (5,1) (being not in accordance with the groups used in Refs. [4, 5]) while the axial point group factors of their full symmetry groups are $\mathbf{D}_{5h}$, $\mathbf{D}_{3h}$, $\mathbf{D}_2$ and $\mathbf{D}_1$, respectively. Consequently, the angular momentum component (along the tube axis) m for these tubes (in precisely the order given above) takes on the values from the intervals $(-5, 5]$, $(-3, 3]$, $(-14, 14]$ and $(-31, 31]$. Besides by m, the irreducible representations are characterized also by the quasi-momentum k and eventually, by the parities: with respect to the two-fold horizontal axis and (for the achiral tubes only) with respect to the vertical and horizontal mirror planes.

RBM and HEM frequencies for the nanotubes (4,0), (4,1), (5,0), (3,3), (4,2), (5,1), (6,0), (4,3) and (5,2) are presented in Fig. 2 and Fig. 3, respectively. The experimentally observed RBMs in the measured Raman spectra (by Z. K. Tang and X. D. Xiao [4]) are denoted by horizontal lines. Obviously, the *POLSym* calculated RBM frequencies of the tubes (3,3), (4,2) and (5,1) match the measured values excellently. Also, RBM

TABLE 1. Interband transitions for the nanotubes (5,0), (3,3), (4,2) and (5,1) in correspondence to the peaks in the calculated polarized optical conductivity (Fig.1): $_kE_m$ and $_kG_m$ denotes two- and four-dimensional irreducible representations, respectively.

E[eV]	(5,0)	(3,3)	(4,2)	(5,1)
1.47				$_kE_{21} \to {}_kE_{21}$
1.65	$_kG_3 \to {}_kG_3$			
2.11	$_kG_4 \to {}_kG_4$			
2.62	$_kG_5 \to {}_kG_5$			
2.69			$_kE_{10} \to {}_kE_{10}$	
2.86				$_kE_{22} \to {}_kE_{22}$
3.20			$_kE_3 \to {}_kE_3$	
3.69		$_kG_2 \to {}_kG_2$		
D[nm]	0.392	0.408	0.414	0.436

frequency of the (5,0) tube is in a fair good agreement with the experimentally observed values. The results for the tubes (3,3) and (4,2) are also in excellent agreement with the previously reported calculations [4].

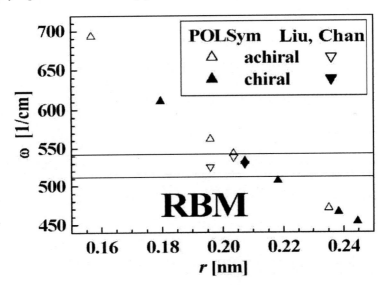

FIGURE 2. RBM frequencies for the nanotubes (4,0), (4,1), (5,0), (3,3), (4,2), (5,1), (6,0), (4,3) and (5,2) calculated by the *POLSym* code and by the first principle calculations [4]. Horizontal lines denotes the measured values.

To resume, we have calculated polarized optical conductivity, RBM and HEM frequencies for the narrow SWCNTs. Contrary to the previously reported *ab initio* calculations of the polarized optical absorption of the 4Å-diameter tubes [2, 4, 5] we find that the presence of the nanotube (5,1) inside porous zeolite can be excluded neither on the basis of the optical response nor on the basis of the RBM frequencies measurements.

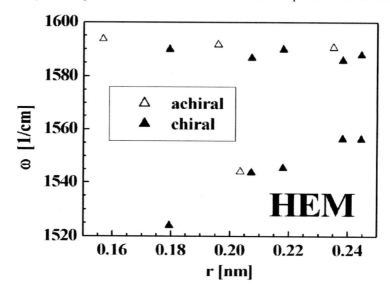

FIGURE 3. HEM frequencies for the nanotubes (4,0), (4,1), (5,0), (3,3), (4,2), (5,1), (6,0), (4,3) and (5,2) calculated by the *POLSym* code.

ACKNOWLEDGMENTS

The authors thank G. Seifert for communicating the DFT data.

REFERENCES

1. Z. K. Tang, H. D. Sun, J. Wang, J. Chen and G. Li, Appl. Phys. Lett. **73**, (1998) 2287; N. Wang, Z. K. Tang, G. D. Li and J. S. Chen, Nature (London) **408**, (2000) 50.
2. Z. M. Li, Z. K. Tang, H. J. Liu, N. Wang, C. T. Chan, R. Saito, S. Okada, G. D. Li, J. S. Chen, N. Nagasawa and S. Tsuda, Phys. Rev. Lett. **87**, (2001) 127401.
3. Y. F. Chan, H. Y. Peng, Z. K. Tang and N. Wang, Chem. Phys. Lett. **369**, (2003) 541.
4. H. J. Liu and C. T. Chan, Phys. Rev. B **66**, (2002) 115416.
5. M. Machón, S. Reich, C. Thomsen, D. Sánchez-Portal, P. Ordejón, Phys. Rev. B **66**, (2002) 155410.
6. I. Milošević, A. Damjanović and M. Damnjanović, Ch. XIV in *Quantum Mechanical Simulation Methods in Studying Biological Systems* ed. D. Bicout and M. Field (Springer-Verlag, Berlin 1996).
7. M. Damnjanović, I. Milošević, T. Vuković and R. Sredanović, Phys. Rev. B **60**,(1999) 2728.
8. M. Damnjanović, T. Vuković and I. Milošević, J. Phys. A **33**, (2000) 6561.
9. D. Porezag, Th. Fraunheim, Th. Kohler, G. Seifert and R. Kaschner, Phys. Rev. B **51**, (1995) 12947.
10. T. Vuković, I. Milošević and M. Damnjanović, Phys. Rev. B **65**, (2002) 045418.
11. I. Milošević, T. Vuković, S. Dmitrović and M. Damnjanović, Phys. Rev. B **67** 1654XX (2003); M. Damnjanović, I. Milošević, T.Vuković, B.Nikolić, E.Dobardžić, Int. Journ. Nanosc. **1** (2002) 313.
12. R. A. Jishi, L. Venkataraman, M. S. Dresselhaus, G. Dresselhaus, Chem. Phys. Lett. **209**, (1993) 77.
13. E. Dobardžić, I. Milošević, B. Nikolić, T. Vuković and M. Damnjanović, Phys. Rev. B (2003, submitted).

Coherent 6 Point Electron Source on the Top of a Caped SWCNT

Leszek Stobinski*^, Ludomir Zommer* and Hong-Ming Lin^

*Institute of Physical Chemistry, Polish Academy of Sciences, 01-224 Warsaw,
Kasprzaka 44/52, Poland (the corresponding author)
^Department of Materials Engineering, Tatung University, Taipei, Taiwan, ROC

Abstract. We have observed the discrete structure of the FEM image in the field emission from the end cap of an individual SWCNT at room temperature. The computer calculations based on the formula of the optical interference can create a similar pattern. Our model suggests that coherent electron emission from the end cap hexagon carbon ring of the SWCNT took place.

Coherent electron emission from many electron sources is causing increasing interest in modern electronics, on account of electron holography and electron microscopy for example. The small radius of curvature of closed-end SWCNT tips (from 0.2 to 1.2 nm) makes them interesting tips for electron field emission. Conventional emission regions are much larger than the area of the SWCNT apex. Since the area of the SWCNT cap is so small, the coherent electronic states near the Fermi level can spread over the emission area. The emitting electrons could therefore interfere with each other. The prospect of inventing a qualitatively new electron source, whose mechanism could be based on electron interference phenomena and could provide a monochromatic and intensive beam, becomes a driving force here. Still, apart from several reports devoted to CNT tips [1,2,3], electron interference phenomena have not been observed during electron emission from metal or semiconductor tips. Usually field emission (FE) patterns have been explained in the terms of anisotropic work functions or local electric field for different emitting micro-regions of the source tip. The reason for this is that the conventional emission areas are much larger than the coherent areas of their electronic states.

The aim of our work is to show that the experimentally obtained FE image consisting of six bright spots, located at the corners of a hexagon ring, with one brighter spot placed in the centre of this hexagon ring, can be explained by our simplified model based on electron interference.

The Basis of Our Model

Let's imagine some sources of electrons, z_j, located on the equipotential surface Ω and equipotential positively charged plane π (here the phosphor screen) on which electron interference patterns are observed. Between these two surfaces the voltage U is applied. We assume that an electron travelling from the j-th source on the Ω surface and reaching the π plane at the point P can be depicted by a spherical de'Broglie wave of the form:

CP685, *Molecular Nanostructures: XVII Int'l. Winterschool/Euroconference on Electronic Properties of Novel Materials,*edited by H. Kuzmany, J. Fink, M. Mehring, and S. Roth
© 2003 American Institute of Physics 0-7354-0154-3/03/$20.00

$$\psi_j = \frac{A}{r_j} e^{i\left(\frac{E}{\hbar}t - kr_j\right)}$$

(1)

where $k=2\pi/\lambda$, $\lambda=h/p$ (h – Planck's constant, p - the electron momentum) and $\overline{r}_j$ is the vector drawn from the j-th source to the point P. An electron reaching the π plane has kinetic energy equal to $p^2/2m=eU$ and its electron wave can be expressed by $\lambda=h/\sqrt{2meU}$ (initial kinetic energy = 0). Electrons interfering at the point $P(x_s,y_s,z_s)$ can be described by the wave function Φ.

$$\Phi(x_s, y_s, z_s) = \sum_j \psi_j = \sum_j \frac{A}{r_j} e^{i\left(\frac{E}{\hbar}t - kr_j\right)}$$

(2)

The probability of finding an electron at the point P and the intensity I of the created image on the screen is proportional to $|\Phi(x_s,y_s,z_s)|^2$:

$$I \propto |\Phi(x_s,y_s,z_s)|^2 = \left|\sum_j \frac{A}{r_j} e^{-ikr_j}\right|^2$$

(3)

where: $r_j = \sqrt{(x_{z_j} - x_s)^2 + (y_{z_j} - y_s)^2 + (z_{z_j} - z_s)^2}$

Experimental Setup

An ultra-high vacuum (1-2×10^{-10} Torr) metal system was used for the field emission experiments. Single-wall carbon nanotubes (Carbo-Lex, as prepared) were deposited on a W loop and were vacuumed and heated for several hours at 1000-1200 C. Applying an electric field, π-electrons can be emitted from the tips of SWCNTs to the vacuum by the quantum mechanical tunneling effect. The electrons are accelerated along the lines of the electrostatic field force and are imaged on a phosphor screen placed about 0.17 m from the single SWCNT tip chosen by the manipulator. The intensity of this image is proportional to the impinging electron beam with the screen. The distance between the closest carbon neighbours in this hexagon ring is about 0.14 nm and the diameter of the hexagon ring at the end cap of the SWCNT is about 0.28 nm.

Results and Discussion

Fig.1 shows the discrete structure of the FEM image obtained from the end cap of individual SWCNT at 700 V. It is important to highlight that we have never observed this kind of pattern at low voltage: 50-300 V, but only one central bright spot.

436

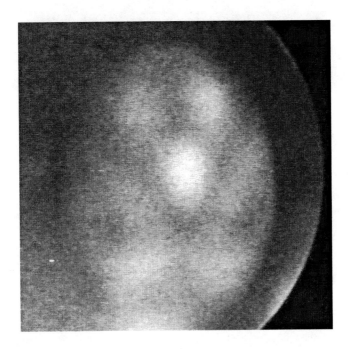

FIGURE 1. The FEM discrete image obtained from the single SWCNT end cap at 700 V.

For the same voltage of 700 V, based on the electron interference model described above, we have simulated the interference electron pattern (see Fig. 2) for six electron sources placed at the corners of a regular hexagon with the same distances (0.14 nm) between carbon neighbors as in the real carbon hexagon ring of the SWCNT end cap. Because each carbon atom in the hexagon ring cannot be treated like a point electron source, in these calculations we have additionally assumed that π-electrons around each of the carbon atoms are partly delocalized [4] according to the Gaussian function.

For voltages below 700 V, like 300 V or lower (for instance 50-100 V) we could not observe a discrete structure as shown in Fig. 1, but only one central bright spot. This is understandable in the light of our model because the de'Broglie wave is longer than that for 700 V and, as shown by our simulation, the created interference picture is too large to be imaged on the screen. Thus, one bright spot was observed in this case.

CONCLUSIONS

The origin of the experimentally obtained fine field emission pattern from an individual capped SWCNT can be interpreted by our model describing the interference of the electron waves emitted from the hexagon carbon ring placed at the end cap of SWCNT. Some delocalization of the π–electrons has to be assumed.

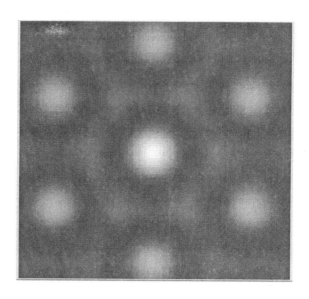

FIGURE 2. The simulated FEM image for the hexagon carbon ring for 700 V.

ACKNOWLEDGMENTS

The FE image (Fig. 1) was obtained at Nanoscience Laboratory, Institute of Physics, Academia Sinica, Nankang, Taipei, Taiwan, ROC, courtesy of Profs. T. T. Tsong and C.S. Chang.

We would like to thank the National Science Council, Republic of China for financial support through Contract Number NSC 91-2216-E-036-017 and NSC 91-2811-E-036-001.

REFERENCES

1. T. Yamashita, K. Mastuda, T. Kona, Y. Mogami, M. Komaki, Y. Murata, C. Oshima, T.Kuzumaki, Y. Horiike, *Surface Science* **514**, 283–290 (2002).
2. C. Oshima, K. Mastuda, T. Kona, Y. Mogami, M. Komaki, Y. Murata, and T. Yamashita, T. Kuzumaki and Y. Horiike*, Phys. Rev. Letters* **88,** 038301-1 - 038301-4 (2002).
3. Y. Saito, K. Hata and T. Murata, *Jpn. J. Appl. Phys.* **39**, L271-L272 (2000).
4. M. Buhl and A. Hirsch, *Chem. Rev.* **101**, 1153-1183 (2001).

Geometrical Effects of Wave Functions of Carbon Nanosystems

Levente Tapasztó[1], Géza I. Márk[1], József Gyulai[1], Philippe Lambin[2],
Zoltán Kónya[3], and László P. Biró[1]

[1] Research Institute for Technical Physics and Materials Science, H-1525 Budapest, P.O. Box 49,
Hungary, E-mail: tapaszto@mfa.kfki.hu
[2] Departement de Physique, FUNDP, 61 Rue de Bruxelles, B–5000 Namur, Belgium
[3] University of Szeged, Applied and Environmental Chemistry Department, H-6720 Szeged, Hungary

Abstract. Networks built from carbon nanostructures and in particular from nanotubes promise exciting nanoelectronic applications. It is thus important to fully understand the quantum mechanical rules of charge propagation through these nanostructures. Some features of the electronic properties of the nanosystems are of purely geometrical origin. These can be investigated in the framework of the jellium model [1]. Wave packet dynamical calculations showed that, when the current is tunneling in a transversal direction through a carbon nanotube "sandwiched" between two electrodes, the energy dependence of tunneling probability shows a plateau in a well defined range. In this work, by solving analytically the stationary Schrödinger equation of a model system, we demonstrate that this plateau is due to electrons being trapped in stationary states of the nanotube. Geometrical features, like diameter dependence of the wave functions and binding energies are studied. Comparison of the results of the widely used zone folding technique with exact jellium wave functions shows a discrepancy at small diameters and an excellent agreement for $d > 1$ nm.

INTRODUCTION

Electron transport through carbon nanosystems is a basic phenomenon for the understanding of the Scanning Tunneling Microscopy (STM) investigation of these structures and also provides the principal operational principles for carbon nanotube based nanoelectronics. Understanding the current flow through a carbon nanotube during the STM measurements is of great interest in the interpretation of experimental STM images, which contain always a mixture of geometrical features and the electronic structure of both the sample and the STM tip. In our former wave packet dynamical simulations [2] following the time evolution of the electron tunneling through the full three-dimensional system the tunneling problem was regarded as a problem in potential scattering theory with a jellium model potential and the time dependent Schrödinger equation was solved numerically. The study of energy dependent transmission of a wave packet showed [3] when the tunneling occurs in a transversal direction through a carbon nanotube "sandwiched" between the STM tip and the support surface, the energy dependence of the tunneling probability shows a constant plateau (instead of typical exponential behavior), in a well defined energy range (Fig.1.left.). This means an increased transition probability from tip to support in that energy range compared to the case when the nanotube is not present between the electrodes. This is a signature of resonant tunneling, i.e. the existence of quasi-stationary states in the mentioned energy range.

CP685, *Molecular Nanostructures: XVII Int' l. Winterschool/Euroconference on
Electronic Properties of Novel Materials,* edited by H. Kuzmany, J. Fink, M. Mehring, and S. Roth

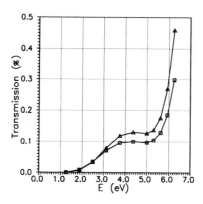 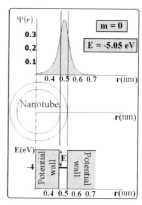

FIGURE 1. Left: Energy dependent transmission of a wave packet tunneling through a nanotube in an STM model for tip positive (triangles) and negative (squares) 1 V bias potential (taken from ref.[3]). Right: The potential model used in analytical calculations, and the stationary solution of the Schrödinger equation corresponding to the zero angular momentum state of a jellium tube.

CALCULATION METHOD AND RESULTS

The stationary states of the system can be found by solving the stationary Schrödinger equation for the given model. Due to the simplicity of the jellium model, this can be done analytically. The potential model is shown in Fig.1.(right). The potential inside the walls of the tube is set to -9.81eV. Outside the walls of the tube the potential is zero. The diameter of the tube is taken 1 nm corresponding to a typical SWCNT value, while the width of the wall is chosen 0.14 nm. See ref. [3] for details. Because of the cylindrical symmetry it was feasible to solve the Schrödinger equation in cylindrical coordinates. The solutions can be written as a combination of first and second order, Bessel, and modified Bessel functions.

$$\psi_m(r) = A_m^i I(r,m) + B_m^i J(r,m) + C_m^i Y(r,m) + D_m^i K(r,m)$$

The coefficients of the Bessel functions were obtained from matching conditions at the boundaries of the three potential regions.

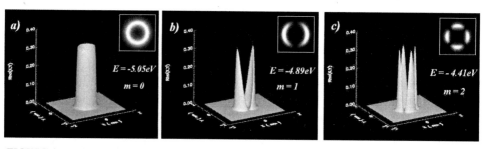

FIGURE 2. Analytical solutions of stationary Schrödinger equation for jellium tube. (x,y) are the cross sectional coordinates in nm while the vertical coordinate corresponds to the probability density. The insets show the ρ(x,y) in nanotubes cross section in grayscale. E is the binding energy for the different angular momenta.

DISCUSSION AND CONCLUSIONS

As shown in Fig.3.(left), for a jellium tube with 1 nm diameter there are six allowed energy states, corresponding to angular momentum quantum numbers $m = 0$, 1, 2, 3, 4, and 5. The energy distribution of the wave packet used in the time dependent simulations is also shown.

As can be seen in Fig.3.(left), the incoming wave packet can excite with significant probability only three states of the jellium tube, for $m = 0$, 1, and 2. This means that the quantum mechanical state of the electrons at every moment can be obtained as the superposition of these three states with different, time dependent, coefficients.

During the time evolution of the system there can occur time intervals when only one of the coefficients of these three states is significant, then we can say that the system is in a quasi-stationary state.

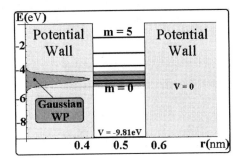 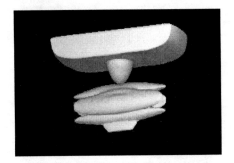

FIGURE 3. Left: The stationary energies and the energy distribution of wave packet used in time dependent simulations. Right: A snapshot taken from simulation at t = 5.4 fs showing a quasi-stationary state.

Following the time evolution of our system (the tunneling process) in simulations, we have found states that shows evident similarity with our analytical results, in the Fig.3.(right) it is shown the system in a quasi-stationary state that corresponds to our analytically calculated stationary state for $m = 2$. The metastability of these states can explain the increased dwell times of the electron (see Ref. [4]) in the tube region.

Knowing the analytical wave functions of the system, the jellium model enables us to investigate the geometrical effects such as diameter dependence of the stationary energies for different angular momentum quantum states. As expected, with increasing diameter, the biding energies for all m values tend to that calculated for the graphene layer using also the jellium potential model (Fig.4.left).

It is worth emphasizing that the zero angular quantum momentum state has inverse diameter dependence.

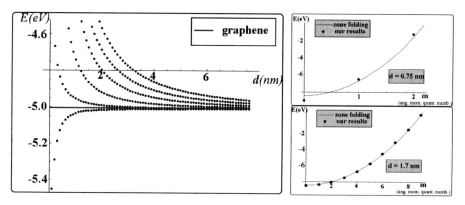

FIGURE 4. Left: diameter dependence of binding energies for electrons with different angular momenta. Right: comparison between the exact results and the results using zone folding approximation, for small and large tube diameters.

A widely used simple approximation is the so-called zone folding method. In tight biding approximation one calculates the wavefunctions and energies for the graphene sheet, than applies a periodic boundary condition in the rolling direction to find out the wavefunctions for the carbon nanotubes. However it is obvious from experimental findings that there are problems applying the zone folding approximation to tubes with small diameters [5]. This error was attributed to the hybridization of the σ and π orbitals of the graphene, that are no more perpendicular to each other when the graphene sheet is rolled up to form a cylinder. As seen in Fig.4.(right) zone folding leads differences when applied to small diameter tubes even *from purely geometrical effects*.

In conclusion our analytical results are in excellent agreement with numerical simulation results, and can explain some interesting features of earlier simulations. Geometrical effects like diameter dependence of binding energies, and geometrical corrections to zone folding approximation were found.

ACKNOWLEDGMENTS

This work was supported by the EU5 contracts NANOCOMP, HRPN-CT-2000-00037 and EU5 Centre of Excellence ICAI-CT-2000 70029 and by OTKA grant T 043685 in Hungary, and the IUAP program P5/01 of the Belgian state OSTC office.

REFERENCES

1. Östling, D., Tománek, D., and Rosen, A., *Phys. Rev. B*: **55**, 13980 (1997).
2. Márk, G.I., Biró, L.P., Koós, A., Osváth, Z., Gyulai, J., Benito, A.M., Thiry, P.A., and Lambin, Ph., in *Electronic Properties of Novell Materials*-2000, edited by H. Kuzmany et. al. AIP Conference Proceedings **591**, New York, 2001, pp. 364-367.
3. Márk, G.I., Biró, L.P., Gyulai, J., Thiry, P.A., Lucas, A.A., and Lambin, Ph., *Phys.Rev.B*:**62**, (2000).
4. Márk, G.I., Gyulai, J., and Biró, L.P., *Phys.Rev.B*:**58**, 12645 (1998).
5. Stojkovic, D., Zhang, P.H., Crespi, V.H., *Phys Rev. Lett:* **87**, 125502, (2001).

Calculating the Structure of the Raman D Band
of Bundles of Single-Wall Carbon Nanotubes

Viktor Zólyomi*, Jenő Kürti* and Hans Kuzmany†

*Department of Biological Physics, Eötvös University Budapest, Pázmány Péter sétány 1/A,
H-1117 Budapest, Hungary
†Institut für Materialphysik, Universität Wien, Strudlhofgasse 4, A-1090 Wien, Austria

Abstract. Measured D bands of bundles of single wall carbon nanotubes have a fine structure. The D band is well reproduced using the double resonance theory for four different laser excitation energies. We provide and discuss a comparison of calculated results obtained from different methods: a simple but effective model phonon dispersion versus a DFT phonon dispersion, and the well known tight binding electron dispersion versus a DFT electron dispersion.

Disordered sp^2 carbon materials show a weak, wide and dispersive band in the 1300-1400 cm^{-1} region, the so-called D band; this was first observed in graphite [1], where a nearly linear upshift of the peak position is observed as the laser excitation energy is increased [2]. This originates form disorder induced double resonant processes, as was explained by Thomsen and Reich [3]. The linear shift is present in bundles of single wall carbon nanotubes (SWCNTs) as well [4], but there is an oscillatory fluctuation superimposed on it, a result of resonances with the van Hove singularities of different SWCNTs [5]. Beyond the behaviour of the peak positions however, the D band of bundles of SWCNTs shows a fine structure, consisting of several subbands [6].

In Ref. [5] we have shown that the intensity of the calculated D band of single wall carbon nanotubes greatly increases whenever the energy of the incoming or the scattered photon matches or nearly matches a van Hove singularity in the joint density of states, and this allows us to easily calculate the *peak position* of the D band by considering only the transitions corresponding to van Hove singularities (the so-called "triple resonant transitions"). But explicit *integration* of the proper perturbation formulas [5] is needed to obtain the *shape* of the D band. We performed all our calculations using zone-folded graphitic dispersion relations.

In our first calculation we used the well known tight binding (TB) approximation for the electron dispersion relation (see Ref. [7]) and the simple model formula for the phonon dispersion relation taken from Ref. [5] (slightly modified so it would behave well at the Brillouin-zone boundary):

$$\omega(\tilde{q}_r, \tilde{q}_\varphi) = \omega_0 + \omega_1 \tilde{q}_r (1 - (\delta - \tilde{q}_r \frac{4\pi\delta - 3}{4\pi^2}) \tilde{q}_r cos(3\tilde{q}_\varphi)) \tag{1}$$

where $\tilde{q}_r$ and $\tilde{q}_\varphi$ are the dimensionless polar coordinates of the wave vector relative to the K point ($\tilde{q}_r$ is in units of $1/(\sqrt{3}a_0)$ where a_0 is the C-C bond length). For the K point frequency $\omega_0 = 1220$ cm^{-1} was used, while ω_1 and the anisotropy parameter δ

CP685, *Molecular Nanostructures: XVII Int'l. Winterschool/Euroconference on Electronic Properties of Novel Materials*, edited by H. Kuzmany, J. Fink, M. Mehring, and S. Roth

were 120 cm^{-1} and 0.06 respectively (as in Ref. [5]). The result is seen on Fig 1 for $E_{laser} = 1.761$ eV (dashed curve: calculated results after a frequency upscaling of 25%, solid curve: measured spectrum; measurements were carried out by A. Grüneis):

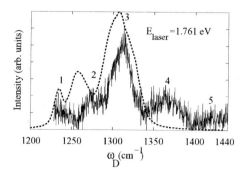

FIGURE 1. Measured (solid) and calculated (dashed) D band spectra for $E_{laser} = 1.761$ eV; calculation was done with low anisotropy model phonon dispersion (see text)

It is seen that the high frequency subbands of the measurements do not appear in the calculated spectrum at all. However if we increase the anisotropy parameter of the model dispersion, then the triple resonant transitions will be scattered on a wider range, thus expanding the frequency interval that the D band spans. After numerous attempts, the values of $\omega_1 = 170$ cm^{-1} and $\delta = 0.30$ finally resulted in a calculated spectrum very similar to the measured one. We get similar results for other laser excitation energies as well. However, the calculated spectra need to be scaled up along the frequency axis relative to the K point frequency, by an average of 6 % [6]. The result can be seen for four selected laser excitation energies on Fig 2. The dotted curves show the measurements, while the dashed ones show the calculations with the increasedly anisotropic model phonon dispersion (and TB electron dispersion).

Note one very important feature of the calculations, the anisotropy factor is five times greater than what we used in Ref. [5], raising the question whether our approach is physically meaningful. We have compared our model phonon dispersion to the density functional theory (DFT) calculations of G. Kresse and O. Dubay [8]. Comparison of equi-phonon frequency contours shows that while in detail they differ, the anisotropy of the two dispersions is about the same, and the model dispersion is in fact quite close to the DFT calculated one (Fig 3).

For further comparison, we repeated the calculations using the DFT phonon dispersion of Kresse and Dubay, still using TB electron dispersion (thin solid curves on Fig 2). Note, that these spectra were not scaled at all along the frequency axis, thus the positions of the subbands are much better reproduced now, however the relative intensities are not always right. This can be caused by the several factors that we can't explicitly include: neglection of curvature effects, intertube-interaction, and wave-vector dependence of Raman matrix elements. However, so far we used a simple TB electron dispersion, and we can at least step beyond this by using a DFT electron dispersion instead of TB.

We repeated the calculations with the DFT phonon dispersion once more, this time using DFT results for the electron dispersion, which we calculated ourselves [9], using the same geometry that Kresse and Dubay used. The results are plotted by the thick

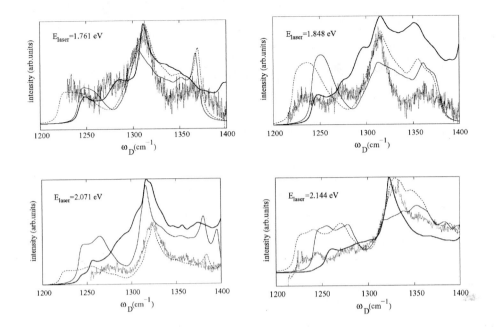

FIGURE 2. Measured spectra (dotted curves) compared with different calculations (dashed: TB electron-, high anisotropy model phonon dispersion; thin solid: TB electron-, DFT phonon dispersion; thick solid: DFT electron-, DFT phonon dispersion) for four different values of E_{laser}

solid curves on Fig 2. The thin solid and thick solid spectra are clearly different, the latter showing more resemblence to the measured spectra. But in detail there are still differences. This shows that using a DFT electron dispersion gives much better results for the D band, than using simple TB approximation, but it also shows that the neglected factors (curvature, intertube-interaction, matrix elements) play a significant role in determining the *exact* lineshape, and including these would certainly improve the calculations even further.

In summary, the fine structure of the D band in the Raman spectrum of bundles of single wall carbon nanotubes was investigated theoretically and experimentally. Our calculations show, that a sufficiently anisotropic model phonon dispersion well reproduces the measured fine structure. This model dispersion compares well with DFT results, so its high anisotropy is physically justified. Repeating the calculations using the DFT phonon dispersion the absolute and relative positions of the subbands are improved, but the relative intensities are not always right. Calculated spectra are further improved significantly by repeating the calculations with the DFT phonon dispersion *and* with our own DFT *electron* dispersion. Remaining discrepancies may be resolved by inclusion of neglected factors (see text)

We acknowledge support from the OTKA T038014 and FKFP-0144/2000 grants in Hungary.

445

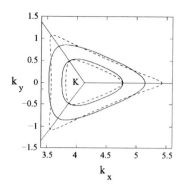

FIGURE 3. Comparison of equi-phonon-frequency contours for the DFT and the model dispersion used (solid curves: model dispersion, dashed curves: DFT dispersion; inner curves: 1300 cm^{-1}, outer curves: 1350 cm^{-1})

REFERENCES

1. Tuinstra, F., and Koenig, J. L., *J. Chem. Phys.* **53**, 1126 (1970).
2. Pócsik, I., Hundhausen, M., Koós, M., and Ley, L., *J. Non-Cryst. Solids* **227-230B**, 1083 (1998).
3. Thomsen, C., and Reich, S., *Phys. Rev. Lett.* **85**, 5214 (2000).
4. Pimenta, M. A., Hanlon, E. B., Marucci, A., Corio, P., Brown, S. D. M., Empedocles, S. A., Bawendi, M. G., Dresselhaus, G., and Dresselhaus, M. S., *Braz. J. Phys.* **30**, 423 (2000).
5. Kürti, J., Zólyomi, V., Grüneis, A., and Kuzmany, H., *Phys. Rev. B* **65**, 165433 (2002).
6. Zólyomi, V., Kürti, J., Grüneis, A., and Kuzmany, H., accepted for publication in *Phys. Rev. Lett.*
7. Saito, R., Dresselhaus, G., and Dresselhaus, M., S., *Physical Properties of Carbon Nanotubes,* Imperial College Press, London, 1998, pp. 25-33.
8. Dubay, O., and Kresse, G., *Phys. Rev. B* **67**, 035401 (2003).
9. DFT calculation for the electron dispersion was performed on the Schroedinger I cluster at the University of Vienna, using the Vienna *ab initio* Simulation Package (VASP)

Thermodynamic calculations on the catalytic growth of carbon nanotubes

Christian Klinke*, Jean-Marc Bonard* and Klaus Kern*†

*Ecole Polytechnique Federale de Lausanne, CH-1015 Lausanne, Switzerland
†Max-Planck-Institut für Festkörperforschung, D-70569 Stuttgart, Germany

Abstract. Based on previous experimental results on the catalytic growth of carbon nanotubes we developed a growth model. For this model we performed calculations and simulations which yielded predictions for the growth, as there are e.g. growth time and growth velocity. The calculated results are compared with experimental data obtained for the growth.

Carbon nanotubes [1] have been studied for more than ten years and are now considered for applications in miscellaneous devices such as tubular lamps [2], flat panel displays [3] and nanometric electronic devices [4, 5]. Such devices implies specific demands on the properties of the tubes, as the length, diameter and electronic properties have a strong influence on the final performance of the device. Consequently nanotube growth has to be controlled and understood. However, the growth mechanism is at present poorly understood, be it for arc discharge, laser ablation, or catalytic growth [6].

Based and inspired by previous results [7, 8] a new model for the growth of carbon nanotubes is proposed. Here, the possibility is demonstrated to calculate the growth of carbon nanotubes for this model by simple thermodynamics equations leading e.g. to the prediction of the growth velocity. Additionally, the heat generation and distribution and the carbon migration in the nanotubes was simulated by means of a finite elements method software. The calculations and simulations are demonstrated here exemplarily for the nanotube growth at 650°C using iron ink, but may easily be adapted to different conditions.

We suppose that the mechanism of the catalytic growth of carbon nanotubes is similar to the one described by Kanzow et al. [9]. Acetylene is thermally stable at temperatures below 800°C and can be dissociated only catalytically on small metal (oxide) particles. In a first step the acetylene reduces the metal oxide grains to pure metal: $Fe_2O_3 + 3C_2H_2 \rightarrow 2Fe + 6C + 3H_2O$, whereas the iron remains on the substrate surface as grain, the carbon diffuses into the metal and the water evaporates. In the following, the catalytic dissociation of acetylene takes presumably place at facets of well-defined crystallographic orientation of those iron particles [10], the resulting hydrogen H_2 is removed by the gas flow whereas the carbon is dissolved in and diffuses into the particle. For unsaturated hydrocarbons this process is highly exothermic. When the particle is saturated with carbon, the carbon segregates on another, less reactive surface of the particle, which is an endothermic process. The resulting density gradient of carbon dissolved in the particle supports the diffusion of carbon through the particle. In order to avoid dangling bonds, the carbon atoms assemble in an sp^2 structure at a less reactive

CP685, Molecular Nanostructures: XVII Int'l. Winterschool/Euroconference on
Electronic Properties of Novel Materials, edited by H. Kuzmany, J. Fink, M. Mehring, and S. Roth
© 2003 American Institute of Physics 0-7354-0154-3/03/$20.00

facet of the particle, which leads to the formation of a nanotube.

In principle the growth can take place with a particle at the top of the nanotube or at the bottom. Both cases work in the same way, but in the second one the particle adheres more to the substrate surface than in the first case. There must be free particle surfaces that are exposed to the gas for the growth to proceed. In the second case the acetylene diffuses from the sides into the particle and the nanotube is constructed from the bottom up, whereas in the first case the gas diffuses from the sides and from the top into the particle (see Fig. 1). The second case seems to be the favored mechanism in our experiments as typically 90 % of the tubes have closed tips without a catalytic particle at the top [7].

The catalytic reaction $C_2H_2 \xrightarrow{Fe} 2C_{graphitic} + H_2$ is highly exothermic and enthalpy driven. This reaction frees at 650°C an energy of about 262.8 kJ/mol (226.7 kJ/mol at 25°C) [11]. The two carbon atoms diffuse at a reactive facet into the catalyst particle and the hydrogen is taken away by the gas flow. The carbon will diffuse through the particle to another less reactive facet where the carbon concentration is smaller and the temperature is lower. Similar models were suggested by different authors [9, 12, 13].

In an extensive study on catalytic particles on top of carbon nanotubes prepared by CO decomposition Audier et al. [10] found that there are relations between the crystallographic structure of the catalyst particles and the attached nanotubes. In the case of a bcc structure of the catalyst particle, the particle is a single crystal with a [100] axis parallel to the axis of the nanotube, and the basal facets of the truncated cone, which appeared free of carbon, are (100) facets. Anderson et al. [14] determined theoretically different activities of decomposition of acetylene on iron facets and Hung et al. [15] mention that with Fe(bcc) a complete decomposition of acetylene takes place at the Fe(100) facets, whereas at the Fe(110) and Fe(111) facets molecular desorption was observed. This may be due to the different surface roughnesses. The diameter of nanotubes is determined by the size of the catalyst particle as already proven in [16].

In order to calculate the heat and the particle diffusion through a catalyst grain we use basic thermodynamic formulas like the Heat Equation:

$$\frac{\partial T}{\partial t} - \kappa \triangle T = 0 \qquad \text{with} \qquad \kappa = \frac{\lambda}{c_q \rho} \tag{1}$$

were λ - heat conductivity, c_q - heat capacity, ρ - density
and Fick's First Law:

$$j_p = -D \nabla c \tag{2}$$

where D - diffusion constant, c - concentration and D is given by

$$D = D_o \cdot exp[-E_a/kT] \qquad \text{Arrhenius equation} \tag{3}$$

where E_a - activation energy, D_o - diffusion factor

Following the iron-carbon-diagram the saturation of carbon in iron without forming any chemical bonds (e.g. Fe_3C) is 65 ppm(weight) at 650°C [17]. This determines the concentration gradient ∇c.

After some considerations it is possible to calculate the growth time for a nanotube as (details can be looked up in [18]):

$$t_{growth} = \frac{\pi \cdot l_{nt}}{4}(d_{out}^2 - d_{in}^2) \cdot \frac{1}{V_{mol}[C]} \cdot \frac{1}{D \cdot |\nabla c|} \cdot \frac{1}{\frac{1}{2} \cdot \pi \cdot d_{particle}^2} \tag{4}$$

$$= \frac{l_{nt} \cdot (d_{out}^2 - d_{in}^2)}{2 \cdot V_{mol}[C] \cdot d_{particle}^2 \cdot D \cdot |\nabla c|} \tag{5}$$

From the experiments we can define a standard nanotube: a hollow cylinder with a length of $l_{nt} = 5$ μm, an inner diameter of $d_{in} = 10$ nm and an outer diameter of $d_{out} = 20$ nm (compare with [7, 8]). For such a standard tube the growth time is:

$$\underline{t_{growth}(\text{Fe}_{bcc}) = 3.794 \text{ s}}$$

$$\hookrightarrow \quad v_{growth}(\text{Fe}_{bcc}) = \frac{l_{nt}}{t_{growth}}$$

$$\hookrightarrow \quad \underline{v_{growth}(\text{Fe}_{bcc}) = 1.318 \frac{\mu\text{m}}{\text{s}}}$$

For the standard tube, the molar volume of graphite $V_{mol}[C]$, $D(\text{Fe}_{bcc}@650°C)$, and $d_{diff} \simeq \frac{1}{2}d_{particle}$ the growth velocity v_{growth} simplifies to:

$$\underline{v_{growth} = 1.976 \cdot 10^{-14} \frac{\text{m}^2}{\text{s}} \cdot \frac{d_{particle}}{(d_{out}^2 - d_{in}^2)}} \tag{6}$$

As the particle size is correlated with the tube diameter like $d_{particle} \simeq d_{out}$ and $d_{particle} \simeq 2 \cdot d_{in}$, v_{growth} is proportional to $1/d_{particle}$. This is similar to the result Baker [12] found experimentally for carbon filaments $v_{growth} \propto 1/\sqrt{d_{particle}}$.

In order to simulate the temperature distribution in the system (catalyst particle - carbon nanotube - silicon substrate) two-dimensional finite element method simulations (FEM) have been employed (FreeFEM+ [19]). A standard setting ($l_{nt} = 5$ μm, $d_{nt} = 10$ nm) reaches a temperature rise of $\Delta T_{top} = 6.474 \cdot 10^{-4}$ K with the particle on top (Fig. 1). Whereas the particle-on-bottom configuration is almost parameter independent. In this configuration the standard setting reaches a temperature rise of $\Delta T_{bottom} = 1.165 \cdot 10^{-5}$ K. As the maximal temperature rise in the particle is very low the diffusion of carbon through the particle can be considered just as driven by the concentration gradient.

Under the considered conditions a growth by surface diffusion is unlikely (relative high pressure, complete surface coverage) and it can not explain the growth of multi-wall nanotubes (growth of several walls with the same velocity, diffusion of carbon through the already created walls). Hung et al. [15] report a carbon diffusion into the bulk at temperatures $T > 773$ K (500°C). This means that multi-wall carbon nanotubes

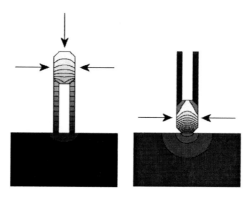

FIGURE 1. *FEM simulation: Penetration of heat at certain facets of the iron particle (due to the decomposition of acetylene) assuming a constant temperature at the silicon sample. Left: particle-on-top, right: particle-on-bottom setting. Arbitrary units for all dimensions (light: high temperature, dark: low temperature).*

can not be created below this temperature. The mobility of carbon in iron will increase with the temperature (diffusion). This explains why the nanotube growth is a thermal process. For the growth a certain dynamic in the catalyst particle must be established which is determined mainly by the diffusion constant. According to Hung et al. [15] C_2H_2 decomposes catalytically to pure C at temperatures T > 400 K (127°C). Thus there is a small window for the growth of pure nanotubes. The lower boundary is 500°C and the upper boundary is about 750°C. At still higher temperature polycrystalline carbon adsorbes on the nanotube surface as mentioned in [7]. In the frame of the experiments discussed here the lowest growth temperature was 620°C. For plasma-enhanced CVD lower deposition temperatures are reported: e.g. Choi et al. reported the experimental growth of nanotubes at 550°C [20].

The calculated values correlate well with the experimental data. In Tab. 1 some experimental time data of the nanotube growth are listed. The calculated growth rate falls well in the range of experimental data, and the calculated growth time is at the lower limit of the experimental data. According to the calculations the growth rate is not influenced by a varying length but just by the diameter of the nanotube like $v_{growth} \propto 1/d_{NT}$. Baker et al. [13] found experimentally a dependence of $v_{growth} \propto 1/\sqrt{d_{NT}}$ for carbon filaments. The difference between the $1/\sqrt{d_{NT}}$ dependence for filaments and the calculated dependence $1/d_{NT}$ for nanotubes for the growth velocity v_{growth} may lie in the fact that carbon nanotubes are hollow graphite-like structures and the carbon filaments consist of monolithic amorphous carbon, which may result in a slower growth.

Even if the reported values apply not exactly, they give an impression and the equations means to calculate and estimate the nanotube generation process.

The small temperature rise $\Delta T = 6.474 \cdot 10^{-4} \, K$ can be explained by the high thermal conductivities of iron, graphite and silicon. The produced heat is distributed in the material very rapidly (diffusive flux through the iron particle, the nanotube, the silicon substrate and out of the considered volume). In the particle-on-bottom setting this flux is led away even more rapidly and the temperature rise smaller (diffusion direct into silicon

TABLE 1. *Comparison of the calculated values for growth time and growth rate with the experimental data.*

Source	Growth time [s]	Growth rate [μm/s]
Calculated	3.8	1.3
Ref. [8]	< 60	> 0.16
Ref. [16]	< 10	> 0.1
Ref. [21]	-	0.9 - 5.1
Ref. [22]	10 - 15	1.3 - 7.2

substrate).

The calculations can easily be repeated with other material parameters (e.g. for CH_4 as carbon source gas and nickel as catalyst particle).

REFERENCES

1. S. Iijima, Nature 354 (1991) 56.
2. J. M. Bonard et al., Appl. Phys. Lett. 78 (2001) 2775.
3. W. B. Choi et al., Appl. Phys. Lett. 75 (1999) 3129.
4. Z. Yao et al., Nature 402 (1999) 273.
5. M. Hirakawa et al., Appl. Surf. Sci. 169 (2001) 662.
6. J. C. Charlier et al., in: M. S. Dresselhaus et al. (Eds.), Carbon Nanotubes (Springer, 2001).
7. C. Klinke et al., Surf. Sci. 492 (2001) 195.
8. C. Klinke et al., J. Phys. Chem. B 106 (2002) 11191.
9. H. Kanzow et al., Chem. Phys. Lett. 295 (1998) 525.
10. M. Audier et al., J. Cryst. Gr. 55 (1981) 549.
11. NIST WebBook (October 21, 2002): http://webbook.nist.gov
12. R. T. K. Baker, Carbon 27 (1989) 315.
13. R. T. K. Baker et al., J. Catal. 26 (1972) 51.
14. A. B. Anderson et al., Surf. Sci. 136 (1984), 398.
15. W. H. Hung et al., Surf. Sci. 339 (1995), 272.
16. J. M. Bonard et al., Nano Lett. 2 (2002) 665.
17. T. B. Massalski, Binary Alloy Phase Diagrams Vol. 1 (1986).
18. C. Klinke, Thesis at the EPFL (2003).
19. FreeFEM+ (v1.2.10): http://www.freefem.org
20. Y. C. Choi et al., Syn. Metals 108 (2000) 159.
21. J. M. Bonard et al., Appl. Phys. Lett. 81 (2002) 2836.
22. J. M. Bonard et al., Phys. Rev. B 67 (2003) 085412.

Interpretation of the Low-Frequency Raman Modes in Multiwalled Carbon Nanotubes

J.P. Buisson*, J.M. Benoit[a], C. Godon, O. Chauvet and S. Lefrant

Institut des Matériaux Jean Rouxel, Nantes, France
[a] on leave at MPI, Stuttgart, Germany

Abstract. Multiwall Carbon Nanotubes (MWNT's) prepared by the electric arc method and purified by oxidation in air, have been studied in Raman spectroscopy. Low frequency vibrational modes are unambiguously identified, thanks to their small internal diameters. We have built a model showing that these modes originate from the radial breathing vibrations of individual tubes which are coupled by Van der Waals interactions between adjacent walls. Using a bond polarization theory, we have derived the relative intensity of these modes, introducing structural characteristics obtained in transmission electron microscopy such as the diameter of the internal tube, as well the average number of tubes constituting the MWNT's.

INTRODUCTION

Multiwall carbon nanotubes (MWNT's) have been studied extensively and characterized by Raman spectroscopy. The low frequency modes are clearly observed in the high quality sample prepared by the electric arc method and purified by oxidation in air. Resonance effects have been also detected[1].

We present a model to interpret these modes based on the Van der Waals[2] coupling of the radial breathing mode of each individual wall of the MWNT. Our method of calculation of the frequencies uses a Lennard Jone potential to describe the carbon-carbon long range interaction, where r is the distance between the two C atoms :

$$V = -\frac{a}{r^6} + \frac{b}{r^{12}} \tag{1}$$

The a and b constants are obtained from graphite data.

We limit the interactions to the first neighboring walls and we consider infinitely long tubes. We assume a continous mass density over the tubes. The tube tube interaction is obtained from the mutual integration of the potential over adjacent walls.

Intensity calculations are carried out as well. Since no reliable data exist on the electronic density of states of MWNT's, we do not take into account any possible resonance effect. The polarizability is expressed as a sum of individual bond polarizabilities, which are obtained using non-resonant bond polarization theory[3] and can thus be written as :

$$P_{\alpha\beta} = \frac{1}{3}\left(\alpha_{\parallel} + 2\alpha_{\perp}\right)\delta_{\alpha\beta} + \left(\alpha_{\parallel} - \alpha_{\perp}\right)\left(\frac{R_{\alpha}R_{\beta}}{R^2} - \frac{1}{3}\delta_{\alpha\beta}\right) \tag{2}$$

CP685, Molecular Nanostructures: XVII Int'l. Winterschool/Euroconference on Electronic Properties of Novel Materials, edited by H. Kuzmany, J. Fink, M. Mehring, and S. Roth

where R is the vector connecting the two atoms linked by the bond, $\alpha_{\parallel}$ and $\alpha_{\perp}$ depend on the distance R between the two atoms.

MODEL

For uncoupled tubes, the normal coordinate of the breathing mode of the tube i is :

$$S_i = \frac{1}{\sqrt{N_i}} \sum_n \lambda_{in} \qquad (3)$$

here N_i is the number of atoms and λ_{in} the reduced radial displacement of atom n relatively to tube i.

The total potential of the multi-walled nanotube is :

$$\phi_o = \frac{1}{2} \sum_i \omega_{io}^2 S_i^2 \quad with \quad \omega_{io}(cm^{-1}) = 111.5 / R_{io}(nm) \qquad (4)$$

ω_{io} is the breathing mode frequency and R_{io} the radius of the tube.

In the continuum model, the tube layers are considered as homogeneous cylindrical surfaces with constant surface mass density. The normal coordinate of the breathing mode can be written :

$$S_i = \sqrt{m_c N_i}(\Delta R_i) = \sqrt{m_c N_i}(R_i - R_{io}) \qquad (5)$$

When the tubes are coupled by Van der Waals interaction, the potential becomes, in the harmonic approximation :

$$\phi = \frac{1}{2} \sum_i \omega_{io}^2 S_i^2 + \sum_i \sum_i K_{ij} S_i S_j \qquad (6)$$

where we have defined :

$$K_{ii} = \sum_{j \neq i} \left(\frac{\partial^2 \phi_{ij}}{\partial S_i^2} \right)_o \quad and \quad K_{ij} = \left(\frac{\partial^2 \phi_{ij}}{\partial S_i \partial S_j} \right)_o \qquad (7)$$

Here ϕ_{ij} is the interaction potential between tubes i and j.

We obtain a system of N coupled harmonic oscillators, N being the number of walls. To resolve this system we diagonalize the dynamical matrix and we obtain the eigenvalues ω_k relatively to the normal coordinates :

$$Q_k = \sum_\ell a_{k\ell} S_\ell \qquad (8)$$

The Raman intensity of the vibrational mode k is proportional to the square of the polarizability derivatives with respect to the normal coordinate Q_k. It turns out that the derivative polarizability tensor of an isolated wall with radius R_{ho} scales with $R_{ho}^{-1/2}$:

$$(\partial[\alpha]/\partial S_h)_o = [\beta]\sqrt{R_{ho}} \qquad (9)$$

where the tensor [β] depends on $\alpha_{\parallel} - \alpha_{\perp}$, $\alpha_{\parallel} + 2\alpha_{\perp}$ and the derivative of the bond polarizability with respect to the bond length $\alpha'_{\parallel} - \alpha'_{\perp}$.

The multiwall nanotube Raman intensity is the sum of scattered intensities of each peculiar mode k :

$$I(\omega) \ \propto \ \sum_k \left\{ \sum_h a_{kh}/\sqrt{R_{ho}} \right\}^2 x \left\{ \frac{1}{1 + ((\omega - \omega_k)/\Delta\omega_k)^2} \right\} \qquad (10)$$

We arbitrarily use a FWHM, $\Delta\omega_k = 4$ cm^{-1}.

RESULTS AND DISCUSSION

The calculated vibrational modes for two-layer tubes are shown in figs. 1 and 2. For small radius tubes (fig. 1) each of the two modes with frequencies 168 and 325 cm^{-1} has the characteristic features of the breathing modes of the isolated layers (162 and 318 cm^{-1}). Mixing of both modes is negligible and the Raman intensities ratio of the two modes is the same as for isolated tubes.

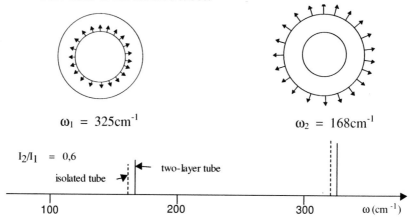

FIGURE 1. Atomic displacements, frequencies and intensities for the two vibrational modes of the two-layer tube with small inner radius, $R_{in} = 0.35$ nm, and with R_2-$R_1 = 0.34$ nm.

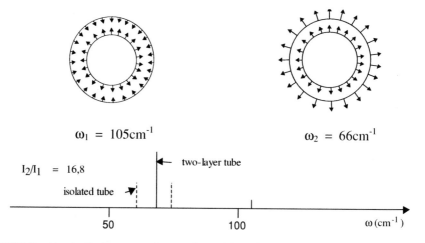

FIGURE 2. Atomic displacements, frequencies and intensities for the two vibrational modes of the two-layer tube with large inner radius, $R_{in} = 1.5$ nm, and with R_2-$R_1 = 0.34$ nm.

454

In fig. 2 we present the results for large radius tubes, where the frequencies ot the two isolated walls (61 and 74 cm^{-1}) are closed. In this case, the low and high frequencies correspond to in-phase and counter-phase modes of the two layers. The frequency up-shift of the counter-phase mode (105 cm^{-1}) is bigger than small radius two-layer tubes, and its intensity is smaller than the in-phase mode.

The combined results of both frequency and intensity calculations of MWNT's[4] are shown in fig. 3. As shown in the right panel of the figure, the modes at highest frequencies originate from the innermost wall, and their intensities become vanishingly small as soon as the inner radius exceeds 1 nm. These modes are only weakly affected by the number of walls (left panel of the fig. 3)

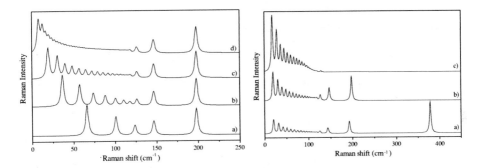

FIGURE 3. Simulated low frequency Raman spectra of MWNT's. The FWHM of each peak is arbitrary fixed to 4 cm^{-1}. Left panel : effect of the number on a R_{in} = 0.6 nm, a) 5; b) 10; c) 20 and d) 50 walls. Right panel : effect of the inner radius on a 20-wall tube : a) 0.3; b) 0.6 and c) 1.2 nm.

CONCLUSION

The first neighbor wall-wall interactions turn out to be independent of the wall diameters with force constants close to the one for two adjacent graphite sheets. The N individual radial breathing modes are coupled through the tube-tube interactions, resulting in N new modes. These new modes are upshifted by comparison with the breathing vibrations of the isolated tubes. The modes at highest frequencies originate from the innermost wall. However, their intensities become vanishingly small as soon as the inner diameter exceeds 2 nm. The smallest frequency mode, originating from the outermost tubes, is associated with an in-phase vibration, inducing a cumulative effect in terms of intensity.

REFERENCES

1. Jantoljak, H., Salvetat, J. P., Forro, L. and Thomsen, C., *Appl. Phys. Letters A : Mater. Sci.Process.* **67**, 113-118 (1998).
2. Henrard, L., Hernandez, E., Bernier, P. and Ruboi, A., *Phys. Rev. B* **60**, 8521-8525 (1999).
3. Guha, S., Menendez, J., Page, J. B. and Adams, G. B., *Phys. Rev. B* **53**, 13106-13114 (1996).
4. Benoit, J. M., Buisson, J. P., Chauvet, O., Godon, C. and Lefrant, S. B., *Phys. Rev. B* **66**, 073417-1-4 (2002).

Theoretical Investigation of Small Diameter Single-Wall Carbon Nanotubes

Jenő Kürti* and Viktor Zólyomi*

*Department of Biological Physics, Eötvös University Budapest, Pázmány Péter sétány 1/A,
H-1117 Budapest, Hungary

Abstract. Density functional theory calculations were carried out to obtain the geometries, electronic band structures and the radial breathing mode frequencies of 23 single wall carbon nanotubes (armchair, zigzag as well as chiral tubes) within the diameter range of 0.3-0.8 nm. These parameters differ from those obtained from simple graphene wrapping and depend not only on the diameter but the chiral angle as well. In particular, the frequencies of the radial breathing mode do not follow a simple 1/d behavior for small diameters, starting from $\approx$0.7 nm.

INTRODUCTION

The properties of single wall carbon nanotubes (SWCNTs) are determined by the well known two integer numbers (wrapping indices), or, equivalently saying, by their diameters and chiral angles. SWCNTs have a diameter which is usually larger than 10 Å. However, nanotubes with a diameter as small as 4 Å can also be prepared, e.g. in zeolites [1] or by vacuum annealing of fullerene peapods [2].

Whereas the electronic and vibrational properties of SWCNTs with large enough diameter can be reasonably well described by folding the dispersion relations of graphene, the behavior of the nanotubes with small diameter can differ significantly from the 'average folded graphitic' behavior.

One important factor is the increasing influence of the curvature on the properties of the SWCNTs with decreasing diameter, mainly due to increasing rehybridization effects. However, in addition to that, the individualities of the tubes become more and more important with decreasing diameter. The properties of tubes with the same diameter but with different chiral angle can differ significantly. We show this on the example of the radial breathing mode (RBM) frequency.

We carried out density functional theory (DFT) calculations within the local density approximation (LDA), using the Vienna *ab initio* simulation package (VASP) [3] in order to determine the optimized geometries, the electronic band structures and the RBM frequencies of 23 SWCNTs (armchair, zigzag as well as chiral tubes) within the diameter range of 3-8 Å.

CP685, *Molecular Nanostructures: XVII Int'l. Winterschool/Euroconference on
Electronic Properties of Novel Materials,* edited by H. Kuzmany, J. Fink, M. Mehring, and S. Roth
© 2003 American Institute of Physics 0-7354-0154-3/03/$20.00

METHOD

DFT calculations were carried out by the Vienna *ab initio* simulation package (VASP) in the LDA approximation, employing a plane-wave basis set, and using ultrasoft pseudopotentials. The cutoff energy was set to 290 eV. A gaussian broadening of the eigenvalues was used with a smearing width of 0.1 eV. In the case of graphite a vacuum of 7 Å between the individual graphene layers was kept fixed. The SWCNTs were arranged in a tetragonal lattice with large enough vacuum between them: the tetragonal lattice constant was fixed to 20 Å. 240 irreducible k-points (30x30x1 Monkhorst-Pack grid) were used in the Brillouin zone of graphene. For the geometry optimization of the nanotubes, 30, 15, 5, and 1 k-points along the tube axis were used, depending on the number of atoms (N) in the unit cell. 31 irreducible Γ-centered k-points (1x1x61 M-P grid) were used in the band structure calculations of tubes for which the number of atoms in the unit cell was less than 56. Strict optimization conditions were used: all forces were less than 0.001 eV/Å in the optimized geometries.

RESULTS AND DISCUSSION

First, the calculations for graphene were carried out. The optimized bond length a_{CC} was found to be 1.407 Å. This was taken as a reference value for obtaining the geometrical parameters of the wrapped graphene sheet.

We optimized the geometries of all 23 SWCNTs within the diameter range of 3-8 Å, for which the number of atoms in the unit cell (N) is less than 200. The results are collected in Table 1. The chiral angles obtained from DFT calculations (θ^{DFT}) are practically the same as obtained from simply wrapping the graphene sheet (θ_G). The DFT diameters (d^{DFT}), however, are larger than those obtained from graphene wrapping (d_G). The deviation increases with decreasing diameter.

The bond lengths (not shown in Table 1) are not uniform in SWCNTs. Achiral (armchair and zigzag) tubes have two, chiral tubes have three different bond lengths – some of them are longer, some of them are shorter than the bond length in graphene. The difference between bond lengths increases with decreasing diameter. As an example, the difference is 0.01 Å and 0.04 Å for (9,0) and (5,0) tubes, respectively. There are two and three different bond angles in achiral and chiral tubes, respectively. At least one of them remains near 120°, but the smallest one decreases significantly with decreasing diameter. Also the lattice parameters along the tube axes are longer than the value calculated from graphene folding, although the effect is very small: a few tenth of a percent.

One additional observation to be emphasized is: whereas there is a general trend that the deviation of geometrical parameters from their graphene counterparts increases with decreasing diameter, this trend can not be described by a smooth curve. The nanotubes exhibit some individualities, their behavior is also influenced by their chiral angles. Interestingly, this is not the case for the folding energy (the difference between the energies of a SWCNT and graphene, per carbon atom): it shows a rather smooth $1/d^2$ behavior with only a very small scattering.

The most important result concerns the radial breathing mode (RBM). It is usually

TABLE 1. Diameter (d^{DFT}), chiral angle (θ^{DFT}), and RBM frequency (ω_{RBM}^{DFT}) of small diameter SWCNTs, together with some other parameters

n	m	N	d_G (Å)	θ_G	θ^{DFT}	d^{DFT} (Å)	$2340/d^{DFT}$ (cm^{-1})	ω_{RBM}^{DFT} (cm^{-1})
4	0	16	3.11	0	0	3.328	703.1	655.8
3	2	76	3.38	23.4	24.4	3.519	664.9	653.4
4	1	28	3.56	10.9	11.5	3.719	629.2	591.1
5	0	20	3.88	0	0	4.028	581.0	546.9
3	3	12	4.03	30	30.6	4.162	562.2	544.2
4	2	56	4.10	19.1	19.8	4.238	552.2	542.0
5	1	124	4.32	8.9	9.3	4.447	526.2	495.2
6	0	24	4.65	0	0	4.781	489.4	465.0
4	3	148	4.72	25.3	25.8	4.818	485.7	483.1
5	2	52	4.85	16.1	16.5	4.954	472.3	454.5
6	1	172	5.09	7.6	7.8	5.200	450.0	441.2
4	4	16	5.37	30	30.4	5.472	427.6	427.9
5	3	196	5.43	21.8	22.2	5.528	423.3	419.9
7	0	28	5.43	0	0	5.528	423.3	412.9
6	2	104	5.59	13.9	14.2	5.686	411.5	407.9
7	1	76	5.85	6.6	6.7	5.952	393.1	376.7
6	3	84	6.15	19.1	19.4	6.244	374.8	368.0
8	0	32	6.20	0	0	6.297	371.6	366.2
5	5	20	6.72	30	30.3	6.800	344.1	340.3
6	4	152	6.76	23.4	23.8	6.843	341.9	338.4
9	0	36	6.99	0	0	7.063	331.3	319.4
8	2	56	7.11	10.9	11.1	7.187	325.6	318.1
7	4	124	7.48	21.1	21.3	7.557	309.6	309.0

assumed in the literature that its frequency is inversely proportional with the diameter [4, 5, 6]. The 8th column of Table 1 shows the 1/d calculated results, using the proportionality constant from Ref. [4]. However, Fig. 1 shows that this is only true for larger diameters. The points corresponding to larger diameters are taken from Ref. [4]. With decreasing diameter, a significant decrease of the RBM frequency, as compared to the 1/d behavior, can be seen, together with a strong chirality dependence. Especially interesting is the comparison of the (7,0) and (5,3) tubes. Their diameters are exactly the same for graphene wrapping, and for three decimals the same also from DFT calculations. However, their RBM frequencies differ by 7 wave numbers.

We calculated also the band structure of the nanotubes, for which the number of atoms in the unit cell is not larger then 76, in the diameter range between 3 and 8 Å. The tubes (9,0) and (8,2) have relatively large diameters and their electronic properties are in accordance with the simple tight binding predictions. Both of them are metallic, as expected. (8,0) is semiconducting, also as expected. In contrast to this, the band structure of tubes with small diameters differs significantly from the tight binding (TB) prediction. For example, (5,0) and (4,2) are not insulators as expected from TB approximation: (4,2) is a very small band gap semiconductor, and (5,0) is a metal. The transition energies for the metallic (3,3) tube are much less then predicted from TB calculations. The reason for the deviation from simple TB behavior is the strong $\sigma - \pi$ rehybridization for small

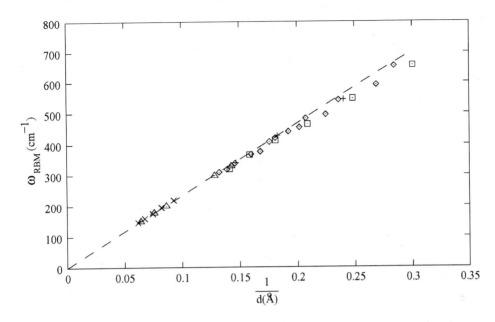

FIGURE 1. Deviation from the 1/d behavior of the RBM frequency calculated by DFT

diameters [7, 8, 9].

ACKNOWLEDGMENTS

We acknowledge financial support from the grants OTKA T038014 and FKFP-0144/2000 in Hungary. DFT calculations were performed on the Schroedinger I cluster at the University of Vienna, using the Vienna *ab initio* simulation package (VASP).

REFERENCES

1. Tang, Z. K., Sun, H. D., Wang, J., Chen, J., and Li, G., *Appl. Phys. Lett.*, **73** (16), 2287 (1998).
2. Bandow, S., *et al.*, *Chem. Phys. Lett.*, **337**, 48 (2001).
3. Kresse, G., and Hafner, J., *Phys. Rev. B*, **48**, 13115 (1993); Kresse, G., and Furtmüller, J., *Comput. Mater. Sci.*, **6**, 15 (1996); *Phys. Rev. B*, **54**, 11169 (1996).
4. Kürti, J., Kresse, G., and Kuzmany, H., *Phys. Rev. B*, **58**, R8869 (1998).
5. Jishi, R. A., Venkataraman, L., Dresselhaus, M. S., and Dresselhaus, G, *Chem. Phys. Lett.*, **209**, 77 (1993).
6. Bachilo, S. M., Strano, M. S., Kittrell, C., Hauge, R. H., Smalley, R. E., and Weisman, R. B., *Science*, **298**, 2361 (2002).
7. Blase, X., Benedict, L. B., Shirley, E. L., and Louie, S. G, *Phys. Rev. Lett.*, **72**, 1878 (1994).
8. Kanamitsu, K., and Saito, S., *J. of Phys. Soc. Jap.*, **71**, 483 (2002).
9. Reich, S., Thomsen, C., and Ordejón, P., *Phys. Rev. B*, **65**, 155411 (2002).

Excitons in Carbon Nanotube Fluorescence Spectroscopy

E. J. Mele and C. L. Kane

Department of Physics and Astronomy
Laboratory for Research on the Structure of Matter
University of Pennsylvania
Philadelphia, PA 19104 USA

Abstract. We investigate the effects of electron hole interaction in the photoexcited states of carbon nanotubes and apply the theory to study the band gap ratios and band gap modulations observed in fluorescence spectroscopy on individual tubes. The former are affected by the long range part of the Coulomb potential, the latter arise from central cell corrections due to anisotropy in the short range part of the Coulomb potential.

Introduction

By isolating single wall carbon nanotubes within micelles O'Connell *et al.* [1] have observed that semiconducting nanotubes fluoresce after optical excitation above the fundamental absorption edge. The photoluminescence efficiency (PLE) at a single emission frequency as a function of the exciting frequency probes the excitations of individual nanotubes. In experiments on solutions containing micelles encapsulating semiconducting tubes of varying radii and chiralities, the observed absorption and emission spectra of the ensemble are quite complex [1,2], but PLE spectra tagged to a single fluorescence frequency show simpler structure and allow one to experimentally assign the fundamental optical excitations of individual tubes [2]. Interestingly the measured spectra contain features that are not anticipated by existing one electron models for the electronic structure. These deviations arise from the residual Coulomb interaction between the excited electron and its valence band hole. Excitonic effects can be quite strong in a one dimensional semiconductor; here the one dimensional structure *and* compact geometry of the nanotube lead to new physical effects.

Failures of the One Electron Theory

There are two aspects of the experimental data in which the effects of the electron hole interaction are particularly apparent.

The "Ratio" Problem: A plot of the ratio of the energy of the second peak in the PLE spectrum to the emission frequency as a function of the inverse tube diameter reveals that the "average" gap ratio is approximately 1.75 and that this ratio persists when the data are extrapolated to large tube radii. In the simplest interpretation of the

CP685, Molecular Nanostructures: XVII Int'l. Winterschool/Euroconference on
*Electronic Properties of Novel Materials,*edited by H. Kuzmany, J. Fink, M. Mehring, and S. Roth
© 2003 American Institute of Physics 0-7354-0154-3/03/$20.00

experiments, this ratio relates the magnitudes of the bandgaps for the first and second pairs of quantized subbands on a semiconducting tube (i.e the gap ratio r = E_{22}/E_{11}). By studying the wrapping of the electronic states on the surface of the tube, one concludes that this ratio should be two. However this theoretical argument is oversimplified, since for small radius tubes tube curvature, s-p hybridization of the electronic bands, and threefold anisotropy in the band structure (trigonal warping) all lead to deviations from r=2. Note however that all of these one electron effects vanish in the limit of large tube radius, and thus one electron theory requires that the ratio should *asymptotically* approach two for large tube radii. This prediction applies quite generally in any model where the quasiparticle self energy can be linearized near the Fermi points. The experimental ratio r≈1.75 extrapolated to large tube radii, and even for the small tube radii measured experimentally, this ratio varies with the tube chirality, but with a mean near r≈1.75.

The "Deviations" Problem: The sizes of the observed gaps (measured by fluorescence at the first subband threshold and PLE at the second subband threshold) depend on tube wrapping and show a "period 3" modulation as a function of the wrapping vector [2]. The signs of these modulations are anticorrelated at the first and second subband edges, so that the primary gap is enhanced for tubes where the second subband gap is reduced and vice versa. This effect has been assigned to the effects of trigonal warping (threefold anisotropy) in the bandstructure [2]. We find that gap correlations of this form are generic to any perturbation that depends on the chiral angle θ following the sin (3θ) law required by threefold symmetry of the graphene sheet [3]. Moreover when the magnitudes of the experimental deviations at the first and second band edges are compared they are found to obey a nonlinear scaling relation while the theory of trigonal warping predicts that they are proportional.

Electron Hole Interactions in the Excited States

Both of these difficulties can be resolved by treating the effects of the electron hole interactions in the photoexcited states. Interestingly, this changes the physical picture of the allowed electronic excitations on the nanotube in a fundamental way.

In the conventional band models, one treats the electrons as occupying well defined azimuthal subbands labeled by a discrete band index *m*, with a conserved crystal momentum *q*, and a well defined quasiparticle energy $E_m(q)$. Excitations energies are then obtained from differences of the *E's* and selection rules require that optical excitations with the exciting field polarized along the tube direction can change neither *m* nor *q*.

In the presence of a residual (e.g. Coulomb) interaction between the electron and hole, the electron and hole can scatter exchanging momentum and this leads to a more complicated picture of the excitations. For example, in the presence of scattering along the tube direction the longitudinal crystal momenta of the electron and hole are not separately conserved, but their difference q_e-q_h is conserved. In a similar way, the azimuthal index *m* is not a sharp quantum number, since by scattering along the circumferential direction the electron and hole can exchange angular momentum, allowing mixing of particle-hole excitations residing in different azimuthal subbands (shown by the decay process in Figure 1(c)). Finally, when a hot electron hole pair is created at sufficiently high energy above the fundamental band edge, it can relax by exciting a second electron hole pair as shown in Figure 1(d-e). By this mechanism the single particle-hole excitations and two particle-hole excitations are mixed by the Coulomb interaction (for very high energy excitations the phase space for mixing with multiparticle excitations becomes even more complex). This configuration mixing is well known in the photophysics of polyenes and other complex molecules and plays an important role in the nanotubes as well.

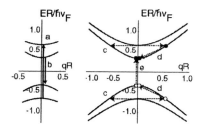

FIGURE 1. Left: Schematic of the measurement of electronic excitations on individual nanotubes by fluorescence spectroscopy. After primary excitation (a) the electron and hole relax to the first azimuthal subband from which they decay radiatively (b). By measuring the efficiency of process (b) as a function of the exciting frequency, the spectrum of primary excitations of individual nanotubes can be studied. Right: two important relaxation channels for interband excitations in higher subbands. In process (c) the second subband electron and hole scatter into a lower azimuthal subband, conserving the total crystal momentum. In process (d-e) they relax by excitation of a second particle hole pair.

Nanotube Excitons and Relaxation Processes

To study the effect of the Coulomb interaction on the photoexcited states it is useful to consider separately the effects of the long range part and short range parts of the Coulomb interaction between charges in the tangent plane of the tube. The long range part of the interaction binds the exciton, and this binding depends on the tube radius and subband index. The short range part of the potential gives a central cell correction, i.e. a perturbation to the binding energy that depends on the tube chirality through its chiral index. The former turns out to be responsible for the ratio problem, and the latter is responsible for the deviations problem.

Our theory treats the long range part of the excitonic interaction using a variational method. All energies in this problem can be expressed in units of the quantization energy on a tube of radius R through the dimensionless energy parameter $\varepsilon = ER / \hbar v_F$ and the interaction strength is given by the dimensionless coupling constant $\tilde{\alpha} = e^2 / (2\pi\kappa\hbar v_F)$ screened by the dielectric constant κ. For an electron hole pair excited into a bound exciton formed from the n-th azimuthal subband $\Phi(z) \propto \exp(-z / \xi_n)$, one finds that $\xi_n = 1.5\kappa R / n$ with a binding energy $\delta\varepsilon = 0.67 \, n\tilde{\alpha} / \kappa$. Thus the long range excitonic effects are twice as strong in the second azimuthal subband ($n=2$) than in the first.

The first and second subband excitons can be distinguished by their relaxation processes. Processes (c) and (d-e) in Figure 1, are allowed for the second subband exciton, but not for the first subband exciton. We have calculated the exciton self energies [4] due to these processes and compute the Green's function for the second subband exciton.

$$G(\varepsilon) = \left(\varepsilon - \varepsilon_0 + iA\tilde{\alpha}^2 + B\tilde{\alpha}^4 / \sqrt{\varepsilon_0^2 - \varepsilon^2} \right)^{-1} \qquad (1)$$

where $A \approx 2.3$ and $B \approx 5.8$ are phase space factors for the single pair and two-pair decay channels. The linewidth and energy shift of the exciton appear at different orders in the dimensionless coupling constant $\tilde{\alpha}$. Solving for the pole of Eqn. (1) we find a shift $\Delta\varepsilon = -3^{1/3} B^{4/3} \tilde{\alpha}^{8/3} / 2$ and an intrinsic width $\gamma = A\tilde{\alpha}^2$. The lineshape obtained from Eqn. (1) is not Lorentzian as shown in Figure 2. Comparison with experiment indicates that the effective coupling constant $\tilde{\alpha} \approx 0.20$, approximately half the value of the unscreened coupling constant.

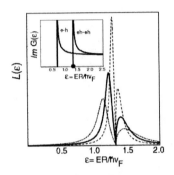

FIGURE 2. Lineshapes for the second subband exciton resonating with various first subband excitations plotted as a function of the dimensionless energy ε. The inset shows the spectral densities for the one particle-hole and two particle-hole continua. The three curves give the lineshapes for three values of the screened coupling constant : $\tilde{\alpha} = 0.15$ (dashed), 0.20 (solid) and 0.25 (dot-dashed).

The period 3 gap modulations observed in the experiments of reference 2 can arise from various physical effects. Anticorrelations of the gap magnitude deviations at the first and second azimuthal band edges occur for any source of anisotropy modulated $\sin(3\theta)$ where θ is the chiral angle. Table 1 lists three candidate sources of this anisotropy that can be distinguished by their dependence on subband index, n, and tube radius, R [3]. In one electron theory the trigonal warping corrections are a strong function of subband index (proportional to n^2) and tube radius (proportional to R^{-2}).

TABLE 1. Scaling of contributions to modulations of subband gap magnitudes.

	θ	N	R
Tube Curvature	$\sin(3\theta)$	n^0	R^{-2}
Trigonal Warping	$\sin(3\theta)$	n^2	R^{-2}
Coulomb Anisotropy	$\sin(3\theta)$	n	$R^{-3.7}$

The third row of Table 1 gives the contribution of Coulomb anisotropy to the deviations. The Fourier components of the Coulomb potential in the first star of reciprocal lattice vectors allow scattering of an electron hole pair between equivalent K points at the Brillouin zone corners. The scattering amplitudes between K points are isotropic for a graphene sheet, but inherit a twofold anisotropy on the tube due to its curvature. This leads to a perturbation to the exciton binding energy that depends explicity on the chiral angle θ, band index, n, and exciton localization length ξ_n through $\delta E = -4\gamma\sin(3\theta)/3\delta_n\xi_n^2$, where γ is a constant that measures the anisotropy of the Coulomb matrix elements[3]. The dependence on the localization length leads an additional nonlinear dependence of the correction on $\sin(3\theta)$. We calculate the perturbation to the exciton binding energy by combining contributions from the trigonal warping with the Coulomb anisotropy

$$\delta E = \pm\left(\frac{A(r)n}{R}\left(1-\frac{\delta E}{E}\right)+B\frac{n^2}{R^2}\right)\sin(3\theta) = \pm\frac{\left(\frac{A(r)n}{R}+B\frac{n^2}{R^2}\right)\sin(3\theta)}{1\pm\frac{A(R)n}{ER}\sin(3\theta)} \qquad (2)$$

where A, B and E are (known) constants that describe the Coulomb anisotropy, trigonal warping and exciton binding energy respectively. In Table 2 we compare the values of these constants obtained from fits of Eqn (2) to the experimental data with estimates from microscopic theory. Equation (2) also predicts a nonlinear scaling relation between the first and second subband deviations, where the nonlinearity has the effect of enhancing the shifts to higher binding energy and suppressing the positive deviations, as seen experimentally. The solution to Eqn (2) shows that as the Coulomb anisotropy perturbs the localization length, it amplifies the effects of trigonal warping in the energies. The fits show that the two make comparable contributions to the observed bandgap deviations.

TABLE 2. Parameters contributing to bandgap deviations.

	Fit	Estimate
A (cm^{-1}-A)	1900	2700
B (cm^{-1}-A^2)	5800	4000
E (cm^{-1})	870	820

ACKNOWLEDGMENTS

This work was supported by the Department of Energy under Grant DE-FG-02-ER01-45118 and by the National Science Foundation under grant DMR-00-79909. We thank Bruce Weisman and Mark Strano for helpful discussions.

REFERENCES

1. M.J. O'Connell *et al.*, Science **297**, 593(2002)
2. S.M. Bachilo *et al.*,Science **298**, 2361 (2002)
3. C.L. Kane and E.J. Mele (to be published)
4. C.L. Kane and E.J. Mele (cond-mat 03-03528, submitted to Physical Review Letters)

Tubular Image States and Light-Driven Molecular Switches

Petr Král[1,2], Brian Granger[2], H. R. Sadeghpour[2], Ioannis
Thanopulos[1], Moshe Shapiro[1] and Doron Cohen[3]

[1] *Department of Chemical Physics, Weizmann Institute of Science, Israel*
[2] *ITAMP, Harvard-Smithsonian Center for Astrophysics, Cambridge, Massachusetts 02138*
[3] *Department of Physics, Ben-Gurion University, Beer-Sheva, Israel*

Abstract. We introduce new tubular image states (TIS) that can be formed around
linear conductors and dielectrics, like metallic carbon nanotubes. These Rydberg-like
molecular states have a very large extent and possess peculiar physical properties. We
also present a two-step light-driven enantiomeric switch, which within 100 ns can turn
a mixture of left and right chiral molecules into a pure enantiomeric form. Molecular
switches with more quasi-stable states can be used as dynamic memories or motors.

Today's nanotechnology operates at the crossroad between physics, chemistry
and biology. The prepared molecular-scale systems fulfill various complex tasks,
like chemical nano-sensing or providing of motoric activity. Coherent transport
regimes can often be present in these systems, since their scale is too small for a
proper relaxation to occur. We discuss two examples of such unique systems.

TUBULAR IMAGE STATES

Large Rydberg states with long lifetimes can be observed in atoms and molecules.
Extended electronic image states can be also observed above bulk conductors or
dielectrics, clusters and liquid He. Their Coulomb-like potential $V(\rho) = -\frac{e^2}{4\rho}\left(\frac{\epsilon-1}{\epsilon+1}\right)$
gives binding energies $E_n = -\frac{13.60}{16n^2}\left(\frac{\epsilon-1}{\epsilon+1}\right)^2$ eV=15-40 meV. Practical application of
these states is usually limited by their picosecond lifetimes, mostly given by their
spatial overlap with the surface states.

Recently, we have suggested a new class of "tubular image states" (TIS), formed
above freely suspended linear molecular conductors or dielectrics [1]. These TIS
have a *non-zero angular momentum l*. The resulting centrifugal forces keeps the
wave functions away from the surface, so the TIS population becomes stabilized
against collapse on the nanotube, rendering it long living at low temperatures.

CP685, *Molecular Nanostructures: XVII Int'l. Winterschool/Euroconference on
Electronic Properties of Novel Materials,*edited by H. Kuzmany, J. Fink, M. Mehring, and S. Roth
© 2003 American Institute of Physics 0-7354-0154-3/03/$20.00

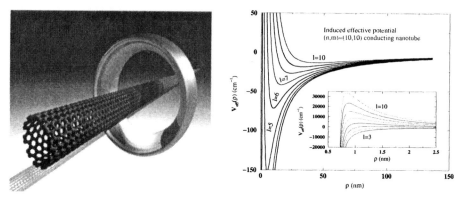

FIGURE 1.

In Fig. 1 (left), we show a visualization of an idealized TIS around a $(10, 10)$ metallic carbon nanotube of radius $a = 0.68$ nm. External electron is *attracted* to the material surface by its image charge and *repulsed* by its angular momentum, so the states are formed in quantum wells separated from the surface (right).

In the first approximation, we can model the linear conductor by a metallic cylinder of radius a. Point charge, at distances $\rho_0 \gg a$ above the cylinder, has a potential Φ_0, that polarizes the tube and induces the potential Φ_{ind}. The total potential $\Phi_{tot} = \Phi_0 + \Phi_{ind}$ vanishes at the nanotube surface, that results in

$$
\Phi_{ind}(\rho, \varphi, z) = -\frac{2q}{\pi} \sum_{m=-\infty}^{m=+\infty} \int_0^\infty dk \, \cos(kz) \, \exp(im\phi)
$$
$$
\times \frac{I_m(ka)}{K_m(ka)} K_m(k\rho_0) K_m(k\rho). \tag{1}
$$

The electron potential energy is given by $V(\rho_0) = \frac{1}{2} q \, \Phi_{ind}(\rho_0, 0, 0)$. At $\rho_0 \gg a$ we can find the attractive potential $(\mathrm{li}(x) \equiv \int_0^x dt/\ln(t))$

$$
V(\rho_0) \sim \frac{q^2}{a} \mathrm{li}\left(\frac{a}{\rho_0}\right) \approx -\frac{q^2}{a} \frac{1}{(\rho_0/a) \ln(\rho_0/a)}, \tag{2}
$$

which interpolates the potential for a charge above a metallic plane, $V \sim -1/\rho$, and a sphere, $V \sim -1/r^2$. The effective potential of the electron has the attractive induced part and the repulsive centrifugal part, $V_{eff}(\rho) = V(\rho) + \frac{(l^2 - \frac{1}{4})}{2m_e \rho^2}$. For $l > 5$, the system develops extended but shallow quantum wells that support bound TIS, as shown in Fig. 1 (right). The collapse of their population on the tube is suppressed by the centrifugal barrier (see the inset).

The total TIS wave functions $\Psi_{n,l,k}(\rho, \varphi, z) = \psi_{n,l}(\rho) e^{il\varphi} \phi_k(z)/\sqrt{2\pi\rho}$ have the eigenenergies $E_{n,l,k} = E_{n,l} + E_k$, where $E_{n,l}$ is related with the radial electron motion and E_k is the kinetic energy for the axial motion along the tube. The radial wave functions $\psi_{n,l}(\rho)$ satisfy the Schrödinger equation

$$
\left(\frac{d^2}{d\rho^2} + 2m_e \left[E_{n,l} - V_{eff}(\rho)\right] \right) \psi_{nl}(\rho) = 0. \tag{3}
$$

In Fig. 2 (left), we show some of the states with $n = 1$. Their maxima are located at the distance $r_{max} \sim l^3$ $(10 - 50$ nm for $l > 5)$ from the tube surface.

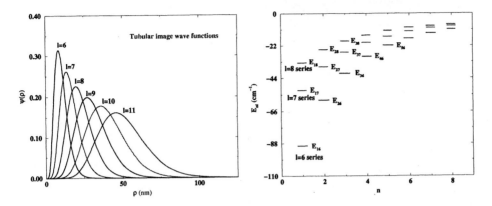

FIGURE 2.

In Fig. 2 (right), we show the eigenenergies $E_{n,l} \sim 1 - 10$ meV related with the radial motion. TIS with the same quantum number n but different angular momentum l are non-degenerate and scale as $E_{n,l} \sim l^{-3}$. They can be filled by radiative recombination, similarly as image states above bulk metals.

Presently, we also study TIS above several tubes and TIS with more electrons. Two TIS electrons spin in the same direction, on opposite sides of the tube circumpherence, and are shifted one from another along the tube to a distance, determined by an additional external potential.

LIGHT-DRIVEN MOLECULAR SWITCHES

Let us now discuss an optical "enantio-selective switch" [2], that, in two steps, turns a ("racemic") mixture of left-handed and right-handed chiral molecules into the enantiomerically pure state of interest. The switch is applied on the (transiently chiral) D_2S_2 molecule, shown schematically in Fig. 3 (left), together with a one-dimensional cut of the ground electronic potential energy surface along the enantio-mutative path, with a few chiral and non-chiral ro-vibronic states.

The enantio-switch is composed of an "enantio-discriminator" and an "enantio-converter" acting in tandem. The enantio-discriminator is based on our "Cyclic Population Transfer" scheme (CPT) [3]. The approach is akin to the Adiabatic Passage (AP), used to *completely* transfer population between quantum states, that are optically coupled as, $|1\rangle \leftrightarrow |2\rangle \leftrightarrow |3\rangle$. In chiral molecules, lacking an inversion center and thus having eigenstates with ill defined parity, $|k\rangle_{L,D} = \frac{1}{\sqrt{2}} (|k\rangle_S \pm |k\rangle_A)$ $(k = 1, 2, 3)$, it is possible to close the "cycle" by introducing a third field which

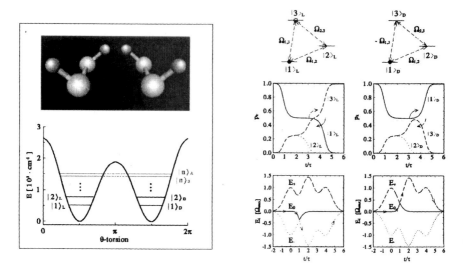

FIGURE 3.

couples the states $|1\rangle \leftrightarrow |3\rangle$ directly (see Fig. 3 (right top)). The interference of one and two-photon transitions along the two paths renders the evolution dependent on the total *phase* φ of the three coupling terms [3]. Since the transition dipoles of the two enantiomers differ in sign, their evolution under the action of the three fields is different, and the two can be *separated* and *converted* one to another.

We denote the levels in each enantiomer by $|i\rangle$, with their energies ω_i ($\hbar = 1$ in atomic units). The external electric field is chosen to be a sum of components, each being in *resonance* with one of the $|i\rangle \leftrightarrow |j\rangle$ transition frequencies of interest, $\mathbf{E}(t) = \sum_{i \neq j} \mathcal{R}_e \left[\hat{\epsilon} \mathcal{E}_{i,j}(t) e^{-i\omega_{i,j}t} \right]$, where $\omega_{i,j} = \omega_i - \omega_j$, and $\hat{\epsilon}$ is the polarization direction. The Hamiltonian of the system in the rotating wave approximation is,

$$H = \sum_{i=1}^{3} \omega_i |i\rangle\langle i| + \sum_{i>j=1}^{3} \left(\Omega_{i,j}(t) e^{-i\omega_{i,j}t} |i\rangle\langle j| + \text{H.c.} \right). \tag{4}$$

It depends on the Rabi frequencies, $\Omega_{i,j}(t) = \mu_{i,j} \mathcal{E}_{i,j}(t)$, where $\mu_{i,j}$ are the transition-dipole matrix elements. Expanding the system wave function in the material states $|i\rangle$ as, $|\psi(t)\rangle = \sum_{i=1}^{N} c_i(t) e^{-i\omega_i t} |i\rangle$, the (column) vector of the slow varying coefficients $\mathbf{c} = (c_1, c_2, c_3)^{\mathsf{T}}$, with T designating the matrix transpose, is the solution of the matrix-Schrödinger equation $\dot{\mathbf{c}}(t) = -i\,\mathsf{H}(t) \cdot \mathbf{c}(t)$, where $\mathsf{H}(t)$ is the effective Hamiltonian matrix

$$\mathsf{H}(t) = \begin{bmatrix} 0 & \Omega_{1,2}^*(t) & \Omega_{1,3}^*(t) \\ \Omega_{1,2}(t) & 0 & \Omega_{2,3}^*(t) \\ \Omega_{1,3}(t) & \Omega_{2,3}(t) & 0 \end{bmatrix}. \tag{5}$$

The phases of the Rabi frequencies $\Omega_{i,j}(t)$ are given by $\phi_{i,j} = \phi_{i,j}^{\mu} + \phi_{i,j}^{E}$, where $\phi_{i,j}^{\mu}$ are the phases of the dipole matrix elements $\mu_{i,j}$, and $\phi_{i,j}^{E}$ are the phases of the electric field components $\mathcal{E}_{i,j}$. The evolution of the system is determined by the *total* phase $\varphi \equiv \phi_{1,2} + \phi_{2,3} + \phi_{3,1}$, which is shifted by π in the two enantiomers, since one or three of their Rabi frequencies $\Omega_{i,j}(t)$ are opposite (see Fig. 3).

The process applies a "dump" pulse $\mathcal{E}_{2,3}(t)$ that couples the $|2\rangle \leftrightarrow |3\rangle$ states and two subsequent "pump" pulses, overlapping with the dump pulse, that couple the $|1\rangle \leftrightarrow |2\rangle$ and the $|1\rangle \leftrightarrow |3\rangle$ states. The Rabi frequencies are $\Omega_{2,3}(t) = \Omega^{\max} f(t)$, $\Omega_{1,2}(t) = \Omega^{\max} f(t-2\tau)$, $\Omega_{1,3}(t) = \Omega_{1,2}(t) + \Omega^{\max} f(t-4\tau) \exp\{-i t \Omega^{\max} f(t-6\tau)\}$, where $\Omega^{\max} = 1$ ns^{-1} and $f(t) = \exp[-t^2/\tau^2]$.

The population in both enantiomers initially follows the eigenstate $|E_0\rangle$ of the zero eigenvalue E_0 of the Hamiltonian (5). As the three Rabi frequencies get closer in magnitude, $|\Omega_{1,2}| = |\Omega_{1,3}| \approx |\Omega_{2,3}| = \Omega$, the eigenvalues E_0 crosses the eigenvalues E_- or E_+, depending on the enantiomer (see Fig. 3 (right bottom)). As a result, their population is *diabatically* transferred to the respective eigenstate $|E_+\rangle$ or $|E_-\rangle$, depending on the enantiomer. Next we adiabatically turn off the pulse $\mathcal{E}_{1,2}(t)$. Therefore, the zero adiabatic eigenstate $|E_0\rangle$ correlates adiabatically with state $|2\rangle$, which thus becomes *empty* after this process, while the occupied $|E_+\rangle$ and $|E_-\rangle$ states correlate to, $|E_{\pm}\rangle \rightarrow (|1\rangle \pm |3\rangle)/\sqrt{2}$.

The *chirp* in the second term of $\Omega_{1,3}(t)$ causes a $\pi/2$ *rotation* in the $\{|1\rangle, |3\rangle\}$ subspace at $t \approx 5\tau$. As a result, state $|E_+\rangle$ goes over to state $|3\rangle$ and state $|E_-\rangle$ goes over to state $|1\rangle$, or vice versa, depending on φ. The net result of the adiabatic passage and the rotation is that one enantiomer returns to its initial $|1\rangle$ state and the other switches over to the $|3\rangle$ state. As shown in the middle panel of Fig. 3, the enantio-discriminator is very *robust*, with all the population transfer processes occurring in a smooth fashion.

We then apply the "enantio-converter" process [2], based on other phase sensitive population transfer methods [4,5]. The process converts the excited $|3\rangle_L$ state to the $|4\rangle_D$ state, while going through two higher excited states as follows $|3\rangle_L \rightarrow \alpha e^{-i\omega_{5S}t} |5\rangle_S + \beta e^{-i\omega_{5A}t} |5\rangle_A \rightarrow |4\rangle_D$. Thus in the end all the molecules have the same symmetry. This methodology could lead to applications in organic chemistry, biochemistry and drug industry. Moreover, we are studying its application to Jahn-Teller molecules with more than two quasi-stable states. Such systems could applied in light-driven molecular motors and multi-state memories.

REFERENCES

1. B. Granger, P. Král, H. R. Sadeghpour and M. Shapiro, Phys. Rev. Lett. **89**, 135506 (2002).
2. P. Král, I. Thanopulos, M. Shapiro and D. Cohen, Phys. Rev. Lett. **90**, 033001 (2003).
3. P. Král and M. Shapiro, Phys. Rev. Lett. **87**, 3002 (2001).
4. P. Král and M. Shapiro, Phys. Rev. A **65**, 043413 (2002).
5. P. Král, Z. Amitay and M. Shapiro, Phys. Rev. Lett. **89**, 063002 (2002).

BIOLOGICAL NANOSTRUCTURES
AND NEW MATERIALS

Molybdenum Disulfide - Amine Nanostructures

V. Lavayen[1], N. Mirabal[1], E. Benavente[1], J. Seekamp[2], C. M. Sotomayor Torres[2], Guillermo González[1]

[1]Department of Chemistry, Faculty of Sciences, Universidad de Chile, Las Palmeras 3425, PO BOX 653, Santiago, Chile

[2]Institute of Materials Science, Dep. Electrical and Information Engineering. University of Wuppertal, 42097 Wuppertal, Germany

Abstract. MoS_2 with intercalated primary amines was synthesized and structurally characterized. The interlaminar distance of the MoS_2 was found to increase from 0.6 nm to 5.1 nm for hexadecylamine. Hydrothermal treatment of these nanocomposites yields nanoparticles. Among them are nanotubes and nanofibres with inner diameters of 100-25 nm and a length ranging 2-12 μm. A first growth mechanism for these tubes and fibers is proposed.

INTRODUCTION

In recent years interest in atomic as well as supramolecular species with structures of nanometer size has steadily increased. A prominent example of these structures are intercalation derivatives of layered solids [1,2,3,4,5,6,7], which can be considered as two-dimensional (2D) structures and various types of nanotubes, representing one-dimensional (1D) structures. Nanotubes were first synthesised from carbon with newer developments yielding nanotubes of metal oxides and chalcogenides. Layered compounds like graphite [9,10] or molybdenum disulfide [11] exhibit a rich intercalation chemistry [12,13], which permits a variety of intercalates to be inserted in the host structures. Intercalation results in novel two dimensional nanometric phases produced in interlaminar spaces. In these low dimensional phases confinement effects often permit the stabilisation of new species, e.g., the generation of trinuclear lithium clusters for the co-intercalation of lithium and diethylamine into MoS_2 [14]. At the same time the nature of the interlaminar phase inevitably alters the behaviour of the host. Thus, the properties of the intercalates and the host could, in principle, be designed. [1,15]. Furthermore, amines intercalated in V_2O_5, e.g., show a marked template activity in the structural conversion of the solids into micro and nano tubular species [8,16].

In this paper we describe the products of the intercalation of long-chain alkyl amines into molybdenum disulfide and its conversion to nanotubes under hydrothermal conditions.

CP685, *Molecular Nanostructures: XVII Int'l. Winterschool/Euroconference on Electronic Properties of Novel Materials*, edited by H. Kuzmany, J. Fink, M. Mehring, and S. Roth

EXPERIMENTAL

Lithium intercalated MoS_2, previously prepared by direct intercalation of lithium using n-BuLi (Li_xMoS_2, $x \approx 1$) [17], is exfoliated by a 24 h moisture treatment. The product is then brought into an aqueous suspension.

Amine-MoS_2 nanocomposites are prepared by addition of an aqueous solution of amine to the suspension of exfoliated MoS_2, followed by stirring the reaction mixture at 50°C for 48 h.

The elemental analysis of samples by means of chemical analysis clearly showed the following composition for $Li_{(x<1)}MoS_2$ (HDA)$_{4.5}$ N, 4.19; C, 64.03; H, 12.36; S, 4.70. This matches the calculated values of the formula: N, 5.06; C, 69,43; H, 12.08; S, 5.15.

The conversion of the laminar species $Li_{(<0.1)}MoS_2(HDA)_{4.5}$ to micro fibers or nanotubes is investigated by treating aqueous suspensions of the nanocomposite in a Teflon-lined autoclave for periods between 12 h and 120 h at temperatures in the range 100-150°C.

Compositions are determined by chemical analysis (SISONS model EA-1108) and atomic absorption spectrometry (UNICAM 929). Further characterisation is performed by X-ray diffraction analysis (SIEMENS D-500, Cu-Kα, λ=1.5418 Å, operation voltage 40kV, and current 30mA); scanning electron microscopy (Phillips M300, XL-30); and transmission electron microscopy (JEOL 100-SX).

RESULTS AND DISCUSSION

The intercalation of hexadecylamine (HDA) or octadecylamine (ODA) into MoS_2, sketched in Figure 1, leads to well-structured and intercalated laminar solids.

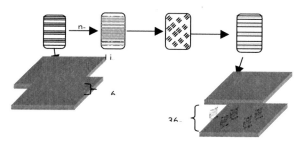

FIGURE 1. Synthesis route for MoS_2-based intercalation compounds

The products were characterised mainly by elemental and powder X-ray diffraction analyses. The diffraction patterns are shown in Figure 2. The narrow and intense Bragg diffraction of the 00n (n$\in$ {1,2,3}) planes (Fig. 2) indicates a highly ordered laminar system and the observed interlaminar distance is 5.1 nm. This matches the length of the amine alkyl-chain. Molecular dynamic calculations [11] indicate that the

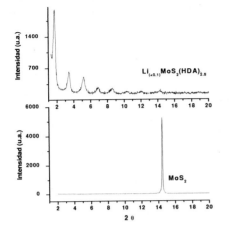

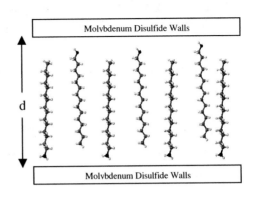

FIGURE 2. Powder X-ray diffraction pattern of $Li_{(x<1)}MoS_2(HDA)_{2.5}$ (top) and a XRD pattern of MoS_2 (bottom)

FIGURE 3. Sketch of alkyl amines intercalated into MoS_2

interlaminar phase is an amine bilayer as sketched in Figure 3. Comparing alkyl amine intercalates with other MoS_2 intercalates, the first show a higher laminar order. This structure agrees with the pattern spontaneously formed by long-chain dipolar amines in a self-assembly process [18].

The treatment of the amine intercalate under hydrothermal conditions used leads to the formation of a variety of nano/micro particles. Transmission electron micrographs of these products are shown in Figure 4.

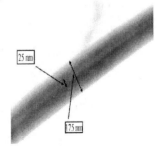

FIGURE 4. TEM image of nanocomposite microfibres and lamellar particles.

FIGURE 5. TEM image of $MoS_2(HDA)_{4.5}$ microfibres

FIGURE 6. TEM image of a MoS_2 nanocomposite nanotube.

475

They have a size dispersion between 1 µm and 10 nm. The aspect ratio depends essentially on the temperature and the reaction time. For a comparatively low temperature (110°C) and short time (12 h) micro/nano fibres are obtained (Fig. 5). Nanotubes evolve for a temperature of 140°C and reaction times of 12-24 h. An example of the latter is shown in Figure 6. It shows a nanotube with an outer diameter of 175 nm and an inner diameter of 25 nm. Closer investigation of particle morphologies points to a possible formation sequence. For lower temperatures around 110°C, the lamellar sheets wrap around a core of pure alkyl amine and form microfibers. With increasing temperature hollow nanotubes form which may grow to microtubes. For a further increase of both reaction time and temperature (5 days, 150°C) the MoS$_2$-alkyl-amine nanocomposites decompose. Further experiments towards a better understanding of the mechanism of the observed structural conversion are in progress.

ACKNOWLEDGMENTS

Deutscher Akademischer Austauschdienst (DAAD), Program ALECHILE, FONDECYT (1010924) and Universidad the Chile DID (PG/93/2002).

REFERENCES

1. D. O'Hare, *Inorganic Materials* D. O'Hare and D. W. Bruce, Eds., John Wiley and Sons: London **1992**.
2. J. L. Atwood, J. E. D. Davies, D. D. MacNicol. *"Inclusion Compounds"* Edited by Academic Press, London **1994**. Vol 1.
3. Byung Hee Hong, Jin Yong Lee, Chi Wan Lee, Jong Chan Kim, Sung Chul Bae and Kwang S. Kim. *J.Am.Chem.Soc.*, *123*, 10748. (**2001**).
4. Thomas D. M.; McCarron E. M. *Mater.Res.Bull. 21*, 945. (**1986**).
5. Nazar L. I.; Wu H.; Power W. P. *J. Mater.Chem 5*, 11. (**1995**)
6. Mujica C.; Duran R.; Llanos J.; Clavijo R. *Mat.Res.Bull. 5*, 483. (**1996**)
7. Yoffe D. *Solid State Ionics 1*, 39. (**1990**)
8. Ghadiri M.R.; Granja J.R.; Milligan R.A.; McRee D.E.; Khazanovich N. *Nature* 366, 324. (**1993**).
9. Scrosati B. *J.Electrochem.Soc. 139*, 2276. (**1992**).
10. Megahed S.; Scrosati B. *Interface 4*, 34. (**1995**)
11. Benavente E.; Santa Ana M. A.; Mendizábal F.; González G. *Coord.Chem. Rev. 224*, 87. (**2002**)
12. Ozin G. A. *Adv.Mater. 4* ,612 (**1992**)
13. Ozin G. A. *Chem. Commun.* 419. (**2000**).
14. Bloise A. C.; Donoso J. P.; Magon C. J. ; Schnider J.; Panepucci H.; Benavente E.; Sanchez V.; Santa Ana M. A.; González G. *J.Phys.Chem.B 106*, 11698. (**2002**).
15. Sanchez V.; Benavente E.; Santa Ana M. A.; González G. *Chem. Mater. 11*, 2296. (**1999**).
16. Niederberger M.; Muhr H. J.; Krumeich F.; Bieri F.; Gunther D.; Nesper R.; *Chem.Mater. 12*, 1995. (**2000**).
17. Joensen P.; Frindt R. F.; Morrison S. R. *Mater.Res.Bull. 21*, 457. (**1986**).
18. Mirabal N.; Sanchez V.; Benavente E.; Santa Ana M. A.; González G. *Mol. Cryst. Liq. Cryst. 374*, 229. (**2002**).

Structure and field emission of C-N nanofibers, formed in High Isostatic Pressure Apparatus

V. D. Blank[1], S.G.Buga[1], E. V. Polyakov[1], D. V. Batov[1],
B. A. Kulnitskiy[1],
U. Bangert[2], A. Gutiérrez-Sosa[2], A. J. Harvey[2], A. Seepujak[2],
Yang-Doo Lee[3], Duck-Jung Lee[3], Byeong-Kwon Ju[3]

[1]Technological Institute for Superhard and Novel Carbon Materials (TISNCM), 7a Centralnaya Street, 142190, Troitsk, Moscow region, Russia
[2]Department of Physics, UMIST, Manchester M60 1QD, UK
[3]Microsystem Research Center / Korean Institute of Science and Technology P.O. Box 131, Cheogryang, Seoul 130-650, Korea

Abstract. The curved intersected inner C-N layers have been observed inside nanofibers, formed in the High Isostatic Pressure (HIP) apparatus. The threshold field strength value of about $1 V/mkm$ for the field electron emission, measured in the present study, falls among best values, found for graphite-like materials. The electron emission depends on peculiarities of the energy band diagram of the material and on the amount of band bending at the surface. Significant curvature of nanocarbons, produced with the participation of nitrogen and containing C-N bonds, is responsible for improved emission properties of our deposit. Close to the spherical shape of many fragments and numerous curved surfaces of our fibers allow not to orient them along the electric field, because such fragments will emit electrons at any orientation.

INTRODUCTION

Many forms of carbon such as nanotubes, nanofibers, diamond, diamond-like carbon, tetrahedral amorphous carbon (ta-C) are known as good electron field emitters [1-3]. There are various electronic devices, which can use emitters, based on carbon. Nanocarbons are capable of emitting high currents (up to 1 A/cm^2) at low fields (~ 5 $V/\mu m$). Aside from pure carbon, C-N nanostructures attract considerable interest as alternative material for cold emitters [4-5]. Nitrogen concentration in C/N nanotubes, formed by different methods, varied from 2 to 5 %, whereas threshold fields varied from 3.8 to 11.5 $V/\mu m$. In this article we investigated structure, nitrogen content and emissive properties of different C-N nanofibers, obtained by HIP apparatus and found that they are competitive with those of pure carbon.

CP685, *Molecular Nanostructures: XVII Int'l. Winterschool/Euroconference on Electronic Properties of Novel Materials*, edited by H. Kuzmany, J. Fink, M. Mehring, and S. Roth
© 2003 American Institute of Physics 0-7354-0154-3/03/$20.00

EXPERIMENTAL

The nitrogen containing carbon nanostructures were formed in the HIP-unit. Details of experimental procedure have been presented in [6]. Carbon evaporation was carried out by direct resistive electrical heating. Carbon deposit was investigated by transmission (TEM) and scanning transmission electron (STEM) microscopy, using JEM-200CX, CM20 Philips TEM, VG 601 UX STEM and Hitachi S-4300. The C-N nanofibers were printed on the active area of cathode plate and then the diode-type flat lamp was assembled for the field emission study. The field emission measurements were performed with the gap between anode and cathode varying in the range of 300 – 900 μm in the vacuum chamber at a pressure of 10^{-6} Torr using F.u.G. Elektronik DC Power Supply.

RESULTS

Carbon-nitrogen nanostructures, formed at conditions of our experiments, could be divided into some groups: bamboo-like fibers, net- and corrugated structures and bead necklace-like fibres. All these structures are characterised by curved carbon layers as a result of the presence of included nitrogen atoms. Different types of formed fibers were studied by EELS-method. We analysed spectra, obtained from different points of

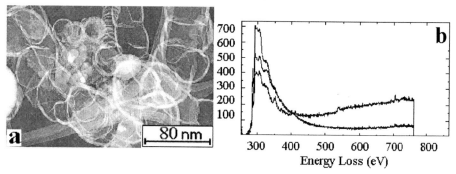

Fig. 1. a). Bead necklace-like fibres, b) EELS spectra for different points of a). Maximal value of the Nitrogen concentration is equal to 0.08 ± 0.01%

as-grown nanocarbons. We can associate the peak occurring at 401 eV with substitutional trivalent nitrogen atoms in a hexagonal lattice. The greatest value of nitrogen concentration of about 8% was found for bead necklace-like fibers, shown in fig.1a. Corresponding EELS spectra are shown in fig.2b. The threshold field strength value of about 1 v/μm for the field electron emission, measured in the present study, falls among the best values, found for graphite-like materials (0.8-6.0 v/μm). Typical current-voltage characteristic for C-N nanofibers is presented on the fig.2.

DISCUSSION

The formation of numerous morphological types of nanocarbons could be explained through the influence of metal-catalysts, and through fluctuations of temperature and carbon and nitrogen concentration in the gas phase. It is known that emission properties of metallic emitter strongly depend on valleys and ridges, existing on its surface. The electric field tension near these points of the surface can achieve many times more values then the mean value of this field. The electron emission

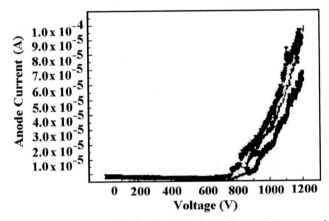

Fig.2. Current-voltage characteristic for C-N nanofibers. The gap between anode and cathode is equal to 700 μm.

depends on peculiarities of the energy band diagram of the material and on any band bending at the surface. Surface irregularities play a key role for emission properties of nanocarbons as well. An emission centre on the surface of a carbon material can represent a single bent atomic layer. A chain of carbon atoms at the end of nanofiber is considered sometimes as responsible for electron emission. Work function depends on changing of configuration of electrons at the end of nanocarbons [7]. A lateral surface of carbon tubes shows electron emission as well. Although a current density of lateral surface is not so high as for the end of tube, it is possible to increase this value by varying its orientation. It is shown as well that the increase of sp^3-bonds content leads to the growth of emission value. It was shown in [8]. that the presence of clusters with diamond-like interatomic bonds in nanostructured graphite-like carbon materials leads to the distortion of curvature of fragments of surface, local decrease in the electron work function in the region of such clusters and the growth of emission. When the sheet bends, some sp^2-bonds transform into sp^3-bonds. It is known that sp^3-hybridisation in diamond is the reason of negative electron affinity on his surface, being responsible for a very low energy threshold for electrons. High value of emission was found for ta-C films. Some sp^3-bonds have been found as well for catalitically grown nanofibers, formed at conditions of high gaseous pressure [9]. The increase of field emission of the CVD-diamond film by nitrogen doping was observed in [10]. Although the role of nitrogen for structure formation as well as for modification of material properties is not completely clear, we believe that the low

479

value of onset field found for our carbon powders, can be explained by features of their structures, caused by the presence of nitrogen. It was shown in [11] that nanocarbons of different configurations were formed under nitrogen pressure. The curved intersected inner C-N layers forming a wavy net-like structure, have been observed inside nanotubes. Nitrogen is assumed to be substitutionally incorporated into the graphitic lattice. The corrugated structure was explained in [12] by formation of pyridine-like bonds between nitrogen and carbon atoms. The structure of the inner layers could be explained by formation of intercrossing bonds, which would produce fullerene-like structures [13]. Nanocarbons of different shapes with significant curvature, containing C-N bonds, are responsible for improved emission properties of our deposit. Considering emission possibilities of carbon-nitrogen structures, produced in the present study, we concluded that our deposit has some advantage over other carbon structures. Close to spherical shape of many fragments and numerous curved surfaces of our fibers allow us not to orient them along the electric field, as at any orientation some of fragments will emit electrons.

REFERENCES

1. Y. Saito, S. Uemura, *Carbon* **38**, 169, (2000).
2. B.S. Satyanarayana, J.Robertson and W.I. Milne, *J. Appl. Phys.*, **81**, 3126, (2000).
3. Y.-H. Lee, Y.-T. Jang, D.-H. Kim, J.-H Ahn and B.-K. *J. Adv. Mater.*, **13**, 479, (2001).
4. R.Kurt, J.M.Bonard, A.Karimi, *Thin Solid Films*, **193**, 398-399 (2001).
5. Y. Zhang, X.C. Ma, D.Y. Zhong, and E.G. Wang , *J. Appl. Phys.* **91**, 9324, (2002).
6. V.D.Blank, I.G. Gorlova, J.L. Hutchison, N.A. Kiselev, A.B. Ormont, E.V. Polyakov, J. Sloan, D.N. Zakharov, S.G. Zybtsev, *Carbon*, **38**, 1217, (2000).
7. A.G.Rinzler, J.H.Hafner, P.Nikolaev et al, *Science*, **269**,1550, (1995).
8. A.N.Obraztsov, A.P.Volkov, A.I.Boronin, S.V.Koshcheev, *JETP*, **93**, 4, 846, (2001)
9. V. D. Blank, B. A. Kulnitskiy, D. V. Batov, U. Bangert, A. Gutiérrez-Sosa, A. J. Harvey, *Diamond and Related Materials*, **11**, 931, (2002).
10. C.-F. Shih, K.-S. Liu, I.-N. Lin, *Diamond and Rel. Mat.*, **9-10**, 1591, (2000).
11. V. D. Blank, E. V. Polyakov, D. V. Batov, B. A. Kulnitskiy, U. Bangert, A. Gutiérrez-Sosa, A. J. Harvey, A. Seepujak, *Diamond and rel. Mat.*, 2003 (to be published).
12. M. Terrones, H. Terrones, N. Grobert, W.K. Hsu, Y.Q. Zhu, J.P. Hare, H.W. Kroto, D.R.M. Walton, Ph. Kohler-Redlich, M. Rühle, J.P. Zhang, A.K. Cheetham, *Appl. Phys. Lett.*, **75**, 3932 (1999).
13 H. Sjöström, S. Stafström, M. Boman, J.E. Sundgren, *Phys. Rev. Lett.* **75** , 1336, (1995).

Polymer and Carbon Nanostructure Dispersions for Optical Limiting

S. O'Flaherty, R. Murphy, S. Hold, A.Drury, M. Cadek, J. N. Coleman and W. Blau

Department of Physics, Trinity College Dublin, Republic of Ireand

Abstract. Experimental measurements of optical limiting of nanosecond laser pulses by two distinctly different polymer and carbon nanostructure composite materials dispersed in solution is reported here. The composites consist of multi walled carbon nanotubes, other clearly defined carbon nanoparticles and polymer. Furthermore, the scattering of high intensity light from the materials was qualitatively probed and its angular dependence investigated

INTRODUCTION

'Optical limiters' are systems that permit the transmission of ambient light levels but which strongly attenuate high intensity, potentially damaging light such as focused laser pulses. In recent years carbon nanotubes [1], both singlewalled (SWNT) and multiwalled (MWNT), fullerenes and amorphous carbon black have been investigated for many potential applications including optical limiting. Mansour *et al.* [2] concluded that optical limiting in carbon black suspensions and carbon black deposited on glass was dominantly due to thermally induced scattering. Similar optical limiting studies of SWNTs [3,4] and MWNTs [5,6] have reported similar conclusions. Riggs *et al.* [6] performed optical limiting experiments on suspended and solubilized full length and shortened SWNTs and MWNTs.

Previously it has been shown that MWNTs can be stably dispersed using the polymer poly(meta-phenylenevinylene-co-2,5-dioctyloxy-para-phenylenevinylene) [7] (PmPV). This has the added advantage of purifying the nanotube soot [8] (Figure 1a), embedding the nanotubes in a polymer matrix which forms a stable dispersion over time and which is easily fabricated into films or other material forms commonly fabricated from polymers. One is left with a suspended composite material solution consisting almost solely of solvent, PmPV and MWNTs. These composite materials have been extensively investigated mechanically [9], electrically [10] and optically [11].

In this paper, application of a new temporally stable, dispersed polymer and carbon nanostructure composite system is presented and compared with the aforementioned system. The second polymer used was the commercially available poly(9,9-di-n-octylfluorenyl-2,7'-diyl) (PFO). [12] It was found that in composites with PFO as the host polymer, MWNTs and other clearly defined carbon nanoparticles were stably

CP685, *Molecular Nanostructures: XVII Int'l. Winterschool/Euroconference on Electronic Properties of Novel Materials,* edited by H. Kuzmany, J. Fink, M. Mehring, and S. Roth

dispersed in the polymer (Figure 1b). [12] Thus, two distinctly different systems are under investigation.

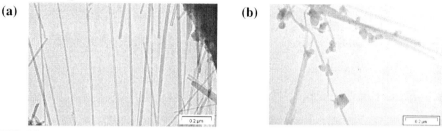

FIGURE 1. TEM images of typical carbon nanostructures held in composites fabricated from (a) PmPV and MWNTs and (b) from PFO, MWNTs and other graphitic nanoparticles. In these images the polymer has been removed using Buchner filtration as described in ref[16]. The nanotubes and graphitic particles can clearly be seen in the images.

EXPERIMENTAL SECTION

The composites were prepared as follows. A 20 g L-1 solution of the polymer (PmPV [13] and PFO (American Dye Source Inc.)) in spectroscopic grade toluene was prepared. To this a mass of Krätschmer arc discharge generated soot equal to half the total mass of the polymer was added. This was ultrasonically agitated using a high power sonic tip (120 W) over 60s and then transferred to a low power sonic bath (60 W) where it was gently agitated for a number of hours. The solution was then left to stand undisturbed for a number of days allowing the sedimentation of the non-nanotube graphitic particles. This suspension was then separated from the sediment by decantation. To vary the mass fraction of nanotubes or nanostructures the composite was blended with a pure polymer solution. The MWNT or carbon nanostructure mass fraction in each composite was measured using thermo-gravimetric analysis, (TGA). [12,14]

The open aperture of a Z-scan experiment for the optical limiting experiments [15], performed using $\approx$ 6 ns Gaussian pulses from a Q switched Nd:YAG laser operated at its second harmonic, 532 nm, with a pulse repetition rate of 10 Hz. All samples were measured in quartz cells with 1 mm path lengths at diluted concentrations of ≈ 1 g L^{-1}.

RESULTS AND CONCLUSIONS

The normalized nonlinear transmission plotted against incident pulse energy density for each of the samples can be seen in Figure 2. All samples, with the exception of pure PFO, exhibited positive nonlinear extinction, or 'optical limiting'. For the PmPV samples (Figure 2a) the limiting performance was similar for 0% and 1.3% and dramatically improved for the MWNT $\approx$ 5.9% sample. At 10 J cm^{-2} the sample with $\approx$ 5.9% MWNT mass has a normalized extinction of $\approx$ 41% while the other two samples (0 and 1.3% MWNT masses) have $\approx$ 0% nonlinear extinction at this energy density. The point at which nonlinear response begins, is reduced by an order of magnitude from approximately 10 J cm^{-2} to less than 1 J cm^{-2}. Similar plots for the PFO composite samples are depicted in Figure 2b. Pure PFO exhibited no response over the energy density that was investigated, and has been omitted from the figure in the interest of clarity. The onset of optical limiting for the other samples (1.1%, 3.8% and 6.6%) is similar in all cases, and occurs approximately at ~ 0.3 J cm^{-2}. The limiting of the samples is mass fraction of carbon nanostructure dependent with the nonlinear effects becoming saturated at carbon nanostructure mass fractions in excess of 3.8%.

(a)   **(b)**

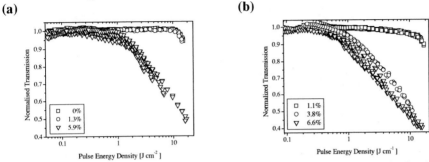

FIGURE 2. Plots of normalized transmission against incident pulse energy density for (a) PmPV composite samples and (b) PFO composite samples.

The scattering from the samples with highest carbon nanostructure contents was qualitatively probed. The approach adopted involved measuring the scattered light from the samples, using a similar approach adopted by Mansour *et al.* [2] for carbon black dispersions. The sample was placed on the focus of the z-scan focusing lens and a photo-detector was placed in the plane of the z-axis of the lens and the laser pulses and was moved about the sample at a fixed radius. The angular dependence of the scattered light from the samples with maximum carbon nanostructure content has been plotted in Figure 3a, where the scattered signal in arbitrary units has been plotted as a function of angular position of the detecting diode. The relative size of the front and back scattering lobes in Figure 3a for both the PFO and PMPV based composite dispersions is in qualitative agreement with the results quoted for carbon black [2] and MWNTs. [5] In these studies the cause of the scattering was reported to be due to thermally induced Mie scattering. It can be seen that there is a difference in the angular scattering profile of the PmPV and the PFO based composite dispersions. In the PmPV composite the carbon nanostructure inclusions are almost exclusively MWNTs and accordingly exclusive Mie regime scattering with the front scattering lobe much larger than the back scattering lobe is observed. However, in PFO based composite dispersions there are many spherical nanoparticles. Spherical particles with dimensions smaller than the wavelength of the incident light would be expected to exhibit Rayleigh scattering. Theoretically the angular profile of Rayleigh scattered light has front and back scattering lobes in equal proportions. Thus, in Figure 3a the PFO composite dispersion angular scattering profile is probably a linear sum of Mie scattering from the MWNTs and Rayleigh scattering from the spherical graphitic nanoparticles. A spectrum of the laser pulse and a spectrum of the scattered light from both composites were collected over 320-850 nm (Figure 3b) confirming that the scattered light was completely due to elastic photon scattering.

In summary composite materials consisting of MWNT and multishelled graphitic particle inclusions stably dispersed in PmPV and PFO hosts have been fabricated, investigated and their optical limiting properties have been experimentally measured.

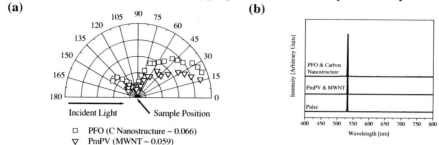

(a)

Incident Light   Sample Position

□   PFO (C Nanostructure ~ 0.066)
▽   PmPV (MWNT ~ 0.059)

(b)

FIGURE 3. Polar plot displaying angular profile of scattered light from the 5.9% MWNT in PmPVsample and the 6.6% carbon nanostructures in PFO and (b) Spectral profile of the scattered light from both composites compared with that of the laser pulse before incidence on the sample.

The MWNT and spherical graphitic particle inclusions could also be differentiated in the scattering data indicating that this method may be an efficient technique that could be developed as a method to measure composite purity.

ACKNOWLEDGMENTS

The authors wish to thank the Irish Higher Education Authority (HEA), Enterprise Ireland and the European Union for financial support

REFERENCES

1. S. Iijima, Nature **354,** 56-58 (1991).
2. K. Mansour, M. J. Soileau, and E. W. Van Stryland, Journal of the Optical Society of America B **9,** 1100-1109 (1992).
3. S. R. Mishra, H. S. Rawat, S. C. Mehendale, K. C. Rustagi, A. K. Sood, R. Bandyopadhyay, A. Govindaraj, and C. N. R. Rao, Chemical Physics Letters **317,** 510-514 (2000).
4. L. Vivien, E. Anglaret, D. Riehl, F. Hache, F. Bacou, M. Andrieux, F. Lafonta, C. Journet, C. Goze, M. Brunet, and P. Bernier, Optics Communications **174,** 271-275 (2000).
5. X. Sun, R. Q. Yu, G. Q. Xu, T. S. A. Hor, and W. Ji, Applied Physics Letters **73,** 3632-3634 (1998).
6. J. E. Riggs, D. B. Walker, D. L. Carroll, and Y. P. Sun, Journal of Physical Chemistry B **104,** 7071-7076 (2000).
7. J. N. Coleman, D. F. O'Brien, A. B. Dalton, B. McCarthy, B. Lahr, A. Drury, R. C. Barklie, and W. J. Blau, Chemical Commmunications, 2001-2002 (2000).
8. R. Murphy, J. N. Coleman, M. Cadek, B. McCarthy, M. Bent, A. Drury, R. C. Barklie, and W. J. Blau, Journal of Physical Chemistry B **106,** 3087-3091 (2002).
9. M. Cadek, B. L. Fougloc, J. N. Coleman, V. Barron, J. Sandler, M. S. P. Shaffer, A. Fonseca, M. v. Es, K. Schulte, and W. J. Blau, Kirchberg/Tirol, Austria, AIP Conferece Proceedings (2002).
10. J. N. Coleman, S. Curran, A. B. Dalton, A. P. Davey, B. McCarthy, W. Blau, and R. C. Barklie, Physical Review B **58,** R7492-R7495 (1998).
11. A. B. Dalton, H. J. Byrne, J. N. Coleman, S. Curran, A. P. Davey, B. McCarthy, and W. Blau, Synthetic metals **102,** 1176-1177 (1999).
12. S. M. O'Flaherty, R. Murphy, S. V. Hold, M. Cadek, J. N. Coleman, and W. J. Blau, Journal of Physical Chemistry B **107,** 958-964 (2003).
13. A. Drury, S. Maier, A. P. Davey, A. B. Dalton, J. N. Coleman, H. J. Byrne, and W. J. Blau, Synthetic Metals **119,** 151-152 (2001).

14. J. N. Coleman, D. F. O'Brien, A. B. Dalton, B. McCarthy, B. Lahr, R. C. Barklie, and W. J. Blau, Journal of Chemical Physics **113,** 9788-9793 (2000).
15. M. Sheik-Bahae, A. A. Said, T.-H. Wei, D. J. Hagan, and E. W. Van Stryland, IEEE Journal of Quantum Electronics **26,** 760-769 (1990).

First-principle DFT modeling of IR spectra of oriented helical HS(CH$_2$CH$_2$O)$_n$CH$_3$ molecules

Lyuba Malysheva* and Alexander Onipko[†]

*Bogolyubov Institute for Theoretical Physics, Kiev, 03143, Ukraine
†Division of Physics, Luleå University of Technology, S-971 87 Luleå, Sweden

Abstract. Only very recently have chemically defined, highly ordered self-assembled monolayers (SAMs) of HS(CH$_2$CH$_2$O)$_n$CH$_3$, $n = 5, 6$, become accessible for systematic infrared spectroscopy analysis [1]. Here we report the results of the first ab initio modeling of reflection-absorption (RA) spectra of specifically oriented non-interacting molecules which represent these SAM constituents. [1] The RA spectra of HS(CH$_2$CH$_2$O)$_{5,6}$CH$_3$, are obtained by DFT methods with gradient corrections and using a variety of basis sets, including 6-311++G**. Positioning and relative intensities of all model spectra are unambiguously identified in all but one intense band in the CH$_2$-stretching region of experimental SAM spectra. Arguments which show that the observed band (which earlier was attributed to symmetric CH$_3$-stretching vibrations) has a distinctively different character are presented.

INTRODUCTION

Oligo(ethylene glycol) (OEG) terminated SAMs are attracting much attention as biomimatic materials, principally (but not only) because of their unique protein selective surface properties due to exposed OEG-segments, see [1, 2] and references therein. Information about the structure of the surface and SAM molecular constituents is mostly inferred from infrared RA and IR–visible sum frequency vibrational spectra [3, 4]. In this type of analysis, the availability of reliable reference spectra is crucially important. The well-established spectral data for ethylene-glycol polymers are not quite suitable and therefore, often insufficient for the purpose of comparison. Additionally, the OEG-containing SAMs studied so far include a large a lkyl portion whose contribution to the observed spectrum makes the identification of the OEG bands difficult and uncertain. For these reasons for a long time one of the most characteristic features in the RA spectra of self-assembled OEG-terminated alkanethiolates was misassigned and identified as the symmetric CH$_2$-stretching vibration of the OEG in the helical conformation [5].

Very recently, highly ordered self-assemblies of OEGs with thiol and methyl termini [HS(CH$_2$CH$_2$O)$_n$CH$_3$, $n = 5,6$] have been obtained and characterized by RA spectroscopy and other methods [1, 2]. This suggests that we have a possibility of straightforward comparison of the first-principle calculations of the OEG structure and

---

[1] The authors thank Dr. D.J. Vanderah for his data files of the reflection-absorption SAM spectra.

CP685, Molecular Nanostructures: XVII Int'l. Winterschool/Euroconference on
Electronic Properties of Novel Materials, edited by H. Kuzmany, J. Fink, M. Mehring, and S. Roth

infrared activity with the relevant experimental spectra of self-assemblies of unidirectional molecules that are precisely defined chemically. Here we present an extensive ab initio RA-spectrum modeling for the $HS(CH_2CH_2O)_{n=5,6}CH_3$ oligomers in their helical conformation as observed in the afore-referenced experiments [1, 2]. Henceforth, these oligomers are referred to as $(EG)_n$, and the corresponding SAMs – as $(EG)_n$-SAMs.

The molecular geometry optimization and calculation of the vibrational spectra were performed with the BP86 exchange-correlation functional utilizing four different basis sets and also with the LSDA/3-21G*, all calculations were made using the G98 suite of programs. The choice of the appropriate method is crucial and always questionable, therefore, a comparison of the results obtained by different methods is of interest.

To model the spectrum for the given orientation of the molecule with respect to the SAM supporting surface, the output of transition dipole moments (TDMs) was recalculated to obtain the TDM components in the molecular coordinates, where z axis coincides with the molecular helical axis. The latter has been assumed to be normal to the substrate surface which means that only z TDM component contributes to the absorption. The model spectra were obtained in the dipole approximation, as the sum of Gaussian-shaped peaks, each centred at the fundamental mode frequency and having the half width at half-maximum (HWHM) indicated in Figs. 1, 2; the peak height is proportional to the squared TDMs z component of the corresponding mode. In the presented results, no frequency scaling or fitting of the relative peak intensity was used.

RESULTS OF THE MODELING AND DISCUSSION

When compared to experimental results, a certain improvement in the appearance of calculated spectra obviously favors the more accurate method (BP86) and a larger basis set (6-311++G**). However, the spectrum description given by a rather crude LSDA/3-21G* method seems to be quite reasonable, taking into account a large reduction of computing. Also we find the use of more elaborate basis sets does not give significantly better results. In fact, by passing from the 6-31G* basis set to 6-311++G**, we obtained only slightly better relative peak positioning. Thus the modeling shows that the BP86/6-31G* method can be regarded as sufficiently reliable for analysis of the infrared spectra of the given oligomer family. In this connection, the BP86/3-21G* method, which is generally believed to be more accurate than the LSDA/3-21G*, is unsuitable in this application, giving a spurious band and generally poor reproduction of the RA spectrum.

Fingerprint region (900 – 1500 cm^{-1}). Figure 1 displays the observed RA spectrum of $(EG)_5$-SAM [1] in the fingerprint region, which is compared with the spectra calculated by DFT LSDA and BP86 methods with different basis sets for a single molecule of $(EG)_5$. The spectra of $(EG)_6$-SAM (experimental) and $(EG)_6$ (calculated) are practically the same except they differ by the expected increase of band intensity.

The shape of the experimental curve is mostly determined by the characteristic bands of ethylene glycol oligomers, which are assigned to modes associated with asymmetric COC stretching + CH_2 rocking (965 cm^{-1}), asymmetric COC stretching (1118 cm^{-1}), and CH_2 twisting (1243 cm^{-1}), wagging (1347 cm^{-1}), and scissoring (1462 cm^{-1}) vibrations [1]. These calculations identify a small peak at 1201 cm^{-1} as a manifestation

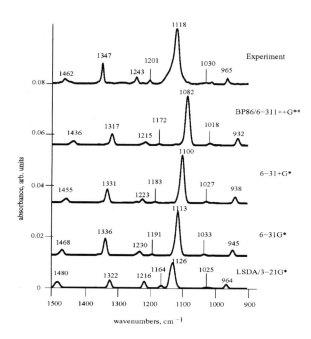

FIGURE 1. Observed and calculated spectra in the fingerprint region. Upper curve represents the RA spectrum of (EG)$_5$-SAM [1]. The single-molecule (EG)$_5$ spectra are calculated with the use of (from bottom to top) LSDA/3-21G* and BP86 method with 6-31G*, 6-31+G*, and 6-311++G** basis set, respectively; HWHM = 5 cm^{-1}.

of local wagging-rocking CH$_3$ vibrations where the wagging component corresponds to the out-of-plane hydrogen atoms. Furthermore, the weak structure centered at 1030 cm^{-1} (experimental curve) corresponds to the asymmetric COC stretching mode at 1033 cm^{-1} (BP86/6-31G* method) and thus, it is of the same nature as the dominant peak in the region. Notice that the fingerprint spectrum of EG$_{15}$-SAM [5] (the SAM of commercial compound HS(CH$_2$)$_2$CONH(CH$_2$CH$_2$O)$_n$CH$_3$, $n \approx 15$) has essentially the same appearance with not more than 1-2 cm^{-1} shift of the band maxima. The only significant distinction observed for EG$_{15}$-SAMs is a prominent shoulder on the high-frequency side of the main peak, the reminiscence of which is observed in the (EG)$_{5,6}$-SAM spectra. The asymmetry mentioned has been attributed to the presence of non-helical conformations which seem to produce minor effects in (EG)$_{5,6}$-SAMs.

CH$_2$-stretching region (2700–3100 cm^{-1}). In the CH$_2$-stretching region, all the methods LSDA/3-21G*, BP86/6-31G*, 6-31+G*, and 6-311++G** give a satisfactory description only of the most intense broad band with the maximum at 2893 cm^{-1} and a pronounced shoulder at 2860 cm^{-1}see spectral curves in Fig. 2. Nearly the same positioning and relative intensities of these features (but without the well-resolved long- and short-wave side maxima near 2816 cm^{-1} and 2979 cm^{-1}, respectively) were observed in the SAMs containing three times longer OEG chains [5]. Modeling the latter experiment,

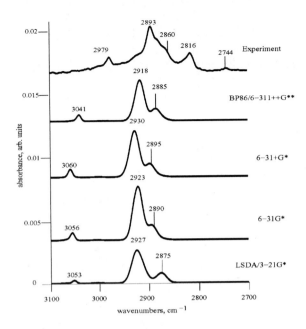

FIGURE 2. The same spectra as in Fig. 1 but for the CH_2-stretching region. The HWHM = 5 cm^{-1} and HWHM = 10 cm^{-1} were used to model the peaks due to the CH_3-stretching modes and the rest of vibrational modes, respectively.

we noticed that the dominant peak represents the asymmetric CH_2-stretching vibrations (in many related studies it was often misassigned and reported to be due to the symmetric vibrational modes). That observation was based on LSDA/3-21G calculations of the infrared spectra of ethylene-glycol oligomers $H(CH_2CH_2O)_nH$ in reference to the EG_{15}-SAM RA spectra. These calculations, which involve much more advanced basis sets, and their direct comparison with the experimental data for (unlike EG_{15}-SAM) chemically well defined and highly ordered self-assemblies of $HS(CH_2CH_2O)_{5,6}CH_3$ molecules, fully support the conclusion that the helical OEG conformation reveals itself in the infrared absorption as two overlapping bands of more intense asymmetric and less intense symmetric methylene vibrational modes. Hence the presented modeling leaves no doubt about the nature of the main band in the CH_2-stretching region.

In the low-frequency part of the region, there is a substantial discrepancy between the ab initio calculations and experimental spectra obtained for $(EG)_{5,6}$-SAMs. A very weak band observed at 2744 cm^{-1} represents combination (twisting+bending) excitations of CH_2 vibrations [6] (these are not described by the present modeling). The most striking discrepancy is the fact that the band at 2816 cm^{-1}, which was tentatively assigned to be due to the methyl symmetric stretching vibration [3, 7], falls into the interval which, according to this extensive modeling, is vibrationally inactive. This cannot be explained as a consequence of the approximations inherent to the methods used, because the detailed analysis does not reveal any symmetric modes which could be associated with stretching vibrations localized in the vicinity of the CH_3 group. The calculations show that all symmetric modes involve collective vibrations of CH_3 and CH_2 groups. (Also, the MP2/6-31* method does not reveal CH_3-localized symmetric vibrations for a shorter molecule of $(EG)_2$.) Moreover, the extended symmetric modes involving only methylene stretching vibrations are lower in energy than the mixed methylene-methyl (CH_3-CH_2) counterparts. In all model spectra obtained by the BP86 method, the contribution from mixed symmetric vibrations forms the low-frequency shoulder of the main band. The low-frequency maximum of the LSDA spectral curve has the same nature, i.e., it is due to collective CH_3-CH_2 vibrations. Only the asymmetric high-frequency modes have a pronouncedly local character, that is, they are localized on the methyl group and are independent of methylene asymmetric stretching vibrations. The in-plain mode manifests itself as a high-frequency peak in Fig. 2. It is present in all model spectra, though the peak is detached from the main band by a noticeably larger distance than that is observed experimentally. Interestingly, the absence of CH_3-localized symmetric vibrations is a distinctive characteristic of the helical OEG conformation only. As shown by our ab initio modeling of all-trans OEG and methoxy hexadecanethiol (not presented here), the vibrational spectra of these molecules contain both symmetric and asymmetric CH_3-localized modes; also this is the case for all-trans alkanethiolates [8].

The above arguments question the existing interpretation of the peak at 2816 cm^{-1} in terms of symmetric vibrations localized on the CH_3 group. However, since the modeling was concentrated on the manifestation of only fundamental vibrational excitations, the presence of overtones, combination-mode excitations, etc., is left beyond the scope of this preliminary report. Some conformational effects also may result in the appearance of the law-frequency band. To clarify this issue, further analysis is in progress.

ACKNOWLEDGMENTS

Computational resources for this work were provided by the Swedish National Allocation Committee for High Performance Computing (SNAC). The authors are grateful to R. Valiokas for his comments to the manuscript.

REFERENCES

1. Vanderah, D. J., Arsenault, J., La, H., Gates, R. S., Silin, V., and Meuse, C. W., *Langmuir*, **19** (2003).
2. Vanderah, D. J., Valincius, G., and Meuse, C. W., *Langmuir*, **18**, 4674–4680 (2002).
3. Zolk, M., Eisert, F., Pipper, J., Herrwerth, S., Eck, W., Buck, M., and Grunze, M., *Langmuir*, **16**, 5849–5852 (2000).
4. Dreesen, L., Humbert, C., Hollander, P., Mani, A. A., Ataka, K., Thiry, P. A., and Peremans, A., *Chem. Phys. Lett.*, **333**, 327–331 (2001).
5. Malysheva, L., Klymenko, Y., Onipko, A., Valiokas, R., and Liedberg, B., *Chem. Phys. Lett.*, **370**, 451–459 (2003).
6. Miyazawa, T., Fukushima, K., and Ideguchi, Y., *J. Chem. Phys.*, **37**, 2764–2776 (1962).
7. Ong, T. H., and Devies, P. B., *Langmuir*, **9**, 1836–1845 (1993).
8. Vanderah, D. J., Meuse, C. W., Silin, V., and Plant, A. L., *Langmuir*, **14**, 6916–6923 (1998).

V_2O_5 nanofiber-based chemiresistors for ammonia detection

U. Schlecht[*], I. Besnard[†], A. Yasuda[†], T. Vossmeyer[†] and M. Burghard[*]

[*]Max-Planck-Institut für Festkörperforschung, Heisenbergstr. 1, D-70569 Stuttgart, Germany
[†]Sony International (Europe) GmbH, Materials Science Laboratories, Hedelfinger Str. 61,
D-70327 Stuttgart, Germany

Abstract. Vanadium pentoxide (V_2O_5) nanofibers with a cross-section of 15 nm^2 were obtained by self-assembly in aqueous solution. These fibers - the length of which can reach 10-15 μm - have been deposited as a thin network on interdigitated electrode structures with 10 μm electrode gap. Exposure to ammonia at room-temperature changes the conductance of the networks, which makes them useful as chemiresistors for gas-sensing applications. By evaluating the concentration dependence of the sensor response, we found that the gas adsorption follows the Langmuir adsorption isotherm.

INTRODUCTION

Classical metal oxide gas sensors have to be operated at elevated temperatures (typically above 250 °C) to achieve sufficient sensitivity [1]. In order to reduce the power consumption, it is desirable to develop gas sensors operating at room-temperature. Recently, advances have been achieved with nanosized materials, due to their large surface/volume ratio [2, 3, 4]. Although carbon-nanotubes could be used as gas sensors [3, 4], they have the disadvantage to be entangled in ropes of metallic and semiconducting tubes, decreasing the surface/volume ratio. Furthermore the conductivity is dominated by the metallic tubes, which are less sensitive to gases than the semiconducting tubes [4]. This encouraged us to investigate vanadium pentoxide (V_2O_5) nanofibers as sensor material, because each fiber shows an identical hopping like conductance [5, 6]. In addition the very large surface/volume ratio of these fibers is comparable to that of an isolated carbon nanotube.

EXPERIMENTAL

V_2O_5 nanofibers were prepared from 0.2 g ammonium(meta)vanadate (NH_4VO_3) and 2 g acidic ion exchange resin (DOWEX 50WX8-100) in 40 ml water, as described previously [5]. Due to a polycondensation process, fibers are formed as an orange colored gel. In order to obtain fibers with a mean length exceeding 5 μm, the solution was aged for several weeks under ambient conditions.

Interdigitated electrode structures with 50 fingers, a separation of 10 μm, and an overlap of 1800 μm were fabricated by optical lithography. The structure was amino-

CP685, *Molecular Nanostructures: XVII Int'l. Winterschool/Euroconference on
Electronic Properties of Novel Materials*, edited by H. Kuzmany, J. Fink, M. Mehring, and S. Roth
© 2003 American Institute of Physics 0-7354-0154-3/03/$20.00

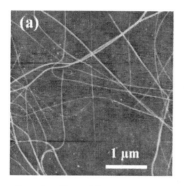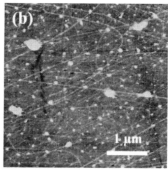

FIGURE 1. Scanning Force Microscopy image of (a) unmodified and (b) with 1 nm gold modified V_2O_5 network.

silanized (using N-[3-(Trimethoxysilyl)propyl]ethylenediamine) prior to dip coating with a 1:10-diluted V_2O_5 solution. Figure 1a shows a typical scanning force microscopy image of these networks, whereas Figure 1b displays a network, modified by evaporating 1 nm gold on top.

The resistances of the networks are on the order of 10 kΩ and were monitored as a function of time in the constant current mode, with the current adjusted to obtain $\sim$100 mV prior gas exposure. Sensors were placed within a teflon cell and a constant gas-flow of 0.4 l/min was supplied by a mass flow system, that also generated the appropriate gas-concentrations. Ammonia was obtained from a gas cylinder containing 1% of the test gas in nitrogen. All experiment were performed under dry conditions at room-temperature, in each case involving three steps: an equilibration time (0-1 min), an exposure phase (1-3 min) and a recovery period (3-8 min).

RESULTS AND DISCUSSION

Figure 2a shows the response $\Delta R/R_{ini}$ (change of resistance ΔR, normalized to initial resistance R_{ini}) of an unmodified V_2O_5 network, when exposed to various concentrations of ammonia. A fast resistance drop over a time scale of $\sim$1 min is observed. In addition, the recovery occurs in less than 5 min. The high sensitivity with a detection limit below 5 ppm and the reversibility at room-temperature, make this material a promising candidate for sensor applications.

It was found that metal evaporation improves the performance of the sensors. For that purpose, nominally 1 nm of gold was evaporated onto the same network, which resulted in clusters of various sizes, as can be seen in the scanning force microscope image of Figure 1b. The response of the sensor was measured prior and after the modification to various concentrations of ammonia, revealing a sensitivity increase by a factor of 7 (at 100 ppm NH_3).

Figure 2b compares the traces of the unmodified (dashed line) and the modified (solid line) sensor in the case of 100 ppm ammonia exposure. Obviously the improved

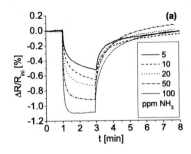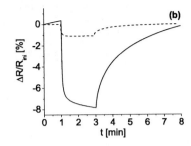

FIGURE 2. (a) Response of an unmodified V_2O_5 sensor to various concentrations of ammonia. (b) Comparison of responses of an unmodified (dashed line) and a gold modified (solid line) V_2O_5 sensor, when exposed to 100 ppm of ammonia.

sensitivity has the disadvantage of an increase in recovery time. For applications, one defines the t_{90}-time, as the time needed to give either 90 % of the response in the case of adsorption or a 90 % recovery in the case of desorption. For an exposure to 100 ppm ammonia, we obtain recovery-values for t_{90} of 100 s and 180 s for the unmodified and the modified V_2O_5 network, respectively. Both, the increased response and the slower recovery indicate a stronger interaction of ammonia with the gold-nanoparticle/V_2O_5-system, compared to the non-modified sample.

To shine more light onto the gas-nanofiber interaction, we analyzed the sensor response measured as a function of the gas-concentration c. Assuming, that the response $\Delta R/R_{ini}$ is proportional to the gas coverage θ on the nanofibers, we can apply the Langmuir adsorption isotherm [7]:

$$\theta_c = \frac{K \cdot c}{1 + K \cdot c}, \qquad (1)$$

where θ_c is the gas-dependent coverage ($0 \leq \theta_c \leq 1$), and K is the binding constant. It is noted that after the exposure time, the sensor response did not fully reach its equilibrium. This is especially true for the lower ammonia concentrations. Nevertheless, Figure 3 shows reasonable agreement between the experimental results (symbols) and the Langmuir fits (lines), for both undoped (triangles) and gold nanoparticle doped (squares) V_2O_5 sensors.

In addition, the observed deviation from ideal Langmuir adsorption behavior can be explained by the presence of different adsorption sites, such as the electrode-fiber contact, on the fiber or on fiber-fiber interconnects.

CONCLUSIONS

V_2O_5 nanofiber networks show a high sensitivity to ammonia at room-temperature. This sensitivity can be improved via the deposition of gold-clusters.

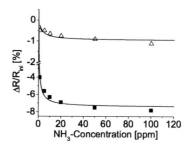

FIGURE 3.   Response of unmodified (triangles) and modified (squares) V_2O_5 network as a function of ammonia-concentration. The solid lines correspond to the Langmuir fits.

The observed resistance decrease is in accordance with the known electron donor property of ammonia and the n-type semiconducting behaviour of V_2O_5 [6]. On this basis, we currently assume that the ammonia predominantly interacts with the gold nanoparticles and changes the conductivity in the underlying V_2O_5 fibers. However, further investigations are required to clarify by which extent the gold particles enhance the e^- transfer from the NH_3 to V_2O_5.

In future experiments, our goal is to achieve a higher selectivity by specific doping, similar to what has been demonstrated for carbon nanotube sensors, where palladium modification introduces a high selectivity to hydrogen [4].

REFERENCES

1. Mandelis, A., and Christofides, C., *Physics, Chemistry and Technology of Solid State Gas Sensor Devices*, John Wiley & Sons, Inc, New York, 1993.
2. Comini, E., Faglia, G., Sbergveglieri, G., Pan, Z., and Wang, Z., *Appl. Phys. Lett.*, **81**, 1869 (2002).
3. Kong, J., Franklin, M., Zhou, C., Chapline, M., Peng, S., Cho, K., and Dai, H., *Science*, **287**, 622 (2000).
4. Kong, J., Chapline, M., and Dai, H., *Adv. Mater.*, **13**, 1384 (2001).
5. Muster, J., Kim, G., Krstic, V., Park, J., Park, Y., Roth, S., and Burghard, M., *Adv. Mat.*, **12**, 420 (2000).
6. Kim, G., Muster, J., Krstic, V., Park, J., Park, Y., Roth, S., and Burghard, M., *Appl. Phys. Lett.*, **76**, 1875 (2000).
7. Langmuir, I., *J. Am. Chem. Soc.*, **30**, 1361 (1918).

Synthesis and characterization of new Mo-Chalcogenide based nanowires

D. Vrbanic[1], R. Dominko[2], V. Nemanic[3], M. Zumer[3], M. Remskar[3], A. Mrzel[3], D. Mihailovic[3], S. Pejovnik[1] , P. Venturini[2]

[1]Faculty of Chemistry and Chemical Technology, Askerceva 5, 1000 Ljubljana, Slovenia
[2]National Institute of Chemistry, Hajdrihova 19, 1000 Ljubljana, Slovenia
[3]Jozef Stefan Institute, Jamova 39, 1000 Ljubljana, Slovenia

Abstract. We report on two interesting applications of new nanostructured transitional- metal chalcogenide (TMC) materials based on the Mo-S-I system. We show that some new Mo-S-I materials can controllably adsorb and release relatively large amounts of Li ions in one–dimensional storage channels, and may form the basis for promising and safe new battery electrode materials. The storage capacity is found to be close to that of MoS_2 NTs, comparing favourably with state-of-the-art graphite electrodes. The field emission (FE) from Mo-S-I nanostructures also appears to show very favourable characteristics compared to carbon nanotubes. The ease of processing of emitters based on TMC materials, together with the relative ease of synthesis and excellent stability suggest that the materials could be useful for e-sources in many very different electronic devices (electron microscopes, flat pannel displays, lamps etc.).

INTRODUCTION

MoS_2 nanotubes (1) appear to be just one of a large number of diverse transitional-metal chalcogenide (TMC) nanostructured materials. Recent experiments show that a very wide range of stoichiometries $Mo_xS_yI_z$ can be achieved by varying synthesis conditions, yielding a variety of very different materials. Selective synthesis of such new nanostructured materials is a promising route for further development of new electrode materials in Li-ion batteries (2) and field emission applications (3). In this paper we show that the new material is capable of reversible lithium intercalation and may be used as a negative electrode in rechargeable lithium batteries with some potential advantages over other systems currently in use. We also report on field emission properties of the new material.

CP685, *Molecular Nanostructures: XVII Int'l. Winterschool/Euroconference on Electronic Properties of Novel Materials,*edited by H. Kuzmany, J. Fink, M. Mehring, and S. Roth
© 2003 American Institute of Physics 0-7354-0154-3/03/$20.00

EXPERIMENTAL

The ropes or bundles of the material, described by the general formula $Mo_xS_yI_z$ with diameters from 20 nm up to 2 μm and length of several millimetres have been synthesized by the chemical transport reaction, which was found to be a succesful technique for the production of different types of a new quasi one dimensional nanostructures. The detailed synthesis and structure of the materials will be reported elsewhere.

 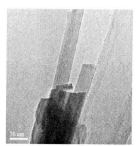

FIGURE 1.: SEM and TEM images of new material.

RESULTS AND DISCUSSION

Li- insertion: The lithium cation is small enough to be inserted into the channels between the bundles. Electrochemical lithium insertion into bundles of new material was carried out in a laboratory-made three-electrode cell (4). The cell body was made of polypropylene, while the electrode holders, which also served as contacts, were made from stainless steel. Metallic lithium was used as a counter and reference electrode. The working electrodes were prepared by spreading as-grown bundles dispersed in sulphonated polyaniline solution onto copper substrate and dried in vacuum at 100 °C. The electrolyte was a 1M solution of $LiClO_4$ in propylene carbonate. Electrochemical measurements were recorded using an EG&G 283 electrochemical interface at constant current, typically 0.025 mA. The geometric surface area of the working electrode was always 0.5 cm^2 containing approximately 1 mg of active material. Lithium insertion into bundles of $Mo_xS_yI_z$ takes place at potentials that are closer to the potential of metallic lithium than in the case of layered MoS_2. The difference between the inserted and extracted lithium in the first cycle is believed to be due to the irreversible reaction with intersitial iodine remaining in the bundles of material after synthesis. (Fig. 2a). Preliminary experiments on material in which iodine was partially removed (by heating material in vacuum), show substantially higher extraction capacity, indicating that iodine may be partly responsible for the irreversible fraction of the doping cycle. (Fig 2b).

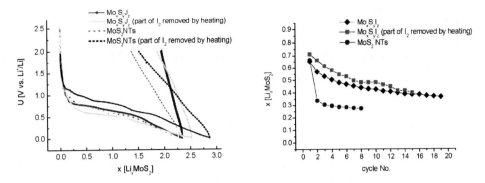

FIGURE 2.: a) The average charge capacity in the first cycle is 2,3 mol of lithium per MoS_2 unit for MoS_2 NTs. Preliminary experiments with bundles of the new material where iodine was partialy removed show even higher capacities. b) The difference between the inserted and extracted lithium in the first cycle is believed to be due to the irrevesible reaction with intersitial iodine remaining in the bundles of material after synthesis.

Field emission: Bundles of $Mo_xS_yI_z$ material were attached by silver paste to a metal tip. Initial checks of field emission (FE) properties were measured from large, but unknown number of sharp emitters, from an area of 0.75 mm². The preliminary measurements were performed in a field emission microscope (FEM), where the emission pattern was observed on a 22 mm diameter aluminized luminiscent screen. The distance from the pin to the screen was set to between 10 and 20 mm. The base pressure was lower than 10^{-7} mbar. The field emission measurements were performed under continuous bias conditions at room temperature at high voltage supplied.

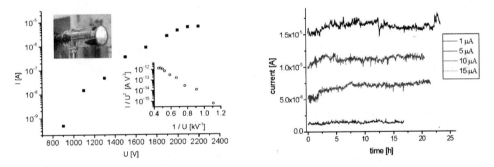

FIGURE 3.: a) The I-V characteristic of the large area emitter. In the insert is the corresponding F-N plot. Emission current caused a patterned interesting image on the FEM screen. b) The emission current stability of the area emitter of material (A= 0,75 mm²) measured at current settings I= 1, 5, 10 and 15 μm respectively.

497

The I-V characteristics of emitters are shown in Fig. 3a. Each point in the diagram represents the current averaged over 15 s. The corresponding Fowler-Nordheim plots are shown in the insert of Fig. 3a. The FE pattern on the screen was observable in the shape of several luminous spots of various shape (insert of Fig. 3a). The emission current shows excellent stability of the emitting site (Fig. 3b). None of tips burnt out at the rated currents from 1 to 15 μA during the duration of the experiment. Because of the risk of damage to the measuring instruments, we did not test the FE tips for burn-out threshold.

CONCLUSIONS

The main advantage would be relatively easy synthesis of this transitional- metal chalcogenide as a nanomaterial. The one-dimensional lithium insertion takes place close to the metallic lithium potential, which suggests that these Mo-S-I material could serve as a potential negative electrode in Li-ion batteries. The material functions well with propylene carbonate electrolyte, which operates at low temperatures, down to −50°C. The Mo-S-I materials could also serve as a design model for further synthesis of electrode materials based on similar nanostructures.
The initial investigations of FE properties of Mo-S-I compounds show that they may represent group of new nanomaterials with promising FE properties. The ease of handling and very good dispersion characteristics in combination with their good FE performances suggests that new material may offer specific advantages in potential applications.

ACKNOWLEDGEMENTS

The financial support from the Ministry of Education, Science and Sport of Slovenia is fully acknowledged.

REFERENCES

1. Remskar, M., Mrzel, A., Skraba, Z., Jesih, A., Ceh, M., Demsar, J., Stadelman, P., Levy, F., Mihailovic, D., *Science* **292**, 479-481 (2001)
2. Dominko, R., Arcon, D., Mrzel, A., Zorko, A., Cevc, P., Venturini, P., Gaberscek, M., Remskar, M., Mihailovic, D., *Advanced Materials* **14**, 1531-1534 (2002)
3. Nemanic, V., Zumer, M., Zajec, B., Pahor, J., Remskar, M., Mrzel, A., Panjan, P., Mihailovic, D., *Applied Physics Letter* (2003) to be published.
4. Gaberscek, M., Bele, M., Drofenik, J., Dominko, R., Pejovnik, S., *Electrochem. and Solid State Lett.* **3**, 171-173 (2000)

Optical Properties of Self-Assembling Molecular Nanowires

A. J. Fleming[°], J. N. Coleman[°], A. Fechtenkötter[‡], K. Müllen[‡], W. J. Blau[°]

[°]*Department of Physics, Trinity College Dublin, Dublin 2, Ireland*
[‡] *Max-Plank-Institut für Polymerforschung, Ackermannweg 10, 55128 Mainz, Germany*

Abstract. The vibronic structure of the luminescence and luminescence-excitation spectra of alkyl substituted hexa-*peri*-hexabenzocoronene (HBC-$C_{8,2}$) and hexa(4-*n*-dodecylphenyl) substituted hexa-*peri*-hexabenzocoronene (HBC-PhC_{12}) are described and explained in terms of the collective behaviour of molecules in a molecular nanowire structure. At medium (10^{-8} M) and high (10^{-6} M) concentration the HBC molecules aggregate into nanowires and a change in the vibronic and electronic structure is observed. The addition of exo-phenyl groups, as in the case of HBC-PhC_{12}, was found to increase the configurational coordinate displacement in the photo-excited state by increasing the intermolecular vibronic coupling. These results correlate well with the observed reduced photo-luminescence efficiency of the HBC-PhC_{12} nanowires.

INTRODUCTION

Intermolecular vibronic coupling in aggregate systems can explain some of the observed spectral differences between the luminescence and luminescence-excitation spectra of isolated molecules and self-assembled molecules of alkyl- and alkyl-phenyl-substituted hexa-*peri*-hexabenzocoronene. The electronic properties of hexabenzocoronene derivatives upon aggregation were previously discussed by Fleming *et al*. [1] and here some of the observed changes to the vibronic structure of the luminescence and luminescence-excitation spectra as a function of concentration are explained.

EXPERIMENTAL

The HBC materials in powder form, as synthesised by Fechtenkötter *et al* [2,3], were weighed and dissolved in spectroscopic grade toluene and 24 solutions of progressively lower concentrations were made up for each material (from 10^{-3} M to approximately 10^{-15}M). Each photo-luminescence and excitation spectra measured was carried out in solution in 1cm quartz sample cuvettes using a Perkin-Elmer LS 50 B Luminescence Spectrometer.

CP685, *Molecular Nanostructures: XVII Int'l. Winterschool/Euroconference on Electronic Properties of Novel Materials*, edited by H. Kuzmany, J. Fink, M. Mehring, and S. Roth
© 2003 American Institute of Physics 0-7354-0154-3/03/$20.00

RESULTS & DISCUSSION

The luminescence and luminescence-excitation of both derivatives at key concentrations are given in figure 1. As the concentration is increased, molecules come closer together and couple electronically. This coupling is described by the resonance interaction β, which was measured[1] to be 0.27 ± 0.02 eV and 0.23 ± 0.02 eV for HBC-C$_{8,2}$ and HBC-PhC$_{12}$ respectively. The splitting of the electronic levels is a result of the dipole-dipole interaction [4,5]. The arrangement into a nanowire structure has stacked all the transition dipole moments into a parallel alignment (an H-aggregate).

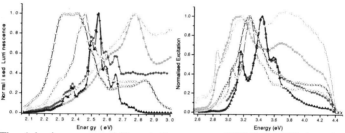

FIGURE 1. The a) luminescence and b) excitation spectra of HBC-C$_{8,2}$ (filled symbols) and HBC-PhC$_{12}$ (unfilled symbols) at low (squares), medium (circles) and high (triangles) concentrations. The excitation peak at 2.85 eV and low energy features are spectrometer related artefacts prominent only at low concentration.

Medium and High Concentration

In the medium (1.7×10^{-8} M) and high (1.4×10^{-6} M) concentration regime a distinct vibronic structure can be observed in the excitation spectra for both derivatives (figure 1(b)). This is due to the collective exchange-narrowing [6] of the inhomogeneously broadened peaks. Plotted in figure 2(a) are the peak positions of the excitation spectra at high concentration (medium concentration peak positions are the same but are broader). Three distinct regions are evident in the plot, each with a different slope (equal to E_{vib}), suggesting that there are three different vibrational components to the aggregates excitation spectrum, unlike the low concentration species, which only has one. The luminescence spectra for the medium and high concentrations also have vibronic structure, and the peak positions are plotted in figure 2(b). The average slopes of the two components are: 0.05 ± 0.01 eV and 0.04 ± 0.02 eV for HBC-C$_{8,2}$ and HBC-PhC$_{12}$ respectively.

Comparing the luminescence of the low concentration species with that found at medium concentration, the vibrational energy has decreased by 66% (from 0.15 eV to 0.05eV). This is because in the photo-excited state the electron orbital can delocalise over several molecules via the intermolecular electronic coupling. This makes the energy of the orbital less dependent on the configurational coordinate and in so doing, decreases the curvature of the harmonic potential. The vibronic levels are therefore brought closer together, giving a gentler slope in the plot (figure 2(c) ϕ_i and ϕ_i^*).

The observed Stokes shift increases by a factor ~2.47 and ~2.12, for HBC-C$_{8,2}$ and HBC-PhC$_{12}$ respectively, from low to medium concentration[1]. The Stokes shift, explained diagrammatically in figure 2(c) by the sum of $\Delta E_i + \Delta E_f$, arises from the geometrical relaxation of the whole system i.e. it extends as far as the electron is delocalised. A photo-excited isolated molecule can only relax its geometry as much as the bonds will allow a redistribution of the nuclear positions. This is limited by the fact that there are only a limited number of bonds available. In an aggregate, the number of available bonds is increased (at least doubled) so one would expect a larger Stokes shift as well as a decrease in E_{vib} (the slightly larger increase in Stokes shift for HBC-$C_{8,2}$ is due to its slightly larger resonance interaction). At high concentration the HBC-PhC_{12} Franck-Condon luminescence maximum shifts to the 4-0 vibronic transition [7] (figure 1(a)). A displacement of the harmonic potential along the configurational coordinate is required for this. This is illustrated in figure 2(c) by ΔQ_a and ΔQ_l. Normally these displacements are equal, as is the case with HBC-$C_{8,2}$ (the Franck-Condon maxima are 0-1 and 1-0 for the absorption and luminescence), but for HBC-PhC_{12} at high concentration $\Delta Q_a < \Delta Q_l$. Changes in the configurational coordinate originate from a displacement of the harmonic oscillator as the charge is accommodated on the molecule. This displacement is static, compared to the vibrating modes of the oscillator, and is a proportional to $\sqrt{\sum_j (\Delta r_j)^2}$, where Δr_j is the individual nuclear displacement. In the photo excited state the vibronic coupling in the HBC-PhC_{12} stack (helped by the exo-phenyl groups) is such that the small displacements of the nuclei sum up resulting in a large ΔQ_l.

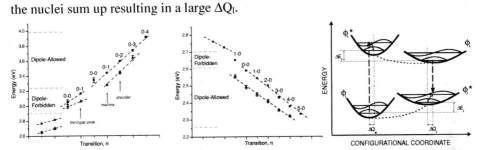

FIGURE 2. Plot of the vibronic structure at high concentration of (a) the excitation spectrum and (b) the luminescence spectrum. HBC-$C_{8,2}$ (squares) and HBC-PhC_{12} (triangles). (c) Configurational coordinate diagram illustrating 0-1 and 1-0 vibronic transitions and harmonic oscillator potential curves.

In the ground electronic state ϕ_i, and at high concentration, the HBC-PhC_{12} excitation Franck-Condon maxima is still the 1-0 transition. This is because the weak vibronic coupling in the ground state results in fewer molecules coupled to a vibronic mode, thus ΔQ is small. However for small changes to Q around $Q = 0$ there is a quick change of the Franck-Condon maximum to the 1-0 transition [8]. This is the situation when two molecules come together (or when weak vibronic coupling limits the range of nearest neighbour interaction) and explains why for medium and high concentration the 0-0 vibronic transition is no longer the Franck-Condon maximum. From figure 2(a), the slope E_{vib} for the dipole-allowed region is the same as that for the isolated molecule i.e. in the ground state the harmonic potential is the same as for the isolated molecule. Aggregation has not perturbed the harmonic potential; otherwise, the vibronic levels would deviate from the linear slope. If there is a perturbation, it is not observable in figures 2 (a) and (b).

CONCLUSIONS

The role of the vibronic coupling in self-assembled nanowires is to increase the number vibronic modes and the number of phonons accompanying an optical transition. By coupling any displacements in the configurational coordinate, the Frank-Condon maximum is shifted to higher vibronic transition. By increasing the phonon density, any excess energy can be redistributed further (limited by β), increasing the geometrical relaxation (Stokes shift) and/or making $\sum_j (\Delta r_j)^2$ increase rapidly. The degree of vibronic coupling will limit the range of the nearest-neighbour interaction. A large effective coupling to the vibronic manifold will also lower the photo-luminescence efficiency. This is exactly what is observed: HBC-PhC$_{12}$ nanowires have a photo-luminescence efficiency that is at least an order of magnitude less [9] than for HBC-C$_{8,2}$. In conclusion, the results and interpretations of the vibronic structure correlate well with the electronic interactions described previously. The discovery of the isolated HBC-C$_{8,2}$ and HBC-PhC$_{12}$ molecular excitation and luminescence spectra is significant as several spectra and results, published previously [10,11] on HBC derivatives, will have to be re-assessed in light of the fact that they were the spectra of aggregates and not isolated molecules.

ACKNOWLEDGEMENTS

A. F. would like to thank Enterprise Ireland and the Higher Education Authority for funding this research.

REFERENCES

1. Fleming, A. J.; Coleman, J. N.; Dalton, A. B.; Fechtenkötter, A.; Watson, M. D.; Müllen, K.; Byrne, H. J.; Blau, W. J.; *J. Phys. Chem. B*, **107**, 37-43, 2003.
2. Fechtenkötter, A.; Tchebotareva, N.; Watson, M.; Müllen, K.; *Tetrahedron*, **57**, 3769, 2001.
3. Fechtenkötter, A.; Saalwächter, K.; Harbison, M. A.; Müllen, K.; Spiess, H. W.; *Angew. Chem. Int. Ed.*, **20**, 3039, 1999.
4. The dipole-dipole model is a simplistic approximation as HBC molecules are large discs (~ 10 Å) closely spaced (~3.5 Å). A computationally heavy supermolecular approach is required for an accurate interpretation.
5. Cornil, J.; Beljonne, D.;Calbert, J. P.; Brédas, J. L.; *Adv. Mater.*, **13**, 1053-1067 , 2001.
6. Abram, I. I.; Hochstrasser, R. M.; *J. Chem. Phys.*, **72**, 3617, 1980.
7. At concentrations just above 1.4×10^{-6} M it the 4-0 transition becomes the sole maximum. The excitation spectra of HBC-C$_{8,2}$ and HBC-PhC$_{12}$ in figure 2(b) are similar in profile, but not identical. The detailed structure of the HBC-C$_{8,2}$ medium concentration luminescence, when matched to its mirror image excitation spectrum, enables the vibronic transitions for each peak in the spectrum to be labelled.
8. Pope, M.; Swenberg, C. E.; *Electronic Processes in Organic Crystals and Polymers*, Oxford Science Publications, 2nd Ed. 1999, p 28.
9. Results not published yet.
10. Hendel, W.; Khan, Z. H.; Schmidt, W.; *Tetrahedron*, **42**, 1127 – 1134, 1986.
11. Biasutti, M. A.; Rommens, J.; Vaes, A.; De Feyter, S.; De Schryver, F. D.; *Bull. Soc. Chim. Belg.*, **106**, 659, 1997.

Low-Temperature Self-Assembly of Novel Compound Nanowires

S. Hofmann*, C. Ducati, J. Robertson

University of Cambridge, Engineering Department, Cambridge CB2 1PZ (UK),
*E-mail: sh315@cam.ac.uk

Abstract. Sulfide nanowires and filled MoS_2 nanotubes are seen to selectively self-assemble from multilayered fullerene and Ni thin films on molybdenum substrates by annealing in a controlled atmosphere at temperatures below 550°C. Patterned, direct growth on transmission electron microscopy grids allowed a detailed electron microscopy study of as-grown structures. The nanorods are less than 100 nm in diameter and more than 10 μm long. Despite the comparatively low synthesis temperature, the structures are of very high crystallinity. The crystal growth is dependent on precursor stacking, annealing atmosphere and temperature. We could not reproduce the synthesis of crystalline arrays of single walled carbon nanotubes of identical chirality, as reported in the literature.

INTRODUCTION

Atomically well defined one-dimensional nanostructures are expected to play a key role in future nanotechnology as well as to provide model systems to demonstrate (quantum) size effects [1]. A great advantage of synthetic nanostructures is their inherent growth in nanometer dimensions without the need of complicated lithography processes. Carbon nanotubes (CNTs) are one example of such nanostructures, being studied in great detail [2]. However, despite considerable progress in their assembly and integration, it is still not yet possible to control the chirality of CNTs, which determines their band gap. Therefore, alternative types of nanotubes and nanowires are of great interest. Presently, a disadvantage of such inorganic nanocrystals, like MoS_2 nanotubes [3], is that they usually require synthesis temperatures of over 900°C and their growth hardly allows selective positioning and alignment, which restricts their full potential for applications.

In the present work a versatile method is presented to synthesize high aspect ratio sulfide nanocrystals and compound filled MoS_2 nanotubes at temperatures below 550°C with a high degree of positional control [4]. The detailed influence of processing conditions and the solid precursor materials is discussed. The result is compared to previous work involving similar pre-patterning that claimed to show crystalline arrays of single-walled CNTs of identical chirality, synthesized by high temperature annealing (950°) in a magnetic field [5].

CP685, *Molecular Nanostructures: XVII Int'l. Winterschool/Euroconference on
Electronic Properties of Novel Materials,*edited by H. Kuzmany, J. Fink, M. Mehring, and S. Roth
© 2003 American Institute of Physics 0-7354-0154-3/03/$20.00

EXPERIMENTAL

The nanocrystals were grown from patterned thin film precursor materials on molybdenum substrates by annealing in a controlled atmosphere. Pillars of alternating C_{60}/C_{70} and metal layers were formed by evaporating C_{60}/C_{70} powder (fullerite, Sigma-Aldrich) and high purity metal wire from tungsten boats mounted on a four turret source in a standard evaporator. The use of four evaporation sources gives a large variety of possible layer-by-layer precursor structures. The fullerite powder was degassed for several minutes before sublimation. The system was allowed to reach its base pressure of 10^{-6} mbar again after the deposition of each individual layer. The films were condensed onto sulfur-containing 300 mesh Mo TEM grids (Agar Scientific) and plain Mo sheets. Figure 1 shows a SEM photograph of such a TEM grid with multilayered pillars of sublimed C_{60}/C_{70}-Ni layers. The patterning on the uneven substrate was achieved by using various Cu TEM grids as disposable shadow masks, the highest mesh (2000) leading to a 10 μm × 10 μm feature size with a 10 μm pitch. The patterning dimensions and the thickness of individual layers were measured by atomic force microscopy (AFM, Digital Instruments Nanoscope III). The as-prepared samples were transferred to a vacuum system fitted with a ceramic heating element (Elstein-

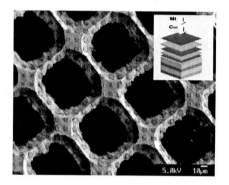

FIGURE 1. SEM photograph of a 300 mesh Mo TEM grid with multilayered pillars of sublimed C_{60}/C_{70} and Ni. The inset shows a schematic of the deposition process.

Werk). After evacuation to below 10^{-6} mbar, the chamber was refilled with nitrogen (5.0 grade) and the samples were typically processed at 5-10 mbar and 510-550°C for up to 1h. Alternatively, some pre-patterned samples were annealed at 950°C in a vacuum furnace (base pressure 10^{-6} mbar) for times varying from a few minutes to an hour. The structure and composition of the nanocrystals were analyzed by a combination of scanning electron microscopy (SEM, Jeol 6340 FEGSEM), high-resolution transmission electron microscopy (HREM, Jeol JEM 4000EX, 400 kV), energy dispersive X-ray spectroscopy (EDX), electron energy loss spectroscopy (EELS, VG HB501 STEM, 100 kV) and Raman spectroscopy (Renishaw RM series).

RESULTS AND DISCUSSION

Figure 2 shows SEM photographs of the edge of a patterned 300 mesh Mo TEM grid processed at 530°C in 5 mbar nitrogen for 30 min. Self-assembled nanostructures originate from the heavily reacted precursor pillars. No structures are seen in between the pillars, demonstrating the possible selective positioning which is important for future applications. The initial pillars were roughly 150 nm high, consisting of nine alternating fullerene-Ni layers. The rod-shaped as-grown structures are less than 100 nm in diameter and more than 10 μm long.

High-resolution TEM investigation revealed that the nanowires are highly crystalline (Figure 3). Although no sulfur was introduced during the patterning and synthesis process, the nanowires were identified as $Ni_xS_yC_z$ compounds by EDX and EELS analysis. Sulfur is a major contaminant of commercial molybdenum grids and sheets, being adsorbed at the metal surface during wet etching. The nanocrystals were observed to be stable and did not suffer beam damage.

Figure 4 shows HREM images of C_{60}/C_{70}-Ni pillars processed at 550°C in 10 mbar nitrogen for 1 hour. Mo is incorporated from the substrate grid forming MoS_2 layers. Fig. 4(a) shows a heterostructure of 14 outer MoS_2 walls (interplanar spacing of 6.2 Å) encapsulating an inner Ni sulfide core. Fig. 4(b) shows a compound crystals with 4 outer MoS_2 layers.

A systematic study of the influence of processing conditions, precursor and substrate materials was performed. A lower nitrogen pressure during annealing gives a lower yield. Neither empty reference Mo grids, nor grids with a single layer of Ni or C_{60}/C_{70} led to crystal formation. The nanowire yield from a multilayered precursor stack critically depends on the thickness of the individual layers. A very thick fullerene layer sandwiched between Ni layers gave a spider web-like carbon network rather than long sulfide crystals. Very thick Ni layers in between fullerene layers blocked growth, whereas a last thick metal layer on top of the precursor pillar left the nanostructures solely emerging from the side, showing the potential for alignment control with this method.

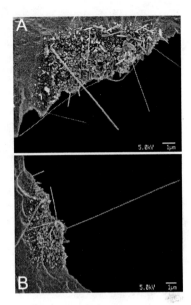

FIGURE 2. SEM photographs of self-assembled structures originating from multilayered pillars at the edge of a 300 mesh Mo TEM grid.

FIGURE 3. HREM image of Ni sulfide nanowire (scale bar 2nm).

As a substitute for the fullerene material, amorphous carbon layer were grown by plasma enhanced chemical vapor deposition from methane. Sandwich structures of up to 9 alternating layers of a-C and Ni have been deposited. The thickness of individual a-C layer was similar to the fullerene films. However, no nanowires could be grown involving the a-C precursor thin films. As a substitute for Ni, other transition metals were included in the precursor pillars. Long, one-dimensional nanostructures self-assembled from C_{60}/C_{70}-Co patterns, similar to Ni. C_{60}/C_{70}-Fe pillars showed strong oxidizing behavior and only short needle-like structures emerged. We are currently analyzing those nano-crystals by EELS elemental mapping to get more detailed information on how the precursor metal is included in the as-grown structures.

Trying to reproduce the reported synthesis of crystalline arrays of single walled CNTs of identical chirality [5], we processed multilayered C_{60}/C_{70}-Ni patterns in a high temperature vacuum furnace at 950°C for up to an hour. Although no magnetic field was applied, we found nanorods emerging perpendicularly from the substrate (Fig. 5). The rods are highly crystalline, up to 1 μm long and typically 100 nm in diameter, very similar to the structures reported by Schlittler et al. However, these structures grow not only on the original pillars but also emerge in between the patterned precursors stacks. This explains why there is no Ni incorporation and the reported structures are found to be Mo-C-O compounds or MoO_2 [6], which tends to form short rod-like structures. Similar to other groups, including some of the original authors, we can not verify the synthesis of single walled CNT crystals with this method [6][7].

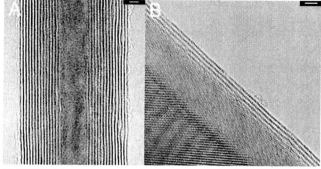

FIGURE 4. HREM images of compound filled MoS_2 nanotube (A) and crystal enwrapped by MoS_2 layers (B) (scale bars 2 nm).

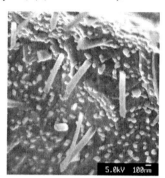

FIGURE 5. SEM photograph of sample processed at 950°C.

In conclusion, a versatile low temperature synthesis route for crystalline compound nanowires and filled MoS_2 nanotubes was presented. Apart from being an interesting system for fundamental studies, we hope that facilitated by selective positioning and alignment these nanocrystals will open new fields of possible applications. This work was supported by the European Union (CARBEN). We thank A. C. Ferrari and D. P. Chu for Raman and AFM measurements, respectively. The help of F. Piazza with the a-C thin film depositions is acknowledged.

REFERENCES

1. Y. Xia, P. Yang, Y. Sun, Y. Wu, B. Mayers, B. Gates, Y. Yin, F. Kim, and H. Yan, Adv. Mat. **15**, 353 (2003).
2. M.S. Dresselhaus, G. Dresselhaus, P. Avouris, *Carbon Nanotubes* (Springer, Berlin, 2000).
3. Y. Feldmann, E. Wassermann, D.J. Srolovitz, and R. Tenne, Science **267**, 222 (1995).
4. S. Hofmann, C. Ducati, and J. Robertson, Adv. Mat. **14**, 1821 (2002).
5. R.R. Schlittler, J.W. Seo, J.K. Gimzewski, C. Durkan, M.S.M. Saifullah, and M.E. Welland, Science **292**, 1136 (2001).
6. C. Durkan, A. Ilie, M.S.M. Saifullah, and M.E. Welland, Appl. Phys. Lett. **80**, 4244 (2002).
7. S.K. Moore, IEEE Spectrum 2, 24 (2002).

Raman Spectroscopy of Silicon Nanowires: Phonon Confinement and Anharmonic Phonon Processes

A. C. Ferrari[1], S. Piscanec, S. Hofmann, M. Cantoro, C. Ducati, J. Robertson

University of Cambridge, Engineering Department, Cambridge CB2 1PZ, UK

Abstract. We calculate the effects of phonon confinement on the Raman spectra of Silicon nanowires. The theoretical predictions are checked by measuring the Raman spectra of SiNWs selectively grown by plasma enhanced chemical vapor deposition (PECVD) employing gold as a catalyst. In order to fully account for the measured spectra and their variation as a function of laser power, the standard confinement theory is extended to include anharmonic phonon processes.

INTRODUCTION

Crystalline nanostructures such as nanotubes and nanowires offer unique access to low dimensional physics, and they can be used as nanotechnology building blocks to reach higher device integration densities than conventional fabrication methods [1,2].

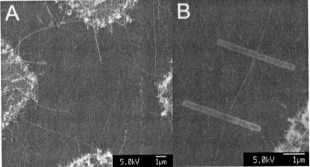

FIGURE 1. (A) SEM image of selectively grown SiNWs on Au patterned substrate. (B) Individual SiNW pinned by Pt metal contacts written by a Ga$^+$ focused ion beam from a metal-organic precursor.

One-dimensional nanomaterials could play a key role in nanotechnology, as well as provide model systems to demonstrate quantum size effects. Nanowires have recently attracted attention as an alternative system to carbon nanotubes (CNTs) [3]. Silicon nanowires (SiNWs) are particularly attractive, due to the central role of the silicon semiconductor industry. The carrier type and concentration in crystalline SiNWs could be controlled by doping, as in bulk Si [4]. Si turns into a direct band gap

---

[1] Email: acf26@eng.cam.ac.uk

semiconductor at nanometer size due to quantum confinement [5] so it could be used in optoelectronics unlike bulk Si.

Raman spectroscopy is a non-destructive analytical instrument, widely employed for semiconductors characterisation. Raman spectroscopy is also one of the main tools to probe CNTs, being able to assess their chirality, diameter and their metallic or semiconducting nature [6,7]. As for carbon nanotubes, Raman spectroscopy can become a standard technique for non-destructive characterisation of SiNWs and a direct probe of quantum confinement effects, the scale for phonon or electron confinements effects being roughly the same [8]. Here we calculate the effects of phonon confinement [9] on the Raman spectra of SiNWs and compare them to the corresponding changes in Si dots. The theoretical predictions are checked by measuring PECVD deposited SiNWs. In order to fully account for the measured spectra and their change as a function of laser power, we improve the confinement theory by including anharmonic phonon processes [10].

RESULTS AND DISCUSSION

The SiNWs used in this study were grown by PECVD using SiH_4 as the Si source and Au as the catalyst, at a substrate temperature of 380 °C. The details of the deposition process are reported elsewhere [11]. Elemental mapping by electron energy loss spectroscopy shows that they consist of a pure Si core surrounded by ~2 nm of SiO_x [11]. CVD allows selective growth. Fig. 1(a) shows as-grown SiNWs from a patterned 1 nm thick Au layer. Fig. 1(b) shows a SiNW on the same sample pinned by two Pt contacts.

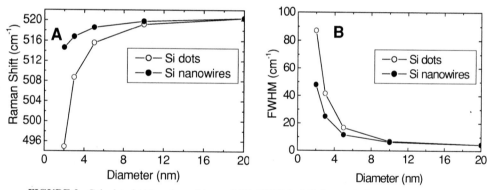

FIGURE 2. Calculated (A) peak position and (B) FWHM of Si dots and SiNWs vs. diameter.

The patterned growth together with pinning and marking of individual SiNWs was used for subsequent Raman analysis. Raman scattering is an inelastic process in which incoming photons exchange energy with the crystal vibrational modes. As the photon momentum is very small on the scale of the Brillouin zone, in an infinite crystal light can only interact with phonons having zero momentum, which gives the fundamental selection rule $q{\sim}0$, where q is the wave vector of the scattered phonon [8]. The selection rule is relaxed for a finite size domain, due to the Heisenberg uncertainty principle, allowing the participation of phonons near the Brillouin zone centre. The phonon uncertainty goes roughly as $\Delta q{\sim}1/d$, where d is the grain dimension or NW diameter. As the optic phonon frequency falls away from the zone centre in Si, confinement causes lower frequency phonons to participate [9]. This gives a downshift of the Si peak and an asymmetric broadening. These features can be predicted and calculated by applying the model of Campbell and Fauchet [9] to a SiNW. Within this framework the Raman intensity $I(\omega)$ is given by:

$$I(\omega) = \int \frac{|C(0,q)|^2}{[\omega - \omega(q)]^2 + (\Gamma_0/2)^2} d^3q \qquad (1)$$

where $C(0,q)$ is the Fourier coefficient of the confinement function, and it is a measure of the number of $q \neq 0$ phonons participating to the scattering process. In our case $|C(0,q)|^2 = \exp(-q^2 d^2/16\pi^2)$. $\omega(q)$ is the phonon dispersion curve. We approximate the dispersion of the Si TO branch by choosing an isotropic $\omega(q) = [A + B\cos(q\pi/2)]^{0.5}$, with $A = 1.714 \times 10^5$ cm^{-2} and $B = 10^5$ cm^{-2}, derived by fitting the experimental trend of the Si TO branch. Γ_0 is the Full Width at Half Maximum (FWHM) of a reference Si as measured with our Raman spectrometer (Renishaw 1000) for 514.5 nm excitation at room temperature. In the case of a quantum dot $d^3q \propto q^2 dq$, whilst for a cylindrical nanowire with length >> diameter $d^3q \propto q dq$. Indeed, wires are not confined along the axis unlike in quantum dots. Thus, the expected Raman frequency and FWHM differ in each case, as shown in Fig. 2(a,b).

Raman spectra of SiNWs have been previously reported in literature [12]. Peak positions of ~500-505 cm^{-1} are reported for wires of ~10-15 nm diameter [12], and it is claimed that this downshift is due to phonon confinement. However, Fig 2 clearly shows that 10-15 nm diameter wires should have a downshift of only ~1 cm^{-1}. To verify this, several spectra were acquired with a 100x objective and laser power of 0.04 to 4 mW. A power of a few mW is common in micro-Raman measurements and

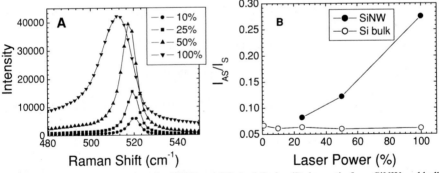

FIGURE 3. (A) Raman spectra of a SiNW and (B) Anti-Stokes/Stokes ratio for a SiNW and bulk Si as a function of laser power. 100% power corresponds to ~4mW on the sample.

is comparable to that used in previous papers [12]. Fig. 3(a) shows how, for increasing laser power, the SiNW spectra broaden and downshift significantly. The spectrum taken at ~4 mW closely resembles previously reported data for wires of similar size [12]. However this is inconsistent with just phonon confinement, as for Fig. 2. Fig. 3(b) shows how the Anti-Stokes/Stokes intensity ratio increases with laser power for a SiNW, whereas it is constant for a bare Si substrate. This ratio increases with local temperature [8]. In order to explain the trends in Fig. 3(b) and to account for possible local heating, we improved the standard confinement theory by including the anharmonic phonon processes resulting from local heating [10].

Within this framework $\omega(q)$ in (1) becomes $\omega(q, T) = \omega_0(q) + \Delta(T)$ and Γ_0 becomes $\Gamma(T)$. An analytical expression for $\Delta(T)$ and $\Gamma(T)$ in bulk Si was given in [10], as a function of 4 anharmonic constants. We calibrated these constants in order to reproduce the experimental trends on bulk Si measured on our Raman spectrometer with a hot-cold Linkam stage. Eq. (1) now depends only on 2 parameters, the local temperature T and the diameter d. For low power measurements the local temperature can be set by the hot-cold stage and the diameter can be measured by SEM or TEM. A thorough study of different wires grown in different conditions, for different local temperatures was conducted. This will be presented in detail elsewhere [13] and shows that our simple model can explain the main features of the SiNWs Raman spectra.

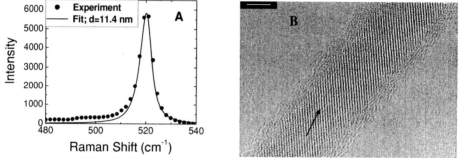

FIGURE 4. (A) Fit to a low power Raman spectrum of a SiNW. The fit gives d~11 nm and T~300K, consistent with the average diameter found by TEM (B). The scale bar in (B) is 3nm

Fig. 4(a) shows a fit to a SiNW spectrum. This gives a diameter of ~11 nm and a local temperature of ~300 K is good agreement with the electron microscopy results, Fig. 4(b). This also demonstrates quantum confinement effects in our PECVD SiNWs.

ACKNOWLEDGMENTS

A. C. F. acknowledges funding from the Royal Society.

REFERENCES

1. S. J. Wind et al. *Appl. Phys. Lett.* **80**, 3817 (2002).
2. Y. Cui, C. M. Lieber, *Science* **291**, 851 (2001).
3. D. Appell, *Nature* **419**, 553 (2002); Y. Xia et al. *Adv. Mater.* **15**, 353 (2003)
4. Y. Cui, X. Duan, J. Hu, C. M. Lieber, *J. Phys. Chem. B* **104**, 5213 (2000).
5. L. T. Canham, *Appl. Phys. Lett.* **57**, 1046 (1990);G. D. Sanders et al. *Phys. Rev. B* **45**, 9202 (1992).
6. A. M. Rao, et al. *Science* **275**, 187 (1997); A. Jorio et al. *Phys. Rev. Lett.* **86**, 1118 (2001)
7. J. Maultzsch, S. Reich, C. Thomsen, *Phys. Rev. B* **65**, 233402 (2002)
8. P. Y. Yu, M. Cardona, *Fundamentals of Semiconductors,* Springer (1999)
9. H. Richter et al. *Solid State Comm.* **39**, 625 (1981); I. H. Campbell et al. ibidem. **58**, 739 (1986)
10. M. Balkanski et al. *Phys. Rev. B* **28**, 1928 (1983);J. Menendez et al. *Phys. Rev. B* **29**, 2051 (1984)
11. S. Hofmann et al., *J. Appl. Phys.,* submitted (2003)
12. S. L. Zhang et al. *Appl. Phys. Lett.* **81**, 4446 (2002); B. Li et al. *Phys. Rev. B* **59**, 1645 (1999); P. C. Eklund, *Proc. of the XVIIIth Int. Conf. Raman Spectr.,*Wiley (2002).
13. S. Piscanec et al. unpublished (2003)

Characterization of Electromigration-induced Gold Nanogaps

S. J. van der Molen, M.L. Trouwborst, D. Dulic, B.J. van Wees

Department of Applied Physics and Materials Science Center,
Rijksuniversiteit Groningen, Nijenborgh 4, 9747 AG Groningen, The Netherlands

Abstract We study the formation and stability of nanometer-sized gaps, created by electromigration, in thin gold wires. After a wire breaks due to a high local current density, nanogaps of random size are found, all exhibiting tunneling behavior at low bias. Surprisingly, we find that small gaps (< appr. 0.5 nm) can be closed again when a voltage of about 2 V is applied. For larger gaps this is not possible, but the gold does become unstable, leading to an apparent negative differential resistance. We relate these effects to field evaporation and field-enhanced diffusion, respectively.

INTRODUCTION

Recently, a considerable amount of scientific effort has been put in the design and optimization of electrodes with nanometer-size separation. The main reason behind this ongoing research is the possibility of using such nanoelectrodes to contact single molecules. Since organic molecules can be connected to gold via a thiol (S) bridge there has been a special focus on gold electrodes. Several methods have been used to obtain nanogaps, including mechanically controllable break junction, scanning tunneling microscopy (STM) and electrodeposition techniques. Another way to create gold nanogaps, based on controlled electromigration, was introduced by H. Park *et al.* [1]. The potential of this method was beautifully demonstrated in recent measurements of the Kondo effect in single molecules. [2,3]. In this contribution, we continue the work initiated by H. Park *et al.*, because we believe that a thorough characterization of gold nanogaps is essential for molecular electronics. Without a good knowledge of the behavior of the contacts, it is difficult to make a separation between the transport properties of the molecules themselves and those of the gold electrodes.

EXPERIMENTAL

We fabricate narrow gold wires on SiO_2/Si substrates using a combination of standard UV optical lithography and electron beam lithography. Two sets of samples have been fabricated. For both sets, first an optical pattern is defined and developed after which a 5 nm Ti and a 25 nm Au layer are deposited, respectively. Subsequently, a 200 nm wide wire containing a constriction with a local width of appr. 50 nm is defined by e-beam lithography. For the first set of samples, a 3 nm Ti and a 13 nm Au

CP685, *Molecular Nanostructures: XVII Int'l. Winterschool/Euroconference on*
Electronic Properties of Novel Materials, edited by H. Kuzmany, J. Fink, M. Mehring, and S. Roth
© 2003 American Institute of Physics 0-7354-0154-3/03/$20.00

layers are deposited, respectively. For the second set, 3 nm of Ti is deposited at an angle of 45° with respect to an in-plane line perpendicular to the wire. After this, 17 nm of Au is deposited at 90° with the sample plane. Therefore, this shadow technique results in the absence of Ti, but presence of Au at the constriction. Figure 1a) shows a scanning electron microscope (SEM) picture of the resulting pattern. Contacting to the outside world takes place via bondings connected to the optically defined Au layer. Our samples are characterized using a simple DC voltage source (Keithley 230) as well as a current meter (Keithley 6517) and voltage sensor (Keithley 195A).

RESULTS AND DISCUSSION

Nanometer-sized gaps are created by linearly increasing a voltage applied to the wire, while measuring the current [1]. After initial ohmic behavior, the wire breaks somewhere within the constriction when a critical voltage V_b (and accompanying current I_b) is applied [see Fig. 1b)]. This results in a sudden drop of the current.

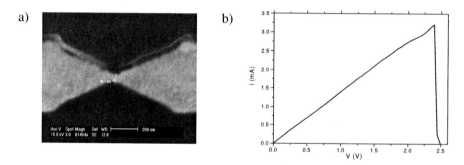

FIGURE 1. a) SEM picture of a typical sample's constriction. b) Slit formation when a voltage V is ramped over the sample and the current I is measured. At breakdown due to electromigration, the current drops, but a small tunnel current is still measured. From the latter the gap d can be determined.

To test if indeed a nanogap has been induced between the contacts, we measure IV curves of the broken junction. In all cases, we observe highly non-linear IV's, that obey Simmons' description of tunneling reasonably well for V < 1V [4]. Zero-bias resistances $R_t \equiv [dV/dI]_{V=0}$ range from $\sim 10^4$ Ω to > 10^{13} Ω. So far we have not been able to find a clear relation between the size of the nanogap and experimental parameters such as V_b, I_b and the ramp speed. Nevertheless, we do observe that samples with a Ti adhesion layer within the constriction generally end up with a relatively low tunnel resistance.

It is rather interesting to find that electromigration, which is often an unwanted effect, can be used advantageously as well. To understand the evolution of slits in the presence of an electric current, we turn to electromigration theory. When an electric field is present, an atom in a metallic sample experiences two forces. First there is the so-called direct force $F_d = Z_d e E$ of the field on the (screened) ion. The size of Z_d, i.e. it being equal to zero or to the atomic valence, has been the subject of fierce debates [5]. In general however, F_d is small compared to the second force an atom experiences, the wind force F_w, which is a result of momentum transfer by electrons to the atom. If there is a net current density j, this transfer is non-zero on average and hence electrons can 'drag' an atom along, with effective force $F_w = K e j$, where K is a constant. Clearly, F_w is minimal for atoms in their exact lattice positions (as these do not scatter Bloch electrons), but for impurities [6] and vacancies [7], F_w is substantial. The transport of vacancies as well as electromigration of surface atoms makes a surface unstable if j exceeds a critical density j_c. Once a slit starts growing, its growth velocity increases further and further due to the decreasing local conductive area and hence increasing current density (at constant I). Finally this process results in complete failure. This mechanism is described thoroughly by Mahadevan et al.[8] and we are currently performing experiments to follow it in greater detail.

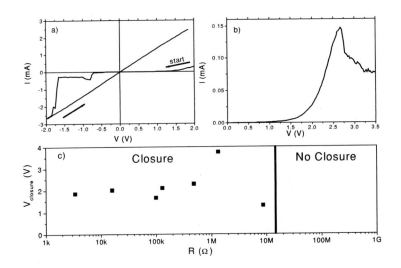

FIGURE 2. a) Closing a small gap by varying V from +2 to –2 V. After initial tunnel behavior the gap is closed at V=V_c and ohmic behavior is found again. b) For larger nanogaps, closure is not observed, but we do see a pseudo negative differential resistance. c) Absolute value of the closure voltage $|V_c|$ as a function of zero bias resistance R_t. If $R_t > 10^7$ Ω, we are not able to close the gap.

To investigate the stability of electromigration induced nanogaps, we once more ramp the voltage, while measuring the current. We find that all of our samples become unstable at some applied voltage V, though two very different kinds of behavior are observed, splitting our samples in two groups. In Fig. 2a) we demonstrate a remarkable effect seen in one group of samples. After initial tunneling behavior, the current starts fluctuating for higher $|V|$, until there is a sudden jump to a much higher value. Continuing the IV scan, the junction proves to have become ohmic again, with a relatively low resistance. Hence, Fig. 2a) shows an interesting inversion: the nanogap that was first generated by a voltage (current), is now closed again by a similar voltage

513

(but much lower current). Note that it is indeed possible to open such a closed junction again by ramping the voltage [as in Fig. 1b)], granting us a pseudo-reversibility.

If the gap is larger than a critical value d_c, the junction cannot be closed again, although it does reorder. An example of this process is given in Fig. 2b). Here we display an anologous IV-experiment for a sample having a higher initial tunneling resistance R_t and hence larger initial gap d. As in Fig. 2a), the IV curve initially shows the expected tunnel behavior, but around 2.7 V, a surprising deviation is observed. The current stops increasing, reaches a maximum value at $V=V_{peak}$ and decreases for higher V. Such an effect could be interpreted as a negative differential resistance (NDR) if it were reversible. Upon cycling between $V= -V_{peak}$ and $V=V_{peak}$ however, we observe a degradation of the peak till finally the IV's are stable again (not shown). Furthermore, the peak is only observed when |V| increases. For decreasing |V| we also find a monotonous decrease in |I|. Hence this phenomenon seems to be a gold surface reordering effect, induced by the electric field. This observation stresses the importance of characterizing nanogaps before applying molecules, so that one can separate molecular transport from gold instabilities. After all, there are also molecules that have been reported to exhibit NDR effects [9].

In Fig. 2c), we plot the (absolute) critical voltage for which closure occurs, $|V_c|$, versus R_t. The gap can only be closed if $R_t \leq 10^7 \Omega$. Using $\phi = 5$ eV, this corresponds to a critical gap $d_{close} \sim 0.5$ nm.[1] For the other samples apparent NDR's are observed.

One can think of several mechanisms that may be responsible for gap closure, implying either a critical field E_c or a critical current I_c. Looking at Fig. 2c) however, it appears that it is not the current $I_c \propto (R_t)^{-1}$, but the voltage V_c that is the critical parameter for closure (in fact, I_c varies by 3 orders of magnitude). Therefore we continue to discuss two field-induced mechanisms that may be responsible for closure. The first is field-induced surface diffusion. According to Méndez et al. [12], gold surface modifications can occur at fields of ~ 2 V/nm due to diffusion of surface gold atoms to the site where the field is highest. One can imagine that such transport results in gap closure. Unfortunately, this would not explain the impossibility of closure for larger gaps (and equal or higher local fields). The only field-induced mechanism that does, has been put forward by Mamin et al. [13] to explain their STM results. Using both a gold tip and substrate, they were able to create small mounds of gold on the substrate by applying a high field between tip and substrate. In their interpretation this is due to field evaporation of small gold clusters from tip to substrate, similar to field evaporation in a field ion microscope (FIM). In a FIM, fields of order 35 V/nm are needed whereas Mamin et al. could do with ~ 4 V/nm. To explain this descrepancy they show that the classical barrier for field evaporation can be substantially lowered when two gold electrodes are in close proximity [13]. Hence, field evaporation is

---

[1] To determine d from R_t, one needs to know how many atoms are taking part in the tunnel process. If tunneling is dominated by one atom only, we have $d_c = 0.3$ nm from $R_t = 10^7 \Omega$, but in a rough gap this is not expected. If 10 % of the gap atoms are equally involved, we would have $d_c \approx 0.6$ nm. Furthermore, $\phi \approx 5$ eV is true for large distances only. As Lang showed [10], ϕ decreases strongly for gaps smaller than appr. 0.5 nm. See e.g. Ref. [11] in which $\phi = 1$ eV is found.

relatively easy for small gaps, but much harder for larger separations. The (soft) boundary between both regimes lies at appr. 0.5 nm [10,13]. It is quite conceivable that closure of our nanogaps takes place by the jump and reordering of a small gold cluster through the gap. The fact that closure is not possible for gaps larger than $d_c \sim$ 0.5 nm, even for higher fields is consistent with this picture. For these larger gaps, however, surface diffusion and reordering remain possible, probably leading to NDR-like effects. To test our interpretation, we are currently performing low temperature measurements and transmission electron microscopy (TEM). The latter technique can also provide information about the role of the Ti in the breaking and repairing process.

CONCLUSIONS

We have studied nanogap formation due to electromigration in thin gold wires. These gaps are created by ramping a voltage over the samples until breakdown. Remarkably, tunnel junctions with small gaps (< appr. 0.5 nm) can be closed again by applying a moderate voltage. Samples with larger gaps cannot. These results are interpreted within Mamin's picture of field evaporation for electrodes in close proximity. Measuring IV's of samples with a larger gap, we find apparent negative differential resistances. Upon cycling, however, this 'NDR' effect disappears, and we conclude it is due to reordering of the gold. This observation stresses the importance of characterizing gold-gold tunnel junctions carefully before applying molecules.

ACKNOWLEDGEMENTS

We thank Bernard Wolfs, Siemon Bakker and Gert ten Brink for their support. This work was funded by the Stiching Fundamenteel Onderzoek der Materie (FOM) and the Dutch Organization for Scientific Research (NWO).

REFERENCES

1. Park, H., *et al.*, *Appl. Phys. Letters* **75**, 301-303 (1999).
2. Park, J. *et al.*, *Nature* **417**, 722-725 (2002).
3. Liang, W. *et al.*, *Nature* **417**, 725-729 (2002).
4. Simmons, J.G., *J. Appl. Phys.* **34**, 1793-1803 (1963).
5. Sorbello, R.S., *Solid State Phys.* **51**, 159-231 (1998).
6. Den Broeder, F.J.A., *et al.*, *Nature* **394**, 656-658 (1998).
7. Van der Molen, S.J., *et al. Phys. Rev. Letters* **85**, 3882-3885 (2000).
8. Mahadevan, M., Bradley, R.M., *Phys. Rev. B* **59**, 11037-11046 (1999).
9. Chen, J. *et al.*, *Appl. Phys. Letters* **77**, 1224-1226 (2000).
10. Lang, N.D., *Phys. Rev. B* **65**, 10395-10398 (1988).
11. Kergueris, C. *et al.*, *Phys. Rev. B* **59**, 12505-12513 (1999).
12. Méndez, J. *et al.*, *J. Vac. Sci. Technol. B* **14**, 1145-1148 (1996)
13. Mamin, H.J. *et al.*, *Phys. Rev. Letters* **65**, 2418-2421 (1990).

APPLICATIONS

Thermal Carrier Injection into Ambipolar Carbon Nanotube Field Effect Transistors

M. Radosavljević, S. Heinze, J. Tersoff, Ph. Avouris

IBM Research Division, T. J. Watson Research Center, Yorktown Heights, New York 10598, USA

Abstract. We demonstrate thermal injection of charge carriers into ambipolar carbon nanotube field-effect transistors with significantly smaller Schottky barrier to the conduction band of the nanotube than to the valence band. The asymmetry of the observed transfer characteristics with drain voltage can be used to find the crossover between the tunneling and the thermal regime. The experimental data can be explained based on a semiclassical model.

INTRODUCTION

Using semiconducting single-walled carbon nanotubes (NTs) as active channels in field effect transistor (FETs) devices is one of the most promising applications for these one-dimensional structures.[1] The last several years have brought much improvement in device characteristics [2] and more detailed understanding of the performance scaling in state-of-the-art CNFETs.[3,4] In particular the scaling rules depend on the fact that the current injection in the turn-on (also called subthreshold) regime is controlled by gate-dependent tunneling through the Schottky barrier (SB) formed at the contact between the metal electrode and the NT.[5] Since the contact quality and SB height vary substantially between different devices, an unambiguous demonstration of thermal injection into the NTs has not been made to date. Thermal current it implies conventional switching via modulation of the bulk (NT) barrier. Bulk switching also has the advantage of the small switching (turn-on) voltage, which is important for applications.

We demonstrate the involvement of thermal injection into NTs by studying ambipolar CNFETs with scaled-down physical dimensions and band line-up that is strongly asymmetric with respect to Fermi level of the electrodes. Thermal current is evident only when the SB height (Φ) is small enough that the thermionic portion of the current is comparable to the tunneling component. The most challenging aspect of determining the transport mechanism from the subthreshold slope ($S = dV_g/d(\log I)$, where V_g is the gate voltage and I is the current in the device) is that its value depends critically on device geometry when tunneling dominates. Moreover, S can approach the thermal limit (60mV/decade) in our scaled devices. For that reason, it is difficult to prove thermal injection based on the absolute value of S. Instead, we use simulations and transport data to show that the thermal regime can be detected by measuring transfer characteristics at various drain voltages (V_d). The asymmetries observed in such measurements give a clear fingerprint of thermal emission.

CP685, *Molecular Nanostructures: XVII Int'l. Winterschool/Euroconference on Electronic Properties of Novel Materials*, edited by H. Kuzmany, J. Fink, M. Mehring, and S. Roth
© 2003 American Institute of Physics 0-7354-0154-3/03/$20.00

METHODS

Carbon nanotube FETs are fabricated in a back-gated geometry using a two-step oxidation process involving small areas of ultra-thin oxide on which NTs are contacted. The device preparation begins with a substrate that consists of 100nm thick silicon nitride on top of degenerately doped silicon wafer. Small, protected areas are defined using electron beam lithography (EBL) and reactive ion etching of the nitride. The exposed silicon is then covered by 120nm thick, thermally grown silicon dioxide. Phosphoric acid is used to dissolve the remaining nitride, and precisely controlled dry oxidation is used to deposit another t_{ox} = 2nm or 5nm of gate oxide dielectric. Carbon nanotubes[6] are dispersed on these substrates, and source/drain connections to NTs located on the ultra-thin oxide areas are defined using standard EBL and lift-off. Scanning electron micrograph (SEM) of typical device is shown in Fig. 1.

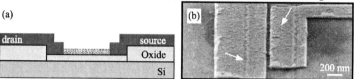

FIGURE 1. (a) Schematic of the CNFET on an oxide dielectric with step profile. (b) Sharp angle SEM micrograph of a completed device with t_{ox} = 2nm. The source-drain separation, L = 60nm is visible in the image; the edge of the ultra-thin oxide area and NT between source and drain are marked by arrows.

Theoretically, we use a semiclassical approach to compute the total current through the device. The current is calculated using the Landauer-Büttiker formalism, assuming ballistic transport in the NT. The energy-dependent transmission through the device is controlled by the SBs at the source and drain contact, and is computed within the WKB approximation. The bandgap of NTs is assumed to be 0.6eV corresponding to the diameter used experimentally.[6] The shape of the SBs depends on the electrostatic potential along the NT. We calculate the potential by solving the Laplace equation for realistic device geometry with a bottom gate at t_{ox} from the NT and a grounded top electrode that is far from the active area of the device (c.f. Fig. 1(a)). A more detailed description of the semiclassical model can be found elsewhere.[3,7]

DISCUSSION

The importance of tunneling through the source SB in CNFETs has been demonstrated by interchanging the source and drain electrodes which often results in different ON-state characteristics.[4] Figure 2(a) shows the difference in the subthreshold (turn-on) region upon swapping of source/drain in a CNFET (t_{ox} = 5nm, L = 300nm). The observed change, $S_1/S_2 \sim 1.7$, cannot be explained by adjusting Φ, because the barrier height affects only the overall current level and has no impact on S (c.f. inset in Fig. 2(b)). However, if the line-up of the metal Fermi level is such that it is closer to one of the bands, the curves (inset in Fig. 2(b)) now feature a steep portion at low V_g which is due to thermal injection into that band, as discussed below.[8] For that reason, the observed change in S (Fig. 2(a)) can only be understood by considering nanoscale variations in the geometry of the CNFET, in particular at the metal-NT contact. The role of these variations is to induce different electric fields at a given potential difference between the contact and the gate. This results in different values of S when source and drain are exchanged. Figure 2(b) shows that e.g. a variation in the dielectric oxide thickness below the source and drain contact can change S significantly. However, other unintentionally introduced geometric variations, such as different contact shapes, have similar effect on the subthreshold characteristics. (A detailed study of different gating schemes will be presented elsewhere.)

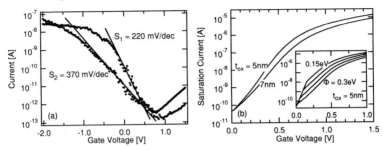

FIGURE 2. (a) The subthreshold slope varies by a factor of 1.7 when exchanging source and drain in a device ($V_d = -0.5V$, $t_{ox} = 5nm$, $L = 300nm$). (b) The change in the slope S is modeled assuming different gate induced potential drop at each electrode, such as by adjusting t_{ox} from 10 nm to 12 nm along the NT. Inset: S does not depend on SB height between 0.15 and 0.3eV (shown in 0.05eV steps).

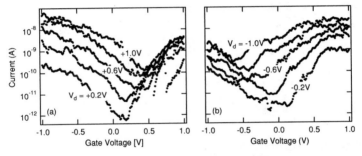

FIGURE 3. Experimental I–V_g curves for a CNFET with $t_{ox} = 2nm$ and $L = 60nm$ taken at (a) positive and (b) negative V_d in steps of 0.2V. Note the steeper subthreshold slope (S = 150mV/decade) for the electron current than for holes in (a). A similar S (at positive V_g) occurs only for the lowest V_d in (b), and it continuously deteriorates by a factor of about 3 as V_d is increased.

We focus now on thermal injection of carriers into ambipolar CNFETs. Figure 3 demonstrates that, depending on the polarity of V_d and V_g, an ambipolar device gives useful information about injection of holes (or electrons) into the NT valence (or conduction) band through either of the two contacts. It is however, important to note that ambipolar devices presented here are not suitable for applications because they suffer from simultaneous charge injection at both source and drain, as we discuss in Ref. [7]. Figure 3(a) shows that the electron current at the source (for V_d, $V_g > 0$) is characterized by S = 150mV/decade, which is consistent with scaling dependence of S on t_{ox} in the case of tunneling injection.[3,4] The observed value of S is also within a factor of three of the thermal limit; however without further information it is difficult to infer if the charge injection proceeds by tunneling through the SB, or thermally over the bulk (NT) barrier.

The main clue in this regard is provided by looking at current injection at the drain, and expanding the definition of subthreshold slope S to include this injection mechanism in addition to more conventional source injection considered above. Figure 3(b) shows that the electron current injected from the drain ($V_d < 0$, $V_g > 0$) is strongly dependent on V_d as S changes by a factor of three (from 190 to 580mV/decade) when V_d is varied from –0.2V to –1.0V. The observed change in S can not be attributed to change in geometry, as discussed above or change in SB height because these data are taken in the same CNFET configuration. However, the kink in the curve for $V_d = -0.2V$ is similar to the change between thermal and tunneling injection for small SB heights shown in the inset of Fig. 2(b).

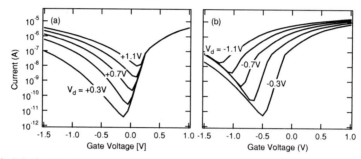

FIGURE 4. Calculated I–V_g curves in the case of asymmetric NT band line-up for (a) positive and (b) negative V_d in steps of 0.2V. In particular, $\Phi = 0.1eV$ for the electrons, and 0.5eV for holes. Electrons are thermally injected over the SB for positive V_d. At negative V_d, larger drain potentials thin the SB and enhance tunneling injection resulting in deteriorating value of S.

For this reason, the CNFET is modeled as a device in which the metal work function has been lowered by 0.2eV compared to the case where there is a midgap line-up for both source and drain.[9] Figure 4 shows that the calculated characteristics for this asymmetric device with SB height of 0.1eV for electron injection (to the conduction band) are in excellent qualitative agreement with data in Fig. 3. More importantly, the simulations allow us to correlate different aspects of the device characteristics with the corresponding band bending profiles for specific V_d and V_g values (see similar profiles in Ref. [7]).

The advantage of studying injection from the drain electrode is that in this regime the gate and the drain potentials independently control the local band bending near this contact. For instance, large negative V_d (at $V_g = 0$) severely thins the drain SB resulting in large tunneling current. The transmission of the drain SB is so high that even as the bands are modulated by V_g the thermal current over 0.1eV barrier is orders of magnitude smaller. For that reason, simulations show a subthreshold slope for electrons at $V_d = -1.1V$ which has similar tunneling value as that for holes despite five-fold difference in the respective SB heights. As V_d is lowered in Fig. 4(b), the drain potential induced thinning of the drain SB is reduced, resulting in smaller tunneling injections. It is in this regime that thermal current becomes a more dominant contribution, at least as long as gate field keeps flattening the drain SB. Indeed, we conclude that the gradual increase in S as V_d is lowered (and near $V_g = 0$) is due to drain induced evolution from tunneling to thermal current. The experimental observation of this transition (Fig. 3(b)) provides compelling evidence that thermally injected electrons from the source are responsible for switching seen in Fig. 3(a). Our conclusions are unaffected by the absolute value of S = 150mV/decade which is still quite far from the thermal limit. We believe that this discrepancy must be related to geometric factors (contact angles, trap densities,...) discussed above (c.f. Fig. 2(b)).

CONCLUSION

We have used CNFETs with scaled-down physical dimensions to emphasize the effect of geometry on subthreshold (turn-on) performance and the exhibit the possibility of thermal injection of charge into these devices. Slight variations in geometry across the device do not allow direct comparison of S with the thermal limit. Instead we rely on V_d dependent change in S for current injected at the drain and show that this change is due to drain induced transition between tunneling and thermal current. The observation of this transition demonstrates that thermal current and bulk switching are possible in these devices.

These findings are complimentary to fabrication of multi-gate CNFETs that can operate as either SB or bulk switching transistors.[10] These new types of devices will lead to direct evaluation of scattering mechanisms in semiconducting NTs.

ACKNOWLEDGMENTS

We thank J. Appenzeller and S. Wind for useful discussions and experimental advice, and J. Bucchignano and B. Ek for expert technical assistance. The devices were prepared in part at ASTL and CSS (T. J. Watson Research Center). S. H. is supported in part by the Deutsche Forschungsgemeinschaft under the Grant number HE3292/2-1.

REFERENCES

1. Tans, S. J., Verschueren, A. R. M., and Dekker, C., *Nature* **393**, 49-52 (1998); Martel, R., Schmidt, T., Shea, H. R., Hertel, T., and Avouris, Ph., *Appl. Phys. Lett.* **73**, 2447-2449 (1998).
2. Wind, S.J., Appenzeller, J., Martel, R., Derycke, V., and Avouris, Ph., *Appl. Phys. Lett.* **80**, 3817-3819 (2002); Rosenblatt, S., et al., *Nano Lett.* **2**, 869-872 (2002); Javey, A., et al., *Nat. Mater.* **1**, 241-246 (2002).
3. Heinze, S., et al., *Phys. Rev. Lett.* **89**, 106801 (2002); Heinze, S., Radosavljević, M., Tersoff, J., and Avouris, Ph., *cond-mat/0302175*.
4. Appenzeller, et al., *Phys. Rev. Lett.* **89**, 126801 (2002).
5. Freitag, M., Radosavljević, M. Zhou, Y. X., Johnson, A. T., and Smith, W. F., *Appl. Phys. Lett.* **79**, 3326-3328 (2001); Martel, R., et al., *Phys. Rev. Lett.* **87**, 256805 (2001).
6. The nanotubes with mean diameter of 1.4nm were grown by laser ablation as described in Thess, A., et al., *Science* **273**, 483-487 (1996). They were used without further processing and dispersed in dichloroethane by a brief exposure to ultrasound.
7. Radosavljević, M., Heinze, S., Tersoff, J., and Avouris, Ph., *cond-mat/0305570*.
8. Knoch, J., and Appenzeller, J., *Appl. Phys. Lett.* **81**, 3082-3084 (2002).
9. Additional difference is that the simulated t_{ox} is increased to account for an experimentally observed weaker response to V_g than expected for t_{ox} = 2nm (c.f. Fig. 2).
10. Wind, S.J., Appenzeller, J., and Avouris, Ph., *cond-mat/0306295* (accepted to *Phys. Rev. Lett.*).

High-Mobility Semiconducting Nanotubes

T. Dürkop, E.Cobas, M. S. Fuhrer

Department of Physics and Center for Superconductivity Research, University of Maryland, College Park, MD 20742-4111, USA

Abstract. Carbon nanotube transistors with channel length exceeding 300 microns have been fabricated. The gate-voltage dependence of carrier transport through these long-channel transistors is similar to short channel (few micrometer) transistors. We place a conservative lower bound for the hole mobility in nanotube transistors at 20,000 cm^2/V·s at room temperature, and offer evidence that the mobility is much greater. This high mobility corresponds with a mean free path for holes of 2.9 μm at a gate voltage of -10 V.

INTRODUCTION

Single wall carbon nanotubes (SWNT) are nanometer-diameter graphite cylinders. Depending on their diameter and chiral angle (the angle of the circumferential vector with respect to the graphite lattice vectors) they may be either metallic or semiconducting. While metallic SWNTs exhibit properties of a 1D-Luttinger liquid with ballistic conductance [1] over distances of several microns, the nature of conductance in semiconducting SWNTs is not completely understood. Recently several publications [2, 3] have established that the behavior of short (channel length less than 1 μm) field-effect transistors (FETs) fabricated from semiconducting SWNTs is governed by the Schottky-barriers between the contacts and the nanotubes. However the intrinsic mobility of the semiconducting nanotubes and the processes that limit it are not fully understood, though a first number for the mobility measured in top-gated SWNT-FETs has been reported to be 3000 cm^2/V·s [4]. To address this question we have fabricated devices with tube lengths of over 300 μm in which we may conservatively estimate a lower bound for the mobility of 20,000 cm^2/ V·s.

DEVICE FABRICATION

Our devices are fabricated using nanotubes grown with chemical vapor deposition (CVD) directly on the substrate (highly doped Si with 500 nm oxide) following a growth process adapted from [5, 6]. We deposit catalyst by first dipping the chips into a solution of Fe(NO$_3$)$_3$ in isopropanol and then in hexane to force the Fe(NO$_3$)$_3$ to precipitate onto the chip. After that we use methane to grow nanotubes at 900°C in a tube furnace. After depositing alignment markers with standard e-beam lithography technique we use a Field-Emission SEM with in-lens detector [7] to find individual nanotubes that are then contacted with Cr/Au-contacts. Figure 1 shows an image of two such devices.

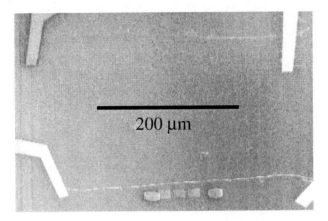

FIGURE 1. FESEM image of two long nanotube-devices. The upper device has a length of 325 μm the lower is 345 μm. The shapes at the bottom of the image are parts of a pattern of alignment markers.

RESULTS AND DISCUSSION

We have measured the electrical properties of devices with different length at different temperatures. Figure 2 shows the behavior of two tubes with lengths of 5 μm (2.7 nm diameter) and 325 μm (3.9 nm) respectively.

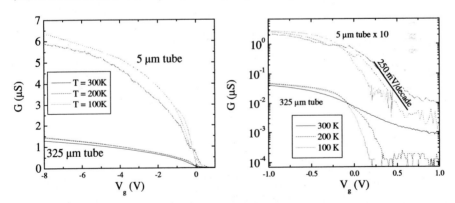

FIGURE 2. Left side: Conductance of two nanotube-FETs for different temperatures in linear scale; Right side: Conductance in log-scale for the same devices emphasizing the subthreshold region, which determines the turn-on behavior of a device. The subthreshold swing is very similar for both devices (250 mV/decade at 200 K).

Comparing the devices shown in Figure 2 one notices that their difference in total conductance is just a factor of five while their lengths differ by almost two orders of magnitude. Furthermore the subthreshold swing S ($S = dV / d\log I$) does not significantly depend on the device length. This agrees with the findings of Avouris, et al. [2, 3], which were interpreted as evidence of Schottky-barrier-dominated transistor behavior in short (< 1 μm) nanotube-FETs. In contradiction to [3] however, the subthreshold swing does show temperature dependence for both tubes and its value of 250 mV/decade is between the limits of an ideal FET (40 mV/decade) and the ideal Schottky-barrier model (1000-2000 mV/decade). Large subthreshold swings may also arise from the filling of localized states which do not contribute to the conductance of the device. Taking into account electrostatic-force-microscopy measurements [8] which show a potential drop along semiconducting nanotube devices we assume that for long tubes the conduction is diffusive and intrinsic tube resistance plays a significant role in determining the device resistance.

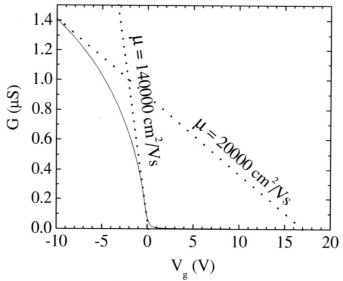

FIGURE 3. Estimate for bounds of the mobility. The intrinsic nanotube-threshold has been estimated from [4] and then rescaled for the backgate in our devices yielding V_{th}=17 V. A lower bound for the mobility is then calculated by drawing a tangent to the $G(V_g)$-curve. The upper bound represents the slope of this curve at its steepest part.

We examine the carrier mobility μ given by the formula:

$$\mu = \frac{L^2}{C_g} \frac{G}{(V_g - V_{th})}$$

where G is the conductance, L is the channel length, C_g the capacitance of the channel to the gate, V_g the gate voltage, and V_{th} the threshold voltage (the gate voltage where the first carriers enter the nanotube channel). The gate capacitance per unit length may be estimated from Coulomb blockade measurements on shorter nanotube devices on the same substrates to be approximately 10 aF/μm.

If we assume that the conductance shown in Figure 3 is the intrinsic conductance of the nanotube channel, then the mobility corresponds to the slope of the $G(V_g)$ curve. The maximum slope of this curve (near $V_g = 0$) would correspond to a mobility of about 140,000 cm^2/Vs. It is possible, however, that the steep portion of the curve is due to the rapid turn-off with increasing V_g of Schottky barriers at the contacts. In this scenario, the nanotube would still have a finite carrier density at $V_g = 0$, and the threshold would occur at positive V_g. We can make a conservative estimate of the positive threshold by estimating the carrier concentration at $V_g = 0$ in the top-gated devices of Javey, et al. [4]. We arrive at an estimate of $V_{th} = +17$ V (this is almost certainly too large, as many of our devices show a finite n-type conduction with an onset at $V_g = 5\text{-}10$ V). However using this estimate, the lowest slope of $G(V_g)$ which does not intersect our measured curve corresponds to a mobility of 20,000 cm^2/V·s. This very conservative lower bound would correspond to the rather artificial case in which the conductance is dominated by Schottky barriers near $V_g = 0$, but then becomes intrinsic (transparent Schottky barriers) at more negative V_g. We conclude that the true mobility of our devices is likely much higher than this estimate.

The two-terminal conductance of the nanotube transistor gives a lower bound for the 1D conductivity of the nanotube $\sigma_{1D} = GL$. At $V_g = -10$ V, the conductivity of the nanotube shown in Figure 3 is 4.6×10^{-8} S·cm. In a 1D conductor, the mean-free-path is given by $l = \sigma/NG_o$, where G_o is the conductance quantum and N the number of 1D channels, which we assume to be 2 (if this nanotube is multiwalled, it is likely the outer wall carries most of the current [9]). We arrive at $l = 2.9$ µm at $V_g = -10$ V and room temperature, a lower bound (a contribution of Schottky barriers to the resistance would increase this number). This indicates that quantum transport will dominate in even micron-length semiconducting nanotube devices at room temperature.

This research was supported by ARDA and the Office of Naval Research through Grant No. N000140110995, and the National Science Foundation through Grant No. DMR-0102950.

REFERENCES

1. Bachtold, A. et al., *Phys. Rev. Letters* **84**, 6082 (1994).
2. Heinze, S. et al., *Phys. Rev. Letters* **89**, 106801 (2002).
3. Appenzeller, J. et al., *Phys.Rev.Letters* **89**, 126801 (2002).
4. Javey, A. et al., *Nature Materials* **1**, 241 (2002).
5. Hafner, J. H., Cheung, C.-L., Oosterkamp, T. H., Lieber, C. M., *J. Phys. Chem. B* **105**, 743 (2001).
6. Kim, W. et al., *Nano Letters* **2**, 703 (2002).
7. Brintlinger, T. et al., *Applied Phys. Letters* **81**, 2454 (2002).
8. Fuhrer, M. S., Forero, M., Zettl, A. and McEuen, P. L., "Ballistic Transport in Semiconducting Carbon Nanotubes" in *Electronic Properties of Molecular Nanostructures*, edited by H. Kuzmany et al., AIP Conference Proceedings 591, New York: American Institute of Physics, 2001, pp. 401-404.
9. Collins, P. G., Arnold, M. S., Avouris, Ph., *Science* **292**, 706 (2001).

Contacting Carbon Nanotubes via AC-Dielectrophoresis

R. Krupke[1], F. Hennrich[1], M. M. Kappes[1,2], H. v. Löhneysen[3,4]

[1]Forschungszentrum Karlsruhe, Institut für Nanotechnologie, D-76021 Karlsruhe
[2]Institut für Physikalische Chemie II, Universität Karlsruhe, D-76128 Karlsruhe
[3]Physikalisches Institut, Universität Karlsruhe, D-76128 Karlsruhe
[4]Forschungszentrum Karlsruhe, Institut für Festkörperphysik, D-76021 Karlsruhe

Abstract. AC-Dielectrophoresis is a powerful method for the selective deposition of single walled carbon nanotubes (SWNTs), individually or as single bundles. Alignment of the deposited tubes is achieved for frequencies between 10 kHz and 10 MHz. Within this frequency range capacitive coupling to the ground becomes important and for gaining control on the number of deposited tubes it is crucial that a good electric contact between tube and metal electrodes is formed during the deposition. For SWNTs with COOH functional groups we have achieved best result with silver as electrode material.

INTRODUCTION

A major problem in the realization of electronic circuits is the difficulty to wire up carbon nanotubes, i.e. to position and contact them in a controlled way. Several methods have been reported so far, including: (a) spraying of nanotubes [1,2] – a method where nanotubes are deposited in a random manner onto silicon prior to or after lithographic structuring of metallic contacts, (b) catalytic growth of nanotubes [3] – a hightemperature process where nanotubes are grown on silicon from predeposited catalyst islands and (c) self-assembling on chemically modified surfaces [4,5] – a process using chemically modified silicon surfaces for the selective deposition of carbon nanotubes. Recently dielectrophoresis has attracted much interest for positioning SWNTs. The first alignment of SWNTs bundles on a surface by alternating electric fields has been observed by Yamamoto and coauthors [6]. Since then alternating current (ac)-dielectrophoresis has been used for the deposition and alignment of large numbers of SWNTs bundles [7], for the assembling of small networks of carbon nanotubes [8] and for the positioning and contacting of small numbers of carbon nanotubes on sub-micron scale electrodes [9]. Recently we have shown that it is possible to selectively deposit and contact a single bundle of SWNTs by controlling the chemical bonding in actually forming the contact between metal and nanotubes [10]. Some of the important aspects of the deposition method will be described in the following.

CP685, *Molecular Nanostructures: XVII Int'l. Winterschool/Euroconference on*
*Electronic Properties of Novel Materials,*edited by H. Kuzmany, J. Fink, M. Mehring, and S. Roth
© 2003 American Institute of Physics 0-7354-0154-3/03/$20.00

EXPERIMENTAL

Single-walled carbon nanotubes were grown in a laser ablation system, purified and finally suspended in N,N-dimethylformamide (DMF) [11,12]. IR-spectra reveal COOH functional groups associated with the acid treatment. For the ac-dielectrophoretic trapping experiments, the suspension was repeatedly sonicated and diluted to the extent that the suspension appears colorless and transparent (SWNT concentration $\approx$ 10 ng/ml).

Electrodes were prepared on thermally oxidized silicon substrates with standard electron-beam lithography and liftoff technique. The thickness of the oxide layer is 600 nm. The electrodes are 20-nm thick, 80–150-nmwide and the electrode distance is of the order of 100 nm. Au and Ag are used as top-electrode materials.

Prior to trapping of tube bundles, the structure has been bonded onto a chip carrier and wired up with a series resistance R_s = 500 MΩ, as sketched in Fig. 1. The circuit is powered by a low-impedance frequency generator with optional dc-offset voltage. During trapping, the electric current is monitored by dc voltage across the series resistance V_{DC}.

After switching on the frequency generator, a drop of nanotube suspension (10 µl) is applied onto the chip with a pipette. After a delay of typically one minute, the drop is blown off the surface with nitrogen gas. Finally, the generator is turned off and the sample is subjected to scanning electron microscopy (SEM) characterization and transport measurement.

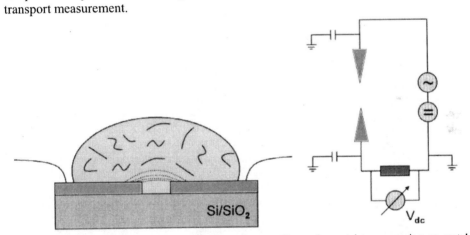

FIGURE 1. Left: Illustration of the experimental setup. Drop of nanotubes suspension on metal electrodes. Dotted lines indicate electric field. Right: Wiring scheme. Ac-voltage is applied for trapping of tubes. Small dc-bias is applied only for detecting the moment of contact forming. Capacitive coupling of the wires to the ground is indicated.

RESULTS AND DISCUSSION

Fig. 2 shows an example of a single carbon nanotube bundle trapped along four Au electrodes using the experimental setup described above. The trapping has been performed with an applied field of V_{rms} = 1 V at the frequency of f = 1 MHz. The chip was exposed to the tube suspension for 20 s. The bundle has been trapped with an excellent alignment and this result is highly reproducible at this frequency, independent of the chosen solvent or electrode material. We have observed in our experiments that nanotubes align only at trapping frequencies above 1 kHz.

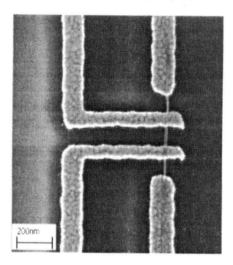

FIGURE 2. Scanning electron micrograph of a single bundle of SWNT deposited onto electrodes via ac-dielectrophoresis.

We assume that dielectrophoretic forces are responsible for the above observations. SWNTs can be strongly polarized in an electric field [13] and acquire a dipole moment which aligns them along the electric field lines. In addition the aligned tubes move along the field gradient of the inhomogeneous electric field towards the central area between the electrodes (Fig. 3). Finally, the tubes attach to the electrodes and bridge them in a straight line.

In Fig. 2 only one bundle is trapped, while with Au electrodes we most often observe more than one. In the following we show that the control over the number of trapped bundles depends decisively on the electrode material. The number of bundles N can be reliably determined with SEM. With Au electrodes, N increases with the time interval the electrodes are exposed to the suspension, and furthermore with the concentration of the suspension. With Ag electrodes, on the other hand, the situation is different. Here, only very few bundles, and quite often just a single one, are trapped independent of time and concentration. This observation goes along with a significant reduction and reproducibility of the two-terminal resistance of the trapped bundles when using Ag instead of Au electrodes. For an understanding it is necessary to consider the experimental setup in some detail. The series resistance is expected to act as a voltage divider and current limiter. As soon as a contact is formed between the electrodes via a nanotube and the resulting resistance R_T is smaller than the series resistance R_S, the applied voltage will mainly drop along the latter. Hence the field between the electrodes will collapse and the trapping of additional tubes will be automatically prevented.

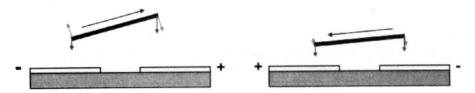

FIGURE 3. Schematic drawing of the forces acting on a tube in an inhomogeneous ac-field during two half cycles (left, right). Induced dipole moment, alignment torque and dielectrophoretic forces indicated by black, red and blue arrows respectively.

This protocol apparently works well only when we use Ag electrodes. The failure of the mechanism with Au electrodes can be understood in the following way: if an alternating voltage is applied, the voltage divider operates only at resistance values much lower than the series resistance, because the unavoidable capacitive reactance Im $Z = \omega C$ of the leads and of the measurement devices short circuits the series resistance with a typical capacitance of $C = 20$ pF. For instance, when attempting trapping at $f = 1$ MHz the accumulation of bundles continues, unless a resistance of the order of $R_T = 10$ kΩ or lower is formed. For lower frequencies R_T is correspondingly shifted to larger values. Hence for preventing the accumulation of many tubes a good electric contact needs to be formed between the first deposited tube and the electrodes, still in the presence of the solvent.

The question arises why nanotubes form an good electric contact with Ag in the presence of a suspension and why they do not with Au. The origin may be found in the strong affinity of the COOH groups of our acid-treated nanotubes to Ag surfaces. For instance, it is known that n-alkanoic acid [CH$_3$(CH$_2$)$_m$ COOH] forms a self-assembled monolayer on native Ag oxide surfaces [14] and that short, COOH-functionalized, nanotubes attach parallel to the surface normal of Ag films [15]. Hence it is quite likely that COOH-nanotube bundles undergo chemical bonding to Ag electrodes even in the presence of solvent. In contrast, COOH groups have no affinity to Au surfaces (Fig. 4).

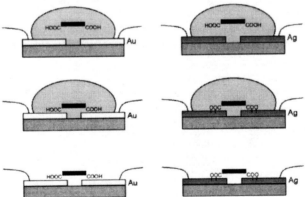

FIGURE 4. Illustration of the process of contact forming (top to bottom) for COOH-functionalized SWNTs on gold and silver electrodes (left, right).

SUMMARY

In summary, we have demonstrated that a single bundle of carbon nanotubes can be selectively trapped onto electrodes with alternating electric fields. The alignment is excellent for frequencies of the order of 1 MHz. Matching of the electrode material with the chemical functional groups of the carbon nanotubes appears to be essential for controlling the number of trapped bundles and for obtaining low contact resistance. Best results for nanotubes with COOH groups have been achieved with Ag electrodes. With Au electrodes the formation of an electrical contact during trapping is inhibited, probably by a layer of solvent. Thus, besides establishing the use of alternating electrical fields to position nanotubes, our results stress the important role of chemical bonding in actually forming the contact between metals and nanotubes.

ACKNOWLEDGMENTS

The authors thank H. B. Weber, O. Hampe and D. Beckmann for helpful discussions.

REFERENCES

1. S.J. Tans, A. Verschueren, C. Dekker, *Nature* **393**, 49 (1998)
2. M. Bockrath, D.H. Cobden, J. Lu, A.G. Rinzler, R.E. Smalley, L. Balents, P.L. McEuen, *Nature* **397**, 598 (1999)
3. J. Kong, H. Soh, A. Cassell, C. Quate, H. Dai, *Appl. Phys. A* **69**, 305 (1999)
4. J. Liu, M.J. Casavant, M. Cox, D.A. Walters, P. Boul, W. Lu, A.J. Rimberg, K.A. Smith, D.T. Colbert, R.E. Smalley: *Chem. Phys. Lett.* **303**, 125 (1999)
5. R. Krupke, S. Malik, H.B. Weber, O. Hampe, M.M. Kappes, H. v. Löhneysen, *Nano. Lett.* **2**, 1161 (2002)
6. Yamamoto K., Akita S., Nakayama Y., *J. Phys. D: Appl. Phys.* **31**, L34-36, (1998)
7. Chen X. Q., Saito T., Yamada H., Matsushige K., *Appl. Phys. Lett.* **78**, 3714-3716 (2001)
8. Diehl M. R., Yaliraki S. N., Beckmann R. A., Barahona M., Heath J. R., *Angew. Chem.* **114**, 363-336 (2002)
9. Nagahara L. A., Amlani I., Lewenstein J., Tsui R. K., *Appl. Phys. Lett.* **80**, 3826-3828 (2002)
10. Krupke R., Hennrich F., Weber H. B., Beckmann D., Hampe O., Malik S., Kappes M. M., v. Löhneysen H., *Appl. Phys. A* **76**, 398-400 (2003)
11. Lebedkin S., Schweiss P., Renker B., Malik S., Hennrich F., Neumaier M., Stoermer C., Kappes M. M., *Carbon* **40** 417-423 (1998)
12. Reflux in 3M HNO3 for 48 h, centrifugation, dispersion/suspension with Triton X-100 in water, ultra filtration, membrane filtering, washing with H2O, acetone, and drying in vacuum at 10-3 mbar for 12 h
13 Benedict L. X., Louie S. G., Cohen M. L., *Phys. Rev. B*, **52**, 8541-8549 (1995)
14 Y.T. Tao: *J. Am. Chem. Soc.* **115**, 4350 (1993)
15 B. Wu, J. Zhang, Z. Wei, S. Cai, Z. Liu: *J. Phys. Chem. B* **105**, 5075 (2001)

Self-Aligned Contacting of Carbon Nanotubes

Maik Liebau*, Eugen Unger, Andrew P. Graham, Georg S. Duesberg,
Franz Kreupl, Robert Seidel, and Wolfgang Hoenlein

Infineon Technologies AG, Corporate Research, 81730 Munich, Germany

Abstract. An important prerequisite for the implementation of carbon nanotubes (CNTs) in microelectronic circuits is the evaluation of their electronic properties. For this, the generation of low-ohmic electrical contacts between CNTs and metallic circuit lines is crucial. We describe a parallel process which does not rely on e-beam lithography and yields self-aligned, low-ohmic contacts on wafer scale. CNTs that are spray deposited on metallic test structures created by optical lithography are subsequently embedded by electroless metallization and annealed. With this method we are able to simultaneously lower the contact resistance of an unlimited number of CNTs. The reduction of the contact resistance depends on the metals involved and the annealing process. This new method enables the statistical evaluation of the CNT quality dependence on synthesis conditions.

INTRODUCTION

Due to their unique electrical and mechanical properties carbon nanotubes (CNTs) have been proposed as nanoscale building blocks, in particular for the construction of molecular electronic devices.[1] The realization of good contacts between metallic microelectrodes and CNTs is especially critical for all applications of CNTs in microelectronics.[2] Whereas the direct deposition of individual CNTs from solution onto predefined electrodes[3] suffers from high contact resistances, better electrical contacts can be achieved when the metal electrodes are deposited on top of the CNTs, e.g. by electron beam lithography.[4] The latter approach is, however, very time consuming and is applicable only for a limited number of contacts. There are also other approaches to improve CNT-metal contacts, e.g. by mechanical manipulation[5] or by soldering of CNTs with an electron beam.[6]

We present a new approach for the highly parallel enclosure of CNTs in microelectrodes by electroless deposition of nickel.[7] Using this technique we demonstrate the improvement of the contact resistances between CNTs and gold electrodes for about 30 individual devices. We show that electroless nickel deposition is generally applicable for the fabrication of mechanical and electrical composites between metals and CNTs on a large scale.

EXPERIMENTAL

Mutli-walled CNTs were synthesized by thermal chemical vapor deposition (CVD) as described elsewhere.[8] The synthesis was optimized for the growth of CNTs that are

CP685, *Molecular Nanostructures: XVII Int'l. Winterschool/Euroconference on Electronic Properties of Novel Materials*, edited by H. Kuzmany, J. Fink, M. Mehring, and S. Roth
© 2003 American Institute of Physics 0-7354-0154-3/03/$20.00

shorter than 5 μm and have diameters of about 15 nm. Further, CNTs were oxidized for 5 min at 700°C under ambient conditions, dispersed in dimethylformamide (DMF), and finally distributed onto patterned gold microelectrodes.

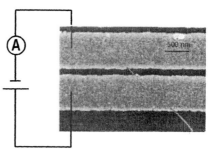

Electrical resistances were determined by four-terminal measurements on five devices and by two-terminal measurements on ca. 25 devices before and after electroless deposition of nickel. An illustration of the measurement arrangement is given in figure 1.

The electroless deposition of nickel was performed in an alkaline solution, containing 21 g/l nickel chloride, 24 g/l sodium phosphinate, and 40 g/l sodium citrate. The pH

FIGURE 1. SEM image of a multi-walled CNT that is embedded in nickel and schematic illustration of the measurement setup.

was controlled by the addition of ammonia solution. The deposition was performed at 80°C and the metal thickness was controlled by time (Fig. 2). After nickel deposition an annealing step was performed for 5 min at 400°C under argon atmosphere.

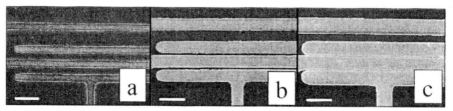

FIGURE 2. SEM images of Au-microelectrodes (a) before, (b) after 15 s, and (c) after 30 s electroless deposition of nickel at 80°C. The error bars are 2 μm long.

RESULTS AND DISCUSSION

In order to study the effect of electroless nickel deposition on the contact resistances between carbon nanotubes and microelectrodes, gold was used as base electrode material. Unfortunately, direct electroless deposition of nickel on metals such as gold and silver is impossible, because of the electrochemical potentials.[9] Thus, the gold electrodes were coated with a thin iron layer (10 nm) before applying to the nickel deposition bath. We also performed studies with an iron layer (10 nm) deposited under the gold electrodes (40 nm) and found no effect on the quality and quantity of electroless nickel deposited. Therefore, it was concluded that iron diffuses quickly into the gold electrode and the electrode was in fact an alloy of iron and gold.

Contact resistances between carbon nanotubes and gold microelectrodes were statistically evaluated: four-terminal measurements were performed on five oxidized, multi-walled CNTs. Since the oxidation procedure generates a number of defects, the intrinsic resistance of CNTs is length dependent. We determined an average intrinsic resistance of oxidized multi-walled CNTs of 41 ± 7 kΩ/μm.

After nanotube deposition and without any further manipulation of the device (Fig. 3a) no electrical contacts could be measured by two-terminal measurements. Annealing at 400°C slightly improved the electrical resistances. We measured resistances of annealed samples, prior to electroless nickel deposition, ranging from 20 MΩ to 2.5 GΩ. After soldering the tubes by nickel deposition dramatic improvements of the electrical resistances were found. Resistances of nickel embedded CNTs are ranging from 25 kΩ (CNT length: 0.8 μm) to 165 kΩ (CNT length: 4 μm).

Based on the hypothesis that the total resistances depend linearly on the length of the nanotubes, which is reliable at least until a length of 5 μm, it is appropriate to plot the uncorrected results versus length (Fig. 3). Using a linear least-square fit a contact resistance of 22 kΩ was calculated for nickel embedded CNTs (filled circles).

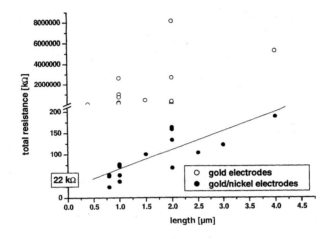

FIGURE 3. Total electrical resistances of multi-walled CNTs prior to nickel deposition (○) compared to resistances of CNTs after the electroless deposition if nickel (●).

The measurements reported here were performed after annealing at 400°C for 5 min. Temperatures below 200°C do not significantly improve the nickel/nanotube/gold contacts, whereas higher temperatures (700°C) can destroy the nanotubes. The deformation of CNTs at elevated temperatures is probably caused by mechanical stress of the material during the formation of alloys. An example of several CNTs that are destroyed after 700°C anneal is given in figure 4.

There are a number of factors which influence the scatter observed in electrical resistances. The lengths of the CNTs were derived from SEM images and involve the assumption that the electrically active contact areas correspond to the metal-covered areas as visible in the SEM images (compare Fig. 1). Although this is most likely, it is possible that the active contact area is smaller and the measured lengths differ by the widths of the contact pads. Furthermore, one has to consider that the intrinsic resistances of the gold microelectrodes are in the range of several hundred ohms. However, this corresponds to only 10 % of the measured contact resistances, which is in the range of the deviation of the uncorrected results (Fig. 3).

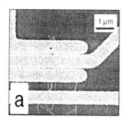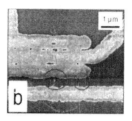

FIGURE 4. SEM images of multi-walled CNTs that cover four microelectrodes. Image (a) shows the situation after nickel deposition. In (a) the CNT appears to be intact whereas after annealing at 700°C (b) the CNT was broken.

Finally, the number of defects, leading to the intrinsic resistance will be different for each individual CNT. For this reason this study was performed only on multi-walled CNTs without any obvious defects (e.g. kinks). Thus, it was reasonable to assume that the intrinsic resistance (41 kΩ) of the CNTs was identical within the standard deviation (17%).

CONCLUSION

Contact resistances between CNTs and microelectrodes were significantly improved by the electroless deposition of nickel. This study focused on gold electrodes but it is already known that the presence of d-states at the Fermi energy affects the coupling between metals and nanotubes.[10] Therefore, gold is a relatively poor contact material in comparison to transition metals with unoccupied d-states that can provide lower contact resistances. Further studies are underway for a number of different electrode metals.

These results have shown that electroless metal deposition is a general method for the improvement of the mechanical and electrical contact between metal electrodes and CNTs, as a prerequisite for the future application of CNTs in microelectronic chips.

REFERENCES

1. Kong P.G. Collins, P. Avouris: Sci. Am. **62**, (December 2000); H. Dai, E. W. Wong, C. M. Lieber: Science, **272**, 523, (1996)
2. A. Bachtold, P. Hadley, T. Nakanishi, C. Dekker: Science, **294**, 1317, (2001); S. Frank, P. Poncharal, Z. L. Wang,, W. A. de Heer: Science, **280**, 1744, (1998)
3. S.J. Tans, M.H. Devoret, H.Dai, A. Thess, R.E Smalley, L.J. Geerligs, C. Dekker: Nature **386**, 474, (1997)
4. H. Dai, J. Kong, C. Zhou, N. Franklin, T. Tombler, A. Cassell, S. Fan, M. Chapline: J. Phys. Chem. B **103**, 11246, (1999)
5. P. A. Williams, S. J. Papadakis, M. R. Falvo, A. M. Patel, M. Sinclair, A. Seeger, A. Helser, R. M. Taylor II , S. Washburn, R. Superfine: Appl. Phys. Lett. **80**, 2574, (2002; J. Cumings, A. Zettl: Science **289**, 602, (2000)
6. F. Banhart: Nano Lett. **1**, 329, (2001); D. Nørgaard Madsen, K. Mølhave, R. Mateiu, A. M. Rasmussen, M. Brorson, C. J. H. Jacobsen, P. Bøggild: Nano Lett. **3**, 47, (2003)
7. F. A. Lowenheim: Modern Electroplating, Chapter 33. Electroless Plating, John Wiley, Inc., New York (1963)
8. F. Kreupl, A. P. Graham, G. S. Duesberg, W. Steinhögl, M. Liebau, E. Unger, W. Hönlein: Microelec. Eng., **64**, 399, (2002); G. S. Duesberg, A. P. Graham, M. Liebau, R. Seidel, E. Unger, F. Kreupl, W. Hönlein: Nano Lett. **3**, 237, (2003)
9. Milazzo et al., Tables of Standard Electrode Potentials, New York: Wiley (1978)
10. J.J. Palacios, A.J. Perez-Jimenez, H. Louis, E. SanFabian, J.A. Verges: Phys. Rev. Lett. **90**, 106801 (2003)

Photophysics and Photovoltaic Device Application of Fullerene Containing Phthalocyanine Dyads

Helmut Neugebauer[a], Maria Antonietta Loi[b], Christoph Winder[a], N. Serdar Sariciftci[a], G. Cerullo[c], Andreas Gouloumis[d], Purificación Vázquez[d], Tomás Torres[d]

[a]LIOS, Physical Chemistry, Johannes Kepler University Linz, Altenbergerstraße 69, A-4040 Linz, Austria
[b]ISMN-C.N.R., Via P.Gobetti, 101, 40129 Bologna, Italy
[c]Dipartimento di Fisica, Politecnico di Milano, Piazza L. da Vinci 32, I-20133 Milano, Italy
[d]Departamento de Quimica Organica, Universidad Autonoma de Madrid Canto Blanco, Madrid, Spain

Abstract. The use of conjugated polymers as antenna systems in photovoltaic devices with phthalocyanine-fullerene dyad compounds is described. With luminescence measurements and time resolved transmission changes, energy transfer processes are studied. The influence on photovoltaic device properties is presented.

INTRODUCTION

Photoinduced electron transfer in organic molecules is an intensively investigated process in natural as well as in artificial systems. Dyad systems composed of electron acceptor molecules covalently linked to photoactive donors are candidates to perform photoinduced electron transfer. Due to their outstanding electronic and optical properties, fullerenes as electron accepting units in combination with phthalocyanines as electron donors appear particularly promising for optoelectronic applications. Recently, charge transfer states with lifetime of 3 ns have been reported for phthalocyanine-fullerene dyad compounds in solution [1], as well as long living photoinduced charge separation in the solid state [2,3]. Besides fundamental interest in the nature of the process, charge collection after charge separation may open a possibility to use such electron transfer systems in organic photovoltaic applications. Photovoltaic devices using films of phthalocyanine-fullerene dyads were built and characterized [2,3].

In addition, energy transfer from conjugated polymers may enhance the spectral range of the dyad for the photocurrent, which improves the spectral mapping to the terrestrial solar spectrum. Photophysical studies on these processes are described in the present paper.

CP685, *Molecular Nanostructures: XVII Int'l. Winterschool/Euroconference on Electronic Properties of Novel Materials*, edited by H. Kuzmany, J. Fink, M. Mehring, and S. Roth
© 2003 American Institute of Physics 0-7354-0154-3/03/$20.00

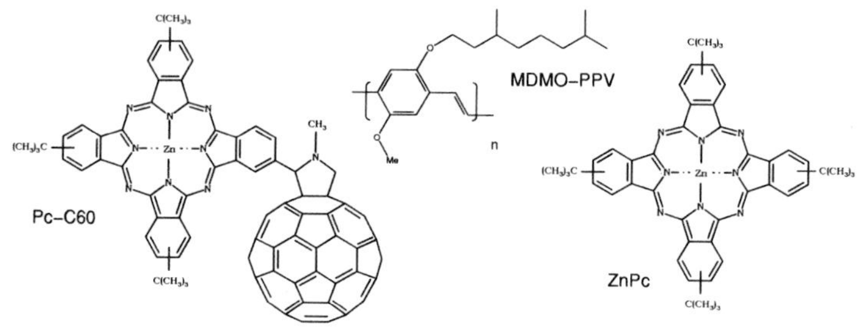

FIGURE 1. Structure of the compounds

EXPERIMENTAL

The synthesis of the dyad, abbreviated as Pc-C60, has been described elsewhere [4]. The structure together with the structure of a model compound ZnPc and the conjugated polymer MDMO-PPV [5] used for energy transfer is shown in Fig. 1.

Thin film samples for optical measurements were prepared by spin coating from toluene solution onto fused silica substrates. Luminescence was measured with excitation at 476 nm. Time resolved spectroscopy has been performed with a sub-10 fs laser system using pump and probe technique. Normalized transient transmission changes $(-\Delta T/T)$ were measured by spectrally filtering the probe beam and combining differential detection with lock-in amplification. The setup of the system has been described elsewhere [6]. Photovoltaic device characteristics were measured under illumination intensity of 80 mW/cm^2 of white light from a solar simulator.

RESULTS AND DISCUSSION

Evidence for energy transfer from MDMO-PPV to phthalocyanine is found from luminescence measurements (Fig. 2). With excitation at 476 nm (where only MDMO-PPV absorbs), the luminescence around 600 nm from MDMP-PPV (left curve) is quenched in a 5% mixture with zinc-phthalocyanine (ZnPc), where only the luminescence of ZnPc around 720 nm is obtained (right curve).

Time resolved transmission changes of the MDMO-PPV:ZnPc mixture are shown in Fig. 3. After exitation of the polymer, the bleaching of the ground state absorption of MDMO-PPV around 500 nm decays within 10 ps, and with the same rate the bleaching of the ground state absorption of ZnPc at 700 nm appears. The inset in Fig. 3 compares time traces of the features at 540 nm and 700 nm for mixtures MDMO-PPV:ZnPc and MDMO-PPV:Pc-C60. The high similarity of the rate constants implies a similar energy transfer process from MDMO-PPV to phthalocyanine in both mixtures.

Utilizing the energy transfer, MDMO-PPV may be used as antenna system for photovoltaic applications of Pc-C60, extending the spectral range by the polymer absorption around 500 nm. Figure 4 shows the IPCE (incident photons converted to electrons)

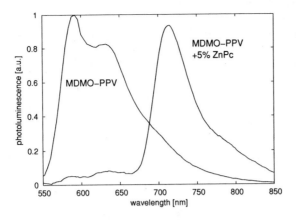

FIGURE 2. Luminescence of MDMO-PPV and of MDMO-PPV + 5% ZnPc, excitation 476 nm

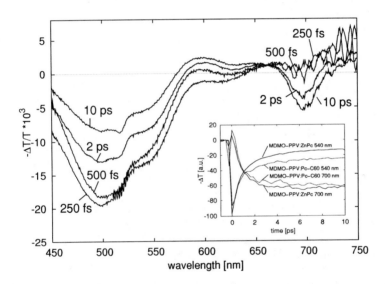

FIGURE 3. Time resolved transmission changes after excitation at 476 nm of MDMO-PPV:ZnPc. Inset: traces at 540 nm and 700 nm for MDMO-PPV:ZnPc and MDMO-PPV:Pc-C60

curves of photovoltaic devices made with Pc-C60 and with a Pc-C60:MDMO-PPV 10:1 mixture. Indeed, the absorption range around 500 nm contributes to the photocurrent. The open circuit voltage with white light illumination is increased to around 0.5 V (inset of Fig. 4). However, the short circuit current in the the mixture is lower (reflected also by the lower current in the maxima of the IPCE curve), indicating charge transport problems within the device.

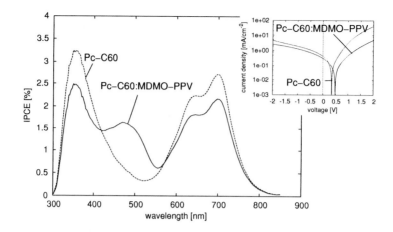

FIGURE 4. IPCE of devices with Pc-C60 and Pc-C60:MDMO-PPV. Inset: I/V characteristics

CONCLUSION

Energy transfer processes from conjugated polymers to phthalocyanine-fullerene dyads have been studied. The extension of the absorption range can be used for a better matching of the photocurrent spectrum to the solar emission spectrum. Devices with MDMO-PPV as antenna system mixed into Pc-C60 dyads show the spectral extension, however with reduction in the obtained current.

ACKNOWLEDGMENT

Supported by EU RTN Project No. HPRN-CT-2000-00127 "EUROMAP"

REFERENCES

1. Guldi, D., Gouloumis, A., Vázquez, P., and Torres, T., *Chem. Commun.* 2056 (2002).
2. Loi, M., Denk, P., Neugebauer, H., Brabec, C., Sariciftci, N. S., Gouloumis, A., Vázquez, P., and Torres, T., in *Structural and Electronic Properties of Molecular Nanostructures*, edited by H. Kuzmany et al., AIP Conference Proceedings 633, Melville, New York, 2002, p. 488.
3. Loi, M. A., Denk, P., Hoppe, H., Neugebauer, H., Winder, C., Meissner, D., Brabec, C., Sariciftci, N. S., Gouloumis, A., Vázquez, P., and Torres, T., *J. Mat. Chem.* **13**, 700 (2003).
4. Gouloumis, A., Liu, S.-G., Sastre, A., Vázquez, P., Echegoyen, L., and Torres, T., *Chem. Eur. J.* **6**, 3600 (2000).
5. Brabec, C., Sariciftci, N. S., and Hummelen, J. C., *Adv. Funct. Mat.* **11**, 15 (2001).
6. Lanzani, G., Cerullo, G., Brabec, C., and Sariciftci, N. S., *Phys. Rev. Lett.* **60**, 047402-1 (2003).

Room-Temperature Fabrication of High-Resolution Carbon Nanotube Field-Emission Cathodes by Self-Assembly

S.J. Oh[1], Y. Cheng[2], J. Zhang[2], H. Shimoda[3], Q. Qiu[3], and O. Zhou[1,2]

[1]*Curriculum in Applied and Materials Sciences, University of North Carolina, Chapel Hill, NC 27599*
[2]*Dept. of Physics and Astronomy, University of North Carolina, Chapel Hill, NC 27599*
[3]*Applied Nanotechnologies, Inc., 308 W. Rosemary St. Chapel Hill, NC 27516*

Abstract. We report a process to assemble carbon nanotubes (CNTs) into patterned and periodic structures by self-assembly at room temperature. Patterns of 10 μm or smaller can be readily deposited on various types of substrates using functionalized CNTs. The self-assembled CNTs have in-plane orientational order, and adhere strongly to the substrates. Under applied electric field they emit electrons with emission characteristics comparable to the CNT cathodes made by other techniques. This room temperature process can be utilized for assembly and integration of nano-structured materials for a variety of devices including field emission displays.

INTRODUCTION

Carbon nanotubes (CNTs)[1] are a promising class of electron field emission materials[2] with a low threshold field for emission and high emission current density[3-5]. They are being investigated for applications in a variety of vacuum electronic devices including field emission display (FED) [6] and x-ray [7]. Although several techniques have been developed to synthesize raw CNT materials, assembly and integration of functional macroscopic CNT structures for device applications remain challenging. Among the synthesis methods developed to date, chemical vapor deposition (CVD) is the only one capable of directly depositing CNT and CNT-like materials on device surfaces. However, the requirements of high temperature and reactive environment have prevented its application from systems with limited thermal and/or chemical stability such as the field emission displays, where low-melting-temperature glass substrates are required [8]. Screen-printing technique that is currently used to fabricate FED cathodes [9] suffers from low-resolution and inefficient use of materials.

Recently we demonstrated that functionalized CNTs can self-assemble into uniform and ordered forms [10]. Here we demonstrate that patterned and periodic structures of CNTs can be readily prepared by this self-assembly process using template substrate. The self-assembled CNT films were investigated for their field emission property.

CP685, *Molecular Nanostructures: XVII Int'l. Winterschool/Euroconference on Electronic Properties of Novel Materials,* edited by H. Kuzmany, J. Fink, M. Mehring, and S. Roth
© 2003 American Institute of Physics 0-7354-0154-3/03/$20.00

EXPERIMENTAL

Single wall carbon nanotubes (SWNTs) synthesized by laser ablation method at UNC [11] were used for this experiment. They were chemically processed to reduce the aspect ratio to around 100 and to have hydrophilic functionality [12]. The Raman excited breathing and tangential modes were unchanged after processing (Figure 1A). The FTIR spectrum showed a strong peak at the C=O stretching mode frequency, suggesting the defects sites were terminated with carboxylic groups [13]. Due to the carboxylic groups, the CNTs dispersed easily in hydrophilic solvents such as water or alcohol and the suspension was stable for over a week without precipitation.

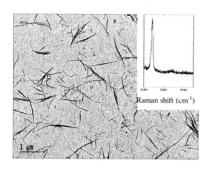
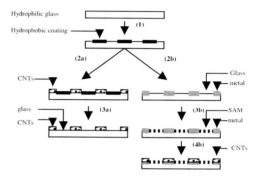

FIGURE 1 (A): TEM image of the etched SWNTs with an average bundle length of 1-2 μm (inset: the characteristic Raman breathing mode from the etched SWNTs)
(B). Schematic illustration of the deposition process: (1) a hydrophilic glass substrate is patterned with a layer of hydrophobic photoresist by photolithography. (2a) SWNTs are deposited on the hydrophilic regions of the substrate (exposed glass surface) by the self-assembly process. (3a) Removal of the hydrophobic polymer coating by washing the SWNT coated substrate in acetone. (2b) Metallization by thermal evaporation of metals (Cr/Al) and photoresist lift-off. (3b) Attachment of OTS molecules on hydroxyl terminated glass surface. (4) Deposition of CNTs on metal strips.

To deposit patterned CNT structures, the substrates were first patterned with periodic alternations of hydrophilic and hydrophobic regions (Figure 1B). Two methods were used which are illustrated in Fig. 2. A homogeneous suspension of the CNTs was then stabilized in de-ionized water at a nanotube concentration up to 1 g/L. The patterned substrate was submersed into the CNT/water suspension at room temperature. The nanotubes assembled along the water-substrate-air triple line on the hydrophilic regions of the substrate. As the triple line moved downward when the water gradually evaporated, a continuous CNT film formed. The nanotubes adhered strongly to the hydrophilic substrates. The self-assembled CNT structure was undamaged under sonication in Acetone for a few minutes, enabling removal of the hydrophobic photoresist coating. The strong adhesion is attributed to the interaction between the hydroxyl groups on the glass substrates and the carboxylic groups terminating CNT defect sites. All samples were annealed at 450C for an hour before further experiments.

The electron field emission characteristics of these self-assembled SWNT films were measured in a vacuum chamber at 5×10^{-7}Torr base pressure. Metal strips evaporated onto the two edges of the CNT film provided the electrical contact to the nanotube films. The current-voltage (I-V) characteristic was measured using a hemispherical tungsten tip (5mm in radius) as the anode. The cathode-anode distance was kept at 168 µm. The emission image was taken using a phosphor coated ITO glass as the anode.

RESULTS AND DISCUSSION

As shown in Fig. 2 patterned CNT films with the line width down to 10µm were obtained by this self-assembly process. Pattern size and edge sharpness depended on the template preparation technique. The template used to obtain the patterns shown in the figure was created using a printed polymer film with a 5µm resolution limit. The film characteristics such as thickness, uniformity, and morphology generally depended on the concentration of the suspension and the solvent evaporation rate. When the temperature of the water/CNT suspension was increased the resulted film thickness decreased, and the uniformity fluctuation was larger than that obtained at room temperature. Above 40C the film was no longer continuous over the entire growth area. When fast evaporating solvent such as ethyl alcohol was used, thin and non-uniform film resulted. In the samples from highly concentrated suspensions (1-2 g/L), dendrite structures of SWNT bundles were observed.

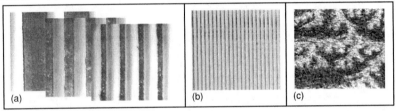

FIGURE 2.(a) Patterned CNT structures. The widths of strips shown above are 100, 40, 10µm, respectively. (b) Uniformly patterned CNT structure over large area (c) The "dendrite" structure found in films from highly concentrated suspension

Fig. 3 shows the emission I-V characteristics of the self-assembled CNT film. Initially the film showed a threshold electric field of 11V/um for 10 mA/cm^2 current density which is substantially higher than the values we previously reported for films fabricated by other techniques. But this value was reduced to 6V/um after an electrical conditioning process. This value is comparable to what we reported for free-standing membranes of long SWNT bundles (4-7V/µm for 10 mA/cm^2) [5]. The emission characteristics of the self-assembled film are comparable with those from the CVD-grown CNTs [14] and cathodes by the screen-printing methods [15]. Figure 3(b) shows the emission image of the self-assembled film at 6V/µm applied field. The field was supplied at pulsed mode (frequency 100 Hz, pulse width 1 ms) to prevent the phosphor screen from damaged by excessive electron bombardment. The two dark lines across the image are dividing lines of three different regions of the sample. This self-assembly technique is versatile in that it can be used for a variety of different substrates such as silicon, glass, ITO coated glass, Al, Cr and polymers.

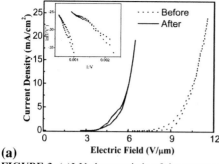

(a)   Electric Field (V/μm)   **(b)**

FIGURE 3. (a)I-V characteristic of the patterned CNTs (stripe width 100 um, pitch 300 um). After the first measurement, a relatively high electrical field was applied between the anode and the cathode for a short time. The threshold field for emission was substantially reduced after the conditioning process. (b) Emission image over 2 x 2 cm.

ACKNOWLEDGEMENT

This work was supported by the Office of Naval Research through a MURI program at UNC (N00014-98-1-0597) and by a grant from NASA (NAG-1-01061). We thank L.E. McNeil for use of the Raman spectrometer, J. Zhang and R.W. Murray for IR measurements, and S. Washburn for using the photolithography facility.

REFERENCES

1. S. Iijima, Nature **354**, 56-58 (1991)
2. I. Brodie and C. A. Spindt, Advanceds in Electronics and Electron Physics **83**, 1-106 (1992)
3. W.A.d. Heer, A. Chatelain, and D. Ugarte, Science **270**, 1179-1180 (1995)
4. P.G. Collins and A. Zettl, Appl. Phys. Lett. **69**(13), 1969-1971 (1996)
5. W. Zhu, C. Bower, O. Zhou, G.P. Kochanski, and S. Jin, Appl. Phys. Lett. **75**(6) 873-875 (1999)
6. W.B. Choi, D.S. Chung, J.H. Kang, H.Y. Kim, Y.W. Jin, I.T. Han, Y.H. Lee, J.E. Jung, N.S. Lee, G.S. Park, and J.M. Kim, Appl. Phys. Lett. **75**, 3129-3131 (1999).
7. G.Z. Yue, Q. Qiu, B. Gao, Y. Cheng, J. Zhang, H. Shimoda, S. Chang, J.P. Lu, and O. Zhou, Appl. Phys. Lett. **81**(2), 355-357 (2002).
8. J. A. Castellano, Handbook of Display Technology (Academic Press, 1992)
9. Y. Huang, X. Duan, Q. Wei, and C.M. Lieber, Science **291**, 630-633 (2001)
10. H. Shimoda, S.J. Oh, H.Z. Geng, R.J. Walker, X.B. Zhang, L.E. McNeil, and O. Zhou, Adv. Mater. **14**(12), 899-901 (2002).
11. X.P. Tang, A. Kleinhammes, H. Shimoda, L. Fleming, C. Bower, S. Sinha, O. Zhou, and Y. Wu, Science **288**, 492-494 (2000)
12. O. Zhou, H. Shimoda, B. Gao, S.J. Oh, L. Fleming, and G.Z. Yue, Acc. Chem. Res. **35**, 1045-1053 (2002)
13. J. Liu, A. Rinzler, H. Dai, J. Hafner, A.R. Bradley, P.Boul, A. Lu, T. Iverson, A.K. Shelimov, C. Huffman, F. Rodriguez-Macias, Y. Shon, R. Lee, D. Colbert and R.E. Smalley, Science **280**, 1253-1256 (1998).
14. X. Xu, G. R. Brandes, Appl. Phys. Lett. **74**(17), 2549-2551 (1999).
15. Y.R. Cho, et al. J. Vac. Sci. Technol. B **19**(3), 1012-1015 (2001)

Attachment of Single Multiwall WS$_2$ Nanotubes and Single WO$_{3-x}$ Nanowhiskers to a Probe

I. Ashiri[1], K. Gartsman[2], S.R. Cohen[3], R. Tenne[1]

[1]*Department of Materials and Interfaces, Weizmann Institute of Science, Rehovot 76100, Israel*
[2]*Electron Microscopy Unit, Weizmann Institute of Science, Rehovot 76100, Israel*
[3]*Surface Analysis Laboratory, Weizmann Institute of Science, Rehovot 76100, Israel*

Abstract. WS$_2$ nanotubes were the first inorganic fullerene-like (*IF*) structures to be synthesized. Although the physical properties of *IF* were not fully studied it seems that the WS$_2$ nanotubes can be suitable for applications in the nanoscale range. An approach toward nanofabrication is simulated in this study. High resolution scanning electron microscope equipped with micromanipulator was used to attach single multiwall WS$_2$ nanotubes and single WO$_{3-x}$ nanowhiskers to a probe, which is an atomic force microscope (AFM) silicon tip in the present case. The imaging capabilities of this nanotube or nanowhisker tip were tested in the AFM. The WO$_{3-x}$ nanowhisker tip was found to be stable, but it has a low lateral resolution (100nm). The WS$_2$ nanotube tips were found to be stable only when its length was smaller than 1μm. The fabrication technique of WS$_2$ nanotube tip and WO$_{3-x}$ nanowhisker tip was found to be controllable and reliable and it can probably be used to various applications as well as for preparation of single nanotubes samples for measurements, like mechanical or optical probes.

INTRODUCTION

WS$_2$ Nanotubes

The synthesis of fullerene-like nanoparticles and nanotubes of WS$_2$ was first reported in 1992 [1]. A number of methods have been developed for the synthesis of WS$_2$ (MoS$_2$) multiwall nanotubes [2]. The preferred procedure for the synthesis of large amounts of WS$_2$ fullerene-like nanoparticles and nanotubes is generally based on the conversion of tungsten oxide particles into WS$_2$ nanoparticles by reacting the oxide in a reducing (H$_2$) and sulfidizing atmosphere (H$_2$S) at elevated temperatures (840°C) [1-7]. The phase composition (fullerene-like nanoparticles or nanotubes) and the size of the WS$_2$ nanoparticles is determined by the synthesis conditions (total flow rate, flow rate of hydrogen, etc.) and the amount of tungsten oxide phase (WO$_3$ nanoparticles, WO$_{3-x}$ needle-like particles) in the reactor.

CP685, *Molecular Nanostructures: XVII Int'l. Winterschool/Euroconference on Electronic Properties of Novel Materials,*edited by H. Kuzmany, J. Fink, M. Mehring, and S. Roth

At present, by using the fluidized bed reactor (FBR) with somewhat modified conditions, up to 100g of fullerene-like nanoparticles enriched by 5% WS_2 multiwall (5-7 layers) nanotubes 20nm in diameter and as long as 0.5mm can be obtained [8]. These long nanotubes have a hollow space occupying about 60% of their volume and they are open ended.

Nanotubes as AFM Tips

A very known and studied application of carbon nanotubes is AFM tips [9]. They combine a range of properties that make them well suited for use as probe tips in atomic force microscope (AFM). Their high aspect ratio opens up the possibility of probing deep crevices that occur in microelectronic circuits. The small effective radius of the nanotube tip significantly improves the lateral resolution beyond what can be achieved using commercial silicon tips. Another characteristic feature of nanotubes is their ability to buckle elastically, which makes them very robust while limiting the maximum force that is applied to delicate organic and biological samples [10]. Carbon nanotube AFM tips were first fabricated by direct mechanical assembly [9].

Inorganic nanotubes can be useful as tips in AFM too. The WS_2 nanotubes by virtue of their stiffness and inertness are likely to serve well in the high resolution imaging of rough surfaces that have features with a large aspect ratio [11].

EXPERIMENTAL

IF-WS_2 powder was crushed on a substrate. Some of the nanotubes were observed to protrude from the substrate, allowing manipulation of a single nanotube at a time. A micromanipulator system was placed, in the scanning electron microscope (SEM) chamber [12-14]. The substrate with nanotubes or nanowhiskers protruding out was placed on the SEM stage. A convential AFM silicon tip was mounted at the end of a probe, which is part of the micromanipulator system. The attachment of the nanotube or nanowhisker to the AFM tip was done by using glue.

Imaging of a well-characterized sample (an evaporated titanium film which forms sharply spiked ridges) was used to test the nanotube/nanowhisker tips in the AFM. Also, the WO_{3-x} nanowhisker tip's ability to image high aspect ratio features was tested by imaging a grid, which is a replica with series of grooves turning parallel to each other in varying distances. [9,11,14,15,16].

RESULTS

The results of the manipulation process can be seen in Fig.1. Fig.1A presents a nanotube before it is being picked by the AFM tip from the substrate, and Fig.1B presents another WS_2 nanotube attached to the AFM cantilever.

FIGURE 1. WS$_2$ nanotube tip mounting on AFM tip

Fig.2 presents a perforated Ti film which was imaged by the WS$_2$ nanotube tip (Fig.2A) and by the WO$_{3-x}$ nanowhisker tip (Fig.2B). The WS$_2$ nanotube tip that scanned this image was 20nm in diameter and 500nm long. Several tips with lengths of 1μm or more were attempted on this sample and they were all unable to produce a noise-free image. An important requirement for AFM tip is its stiffness. Stiffness is a function of the material's Young's modulus and its dimensions. Lateral stiffness varies inversely as the 3rd power of the nanotube length [35]. Since all the WS$_2$ nanotube tips have almost the same diameter, increasing its length by a factor of two reduces its stiffness by nearly an order of magnitude. The nanotube tip that led to the highest quality of a scanned image was relatively short and stiff, illustrating the importance of lateral rigidity in obtaining faithful scan.

The WO$_{3-x}$ nanowhisker tip (100nm in diameter, 6μm long) provides a very noisy image that does not present faithfully the image of the surface, in comparison with the WS$_2$ nanotube tip.

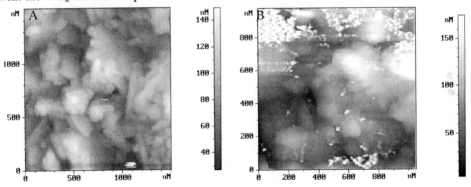

FIGURE 2. Sample of Ti film that was imaged with the WS$_2$ nanotube tip (left) and the WO$_{3-x}$ nanowhisker tip (right)

Fig.3 presents an image of the grid with grooves 800nm long and 400nm wide, scanned by the WO$_{3-x}$ tip. The image and the profile represent the grid faithfully. The image is clear except for the noise at the right wall. This noise can be possibly ascribed to the lateral interaction between the tip and the grid. The small deviation of the nanotube from perpendicular orientation with respect to the sample could lead to enhance interaction of the tip with one side of the step.

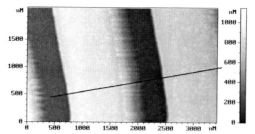

FIGURE 3. A grid imaged by the WO_{3-x} tip

DISCUSSION

The process of manipulation and attachment of single multiwall WS_2 nanotubes and single WO_{3-x} nanowhiskers to an AFM tip inside the SEM was found to be controllable and reliable.

Preliminary results with WS_2 nanotubes and WO_{3-x} nanowhiskers as AFM tips were demonstrated. The WO_{3-x} nanowhisker tip was very stable and stiff; these properties are essential for a good AFM tip, still the lateral resolution was quite low (100nm) compared to commercial silicon tip (5nm). The long WS_2 nanotube tips (1.5-2µm) were not sufficiently stiff to permit scanning of a reliable image, but when a shorter WS_2 tip (0.5 µm in length) was tested it was found to be sufficiently stiff and provide improved resolution (20nm) compared with the WO_{3-x} nanowhisker tip.

This study demonstrates the nanofabrication capabilities of using SEM with micromanipulator. This system can be applied to nanotubes of various sorts, it can be used to move nanotubes or nanoparticles, the nanotube can be attached to other objects (not just to an AFM tips), etc.

Nanotubes attached to AFM tips can be used not only as probes for imaging but also can be used as a samples for measuring the physical properties of single nanotubes [17, 18].

The technique used in the present study will be further used to investigate the physical properties of WS_2 nanotubes and to fabricate various nanodevices.

REFERENCES

1. R. Tenne, L. Margulis, M. Genut, G. Hodes: Nature, **360**, 444, (1992).
2. C.N.R. Rao, M. Nath: Dalton Trans., **1**, 1, (2003).
3. Y. Feldman, G.L. Frey, M. Homoyonfer, V. Lyakhovitskaya, L. Margulis, H. Cohen, G. Hodes, J.L. Hutchison, R. Tenne: J. Am. Chem. Soc., **118**, 5326, (1996).
4. A. Rothschild, G.L. Frey, M. Homoyonfer, R. Tenne, M. Rappaport: Mater. Res. Innovat., **3**, 145, (1999).
5. A. Rothschild, R. Popovitz-Biro, O. Lourie, R. Tenne: J. Phys. Chem. B, **104**, 8976, (2000).
6. A Rothschild, J. Sloan, R. Tenne: J. Am. Chem. Soc, **122**, 5169, (2000).
7. R. Tenne: Chem. Eur. J., **23**, 5296, (2002).

8. R. Rosentsveig, A. Margolin, Y. Feldman, R. Popovitz-Biro, R. Tenne: Chem. Mater., **14**, 471, (2002).
9. H. Dai, J.h. Hafner, A.G. Rinzler, D.T. Colbert, R.E. Smalley: Nature, **384**, 147, (1996).
10. S.S. Wong, E. Joselevich, A.D. Woolley, C.L. Cheung, C.M. Lieber: Nature, **394**, 52, (1998).
11. A. Rothschild, S.R. Cohen, R. Tenne: Appl. Phys. Lett., **75**, 4025, (1999).
12. J. Hafner: http://cnst.rice.edu/mount.html
13. H. Nishijima, S. Kamo, S. Akita, Y. Nakayama, K.I. Hohmura, S.H. Yoshimura, K. Takeyasu: Appl. Phys. Lett., **74**, 4061, (1999).
14. H. Nishijima, S. Akita, Y. Nakayama: Jpn. J. Appl. Phys., **38**, 7247, (1999).
15. R.M.D. Stevens, N.A. Frederik, B.L. Smith, D.E. Morse, G.D. Stucky, P.k. Hansma: Nanotechnology, **11**, 1, (2000).
16. C.V. Nguyen, K.J. Chao, R.M.D. Stevenes, L. Delzeit, A. Cassell, J. Han, M. Meyyappan: Nanotechnology, **12**, 363, (2001).
17. E.W. Wong, P.E. Sheehan, C.M. Lieber: Science, **277**, 1971, (1997).
18. N.R. Wilson, D.H. Cobden, J.V. Macpherson: J. Phys. Chem. B, **106**, 13102, (2002).

Field emission of aligned grown carbon nanotubes

Karl Bartsch and A. Leonhardt

Leibniz-Institut of Solid State and Materials Research Dresden, Helmholtzstr. 20,
D-1069 Dresden, Germany

Abstract: Tungsten carbide/cobalt hard metals were coated with multi-walled carbon nanotubes (CNT) using microwave assisted chemical vapor deposition (MWCVD). The aspect ratio of the tubes and their packing density were changed by the deposition conditions and by wet-chemical pre-treatments of the substrate surfaces. The influence of the morphology of the nanotube layers on the electron field emission was investigated. The best layers yielded field enhancement factors of 1500 - 2000 and current densities up to 0.1 A/cm^2.

INTRODUCTION

The high efficiency of CNT for electron field emission is of particular relevance with respect to their application as cold cathodes in vacuum micro-electronic devices, as for example flat-panel displays, parallel a-beam lithography systems, microwave power amplifiers, and X-ray sources. In order to achieve optimal performance, controlling both the alignment, and the distribution, and perhaps also the structure of the nanotubes is desirable. Among the different methods for manufacture of nanotubes, catalytical CVD is the most promising one for the applications mentioned above. But a lot of unsolved issues have still to be overcome. In this contribution, some relations between the aligned growth of multi-walled CNT using MWCVD and the electron field emission behavior are investigated. Obtaining a good structure of the CNT requires at least a deposition temperature of approximately 700 °C, which is quite high and results in chemical interactions between the catalyst metal (e.g. Fe, Co, Ni) needed for the growth and the commonly used silicon substrates. Because the use of thin inter-layers, like SiO_2 barriers, is not indicated due to the high surface resistance, which does not enable field emission measurements, the deposition on hard metal substrates was tried.

EXPERIMENTAL

The deposition of the multi-walled CNT was performed in a MWCVD set-up equipped with a quartz tube as a deposition chamber. Three slightly different procedures were used, which are denoted as sample 1 - 3 in the following. The reaction gas was a H_2 (100 sccm) - CH_4 (4.8 sccm) mixture. Using an additional bias voltage ($\approx$ - 400 V) the CNT grew perpendicular to the substrate surface. WC/6% Co hard metal substrates, consisting of irregularly shaped WC particles embedded in a thin cobalt matrix, were

CP685, *Molecular Nanostructures: XVII Int'l. Winterschool/Euroconference on*
*Electronic Properties of Novel Materials,*edited by H. Kuzmany, J. Fink, M. Mehring, and S. Roth
© 2003 American Institute of Physics 0-7354-0154-3/03/$20.00

directly heated up to about 1000 °C by the hydrogen plasma. The chamber pressure was amounted to 20 mbar (sample 1 and 2) and 2 mbar (sample 3), respectively. In the case of sample 2 Co was partly removed from the substrate surface by leaching with diluted nitric acid. The CNT were characterized by scanning and transmission electron microscopy (SEM, TEM, HRTEM). The field emission measurements were carried out using a hemispherical tungsten anode with a diameter of 1 mm (sample 1 and 2) and a spherical stainless steel and anode with a diameter of 2.5 mm (sample 3). The position of the anode over the substrate could be controlled using an x-y-z manipulator. The pressure amounted to $1-3*10^{-9}$ mbar during the measurements..

RESULTS AND DISCUSSION

Fig. 1 shows SEM micrographs of the samples 1 - 3 in different magnifications, which are characteristic of regions where the field emission was measured.

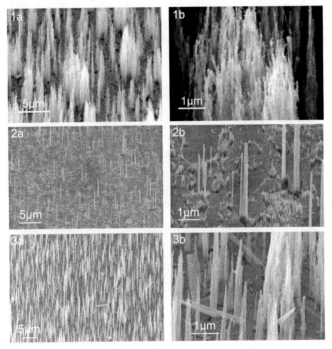

FIGURE 1. SEM micrographs of differently grown CNT (1a,b and 3a,b) and CNF (2a,b), 1 and 3 tilted by 40°, 2 tilted by 60°

The overview images (left column) demonstrate a different growth density and distribution of the tubes. Under greater magnification (right column) the images reveal that in the case of sample 1 and 3 the CNT form cone-like bunches, whereas sample 2 is characterized by single free standing pillars or groups of few pillars. As we do not know the structure of the pillars, the term CNF (carbon nanofiber) is used for them in the following text. The structure of the CNT forming the bundles was identified as the so-called bamboo-like structure [1] by HRTEM. It is distinguishable from the SEM micrographs that the depth of the bunches and their packing density are different for sample 1 and 3.

By visual inspection of SEM micrographs (employing the perpendicular view on the surface in the case of sample 1 and 3, not shown) the packing density of the bunches and the CNF was estimated to be about $2.5*10^7$ cm^{-2} for sample 1, $6*10^6$cm^{-2} for sample 2, and $2.6*10^8$ cm^{-2} for sample 3. Even if the uncertainty of these values is quite high (at least 10 % error), one can assume that the differences between the samples are significant. The images shown in fig. 1 also reveal differences in the length and the diameter of the CNT and CNF. A preliminary estimation of their aspect ratios (length/diameter) from additional SEM and TEM micrographs yielded averaged values of about 150 for the CNT and 20 for the CNF. Apart from the spread of the individual dimensions of the CNT/CNF, the higher aspect ratio of the CNT forming the bunches is apparent. Assuming that smaller Co particles are more quickly removed from the sample surface than larger one during the acid treatment, a higher diameter of the fibers is to be expected in the case of sample 2, because the particle size governs the diameter of the CNT/CNF [2].

The results of the field emission measurements are presented in fig. 2 and 3. The plot shown in fig. 2 exhibits the current density versus the macroscopic field strength. The emission areas used for the calculation of the current density were estimated using the method reported by Zhu et al. [3], and they are given in the figure caption.

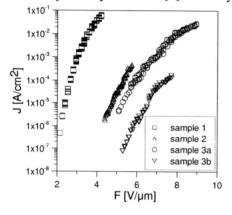
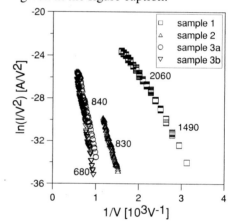

FIGURE 2. Current density vs. field strength (emission areas in cm^2: 1: $3.7*10^{-4}$, 2: $1.8*10^{-4}$, 3a: 10^{-3}, 3b: $7.6*10^{-3}$

FIGURE 3. F-N plots corresponding to figure 2

The best values were obtained for sample 1, though with respect to electrostatic shielding, the packing density of the CNT is not optimal for this sample. According to the rule, that the tube distance should be twice their length [4], the packing density is at least one order of magnitude too high (regarding the packing density of the bunches given above). Otherwise, the averaged packing density of sample 2 is nearly optimal, but the measured current density is clearly lower . Keeping in mind that only a few CNT/CNF with the highest aspect ratios contribute to the emission [5], this discrepancy can be explained as follows. On the one hand, the aspect ratio of sample 1 is higher than that of sample 2, and on the other hand there is also a greater probability that tubes with higher aspect ratios occur at sample 1. With sample 3, measurements were performed at two different positions (a and b). Curve 3a is comparable with curve 2. Because the averaged aspect ratio of sample 1 and 3 is comparable, the decreased current density of sample 3 is assumed to be caused by the high packing density of this sample. Curve 3b was obtained after loading the emission spot with a higher current of $3*10^{-4}$ A. An inspection of the sample by SEM after the measurements revealed a serious degradation of the emission area.

Fig. 3 present the Fowler-Nordheim plots of the measurements corresponding to fig. 2. The figures given in the plot are the field enhancement factors β, which were estimated by the commonly adopted procedure from the slope of the F-N plot ($\ln I/V^2 \propto (-6.83*10^3$ $d\phi^{3//2}/\beta)(1/V)$, d - distance of electrodes in µm, work-function ϕ of the CNT $\cong 5$ eV). The curve for sample 1 changes their slope, thus two ranges were fitted. Similar behavior was also observed by Sveningsson et al. [6]. The cause of this effect is not understood. Sveningsson et. al. assumed a structure change to be responsible for this behavior. The field enhancement factors of sample 2 and 3 are comparable, even if their aspect ratios differ remarkably. Apparently, the unfavorable packing density of sample 3 counteracts the positive effect of the increased aspect ratio. The reduced β related to measurement 3b can be explained by a reduced number of emission sites after the high current load.

CONCLUSIONS

The deposition of multi-walled CNT on heterogeneous catalytic surfaces, such as hard metal surfaces, results in the formation of CNT bunches containing tubes of different length and diameter. The depth and packing density of the bunches can be controlled to a certain extent by the deposition conditions. By partly removing the catalyst metal from the surface, the number of the CNF can be effectively decreased. However, simultaneously an increase of the CNF diameter cannot be avoided. The tendencies of the field emission measurements can be immediately related to the morphological features of the samples, such as aspect ratio and packing density. By improving the CNT adhesion, cold cathodes yielding high emission current densities could be realized on the basis of CNT coated hard metals.

REFERNCES

1. Cui, H., Zhou, O., Stoner, B.R., J. Appl. Phys. **88**, 6072-6074 (2000)
2. Bartsch, K., Leistikow, A., Graff, A., Leonhardt, A.,in Structural and Electronic Properties of Molecular Structures, ed. H. Kuzmany et al., AIP Conf. Proceed. Vol. **633** 2002 pp. 174-177
3. W. Zhu, C. Bower, O. Zhou, G. Kochanski, S. Jin, Appl. Phys. Lett. **75**, 873 (1999)
4. L. Nilsson, O. Groening, C. Emmenegger, O. Kuettel, E. Schaller, L. Schlappbach, Appl. Phys. Lett. **76**, 2071 (2000)
5. L. Nilsson, O. Groening, P. Groening, O. Kuettel, and L. Schlappbach, J. of Appl. Phys. **90**, 768 (2001)
6. Sveningsson, M., Morjan, R.E., Nerushev, O.A., Sato, Y., Bäckström, J., Campbell, E.E.B., Rohmund, F., Appl. Phys. **A 73**, 409-418 (2001)

Transparent CNT Composites

M. Kaempgen and S. Roth

Max-Planck-Institut für Festkörperforschung,
Heisenbergstr. 1, D-70569 Stuttgart, Germany

Abstract. Thin networks of carbon nanotubes not far away from its percolation threshold are used as transparent electrode. At a transmission of about 90% the surface resistance is about 1 kΩ cm and too high to replace ITO. However, it is conducting enough to be used as electrode in electrochemistry. Here the carbon nanotube network is used as conductive backbone structure to deposit another material. In case of conducting polymers the resistance is increasing by about 50-70% for polyaniline and polypyrrol respectively. These CNT/conducting polymer composites are still transparent but now more wavelength selective. The electrical properties are dominated by the CNT whereas the transmission characteristics by the polymer.

INTRODUCTION

In macroscopic scales the direct CVD growth of CNT is the only way to get reproducible aligned patterns in larger scales. This method is already used for field emission applications [1].

Randomly orientated networks is what we get if we work with in macroscopic scales and if we start from CNT suspensions [2]. However, such networks can be directly used as a conductive layers and are already investigated as thin film transistors [3] or as an artificial muscle made from bucky paper [4].

In this work we used an extremely thin carbon nanotube networks not far from its percolation threshold. It is highly transparent and conductive. Therefore it can be used as transparent electrode. This could be an alternative to ITO in several applications e.g. electro chromic devices, light emitting devices or touch sensitive displays [5] since carbon nanotubes have a similar work function compared to ITO [6]. Large scale carbon nanotube networks can be easily produced and tuned between transmissions of 100% and 0% in the UV/VIS/near IR region by air brush technique which could replace the expensive and labour intensive ITO process. The simple air brush technique allows us to create a conductive network on nearly every solid surface and we made good experience on plastic substrates where the adhesion is strong enough to get mechanically stable films even in liquids. Therefore its conductive surface can be used for electrochemical experiments. In the present work we tried out the deposition of conductive polymers like polyaniline to modify electrical and optical properties. We characterized these types of networks or composites by means of conductivity, transparency, SEM, AFM and cyclic voltammetry.

CP685, *Molecular Nanostructures: XVII Int'l. Winterschool/Euroconference on Electronic Properties of Novel Materials,* edited by H. Kuzmany, J. Fink, M. Mehring, and S. Roth
© 2003 American Institute of Physics 0-7354-0154-3/03/$20.00

EXPERIMENTAL

A SDS suspension of SWNT obtained in HiPco process and delivered from CNI was sprayed with an air brush pistol on transparent plastic (PVC) substrates under careful heating. The size of the samples is about 2 cm² but can easily scaled up. After dipping and shaking the sample in pure water in order to remove SDS the CNT network remains in desirable thickness.

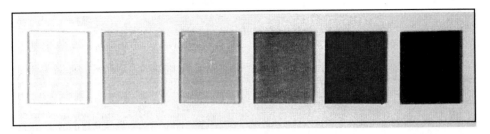

Figure 1. Photos of sprayed CNT networks. The thickness can be tuned between transmissions of 100% and 0%.

DC conductivity was measured in a four probe set up at room temperature with a Keithley 197 DMM. Cyclic voltammetry (CV) and electrochemical deposition was carried out in 1m H_2SO_4 and 0,1m Aniline in 1m H_2SO_4 vs. SCE respectively with a Jaissle IMP 83 PC potentiostat and at a scan rate of 50 mV/s. Polypyrrol was grown from acetontrile with 0,1m pyrrole, 0,2m $LiClO_4$ and 1% H_2O. SEM images are taken with a Hitachi S800 with field emission electron source and the pictures with a digital Olympus C 3040 camera.

RESULTS

We focussed our work on highly transparent networks with a resistance of about 1 $k\Omega cm$ and a transmission in the visible region of about 90 %. Since the resistance is too high compared with ITO (10 Ωcm) we tried to grow a conducting polymer electrochemically on this layer to improve conductivity. The CNT network was used as anode. In this set up the polymer grows along the tubes only and does not fill up the space between which would decrease the transparency without a positive effect on the conductivity. We started with polyaniline because it is well known and simple to produce from aqueous solutions.

First of all both deposition with cycling and constant potential was tried out. It turned out that due to the high resistance a potential drop along the depth of dipping occurs and the polymer grows mostly underneath the solution surface.

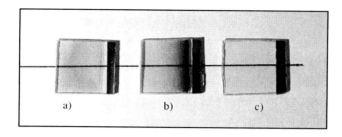

Figure 2. Transparent polyaniline CNT composites, where polyaniline was grown under a) constant and b) sweeping potential, c) pure CNT network for reference (90% Transmission).

SEM images before and after electrochemical polyaniline deposition are shown of the carbon nanotube network in figure 2.

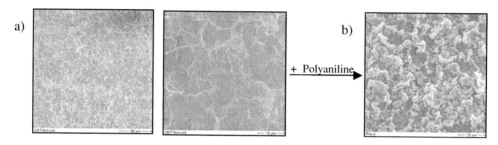

Figure 3. a) SEM images of large scale thin transparent CNT networks on plastic substrates showing a percolating network of bundled singled wall carbon nanotubes. b) Polyaniline grows along the carbon nanotubes.

The SEM images show clearly that the Carbon nanotubes are forming a percolating network of bundled single wall carbon nanotubes. Since this network is the only current path on the surface polyaniline starts growing along the carbon nanotubes (Figure 3).

For further experiments we used potentiostatic method only in order to get a homogenous polymer distribution. After deposition, washing and drying polyaniline stays in its conductive state called polyemeraldine [7] showing a green colour. The hole CNT polyaniline composite is still transparent and occurs now as a transparent more or less green layer depending on the thickness of the polymer. So the transmission characteristics of such composites become more wavelength selective and could be used as electro chromic devices if one could stabilize the different coloured oxidations states of polyaniline.

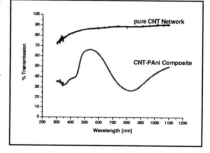

Figure 4. Transmission spectra of CNT and CNT Pani Composite

The conductivity of carbon nanotube networks are limited by the contacts between the CNT. Therefore we hoped that a conducting polymer wrapping around the carbon nanotubes could lower the contact resistance. However, due to the high surface resistance a potential drop could occurs during growth. This leads to growth conditions dependent on the distance to the outer electrical contact to the potentiostat. Therefore different thickness of polymer can be expected. We took this into account and measured the resistance before and after growth at different distances up to 6 cm in each direction from the central contact as shown in figure 5.

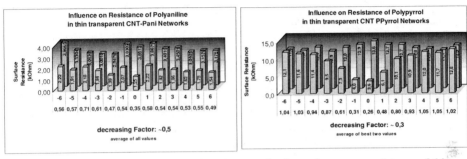

Figure 5. Influence of electrochemically grown conducting polymers on resistance of thin CNT networks. First row: after growth, second row behind: before growth
x-axis: first row: distance to outer contact - second row: factor of changed value for resistance

In case of polyaniline the resistance is decreasing of about 50% but a significant dependence on the thickness was not observed. In case of polypyrrol the resistance is decreasing only near to the outer contact meaning at thicker layers. At distances more than 3 cm the resistance has not changed. The average of the best three values decreases the resistance of about 70%.

We think this is not only due to the different types of the polymers but maybe also due to an influence of the organic solvent. For polypyrrol acetonitrile was used. The wetting behaviour of this organic solvent maybe better and could lead to better growth between the CNT.

The modified networks should lead to characteristic changes in the electrochemical behaviour. Cyclic voltammetry before and after polyaniline deposition was used to indicate these changes. The I/V curves are shown in figure 5.

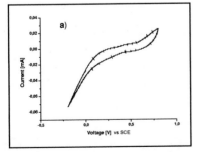 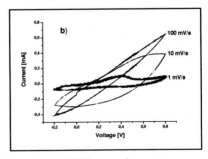

Figure 6. CV in sulphuric acid of a   a) pure thin transparent CNT network
b) CNT Polyaniline composite after growth at different scan rates.

557

The electrochemical behaviour of carbon nanotubes is dominated by capacitive double layer charging resulting in a featureless I/V curve (figure 6a) which are also observed in other groups [8, 9]. Since the CNT/PAni composite is not high conductive the layer behaves at higher scan rates like an capacitor as well and does not show any typical polyaniline features. At lower scan rates these features appear and at a scan rate of 1mV/s we are able to observe the typical CV of polyaniline.

CONCLUSIONS

We have shown that thin transparent carbon nanotube networks can be used as transparent electrodes. The resistivity is too high to replace the well known ITO. However, as shown the material is conductive enough to be used as electrode in electrochemistry. The carbon nanotubes can then used as backbone structure to deposit another material. In case of a CNT/conducting polymer composite the electrical behaviour is dominated by the carbon nanotube network whereas the transmission properties by the deposited polymer.

REFERENCES

1. K. Teo, et al., *Appl. Phys. Lett.*, 79, 1534 (2001).

2. G. Duesberg et al., Appl. Phys. A, 67, 117 (1998)

3. E. S. Snow et al., Appl. Phys. Lett., 82, 2145 (2003).

4. R. H. Baughman et al., Science 284, 1340 (1999).

5. K. L.Chopra et al., Thin Solid Films, 102, 1 (1983)

6. M.Shirashi et al., Carbon 39, 1913 (2001)

7. Handbook of Organic Conductive Molecules and Polymers. Vol. 3

8. J. H. Chen et al., Carbon, 40, 1193 (2002)

9. K.H.An et al., J. Electrochem. Soc., 149, A1058 (2002)

Optical Limiting Properties of Suspensions of Single-Wall Carbon Nanotubes

L. Vivien[a], N. Izard[a,b], D. Riehl[a], F. Hache[c], E. Anglaret[b]

[a]Centre Technique d'Arcueil, DGA, Arcueil, France
[b]GDPC, UMR CNRS 5581, Université Montpellier II, Montpellier, France
[c]Laboratoire d'Optique et Biosciences, Ecole Polytechnique, Palaiseau, France

Abstract: Suspensions of single wall carbon nanotubes are good candidates for optical limiting. We compare their performances with those of other carbonaceous materials. Z-scan and pump-probe experiments are used to identify and investigate nonlinear scattering as the main origin of optical limiting. In suspensions of SWNT, non linear scattering is due both to heat transfer from particles to solvent, leading to solvent bubble formation, and to sublimation of carbon nanotubes.

INTRODUCTION

The development of high power laser sources has motivated extensive research for the design of auto-activated, so-called optical limiting, systems for eye and sensor protection [1-13]. Optical limiting can be due to several different non linear light-matter interactions, especially nonlinear absorption, refraction and scattering. Non linear absorption leads to optical limiting via reverse saturable absorbption (RSA) [1] or via multi-photon absorption (MPA) [2]. The best materials for RSA are molecules of the phtalocyanines, naphtalocyanines or fullerene families. MPA is observed in some solid semiconductors (GaAs, ZnSe...) or in some organic dyes. Non linear refraction can be due to thermal effects, *i.e.* dilatation of the solvent [3] or to optical Kerr effet (CS_2) [4]. Finally, non linear scattering is observed when a laser beam induces an index mismatch in a biphasic media [5], or a growth of gazeous scattering centers due to heating of absorbing particles [6]. This latter effect is the dominating effect for metallic nanoparticles or carbon black suspensions (CBS). The optical limiting properties of suspensions of carbon nanotubes were demonstrated and investigated since 1998 [7-13]. In this paper, we compare the performances of singlewall carbon nanotube suspensions (SWNT) and other carbonaceous materials. From Z-scan and pump-probe studies of SWNT, we show that nonlinear scattering is responsible for limiting and we discuss the thermodynamical origins of the phenomenon. Finally, we discuss possible routes for improving the optical limiting performances of SWNTS and we show that nanotubes are among the best candidates for efficient protection over broad spectral and temporal ranges.

CP685, *Molecular Nanostructures: XVII Int'l. Winterschool/Euroconference on Electronic Properties of Novel Materials,*edited by H. Kuzmany, J. Fink, M. Mehring, and S. Roth
© 2003 American Institute of Physics 0-7354-0154-3/03/$20.00

EXPERIMENTAL

SWNT were prepared by the electric arc discharge and purified in a three-steps procedure [10]. From scanning electron microscopic measurements, we estimate that this purified material contains about 90 vol% of SWNT. Nanotubes suspensions were prepared in water with the help of a surfactant (Triton X100) or in chloroform. The linear transmittance was adjusted to be close to 70%.

The optical limiting curves were measured in a f/30 focusing geometry using a Q-switched Nd:YAG laser generating 5 ns pulses at 532 or 1064 nm and a Titanium:Sapphite laser generating 80 ns pulses at 1000 nm. Pump-probe experiments were carried out using the Nd:YAG laser as the pump beam and the 632.8 nm line of a continuous He-Ne laser as the probe beam.

RESULTS AND DISCUSSION

As underlined above, many carbonaceous materials exhibit optical limiting. Figure 1 compares the performances of C_{60}, C_{70} (in toluene solutions), CBS (in water/surfactant suspensions), multi-wall carbon nanotubes (MWNT) prepared byt he electric arc and SWNT (in water/surfactant suspensions). Both MWNT and SWNT appear to be competitive with respect to other limiters. Below, we will focus only on SWNT. We will especially discuss –how their performances can be improved in the future. In this section, we first address the origin of optical limiting in suspensions of SWNT.

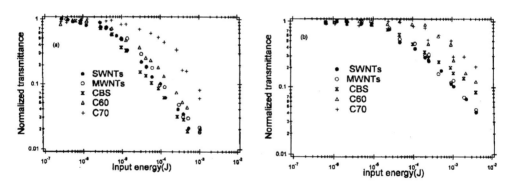

FIGURE 1 : Comparison of optical limiting by solutions of fullerenes in toluene, CBS and suspensions of MWNT or SWNT, for 5 ns pulses, at 532 nm (left) and 1064 nm (right).

First, z-scan experiments were used to identify the phenomenon responsible of limiting in SWNT. In both water and chlorofom, for nanosecond pulses, one observes the signature of an auto-defocalisation effect as well as a very strong non linear scattering [8-9]. The auto-defocalisation effect is associated with a negative nonlinear index, characteristic of a thermal effect related to a heat-induced dilatation of the solvent. A detailed study of the dominating non linear scattering contribution was achieved with the help of pump-probe experiments [10-11]. In such experiments, the non linear effect is excited by a pulsed laser "pump" and studied by measuring the transmittance of a continuous laser "probe". Typical results are presented in figure 2 for 5 ns pump pulses. At low input fluences, the perturbation of the probe beam occurs a few nanoseconds after the pump pulse, both in water (fig. 2a and 2c) and in chlorofom (fig. 2b and 2d), for pump beams at 1064 nm (fig. 2a and 2b) and 532 nm (fig. 2c and 2d). Such a slow process rules out any non linear absorption phenomenon. When the input fluence increases, the perturbation of the probe beam shifts towards small timescales and develops much faster. The optical limiting threshold corresponds roughly to a loss of transmittance of the probe at the top of the pump pulse. This suggests that two different mechanisms are responsible for the probe perturbation. We assign the first (slow) mechanism to the formation of solvent vapor bubbles due to heat transfer from the nanotubes to the surrounding liquid. The second (fast) mechanism occurs at much larger input fluences. We assign it to the sublimation of the nanotubes themselves leading to the formation of carbon vapor bubbles. The thresholds for both mechanisms are much smaller for chloroform than for water. This can be understood easily by considering the thermodynamic properties of the solvents : the heat conductivity, calorific capacity, boiling point and vaporisation energy of chloroform are all much smaller than those of water, which makes much easy both its heating and vaporization. One can also underline that the threshold for mechanism 1 is much smaller for SWNT than for CBS. This may be due to a better heat conductivity of the former and also possibly to a larger specific area. This makes SWNT good candidates for optical limiting for longer pulses. As a matter of fact, we checked that the limiting performances of SWNT suspensions were much better for 80 ns pulses.

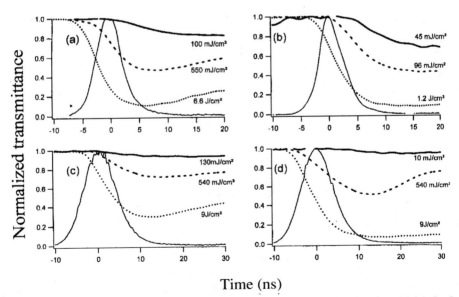

Time (ns)

FIGURE 2 : Pump-probe results for SWNT suspensions in water (left) and in chloroform (right), for 5 ns pulses at 1064 nm (top) and 532 nm (bottom). The peak in solid line corresponds to the pump pulse.

Figure 3 compares typical limiting curves for pulses of 5 and 80 ns, for water and chloroform suspensions. The limiting threshold is significantly smaller at 80 ns for water and much smaller for chloroform. In addition, one can distinguish two regimes of limiting for long pulses and only a single one for short pulses (indicated by straight lines in figure 3). We assign the first regime to scattering by solvent vapor bubbles and the second one to scattering by both solvent and carbon bubbles. As stated by the pump-probe measurements, only this second mechanism can be effective for short pulses. The optical limiting performances were measured for various wavelengths (from 430 nm to 1064 nm) and pulse durations (from 2 ns to 100 ns) [11]. Limiting is effective all over these broad ranges, with better performances for shorter wavelengths, as expected for a scattering process.

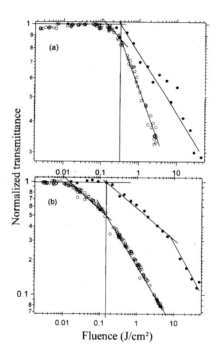

FIGURE 3 : Comparison of optical limiting by suspensions of SWNT, for 5 ns pulses at 1064 nm (full circles) and 80 ns pulses at 1000 nm (open circles) for SWNT suspensions in water (top) and in chloroform (bottom).

CONCLUSIONS AND PERSPECTIVES

Carbon nanotubes are broadband optical limiters from visible to near infrared and for pulse durations from a few nanoseconds. Limiting is due to a strong nonlinear scattering, due to formation of solvent and carbon bubbles. The performances strongly depend on the thermodynamical properties of the solvent and possibly on the structure/specific area of tubes and bundles. In the future, the availability of nanotubes of different structure [13] and the development of chemical modifications through functionalization will probably open many routes to improve the performances of SWNT-based limiters.

REFERENCES

1. Perry, J.W., in *Organics and metal-containing reverse saturable absorbers for optical limiting*, ed by H.S. Nalwa and S. Miyata, CRC Press, Orlando, 1997, p3.
2. Said A.A., Sheik-Bahae M., Hagan D.J., Wei T.H., Wang J., Young J., Van Stryland E.W., J. Opt. Soc. Am. B **9**, 405 (1992).
3. Justus B.L., Huston A.L., Campillo A.J., Appl. Phys. Lett. **63**, 1483 (1993).
4. Boyd R.W., *Non linear optics*, Academic Press, New-York, 1992.
5. Boudrier V., Bourdon P., Hache F., Flytzanis C., Appl. Phys. B **70**, 105 (2000).
6. Mansour K., Soileau M.J., Van Stryland E.W., J. Opt. Soc. Am. **9**, 1100 (1992).
7. Chen P., Wu X., Sun X., Lin J., Ji W., Tan K.L., Phys. Rev. Lett. **82**, 2548 (1999).
8. Vivien L., Anglaret E., Riehl D., Hache F., Bacou F., Andrieux M., Lafonta F., Journet C., Goze C., Brunet M., Bernier P., Opt. Comm. **174,** 271 (2000).
9. Vivien L., Riehl D., Anglaret E., Hache F., IEEE J. Quant. Electron. **36**, 680 (2000).
10. Riggs J.E., Walker D.B., Caroll D.L., Sun Y.P., J. Phys. Chem.B **104**, 7071 (2000).
11. Vivien L., Delouis J.F., Delaire J., Riehl D., Hache F. and Anglaret E., J. Opt. Soc. Am. B **19**, 208 (2002).
12. For a review, see Vivien L., Lançon P., Riehl D., Hache F., Anglaret E., Carbon **40**, 1789 (2002).
13. N. Izard et al, these proceedings.

Progress in actuators from individual nanotubes

J. Meyer*, J.-M. Benoit*†, V. Krstic** and S. Roth*

*Max-Planck Institute for solid state research, Stuttgart (Germany)
†IMN-LPC, CNRS/Nantes University, Nantes (France)
**Grenoble High Magnetic Field Laboratory, MPI-FKF/CNRS, Grenoble (France)

Abstract. Electrochemical double layer charge injection in macroscopic sheets of carbon nanotubes (buckypaper) has been shown to convert electrical energy into mechanical energy. We investigate actuation for single nanotube bundles in order to obtain understanding that may be important for optimizing nanotube arrays (like buckypaper), understanding mechanical effects in electric devices made from nanotubes, and possibly using individual nanotubes as nanoscale actuator devices. Atomic Force Microscopy (AFM) investigations were done on individual suspended single-walled carbon nanotube (SWNT) bundles, showing the possibilites and limits of this technique.

INTRODUCTION

Carbon nanotubes are expected to show, among many other unique properties, charge-induced length changes [1, 2]. Various materials show bond-length changes as a mechanical response to charge-injection [3], however the combination with a high strength and a high young's modulus (in the order of $1\,TPa$) make carbon nanotubes the ideal candidates for nano-actuators. Electromechanical actuation has been shown in networks from single-walled carbon nanotubes [4]. Although these networks have a much lower strength than single nanotubes due to the randomly oriented nanotubes, they already generate higher stresses than natural muscles and higher strains than high modulus ferroelectrics.

Although artificial muscles from carbon nanotube networks have a wide potential field of applications, the underlying mechanism needs to be better understood. One difficulty for previous measurements of macroscopic nanotube arrays (sheets and fibers) is that the responses from many different types of nanotubes are averaged. Theoretical predictions [1, 2] suggest that the sign and size of the actuation of an individual single-walled carbon nanotube depends on its diameter and helicity. Thus investigations on individual nanotubes are crucial for understanding the actuation in macroscopic actuators. Moreover, there will be charge-induced length changes in any carbon nanotube incorporated in an electronic device. Devices like the all-nanotube transistor [5], which consists of a T-like nanotube junction, may actually have more in common with a mechanical relay than a transistor: It has been suggested that the switching behaviour is caused by mechanical deformations of the source-drain nanotube caused by an expanding gate tube. Finally, one can imagine nanoscale mechanical devices driven by expanding and contracting nanotubes.

CP685, Molecular Nanostructures: XVII Int'l. Winterschool/Euroconference on
Electronic Properties of Novel Materials, edited by H. Kuzmany, J. Fink, M. Mehring, and S. Roth
© 2003 American Institute of Physics 0-7354-0154-3/03/$20.00

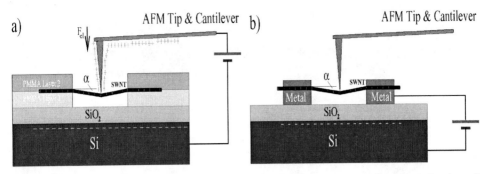

FIGURE 1. a) Nanotube embedded between two layers of PMMA is insulated and has to be charged via the AFM tip, resulting in an electrostatic force on the cantilever. b) A nanotube embedded in metal contacts can be manipulated with a non-conductive AFM tip, and charged through the contacts.

SETUP

Our approach for investigating length changes in an individual nanotube is to measure the deflection of an AFM cantilever placed on a freestanding tube. The ends of the tube can be either embedded between two layers of polymethylmetacrylate (PMMA), or in metallic contacts, as described in [6].

Rather than electrochemically injecting charge into the nanotubes, like in a super capacitor, we will here induce charge using ordinary dielectric charging. The advantage is elimination of the need for an electrolyte; the disadvantage is that the degree of charge injection (and corresponding actuation) is reduced. Experimentally, the charge required to induce length changes in the tube can be induced in two ways: A nanotube embedded in metal can be charged through the contacts. If the tube is isolated, it can be charged through a conducting AFM tip. This latter method has been previously used [7, 8], but, as shown below, a strong electrostatic attraction arises between the conducting cantilever and the backgate which can be larger than the actuation due to bond length changes.

Expected actuation

From the predicted charge-induced length changes and the geometry of our setup, one can calculate the expected deflection of the AFM cantilever by an actuation of the nanotube. To calculate the charge on the nanotube, an estimation of the nanotube-backgate capacity is neccessary. We approximate the setup by an infinitely long conducting cylinder above a planar surface, for which the capacity per unit length can be calculated analytically. We consider a cylinder of radius r at a distance d from the surface. The formula can be simplified with the assumption that $r \ll d$ (meaning that the radius of the nanotube is small compared to its distance to the backgate). The resulting capacity C per unit length l is $\frac{C}{l} = \frac{2\pi\varepsilon_0}{\ln\left(\frac{2d}{r}\right)}$. Using reasonable values for our setup ($l = 500nm$, $d = 100nm$, and $r = \frac{1}{2} \cdot 1.2nm$ for an individual SWNT) the nanotube-backgate capacity

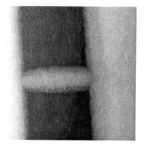

FIGURE 2. SEM (left) and AFM (right) of a freestanding SWNT bundle. The width of the trench is 500nm.

can be estimated as $C \approx 5 \cdot 10^{-18}F$, corresponding to just 30 electrons per volt. For an applied potential of 1.5V, the predicted actuation is a relative length change of up to $2 \cdot 10^{-4}$, depending on the helicity of the nanotube.

Although the capacity increases for larger nanotubes and bundles, the important quantity determining the actuation is the charge *per carbon atom* (or per bond). Since the number of atoms increases faster with diameter than the capacity, the charge per atom strongly decreases with increasing tube or bundle diameter. To investigate actuation in the suggested way, it is therefore required to create a single freestanding SWNT, or at least a very small bundle.

Electrostatic effects

Freestanding bundles of SWNTs embedded between two layers of PMMA have been successfully created (Fig. 2). A conducting AFM tip was placed on the freestanding bundle, and a sinusoidal potential was applied (Fig. 3). Since electrostatic attraction is proportional to the square of the applied potential, the cantilever response corresponding to it must appear with twice the applied frequency. In this way, actuation can be distinguished from electrostatic attraction.

The measurement shows that there is a non-negligible signal due to the electrostatic attraction. It is two orders of magnitude larger than the signal at the applied frequency (where the actuation peak is expected), which is hardly above the noise. Since this is a nanotube bundle, the small actuation due to bond length changes is not surprising.

Taking the relative length changes predicted above for a single tube however, the length change of a 500nm long nanotube would be around 1Å. Due to the geometry of the setup, the deflection of the AFM tip will change by a much larger value than this: The AFM pushes down the freestanding tube so that it is bent by an angle α (fig. 1). With the tip pushing with a (nearly) constant force, a small length change ΔL along the tube is translated into an AFM tip movement $\Delta z = \frac{\Delta L}{2\alpha}$ for small angles α. The effect for an individual SWNT will still be smaller than the electrostatic effect, so care must be taken to distinguish these effects, but should be detectable with the AFM.

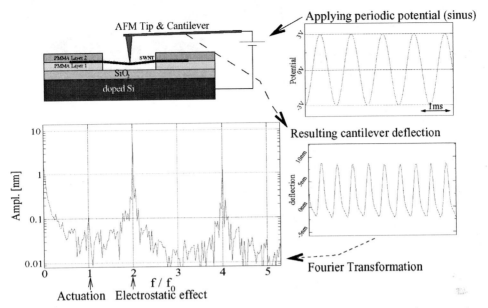

FIGURE 3. An actuation measurement. Charge is injected into the freestanding carbon nanotube, and the resulting cantilever deflection is measured. The different dependence of the actuation and the electrostatic attraction from the applied potential is exploited to distinguish these two effects.

CONCLUSIONS

Our measurements show that it is neccessary to take into account electrostatic attraction in the actuation measurements on nanotubes charged via the AFM tip, which was not considered before. An alternative would be to charge the tube via metallic contacts, and use an insulating AFM tip. For electrostatically charged tubes, in any case, it is required to have an individual SWNT in order to achieve a measurable charge-induced length change. Then, the signal should be detectable; the noise level in our measurements show the sensitivity of the AFM: It is not sufficient to measure actuation in freestanding nanotube bundles, but should be good enough to measure actuation in individual tubes. Work is in progress to achieve freestanding individual single-walled nanotubes in different configurations.

ACKNOWLEDGEMENTS

This work is supported by DARPA contract MDA972-02-C-005, COMELCAN, and a stipend of the Studienstiftung des deutschen Volkes.

REFERENCES

1. Yu N. Gartstein, A. A. Zakhidov, and R. H. Bauhgman. Charge-induced anisotropic distortions of semiconducting and metallic carbon nanotubes. *Phys. Rev. Lett.*, 89, 2002.
2. G. Sun, J. Kurti, M. Kertesz, and R. H. Baughman. Dimensional changes as a function of charge injection in single-walled carbon nanotubes. *J. Am. Chem. Soc.*, 2002.
3. G. Sun, J. Kurti, and M. Kertesz. Dimensional changes as a function of charge injection for trans-polyacetylene: A density functional theory study. *J. Chem. Phys*, 117, 2002.
4. R. H. Baughman, Ch. Cui, A. A. Zakhidov, Z. Iqbal, J. N. Barisci, G. M .Spinks, G. G. Wallace, A. Mazzoldi, D. De Rossi, A. G. Rinzler, O. Jaschinzki, S. Roth, and M. Kertesz. Carbon nanotube actuators. *Science*, 284:1340–1344, 1999.
5. P. W. Chiu, M. Kaemppgen, U. Dettlaff, and S. Roth. All-carbon transistors. 2003. To be published.
6. G.-T. Kim, G. Gu, U. Waizmann, and S. Roth. Simple method to prepare individual suspended nanofibers. *Appl. Phys. Lett.*, 80, 2002.
7. J. M. Benoit, G. Gu, G. T. Kim, A. Minett, R. Baughman, and S. Roth. Actuators from individual carbon nanotubes. *Structural and Electronic Properties of Molecular Nanostructures*, 2002.
8. S. Roth and R. H . Baughman. Actuators or individual carbon nanotubes. *Current Applied Physics*, 2, 2002.

568

Excitation spectrum of hydrogen adsorbed onto carbon nanotubes

B. Renker[1], H. Schober[2], P. Schweiss[1], S. Lebedkin[3] , and
F. Hennrich[4]

[1]Forschungszentrum Karlsruhe, IFP, D-76021 Karlsruhe, Germany
[2]Institut Laue-Langevin, F-38042 Grenoble Cedex, France
[3]Forschungszentrum Karlsruhe, INT, D-76021 Karlsruhe, Germany
[4]Institut für Physikalische Chemie, Universität Karlsruhe, D-76128 Karlsruhe, Germany

Abstract. We report inelastic neutron scattering results on hydrogen physisorbed onto an inner layer in open SWNT's. Quantum rotations and phonon spectra of hydrogen are studied in detail.

Public interest in high performance storage devices has triggered many investigations of hydrogen adsorbed onto carbon nanotubes. Despite this effort results concerning the relevant adsorption processes remain quite controversial. Due to the exceptionally high incoherent cross-section of hydrogen neutron scattering promises to aid some of the missing clues.

In a recent paper [1] we reported that acid treated single walled carbon nanotubes (open SWNTs) can adsorb 3 to 4 times more hydrogen than as prepared material where the tubes are closed by caps. It was found that much of the excess hydrogen, i.e. $\approx$ 0.6 wt% is readily released between 150 K and 300 K by heating. From simple geometrical considerations it was concluded that most of this hydrogen is weakly bond as a molecule to the inner walls of the tubes.

The H_2 molecule is a quantum rotator where the rotational energies are distributed according to $E_J = B\,J(J + 1)$ with a rotational constant $B = 7.35$ meV in free space. Molecular quantum rotations directly reflect interactions with the SWNTs matrix. The rotational spectrum can be suitably studied by inelastic neutron scattering. Since the molecule is composed of two indistinguishable protons the nuclear coordinate and spin wave functions are strictly coupled for a given electronic state. As shown by Tab. 1 there are two modifications – para- and ortho- hydrogen. At higher temperatures the two states are in thermal equilibrium with 3/4 of the molecules in the ortho-state possessing an odd rotational angular momentum J. If the sample is rapidly cooled these molecules will be trapped in the J = 1 ortho state since a transition to the energetically more favorable ground state with J = 0 would require a flip of the nuclear spin which in the absence of magnetic impurities is highly improbable. Neutrons due to their magnetic moment can catalyze the flip of the nuclear spin. They thus allow observing this particular

CP685, *Molecular Nanostructures: XVII Int'l. Winterschool/Euroconference on Electronic Properties of Novel Materials,*edited by H. Kuzmany, J. Fink, M. Mehring, and S. Roth
© 2003 American Institute of Physics 0-7354-0154-3/03/$20.00

TABLE 1. Symmetry properties of the hydrogen molecule. I = nuclear spin, J = molecular rotational quantum number

H_2 molecule	I	J	Nuclear weight	
	0	0, 2, 4, ...	1	Para-hydrogen
	1	1, 3, 5, ...	3	Ortho-hydrogen

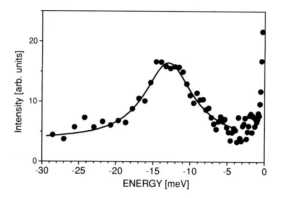

Figure 1. Neutron energy gain spectrum of the J = 1 → 0 rotational transition (ortho → para) measured for H_2 molecules adsorbed onto open SWNTs at 2 K. The result is an average over a larger region of momentum transfers: $1.2l \le Q \le 3.6$ Å^{-1}. The energy expected for unhindered rotations is 14.7 meV.

transition in energy gain. Working in this particular mode has the advantage that all phonon contributions are frozen out at low temperatures. The J = 1 → 0 transition can therefore be studied with high selectivity. In Fig. 1 we show a spectrum at 2 K where the J = 1 → 0 transition is clearly visible.

For our experiment we have used SWNTs produced by laser ablation. The material was acid treated in order to create openings in the tube walls and loaded with hydrogen at 300 K and 200 bar. After quick removal from the pressure cell the samples were stored at 77 K in thin Al sample holders foreseen for the neutron scattering experiment. Immediately before the measurement the SWNTs were warmed up to 300 K, a small hole that allowed for an escape of H_2 gas was opened and the sample was placed into a precooled cryostat featuring dynamic vacuum. For the inelastic scattering we used a low incident neutron energy of 4.8 meV which allowed for a good energy resolution of $\Delta E_{el} = 0.2$ meV and a large region of scattering angles which enabled us to study the J = 1 → 0 transition for different momentum transfers up to $Q_{inel.} \le 3.6$ Å^{-1}.

In Fig. 1 the contributions of all detectors distributed over a large solid angle between 14^0 and 114^0 have been summed up for intensity reasons. More information is obtained from the Q-dependence of the scattering. In Fig. 2 we have plotted the rotational line as a function of Q by summing up intensities for

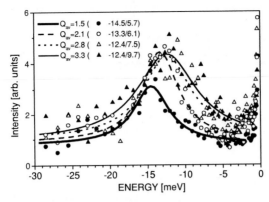

Figure 2. Neutron energy gain spectrum of the $J = 1 \rightarrow 0$ rotational transition as a function of momentum transfer at 2 K. Phonon contributions are frozen out. The lines are fits to the data by Lorentzians (brackets:E_{rot}/Γ (FWHM)). The significant changes are attributed to a recoil effect if the neutrons are scattered by weakly bond H_2 molecules.

groups of detectors. With increasing Q a significant increase in line width and a shift to smaller energies is registered. If we extrapolate the transition energy to Q = 0 we find an energy of ≈ 15 meV that is close to the value of 14.7 meV for a rotationally free H_2 molecule. The observed line width contains two contributions - an instrumental part due to the finite resolution of the spectrometer, a possible lifting of the degeneracy of the J = 1 level by local fields at different adsorption sites and – a Q-dependent part where recoil effects and a coupling to translational vibrations become important. The instrumental resolution at this energy is estimated as $\Delta E \approx 1.5$ meV. Neglecting the splitting we can deduce the mean kinetic energy of the H_2 molecules from the line widths σ via the impulse approximation (IA): $\langle E_K \rangle = 3/2 \, M/h \, (\sigma/Q)^2$ [2] and obtain from the slope σ/Q a value of $\langle E_K \rangle \approx 55$ K. This value is comparable to $\langle E_K \rangle = 77$ K obtained for H_2 adsorbed onto interstitial sites between closed tubes in a bundle [2] and is somewhat lower than that for H_2 adsorbed onto commensurate sites in Grafoil (115 K) [3]. It should be said that at the low incident energies the IA is an only poor approximation. Despite the fact that an exact treatment of the rotational translational coupling is difficult, it is a common result for all the investigated systems that the adsorbed molecules are translationally not free but fixed onto their adsorption sites.

Further information is obtained by warming up the sample. With increasing temperature the observed spectrum becomes dominated by phonon contributions following Boltzman statistics. For temperatures T > 150 K we register a significant drop in the elastic scattering intensity due to an important desorption of H_2 (sample in dynamic vacuum) up to ≈ 300 K. By taking the difference of the dynamic susceptibility extracted from spectra recorded at 150 K and 300 K we

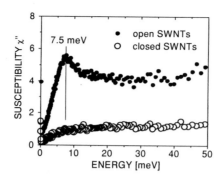

Figure 3. Susceptibility of the H_2 fraction which desorbs between 150 K and 300 K. A significantly higher amount of H_2 is adsorbed by open SWNTs. The particular excitation around 7.5 meV is attributed to H_2 molecules adsorbed onto the inner walls.

isolate the translational vibration spectrum originating from the H_2 fraction, which desorbs easily. The result is shown in Fig. 3. We observe a broad maximum at 7.5 meV, which resembles very much the spectrum observed for H_2 adsorbed on Grafoil [4]. There this excitation could be identified with lateral movements of the adsorbed molecules and a weak force constant of 0.180 N/m of the corresponding bond was determined.

In conclusion, the system of opened SWNTs has been investigated for the first time with neutrons and we find that the results compare favorably with similar information that had been obtained before for H_2 adsorbed onto Grafoil or by bundles of closed SWNTs [5]. The observation of the ortho – para transition for H_2 adsorbed onto SWNTs as well as the particular translational excitation at 7.5 meV in open SWNTs are new and might be specific for this system. Although we used the IA to obtain parameters for the kinetic energy the reader should be aware that a full investigations of the momentum distributions would require much larger neutron energies as the recoil energy should be larger than a typical phonon energy for the IA to be well fulfilled.

REFERENCES

1. B. Renker, H. Schober, F. Hennrich, and S. Lebedkin, AIP Conf. Proc. Of the XVI Int. Winterschool in Kirchberg, Tirol, Austria, 2-9. March 2002, Vol. **633** p. 322 .
   see also: B. Renker, H. Schober, P. Schweiss, S. Lebedkin, and F. Hennrich, **cond-mat/0301266**
2. D. G. Narehood, M. K. Kostov, P.C. Eklund, M. W. Cole, and P. E. Sokol, PRB 65, 233401 (2002)
3. G.J. Kellogg, P.E. Sokol, S.K. Sinha, and D.L. Price, PRB **42,** 7725 (1990)
4. V. L. P. Frank, H. J. Lauter, and P. Leiderer, PRL **61**, 436 (1988).
5. C.M. Brown,T. Yildirim,D.A. Neumann, M.J. Heben, T. Gennett, A.C. Dillon, J.L. Alleman,J.E. Fischer, Chem. Phys. Lett. **329**, 311 (2000)

Recent Progress in hydrogen adsorption in single-walled carbon nanotube systems

Masashi Shiraishi[1], Taishi Takenobu[2], H. Kataura[3]
and Masafumi Ata[1]

[1] *Materials Laboratories, SONY Corporation,*
240-0036 Shin-Sakuragaoka 2-1-1, Hodogaya-ku, Yokohama, Japan
[2] *Institute of Materials Research, Tohoku Univ.*
[3] *Department of Physics, Tokyo Metropolitan Univ.*

Abstract. Hydrogen adsorption properties in well-purified single-walled carbon nanotubes (SWNTs) and peapods (C_{60}@SWNTs) were analyzed and it was elucidated that hydrogen is physisorbed in inter-tube pores. The adsorption potential of the inter-tube pore was determined to be about –0.21 eV, and this large adsorption potential induces the hydrogen physisorption at room temperature.

INTRODUCTION

The interactions between gases and SWNTs attract much experimental and theoretical interest being stimulated by fundamental research in one-dimensional (1D) nano-ordered materials as well as practical applications for fuel cells. The present study is mainly focused on the physical realization of this unique 1D system. Up to now, much effort has been done to analyze a mechanism of hydrogen adsorption, however this is still controversial. In this work, the adsorption mechanism was investigated in detail and it was clarified that sub-nanometer ordered pores which have large adsorption potential are indispensable to hydrogen physisorption at room temperature.

EXPERIMENTS

SWNTs were synthesized by a laser ablation method using a Ni/Co catalyst. They were purified by H_2O_2 and aqueous NaOH solution to remove amorphous carbon nano-particles. The detailed procedure and the effect of this purification process were summarized in the literature [1]. Peapods were synthesized by using the same batch of the SWNTs. The filling rate was estimated to be about 85 % [2]. As a pretreatment, the SWNTs and the peapods were heated up to 723 K at 10^{-4} Pa for 1hr. Hydrogen was introduced at 6 MPa and 0.1 MPa at RT, respectively. The hydrogen desorption properties were analyzed by a temperature programmed desorption (TPD) method.

CP685, *Molecular Nanostructures: XVII Int'l. Winterschool/Euroconference on*
Electronic Properties of Novel Materials, edited by H. Kuzmany, J. Fink, M. Mehring, and S. Roth
© 2003 American Institute of Physics 0-7354-0154-3/03/$20.00

RESULTS AND DISCUSSION

The inset of Figure 1 shows the typical hydrogen desorption peak from SWNTs observed by the TPD method. This hydrogen desorption can be observed both from the SWNTs and the peapods only when 6 MPa of hydrogen was introduced and when NaOH process was applied. In addition, when the applied high-pressure gas was replaced to deuterium, the desorption peak of deuterium can be observed. From these results, it can be concluded that the desorbed hydrogen (and deuterium) was introduced extrinsically to the samples. Because hydrogen was charged at RT, it can be said that room temperature adsorption of hydrogen was achieved.

A water desorption peak was also investigated, and the peak was observed only from the SWNTs. These differences of desorption between hydrogen and water can be of help to identify adsorption sites. As previously reported, an outer surface and inside of a SWNT cannot adsorb hydrogen at room temperature because adsorption potential is too small [3]. The inside of the peapods are occupied by C_{60}s, thus only possible site is an inter-tube pore, of which size is typically 0.3 nm, and hydrogen is adsorbed in the inter-tube pores. , The reason why the inter-tube pores can adsorb

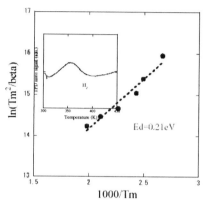

FIGURE 1. TPD signals

hydrogen at room temperature is discussed in the following paragraphs. In SWNTs, water is adsorbed inside of the SWNTs, whereas hydrogen is in the inter-tube pores. The adsorbed water inside of the SWNTs can form ice nanotubes as reported by Maniwa et al. [4].

Kissinger's plot analysis, as shown in Fig.1, was carried out to determine the desorption energy of hydrogen from the inter-tube pores, and the value was estimated to about 0.21 eV. This result directly shows that hydrogen is physisorbed. If hydrogen is chemisorbed, the value should be about ten times larger (~2 eV). Here, because surfaces of the SWNTs are clean enough, the adsorption potential of the inter-tube pores is equal to –0.21 eV. Figure 2 shows the relationship between the adsorption potential of the inter-tube pores and chemical potential of hydrogen. In hydrogen physisorption, the adsorption mechanism can be described by Langmuir adsorption isotherm,

$$f = \cfrac{1}{1 + \exp(\cfrac{\varepsilon - \mu}{kT})} \quad , \tag{1}$$

where ε is adsorption potential, μ chemical potential, k Boltzmann factor, f coverage of surfaces, T temperature.

In this work, the inter-tube pores is activated because of NaOH treatment, the pores with large adsorption potential are available. On the other hand, surfaces and inside of SWNTs have small ε (-0.03~-0.09 eV), they cannot be used for physisorption at room temperature.

The measure ε value is about 2.3 times as large as that of the inside of SWNTs when the latter potential is assumed to be -0.09 eV. A similar ratio, 1.8, has been reported by a recent theoretical calculation [5]. The amount of adsorbed hydrogen was

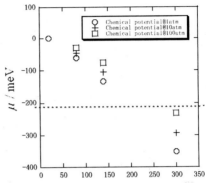

FIGURE 2. ε of the inter-tube pores and μ of hydrogen.

evaluated to be about 0.3 wt% at 10 MPa at room temperature. The hydrogen coverage of the inter-tube pores at 300 K and 10 MPa, f =0.31 as estimated from the μ value in Fig. 2 and experimental ε value, is comparable with the measured value, 0.38, because the theoretical limit of the adsorbed hydrogen in the inter-tube pores is 0.8 wt% [6]. This similarity also corroborates our experimental results and the model used in the present study. The corresponding coverage for the inside of SWNTs, f =0.0044, is smaller by two orders of magnitude, though the adsorption potential of the inter-tube pores is at most 2-3 times as large as that of the intra-tube pores. In other words, only a small difference in ε induces a major difference in f. Therefore, fabrication of minute pores, sized 0.2-0.3 nm and with large ε, is indeed a crucial factor to achieve hydrogen physisorption at ambient temperature [7].

CONCLUSION

In this work, hydrogen adsorption mechanisms were characterized by using TPD method. Because the SWNTs were highly purified, the inter-tube pores, of which size is about 0.3 nm, were activated and hydrogen can be physisorbed in the pores even at room temperature, because such minute pores have large adsorption potential (-0.21 eV).

Hydrogen physisorption can be described by Langmuir adsorption isotherm, and the experimental results have a good accordance with theoretical calculations. This fact corroborates our results and models.

REFERENCES

1. Shiraishi, M. et al. Chem. Phys. Lett. **338**, 213 (2002).
2. Kataura, H. et al. Appl. Phys. A **74**, 349 (2002).
3. Stan, G. et al. J. Low Temp. Phys. **110**, 539 (1998).
4. Maniwa, Y. et al. J. Phys. Soc. Jpn. **71**, 2863 (2002).
5. Murata, K. et al. J. Phys. Chem. B **105**, 10210 (2001).
6. Fujiwara, A. et al., Chem. Phys. Lett. **336**, 205 (2001).
7. Shiraishi, M. et al. Chem. Phys. Lett. **367**, 633 (2003).

Mechanics of nanotubes and nanotube-based devices

S.J. Papadakis*, P.A. Williams*, M.R. Falvo[†], R. Superfine** and S. Washburn**

*Department of Physics and Astronomy, University of North Carolina at Chapel Hill, Chapel Hill, NC 27599, USA
[†]Curriculum in Applied and Materials Science, University of North Carolina at Chapel Hill, Chapel Hill, NC 27599, USA
**Physics and Astronomy, Applied and Materials Sciences, Computer Science, University of North Carolina at Chapel Hill, Chapel Hill, NC 27599, USA

Abstract. We discuss the mechanical properties of carbon nanotubes and devices incorporating carbon nanotubes. We demonstrate novel measurement and force application techniques using an atomic force microscope coupled to a unique computing environment that simplifies manipulations. We report on results from measurements of the mechanical and electronic interactions between nanotubes and graphite surfaces. We also fabricate nanometer-scale electromechanical devices which incorporate nanotubes as springs, and discover a remarkable stiffening behavior of the nanotubes.

INTRODUCTION

Carbon nanotubes (CNTs) are the subject of many investigations due to their small size and remarkable mechanical and electrical properties [1]. They naturally form as nanometer-scale objects, allowing for straightforward fabrication of devices with nanometer-scale features. Rapid progress is being made, for example, in the creation of field-effect transistors using CNTs as the conducting channel [2, 3, 4, 5]. Their small size also means that quantum effects are often observed [5, 6, 7] in such simply-fabricated devices. CNTs are also remarkably robust mechanically [8, 9], which allows them to be mechanically manipulated without damage [5, 8, 10]. In this paper we report on the development of new tools which are particularly useful in the manipulation of CNTs, and experimental results on the fundamental mechanical and electrical properties of CNTs.

NANOMANIPULATOR

All of the measurements we discuss involve the atomic force microscope (AFM). This is an instrument which scans a sharp tip over a sample in order to produce topographical data about the sample. Furthermore, the AFM can make local measurements of sample properties. Forces can be applied to particular locations, and the displacement of the sample measured. With most AFM systems, both vertical (perpendicular to the sub-

CP685, *Molecular Nanostructures: XVII Int'l. Winterschool/Euroconference on Electronic Properties of Novel Materials*, edited by H. Kuzmany, J. Fink, M. Mehring, and S. Roth
© 2003 American Institute of Physics 0-7354-0154-3/03/$20.00

FIGURE 1. The nanomanipulator (nM) in use. The user holds the haptic-feedback stylus in his right hand. He can directly control the movement of, and feel the forces on, the AFM tip (upper right). A three-dimensional representation of the sample is shown on the screen (closeup in the upper left), with an icon representing the AFM tip also visible. This allows the user to place and move the AFM tip precisely to perform local force measurements.

strate) and lateral (parallel to the substrate) forces can be applied and measured. With a conductive tip, electronic properties can also be probed [11, 12, 13].

An AFM, however, has a significant limitation: scanning must be stopped in order to make a local measurement. With the thermal drift typical of instruments operated in ambient conditions, this makes it difficult to be certain of the measurement position. One can reduce the effects of thermal drift by operating in vacuum, and to a lesser extent in a controlled atmosphere. We have developed another way to overcome this limitation with the use of haptic feedback, incorporated into a system called the nanomanipulator (nM) [14, 15]. Using a haptic feedback stylus, the scientist takes real-time control of the tip and can feel (eg) the forces on the tip (Fig. 1). The system records all microscope settings during the run, all tip motion, and all responses from a set of sensors that typically include the topography and the vertical and lateral forces, and which may include electrical conductance or temperature, etc. Since local measurements are usually done at topographic features on the sample, the user can position the tip by feel onto or next to the feature to be probed, and then start the measurement in the desired direction. This is accomplished with natural hand motions very similar to those used in everyday life to explore the world at centimeter scales. The haptic feedback provides confidence that the tip is making contact with the sample at the desired location, which greatly facilitates measurements.

SLIDING, ROLLING, AND CONTACT RESISTANCE OF MULTIWALLED NANOTUBES

The nM was applied to measurements of the frictional properties of multiwalled CNTs (MWNTs) on highly-oriented pyrolytic graphite (HOPG) substrates, and also on atomically flat mica, MoS_2, SiO_2, Si_3N_4. Arc-grown MWNTs [16] were dispersed on the substrates from suspension. The nM was used to take topographical images of and apply lateral forces to the MWNTs [17, 18].

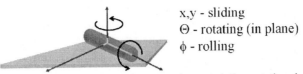

x,y - sliding
Θ - rotating (in plane)
φ - rolling

FIGURE 2. Motion of a cylinder on a flat surface comprises translation, rotation about a vertical axis, and rolling of the cylinder around its axis of symmetry.

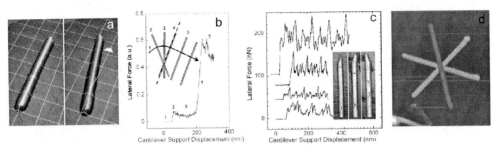

FIGURE 3. a) A MWNT before (left) and after (right) lateral force has been applied. It has both translated and rotated about a vertical axis. b) Periodic force response corresponding to the MWNT circumference. The inset shows a MWNT with an asymmetric tip rolling. c) MWNT force response as it slides on the surfaces until coming into atomic registry, after which it rolls. d) Superposition of three images showing the angles at which the MWNT locks into registry.

With the precision provided by haptic feedback, we were able to apply lateral force both perpendicular and parallel (end-on) to the MWNTs. With the MWNTs on mica, where the force was applied end-on, the MWNTs exhibited an initial static-friction peak followed by relatively smooth sliding. On HOPG, the MWNTs demonstrated repeated stick-slip behavior after they had started moving.

The particularly interesting results appear when the MWNTs are pushed from the side, perpendicular to their length. There are three generic modes of motion for such manipulations (Fig. 2): (1) translation, (2) rotaton of the MWNT about a vertical axis, and (3) rolling of the MWNT about its own axis. Which of these behaviors is observed depends on the interaction, at the atomic level, between the outer shell of the MWNT and the upper layer of atoms of the substrate. In some cases, the MWNT is pushed from the side, and the lateral force data shows a simple jump to a higher level when the AFM tip contacts the MWNT. Scanning the sample again after the manipulation reveals that the MWNT has both translated over the surface and rotated about a vertical axis (Fig. 3a). The motion of the MWNT is described well by modeling the system as a rod lying on a flat surface with uniform rod-surface friction along its length. The results imply that the MWNTs are sliding along the surface. If they were rolling, a uniform friction model would not fit. This type of behavior is observed on both mica and HOPG substrates.

In other cases, when lateral force is applied perpendicular to the MWNT a periodic force response is observed, qualitatively similar to stick-slip behavior. The periodicity of this response corresponds to the circumference of the MWNT. In these cases, topographical scans after the manipulation show that the MWNT has *not* rotated about a vertical axis, but has simply been translated. These characteristics imply that the MWNT is rolling instead of sliding on the surface. Detailed scans of MWNTs with a recognizable feature such as an asymmetrically tapered end confirm the rolling behavior (Fig. 3b) [17]. Rolling is observed only on HOPG substrates.

Further experiments reveal why the MWNT rolls in only certain cases. When the MWNT is lying on the HOPG substrate in a way that puts the outer shell of the MWNT in atomic registry with the upper carbon layer of the HOPG, the MWNT is resistant to sliding and rolls when lateral force is applied perpendicular to it. If the MWNT is lying such that its outer shell is not in registry with the HOPG surface, when force is applied it slides. If the force is applied away from the geometric center, it slides and rotates about a vertical axis (as described above) *until* it rotates into registry (Fig. 3c), after which it starts rolling [18]. The MWNT can be forced out of registry by applying the lateral force far enough away from the center of the MWNT or by applying end-on force. When these manipulations are performed repeatedly on the same MWNT, we find that the MWNT locks into the surface and starts rolling in three different orientations, each separated from the other by 60 degrees (Fig. 3d). This is consistent with the hexagonal configuration of the graphene sheets that make up the MWNT and the HOPG.

It is important to note that the force required to roll the MWNT when it is in registry is much *greater* than the force required to slide it when it is out of registry. This force rises instantaneously within our measurement resolution of ± 1 degree. This is qualitatively different from the behavior of macroscopic objects, which are typically easier to roll than to slide. It is more evidence of the atomic nature of the interaction between the MWNT and substrate. When the MWNT is in registry with the substrate, it is individual atomic interactions between carbon atoms that cause the increase in required force. When it is out of registry, even though on average the atoms in the MWNT are the same distance from the atoms in the substrate, the force required is much less because they are not in the ideal positions for bonding.

It is also interesting to look at the transport properties of this system as these manipulations are performed [19]. A metal-coated AFM tip is used, and the substrate is grounded. With a small voltage applied to the AFM tip, we pass current through the sample and measure the resistance R. This is a two-point measurement, so we measure three resistances in series: the AFM tip-MWNT contact resistance, the resistance across the MWNT's diameter, and the MWNT-HOPG contact resistance.

To begin the experiment we manipulate a MWNT to put it in registry and define its orientation as $\Theta = 0$. We use the nM to place the AFM tip on top of the MWNT and measure R through the MWNT to the substrate. We then rotate the MWNT on the substrate and repeat the R measurement. Figure 4 shows data from a MWNT rotated over a range of 180 degrees. The data show that R is small when the MWNT is in registry, and large when it is out of registry. The period of 60 degrees, and the repeatability from one period to the next, is clear and again reflects the hexagonal structure of the graphene sheets.

It is worth noting the qualitative difference between the dependence of MWNT-HOPG friction on Θ and of R on Θ. R changes gradually, being the result of interactions between allowed momentum states of itinerant electrons, which are determined by the boundary conditions (ie the sizes of the MWNT and substrate) as well as the crystal structure [19]. The friction, however, shows a very sudden decrease when shifted out of registry,

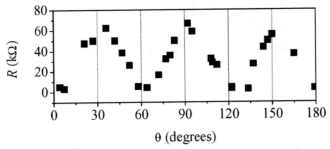

FIGURE 4. Electrical resistance through the MWNT as it is rotated on the substrate about a vertical axis.

because it depends only on the sum of the individual interactions between atoms at the interface.

The friction and transport results taken together show that MWNTs hold promise for use in future electronic and electromechanical devices. The fact that these measurements were all done in ambient conditions and still manifest the effects of atomic-scale interactions in both friction and current measurements demonstrates that these are very robust effects. Until recently, such results were only seen in ultra-high-vacuum systems, where the ultimate in cleanliness is available. Furthermore, the fact that we see gear-like behavior and tunable electronic resistance in objects only a few nanometers in diameter suggests that we may ultimately be fabricating useful devices with dimensions very near the atomic scale. The experiments also demonstrate that the control over assembly of these devices must be exquisite, as sub-nanometer misalignments may create order-of-magnitude variations in the mechanical and electrical properties.

NANOMETER-SCALE DEVICES

We have also begun incorporating MWNTs into nanometer-scale devices. We are interested in structures which first allow us to perform new experiments and to learn about the fundamental mechanical and electrical properties of CNTs, and which might serve ultimately as sensors (for example). In particular, we are fabricating paddle oscillator structures [20, 21, 22] using MWNTs as torsion springs. We perform measurements on these structures by applying forces to them with an AFM which operates inside the chamber of a scanning electron microscope (SEM). The AFM is oriented such that the the point of interaction between the tip and sample is visible in the SEM image. The system allows us to study suspended structures which are too fragile to be scanned with an AFM. The SEM images can be used to position the tip onto the structure to be tested without the need for a topographical scan by the AFM. All measurements are done in a vacuum of order 10^{-6} Torr.

The samples are fabricated by a combination of lithography, metal deposition, and etching. First, the MWNTs are dispersed from suspension onto an Si substrate with 500 nm of SiO_2 and predefined device areas. Electron beam lithography followed by thermal

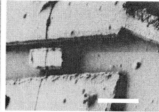

FIGURE 5. Completed paddle devices. The scale bars correspond to 1 micron.

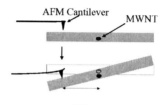

FIGURE 6. Schematic of the paddle deflection when a vertical force-distance curve is performed. The MWNT can be both twisted and deflected downward.

evaporation of 10 nm Cr and 100 nm Au is used to form a small stripe of metal over the center of each MWNT, which acts as a paddle, and large metal pads over the ends of each MWNT which anchor the ends. The sample is then etched in HF long enough to completely undercut the paddle but not long enough to undercut the anchors. This is followed by a supercritical drying step which leaves each paddle suspended above the substrate by an individual MWNT [23].

Figure 5 shows some completed devices. We perform measurements on devices such as these by using the AFM to apply forces to the paddles and to measure their responses. We use Si AFM cantilevers with a spring constant K_c of approximately 1 N/m for all measurements. Before starting the measurements, the AFM is calibrated so that we can quantify both the force on and displacement of the paddles [24]. Vertical force-distance (F-D) traces are performed on the paddle. The AFM tip is initially a small distance above the paddle, and is moved towards the paddle by the scan tube (Z-piezo) of the AFM. When the AFM tip touches the paddle, both the paddle and the cantilever deflect (Figs. 6,7). The AFM senses and records the cantilever deflection in the form of a photodiode signal. It is the slope of this trace (S_{pad}) that is related to the stiffness of the MWNT.

A simple statics analysis shows that there are two modes of deflection we might be interested in (Fig 6). When we push down on the paddle, there is both torsion and a vertical force on the MWNT. Therefore we are interested in both the torsional κ and vertical K_Z spring constants. The spring constants are related to the S_{pad} measured by

FIGURE 7. A vertical force-distance trace in progress. The SEM resolution is reduced when the AFM is in operation. The dashed white line in the first frame shows the location of the MWNT. The solid white line is provided as a guide to the eye, and is at the same angle and location in all frames.

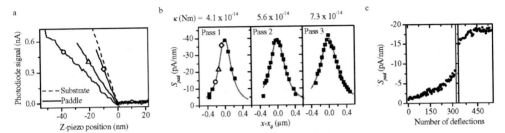

FIGURE 8. a) Data from a vertical force-distance trace on the bare substrate and from three consecutive traces at different $x - x_0$ on paddle C. The slopes (S_{pad}) of the three curves appear, with the same symbols, as points in b. b) Force-distance data from paddle C, shown with fits of Eqn. 1. The panels from left to right show three consecutive sets of data taken on one paddle. c) Repeated force curves taken on paddle D, showing a large increase in stiffness. The AFM tip was kept in one place for curves 1 to 330, after which its position on the paddle was changed by about 50 nm where it remained for subsequent measurements.

the AFM [24]:

$$S_{pad} = \frac{S_{sub}}{1 + \frac{K_c}{K_z} + \frac{K_c}{\kappa}(x - x_0)^2} \tag{1}$$

where S_{sub} is the slope of a force trace taken on the bare substrate and $x - x_0$ is the distance along the paddle measured from its attachment to the MWNT to the point where the AFM tip is touching the paddle.

We perform a series of force curves along the length of the paddle (Fig. 8a). The AFM has nanometer precision in relative positioning of the AFM tip, so that we know the relative $x - x_0$ accurately. Since we perform force curves on both sides of the paddle relative to the MWNT, we get the absolute $x - x_0$ from the symmetry of the data. We then fit Eqn. 1 to the data in order to deduce the spring constants. We perform multiple passes on each paddle that we measure, depicted left to right for paddle C in Fig. 8b. We find that κ gets larger with each pass. We observe this rise in κ for all paddles that we measure, and the percent change in κ from one pass to the next is roughly correlated with the number of F-D curves performed in the last pass. In order to observe this behavior more directly, we perform repeated force curves at one location on Paddle D (Fig. 8c). There is a break in the data because after about 330 force curves, the AFM tip was shifted slightly due to drift in the microscope. The trend in this data is clear, the paddle stiffens continuously as more force curves are performed. Especially of note is that the stiffness saturates after several hundred force curves.

While in principle Eqn. 1 suggests that we should be able to also get K_Z from our fits, in practice we are unable to measure K_Z because K_Z is much greater than K_c (which is chosen to access the force range for κ). The paddle is very long compared to the radius of the MWNT, so the AFM tip has significant leverage to twist the MWNT when it is positioned near the end of the paddle. If we use an AFM tip with K_c small enough to measure forces accurately when it is near the end of the paddle, it is too weak to deflect the paddle downward significantly, which is required to measure K_Z.

In order to compare data from the various paddles, we use the continuum mechanics model to calculate the shear modulus of the MWNT from the κ. One would expect the shear modulus to be the same for all MWNTs. We model the MWNT as a solid rod and calculate an effective shear modulus G_e. Since the shear modulus calculated for a circular rod of radius r from its torsional spring constant goes as $1/r^4$, neglecting the hole in the center of the MWNT is a reasonable approximation. The dominant uncertainty comes from our measurement of the outer radius of the MWNT, which we estimate to have an uncertainty of about 20%. This corresponds to a factor of two uncertainty in G_e. Table 1 shows that the G_e range over more than an order of magnitude. The theoretically predicted value for the shear modulus of MWNTs is $\sim$550 GPa, similar to diamond or basal-plane graphite [25]. Taking into account our expected uncertainty in G_e, and concentrating for the moment on only the first pass on each device, Table 1 shows that the MWNTs have values from much less than the predicted value to near the theoretically predicted value.

Our hypothesis to explain this is that in many cases the inner shells of the MWNT slide with little resistance inside the outer shells. Since the metal of the anchors and paddles clamps only the outer shell, only it is deflected. The conclusion that in some MWNTs the inner shells do not have a strong mechanical coupling to the outer shells has been reached by other investigators [9, 26]. If this is the case, we should be able to get an upper limit on the real shear modulus by assuming that only the outer shell bears the load, and calculating a shell shear modulus G_s. When this calculation is performed for the initial passes on those devices which initially appeared much softer than expected, we find that G_s is consistent with the predicted MWNT shear modulus of $\sim$550 GPa. Especially worth noting is paddle D (Fig. 8c). This device shows a saturation of its stiffness at a κ that leads to a G_e consistent with the theoretical value. Its initial G_s is very similar to the final G_e. Since the major source of error in our shear modulus calculations comes from outer radius, the values for a given device are expected to be self-consistent. The data taken together imply that the reason the MWNTs are stiffening is that the shells are slowly becoming linked mechanically as deflections are performed.

We have two hypotheses as to why this stiffening may be occurring. One is that the repeated deflections cause defects in the tubes that link the shells together. Another is that the deflections allow the individual tubes to shift relative to each other and thus improve their atomic registry. As the data in Sec. 3 show, improving the registry would be expected to increase the mechanical coupling between shells. We feel the second hypothesis is more likely because we are applying strains of between 0.5 and 1 % during

584

TABLE 1. Summary of MWNT torsional spring constants (κ), and effective (G_e) and shell (G_s) shear moduli (See the text).

Device/pass	κ $(10^{-14}$ Nm)	G_e (GPa)	G_s (GPa)
A/1	15	600	
B/1	2.4	60	830
B/2	4.5	120	
B/3	4.6	120	
B/4	10	280	
B/5	22	590	
B/6	46	1200	
C/1	2.5	15	210
C/2	3.4	20	
C/3	4.4	26	
D/i	1.4	30	430
D/f	17	400	

our measurements. Strains of this size are not expected to cause the bonds in a MWNT to break and reform between shells.

SUMMARY

In summary, we have performed experiments on CNTs and their interactions with surfaces. We find that MWNTs on a graphite substrate adhere preferentially when they are in atomic registry, along three directions sixty degrees apart consistent with the crystal structure of the MWNT and graphite. If they are not in atomic registry, when lateral force is applied they slide along the surface, obeying well a model of a rod on a smooth surface with uniform friction. If they are in atomic registry, it takes a larger lateral force to move them, and when they move they roll along the surface instead of sliding. Furthermore, the electrical contact resistance between the MWNT and the substrate is at a minimum when they lie in registry. The results show that even in ambient conditions, the surfaces of graphite and MWNTs remain clean enough that interactions between the carbon atoms dominate its behavior.

We have also fabricated nanometer-scale devices that allow us to apply torsional strains directly to the CNTs. This method suspends the CNT above the substrate, so only the CNT, and not its interactions with the substrate, are studied. The technique will be useful for both mechanical and electrical measurements on CNTs or other rod-like objects. We perform mechanical measurements on MWNTs and find that as the MWNTs are twisted repeatedly, they become stiffer. This is because the mechanical interactions between shells are increased by the repeated deflections.

ACKNOWLEDGMENTS

The authors would like to thank the ONR and NSF for funding for these projects.

REFERENCES

1. See, e.g., Top. Appl. Physics **80**, (2001); C. Dekker, Phys. Today **52**, 92 (1999); and M. Dresselhaus and G. Dresselhaus, *Science of Fullerenes and Nanotubes*, (Academic, San Diego, 1996).
2. R. Martel, T. Schmidt, H. R. Shea, T. Hertel, and P. Avouris, Appl. Phys. Lett. **73**, 2447 (1998).
3. S. J. Tans, A. R. M. Verschueren, and C. Dekker, Nature **393**, 49 (1998).
4. P. G. Collins, M. S. Arnold, and P. Avouris, Science **292**, 706 (2001).
5. H. W. C. Postma, T. Teepen, Z. Yao, M. Grifoni, and C. Dekker, Science **293**, 76 (2001).
6. M. Bockrath, D. H. Cobden, P. L. McEuen, N. G. Chopra, A. Zettl, A. Thess, and R. E. Smalley, Science **275**, 1922 (1997).
7. A. Bachtold, C. Strunk, J.-P. Salvetat, J.-M. Bonard, L. Forró, T. Nussbaumer, and C. Schönenberger, Nature **397**, 673 (1999).
8. M. R. Falvo, G. J. Clary, R. M. Taylor II, V. Chi, F. P. Brooks Jr., S. Washburn, and R. Superfine, Nature **389**, 582 (1997).
9. M.-F. Yu, O. Lourie, M. J. Dyer, K. Moloni, T. F. Kelly, and R. S. Ruoff, Science **287**, 637 (2000).
10. H. W. C. Postma, M. de Jonge, Z. Yao, and C. Dekker, Phys. Rev B **62**, R10653 (2000).
11. A. Bachtold, M. S. Fuhrer, S. Plyasunov, M. Forero, E. H. Anderson, A. Zettl, and P. L. McEuen, Phys. Rev. Lett. **84**, 6082 (2000).
12. S. H. Tessmer, P. I. Glicofridis, R. C. Ashoori, L. S. Levitov, and M. R. Melloch, Nature **392**, 51 (1998).
13. M. A. Topinka, B. J. LeRoy, R. M. Westervelt, S. E. J. Shaw, R. Fleischmann, E. J. Heller, K. D. Maranowskik, and A. C. Gossard, Nature **410**, 183 (2001).
14. Taylor, R. M. *et al.*, *SIGGRAPH '93* (ACM SIGGRAPH, New York, 1993).
15. Finch, M. *et al.*, *ACM Symposium on Interactive Computer Graphics* (ACM SIGGRAPH, Monterey, California, 1995).
16. T. W. Ebbesen and P. M. Ajayan, Nature **358**, 16 (1992).
17. M. R. Falvo, R. M. Taylor II, A. Helser, V. Chi, F. P. Brooks Jr., S. Washburn, and R. Superfine, Nature **397**, 236 (1999).
18. M. R. Falvo, J. Steele, R. M. Taylor II, and R. Superfine, Phys. Rev. B **62**, R10665 (2000).
19. S. Paulson, A. Helser, M. Buongiorno Nardelli, R. M. Taylor II, M. Falvo, R. Superfine, and S. Washburn, Science **290**, 1742 (2000).
20. A. N. Cleland and M. L. Roukes, Nature **392**, 160 (1998).
21. S. Evoy, D. W. Carr, L. Sekaric, A. Olkhovets, J. M. Parpia, and H. G. Craighead, Jour. Appl. Phys. **86**, 6072 (1999).
22. D. W. Carr, S. Evoy, L. Sekaric, H. G. Craighead, and J. M. Parpia, Appl. Phys Lett. **77**, 1545 (2000).
23. P. A. Williams, S. J. Papadakis, A. M. Patel, M. R. Falvo, S. Washburn, and R. Superfine, Appl. Phys Lett. **82**, 805 (2003).
24. P. A. Williams, S. J. Papadakis, A. M. Patel, M. R. Falvo, S. Washburn, and R. Superfine, Phys. Rev. Lett. **89**, 255502 (2002).
25. J. P. Lu, Phys. Rev. Lett. **79**, 1297 (1997).
26. J. Cumings and A. Zettl, Science **289**, 602 (2000).

Towards the integration of carbon nanotubes in microelectronics

A.P. Graham, G. S. Duesberg, F. Kreupl, R. Seidel, M. Liebau, E. Unger, W. Hönlein

Infineon Technologies AG, Corporate Research, Otto-Hahn-Ring 6, D-81739 Munich, Germany.

Abstract: The outstanding properties of carbon nanotubes (CNTs) makes them particularly interesting for microelectronic applications such as interconnects and devices. The parallel integration of CNTs using compatible processes is critical for their future application and implies the simultaneous processing of millions of CNTs. In this paper, we present concepts for CNT-based interconnects for large-scale integration. In order to realize this we demonstrate the creation of vertical interconnects (vias) consisting of multi-walled nanotubes and show, for the first time, the growth of single, isolated CNTs at lithographically defined locations.

INTRODUCTION

The continuing miniaturization of the devices in silicon chips is anticipated to lead to sub-25nm devices by 2016. This has prompted the investigation of a range of novel concepts and materials to replace the current silicon-based transistors and metal wires. One material in particular, namely carbon nanotubes (CNTs), has attracted a great deal of interest primarily due to the exciting properties and applications for nanoelectronic devices that appear to be possible. The dearth of research in this area has resulted in the demonstration of CNT-based transistors with very promising characteristics [1] and ballistic transport, i.e., current flow in CNTs with no resistance except at the contacts [2].

A cursory glance at the cross-section of a modern high-performance microprocessor immediately shows that chips consist mostly of wires, or interconnects, joining the active devices (transistors) and blocks of devices with each other (Fig. 1). At present up to 9 layers of wiring are required and this is expected to increase as the area per transistor decreases. This increase brings with it a host of problems including higher current densities that can lead to electromigration failure, higher resistances and local heating causing thermal breakdown not to mention integration difficulties. Hence, one possible application of the ballistic transport properties of CNTs is as on-chip interconnects; Also important for this application are the excellent thermal transport and structural stability of the CNTs. A further application of CNTs is as nano-scale transistors with transconductivities higher than the silicon equivalent and excellent switching characteristics.

There are several major hurdles to be overcome before integrated CNTs can compete with the existing silicon technology. Hand-picked demonstrators derived from

CP685, *Molecular Nanostructures: XVII Int'l. Winterschool/Euroconference on Electronic Properties of Novel Materials,*edited by H. Kuzmany, J. Fink, M. Mehring, and S. Roth
© 2003 American Institute of Physics 0-7354-0154-3/03/$20.00

randomly deposited CNTs that prove device functionality, e.g., the CNT-FET, are essential for the study of the physical properties of the CNTs and for an understanding of device function. However, ultimately they cannot compete with state-of-the-art nano-scale silicon transistors due to the large footprint and the random nature of the production process. Thus, processes need to be developed to integrate the CNTs in a space saving and silicon-technology compatible way. In this article we present an overview of the work we are currently undertaking towards CNT integration.

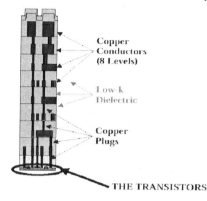

Copper
Conductors
(8 Levels)

Low k
Dielectric

Copper
Plugs

THE TRANSISTORS

Figure 1: Schematic cross-sectional view through a modern high-performance microprocessor. The active devices (transistors) that are anchored to the crystalline silicon substrate are found only at the bottom of the very high stack of metal wires connecting them together.

CARBON NANOTUBE WIRING

As mentioned above, one of the first microelectronics applications of CNTs which can be realized is on-chip wiring. For the integration of CNTs into a wiring scheme several processes are crucial including (a) accurate CNT placement (b) control over the electrical properties of the CNTs and (c) low resistance contacting. We will deal with these problems one at a time:

(a) Although arc-discharge and laser-ablation produce the highest quality CNTs with very few defects and exceptional electrical properties, the contamination of the CNTs with sometimes high quantities of graphite particles and the randomness of the deposition process using solvents, make these processes unsuitable for integration. Rather what is required is the defined placement of the CNTs available using, for instance, chemical-vapor-deposition (CVD) based methods. In CVD and plasma enhanced (PE-) CVD processes, CNTs are grown from arranged catalyst particles using a gas-phase feedstock, usually methane, ethylene or acetylene. Thus, CNTs are produced where the catalyst particles are positioned, a process which can be defined using standard lithographic techniques [3]. One disadvantage of the CVD or PECVD method is the lower quality of the CNTs that are produced. This is mainly due to the lower process temperatures, which are typically between 450°C and 950°C, rather than the 2000-2500°C produced in an arc-discharge. Further, the integration requirements of microelectronics strongly favors the lower end of the CVD temperature range, i.e., T<600°C, due to diffusion in other components.

(b) Controlling the electrical properties of single-walled (SW-) CNTs has proved to be a very difficult goal. Experiments have shown that about two thirds of all SWCNTs are semiconducting while the remaining third are metallic [4]. For microelectronics, either all semiconducting CNTs with defined band-gaps are required for transistor applications, whereas all metallic CNTs are the prerequisite for on-chip wiring. Fortunately, in contrast to the SWCNTs, it would appear that all CVD multi-walled (MW-) CNTs have metallic character, i.e., Ohmic I-V curves [5]. Thus, MWCNTs would appear to be the ideal material for the first step towards CNTs wiring.

(c) Control over the contact region between metals, with essentially free electrons, and CNTs with well-defined quantum states is critical for all applications of CNTs in microelectronics. For transistors, the switching characteristics of the device itself may be good, i.e., high transconductance (=source-drain conductivity for a given gate voltage), but if the contacts have a high resistance, the overall conductivity will be low. For wiring applications it is essential to approach the quantum resistance of the CNTs [2] in order to compete with scaled-down versions of existing metallization schemes. The quantum resistance is given as $1/G_0 = h/2e^2 = 12.9$ kΩ per state where the factor 2 comes from the spin up and down states of the electron. Metallic nanotubes have two conduction levels per shell, leading to a fixed resistance of 6.45 kΩ per shell. Further, if all layers of a metallic MWCNT can be contacted, the resistance will fall since each shell contributes the same parallel resistance of 6.45 kΩ, i.e., a 10 shell MWCNT has a theoretical resistance of only 645 Ω independent of the length in the ballistic transport regime. For comparison, a 10nm diameter copper damascene wire 150nm long is expected to have a resistance of 800-1000 Ω due to size effects related to grain and boundary scattering [6]. Enormous progress has been made in the last few years towards the optimization of metal-CNT contact resistances, particularly the use of carbide formation to contact all layers of a MWCNT [7].

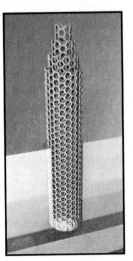

Figure 2: CVD growth of a multi-walled carbon nanotube (MWCNT) from a catalyst particle placed on a metal conductor at the bottom of a hole in an insulating layer. Contacting the top of the MWCNT with a suitable metal provides the electrical contact between the two metal layers.

VERTICAL NANOTUBE CONNECTORS

For the purposes of integration, we have focussed on the production of vertical CNT on-chip wiring. Vertical wires, also known as vias, are particularly prone to electromigration and stress failure and are, consequently, the most promising wiring components for replacement with CNTs. The main idea behind CNT vias is the CVD growth of a MWCNT from a catalyst particle in contact with a bottom electrode (Fig. 2) followed by definition of a top contact electrode. As reported previously, MWCNTs can be grown using a CVD process from iron-based catalysts on metallic layers such as Ti, Ta or TaN [8]. For the production of a vertical wire consisting of only one MWCNT, Fe and Ta were deposited on a TaN covered Si wafer structured using i-line lithography (350-400nm) and covered with 150nm Si_3N_4 in a plasma CVD process. Arrays of holes with diameters between 30nm and 50nm were etched through the nitride layer to the catalyst/support layer using ion beam lithography (direct writing). MWCNTs were then grown from the buried catalyst in a CVD process for 20 minutes using acetylene and hydrogen at 700°C [3], as shown in Fig. 3.

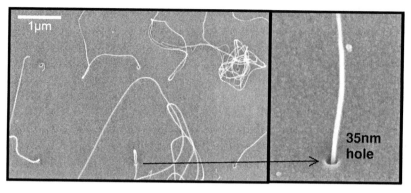

Figure 3: SEM images of MWCNT arrays grown through 35nm diameter holes. The holes were etched through a 150nm thick Si3N4 layer to an iron catalyst layer on a Ta/TaN support. The MWCNTs have diameters of about 15nm.

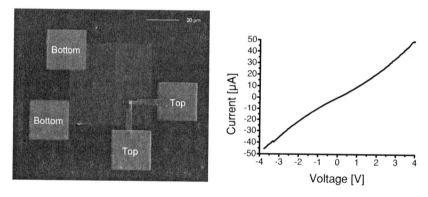

Figure 4: (left) Tungsten contacts deposited with electron and ion beams to form 4-point electrical contacts to a single CNT vertical connector (via). (right) First 4-point current-voltage measurement of a 150nm long MWCNT via. The resistance is about 80 kΩ and the curve shows deviations from a linear ohmic behavior, indicating tunnel or Schottky barriers.

The CVD MWCNTs grown from the buried catalyst layer have diameters around 15 nm, smaller than the diameter at the top of the hole of about 35 nm indicating that the catalyst particles at the bottom of the holes have diameters in the same range. The Fe catalyst and Ta support form the bottom electrode of the nanotube via. Tungsten top contacts were written using electron and ion beam deposition for 4-point resistance measurements, as shown in Fig. 4. As shown by the current-voltage behavior of a single CNT via the resistance is about 80 kΩ and the curve shows significant deviations from a linear ohmic response. Further, currents in excess of 70 µA could be passed through the 15nm diameter MWCNT corresponding to a current density of $4 \cdot 10^7$ A/cm^2 averaging over the whole nanotube diameter.

The deviations from a perfect linear behavior indicate that there are tunnel or Schottky barriers between the contact electrodes and the CNT. Sources of these barriers include oxidation of the catalyst particles and the top few atomic layers of the Ta bottom electrodes and poor e-beam deposited W top electrodes. Further, it is not expected that the W top electrode contacts more than the outer shell of the MWCNT, so only one layer of the CNT probably contributes to the conduction. Hence, the lowest resistance for the present configuration is anticipated to be only $1/2G_0=6.45$ kΩ. Improvement to the bottom electrode contact may be possible by high temperature annealing and reduction or electroless metal deposition (see paper by M. Liebau in these proceedings). Better top electrodes can be obtained using Ti or other carbide forming metals [7] or etching of the MWCNT to expose all shells [9].

CONCLUSIONS

We have presented a concept for the integration of carbon nanotubes into microelectronics in the form of vertical inter-layer connectors (vias). Measurements of the first CNT vias were also shown with a total resistances of 80 kΩ for non-optimized contacts. Currents of over 70µA could be carried by the vias corresponding to current densities in excess of $4 \cdot 10^7$ A/cm^2. Higher current densities and lower resistances are expected for optimized contact materials and processing.

One clear progression from the creation of a vertical CNT interconnect is to replace the metallic MWCNT with a semiconducting SWCNT and include a gate in the dielectric layer between top and bottom electrodes. The first steps towards this vertical VCNT-FET are already underway, as shown by R. Seidel in these proceedings.

REFERENCES

1. Wind, S.J.; Appenzeller, J.; Martel, R.; Derycke, V.; Avouris, Ph.; *Appl. Phys. Lett.* **2002**, 80, 3817-3819
2. Frank, S.; Poncharal, P.; Wang, Z.L.; de Heer, W.A.; *Science*, **1998**, 280, 1744-1746
3. Duesberg, G.S.; Kreupl, F.; Graham, A.; Liebau, M.; Unger, E.; Gabric, Z.; Hönlein, W.; in *Electronic Properties of Molecular Nanostructures*; Kuzmany, H.; Fink, J.; Mehring, M.; Roth, S.; Eds.; AIP 633 New York **2002**; p 157-160.
4. Javey, A.; Wang, Q.; Ural, A.; Li, Y.; Dai, H.; *Nano Lett*, **2002**, 2, 929-932.
5. Kreupl, F.; Graham, A.P.; Duesberg, G.S.; Steinhögl, W.; Liebau, M.; Unger, U.; Hönlein, W.; *Microelec. Eng.* **2002**, 64, 399-408.
6. Steinhögl, W.; Schindler, G.; Steinlesberger, G.; Engelhardt, M.; *Phys. Rev. B.* **2002**, 66, 075414.
7. Martel, R.; Derycke, V.; Lavoie, C.; Appenzeller, J.; Chan, K.K.; Tersoff, J.; Avouris, Ph.; *Phys. Rev. Lett.* **2001**, 87, 256805
8. Kreupl, F.; Graham, A.; Hönlein, W.; *Solid State Technol.* **2002**, 45, s9-16.
9. Huang, S.; Dai, L.; *J. Phys. Chem. B* **2002**, 106, 3543-5

Carbon Nanotubes For Field Emission Displays

J. Dijon, A. Fournier, T. Goislard de Monsabert, B. Montmayeul,
D.Zanghi

CEA -DRT-LETI/DOPT -CEA/GRE 17 rue des martyrs 38054 Grenoble cedex 9 France

Abstract. Display structures suitable for direct growth of CNT films by thermal CVD are described. The CNT film specifications for TV application are the following: a threshold field for emission higher than 3 V/μm, a current density of 140mA/cm^2 and an emission site density around 10^7 cm^{-2}. The major difficulty encountered with the deposited films is to reach the specified emission site density. These films present an exponential distribution of the amplification factor controlling the current emission. Small 1 cm^2 working displays with CNTs grown by using the decomposition of C$_2$H$_2$ over Ni catalyst have been done. These samples show that the three major locks of conventional FED technology, namely large size compatibility, simple low cost cathode and good reliability, are now opening thanks to CNTs.

1 INTRODUCTION

One of the most promising short term application of Carbon Nanotubes (CNTs) is probably the field emission devices. Thanks to their geometry and to the intrinsic chemical stability of carbon, the CNTs can be considered as almost perfect electron emitters. To be more than attractive laboratory demonstrations, a lot of work have to be done to trigger a new wide industrial interest around field emission displays (FED).

In the nineties, wide industrial developments have been done around molybdenum micro tips displays[1]. While successful, the industrialisation of these displays have been stopped for reasons which are more economical than technical. The main one is the difficulty to extend at low cost the technology to large size FED. This extension is needed to ultimately access the wide TV market. This difficulty lies on two facts. Firstly the basic process requested to master the small micron size gate holes of the so called spindt cathodes[2] is a microelectronic one. Secondly the fabrication of the tips themselves needs hudge deposition chambers in order to be scaled up. The extension of these processes to large sizes, while possible, is thus very costly. Furthermore reliability was difficult to achieve on molybdenum tips due to the ease of formation of molybdenum oxide on the tips during emission in a display environment[3]. Nevertheless field emission is a very attractive emissive technology thanks to its low power consumption, high brightness, high contrast, high viewing angle, high speed and image looking like CRT ones. CNTs offer a unique possibility to overcome the FED limitations just discussed. The purpose of this paper is to present the LETI (Laboratory of Electronic and Information Technology) state of the art in that direction.

CP685, *Molecular Nanostructures: XVII Int'l. Winterschool/Euroconference on Electronic Properties of Novel Materials,* edited by H. Kuzmany, J. Fink, M. Mehring, and S. Roth
© 2003 American Institute of Physics 0-7354-0154-3/03/$20.00

In the first part a cathode structure compatible with low cost photolithography and suitable for large area displays is presented. The growth techniques compatible with glass substrate chosen to be developed by our team is also described. The emission performances of CVD grown CNT films is the subject of the second part of the paper. The CNT film specifications are first given, then the achieved emission performances measured in diode mode are discussed. Finally the integration of the CNTs inside the device structure allows to present the performances of real CNT FED.

2 CATHODE STRUCTURE FOR CARBON NANOTUBES FIELD EMISSION DISPLAY

Cathode design

Field emission is the extraction of electrons from a solid by tunneling through the surface barrier by applying a strong electric field. The order of magnitude of the field to extract electrons from materials is around 3000V/µm. To reach such a high field one usually use very sharp objects which amplifies the applied field by a factor β. The so called amplification factor β depends on the geometry of the object and typically, for a long cylindrical object like CNT, is proportional to the aspect ratio h/r, where h is the tube height and r the tube radius. This means that the local field E_{local} at the tube tip is expressed by $E_{local} = \beta E$ where E=V/d is the external applied field. The typical aspect ratio of CNT can be higher than 1000. This explains the low threshold of emission (around few volts/µm) reported in the literature for CNTs[4]. For field emission displays such a low threshold means that either it is possible to drive the panel at low applied voltage V if d is just few microns or, more interestingly, that it is possible to design remote structures to applied the field if one accepts to use higher voltage (between 50V and 100V) still compatible with low cost electronic drivers. In that case the typical dimension between the emitters and the gate electrodes is around 10µm. Such possibility is one of the major interest of CNTs for displays. Indeed, it allows to overcome one of the major difficulties encountered during the development of micro tips displays which is the mastering of micron size photolithography on large area substrates (above 15 inches). This opens the possibility to built low cost cathode structures without loosing emission performances and thus, ultimately, to develop large size CNT FED for TV application. The triode structures developed at LETI are examples of what can be achieved from these concepts.

For FED, a triode structure is needed in order to control the emission by a gate electrode without switching the high voltage and power provided by the anode. The basic materials arrangement as well as the typical geometrical dimensions of the cathode are presented figure 1. The substrate material made of glass is covered by a first metallic layer patterned to form the cathode lines. The cathode lines are then covered by a silicon layer which thickness is typically 1.4µm. Above this ballast layer an insulating layer made of silica and a second conducting layer are then deposited. These two materials are patterned in order to form large trenches with typical dimensions of 15µm at a pitch of 25µm. In the middle of each trench the CNTs are located on catalyst pads which are used to selectively grow the tubes by using thermal CVD process. The dimensions of these catalyst pads are 6µm by 10µm. This

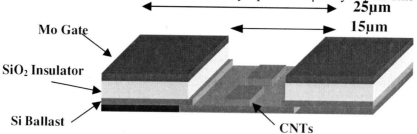

FIGURE 1. Triode structure developed for CNTs FED. The smallest geometrical dimension is the catalyst pad width (6µm). The gate pitch is 25µm and the trench in between the gates is 15µm wide. The whole structure is deposited on glass substrate.

basic structure is used to make a display with 350µm*350µm pixels. Each individual pixel is composed of three color sub pixels with dimensions of 350µm*117µm. In this sub pixel area there are four gate lines with three trenches and 57 catalyst pads. About 8.4% of the display area is covered by catalyst and thus by CNTs. This structure is made by using just three low resolution photolithographic masks. It is compatible with low cost large area cathodes. Small 1 cm² displays have been used to develop the CNTs growth process onto the device and to test the display characteristics.

Direct Growth of Carbon Nanotubes

The CNTs growth is performed directly on the device and is the last technological step to make the cathode. To be compatible with large size and industrial development, simple thermal CVD process have been considered. The used substrate is a borosilicate glass with an annealing point of 680°C. So the growth temperature is typically around 600°C to avoid glass melting or glass deformation. At such a low temperature, the growth is largely controlled by the characteristics of the catalyst particles. These particles are obtained by fragmentation of the catalyst layer during the rising of the temperature [5] before the introduction of the reactive atmosphere. The particle sizes is controlled by the thickness of the catalyst layer [6]. The nickel catalyst is deposited at room temperature by e-beam evaporation with a thickness of 3nm or 10nm. The characteristics of the catalyst droplet population obtained after annealing at 600°C with these two kinds of layer are summarized table 1. The droplets diameter follows a log-normal distribution (figure 2):

$$P(d) = \frac{1}{d\, sig\sqrt{2\pi}} \exp(-\frac{1}{2}\left(\frac{\ln(d)-\ln(do)}{sig}\right)^2) \qquad (1)$$

TABLE 1. Catalyst droplets population characteristics.

Characteristics of the droplets	Catalyst layer of 3nm	Catalyst layer of 10nm
Droplets density (1/µm2)	70	10
Averaged diameter do (nm)	55 ± 5	135 ± 10
Standard deviation sig	0.47 ± 0.01	0.69 ± 0.03

As can be seen on the left part of figure 2 the droplets stay on the pad after their formation.

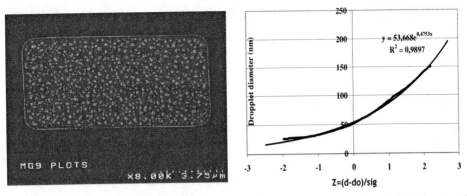

FIGURE 2. Left: droplets of catalyst on the pad after rising of the temperature at 600°C. Right: normal probability plot of the droplet diameter measurements for 3nm catalyst layer. This chart presents the probability to have a droplet of diameter D smaller or equal to d. The measured probability is connected

to z by $P(D \leq d) = \dfrac{1}{\sqrt{2\pi}} \int_{-\infty}^{z} e^{-t^2/2} \, dt$ where sig^2 is the variance of the distribution and do the mean

diameter of the droplets. In this diagram, the exponential behavior of the data indicates that the distribution is a Log-Normal one. The parameters of the exponential gives the mean value and the standard deviation of the distribution.

The growth is performed by using the simplest possible thermal CVD process. The reactive gas is pure acetylene at a typical pressure of 150mT. The growth time is very short, in between one and ten minutes. The CNTs obtained with these conditions are either MWNT or carbon fibers with a diameter around 30nm (figure 3). On each pad a forest of CNTs is obtained

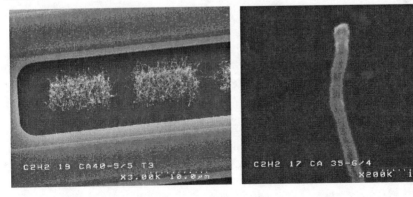

FIGURE 3. CNTs grown in the display at 600°C. Left: CNTs forest in the trench in between the gates. Right typical CNT grown with C_2H_2, the diameter of the tube is about 30nm.

Our purpose is to optimize the growth conditions to obtain the emission characteristics requested by display applications. In the next section we will define these specifications.

3 FIELD EMISSION OF CARBON NANOTUBE FILMS

Film requirements for FED

The emission of individual CNT follows the classical Fowler Nordheim (F-N) theory [7]. Thus the emitted current for one tube can be expressed by

$$I_{CNT} = S_t a(\phi)(F\beta)^2 \exp(-\frac{b(\phi)}{F\beta})$$ (2)

where β is the amplification factor which takes into account the aspect ratio of the tube as already mentioned, St is an equivalent area of emission for a single tube, Φ is the work function of the carbon emitters (typically 5eV [8]) and a(Φ), b(Φ) are elliptic functions[9] which can be considered as constant. In our case a= 2.25 10^{-5} and b=7.26 10^{10} V/m.

In the case of CNT films, one major departure from expression 2 comes from the fact that the geometrical parameters of the tubes and thus β are distributed. So

$$J_{film} = \frac{I_{film}}{S} = \int No S_t a(\phi)(F\beta)^2 \exp(-\frac{b(\phi)}{F\beta})D(\beta)d\beta$$ (3)

where S is the area of current measurement, No is the emission site density (m^{-2}) and D(β) is the distribution function of the amplification factor.

For display applications, requirements are needed on some of these parameters.

The average current density is related to the brightness of the display. One have

$$B = \frac{\eta UI}{\pi A}$$ (4)

where B is the brightness of the display in Cd/m^2 (typically 3000 Cd/m^2 for the green phosphor), U is the anode voltage (3000V in our case), A is the area of the display and η is the luminous efficiency of the phosphor (typically 12 lumen/W with green phosphor).

With these values the average current density is roughly 25 10^{-6} A/cm^2. Due to multiplexing of the rows, the requested peak current that must be delivered by the CNT film is multiplied by the multiplexing ratio. For 1000 rows suitable for TV display it is 12.5 mA/cm^2. For the CNT films, taking into account the filling factor of the cathode, it means that a current density around 140mA/cm^2 have to be reached. Such a high current density have already be reported on CVD grown materials [10].

The uniformity of the emission is partly governed by the density of emissive site (number of CNT really emitting per surface unit). This density must be close to 10^{10} m^{-2} which is the density of Mo micro tips in conventional FED [11].

Finally the emission must be controlled by the gate field. Electrostatic simulations of the field in the display structure previously described [12] show that the applied field at the tips of the CNT is a linear combination of the gate field and of the anode field. The cell gap between the anode and the cathode is 1mm, the applied voltage on the anode is 3kV to achieve a good luminous efficiency. So it produces a field of 3V/µm which must not induce significant current emission. This means that the threshold of emission of the film must be higher than this value. We consider that too low threshold film are not requested for display applications while a high density of emissive site is absolutely needed. It is of prime interest to assess this density experimentally and to optimize the CNT films in that particular direction. This is the subject of the next sections.

Threshold of emission and field amplification factor

From the basic point of view, there is no threshold of emission (see equation 2). In order to make a comparison between different emissive materials authors use two kinds of experimental parameters which are the turn on field (Fto) and the threshold field (Fth), respectively defined as the electrical field needed to obtain a current density Jto of 10µA/cm^2 and Jth of 10mA/cm^2. These parameters are related with the more physical ones which are β and No. Indeed, starting from equation (2), thus considering in a first step that the distribution function D(β) is rather sharp and can be approximate by a dirac function, we have

$$\ln(\frac{J_{th}}{NoSta}) = 2\ln(F_{th}\beta) - \frac{b(\phi)}{F_{th}\beta} \tag{5}$$

Excepted a small logarithmic dependence versus No the solution of this equation is

$$F_{th}\beta = Cte \tag{6}$$

This relation is well followed on our experimental set up: With a 4mm spherical ball anode at 100µm from the tested film [13] we have $F_{th}\beta = 4960$ for a wide range of

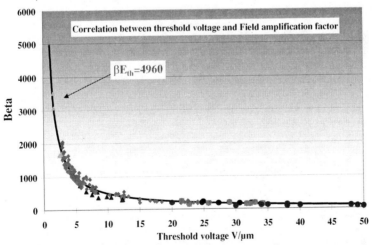

FIGURE 4. Correlation between the threshold field defined for a current density of 10µA/cm^2 and the field amplification factor. The measurements are performed on different CNT materials as well as on textured graphitic materials. This result also means that for carbon emitters from different suppliers the work function φ is almost constant.

CNT films or for other textured graphitic materials (figure 4). So, our requirements on threshold of emission (upper than 3V/μm) is thus transformed in a requirement on β (lower than 1500). Thus, in the following, instead of considering the threshold of emission, we will discuss on the field amplification factor β which have a more direct meaning related with tube geometry.

Distribution function of the amplification factor

The determination of $D(\beta)$ on our films is of prime interest. To do that, both the distribution function of h and the distribution function of r have been measured: on one particular sample (called Di72), 10 contiguous SEM pictures have been taken on the sample tilted at 90° with a magnification of 2.5K. This CNT film has been grown from catalytic decomposition of CO over 3nm thick Ni layer. The magnification has been chosen as a compromise to have enough field of view and enough resolution to see the CNT. The images have been processed to count the number of intersections in between a virtual sliding bar located at a height h above the sample surface and the CNT. With these data a cumulated distribution function N(h) have been built (see figure 5).

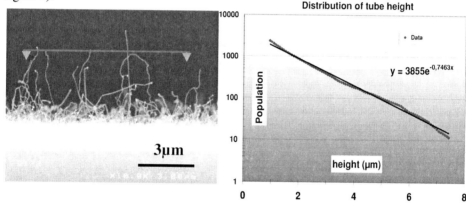

FIGURE 5. Determination of the distribution function of the tube height. On the right an image of the CNTs (magnification 10K) on this sample. The number of tubes at the altitude h is given by the number of intersections between the horizontal bar and the CNTs. The obtained distribution function is on the right. The slope of the exponent is 1/ho=0.746μm⁻¹ . The shown distribution is the cumulated one, due to the exponential shape the slope is the same as the distribution of tube height.

The obtained distribution presents an exponential decay on more than two decades with a characteristic height of 1/0.746=1.35μm.

On the same sample the distribution of diameters has been measured from the same images. It is a gaussian one with an averaged radius r_o of 30nm and a standard deviation of 6nm.

An exponential distribution of the amplification factor is thus expected from these geometrical considerations to be approximately :

$$D(\beta) \propto \exp(-\frac{h/r_o}{ho/r_o}) = \exp(-\frac{\beta}{\beta_o}) \tag{7}$$

with $\beta o = 1.35/0.03 = 45$. The direct numerical convolution of the two experimental distributions confirms this behavior with a more realistic βo coefficient of 65.

An other way to assess the distribution of beta is to perform I(V) measurements. Such exponential distribution have already been proposed and determined by statistical measurements of localized I(V) [14,15,16] but to our knowledge it is the first time that it is directly determined geometrically.

We also performed direct electrical determination of βo. Indeed with $D(\beta)$ specified, it is possible to obtain analytic expression for equation (3) [13] and thus to determine directly by I(V) measurements the slope βo of the distribution. In that case, the shape of the emission current is expected to present a slight upward curvature in the classical F-N plot $\log(J/F^2)$ versus $1/F$. For sample Di72 such a shape is actually observed. The electrical determination of this parameter gives $45 < \beta o < 75$ in agreement with the geometrical determination. This kind of value for βo is also found by Nilsson [17]. Thus by using equation (3) and considering the distribution function of geometrical tube parameters a close agreement is observed between the geometrical determination of the amplification factor and the electrical one.

So with an exponential distribution of β it is βo which govern the slope of the current. If $D(\beta)$ is not considered in equation (3) the obtained amplification factor is an equivalent fit parameter with a value of 800 in the case of sample Di72, far from any geometrical parameter reasonably determined from SEM images (see figure 5). We think that the discrepancies reported in the literature for films [18] are due to the use of simplified versions of the emission current expression which do not take into account the distribution of tubes parameters. This is the reality of CNT films grown without any particular way to deeply modify this distribution. The tuning of the threshold is achievable by the tuning of the geometrical parameters of the CNT length or tube diameter.

Emission site density

The emission site density is a key factor for display applications. To have a fast loop between growth and film optimization it is very likely to be able to assess the site density from direct diode measurements on film deposited onto simplified cathode structure before integration inside the real device. With F-N plot of the emission current it is possible to extract two parameters: β and the product NoSt. To go further and try to quantify the number of emissive site it is necessary to, at least, estimate St. To do it experimentally, triode structures without ballast layer have been used. In that case, the emission current as well as the optical characterization of the emission have been performed simultaneously. From optical measurements a minimum value of the emission site density can be assessed. With a ballast layer in the triode, the emission is smoothed [12] and it is no more possible to determine properly No by an optical way.

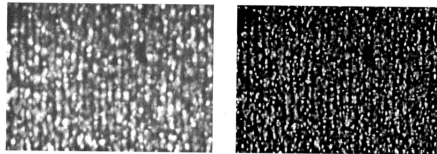

FIGURE 6. Left image: Emission from triode structure without ballast at Vgate=75V Vanode=1000V. The anode covered with phosphor is located at 1mm from the cathode. Right image: after deconvolution by using a gaussian spreading function to take into account the divergence of the electron beam between the anode and the cathode. The image is made with a 12 bits CCD camera and an optical lens assembly which provides a spatial resolution of 6.5µm on the display. The image size is 6*9mm^2

The rough images are processed (figure 6) in order to extract directly by counting the density of sites on 1cm^2 of display. The same work has been done on a conventional µtip display where the number of tips is known (10^{10} m^{-2}) and the percentage of working tips is estimated to be around 30% [19] to check the method (figure 7).

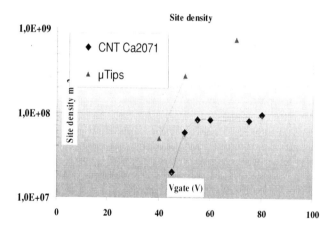

FIGURE 7. Evaluation of the emission site density for two FED displays. The triangles are for µ Tip based FED while diamond are for CNT based FED. With µ Tips around 3 10^9 sites are expected at Vg=90V. It was not possible to determine the number of sites at this gate voltage. Extrapolation of the data suggest that the site number is underestimated by a factor of 2 or 3. The saturation of the CNT curves is related with tube degradation with triode structure without ballast[12,13].

On this display the site density saturate around 9 10^7 m^{-2} which leads to a site density of 2 10^9 m^{-2} for the CNT film taking into account the filling factor of the triode structure (which was 4.7% in that particular display). From current emission the product NoSt is found to be 1.4 10^{-10}. Thus the emission area for one tube is around 1.5 10^{-18} m^2. This value is purely indicative but it is almost two order of magnitude larger than the area obtained the same way for µtip emitters which is typically around 10^{-20} m^{-2}. On the particular display studied the pad density is 1.86 10^9 m^{-2} so less than

10% of the pads emit and considering the tubes number on the pad it means that the probability for a CNT to be an emitter is around 10^{-4}. Hopefully, the addition of a ballast layer in the triode structure improves the situation.

With the above determination of St, different films have been analyzed in term of No and beta. The measurements have been performed in diode mode and the results are reported on the graph figure 8. A lot of comments can be made on these data.

-First the spread in No for a given beta is about 4 orders of magnitude independently of the beta value.

-Second the fit of these data by a power law in beta gives a dependence which is

$$N_o = 410^{15}\,\beta^{\,-2.07}$$

This experimental power dependence is very close to an exponent of minus two. Such value of the exponent is expected when there is a screening effect between CNT [20]. Indeed the optimum distance between tubes is proportional to h [15,20], thus roughly to beta, and the optimum density is proportional to β^{-2}.

-Third the data suggest that there is a limiting achievable density which follow the same behavior as the average value. Our targeted value, around 10^{11} m^{-2} for No, and between 1000 and 1500 for beta, is uncomfortably close to this limit. Without a positive action to control the tube density, standard films have site density between one and two decades from the optimum, leading to an acceptable but still grainy aspect of the emission. To optimize the films it is clear that sparse growth has to be mastered. Our main effort to reach this goal are done on the control of the catalyst properties.

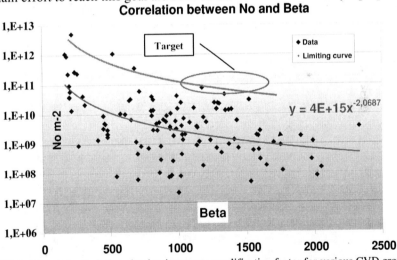

FIGURE 8. Plot of the Emission site density versus amplification factor for various CVD grown CNT films. The emission measurements are done in diode mode with a spherical ball lying at 100μm from the film surface.

Results On Triodes

Typical results achieved on small 1cm^2 displays are summarized table 2 and figures 9, 10.

TABLE 2. Characteristics of 1cm² CNT FED

Number of pixels	32*32*(3)
Pixel size	350μm*350μm
Peak current density before sealing (Vg=120V Va=3kV)	4.3mA/cm²
Peak current density after sealing and pre ageing (Vg=120V Va=3kV)	4.5mA/cm²
Brightness	1200 Cd/m²

On that particular display, the achieved current is about half our ultimate target. This current density allows to achieve at 3kV a brightness of 1200Cd/m². The current voltage characteristic given figure 9 shows that the gate voltage is 20V too high for driving the panel with low cost electronic (typically working below 100V). Nevertheless this is a preliminary results and by tuning the tube geometry it is possible to substantially decreases this voltage as shown on figure 9. The uniformity of the emission is already quite correct while not yet sufficient on the standard of the display application. The remaining grainy aspect is directly related with the problem of emission site density discussed previously. The benefit of the ballast layer on the emission quality is very important as can be seen by comparing figure 6 and 9.

Finally a very important result concerning the long term stability of the emission is demonstrated. In sealed display working at full brightness the emission is very stable on more than 800 hours (figure 10).

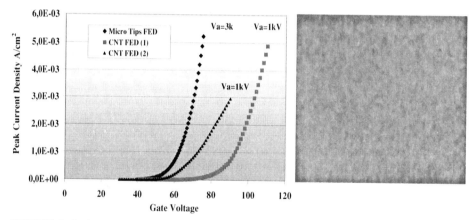

FIGURE 9: Left: comparison of the I(V) characteristics for μTips triode and CNT triodes. The diamonds are from a conventional Mo tips FED. The squares are the characteristic of the display presented on the right with 1kV on the anode. The triangles show the shift of the characteristic achieved on an other display after tuning the growth. In that case the slope is not very steep due to a very high value of the ballast layer. Right: close view of the 1cm² display working at Vg=120V Va=3kV. While correct, the uniformity is not yet perfect due to the lack of emission sites. Compare with figure 6 to appreciate the uniformity improvement induced by the ballast layer.

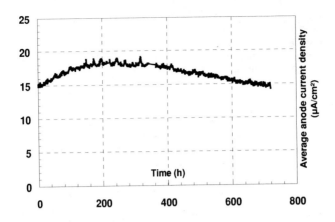

FIGURE 10. Preliminary results on long term ageing of the CNTs display presented figure 9. The display is in full brightness mode with an anode voltage of 3kV, a gate voltage of 120V and a duty ratio of 1/300. The current indicated is the averaged current density in $\mu A/cm^2$ on the anode.

4 CONCLUSION

The CNT FED technology is now credible and close to expectations for display applications. It is moving from concept to feasibility demonstration. Indeed low cost large size compatible triode structures have been successfully developed and integration of thermal CVD growth on the devices have been done. The to date performances are close to the specifications concerning the current density, the brightness and the achieved uniformity of the display. Three significant locks of the FED technology which are large size compatibility, low cost cathodes and reliability are opened by CNTs. To go further in that direction it is necessary to improve the number of CNTs working as emitters on the cathode, still preserving the intrinsic simplicity and low cost of the chosen solutions for the technology. This is the challenge to meet if ones want to see CNTs inside our future flat TV displays.

5 ACKNOWLEDGMENTS

Part of the results shown here have been funded by the European IST CANADIS project under contract number 1999-20590. Thanks are due to our partners of CANADIS IST project particularly to E. Broitman and H.J. Scheibe for providing us textured carbon materials grown by PVD and mentioned in figure 4.
Thanks are also due to P. Gadelle, O. Manfroi, J.P. Pinheiro, M.C. Schouler from INPG for providing us CNT samples grown by CVD with CO, and to E. Rouviere T. Krebs from CEA/DRT/DTEN for making C_2H_2 growth on some triode structures and to P. Smutek from S.A. PLASSYS.

6 REFERENCES

1. R. Meyer Proceeding IDW97 pp 23-26
2. C.A. Spindt J. Appl. Phys 1968, Vol39 pp3504-3505
3. H.H. Busta in "Vacuum micro-electronics" edited by Wei Zhu John Wiley&Son New York 2001
4. A. M. Rao, D. Jacques, R. C. Haddon, W. Zhu, C. Bower, and S. Jin Appl. Phys. Lett 2000, 76, pp3813-3815
5. M. Yudasaka, R. Kikuchi, Y.Ohki, E. Ota, S. Yoshimura Appl. Phys. Lett 1997, 70 (14), pp1817-1818
6. M. Chhowalla, K. B. K. Teo, C. Ducati, N. L. Rupesinghe, G. A. J. Amaratunga, A. C. Ferrari, D. Roy, J. Robertson, and W. I. Milne , J.A.P. 2001 vol 90 (10), pp5308-5317
7. R.H. Fowler, L. Nordheim Proc. Roy. Soc. London,1928, A119, p173
8. J.M. Bonard, H.Kind, T.Stöckli, L.Nilsson Solid State Elect.45 (2001) pp893-914
9. R.H good, E.W. Muller Field emission, Handbuch der physik 1956, Bd21 pp177-231
10. X.Xu, G.Brandes Appl. Phys. Lett 1999, 85 p 2549
11. A. Ghis, R. Meyer, P. Rambaud, F.Levy, T. Leroux IEEE Trans.On Elec. Dev. 1991,Vol38 N°10, pp2320-2322
12. J.Dijon, A.Fournier, B.Montmayeul, D.Zanghi, B. Coll , J.M.Bonard, A.M. Bonnot Proceedings Eurodisplay October 2002 pp821-824
13. J.Dijon, J.F. Boronat, A.Fournier, D.Sarrasin Proceedings Asia Display IDW01, 2001 pp1205-1208
14. J.D. Levine J. Vac. Sci. Technol. B 13, 553 (1995)
15. L. Nilsson, O. Gröning, C. Emmenegger, O.Küttel, E.Schaller, L. Schlapbach,H. Kind, J.M. Bonard K. Kern, Appl. Phys. Lett. 76, 2071 (2000)
16. O. Gröning, O.M. Küttel,C. Emmenegger, P. Gröning, L. Schlapbach, J. Vac. Sci.Technolol. B 18, 665 (2000)
17. L.O. Nilsson PhD Thesis N°1337 University of Freiburg 2001
18. J.M. Bonard, M. Croci, I. Arfaoui, O. Noury, D. Sarangi, A. Châtelain Dianond and relat. Mat. 11, (2002), pp763-768
19. C. Constancias PhD Thesis UJF Grenoble 19 january 1998
20. J.M.Bonard, N. Weiss,H. Kind, T. Stöckli, L. Forro, K. Kern, A. Châtelain Adv. Mater. 2001, 13 No3, pp184-188

Growth of Aligned Multiwall Carbon Nanotubes and the Effect of Adsorbates on the Field Emission Properties

W.I.Milne[1], K.B.K.Teo[1], S.B.Lansley[1], M.Chhowalla[1], G. A. J. Amaratunga[1], V. Semet[2], Vu Thien Binh[2], G. Pirio[3], and P. Legagneux[3]

[1]Cambridge University Engineering Department, United Kingdom
[2]University of Lyon, France
[3]Thales Research and Technology, France

Abstract. In attempt to decipher the field emission characteristics of multiwall carbon nanotubes (MWCNTs), we have developed a fabrication method based on plasma enhanced chemical vapour deposition (PECVD) to provide utmost control of the nanotube structure such as their alignment, individual position, diameter, length and morphology. We investigated the field emission properties of these nanotubes to elucidate the effect of adsorbates on the nanotubes. Our results show that although the adsorbates cause an apparent lowering of the required turn on voltage/field of the nanotubes, the adsorbates undesirably cause a saturation of the current, large temporal fluctuations in the current, and also a deviation of the emission characteristics from Fowler-Nordheim like emission. The adsorbates are easily removed by extracting an emission current of 1uA per nanotube or using a high applied electric field (~25V/um).

INTRODUCTION

There is considerable interest in the use of carbon nanotubes as field emission electron sources because of their whisker-like shape, high aspect ratio, high conductivity, resistance to electromigration and thermal stability [1]. In the literature today, however, it is often difficult to correlate the field emission I-V characteristics of the nanotubes to the morphology of the nanotubes under study because the nanotubes are often in a random network resembling a forest/grass/spaghetti.

In attempt to decipher the field emission characteristics of carbon nanotubes, this paper reviews the work carried out in our laboratory to develop a fabrication technology to provide ultimate control of the nanotube structure; such as their alignment, individual position, diameter, length and shape. Using controlled arrays, it is possible to study the emission from inidividual nanotubes. In this work, we are particularly interested in the effect of adsorbates on the field emission properties of our nanotubes. Such studies have been carried out on individual single wall nanotube emitters by Dean et al [2]. In this work, we look at individual multiwall nanotube emitters as well as an array of emitters. We show that in general, the same effects of apparent turn on voltage lowering, current saturation and current instability are observed on our arrays. Thus, there is a need to remove the adsorbates and we show

CP685, *Molecular Nanostructures: XVII Int'l. Winterschool/Euroconference on Electronic Properties of Novel Materials,*edited by H. Kuzmany, J. Fink, M. Mehring, and S. Roth
© 2003 American Institute of Physics 0-7354-0154-3/03/$20.00

that this is possible by extracting a current of 1uA per nanotube or using an electric field stress of 25V/um.

Experimental Detail

The multiwalled carbon nanotubes were grown in a dc plasma enhanced chemical deposition system described in detail elsewhere [3]. Acetylene and ammonia gas mixtures were added to the PECVD chamber at temperatures between 550C and 900C and a sputtered nickel thin film was used as the catalyst. Examples of the types of nanotubes grown at 700C are shown in the first figure. In the absence of a plasma, spaghetti like carbon nanotubes are produced (figure 1a). When the dc plasma is generated with different bias voltages, the nanotubes begin to gradually align as shown in figure 1b and 1c, with the highest applied voltage giving the best/straightest alignment. The alignment of the nanotubes is due to the presence of an electric field in the plasma sheath [3,4].

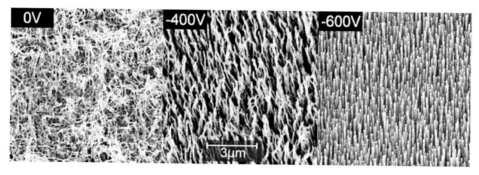

FIGURE 1. Growth of MWCTS with and without Plasma at Substrate Temp of 700C

The "grass like" arrays of fig 1 (b) and (c) look as though they are ideal for field emission applications because of their high aspect ratio and "whisker like" appearance. However as has been reported previously [5], in order to reduce the effects of field screening, the nanotubes have to be placed approximately twice their height apart. In fact, the problem is much worse than this simple analysis would indicate. When one makes a cross section of a random CNT film, one always find long spurious nanotubes which are responsible for the very high beta factors that people often measure from random CNT films. If we now try to make a device by integrating a gate, ideally we want the gate to be as close as possible so we can address the majority of the emitters. This also ensures a low operating voltage, since E=V/d. However the long spurious CNTs often cause short circuits to our gate. Thus an obvious solution would be to place the gate much further away from the film but then only the long spurious CNTs are addressed and we lose the use of the majority of the CNTs. So ideally, what we need is a controlled method of growing the CNTs in order to eliminate these problems. The height of the tubes is controlled by the deposition time and patterned arrays of such tubes can be obtained by e-beam lithographically defining the Ni prior to CNT growth as described in [6]. Examples of such arrays of MWCNTs are shown in figure 2.

FIGURE 2. Ordered array of MWCNTs grown at 700C from e-beam lithgraphically defined 100 nm Ni dots

For numerous field emission applications it will be necessary to produce such arrays over large areas and e-beam lithography is not an effective method for this. Thus alternative technologies are being considered in order to produce such arrays. Laser Interferometry, co-polymer block lithography and nanosphere lithography [7] are all being currently tested.

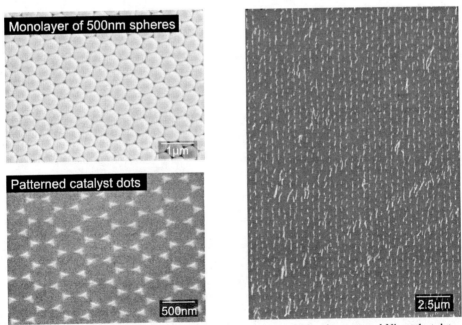

FIGURE 3 SEM image of a monolayer of 500nm polystyrene spheres, the patterned Ni catalyst dots and the arrays of MWCNTS produced at 700°C from the dots

Examples of such arrays produced using 500nm diameter polystyrene nanospheres as the masking layer for subsequent Ni evaporation are shown in figure 3. As can be seen, much work is still required to optimise this process.

Field Emission Results from Individual MWCNTs

Large area measurements on arrays of CNTs provide information on the best emitters whereas information on emission from individual nanotubes can only be obtained by using a scanning anode field emission system as reported by Semet et al [8]. A 100 micron diameter Pt/Ir probe, biased positively, is scanned above an array of individual nanotubes which are spaced 100 microns apart. The positions of the emitters are then located as these would correspond to regions of high emitted current. The probe is then placed directly above an emitter. The typical emitted current as a function of applied voltage (for 3 ramps) is shown in figure 4.

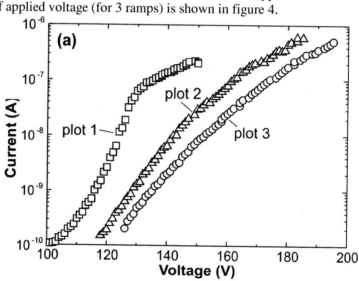

FIGURE 4. Emission Current as Function of Applied Voltage for a single MWCNT

The I-V plot in the initial ramp up (plot 1) always exhibits a saturation effect at high currents. However, the second and third ramp ups of the voltage lead to removal of the saturation effect and the I-V plot becomes reproducible and follows Fowler-Nordheim type behaviour. This is similar to the effects observed by Dean et al [2] in their measurements on SWCNTs. They attributed the behaviour to polar adsorbates and proved it by observing the field emission patterns on a phosphor screen from adsorbate free and adsorbate covered tubes. In an attempt to verify that the effects that our our multiwall tubes also exhibited were due to adsorbates we used a triode FE measurement system with a large anode to grid spacing (~ 1 cm) to ensure that all adsorbed gases can be pumped out from the sample during emission. We tested an array of 100x100 CNTs to provide sufficient signal and ramped up the voltage whilst at the same time sampling gas emissions using a mass spectrometer. At fields of ~15V/um we find that there is an increase in base pressure in the system and the mass spec signal (with the background signal removed), at an applied field of 24V/micron, from this array is shown in the next figure.

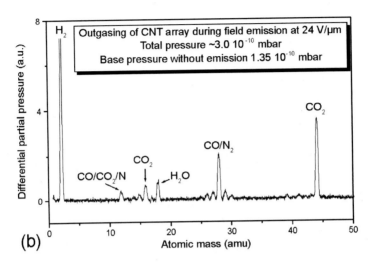

FIGURE 5. Mass Spectrometer Output (minus background signal) measured at 24V/micron field.

As can be seen, as is in Dean's case, we have a significant amount of polar adsorbates in the outgassing spectrum. However, after leaving the field at ~25V/micron for 30 minutes, the outgassing ceases (ie. the nanotubes are cleaned of the adsorbates) and the field emission from this array is also observed to become Fowler-Nordheim like without saturation [9], very similar to the characteristics shown in Figure 4. This shows that with an applied electric field stress, it is also possible to clean the adsorbates from the nanotube emitters.

Lastly, we investigated the stability of the emission current from a single emitter with adsobates and in its clean state. In the as-deposited (ie. with adsorbates) state, there is significant fluctuation in the emission current as shown in figure 6(a). After running a cleaning I-V cycle, the current stability of the 'cleaned' nanotube is much improved as shown in Figure 6(b).

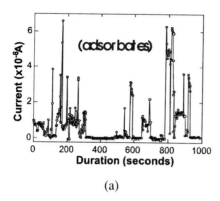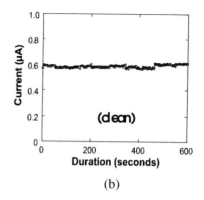

(a)   (b)

FIGURE 6. Emission Current as a function of time (a) with adsorbates and (b) after cleaning [8].

Discussion

The adsorbates are thought to cause a lowering of the effective work function or have resonant tunneling states which give rise to emission at lower fields [2,8]. However, these adsorbates do not support the conduction of large currents and thus cause the observed current saturation. At high current/field, these adsorbates are field-evaporated to yield the clean nanotube which emits with Fowler-Nordheim like behaviour without current saturation. Thus, it is necessary to remove the adsorbates (just by using an I-V cycle or high field/current stress) from nanotube emitters in order for reliable analysis and reproducible, stable measurements to be made.

Conclusions

The growth of controlled arrays of multiwall carbon nanotubes using the PECVD process has been reported. Emission from individual nanotubes has been measured using a scanning probe system. Initial I-V curves exhibit saturation effects but upon "cleaning" with I-V ramps, the nanotubes show Fowler-Nordheim type behaviour. The cleaning process is associated with the removal of adsorbates by means of current (~1uA/emitter) or field (~25V/um). By cleaning the nanotubes with this process, the emission current no longer shows saturation and exhibits stability and Fowler Nordheim like behaviour which is desirable for field emission applications.

ACKNOWLEDGMENTS

This work was funded in part by the EC NANOLITH IST-FET programme and by the EPSRC CBE project.

REFERENCES

1. Purcell, S. T., Vincent, P., Journet, C. and Vu Thien Binh, *Phys. Rev. Lett.*, **88**, 105502 (2002).
2. Dean, K. A. and Chalamala, B. R. *Appl. Phys. Lett.* **76**, 375 (2001); Dean, K. A., von Allmen, P. and Chalamala, B. R., *J. Vac. Sci. Tech. B,* **17**, 1959 (1999).
3. Chhowalla, M., Teo, K.B.K., Ducati, C., Rupesinghe, N.L., Amaratunga, G.A.J., Ferrari, A.C., Roy, D., Robertson, J.and Milne, W.I., *J. Appl. Phys.* **90**, 5308 (2001).
4. Bower, C., Zhu, W., Jin, S. and Zhou, O., *Appl. Phys. Lett.* **77**, 830 (2000).
5. Groening, O., Kuettel, O.M., Emmenegger, Ch., Groening, P. and Schlapbach, L., *J. Vac. Sci. Tech. B*, **18**, 665 (2000).
6. Teo, K.B.K., Chhowalla, M., Amaratunga, G.A.J. , Milne, W.I., Hasko, D.G., Pirio, G., Legagneux, P., Wyczisk, F. and Pribat, D., *Appl. Phys. Lett.*, **79**, 1534 (2001).
7. Huang, Z. P., Carnahan, D. L., Rybczynski, J., Giersig, M., Sennett, M., Wang, D. Z., Wen, J. G., Kempa, K. and Huang, Z. F., *Appl. Phys. Lett.*, **82**, 460 (2003).
8. Semet, V., Thien Binh, Vu, Vincent, P., Guillot, D., Teo, K.B.K., Chhowalla, M., Amaratunga, G. A. J., Milne, W. I., Legagneux, P. and Pribat, D., *Appl. Phys. Lett,.* **81**, 343 (2002).
9. Pirio, G. et al, unpublished results.

Nanowicks: Nanotubes as Tracks for Mass Transfer

B. C. Regan, S. Aloni, B. Huard, A. Fennimore, R. O. Ritchie, A. Zettl

University of California, Berkeley
and
Lawrence Berkeley National Laboratory
Berkeley, CA 94720 U.S.A

Abstract. We have used a manipulation stage to electrically contact individual nanotube bundles coated with metal nanoparticles for *in-situ* studies in a transmission electron microscope. When electrical current is passed through a bundle, unusual mass transport is observed along that bundle. Nanocrystals melt and disappear from a given section, with a correlated growth of similar nanoparticles further along the bundle. This unusual phenomenon, termed *nanowicking*, may provide a method for controlled nanoscale mass transport.

True nanoscale device fabrication presents a formidable challenge, and the tools presently available are quite primitive in comparison to the requirements of an economically useful technology [1]. Fundamental to any construction project is the ability to deliver appropriate quantities of required building material to the assembly site. In this work we describe the observation of directed movement of nanoscale quantities of metals along multiwall carbon nanotube templates or tracks, which we term *nanowicks*. Fully realized, this method of material transport may prove to be useful in many applications, including the formation 'by-design' of electrical and mechanical joints in complex nanostructures.

Our experiments are performed in a Philips CM200 transmission electron microscope (TEM) located at the National Center for Electron Microscopy at Lawrence Berkeley National Laboratory. The sample consists of a boule of multiwalled carbon nanotubes rigidly mounted with silver paint to the sample stage. The boule typically has 25 nm each of indium and tin evaporated onto it, and has been annealed overnight at 120 degrees Celsius. As a result of the annealing, the tubes are covered with nanocrystals of indium/tin alloy with a relatively low (bulk) melting point. From a practical standpoint, this alloy resembles (but is not derived from) commercially available lead-free solder. The sample stage includes a piezo-driven XYZ nanomanipulator. Mounted on the manipulator and opposing the sample is a conducting tip which can be driven to approach a nanotube of our choosing. After a successful approach, electrical contact is made between the tip and sample. Driving a current through the nanotube introduces Joule heating (presumably concentrated at the resistive tip/sample contact). One application is to use the local heating effect together with the locally available indium/tin to solder the nanotube to the tip, forming a

CP685, *Molecular Nanostructures: XVII Int'l. Winterschool/Euroconference on Electronic Properties of Novel Materials,* edited by H. Kuzmany, J. Fink, M. Mehring, and S. Roth
© 2003 American Institute of Physics 0-7354-0154-3/03/$20.00

reliable mechanical and electrical junction. As we demonstrate below, however, a much more interesting phenomenon is also observed.

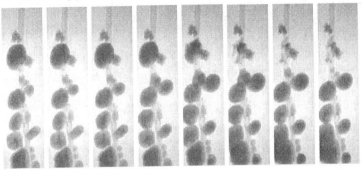

FIGURE 1. TEM image time series (time advances to the right). The field of view is approximately 360 × 1400 nm.

Figure 1 shows a time series of TEM video images taken from one of our experiments. A multiwalled carbon nanotube bundle has been approached from outside the field of view past the top edge of the images. The number of tubes in the bundle varies with position, but we estimate that there are about 6 tubes with characteristic diameters of 20 nm here. During this sequence, which spans about a minute, approximately 10 microamperes are passing through the nanotube bundle, and the total circuit resistance is ~ 30 kΩ. As time passes the mass on the bundle is observed to redistribute, as some globules shrink and others grow. Most striking is the near complete disappearance of a globule near the top of the images, and with corresponding enlargement of the globules directly beneath it. This behavior is repeated multiple times along the bundle. With a given power the metal will redistribute itself over a timescale of about a minute. After the movement has ceased, the applied voltage is increased and the transport is observed to continue from a new location further down the bundle. For this experiment the power dissipated varied from 5 to 200 μW.

We make four immediate observations from this experiment. First, mass always moves away from the tip-sample contact. Second, mass only appears where there is an obvious nucleation site, i.e. a preexisting crystallite. Third, although the TEM can not accurately measure mass or volume, it appears that mass is not lost. Using volume (derived from projected area, assuming spheres) and opacity as test metrics, we find that total nanoparticle mass is approximately conserved. At a minimum 80% of the mass that disappears from one site reappears elsewhere in our field of view. This observation implies directly that the mass is not moving around as vapor, nor even moving symmetrically up and down. Finally, the interesting movement is always localized in one place, not all along the bundle. Taking this transport process to be thermally activated, we conclude that there is a temperature gradient along the nanotube bundle. The most likely thermal source is localized heating at the contact point between the tip and tubes. Any Joule heating along the tube seems to be negligible in comparison.

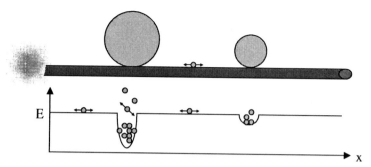

FIGURE 2. Toy model of the mass transport process.

The movement of molten metal globules immediately suggests the Young-Laplace equation as applied to two fluid spheres connected by a channel. The energy E associated with each sphere has a volume term and a surface term:

$$E \sim -aR^3 + bR^2 , \qquad (1)$$

where R is the radius of the sphere, and a and b are constants. This equation implies that such connected spheres, if similarly-sized, are not in mechanical equilibrium with respect to size changes. In order to minimize the total energy the smaller bubble will tend to shrink and the larger one will grow. Extending this argument gives the equation of Young and Laplace, which relates the pressure differential to surface tension and bubble radius: $\Delta P = 2\gamma/R$, where γ is the surface tension and R is the bubble radius. However, this mechanism does not agree even qualitatively with our observations. We see large globules shrink and small ones grow. In our experiments, the critical parameter appears to be position (i.e. local temperature), not globule radius.

The toy model shown in Fig. 2 thus presents itself. A nanotube bundle is locally heated on the left, with two representative liquid metal spheres distributed along its length. Each sphere, or nucleation site, represents a local minimum in the atomic potential. However, the minima are not always deep compared to kT. In the steady state, which obtains quickly, and without distributed heating, a linear temperature profile is expected. At some point along the bundle kT is comparable to the depth of the local minimum, and atoms are excited out of their bubble onto (or perhaps into the troughs between) the carbon nanotubes. There metal constituents randomly walk until they find the next spot where kT is small enough that they can condense.

Surprisingly, this primitive model can be used to extract an estimate for the thermal conductivity of our nanotube bundle. According to theoretical predictions, carbon nanotubes ought to be excellent thermal conductors, and much experimental work has been dedicated to confirming this technologically important property. While we do not know the details of the condensation potential, we can estimate that thermal energy differential $k\Delta T$ between a shrinking sphere and a growing one must be of the same order as the Young-Laplace energy $P\Delta V$ per atom. Furthermore, it follows from the heat equation that the temperature gradient $\Delta T/\Delta x$ is the heat flux Q divided by the thermal conductivity of the bundle κ. Thus, by combining our experimental

parameters (power dissipation $\sim 30~\mu W$, bubble separation $\Delta x \sim 100$ nm, $R \sim 100$ nm, etc.) with the known characteristics of tin-indium alloy (density $\rho \sim 7$ g/cm^3, surface tension $\gamma \sim 500$ dyne/cm), we arrive at an estimate for κ. We find $\kappa \sim 100$ W/K m, which compares well with other values available in the literature on multiwalled carbon nanotube bundles [2-4]. Such a thermal conductivity implies a temperature differential of about 10 K between a site losing mass and one gaining mass, which is physically reasonable.

In conclusion, we report the discovery of nanowicks. Local temperature gradients, produced in our case by Joule heating, can be used to move or remove metal adsorbates on multiwalled carbon nanotube bundles. Possible applications of this process include the construction of electrical and mechanical joints in more sophisticated nanotube devices. Furthermore, we see that these bundles can support 'substantial' temperature gradients. By 'substantial' we mean that a temperature-driven transition between positive and negative mass divergence occurs over length scales of 100 nm or less. Finally, a toy model has been developed that provides an estimate of the bundle thermal conductivity in rough agreement with published values.

ACKNOWLEDGMENTS

This research was supported in part by the National Science Foundation and by the Office of Science of the U.S. Department of Energy under the sp^2 and Interfacing Nanostructures Initiatives.

REFERENCES

1. R.Compañó ed., *Technology Roadmap for Nanoelectronics*, European Commission IST Programme on Future and Emerging Technologies (2000).
2. W. Yi, L. Lu, Zhang Dian-lin, Z. W. Pan, and S. S. Xie, Phys. Rev. B **59**, R9015 (1999).
3. P. Kim, L. Shi, A. Majumdar, and P. L. McEuen, Phys. Rev. Lett. **87**, 215502 (2001).
4. Da Jiang Yang, Qing Zhang, George Chen, S. F. Yoon, J. Ahn, S. G. Wang, Q. Zhou, Q. Wang, and J. Q. Li, Phys. Rev. B **66**, 165440 (2002).

Interaction of Single-Wall Carbon Nanotubes with Gas Phase Molecules

L. Petaccia*, A. Goldoni*, S. Lizzit*, A. Laurita[§], and R. Larciprete*[¶]

* Sincrotrone Trieste, S.S. 14 Km 163.5, I-34012 Trieste, Italy
[§] Centro Interdipartimentale di Microscopia, Università degli Studi della Basilicata, Potenza, Italy
[¶] CNR-IMIP, Zona Industriale, I-85050 Tito Scalo-Potenza, Italy

Abstract. The interaction of single-wall carbon nanotubes (SWCNTs) with various gas molecules has been studied by synchrotron radiation photoemission spectroscopy. Using also x-ray emission microscopy, we found that chemical elements other than C are present in commercial purified SWCNT *bucky paper*. Removal of these residual contaminants made the electronic spectra insensitive to O_2, H_2O, N_2, and CO, while a strong and reversible effect to NO_2, SO_2, NO and NH_3 was observed.

INTRODUCTION

Carbon nanotubes have become increasingly important during the last years both for their physical and chemical properties and for their potential applications [1]. Recently, experimental investigations have shown that the transport and electronic properties of single-wall carbon nanotubes (SWCNTs) might severely change upon exposure to gas molecules such as O_2, NH_3 and NO_2 [2,3]. This has important consequences in practical applications of carbon nanotubes as gas sensors, but it also suggests that the atmospheric environment may strongly influence the performances of the novel carbon-based electronic devices. Gas molecules could affect directly the SWCNT properties by binding to the tubes, but could have also an indirect effect by interacting with donors and/or acceptor centers already bounded to the tubes and due to contaminants. In fact, up to now the purity of SWCNT samples is limited by the extensive use of metal catalysts in the growth process and of chemicals during tube purification and dispersion. The last process requires surfactants to entangle the bundles and isolate the tubes. Unfortunately, in spite of accurate protocols for chemicals elimination, traces remain in the sample as undesired and often unrevealed contaminants, which might interfere with the nanotube properties and simulate exotic and unpredictable behaviors.

Using x-ray emission microscopy and high-resolution photoemission spectroscopy, we investigated the presence of contaminants in commercial purified SWCNTs in the form of *bucky paper*, to determine their cleaning procedure in ultra high vacuum (UHV) conditions and to study how impurities may influence the interaction between SWCNTs and gas phase molecules. Several contaminants have been identified in the pristine *bucky paper*. They were fully removed after prolonged (> 2 h) annealing of the sample in UHV at 1250 K. Indeed, some S and Si impurities remain in the *bucky*

CP685, *Molecular Nanostructures: XVII Int'l. Winterschool/Euroconference on Electronic Properties of Novel Materials*, edited by H. Kuzmany, J. Fink, M. Mehring, and S. Roth
© 2003 American Institute of Physics 0-7354-0154-3/03/$20.00

paper after this cleaning procedure, but they form just localized clusters that do not interact with the SWCNTs and do not interfere with their properties [4]. The exposure of this clean SWCNT *bucky paper* to O_2, H_2O, CO and N_2 does not modify the C 1s photoemission spectra, while strong changes take place when NO_2, SO_2, NO and NH_3 are adsorbed. This confirms the possible application of single-wall carbon nanotubes as high sensitivity chemical sensors of these toxic gases [3].

PRISTINE BUCKY PAPER

The sample used in our experiment was a commercial *bucky paper* (Carbolex, purity > 90% vol.) made of SWCNTs. According to the producer, the tubes were grown by electric arc technique using Ni catalyst particles and, after acid treatments for purification, dispersed in a solution containing sodium lauryl sulfate as surfactant. An example of this purification procedure is described in ref. [5] for their sample #3.

The pristine (as-received) *bucky paper* was examined by microscopy analysis using a XL30 ESEM PHILIPS FEI equipped with an energy dispersion detector for x-ray microanalysis. Scanning electron microscopy images of the sample show CNT bundles sticking out from the edges of the sample. However, the x-ray emission spectra (not reported) taken in different points of the sample attest the presence of chemical elements other than C, that is O, Ni, Na, Si, S, and Ca, inhomogeneously dispersed in the sample. Fig. 1 shows the chemical maps taken in correspondence of the Si Kα (1740 eV), Na Kα (1041 eV), and Ni Kα (7478 eV) x-ray emission peaks, that visualize the lateral distribution of these contaminants in the *bucky paper*.

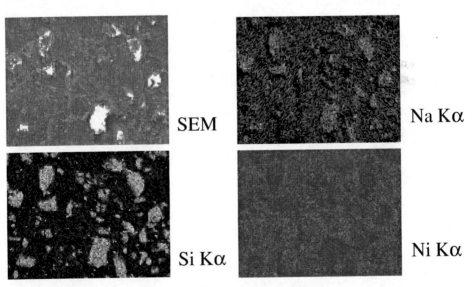

FIGURE 1. $26 \times 20 \ \mu m^2$ low resolution SEM image of SWCNT *bucky paper* and lateral distribution of contaminants in the sample as obtained by x-ray microanalysis. The chemical maps are measured on the Si Kα (1740 eV), Na Kα (1041 eV), and Ni Kα (7478 eV) x-ray emission peaks.

The area chosen for this analysis was among the most contaminated in the sample. The morphology of the sample area, reported in the upper-left image, indicates the presence of contaminating fragments, which according to the Si Kα map turned out to be made of glass. This micro-sized particles, probably coming from ultrasonication treatments in glass beakers during nanotube purification, are also the source of O and Ca contamination. A different distribution in the sample is observed for Ni, whose map shows an inverted contrast with respect to the Si one, attesting the localization of the Ni catalyst particles only in the nanotubes. Instead, Na is detected both inside and outside the glass particles, which demonstrates that Na contamination resides in the NT bundles as well. Although Na (and S) are commonly found in commercial glass, we attribute their presence in the nanotube material to residuals of the chemicals used during purification and dispersion, particularly to sodium hydroxide and sodium lauryl sulfate baths. Of course, it is indispensable to remove the observed contamination before studying the interaction of nanotubes with gas phase molecules.

CLEANING IN UHV BY THERMAL ANNEALING

The SWCNT sample was inserted in the UHV experimental chamber (base pressure 10^{-10} mbar) of the SuperESCA beamline at the ELETTRA synchrotron facility in Trieste, Italy. The sample, mounted on a manipulator that allows cooling to 150 K and annealing up to 1800 K, was examined by photoemission spectroscopy with an overall energy resolution of 100 meV. A first treatment to remove the contaminants consisted in flashing the *bucky paper* at increasing temperature up to 1800 K. This determined the complete desorption of Ni from the sample, and the removal of O and Na from the near surface region as seen by photoemission spectroscopy. Prolonged exposure to the x-ray beam caused the appearance of the O 1s and Na 2s peaks, due to diffusion of these contaminants from the deep layers of the sample to the surface [6]. Complete removal of these species was obtained only after heating the sample at 1250 K for at least two hours in a pressure better than 10^{-8} mbar. S and Si, instead, could not be completely eliminated from the sample. However, after this thermal treatment such impurities do not affect the electronic properties of SWCNTs since they are not bonded to the carbon atoms but form solely isolated clusters [4]. This sample was then considered to be clean. The presence of the SWCNTs in the *bucky paper* after these annealing cycles was checked *in-situ* by electron energy loss spectroscopy and *ex-situ* by Raman spectroscopy [4].

ADSORPTION OF GAS PHASE MOLECULES

We investigated the interaction of SWCNTs cooled at 150 K with gas phase molecules like O_2, H_2O, N_2, CO, NO, NO_2, SO_2, and NH_3.

In the case of O_2, H_2O, N_2, and CO no effects on the C 1s core level photoemission features were observed using a gas partial pressure of the order of 10^{-6} mbar and exposures of several hundreds of Langmuir ($1L=1\times10^{-6}$ Torr s). In Figure 2a we report the C 1s core level of the clean *bucky paper* before and after the exposure to 10^5 L of

O_2. The carbon spectrum remains absolutely the same and no emission in correspondence of the O 1s core level peak was observed after exposure to oxygen. The same happened for the exposure to N_2 and CO [6,7]. In the case of H_2O obviously ice formed at 150 K as also witnessed by the presence of a clear peak in the O 1s photoemission region. However, as shown in Fig. 2b, the C 1s core level remained essentially unaffected. This was also observed after air exposure at atmospheric pressure at both room temperature and about 150 K. These observations indicate that all these gases interact with clean SWCNTs via Van der Waals forces only and do not affect the electronic spectra and properties of nanotubes. In the particular case of oxygen, this observation contrasts with the experiment of Collins and coworkers [2] and with some theoretical results [8], but agrees with more recent experiments [9] and calculations [10]. The possible explanation is that the observed strong sensitivity of SWCNT properties to O_2 reported in ref. [2] was induced by the presence of contaminants, as we observed in oxygen up-takes on the *bucky paper* before the complete removal of Na-containing molecules [6,7]. A further possibility is that oxygen may be chemisorbed at defects sites and open tube caps, which are strongly reduced after the UHV high-temperature annealing [5].

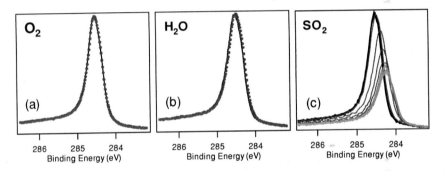

FIGURE 2. C 1s core level photoemission spectra of the clean SWCNT *bucky paper* at 150 K measured: (a) before (solid line) and after (dots) exposure to 100 KL of oxygen; (b) before (solid line) and after (dots) exposure to 1 KL of water; (c) while exposing the sample to sulfur dioxide up to 970 L.

On the contrary, clear changes in the carbon photoemission features were observed by exposing the clean SWCNT sample to NO, NO_2, SO_2, and NH_3. Figure 2c shows the C 1s core level measured in real-time with fast-photoemission (40 s *per* spectrum) during the exposure to SO_2. The C 1s peak changes shape and shifts toward lower binding energy as the exposure time increases. The same happened for the exposure to nitrogen oxides, with comparable binding energy shift. In the case of NH_3 the shift was smaller and in the opposite direction, i.e. towards higher binding energies. If these shifts are ascribed to the nanotube chemical potential, the NO, NO_2 and SO_2 molecules bound to SWCNTs act as charge acceptors, while NH_3 acts as a charge donor. Finally, we note that the changes in the photoemission spectra of SWCNTs exposed to NO, NO_2, SO_2, and NH_3 are completely reversible by annealing the sample above 800 K in UHV. These results reproduce closely those reported in ref. [3] for SWCNTs exposed to ammonia and nitrogen dioxide.

CONCLUSIONS

Our experimental results indicate that chemical processing necessary to purify SWCNTs leads to sample contamination which is not trivial to remove. Prolonged annealing in UHV at high temperatures is necessary to obtain clean samples. The electronic structure as reflected in the photoemission spectra of clean SWCNT *bucky papers* is not influenced by the exposure to molecular O_2, H_2O, N_2, CO and air. However, the interaction between SWCNTs and O_2 may be strongly altered by the presence of residual contaminants and defects. Instead, a significant and reversible effect is observed in the case of SO_2, NO, NO_2 and NH_3 absorption on clean *bucky paper*.

These results highlight the extreme importance of using sensitive diagnostics to state the purity of samples treated with chemical processing, especially when the aim is the investigation of the intrinsic nanotube properties. Moreover, they confirm that SWCNTs could find use as chemical gas sensors of toxic molecules, and strongly suggest that gas molecules present in air weakly interact with SWCNTs.

ACKNOWLEDGMENTS

We are grateful to M. Barnaba for his useful technical assistance.

REFERENCES

1. Baughman, R. H., Zakhidov, A. A., and de Heer, W. A., *Science* **297**, 787-792 (2002).
2. Collins, P. G., Bradley, K., Ishigami, M., and Zettl, A., *Science* **287**, 1801-1804 (2000).
3. Kong, J., Franklin, N.R., Zhou, C., Chapline, M.G., Peng, S., Cho, K., and Dai, H., *Science* **287**, 622-625 (2000).
4. Larciprete, R., *et al.*, unpublished.
5. Monthioux, M., Smith, B. W., Burteaux, B., Claye, A., Fischer, J. E., Luzzi, D. E., *Carbon* **39**, 1251-1272 (2001).
6. Goldoni, A., Larciprete, R., Petaccia, L., and Lizzit, S., unpublished.
7. Larciprete, R., Goldoni, A., and Lizzit, S., *Nuclear. Instrum. and Methods B* **200**, 5-10 (2003).
8. Jhi, S. –H., Louise, S. G., Cohen, M. L., *Phys. Rev. Lett.* **85**, 1710-1713 (2000).
9. Ulbricht, H., Moos, G., and Hertel, T., *Phys. Rev. B* **66**, 075404 (2002).
10. Giannozzi, P., Car, R., Scoles, G., *J. Chem. Phys.* **118**, 1003-1006 (2003).

Surface Modification of Aligned Carbon Nanotubes

Liming Dai* and Ajeeta Patil

Department of Polymer Engineering
College of Polymer Science and Polymer Engineering
The University of Akron, Akron, OH 44304-2909, USA

Richard A. Vaia

Air Force Research Laboratory
Materials and Manufacturing Directorate
AFRL/MLBP, Bldg 654, 2941 P St. Wright-Patterson AFB, OH 45433-7750

Abstract. The excellent optoelectronic, mechanical, and thermal properties of carbon nanotubes have made them very attractive for a wide range of potential applications. However, many applications require the growth of aligned carbon nanotubes with desirable surface characteristics. We have developed a simple pyrolytic method for large-scale production of aligned carbon nanotube arrays *perpendicular* to the substrate. These aligned carbon nanotube arrays can be transferred onto various substrates of particular interest (*e.g.* conducting substrates for optoelectronic devices) in either a patterned or non-patterned fashion. The well-aligned structure further enables us to prepare aligned coaxial nanowires by electrochemically depositing a concentric layer of an appropriate conducting polymer onto the individual aligned carbon nanotubes and to develop a facile approach for modification of carbon nanotube surfaces via plasma activation, followed by chemical reactions characteristic of the plasma-induced functionalities. These surface modification methods are particularly attractive, as they allow surface characteristics of the aligned carbon nanotubes to be tuned for specific applications while their aligned structure can be largely retained. The surface-modified aligned carbon nanotubes are of great significance to various practical applications, ranging from highly efficient electron emitters to highly sensitive sensors.

1. INTRODUCTION

The recent discovery of carbon nanotubes [1] has opened up a new era in material science and nanotechnology. Carbon nanotubes consist of carbon hexagons arranged in a concentric manner. They may be viewed as a graphite sheet that is rolled up into a nanoscale tube form (single-walled carbon nanotubes, SWNTs) or with additional

CP685, *Molecular Nanostructures: XVII Int'l. Winterschool/Euroconference on Electronic Properties of Novel Materials,* edited by H. Kuzmany, J. Fink, M. Mehring, and S. Roth
© 2003 American Institute of Physics 0-7354-0154-3/03/$20.00

graphene tubes around the core of a SWNT to form the so-called multi-walled carbon nanotubes (MWNTs) [2-4]. Usually being capped by fullerene-like structures containing pentagons, these elongated nanotubes have a diameter ranging from a few Ångstrom to tens nanometers and a length of up to several centimeters. Because the graphene sheet can be rolled up with varying degrees of twist along its length, carbon nanotubes have a variety of chiral structures [2-4]. Depending on their diameter and the helicity of the arrangement of graphite rings, carbon nanotubes can exhibit semiconducting or metallic behavior. Dissimilar carbon nanotubes may even be joined together to form molecular wires with interesting electronic, magnetic, nonlinear optical, and mechanical properties [2-7]. As a result, carbon nanotubes are very attractive for many potential applications. Examples include the use of them as new materials for electron field emitters in panel displays [8], single-molecular transistors [9], scanning probe microscope tips (see, for example: [10-13]), gas and electrochemical energy storage [14], catalyst and protein/DNA supports [15, 16], molecular-filtration membranes [17], and artificial muscles [18].

For most of the above-mentioned, and many other, applications, it is highly desirable to prepare aligned/micropatterned carbon nanotubes and to modify their surfaces. Although a large number of solution methods have recently been reported to modify nanotube tips and/or sidewalls through either covalent or non-covalent chemistries [19-21], the simple application of the solution chemistry to the aligned carbon nanotubes could cause a detrimental damage to the alignment structure. To overcome this difficulty, we have developed approaches for chemical modification of aligned carbon nanotubes whilst largely retaining the aligned structure. In particular, we have prepared large-scale aligned carbon nanotube arrays *perpendicular* to the substrate surface by pyrolysis of iron (II) phthalocyanine onto the pristine quartz glass plates [22]. Subsequently, we have developed microfabrication methods for patterning the aligned carbon nanotubes with a sub-micrometer resolution and for patterned/non-patterned transferring such nanotube arrays to various other substrates of particular interest (*e.g.* metal substrates for electrochemistry) [22-26]. Based on these aligned/micropatterned carbon nanotubes, we have also developed several very effective surface modification methods through the electrochemical deposition of a concentric layer of an appropriate conducting (nonconducting) polymer onto the individual aligned carbon nanotubes [27] or by plasma activation of the nanotube surface followed by chemical grafting of functional polymer chains onto the plasma-induced surface groups [28].

Polymer-nanotube hybrid composites constitute a new class of nanomaterials, which could show properties characteristic of both constituent components with potential synergetic effects. For instance, nonaligned carbon nanotubes have been used as electrodes in rectifying heterojunctions with conjugated polymers, showing a significantly reduced onset voltage for nonlinear current injection due to an enhancement of the local field at the tip of the nanotubes [29]. Likewise, light-emitting diodes (LEDs) based on conjugated polymer/carbon nanotube composites have been shown to exhibit lower current densities and better thermal stabilities than the corresponding pure polymer devices, as carbon nanotubes enhance the conductivity and act as nanometric heat sinks [30]. The use of *aligned/micropatterned* carbon nanotubes for making aligned polymer-carbon nanotube coaxial nanowires [27, 28] should provide additional advantages. In this article, we summarize our recent work on microfabrication and chemical modification of aligned carbon nanotubes, along with some potential applications for the modified aligned carbon nanotube arrays.

2. SYNTHESES AND MICROPATTERNING OF ALIGNED CARBON NANOTUBES

Although carbon nanotubes synthesized by most of the common techniques, such as arc discharge and catalytic pyrolysis often exist in a randomly entangled state (Figure 1a), aligned carbon nanotubes have been prepared either by post-synthesis fabrication or by synthesis-induced alignment [2-5, 31]. In this regard, we have prepared aligned carbon nanotubes by pyrolyzing FePc, which has been described in details elsewhere [22, 32]. Without repetition of detailed discussions on the synthesis and structural characterization, a typical SEM image of the FePc-generated aligned carbon nanotube array after having been transferred onto a gold foil is shown in Figure 1b. As can been seen, the constituent nanotubes in the perpendicularly-aligned carbon nanotube array have a fairly uniform length (20 μm) [33].

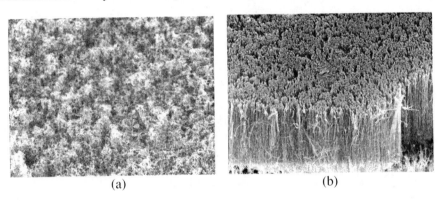

(a) (b)

Figure 1. (a) Typical SEM images of (a) a non-aligned carbon nanotube sample and (b) aligned carbon nanotube arrays prepared by pyrolysis of FePc.

On our further investigation of the aligned carbon nanotubes produced by the pyrolysis of FePc, we, among others [34-38], have recently developed an effective method for photolithographic generation of the perpendicularly aligned carbon nanotube arrays with resolutions down to a micrometer scale [23]. Our method allows not only the preparation of micropatterns and substrate-free films of the perpendicularly-aligned nanotubes but also their transfer onto various substrates, including those which would otherwise not be suitable for nanotube growth at high temperatures (*e.g.* polymer films) [22]. Figure 2a shows the steps of the photolithographic process. In practice, we first photolithographically patterned a positive photoresist film of diazonaphthoquinone (DNQ)-modified cresol novolak (Scheme 1a) onto a quartz substrate. Upon UV irradiation through a photomask, the DNQ-novolak photoresist film in the exposed regions was rendered soluble in an aqueous solution of sodium hydroxide due to photogeneration of the hydrophilic indenecarboxylic acid groups from the hydrophobic DNQ via a photochemical Wolff rearrangement [39] (Scheme 1b). We then carried out the pyrolysis of FePc, leading to region-specific growth of the aligned carbon nanotubes in the UV exposed regions (Figure 2b). In this case, the photolithographically patterned photoresist film, after an appropriate carbonization process, acts as a shadow mask for the patterned growth of the aligned nanotubes. This method is fully compatible with existing photolithographic processes [40, 41].

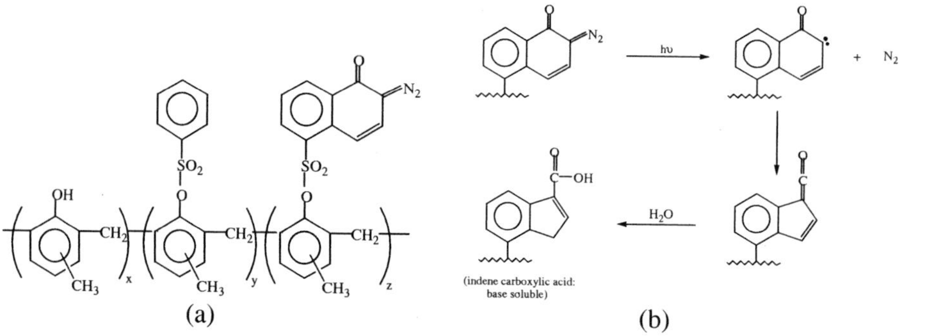

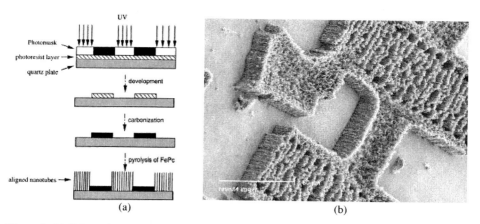

(a)

(b)

Scheme 1 (a) Molecular structure of the DNQ-Novolak photoresist and (b) photochemical reactions of the DNQ-Novolak phoresist [23].

(a)

(b)

Figure 2. (a) Schematic representation of the micropattern formation of aligned carbon nanotubes by photolithographic process. (b) Typical SEM micrographs of patterned films of aligned nanotubes prepared by the pyrolysis of FePc onto a photolithographically prepatterned quartz substrate.

624

Subsequently, we have also used the micro-contact printing (μCP) and micro-molding techniques to prepare micropatterns of carbon nanotubes aligned in a direction normal to the substrate surface [24]. In both cases, a thin layer of polymer film was region-specifically transferred onto a quartz substrate, followed by carbonization of the polymer patterns into carbon black for region-specific growth of the aligned nanotubes in the polymer-free regions by pyrolysis of iron(II) phthalocyanine (FePc) under Ar/H$_2$ atmosphere at 800-1100°C, as is the case for the above-mentioned photolithographic patterning.

As can be seen from above discussion, both the photolithographic and soft-lithographic patterning methods involve a tedious carbonization process prior to the aligned nanotube growth. In order to eliminate the carbonization step, we have also developed a versatile plasma method for making patterns of aligned carbon nanotubes [25]. In this case, we first polymerized certain plasma polymers in a region-specific fashion using a TEM grid as the physical mask. The highly-crosslinked structure of plasma-polymer films [42, 43] could ensure the integrity of the plasma polymer layer to be maintained, even without carbonization, at high temperatures necessary for the nanotube growth from FePc [25]. Therefore, the carbonization processes involved in our previous works on photolithographic [24] and soft-lithographic [25] patterning of the aligned carbon nanotubes can be completely eliminated in the plasma patterning process. Owing to the generic nature characteristic of the plasma polymerization, many other organic vapors could also be used equally well to generate plasma polymer patterns for the patterned growth of the aligned carbon nanotubes. These lithographic/plasma-patterning methods, coupled with the ease with which polymer chains can be chemically and/or electrochemically attached onto the aligned carbon nanotube wall (*vide infra*), facilitated the formation of aligned/micropatterned polymer-nanotube coaxial nanowires for constructing various on-tube optoelectronic devices and sensor arrays, as to be discussed below.

3. SURFACE MODIFICATION OF ALIGNED CARBON NANOTUBES

3.1 Plasma-Coating of Aligned Carbon Nanotubes

The well-aligned structure will not only make the aligned/micropatterned carbon nanotubes very attractive for many potential applications (*e.g.* as electron field emitters in flat panel displays, membranes for molecular separation) but also allow the chemical modification of individual aligned carbon nanotubes without any destruction of the alignment. Unlike most of the solution chemistries that often cause the opening of the nanotube tips and/or detrimental damage to their sidewalls, plasma methods are particularly attractive for the chemical modification of aligned carbon nanotubes. In this context, we have carried out the plasma polymerization on the aligned carbon nanotubes. Figure 3a shows a typical SEM micrograph for the aligned carbon nanotubes *as-synthesized* on a Si wafer or those transferred onto a Scotch tape from the quartz surface [28]. By plasma polymerization (*e.g.* acetaldehyde), a concentric layer of (acetaldehyde) plasma polymer film was homogeneously deposited onto each of the constituent aligned carbon nanotubes (Figure 3b). The SEM image for the plasma-polymer coated nanotubes given in Figure 3b shows the similar features as the aligned nanotube array of Figure 3a, but with a larger tubular diameter and smaller inter-tube distance due to the presence of the plasma coating. The TEM images of the constituent nanotube before and after the plasma treatment are given in the insets of Figures 3a&b, respectively. Comparing the inset of Figure 3b with its counterpart in Figure 3a clearly shows the presence of a homogenous plasma coating along the nanotube sidewall. The coating thickness was determined from the TEM image shown in the inset of Figure 3b to be 20-30 nm for this particular sample, but it can be varied in a controllable fashion by changing the plasma polymerization conditions (*e.g.* treatment time).

(a)   (b)

Figure 3. SEM micrographs of the aligned carbon nanotubes (a) before and (b) after the plasma polymerization of acetaldehyde. The insets show TEM images of an individual nanotube (a) before and (b) after being coated with a layer of the acetaldehyde-plasma-polymer. Note that the micrographs shown in (a) and (b) were not taken from the same spot due to technical difficulties [28].

To characterize the plasma coating, we carried out XPS spectroscopic and air/water contact angle measurements. Figures 4a&b show the XPS survey spectra of the aligned carbon nanotube film before and after the plasma polymerization of acetaldehyde, respectively. As can be seen, Figure 4a shows the expected peak of the graphitic C at 284.7 eV, along with weak signals for N 1s (399 eV), O 1s (531 eV), and Fe 2p (708 eV). The low atomic ratio of O/C = 0.016 deduced from the survey spectrum could indicate an incorporation of a trace amount of oxygen into the nanotube structure. However, the possibility with physically adsorbed oxygen cannot be ruled out since carbon nanotubes are known to be susceptible to oxygen adsorption even at pressures as low as 10^{-8} to 10^{-10} Torr [44], typical for the XPS measurements.

In fact, the C 1s spectrum shown in the inset of Figure 4a is very similar to that of graphite (HOPG) [45], suggesting that the nanotube structure is free from oxygen. On the other hand, previous studies have demonstrated the presence of N in nanotubes prepared by pyrolysis of N-containing molecules (*e.g.* NiPc) [46]. Figure 4b shows a large increase in the atomic ratio of O/C from 0.016 to 0.250. The absence of a signal from N in Figure 4b indicates that the acetaldehyde plasma polymer coating is pinhole-free and thicker than the XPS probe depth (*i.e.* > 10 nm) [47]. The corresponding high-resolution C 1s spectra given in the insets of Figures 4a&b show a significant increase in intensity at 288.0 eV attributable to –C=O groups upon the plasma polymerization of acetaldehyde, clearly indicating the introduction of *aldehyde* surface groups. An important check provided by the XPS C 1s spectra is that the amount of carboxylic groups (at 289 to 289.5 eV), if any, in the plasma layer is very low, which is consistent with previous findings under optimized conditions [48]. The plasma-induced aldehyde surface groups could be used for the surface immobilization of amino-dextran chains onto the aligned carbon nanotubes, as we shall see later.

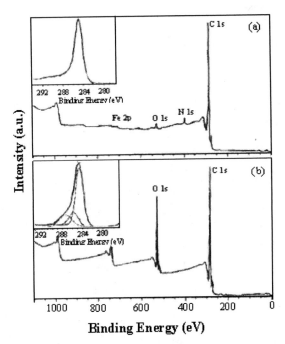

Figure 4. XPS survey and C 1s (inset) spectra of (a) an untreated carbon nanotube film on a Si wafer, (b) after the acetaldehyde plasma polymerization [28].

Recently, we have carried out the investigation of the effect of plasma coating on the electron emission performance of the aligned carbon nanotubes.

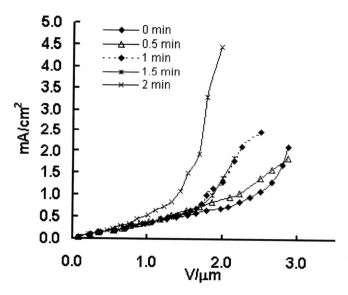

Figure 5 *I-E* plots of the gold-supported aligned carbon nanotubes before and after an *n*-hexane plasma treatment at 250 kHz, 30 W, and monomer pressure 0.65 Torr for different treatment time.

Figure 5 shows changes in the *I-E* curves for the gold-supported aligned carbon nanotubes upon hexane plasma treatment. As can be seen in Figure 5, hexane-plasma coating reduces the turn-on electric field E_{to} [49], coupled with a concomitant increase in the emission current at a constant *V*. We believe that the enhanced field emission by the hexane-plasma treatment was originated from the electrical insulating of the nanotubes along their length, leading to an enhanced field focusing at the nanotube tips.

3.2. Grafting Polymers onto The Plasma-Activated Aligned Carbon Nanotubes

On the basis of the plasma polymerization discussed above, we have developed approaches for chemical modification of aligned carbon nanotubes by carrying out radio-frequency glow-discharge plasma treatment and subsequent reactions characteristic of the plasma-induced surface groups [28]. For instance, we have successfully immobilized NH_2-containing polysaccharide chains onto the acetaldehyde plasma activated carbon nanotubes through the Schiff-base formation, followed by reductive stabilization of the Schiff-base linkage with sodium cyanoborohydride (Scheme 2).

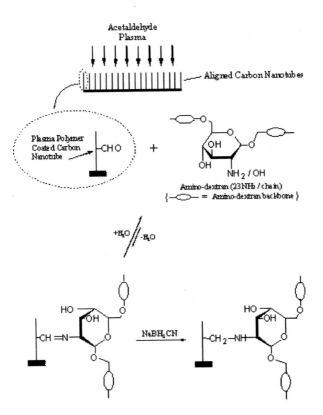

Scheme 2 Grafting polysaccharide chains onto the plasma activated aligned carbon nanotubes through the Schiff-base formation, followed by reductive stabilization of the Schiff-base linkage with sodium cyanoborohydride [28].

Figure 6 shows the XPS survey spectrum of the amino-dextran-immobilized carbon nanotubes. Also included in the inset of Figure 6 is the corresponding XPS C 1s spectrum. Comparing with Figure 4b, Figure 6 shows an increase in the O/C ratio from 0.250 to 0.367. This, together with the reappearance of the N signal in the survey spectrum of Figure 6, clearly indicates the presence of the oxygen-rich amino-dextran coating. Numerical results from the curve-fitted XPS C 1s spectra in Figures 5&6 indicated that the intended grafting reaction was manifested by an increase in the percentage content of C-O from 17.23 to 33.16%, together with the concomitant decrease of the corresponding content for aldehyde group (*i.e.* -C=O) from 16.30 to 9.65%, upon grafting with the amino-dextran chains [28]. The intended covalent interfacial bonding was further tested by autoclaving, a process that has been proven a specific test for distinguishing the reduced Schiff-base linkage from physical adsorption [50]. The XPS measurements on amino-dextran treated nanotubes with insufficient or no NaBH$_3$CN reduction showed part or complete loss of the polysaccharide chains upon autoclaving in water at 121°C for 20 min, whereas similar treatment for the NaBH$_3$CN stabilized samples did not cause any obvious change in the XPS spectra.

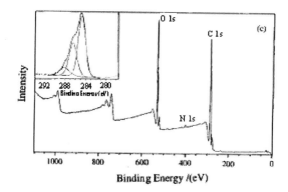

Figure 6. The acetaldehyde-plasma-treated aligned carbon nanotubes after the surface immobilization of amino-dextran chains [28].

The resulting amino-dextran grafted nanotube film showed zero air/water contact angles. The (acetaldehyde) plasma treated carbon nanotube film gave relatively low advancing (90°), sessile (78°), and receding (45°) air/water contact angles comparing to the advancing (155°), sessile (146°), and receding (122°) angles for an untreated sheet of aligned carbon nanotubes. The glucose units within the surface-grafted amino-dextran chains (Scheme 2) can be further converted into dialdehyde moieties by periodate oxidation [50], thereby, providing considerable room for further chemical modification.

3.3. Plasma Etching for Opening of Carbon Nanotubes

We have previously reported that H$_2$O-plasma can be used to effectively etch many substrates including even mica sheets [42]. We found that the H$_2$O-plasma etching technique can also be used to selectively open the top end-caps of the aligned carbon nanotubes without any significant structural change for the sidewalls under appropriate plasma conditions [51]. Figure 7a shows a typical closed structure of the aligned carbon nanotubes with encapsulated Fe rods at their tips. In contrast, Figure 7b clearly shows the removal of the top end-caps from the aligned nanotubes by the H$_2$O-plasma, providing the means to modify even the inner nanotube wall.

(a)   (b)

Figure 7 SEM images of aligned carbon nanotube film before and after the H_2O-plasma etching (a) aligned nanotubes capped by Fe nanorods before plasma treatment. (b) after the H_2O plasma etching for 80 minutes at 250 KHz, 30 W, and 0.62 Torr [51].

3.4. Aligned Carbon Nanotube and Conducting Polymer Coaxial Nanowires

In addition to the chemical grafting of polymer chains onto the carbon nanotube surface, we have also used the aligned carbon nanotubes as nanoelectrodes for making conducting coaxial nanowires by electrochemically depositing a concentric layer of an appropriate conducting polymer uniformly onto each of the constituent aligned nanotubes to form the aligned coaxial nanowires of conducting polymer coated carbon nanotubes (CP-NT) [27] (Figure 8A).

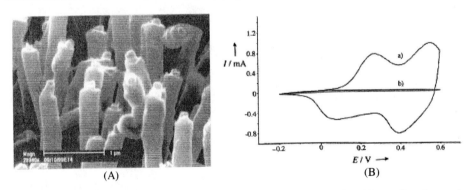

(A)   (B)

Figure 8. (A) Typical SEM images of the CP-NT coaxial nanowires produced by cyclic voltammetry on the aligned carbon nanotube electrode, showing a thin layer of conducting polymer (polypyrrole) coating surrounding each of the constituent aligned carbon nanotubes. (B) Cyclic voltammograms of (a) the polyaniline-coated CP-NT coaxial nanowires and (b) the bare aligned carbon nanotubes. Measured in an aqueous solution of 1M H_2SO_4 with a scan rate of 50 mV/s [27].

631

The electrochemical performance of the aligned CP-NT coaxial nanowires was evaluated by carrying out cyclic voltammetry measurements. As for polyaniline films electrochemically deposited on conventional electrodes, the cyclic voltammetric response of the polyaniline coated nanotube array in an aqueous solution of 1M H_2SO_4 (Figure 8B(a)) shows oxidation peaks at 0.33 and 0.52 V (but with much higher current densities) [52]. This indicates that polyaniline films thus prepared are highly electroactive. As a control, the cyclic voltammetry measurement was also carried out on the bare aligned nanotubes under the same conditions (Figure 8B(b)). In the control experiment, only capacitive current was observed with no peak attributable to the presence of any redoxactive species. The above observed strong redox response suggests that the CP-NT coaxial nanowires are promising for sensing applications.

To demonstrate potential applications of the CP-CNT coaxial nanowires as sensors, we have also immobilized glucose oxidase onto the aligned carbon nanotube substrate by electropolymerization of pyrrole in the presence of glucose oxidase [53, 54]. The glucose oxidase-containing polypyrrole-carbon nanotube coaxial nanowires were used to monitor the concentration change of hydrogen peroxide (H_2O_2) generated from the glucose oxidation reaction by measuring the increase in the electro-oxidation current at the oxidative potential of H_2O_2 (*i.e.* the amperometric method). The amperiometric response was found to be much higher than that of more conventional flat electrodes coated with glucose oxidase-containing polypyrrole films under the same conditions. As shown in Figure 9, a linear response of the electrooxidation current to the glucose concentration was obtained for the CP-NT nanowire sensor. The linear relationship extends to the glucose concentration as high as 20mM, which is higher than 15mM typically for the detection of blood glucose in practice [55]. Furthermore, the amperiometric response was found to be about ten orders of magnitude higher than that of more conventional flat electrodes coated with glucose oxidase-containing polypyrrole films under the same conditions [55].

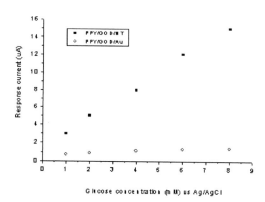

Figure 9. The dependence of electrooxidation current at the oxidative potential of H_2O_2 on the glucose concentration for the CP-NT coaxial nanowire sensor (solid squares) and the conventional polypyrrole sensor on a flat electrode under the same conditions (open circles) [53].

The strong response from the CP-NT sensors can be attributed to the large surface/interface area obtained for the nanotube-supported conducting polymer layer. In addition, the coaxial structure of the aligned CP-NT nanowires allows the nanotube framework to provide mechanical stability [56] and efficient thermal/electrical conduction [57, 58] to and from the conducting polymer layer. Therefore, the CP-CNT nanowires are very useful for making new glucose sensors with a high sensitivity and good thermal/mechanical stability.

4. CONCLUSION

We have demonstrated that aligned carbon nanotubes are attractive for many potential applications. The aligned structure also allowed surface modification of the aligned nanotubes without any destruction of the aligned structure. The polymer and aligned carbon nanotube coaxial nanowires produced both by electrochemically polymerizing a concentric layer of an appropriate conducting polymer onto the individual aligned carbon nanotubes or by chemical grafting of functional polymer chains onto plasma-treated aligned carbon nanotubes are particularly useful for a large variety of device applications, ranging from the improved electron emitters to highly sensitive biosensors. With so many advanced surface functionalization and device fabrication methods already reported and more to be developed, the examples discussed in this paper will surely be only the first few of many potential applications involving the surface-modified aligned carbon nanotubes.

5. REFERENCES

1. Iijima S 1991 *Nature* **354** 56.
2. Dai L 1999 *Polym. Adv. Technol.* **10**, 357, and references cited therein.
3. Dai L, Soundarrajan, P and Kim T 2002 *Pure Appl. Chem.* **74**, 1753.
4. Dresselhaus M S, Deesselhaus G and Eklund P 1996 *Science of Fullerenes and Carbon Nanotubes*; Academic Press: New York.
5. Ebbsen T 1997 *Carbon Nanotubes*; CRC Press: Boca Raton, FL
6. Saito R, Deesselhaus G and Dresselhaus M S 1998 *Physical Properties of Carbon Nanotubes*: Imperial College Press: London.
7. Terrones M, Hsu W K, Hare J P, Kroto H W, Terrones H and Walton D R M 1996 *Phil. Trans. R. Soc. Lond. A* **354** 2025.
8. de Heer W A, Bonard J M., Fauth, K., Châtelain, A., Forró, L., Ugarte, D., 1997. *Adv. Mater.* **9** 87, and references cited therein.
9. Frank S, Poncharal P, Wang Z L and de Heer W A 1998 Science **280** 1744.
10. Dai H, Hafner J H, Rinzler A G, Colbert D T and Smalley R E 1996 *Nature* **384** 147.
11. Harriso J A, Stuart S J, Robertson D H and White C T 1997 *J. Phys. Chem. B* **101** 9682.
12. Wong S S, Harper J D, Lansbury Jr P T and Lieber C M 1998 *J. Am. Chem. Soc.* **120** 603.
13. Wong S S, Joselevich E, Woolley A T, Cheung C L and Lieber C M 1998 *Nature* **394** 52.
14. Che G, Lakshmi B B, Fisher E R and Martin C R 1998 *Nature* **393** 346.
15. Planeix J M, Coustel N, Coq B, Brotons V, Kumbhar P S, Dutartre R, Geneste P, Bernier P and Ajayan P M 1994 *J. Am. Chem. Soc.* **116** 7935.
16. Guo Z, Sadler P J and Tsang S C 1998 *Adv. Mater.* **10** 701 and references cite therein.

17. Jirage K B, Hulteen J C and Martin C R 1997 *Science* **278** 655.
18. Baughman R H, Cui C, Zakhidov A A, Iqbal Z, Barisci J N, Spinks G M, Wallace G G, Mazzoldi A, De Rossi D, Rinzler A G, Jaschinski O, Roth S and Kertesz M 1999 *Science* **284** 1340.
19. Ajayan P M and Ebbesen T W 1997 *Rep. Prog. Phys.* **60** 1026.
20. Hirsch, A. 2002 *Angew. Chem. Int. Ed.* **41**, 1853.
21. Jeffrey L and Tour J M 2002 *J. Mater. Chem.* **12**, 1952.
22. Huang S, Dai L and Mau A W H 1999 *J. Phys. Chem. B* **103** 4223.
23. Yang Y, Huang S, He H, Mau A W H and Dai L 1999 *J. Am. Chem. Soc.* **121** 10832.
24. Huang S, Mau A W H, Turney T W, White P A and Dai L 2000 *J. Phys. Chem. B* **104** 2193.
25. Chen Q and Dai L 2000 *Appl. Phys. Lett.* **76** 2719.
26. Chen Q and Dai L 2001 *J. Nanosci. Nanothchnol.* **1** 43.
27. Gao M, Huang S, Dai L, Wallace G, Gao R and Wang Z 2000 *Angew. Chem. Int. Ed.* **39** 3664.
28. Chen Q, Dai L, Gao M, Huang S and Mau A W H 2001 *J. Phys. Chem. B* **105** 618.
29. Romero D B, Carrard M, de Heer W and Zuppiroli L 1996 *Adv. Mater.* **8** 899.
30. Curran S A, Ajayan P M, Blau W J, Carroll D L, Coleman J N, Dalton A B, Davey A P, Drury A, McCarthy B, Maier S and Strevens A 1998 *Adv. Mater.* **10** 1091.
31. Harris, P J F 1999 Carbon Nanotubes and Related Structures: New Materials for the Twenty-first Century, Cambridge University Press: Cambridge.
32. Dai L and Mau A W H 2001 *Adv. Mater.* **13** 899.
33. Li D-C, Dai L, Huang S, Mau A W H and Wang Z L 2000 *Chem. Phys. Lett.* **316** 349.
34. Fan S, Chapline M G, Franklin N R, Tomber T W, Cassell A M and Dai H 1999 *Science* **283** 512.
35. Li J, Papadopoulos C, Xu J M 1999 *Appl. Phys. Lett.* **75** 367.
36. Ren Z F, Huang Z P, Wang D Z, Wen J G, Xu J W, Wang J H, Calvet L E, Chen J, Klemic J F and Reed M A 1999 *Appl. Phys. Lett.* **75** 1086.
37. Wang X, Liu Y, Zhu D, 2002 *Adv. Mater.* **14**, 165.
38. Suh J S, Lee J S and Jin S 1999 *Appl. Phys. Lett.* **75** 2047.
39. March J 1992 *Advanced Organic Chemistry*, 4th ed, John Wiley, New York.
40. Moreau W M 1988 *Semiconductor Lithography: Principle and Materials,* Plenum, New York.
41. Wallraff G M and Hinsberg W D 1999 *Chem. Rev.* **99** 1801.
42. Dai L, Griesser H J and Mau A 1997 *J. Phys. Chem. B* **101** 9548.
43. H J R and Bell A T 1974 *Techniques and Applications of Plasma Chemistry,* Wiley, New York.
44. Collins PG, Bradley K, Ishigami M and Zettl A 2000 *Science* **287**, 1801.
45. Yudasaka M, Kikuchi R, Ohki Y and Yoshimura S 1997 *Carbon* **35**, 195.
46. Terrones M, Grobert N, Olivares J, Zhang J P, Terrones H, Kordatos K, Hsu, W K, Hare P P, Townsend P D, Prassides K, Cheetham A K, Kroto H W and Walton D R M 1997 *Nature* **388**, 52.
47. Walls J M (Ed.) 1990 *Methods of Surface Analysis: Techniques and Applications*; Cambridge University Press: Cambridge.
48. Gong X, Dai L, Griesser H J, Mau A W H 2000 *J. Polym. Sci. Part B, Polym. Phys.* **38**, 2323.
49. Bonard J M, Salvetat J P, Stöckli T, de Heer W A, Forró L and Chatelain A 1998 *Appl. Phys. Lett.* **73**, 918.
50. Dai L, StJohn H, Bi J, Zientek P, Chatelier R and Griesser H J 2000 *Surf. Interf. Analysis* **29**, 46.
51. Huang S and Dai L 2002 *J. Phys. Chem.* **106**, 3543.
52. Sazou D and Georgolios C 1997 *J. Electroanal. Chem.* **429** 81.
53. Gao M, Dai L and Wallace G 2003 *Electranayl.* (in press).
54. Soundarrajan P, Patil A and Dai L 2003 *J. Vac. Sci. Technol.* (in press).
55. Yasuzawa M and Kunugi A 1999 *Electrochem. Commun.* **1** 459.
56. Gao R, Wang Z L, Bai Z, de Heer W A, Dai L and Gao M 2000 *Phys. Rev. Lett.* **85** 622.
57. Frank S, Poncharal P, Wang Z L and de Heer W A 1998 Science **280** 1744.
58. Odom T W, Huang J-L, Kim P and Lieber C M 2000 *J. Phys. Chem. B* **104** 2794 and references cited therein.

Acknowledgement: We are grateful for financial support from the US AirForce through the AirForce/UAkron Collaborative Center on Polymer Photonics.

Hoenlein, W., 533
Hofmann, S., 503, 507
Hold, S., 481
Holzinger, M., 87, 112
Holzweber, M., 297, 306
Hönlein, W., 95, 587
Huang, S., 103
Huard, B., 612
Huczko, A., 73
Hug, G., 91
Hulman, M., 143, 207, 370
Humbert, B., 181

I

Iijima, S., 318
Ishigami, M., 389
Ishii, H., 139
Iwasa, Y., 77
Izard, N., 235, 559

J

Jacob, A., 19
Jaglicic, Z., 411
Jansen, M., 15, 37
Jeglič, P., 19
Jorio, A., 177, 219
Ju, B.-K., 477

K

Kaempgen, M., 273, 554
Kaiser, A. B., 160
Kalenczuk, R. J., 374, 378
Kamarás, K., 54
Kanai, M., 41
Kane, C. L., 460
Kappes, M. M., 112, 120, 148, 189, 197, 230, 291, 528
Karachevtsev, V. A., 202
Kataura, H., 139, 156, 211, 297, 306, 310, 314, 344, 349, 573
Kavan, L., 344
Kazaoui, S., 257
Keogh, S. M., 261, 265
Kern, K., 169, 447

Khlobystov, A. N., 41
Kikuchi, K., 349
Kiricsi, I., 253
Kishita, K., 302
Kittrell, C., 241, 246
Klinke, C., 447
Klupp, G., 54
Knupfer, M., 127, 361, 374, 378
Kobayashi, Y., 219
Koch, E., 66
Kodama, T., 139, 349
Kondo, T., 302
Kónya, Z., 253, 439
Koós, A. A., 253
Kováts, É., 46
Král, P., 465
Kramberger, C., 297, 306, 310, 314
Krause, M., 50
Kresse, G., 207
Kreupl, F., 95, 533, 587
Krstić, V., 169, 564
Krupke, R., 148, 230, 528
Kukovecz, Á., 11, 185, 211, 253, 306
Kulnitskiy, B. A., 477
Kurashima, K., 366
Kürti, J., 443, 456
Kutner, W., 3
Kuzmany, H., 11, 185, 207, 215, 297, 306, 310, 314, 370, 443
Kuznetsov, V. L., 82, 215

L

Lambin, P., 439
Lange, H., 73
Lansley, S. B., 605
Larciprete, R., 616
László, I., 423
Lathe, C., 164
Laurita, A., 616
Lavayen, V., 473
Lebedkin, S., 148, 230, 569
Lee, D.-J., 477
Lee, Y.-D., 477
Lefrant, S., 181, 384, 452
Legagneux, P., 605
Le Lay, M., 302
Leonhardt, A., 550
Leontiev, V. S., 202

639